LIST OF ELEMENTS WITH THEIR SYMBOLS AND ATOMIC WEIGHTS

Element	Symbol	Atomic number	Atomic weight [a]
Actinium	Ac	89	227.0278
Aluminum	Al	13	26.98154
Americium	Am	95	(243)
Antimony	Sb	51	121.75
Argon	Ar	18	39.948
Arsenic	As	33	74.9216
Astatine	At	85	(210)
Barium	Ba	56	137.33
Berkelium	Bk	97	(247)
Beryllium	Be	4	9.01218
Bismuth	Bi	83	208.9804
Boron	B	5	10.81
Bromine	Br	35	79.904
Cadmium	Cd	48	112.41
Calcium	Ca	20	40.078
Californium	Cf	98	(251)
Carbon	C	6	12.011
Cerium	Ce	58	140.12
Cesium	Cs	55	132.9054
Chlorine	Cl	17	35.453
Chromium	Cr	24	51.996
Cobalt	Co	27	58.9332
Copper	Cu	29	63.546
Curium	Cm	96	(247)
Dysprosium	Dy	66	162.50
Einsteinium	Es	99	(254)
Erbium	Er	68	167.26
Europium	Eu	63	151.96
Fermium	Fm	100	(257)
Fluorine	F	9	18.998403
Francium	Fr	87	(223)
Gadolinium	Gd	64	157.25
Gallium	Ga	31	69.72
Germanium	Ge	32	72.61
Gold	Au	79	196.9665
Hafnium	Hf	72	178.49
Helium	He	2	4.00260
Holmium	Ho	67	164.9304
Hydrogen	H	1	1.0079
Indium	In	49	114.82
Iodine	I	53	126.9045
Iridium	Ir	77	192.22
Iron	Fe	26	55.847
Krypton	Kr	36	83.80
Lanthanum	La	57	138.9055
Lawrencium	Lr	103	(260)
Lead	Pb	82	207.2
Lithium	Li	3	6.941
Lutetium	Lu	71	174.967
Magnesium	Mg	12	24.305
Manganese	Mn	25	54.9380
Mendelevium	Md	101	(258)
Mercury	Hg	80	200.59
Molybdenum	Mo	42	95.94
Neodymium	Nd	60	144.24
Neon	Ne	10	20.179
Neptunium	Np	93	237.0482
Nickel	Ni	28	58.69
Niobium	Nb	41	92.9064
Nitrogen	N	7	14.0067
Nobelium	No	102	(259)
Osmium	Os	76	190.2
Oxygen	O	8	15.9994
Palladium	Pd	46	106.42
Phosphorus	P	15	30.97376
Platinum	Pt	78	195.08
Plutonium	Pu	94	(244)
Polonium	Po	84	(209)
Potassium	K	19	39.0983
Praseodymium	Pr	59	140.9077
Promethium	Pm	61	(145)
Protactinium	Pa	91	231.0359
Radium	Ra	88	226.0254
Radon	Rn	86	(222)
Rhenium	Re	75	186.207
Rhodium	Rh	45	102.9055
Rubidium	Rb	37	85.4678
Ruthenium	Ru	44	101.07
Rutherfordium [b]	Rf	104	(261)
Samarium	Sm	62	150.36
Scandium	Sc	21	44.9554
Selenium	Se	34	78.96
Silicon	Si	14	28.0855
Silver	Ag	47	107.8682
Sodium	Na	11	22.98977
Strontium	Sr	38	87.62
Sulfur	S	16	32.06
Tantalum	Ta	73	180.9479
Technetium	Tc	43	(98)
Tellurium	Te	52	127.60
Terbium	Tb	65	158.9254
Thallium	Tl	81	204.383
Thorium	Th	90	232.0381
Thulium	Tm	69	168.9342
Tin	Sn	50	118.71
Titanium	Ti	22	47.88
Tungsten	W	74	183.85
Unnilennium [b]	Une	109	(266)
Unnilhexium [b]	Unh	106	(263)
Unniloctium [b]	Uno	108	(265)
Unnilpentium [b]	Unp	105	(262)
Unnilquadium [b]	Unq	104	(261)
Unnilseptium [b]	Uns	107	(262)
Uranium	U	92	238.0289
Vanadium	V	23	50.9415
Xenon	Xe	54	131.29
Ytterbium	Yb	70	173.04
Yttrium	Y	39	88.9059
Zinc	Zn	30	65.39
Zirconium	Zr	40	91.22

[a] Numbers in parentheses are mass numbers of the most stable or best-known isotope of radioactive elements.

[b] The official name and symbol have not been agreed to. The names for elements 106, 107, 108, and 109 represent their atomic numbers, as in un (1) nil (0). hex (6) = unnilhexium (Unh) for element 106.

Fundamentals of Organic and Biological Chemistry

John McMurry
Cornell University

Mary E. Castellion
Norwalk, Connecticut

Prentice Hall, Englewood Cliffs, New Jersey 07632

Library of Congress Cataloging-in-Publication Data

McMurry, John.
    Fundamentals of organic and biological chemistry / John
McMurry,
Mary E. Castellion.
            p.   cm.
      Includes index.
      ISBN 0-13-293085-4
      1. Chemistry.   2. Chemistry, Organic.   3. Biochemistry.
I. Castellion, Mary E.   II. Title.
QD31.2.M3878   1994
547—dc20   93-39094
                                            CIP

Editor-in-Chief: Tim Bozik
Acquisitions Editor: Paul Banks
Marketing Manager: Kelly McDonald
Design Director: Florence Dara Silverman
Cover and Interior Designer: Bruce Kensellaer, Meryl Poweski
Manufacturing Buyer: Trudy Pisciotti
Supplements Editor: Mary Hornby
Production Editor: Susan Fisher
Illustrations by Vantage Art
Photo Research Coordinator: Lorinda Morris-Nantz
Photo Researcher: Tobi Zausner
Cover photo by Dr. R. Clark & M. Goff/Photo Researchers

COVER PHOTO is an infrared photo, or thermogram, of hands demonstrating blood flow as reflected in heat given off. Areas of normal blood flow are red and yellow, while cooler areas of diminished blood flow are blue.

Photo credits and acknowledgments appear on pages A–48 and A–49, which constitute a continuation of the copyright page.

 © 1994 by Prentice-Hall, Inc.
A Paramount Communications Company
Englewood Cliffs, New Jersey 07632

Printed in the United States of America
10   9   8   7   6   5   4   3   2   1

ISBN 0-13-293085-4

Prentice-Hall International (UK) Limited, *London*
Prentice-Hall of Australia Pty. Limited, *Sydney*
Prentice-Hall Canada Inc., *Toronto*
Prentice-Hall Hispanoamericana, S.A., *Mexico*
Prentice-Hall of India Private Limited, *New Delhi*
Prentice-Hall of Japan, Inc., *Tokyo*
Simon & Schuster Asia Pte. Ltd., *Singapore*
Editora Prentice-Hall do Brasil, Ltda., *Rio de Janeiro*

Contents

CHAPTER 2
Alkenes, Alkynes, and Aromatic Compounds   *33*

CHAPTER 3
Some Compounds with Oxygen, Sulfur, or Halogens   *66*

CHAPTER 4
Amines   *95*

CHAPTER 5
Aldehydes and Ketones   *119*

CHAPTER 6
Carboxylic Acids and Their Derivatives   *146*

CHAPTER **9**
The Generation of Biochemical Energy *247*

CHAPTER **10**
Carbohydrates   *276*

CHAPTER **11**
Carbohydrate Metabolism   *304*

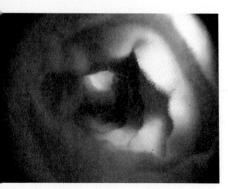

CHAPTER 12
Lipids *330*

CHAPTER 13
Lipid Metabolism *359*

CHAPTER 14
Protein and Amino Acid Metabolism *382*

CHAPTER 15
Nucleic Acids and Protein Synthesis *400*

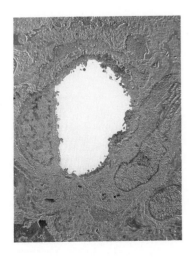

CHAPTER 16
Body Fluids *430*

Appendices

Applications and Interludes

APPLICATIONS

INTERLUDES

Preface

To provide an introduction to organic chemistry and to the chemistry of living things—that is the goal of this textbook. The writing style, content, and organization are directed toward students with career goals in the allied health sciences and toward students seeking to know something about chemistry's role in our complex society.

Teaching chemistry is a challenging activity, just as learning chemistry is challenging. Teaching chemistry all the way from What is an atom? to How do we get energy from glucose? is especially difficult. Conversations with many teachers who face this challenge show that there are just about as many approaches to it as there are teachers. This textbook is designed for the one-semester course in organic and biological chemistry for students who have already had an introduction to general chemistry.

The chapters here are drawn from the organic and biological chemistry portions of *Fundamentals of General, Organic, and Biological Chemistry,* our textbook for the two-semester course. By varying the topics covered and the time devoted to them, each teacher can change the focus of the one-semester course to meet their students individual needs. Our unique biochemistry sequence, described below, allows for an unusual degree of flexibility with this material.

Because an understanding of bonding and acid-base chemistry is so important to success with this material, we have provided appendices on these subjects (Appendix A, Chemical Bonds; Appendix B, Acids and Bases). Marginal notes direct students to the appendices for independent review. The two appendices are written in such a manner that, as an alternative, they can be the basis for class lectures on these topics.

In addition, about 50 new marginal notes have been added to this one-semester text. The purpose of these notes is to provide support for students who may need reminders of essential concepts from their introduction to chemistry.

Most students in this course have their sights set well beyond academic concerns and the laboratory bench. They want to know why: Why must I study organic chemistry?

Why are molecular shapes important for me as a nurse, a farmer, or an informed citizen? We have therefore endeavored at every step along the way to place chemistry in the context of applications and everyday life. To meet this challenge, we have written about these matters in the mainstream of the text as well as in the Application and Interlude sections. With the intent of gaining student confidence, our writing style is relaxed and friendly, and we have included many visual and verbal study aids.

ORGANIZATION

The six *organic chemistry* chapters provide a concise overview of the subject and focus on what students must know in order to get on with the study of biochemistry. Nomenclature rules are included with the introduction to hydrocarbons (Chapters 1 and 2) and thereafter are kept to a minimum. The functional group chapters (Chapters 3–6) emphasize the structures and properties relevant to biomolecules and profile commonly encountered compounds. The chapter on amines presents their numerous biological roles and, by preceding the carboxylic acids chapter, allows amide chemistry to be covered with that of related carboxylic acid derivatives.

Our effort to allow for flexibility is especially evident in the *biological chemistry* chapters, where *structure and function are integrated*. Protein structure (Chapter 7) is followed by enzyme chemistry and biochemical energy (Chapters 8 and 9). Then come carbohydrates and their metabolism, lipids and their metabolism, and protein metabolism. If your time for biochemistry is limited, stop with Chapter 9 on biochemical energy, and your students will have an excellent preparation in the essentials of metabolism. To carry the metabolism story further, cover the next two chapters on carbohydrates and their metabolism (Chapters 10 and 11). And if you want to cover all classes of biomolecules and their metabolism, you will find a thorough, integrated treatment in Chapters 7–14. Nutrition is not treated as yet another separate subject, but is integrated with the discussion of each type of biomolecule. In the last two chapters, we cover subjects that we've found are an essential part of the course to many of you, but optional to others—Nucleic Acids and Protein Synthesis in Chapter 15 and Body Fluids in Chapter 16. Throughout, we have made every effort to provide up-to-date coverage, recognizing that biological chemistry is a most rapidly advancing area of science.

KEY FEATURES

Applications and Interludes   A wide variety of special topics are covered in over 50 Application and Interlude sections. These sections provide thorough coverage—whether the topics are just assigned for additional reading or incorporated in the course, students can gain a reasonable understanding of each topic. Representative subjects (a complete list precedes the Preface) include *everyday chemistry* (Detergents; Sweetness; The Biochemistry of Running), *environmental and societal issues* (Chlorofluorocarbons and the Ozone Layer; Is It Poisonous or Isn't It?), *health and medical applications* (Barbiturates; Ethyl Alcohol as a Drug and Poison; Glucose Tolerance Test), and *modern applied chemistry* (Magnetic Resonance Imaging; Antioxidants; Prodrugs). Questions on the Applications and Interludes are provided in a separate section at the end of each chapter.

Color Photos   Each chapter opens with a photo chosen to generate curiosity about the chapter subject. Then, throughout each chapter, photos are used to enhance understanding and appreciation for the subject matter.

Graphics Molecular structures, chemical equations, and charts and diagrams have been highlighted with color to emphasize their meaning. Many topics in organic and biological chemistry become much clearer when the reacting parts of molecules are color coded. For example, color has been used consistently to highlight phosphate groups in biomolecules and to distinguish energy-rich forms of such molecules as ATP and reduced coenzymes (red) from their lower energy counterparts, ADP and oxidized coenzymes (blue).

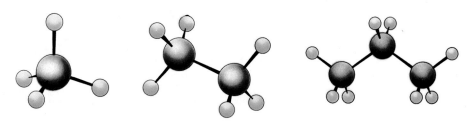

Flavin mononucleotide (FMN)

Nowhere are graphics more important than in clarifying the complexities of biochemical pathways and we believe our graphics are the best ever offered to students in this course. Our overall metabolism diagram has been adapted as a logo and is repeated as appropriate with the topic under discussion highlighted.

Computer-Generated Structures Computer-generated molecular models are used extensively in the organic and biological chapters, both for their accuracy in portraying the three-dimensional structures of molecules and for their visual appeal.

Problem Solving The problem solving skills essential to this course are illustrated with clearly explained and worked out Solved Problems. Solved Problems are all followed by related Practice Problems, and most text sections include additional practice problems to provide an immediate test of understanding. Practice Problems are answered at the back of the book.

PEDAGOGY

Introduction and Goals Each chapter begins with a brief introductory overview, followed by a list of goals for the student to keep in mind while studying.

Marginal Definitions Key terms are boldfaced in the text on their first use, and the definition of each term is provided in the margin nearby for easy review.

Marginal Notes Brief marginal notes provide reminders of essential concepts from introductory chemistry.

Glossary   The definitions of all key terms are collected in alphabetical order in the Glossary at the back of the book.

Summaries   Certain complex topics are summarized immediately after their presentation in bulleted statement lists. Each chapter ends with a clear, concise summary that reviews key points, with essential terms highlighted in boldface type.

Review Problems   The end-of-chapter Review Problems provide approximately 1000 questions and problems: 400 on organic chemistry, and 600 on biological chemistry.

Appendices and Reference Tables   The first two appendices provide review of essential topics from introductory chemistry: Chemical Bonds, and Acids and Bases. Other Appendices provide a review of exponential notation and useful conversion factors. The reference tables inside the cover display for easy reference the periodic table, an alphabetical list of elements, the structural features of important families of organic molecules, and a list of important tables and diagrams in the text.

Index   The index is designed to be especially useful by including both general and specific citations and by the absence of cross reference entries without page numbers.

SUPPLEMENTS

Study Guide and Solutions Manual, by Susan McMurry.   This companion volume answers all in-text and end-of-chapter problems and explains in detail how the answers are obtained. The solutions and data have been carefully prepared and reviewed for accuracy and coordination with the textbook. Chapter summaries, study hints, and self-test materials for each chapter are included.

Instructor's Resource Manual, by Theodore Sakano, and **Prentice Hall Test Manager**   The *Manual* includes chapter overviews, lecture outlines, learning objectives, suggested readings, and 1500 multiple choice test questions. The *Test Manager* provides these questions on disk in either IBM® or MacIntosh® format and includes an editing feature that allows questions to be added or changed.

Laboratory Manual and Instructor's Manual to Laboratory Manual, by Scott Mohr and Susan Griffin   The *Laboratory Manual* provides 24 laboratory experiments adaptable to either two- or three-hour laboratory periods. The *Instructor's Manual* includes detailed descriptions of all necessary chemicals, supplies, and equipment, as well as answers to pre-lab questions, typical student results, and completed report forms.

How to Study Chemistry, by Vernon Burger   This free supplement contains problem-solving strategies, helpful hints for learning and achieving success in chemistry, and a mathematics review.

Transparencies   A set of 100 two-color and four-color transparencies from this and other Prentice Hall chemistry texts is available.

Additional resources   A collection of timely news stories from *The New York Times,* described elsewhere in the opening pages of this book, and several video packages are available upon adoption. For further information please contact your local Prentice Hall sales representative.

ACKNOWLEDGMENTS

It is a pleasure to thank the many people whose help and suggestions were so valuable in preparing *Fundamentals of General, Organic, and Biological Chemistry,* on which this book is based. We especially thank Leslie Kinsland, University of Southwestern Louisiana, for her assistance with questions and problems. The persons listed below provided many excellent suggestions after reviewing all or part of the manuscript. In particular, John M. Daly, Leland Harris, Larry Jackson, Gloria G. Lyle, and Les Wynston travelled across the country to make significant contributions for which we are very grateful, and Larry Jackson helped out with the special topic boxes for Chapter 16.

James N. Beck
McNeese State University

Richard E. Beitzel
Bemidji State University

Rodney Buyer
Hope College

John M. Daly
Bellarmine College

Lindsley Foote (Retired)
Western Michigan State University

Leland Harris
University of Arizona

Kenneth I. Hardcastle
California State University, Northridge

Merrill Hugo
Shasta College

Larry L. Jackson
Montana State University

Gloria G. Lyle
University of Texas, San Antonio

Frank R. Milio
Towson State University

Danny V. White
American River College

Karen Wiechelman
University of Southwestern Louisiana

Donald W. Williams
Hope College

Leslie Wynston
California State University, Long Beach

In addition, the book benefited from the careful reading of galley proofs by Clyde Metz, College of Charleston, and by Leland Harris, University of Arizona. During production, the persistence and professionalism of John Morgan, Production Editor, Prentice Hall; Tobi Zausner, Photo Researcher; and Diane Koromhas, Layout Artist were greatly appreciated. We further extend our thanks to Susan Fisher, Production Editor for this book, and to everyone on the capable staff of Prentice Hall, both those named on the copyright page and the many others who worked and continue to work for the success of our books.

A Note to the Student

Here you are, about to study organic and biological chemistry, perhaps for the first time. The topics you are about to study will be useful in all health-related professions and in many business endeavors. The chemistry you are introduced to in this book will also be useful in exercising judgment in everyday life. Newspapers and magazines are filled with chemistry-related stories about protecting the environment, about new materials designed to improve the quality of life, and about drugs that promise to revolutionize medical care. The better you understand such matters, the better you will be able to function in today's society.

The following suggestions should prove helpful in your study:

Don't read the text immediately. As you begin each new chapter, look it over first. Read the introductory paragraphs and familiarize yourself with the chapter goals. Find out what topics are covered, and take a look at the illustrations—to get a feel for the topics at hand. Then turn to the end of the chapter and read the summary. You'll be in a much better position to learn new material if you first have a general idea of where you're going.

Work the problems. The problems are designed to give you practice in the skills necessary to understand and use chemistry. There are no shortcuts here. The sample problems illustrate the skills, the in-chapter practice problems provide immediate practice, and the end-of-chapter problems provide additional drill. Brief answers to in-chapter practice problems and most even-numbered review problems are given at the end of this book.

Use the study guide. Complete answers and explanations for all problems, along with chapter outlines, additional study hints, and self-tests, are given in the *Study Guide and Solutions Manual* that accompanies this text. The *Study Guide* can be extremely useful when you're working problems and when you're studying for an exam. Investigate what's there now so you'll know where to find help when you need it.

Ask questions.   Faculty members and teaching assistants are there to help you learn. Don't hesitate because you think a question might be stupid or silly. If it's something you need to know to get on with understanding chemistry, it's always a good question.

Many of the words and symbols that lie ahead in this book may at first seem strange to you. We urge you not to let their unfamiliarity cause you to lose sight of your goals: to learn about the amazing kinds of chemistry that keep us all alive and well, and to understand the impact of chemistry on everyday life.

John McMurry
Mary E. Castellion

The New York Times Program

The New York Times and Prentice Hall are sponsoring a THEMES OF THE TIMES, a program designed to enhance student access to current information of relevance in the classroom.

Through this program, the core subject matter provided in the text is supplemented by a collection of time-sensitive articles from one of the world's most distinguished newspapers, *The New York Times*. These articles demonstrate the vital, ongoing connection between what is learned in the classroom and what is happening in the world around us.

To enjoy the wealth of information of *The New York Times* daily, a reduced subscription rate is available in deliverable areas. For information, call toll-free: 1-800-631-1222.

Prentice Hall and *The New York Times* are proud to co-sponsor THEMES OF THE TIMES. We hope it will make the reading of both textbooks and newspapers a more dynamic, involving process.

Fundamentals of Organic and Biological Chemistry

C H A P T E R

1

Introduction to Organic Chemistry: Alkanes

Oil—to some it's a profitable investment, to others it's a source of energy, to still others it's a source of political power. To chemists, it's the principal raw material for creating useful organic compounds. In this chapter, you'll learn about the products of oil refineries like the one shown here.

As knowledge of chemistry slowly evolved in the 1700s, mysterious differences were noted between compounds obtained from animals and those obtained from minerals. Chemicals from animal sources were often more difficult to isolate, to purify, and to work with than those from mineral sources. To express this difference, the term *organic chemistry* was introduced to mean the study of compounds from living organisms, while *inorganic chemistry* was used to refer to the study of compounds from minerals.

Today we know there aren't any fundamental differences between organic and inorganic compounds: The same scientific principles are applicable to both. The only common characteristic of compounds from living sources is that they contain the element carbon.

Why is carbon special? Carbon atoms have the unique ability to bond together, forming long chains and rings. Of all the elements, only carbon is able to form such an immense array of compounds. In this and the next five chapters, we'll look at the chemistry of organic compounds, beginning with an exploration of these topics:

1. *How are organic molecules classified?* The goal: Be able to classify organic molecules into functional-group families.

2. *What are the structures of organic molecules?* The goal: Be able to recognize the main carbon chain in a molecule and identify constitutional isomers.

3. *How are organic molecules drawn?* The goal: Be able to convert between structural formulas and condensed or line structures.

4. *How are alkanes and cycloalkanes named?* The goal: Be able to name an alkane or cycloalkane from its structure or write the structure, given the name.

5. *What are the general properties of alkanes?* The goal: Be able to describe such properties as polarity, water solubility, flammability, toxicity, and chemical reactivity.

6. *What are the major chemical reactions of alkanes?* The goal: Be able to describe the products formed in combustion and halogenation of alkanes.

1.1 THE NATURE OF ORGANIC MOLECULES

Let's review some fundamental ideas about the structures of organic molecules:

● **Carbon is always tetravalent; it always forms four bonds.** In methane, carbon is connected to four hydrogen atoms:

$$\text{Methane, CH}_4 \qquad \text{H}-\overset{\overset{\displaystyle\text{H}}{|}}{\underset{\underset{\displaystyle\text{H}}{|}}{\text{C}}}-\text{H}$$

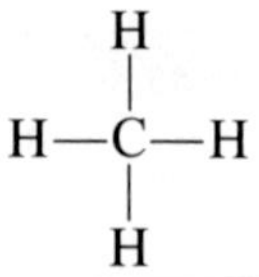

● **Organic molecules contain covalent bonds.** In ethane, the bonds result from the sharing of two electrons, either between C and C atoms or between C and H atoms:

REVIEW   Covalent bonding is reviewed in Appendix A.

$$\text{Ethane, C}_2\text{H}_6 \qquad \text{H} \overset{\displaystyle\text{H}\ \ \text{H}}{\underset{\displaystyle\text{H}\ \ \text{H}}{\text{C} \text{C}}} \text{H} \ = \ \text{H}-\overset{\overset{\displaystyle\text{H}}{|}}{\underset{\underset{\displaystyle\text{H}}{|}}{\text{C}}}-\overset{\overset{\displaystyle\text{H}}{|}}{\underset{\underset{\displaystyle\text{H}}{|}}{\text{C}}}-\text{H}$$

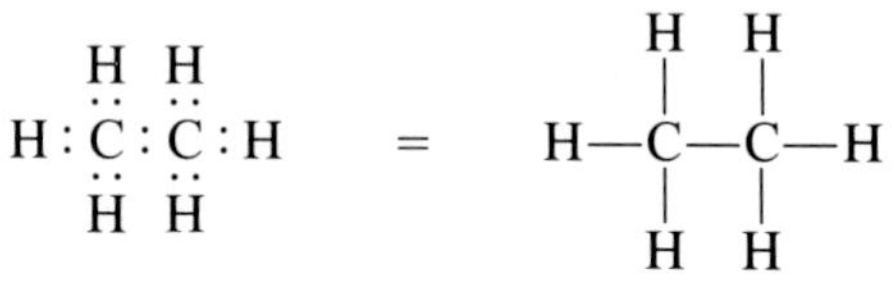

● **Organic molecules contain polar covalent bonds when carbon bonds to an element on the far right or far left of the periodic table.** In chloromethane, the electronegative chlorine atom attracts electrons more strongly than carbon, resulting in polarization of the carbon–chlorine bond so that carbon has a partial positive charge, δ^+:

REVIEW   The more electronegative an atom, the more strongly it attracts electrons in polar bonds (Appendix A.8).

$$\text{Chloromethane, CH}_3\text{Cl} \qquad \text{H} \overset{\displaystyle\text{H}}{\underset{\displaystyle\text{H}}{\text{C}}} \text{Cl} \ = \ \text{H}-\overset{\overset{\displaystyle\text{H}}{|}}{\underset{\underset{\displaystyle\text{H}}{|}}{\text{C}^{\delta+}}}-\text{Cl}^{\delta-}$$

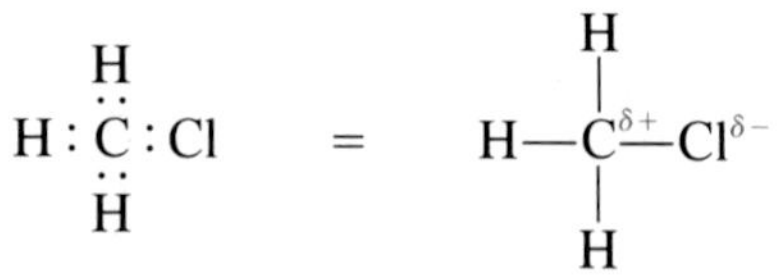

● **Carbon can form multiple covalent bonds by sharing more than two electrons with a neighboring atom.** In ethylene, the two carbon atoms share four electrons in a double bond; in acetylene, the two carbons share six electrons in a triple bond:

$$\text{Ethylene, C}_2\text{H}_4 \qquad \overset{\displaystyle\text{H}}{\underset{\displaystyle\text{H}}{\text{C}}} :: \overset{\displaystyle\text{H}}{\underset{\displaystyle\text{H}}{\text{C}}} \ = \ \overset{\text{H}\qquad\text{H}}{\underset{\text{H}\qquad\text{H}}{\text{C}=\text{C}}}$$

$$\text{Acetylene, C}_2\text{H}_2 \qquad \text{H}:\text{C}:::\text{C}:\text{H} \ = \ \text{H}-\text{C}\equiv\text{C}-\text{H}$$

● **Covalently bonded molecules have specific three-dimensional shapes.** When carbon is bonded to four atoms as in methane, CH_4, the bonds are oriented toward the four corners of an imaginary tetrahedron with carbon in the center:

REVIEW   The shapes of molecules are predicted by the VSEPR model, which assumes that each pair of bonding or nonbonding electrons on a central atom repels the others (Appendix A.7).

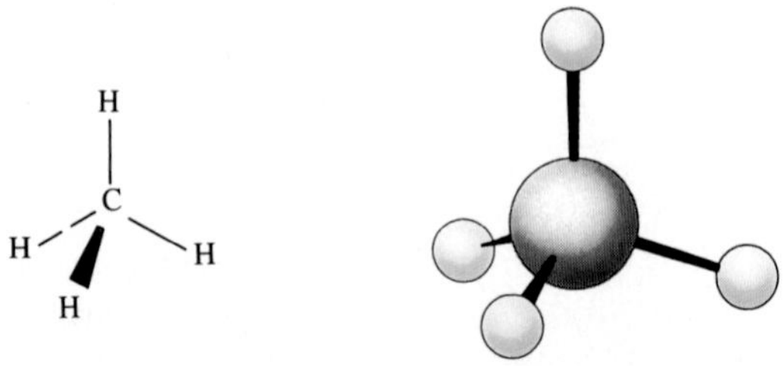

REVIEW   By sharing their
outermost, or valence,
electrons, carbon atoms can
form 4 covalent bonds;
nitrogen atoms can form 3;
oxygen atoms can form 2;
and hydrogen atoms can
form 1 (Appendix A.3, A.4).

REVIEW   Intermolecular
forces of attraction act
between molecules with
unsymmetrical electron
distributions that result in
regions of positive and
negative charge.

REVIEW   Some typical
inorganic salts (Appendix
A.1): sodium chloride (NaCl),
magnesium sulfate (MgSO$_4$),
calcium chloride (CaCl$_2$).

● **Hydrogen, nitrogen, and oxygen are the elements most often present in organic molecules in addition to carbon.** Nitrogen and oxygen can form both single and multiple bonds to carbon.

$$C-N \quad C-O \quad C-H$$
$$C=N \quad C=O$$
$$C \equiv N$$

As a result of their covalent bonding, organic compounds have properties quite different from those of many inorganic compounds. For example, inorganic salts have high melting points and boiling points because they consist of large collections of oppositely charged ions held together by strong electrical attractions. Organic compounds, by contrast, consist of atoms joined by covalent bonds in individual molecules. Thus intermolecular forces, which are fairly weak, have an important influence on the properties of organic compounds. Organic compounds generally have lower melting and boiling points than inorganic salts. In fact, many simple organic compounds are liquids at room temperature, and a few are gases.

Solubility and electrical conductivity are other important differences between organic and inorganic compounds. Whereas many inorganic compounds dissolve in water to yield ions in solutions capable of conducting electricity, most organic compounds are insoluble in water and do not conduct electricity. Only small polar organic molecules, such as glucose and ethyl alcohol (both of which contain —OH groups), or large molecules with many polar groups, such as some proteins, dissolve in water.

The lack of water solubility of organic compounds has many important practical consequences, such as the difficulty in cleaning up greasy dirt and oil spilled in the ocean (Figure 1.1). In digestion, fats and other nonpolar organic molecules consumed in food must be made soluble in water before they can be used by the body.

Figure 1.1
Result of an oil spill in the
Black Sea in February 1991
during the Gulf War between
Iraq and a coalition of
Western and Arab states.
The oil washing up at Jubail,
Saudi Arabia, was spilled in
what was described by some
as an act of "environmental
terrorism."

1.2 FAMILIES OF ORGANIC MOLECULES: FUNCTIONAL GROUPS

REVIEW Physical properties can be observed without changing the identity of a substance; chemical properties involve chemical reactions, which change the identity of one or more substances.

Atoms of a relatively few different kinds of elements can bond together in a great many different ways in organic compounds. At last count, there were more than 10 million organic compounds described in the scientific literature. Each has unique physical properties such as melting point and boiling point, and each has unique chemical properties. The situation isn't as hopeless as it sounds, however, because chemists have learned through experience that organic compounds can be classified into families according to their structural features and that the chemical behavior of the members of a family is often predictable. Instead of 10 million compounds with random chemical reactivity, there are just a few general families of organic compounds whose chemistry is predictable.

Functional group   A part of a larger molecule composed of an atom or group of atoms that has characteristic chemical behavior.

The structural features that allow us to class compounds together are called functional groups. A **functional group** is a part of a larger molecule and is composed of an atom or group of atoms that has characteristic chemical behavior. A given functional group undergoes the same reactions in every molecule it's a part of. For example, the carbon–carbon double bond is one of the simplest functional groups. Ethylene (C_2H_4), the simplest compound with a double bond, undergoes many chemical reactions similar to those of cholesterol ($C_{27}H_{46}O$), a far larger and more complex compound. Both, for example, react with hydrogen in the same manner, as shown in Figure 1.2. These identical reactions with hydrogen are typical: The chemistry of an organic molecule, regardless of size and complexity, is determined by the functional groups it contains.

Table 1.1 lists some of the most important families of organic molecules and their distinctive functional groups. All compounds that contain an —NH_2 group, for example, are known as amines and are members of the amine family.

Figure 1.2
The reactions of (a) ethylene and (b) cholesterol with hydrogen. The carbon–carbon double bond functional groups adds two hydrogen atoms in both cases, regardless of the complexity of the rest of the molecule.

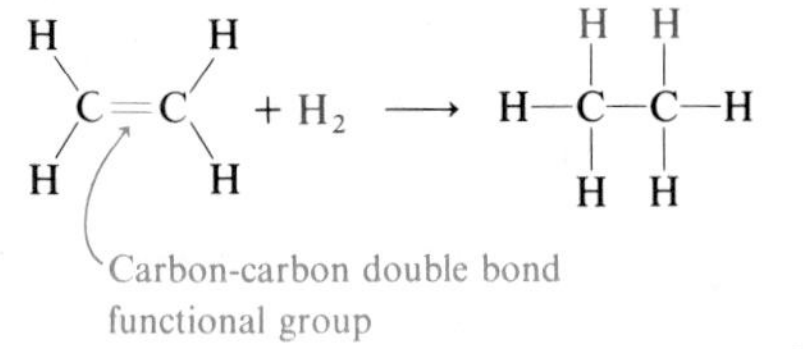

(a) Ethylene

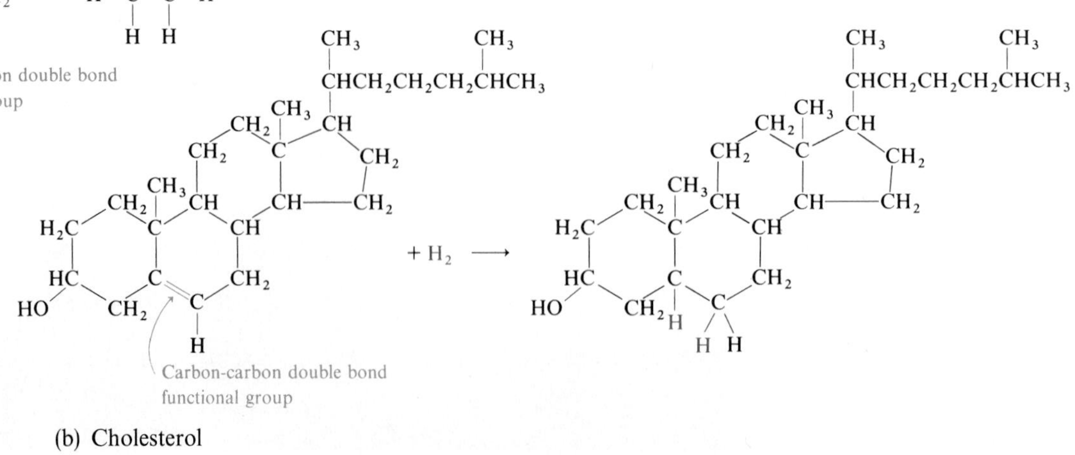

(b) Cholesterol

Table 1.1 Some Important Families of Organic Molecules

Family Name	Functional Group Structure[a]	Simple Example	Name Ending
Alkane	(contains only C—H and C—C single bonds)	CH_3CH_3 ethane	*-ane*
Alkene	C=C	$H_2C=CH_2$ ethylene	*-ene*
Alkyne	—C≡C—	H—C≡C—H acetylene (ethyne)	*-yne*
Arene	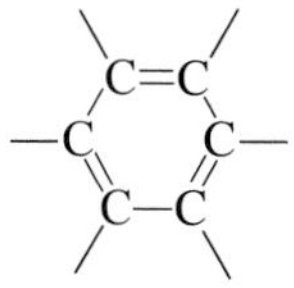 (benzene ring structure)	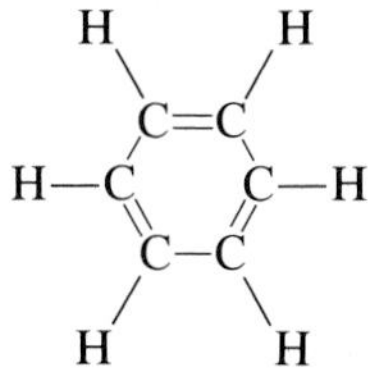 benzene	none
Alkyl halide[b]	—C—X	CH_3—Cl methyl chloride	none
Alcohol	—C—O—H	CH_3—OH methyl alcohol (methanol)	*-ol*
Ether	—C—O—C—	CH_3—O—CH_3 dimethyl ether	none
Amine	—N—H, —N—H, —N—	CH_3—NH_2 methylamine	*-amine*
Aldehyde	—C(=O)—H	CH_3—C(=O)—H acetaldehyde (ethanal)	*-al*
Ketone	—C—C(=O)—C—	CH_3—C(=O)—CH_3 acetone	*-one*
Carboxylic acid	—C(=O)—OH	CH_3—C(=O)—OH acetic acid	*-ic acid*
Anhydride	—C(=O)—O—C(=O)—	CH_3—C(=O)—O—C(=O)—CH_3 acetic anhydride	none
Ester	—C(=O)—O—	CH_3—C(=O)—O—CH_3 methyl acetate	*-ate*
Amide	—C(=O)—NH_2, —C(=O)—N—H, —C(=O)—N—	CH_3—C(=O)—NH_2 acetamide	*-amide*

[a] The bonds whose connections aren't specified are assumed to be attached to carbon or hydrogen atoms in the rest of the molecule.
[b] X = F, Cl, Br, or I.

Hydrocarbon   A compound that contains only carbon and hydrogen.

In this and the next chapter, we'll describe the chemistry of **hydrocarbons,** organic compounds containing only carbon and hydrogen. Members of the first four families listed in Table 1.1—the alkanes, alkenes, alkynes, and arenes—are hydrocarbons when no other functional groups are present in their molecules. Alkanes are the one family of organic compounds that contain no functional groups, for alkanes are constructed entirely of carbon and hydrogen atoms joined by single bonds. As you'll see, the absence of functional groups makes alkanes relatively unreactive.

Every member of the other families is essentially an alkane in which one or more hydrogen atoms have been replaced by functional groups. Some examples of compounds formed by replacement of a hydrogen atom in methane, the simplest alkane, are

$$CH_4 \qquad CH_3OH \qquad CH_3NH_2 \qquad CH_3\overset{\overset{\displaystyle O}{\|}}{C}OH$$

A hydrocarbon   An alcohol   An amine   A carboxylic acid

Much of the chemistry discussed in this and the next five chapters is the chemistry of the families listed in Table 1.1, so it's best to memorize their names and structures now. Note that they fall into three groups: (1) the hydrocarbons of the first four families; (2) those whose functional groups contain only single bonds (alkyl halides, alcohols, ethers, and amines); and (3) those whose functional groups contain a carbon–oxygen double bond (aldehydes, ketones, carboxylic acids and anhydrides, esters, and amides).

Practice Problems

1.1 Locate and identify the functional groups in these molecules:
(a) lactic acid, from sour milk

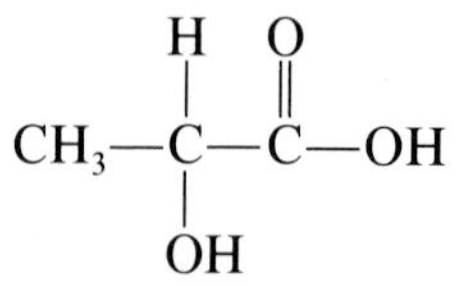

$$CH_3-\overset{\overset{\displaystyle H}{|}}{\underset{\underset{\displaystyle OH}{|}}{C}}-\overset{\overset{\displaystyle O}{\|}}{C}-OH$$

(b) methyl methacrylate, used in making Lucite and Plexiglas

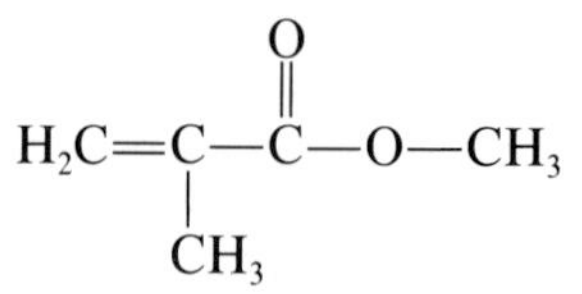

$$H_2C=\overset{\overset{\displaystyle}{}}{\underset{\underset{\displaystyle CH_3}{|}}{C}}-\overset{\overset{\displaystyle O}{\|}}{C}-O-CH_3$$

1.2 Propose structures for molecules that fit these descriptions:
(a) C_2H_4O containing an aldehyde functional group
(b) $C_3H_6O_2$ containing a carboxylic acid functional group

1.3 THE STRUCTURE OF ORGANIC MOLECULES: ALKANES AND THEIR ISOMERS

Alkane A compound that contains only carbon and hydrogen and has only single bonds.

Molecules that contain only carbon and hydrogen and that have only single bonds belong to the family of organic molecules called **alkanes.** If we imagine ways that one carbon and four hydrogens can combine, there is only one possibility: methane, CH_4. If we imagine ways that two carbons and six hydrogens can combine, only ethane, CH_3CH_3, is possible; and if we imagine the combination of three carbons with eight hydrogens, only propane, $CH_3CH_2CH_3$, is possible.

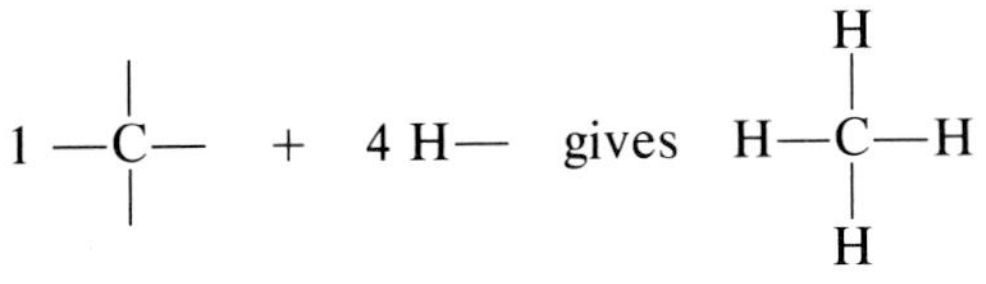

$$1 \ -\overset{\textstyle |}{\underset{\textstyle |}{C}}- \ + \ 4 \, \text{H}- \ \ \text{gives} \quad \text{H}-\overset{\textstyle \text{H}}{\underset{\textstyle \text{H}}{\text{C}}}-\text{H} \qquad \text{Methane}$$

$$2 \ -\overset{\textstyle |}{\underset{\textstyle |}{C}}- \ + \ 6 \, \text{H}- \ \ \text{gives} \quad \text{H}-\overset{\textstyle \text{H}}{\underset{\textstyle \text{H}}{\text{C}}}-\overset{\textstyle \text{H}}{\underset{\textstyle \text{H}}{\text{C}}}-\text{H} \qquad \text{Ethane}$$

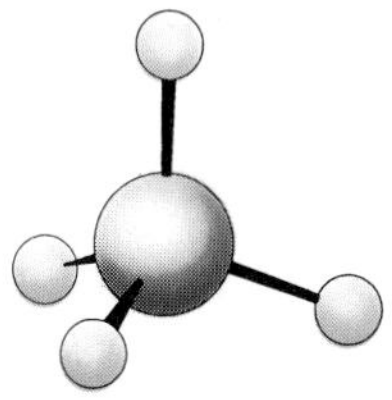

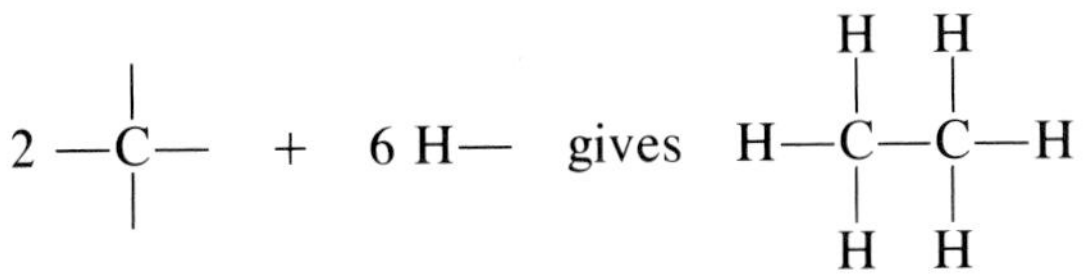

$$3 \ -\overset{\textstyle |}{\underset{\textstyle |}{C}}- \ + \ 8 \, \text{H}- \ \ \text{gives} \quad \text{H}-\overset{\textstyle \text{H}}{\underset{\textstyle \text{H}}{\text{C}}}-\overset{\textstyle \text{H}}{\underset{\textstyle \text{H}}{\text{C}}}-\overset{\textstyle \text{H}}{\underset{\textstyle \text{H}}{\text{C}}}-\text{H} \qquad \text{Propane}$$

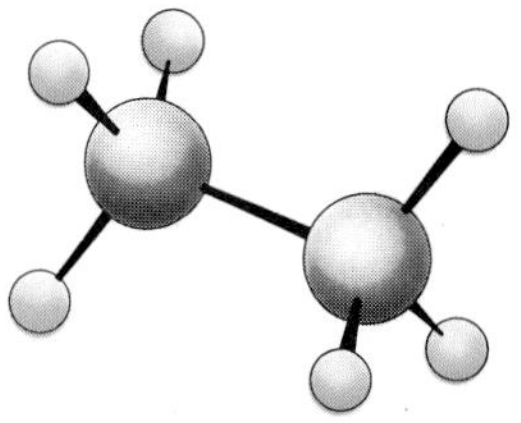

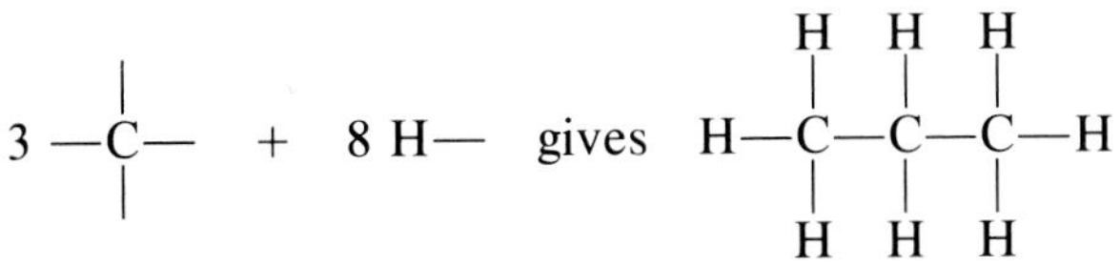

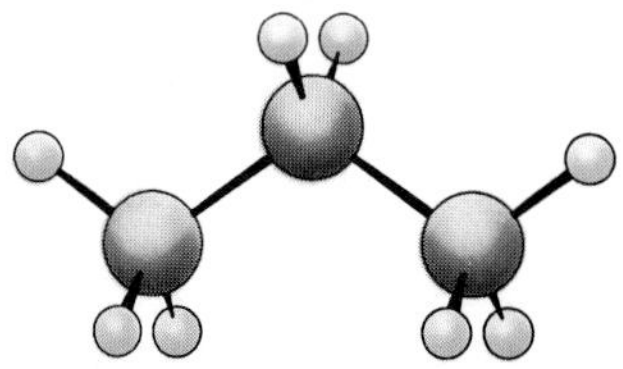

Straight-chain alkane An alkane that has all its carbon atoms connected in a row.

Branched-chain alkane An alkane that has a branching connection of carbon atoms along its chain.

If larger numbers of carbons and hydrogens combine, *more than one kind of molecule can be formed.* There are two ways in which molecules with the formula C_4H_{10} can be formed: The four carbons either can be in a row or can have a branched arrangement. Similarly, there are *three* ways in which molecules with the formula C_5H_{12} can be formed; and so on for larger alkanes. Compounds with all their carbons connected in a row are called **straight-chain alkanes,** whereas those with a branching connection of carbons are called **branched-chain alkanes.** Note that in a straight-chain alkane, you can draw a line through all the carbon atoms without lifting your pencil from the paper. In a branched-chain alkane, however, you must either lift your pencil from the paper or retrace your steps in order to draw a line through all the carbons.

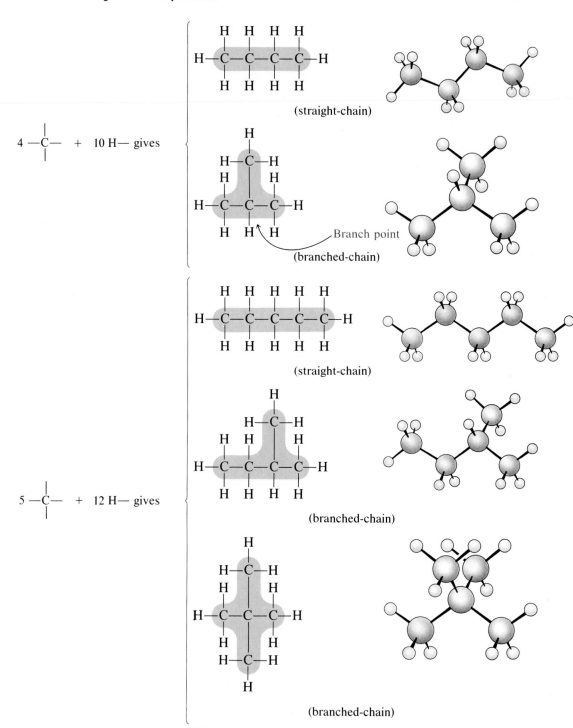

The two different compounds with the formula C_4H_{10} and the three different compounds with the formula C_5H_{12} are called *isomers*. Specifically, these are **constitutional isomers,** compounds with the same molecular formula but with their atoms connected in different arrangements. As Table 1.2 shows, the number of possible alkane isomers grows rapidly as the number of carbon atoms increases.

Isomers, constitutional
Compounds with the same molecular formula but with different connections between atoms.

Table 1.2   Numbers of Possible Alkane Isomers

Formula	Number of Isomers	Formula	Number of Isomers
C_6H_{14}	5	$C_{10}H_{22}$	75
C_7H_{16}	9	$C_{20}H_{42}$	366,319
C_8H_{18}	18	$C_{30}H_{62}$	4,111,846,763
C_9H_{20}	35	$C_{40}H_{82}$	62,491,178,805,831

It's important to realize that different constitutional isomers are completely different chemical compounds. They have different structures, different physical properties such as melting point and boiling point, and potentially different physiological properties. For example, ethyl alcohol and dimethyl ether both have the formula C_2H_6O. Yet ethyl alcohol is a liquid central nervous system depressant with a boiling point of 78.5°C, whereas dimethyl ether is a gaseous substance with a boiling point of -23°C (Table 1.3). You can see why molecular formulas such as C_2H_6O aren't very useful in organic chemistry.

Practice Problems   **1.3**   Draw the straight-chain isomer with the formula C_7H_{16}.

1.4   Draw two branched-chain isomers with the formula C_7H_{16}.

Table 1.3   Some Properties of Ethyl Alcohol and Dimethyl Ether

Name and Molecular Formula	Structure	Boiling Point	Melting Point	Physiological Activity
Ethyl alcohol C_2H_6O	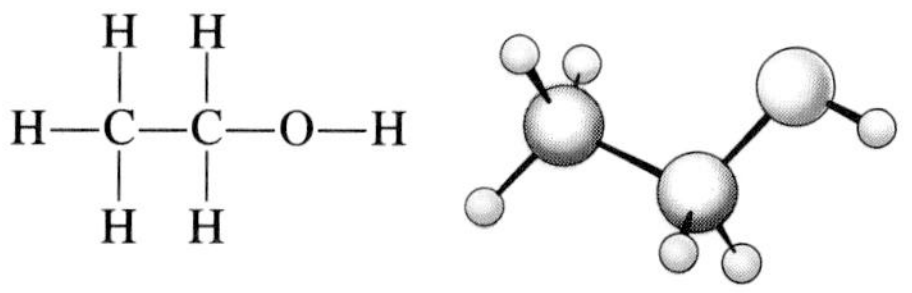	78.5°C	-117.3°C	Central-nervous-system depressant
Dimethyl ether C_2H_6O	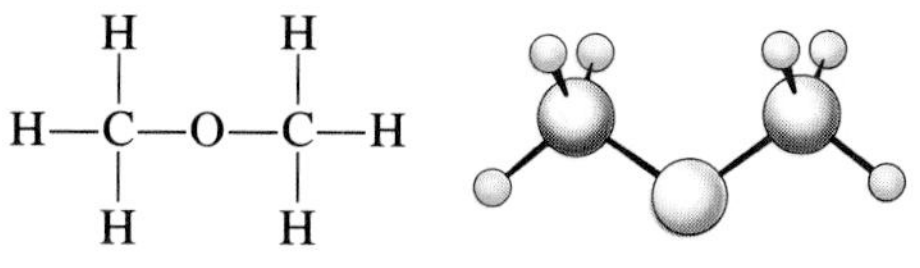	-23°C	-138.5°C	—

AN APPLICATION: NATURAL VS. SYNTHETIC

Before organic chemistry, we had only substances from plants and animals—*natural products*—for treating diseases, perfuming ourselves, flavoring foods, gluing things together, and hundreds of other daily applications. *Essential oils,* for example, are water-insoluble mixtures distilled or extracted from plants. Their names are romantic and alluring: oil of bergamot, oil of sweet bay, oil of rose, and oil of lavender, all of which are used in making perfume. Oil of thyme, oil of sweet almond, and oil of pine needles have medical applications.

Many natural products were first used without any knowledge of their chemical compositions. Sometimes this was a mistake: Oil of bitter almond that's not purified contains hydrogen cyanide. As organic chemistry developed, chemists learned how to work out the structures of the compounds in natural products. The disease-curing miracle of penicillin, the first widely used antibiotic, was discovered in a mold in 1928, but its chemical structure was not determined until about 20 years later. Today there is a revival of interest in folk medicines and an effort to identify the chemical compounds responsible for their effectiveness.

Once a structure is known, organic chemists seek to determine how the compound might be synthesized. If the starting materials are inexpensive enough and the process is simple enough, it may be more economical to manufacture a compound than to isolate it from a natural substance. In the case of penicillin, a complete synthesis was achieved in 1958, and many structurally similar compounds that are also good antibiotics have been made in the laboratory. It is not yet practical, however, to manufacture any of the penicillins by complete synthesis. Instead, we have *semisynthetic* penicillins in which the molecular structure of a mold product is controlled by adding appropriate chemicals to the medium in which the mold grows.

1.4 DRAWING ORGANIC STRUCTURES

Structural formula A formula that shows how atoms are connected to each other.

Condensed structure A structure in which central atoms and the atoms connected to them are written as groups, e.g., $CH_3CH_2CH_3$.

Drawing **structural formulas** like those in Table 1.3 for ethyl alcohol and dimethyl ether is both time-consuming and awkward, even for relatively small molecules. **Condensed structures** are a convenient compromise because they're simpler but show the essential information about which functional groups are present and how atoms are connected. In condensed structures, carbon–hydrogen and carbon–carbon single bonds aren't shown; rather, they're understood to be there. If a carbon atom has three hydrogens bonded to it, we write CH_3; if the carbon has two hydrogens bonded to it, we write CH_2; and so on. For example, the four-carbon straight-chain compound (called *butane*) and its branched-chain isomer (called *2-methylpropane*) can be written as the following condensed structures:

$$\text{H}-\overset{\overset{\displaystyle\text{H}}{|}}{\underset{\underset{\displaystyle\text{H}}{|}}{\text{C}}}-\overset{\overset{\displaystyle\text{H}}{|}}{\underset{\underset{\displaystyle\text{H}}{|}}{\text{C}}}-\overset{\overset{\displaystyle\text{H}}{|}}{\underset{\underset{\displaystyle\text{H}}{|}}{\text{C}}}-\overset{\overset{\displaystyle\text{H}}{|}}{\underset{\underset{\displaystyle\text{H}}{|}}{\text{C}}}-\text{H} = CH_3CH_2CH_2CH_3$$

Butane

$$\text{H}-\overset{\overset{\displaystyle\text{H}}{|}}{\underset{\underset{\displaystyle\text{H}}{|}}{\text{C}}}-\overset{\overset{\overset{\overset{\displaystyle\text{H}}{|}}{\text{H}-\text{C}-\text{H}}}{|}}{\underset{\underset{\displaystyle\text{H}}{|}}{\text{C}}}-\overset{\overset{\displaystyle\text{H}}{|}}{\underset{\underset{\displaystyle\text{H}}{|}}{\text{C}}}-\text{H} = CH_3CHCH_3$$

2-Methylpropane

Large quantities of many of nature's chemicals, however, such as vitamin C and caffeine, are produced synthetically each year, as are large quantities of medicines nature has never produced. Some individuals demand vitamins only from natural sources, assuming that "natural" is somehow better. Do you think it would be possi- ble to devise a chemical test to distinguish between two pure samples of vitamin C, one isolated from rose hips and one made by chemical synthesis? Does anyone ever worry about whether the caffeine in their cola drink came from tea leaves or from a chemical manufacturing plant?

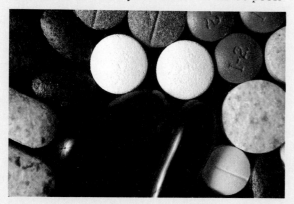 

Two sources of the chemicals known as vitamins.

Note that the horizontal bonds between carbons aren't shown—the CH_3 and CH_2 units are simply placed next to each other—but that the vertical bond in 2-methylpropane is shown for clarity. Occasionally, as a further simplification, a row of CH_2 groups is shown by parentheses, with a subscript equal to the number of groups:

$$CH_3CH_2CH_2CH_2CH_2CH_3 = CH_3(CH_2)_4CH_3$$

Practice Problem **1.5** Draw the three isomers of C_5H_{12} as condensed structures.

(a) 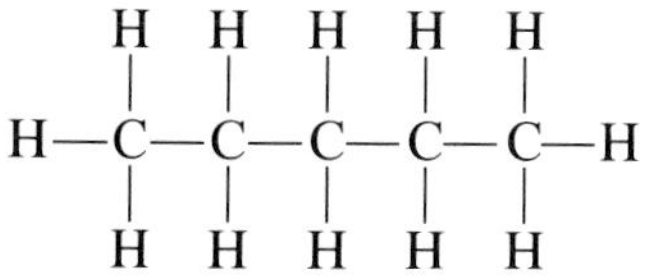 Pentane

(b) 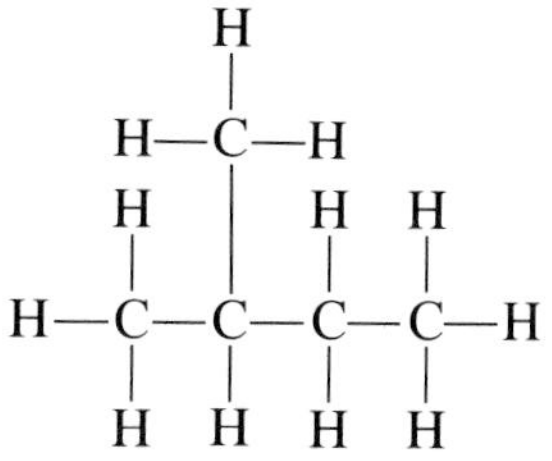 2-Methylbutane

(c) 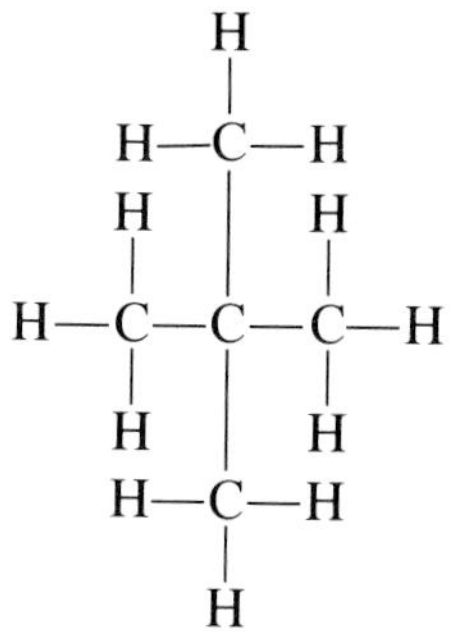 2,2-Dimethylpropane

1.5 THE SHAPES OF ORGANIC MOLECULES

Although every carbon atom in an alkane has its four bonds pointing toward the four corners of a tetrahedron, chemists don't usually worry about exact three-dimensional shapes when writing condensed structures. For example, the straight-chain four-carbon alkane, butane, might be represented by any of the structures shown below. These structures don't imply any particular three-dimensional shape for butane; they only show that butane has a continuous chain of four carbon atoms and indicate the *connections* between atoms without specifying geometry.

Some drawings of butane, C_4H_{10}

$$\underset{\overset{\displaystyle |}{CH_2}\ \ \underset{}{CH_3}}{\overset{CH_3\ \ CH_2}{\diagdown\diagup\diagdown}} \quad \text{or} \quad \underset{\overset{\displaystyle |}{CH_2CH_2CH_3}}{CH_3} \quad \text{or} \quad \underset{\overset{\displaystyle |}{CH_2CH_3}}{CH_2CH_3} \quad \text{or} \quad \underset{\overset{\displaystyle |}{CH_3}}{CH_3CH_2CH_2}$$

> **Conformation** The exact three-dimensional shape of a molecule at any given instant.

In fact, butane has no one single shape because *rotation* is possible around carbon–carbon single bonds. The two parts of a molecule joined by a carbon–carbon single bond are free to spin around the bond, giving rise to an infinite number of possible three-dimensional structures, or **conformations.** A given butane molecule might be fully extended at one instant (Figure 1.3a) but be twisted an instant later (Figure 1.3b). An actual sample of butane contains a great many molecules that are constantly changing shape. At any given instant, however, most of the molecules have the less crowded extended conformation shown in Figure 1.3a. The same is true for all other alkanes: At any given instant most molecules are in the least-crowded conformation.

So long as any two structures show identical connections between atoms, they represent identical compounds, no matter how the structures are drawn. Sometimes you have to mentally rotate structures to see whether they're the same or different. To ''see'' that the following two structures represent the same compound rather than two isomers, picture one of them turned end for end (picture the red CH_3 groups on the same end).

$$\underset{\overset{\displaystyle |}{\underset{\overset{\displaystyle |}{OH}}{CH_2}}}{CH_3CHCH_2CH_2CH_3} \qquad \underset{\overset{\displaystyle |}{\underset{\overset{\displaystyle |}{OH}}{CH_2}}}{CH_3CH_2CH_2CHCH_3}$$

Figure 1.3
Some possible conformations of butane. There are many other conformations as well.

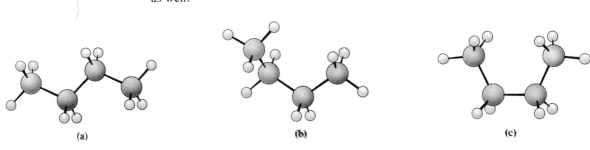

(a)   (b)   (c)

Solved Problem 1.1   The following molecules have the same formula, C_7H_{16}. Which of the structures represent the same molecule?

$$\overset{\displaystyle CH_3}{\underset{\displaystyle |}{}}$$

(a) $CH_3\overset{|}{C}HCH_2CH_2CH_2CH_3$ (b) $CH_3CH_2CH_2CH_2\overset{|}{C}HCH_3$

$$\overset{\displaystyle CH_3}{\underset{\displaystyle |}{}}$$

(c) $CH_3CH_2CH_2\overset{|}{C}HCH_2CH_3$

Solution   The important point in determining whether two structures are identical is to pay attention to the order of connection between atoms. Don't get confused by the apparent differences caused by writing a structure right to left versus left to right. In this example, molecule (a) has a straight chain of six carbons with a —CH_3 branch on the second carbon from the end. Molecule (b) also has a straight chain of six carbons with a —CH_3 branch on the second carbon from the end and is therefore identical to (a). The only difference between (a) and (b) is that one is written "forward" and one is written "backward." Molecule (c), by contrast, has a straight chain of six carbons with a —CH_3 branch on the *third* carbon from the end and is therefore an isomer of (a) and (b).

Solved Problem 1.2   Are the following pairs of compounds isomers, the same compound, or different compounds?

$$\overset{\displaystyle CH_3}{\underset{\displaystyle |}{}} \qquad \overset{\displaystyle CH_3}{\underset{\displaystyle |}{}}$$

(a) $CH_3\overset{|}{C}HCH_2CH_2$ $CH_3\overset{|}{C}HCH_2CH_2CH_3$

$$\underset{\displaystyle CH_3}{\overset{\displaystyle |}{}}$$

$$\overset{\displaystyle CH_2CH_3}{\underset{\displaystyle |}{}}$$

(b) $CH_3CH_2\overset{|}{C}HCH_3$ $CH_3\overset{|}{C}HCH_2$

$$\underset{\displaystyle CH_2CH_3}{\overset{\displaystyle |}{}} \qquad \underset{\displaystyle CH_3}{\overset{\displaystyle |}{}}$$

$$\overset{\displaystyle O}{\underset{\displaystyle ||}{}}$$

(c) $CH_3CH_2OCH_3$ $CH_3CH_2\overset{||}{C}H$

Solution   (a) First, compare molecular formulas to determine whether the compounds are the same or different. In this case, both compounds have the same molecular formula (C_6H_{14}). Next, examine their structures to see if they are the same compound or isomers. To do this, locate the longest straight carbon chain in each and then compare the locations of the **substituents.**

Substituent   A group attached to a root compound.

$$\overset{\displaystyle CH_3}{\underset{\displaystyle |}{}} \qquad \overset{\displaystyle CH_3}{\underset{\displaystyle |}{}}$$

$CH_3\overset{|}{C}HCH_2CH_2$, $CH_3\overset{|}{C}HCH_2CH_2CH_3$

$$\underset{\displaystyle CH_3}{\overset{\displaystyle |}{}}$$

Since the CH_3— group is on the second carbon from the end of a five-carbon chain in both cases, these compounds are identical.

(b) Both compounds have the same molecular formula (C_6H_{14}), and the longest chain in each is five carbon atoms. A comparison shows, however, that the CH$_3$ group is on the middle carbon atom in one structure and on the second carbon atom in the other. These compounds are isomers.

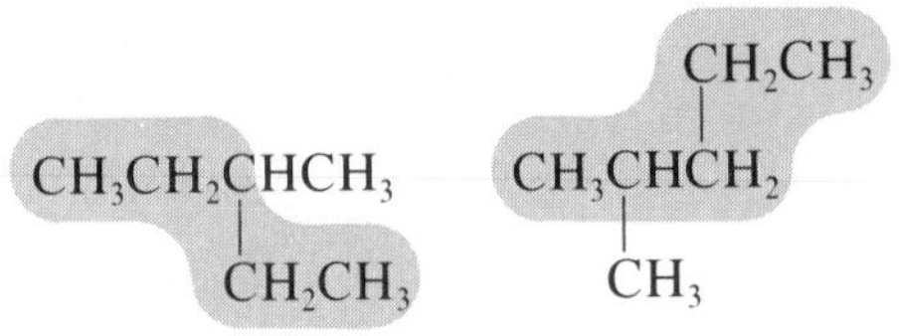

$$CH_3CH_2CHCH_3$$
$$\overset{|}{CH_2CH_3}$$

$$\overset{CH_2CH_3}{\overset{|}{CH_3CHCH_2}}$$
$$\overset{|}{CH_3}$$

(c) Each of these compounds has three carbon atoms and one oxygen atom, but the numbers of hydrogen atoms differ (eight H atoms and six H atoms, respectively). These are different compounds, but they are not isomers.

AN APPLICATION: DISPLAYING MOLECULAR SHAPES

Molecular shape is critical to the proper functioning of all biological molecules. The tiniest difference in shape between two compounds can cause them to behave differently or to have different physiological effects in the body. It's therefore critical that chemists have techniques available both for determining molecular shapes with great precision and for visualizing these shapes in useful and manageable ways.

Three-dimensional shapes of molecules are determined by *X ray crystallography*, a technique that allows us to "see" molecules in a crystal using X ray waves rather than light waves. The molecular "picture" obtained by X ray crystallography looks at first like a series of regularly spaced dark spots on a photographic film. After computerized manipulation of the data, however, recognizable molecules can be drawn. Relatively small molecules like morphine are usually displayed on paper, but enormous biological molecules like immunoglobulins are best displayed on computer terminals where their structures can be enlarged, rotated, and otherwise manipulated for the best view.

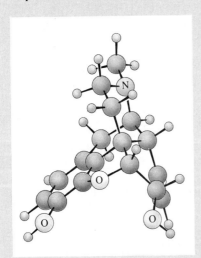

(a)

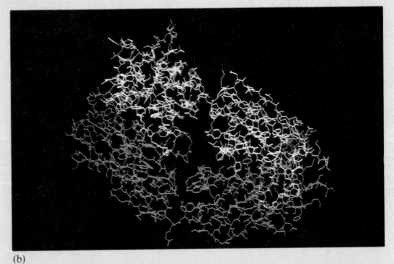

(b)

Computer-generated shapes of (a) morphine and (b) an immunoglobulin, one of the antibodies in blood that protect us from harmful invaders such as bacteria and viruses.

Practice Problems ***1.6*** Which of the following three structures are identical?

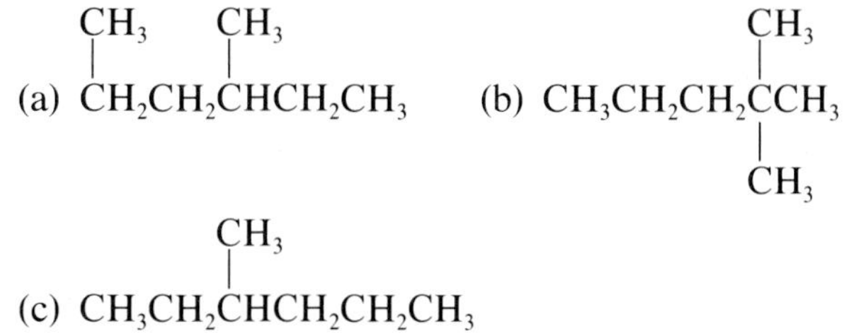

$$\underset{\substack{| \\ \text{(a)} \ \text{CH}_2\text{CH}_2\text{CHCH}_2\text{CH}_3}}{\overset{\substack{\text{CH}_3 \qquad \text{CH}_3}}{}}$$

(a) CH₂CH₂CHCH₂CH₃   (b) CH₃CH₂CH₂CCH₃

(c) CH₃CH₂CHCH₂CH₂CH₃

1.7 According to Table 1.2, there are five isomers with the formula C₆H₁₄. Draw as many as you can.

1.8 Using toothpicks for bonds and two marshmallows to represent the carbon atoms, build a model of an ethane molecule, CH₃CH₃. (Hydrogen atoms can be "understood" to be at the ends of the toothpicks coming out of the marshmallows.) Rotate the model around a C—C bond and note the changing relationships between hydrogens on neighboring carbons.

1.6 NAMING ALKANES

In earlier times when relatively few pure organic chemicals were known, new compounds were named at the whim of their discoverer. Thus, urea is a crystalline substance first isolated from urine, and morphine is a plant extract that's a sedative and is named after Morpheus, the Greek god of dreams. As more and more compounds became known, however, the need for **nomenclature**—a systematic method of naming compounds—became apparent.

Nomenclature A system for naming chemical compounds.

The nomenclature now used is that devised by the International Union of Pure and Applied Chemistry: the IUPAC system (pronounced **eye**-you-pac). In the IUPAC system for organic compounds, a chemical name has three parts: prefix, root, and suffix. The root specifies the overall size of the molecule by telling how many carbon atoms are present in the longest continuous chain, the suffix identifies what family the molecule belongs to, and the prefix specifies the location of functional groups and other substituents on the chain:

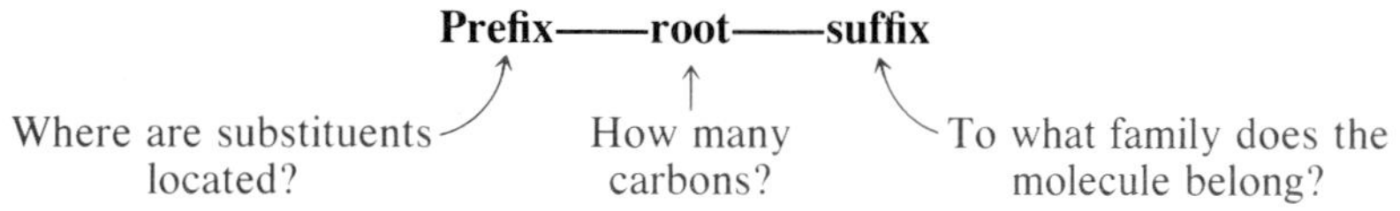

Prefix——root——suffix

Where are substituents located?   How many carbons?   To what family does the molecule belong?

Straight-chain alkanes are named simply by counting the number of carbon atoms in the chain and adding the family suffix *-ane* to the root name to give the name of the parent compound. With the exception of the first four compounds—*meth*ane, *eth*ane, *prop*ane, and *but*ane—whose root names have historical origins, the alkanes are named from Greek numbers according to the number of carbons present. Thus, *pent*ane is the five-carbon alkane, *hex*ane is

Alkyl group The part of an alkane that remains when one hydrogen atom is removed.

Methyl group —CH_3, the alkyl group derived from methane.

Ethyl group —CH_2CH_3, the alkyl group derived from ethane.

the six-carbon alkane, and so on. The roots shown by italics in Table 1.4 are used in naming many types of organic compounds. These first 10 alkane names should be memorized.

Substituents that branch off a main chain are called **alkyl groups.** Each different alkyl group can be thought of as the part of an alkane that remains when one hydrogen atom is removed. For example, removal of one hydrogen from methane gives the **methyl group,** —CH_3, and removal of one hydrogen from ethane gives the **ethyl group,** —CH_2CH_3. Notice that these alkyl groups are named simply by replacing the -*ane* ending of the parent alkane with a -*yl* ending.

Methane

$$H-\underset{\underset{H}{|}}{\overset{\overset{H}{|}}{C}}-H \quad \xrightarrow{\text{remove one H}} \quad -\underset{\underset{H}{|}}{\overset{\overset{H}{|}}{C}}-H \quad = \quad -CH_3 \quad \text{(A methyl group)}$$

Ethane

$$H-\underset{\underset{H}{|}}{\overset{\overset{H}{|}}{C}}-\underset{\underset{H}{|}}{\overset{\overset{H}{|}}{C}}-H \quad \xrightarrow{\text{remove one H}} \quad -\underset{\underset{H}{|}}{\overset{\overset{H}{|}}{C}}-\underset{\underset{H}{|}}{\overset{\overset{H}{|}}{C}}-H \quad = \quad -CH_2CH_3 \quad \text{(An ethyl group)}$$

Methane and ethane are special because each has only one kind of hydrogen. It doesn't matter which of the four methane hydrogens is removed because all four are equivalent. Thus, there is only one possible kind of methyl group. Similarly, it doesn't matter which of the six ethane hydrogens is removed, and only one kind of ethyl group is possible.

The situation is more complex for larger alkanes that contain more than one kind of hydrogen. For example, propane has two different kinds of hydrogens. Removal of any one of the six hydrogens attached to an end carbon yields a straight-chain propyl group called *n*-**propyl,** whereas removal of one of the two hydrogens attached to the central carbon yields a branched-chain propyl group called **isopropyl.** (The "*n*" prefix in *n*-propyl stands for *normal*, meaning straight-chain, and is used with other alkyl groups, too.)

***n*-Propyl group**
—$CH_2CH_2CH_3$, the alkyl group derived by removing a hydrogen atom from an end carbon of propane.

Isopropyl group
—$CH(CH_3)_2$, the alkyl group derived by removing a hydrogen atom from the central carbon of propane.

Table 1.4   Names of Straight-Chain Alkanes

No. of Carbons	Structure	Name
1	CH_4	*Meth*ane
2	CH_3CH_3	*Eth*ane
3	$CH_3CH_2CH_3$	*Prop*ane
4	$CH_3CH_2CH_2CH_3$	*But*ane
5	$CH_3CH_2CH_2CH_2CH_3$	*Pent*ane
6	$CH_3CH_2CH_2CH_2CH_2CH_3$	*Hex*ane
7	$CH_3CH_2CH_2CH_2CH_2CH_2CH_3$	*Hept*ane
8	$CH_3CH_2CH_2CH_2CH_2CH_2CH_2CH_3$	*Oct*ane
9	$CH_3CH_2CH_2CH_2CH_2CH_2CH_2CH_2CH_3$	*Non*ane
10	$CH_3CH_2CH_2CH_2CH_2CH_2CH_2CH_2CH_2CH_3$	*Dec*ane

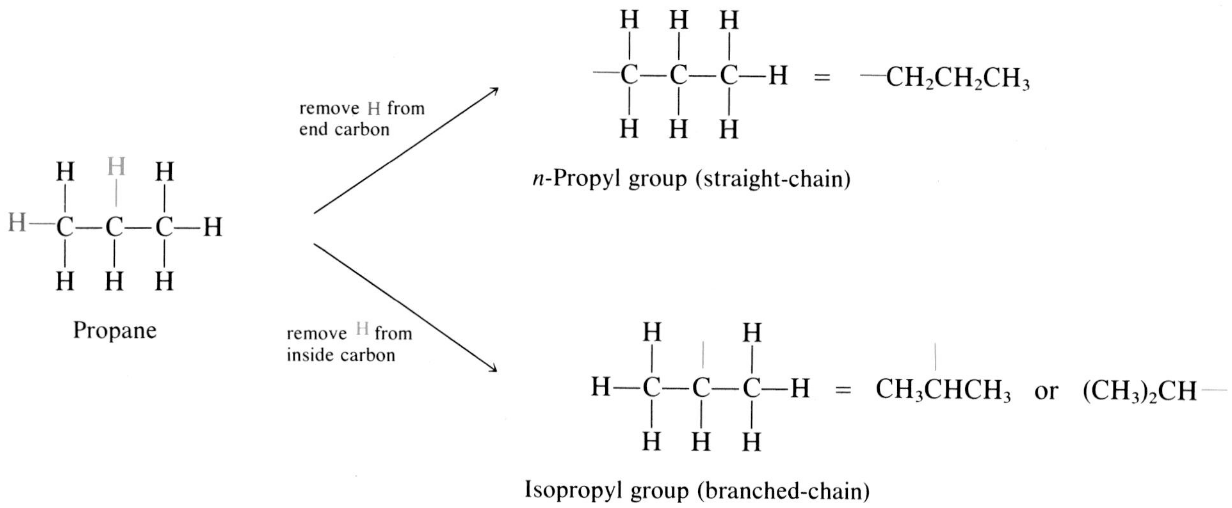

$$\begin{array}{c}\text{remove H from}\\\text{end carbon}\end{array} \nearrow \quad \overset{\overset{\displaystyle H}{|}}{\underset{\underset{\displaystyle H}{|}}{-C}}-\overset{\overset{\displaystyle H}{|}}{\underset{\underset{\displaystyle H}{|}}{C}}-\overset{\overset{\displaystyle H}{|}}{\underset{\underset{\displaystyle H}{|}}{C}}-H \;=\; -CH_2CH_2CH_3$$

n-Propyl group (straight-chain)

$$H-\overset{\overset{\displaystyle H}{|}}{\underset{\underset{\displaystyle H}{|}}{C}}-\overset{\overset{\displaystyle H}{|}}{\underset{\underset{\displaystyle H}{|}}{C}}-\overset{\overset{\displaystyle H}{|}}{\underset{\underset{\displaystyle H}{|}}{C}}-H$$

Propane

$$\begin{array}{c}\text{remove H from}\\\text{inside carbon}\end{array} \searrow \quad H-\overset{\overset{\displaystyle H}{|}}{\underset{\underset{\displaystyle H}{|}}{C}}-\overset{\displaystyle H}{\underset{\underset{\displaystyle H}{|}}{C}}-\overset{\overset{\displaystyle H}{|}}{\underset{\underset{\displaystyle H}{|}}{C}}-H \;=\; CH_3CHCH_3 \;\text{ or }\; (CH_3)_2CH-$$

Isopropyl group (branched-chain)

It's important to realize that alkyl groups themselves are not compounds and that the "removal" of a hydrogen from an alkane is just a way of looking at things, not a chemical reaction. The names of some common alkyl groups are listed in Table 1.5.

Branched-chain alkanes are named by following four steps:

Step 1. Name the main chain. Find the *longest continuous chain of carbons* present in the molecule and name the chain using the root for the number of carbons it contains. The longest chain may not always be obvious from the manner of writing; you may have to "turn corners" to find it.

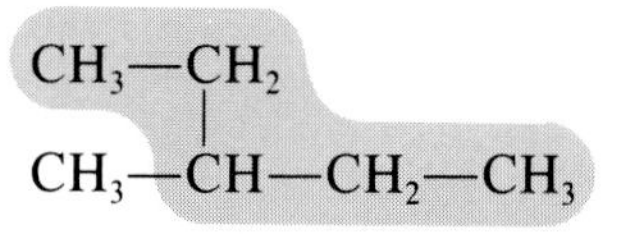

$$\begin{array}{l} CH_3-CH_2 \\ \qquad\qquad | \\ CH_3-CH-CH_2-CH_3 \end{array}$$

Name as a substituted pentane, not as a substituted butane, because the *longest* chain has five carbons.

Step 2. Number the carbon atoms in the main chain. Beginning at the end nearer the first branch point, number each carbon atom in the main chain:

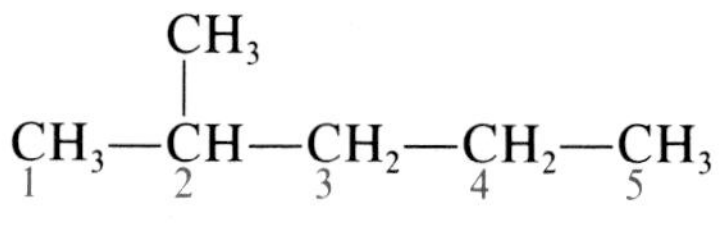

$$\begin{array}{c} CH_3 \\ | \\ \underset{1}{CH_3}-\underset{2}{CH}-\underset{3}{CH_2}-\underset{4}{CH_2}-\underset{5}{CH_3} \end{array}$$

The first (and only) branch occurs at C2 if we start numbering from the left, but would occur at C4 if we started from the right by mistake.

Table 1.5   Some Common Alkyl Groups[a]

			CH₃
CH₃—	CH₃CH₂—	CH₃CH₂CH₂—	CH₃CH—
Methyl	Ethyl	*n*-Propyl	Isopropyl
		CH₃	
CH₃CH₂CH₂CH₂—	CH₃CHCH₂CH₃	CH₃CHCH₂—	CH₃CCH₃
			CH₃
n-Butyl	*sec*-Butyl	Isobutyl	*tert*-Butyl

[a]The red bond shows the connection to the rest of the molecule.

Step 3. Identify and number the branching substituents. Assign a number to each branching substituent on the main chain according to its point of attachment.

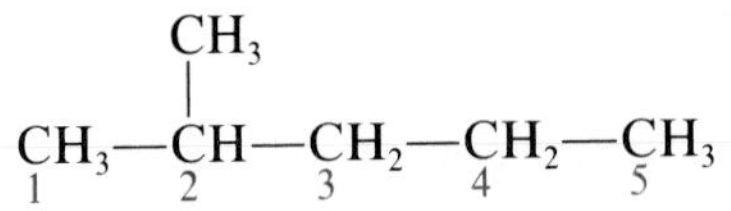

$$
\begin{array}{c}
\quad\quad\quad CH_3 \\
\quad\quad\quad | \\
CH_3{-}\underset{2}{CH}{-}\underset{3}{CH_2}{-}\underset{4}{CH_2}{-}\underset{5}{CH_3} \\
\underset{1}{}
\end{array}
$$

The main chain is a pentane. There is one —CH$_3$ substituent group connected to C2 of the chain.

If there are two substituents on the same carbon, assign the same number to both of them. There must always be as many numbers in the name as there are substituents.

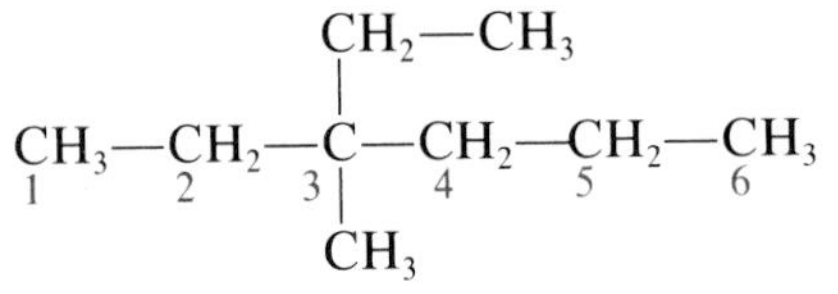

$$
\begin{array}{c}
\quad\quad\quad CH_2{-}CH_3 \\
\quad\quad\quad | \\
CH_3{-}\underset{2}{CH_2}{-}\underset{3}{C}{-}\underset{4}{CH_2}{-}\underset{5}{CH_2}{-}\underset{6}{CH_3} \\
\quad\quad\quad | \\
\quad\quad\quad CH_3
\end{array}
$$

The main chain is a hexane. There are two substituents, a —CH$_3$ and a —CH$_2$CH$_3$, both connected to C3 of the chain.

Step 4. Write the name as a single word. Use hyphens to separate the numbers from the different prefixes and use commas to separate numbers if necessary. If two or more different substituent groups are present, cite them in alphabetical order. If two or more identical substituents are present, use one of the prefixes *di-, tri-, tetra-,* and so forth, but don't use these prefixes for alphabetizing purposes.

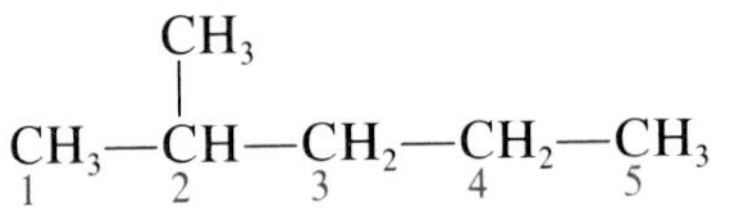

$$
\begin{array}{c}
\quad\quad CH_3 \\
\quad\quad | \\
CH_3{-}\underset{2}{CH}{-}\underset{3}{CH_2}{-}\underset{4}{CH_2}{-}\underset{5}{CH_3} \\
\underset{1}{}
\end{array}
$$

2-Methylpentane (a five-carbon main chain with a 2-methyl substituent)

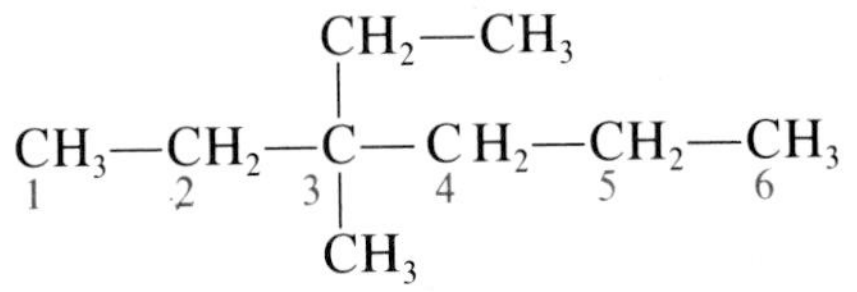

$$
\begin{array}{c}
\quad\quad\quad CH_2{-}CH_3 \\
\quad\quad\quad | \\
CH_3{-}\underset{2}{CH_2}{-}\underset{3}{C}{-}\underset{4}{CH_2}{-}\underset{5}{CH_2}{-}\underset{6}{CH_3} \\
\quad\quad\quad | \\
\quad\quad\quad CH_3
\end{array}
$$

3-Ethyl-3-methylhexane (a six-carbon main chain with 3-ethyl and 3-methyl substituents cited alphabetically)

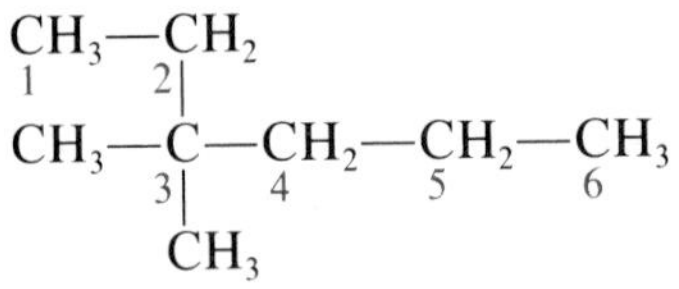

$$
\begin{array}{c}
\underset{1}{CH_3}{-}\underset{2}{CH_2} \\
\quad\quad | \\
CH_3{-}\underset{3}{C}{-}\underset{4}{CH_2}{-}\underset{5}{CH_2}{-}\underset{6}{CH_3} \\
\quad\quad | \\
\quad\quad CH_3
\end{array}
$$

3,3-Dimethylhexane (a six-carbon main chain with two 3-methyl substituents)

Primary (1°) carbon A carbon atom that is bonded to one other carbon atom.

Secondary (2°) carbon A carbon atom that is bonded to two other carbons.

Tertiary (3°) carbon A carbon atom that is bonded to three other carbons.

Quaternary (4°) carbon A carbon atom that is bonded to four other carbons.

One further word of explanation about alkanes and alkyl groups: It's sometimes useful to think about the *number of other carbon atoms attached* to a given carbon atom. There are four possible substitution patterns for carbon, called *primary, secondary, tertiary,* and *quaternary.* As indicated by the following structures, a **primary (1°) carbon atom** has one other carbon attached to it, a **secondary (2°) carbon atom** has two other carbons attached to it, a **tertiary (3°) carbon atom** has three other carbons attached to it, and a **quaternary (4°) carbon atom** has four other carbons attached to it.

$$\underset{\substack{| \\ H}}{\overset{\substack{H \\ |}}{R-C-H}}$$

Primary carbon (1°) has one other carbon attached.

$$\underset{\substack{| \\ H}}{\overset{\substack{R \\ |}}{R-C-H}}$$

Secondary carbon (2°) has two other carbons attached.

$$\underset{\substack{| \\ R}}{\overset{\substack{R \\ |}}{R-C-H}}$$

Tertiary carbon (3°) has three other carbons attached.

$$\underset{\substack{| \\ R}}{\overset{\substack{R \\ |}}{R-C-R}}$$

Quaternary carbon (4°) has four other carbons attached.

R— The general symbol for an alkyl group.

The symbol **R** is used here and in later chapters to represent a *generalized* alkyl group, where R may stand for any alkyl group. For example, the generalized formula R—OH for an alcohol might refer to CH_3OH, CH_3CH_2OH, or any of a great many other possibilities.

Removal of the hydrogen atom from a secondary or a tertiary carbon atom produces a branched alkyl group. Note in Table 1.5 that two of the four-carbon alkyl groups are distinguished from each other by prefixes showing that one bonds through a secondary (*sec-*) carbon atom and the other through a tertiary (*tert-*) carbon atom.

Solved Problem 1.3   What is the IUPAC name of this alkane?

$$\underset{}{CH_3-\overset{\overset{\displaystyle CH_3}{|}}{C}H-CH_2-CH_2-\overset{\overset{\displaystyle CH_3}{|}}{C}H-CH_2-CH_3}$$

Solution   First, find and name the longest continuous chain of carbon atoms (in this case seven carbons, or *hept*ane). Second, number the main chain beginning at the end nearer the first branch (on the left in this case).

$$\underset{1234567}{CH_3-\overset{\overset{\displaystyle CH_3}{|}}{C}H-CH_2-CH_2-\overset{\overset{\displaystyle CH_3}{|}}{C}H-CH_2-CH_3}$$

Name as a heptane.

Third, identify and number the substituents (a 2-methyl and a 5-methyl in this case). Fourth, write the name as one word using the prefix *di-* since there are two methyl groups. Separate the two numbers by a comma and use a hyphen between the numbers and the word:

$$\underset{1234567}{CH_3-\overset{\overset{\displaystyle CH_3}{|}}{C}H-CH_2-CH_2-\overset{\overset{\displaystyle CH_3}{|}}{C}H-CH_2-CH_3}$$

Substituents: 2-methyl, 5-methyl   Name: 2,5-dimethylheptane

Solved Problem 1.4   Identify each of the carbon atoms in this molecule as primary, secondary, tertiary, or quaternary.

$$\underset{\underset{\displaystyle CH_3}{|}}{CH_3CHCH_2CH_2\overset{\overset{\displaystyle CH_3}{|}}{C}CH_3}$$

with CH_3 above the CH as well.

Solution   Look at each carbon atom in the molecule, count the number of other carbon atoms attached, and make the assignment.

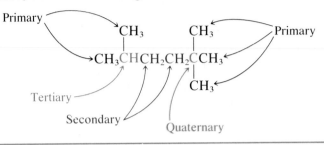

Practice Problems

1.9   What are the IUPAC names of the following alkanes?

$$\text{(a) } CH_3-\underset{\overset{|}{\underset{CH_2-CH_3}{}}}{CH}-CH_2-CH_2-CH_2-\underset{\overset{|}{CH_3}}{CH}-CH_3$$

$$\text{(b) } CH_3-CH_2-CH_2-CH_2-\underset{\overset{|}{C\,H_2-CH_3}}{\overset{\overset{C\,H_2-CH_3}{|}}{C}}-CH_2-CH_3$$

1.10   Draw structures corresponding to these IUPAC names:
(a) 3-methylhexane   (b) 3,4-dimethyloctane   (c) 2,2,4-trimethylpentane

1.11   Identify the carbon atoms in the molecules shown in Problem 1.9 as primary, secondary, tertiary, or quaternary.

1.12   Draw and name alkanes that meet these descriptions:
(a) an alkane with a tertiary carbon atom
(b) an alkane that has both a tertiary and a quaternary carbon atom

1.7   PROPERTIES OF ALKANES

REVIEW   London forces result from momentary unsymmetrical electron distributions in adjacent molecules of any type.

Alkanes contain only nonpolar carbon–carbon and carbon–hydrogen bonds, so the only intermolecular forces influencing their properties are the weak London forces. The effect of these forces is shown in the regularity with which the melting points and boiling points of straight-chain alkanes increase with molecular size (Figure 1.4). The first four alkanes—methane, ethane, propane, and butane—are gases at room temperature and pressure. Alkanes with from 5 to 15 carbon atoms are liquids, and those with 16 or more carbon atoms are generally low-melting, waxy solids.

REVIEW   A volatile substance evaporates easily.

In keeping with their low polarity, alkanes are insoluble in water but soluble in nonpolar organic solvents, including other alkanes. Since alkanes are generally less dense than water, they float on its surface. The liquid alkanes are volatile and must be handled with care because their vapors are flammable, as are the gaseous alkanes. Mixtures of alkane vapors and air can explode when detonated by a single spark.

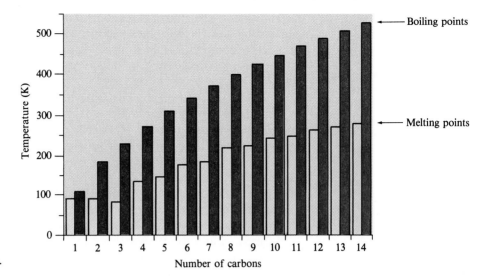

Figure 1.4
Boiling and melting points for the C_1–C_{15} straight-chain alkanes. There is a regular increase with molecular size.

The physiological effects of alkanes are limited. Methane, ethane, and propane gases are nontoxic, and the danger of inhaling them lies in suffocation due to lack of oxygen. Breathing the vapor of higher alkanes in large concentrations can induce sleep followed by loss of consciousness. There's also a danger in breathing droplets of liquid alkanes because they dissolve nonpolar substances in lung tissue and cause pneumonia-like symptoms.

Petrolatum (a jelly) and mineral oil (a thick liquid) are both mixtures of higher alkanes produced directly from petroleum. They're so harmless to tissue that both are approved for use in eye ointments and are used as vehicles for medications of many kinds. Petrolatum (sold as Vaseline) softens, lubricates, and protects the skin. Mineral oil is sometimes used as a laxative; it passes through the body unchanged.

Properties of Alkanes

● Odorless or mild odor, colorless, tasteless
● Nonpolar; insoluble in water and soluble in nonpolar organic solvents; less dense than water
● Flammable
● Regular increase in melting and boiling points with molecular weight; those with up to four C atoms are gases
● Not very reactive.

1.8 CHEMICAL REACTIONS OF ALKANES

REVIEW Polar molecules, which have permanent unsymmetrical electron distributions (Appendix A.8), generally dissolve in polar solvents. Nonpolar molecules, which have symmetrical electron distributions, generally dissolve in nonpolar solvents. Water is highly polar.

Paraffin A mixture of waxy alkanes having 20 to 36 carbon atoms.

If you've ever done any home canning, you might have used paraffin wax to seal the jars. Chemically, what you've used is a mixture of straight-chain alkanes in the C_{20}–C_{36} range. The word **paraffin,** derived from the Latin *parum affinis* meaning "slight affinity," is often applied to alkanes because of their lack of chemical reactivity. Alkanes show little chemical affinity for other molecules and are inert to acids, bases, and most other common laboratory reagents. Their major reactions are those with oxygen and the halogens.

Combustion A chemical reaction in which heat and often light are produced; usually refers to burning in the presence of oxygen.

Combustion   The chemical reaction of alkanes with oxygen occurs during the **combustion** of fuels in engines and furnaces. Carbon dioxide and water are the products of complete combustion of any hydrocarbon (or any organic compound containing carbon, hydrogen, and oxygen), and a large amount of heat is released.

When hydrocarbon combustion is less than complete due to faulty engine or furnace performance, carbon monoxide (CO) and carbon-containing soot are among the combustion products. Carbon monoxide is a highly toxic and dangerous substance. It's especially hazardous because it has no odor and can go undetected. Breathing air containing as little as 2% carbon monoxide for 1 hr can cause either respiratory and nervous system damage or death. The supply of oxygen to the brain is cut off by carbon monoxide because it binds strongly to hemoglobin at the site where oxygen is normally bound. By contrast with carbon monoxide, carbon dioxide is nontoxic and causes harm only by suffocation when it replaces oxygen in the atmosphere.

Practice Problem   1.13   Write a balanced equation for the complete combustion of ethane.

Halogenation, alkane The substitution of one or more hydrogen atoms in an alkane by halogen atoms.

Substitution reaction An organic reaction in which two reactants exchange atoms or groups, $AB + XY \rightarrow AY + XB$.

Halogenation   The second notable reaction of alkanes is the replacement of hydrogen atoms by chlorine or bromine atoms, known as **halogenation.** Such a reaction in which a hydrogen atom or other substituent in an organic compound is replaced by a different substituent is called a **substitution reaction.** Halogenation of alkanes takes place at a useful rate only when initiated by heat or light. Complete chlorination of methane, for example, yields carbon tetrachloride

$$CH_4 + 4\ Cl_2 \xrightarrow{\text{heat or light}} CCl_4 + 4\ HCl$$

REVIEW The halogens make up Group 7A of the periodic table:
F   Cl   Br   I

Although we've written a neatly balanced equation for the reaction of methane with chlorine, it doesn't represent well what actually happens because this reaction, like many organic reactions, usually yields a mixture of products:

$$CH_4 + Cl_2 \longrightarrow CH_3Cl + HCl$$
$$\downarrow Cl_2$$
$$CH_2Cl_2 + HCl$$
$$\downarrow Cl_2$$
$$CHCl_3 + HCl$$
$$\downarrow Cl_2$$
$$CCl_4 + HCl$$

CH_3Cl, methyl chloride
CH_2Cl_2, dichloromethane
$CHCl_3$, chloroform
CCl_4, carbon tetrachloride

REVIEW Equations for inorganic reactions are almost always balanced, that is, the numbers of atoms of each type shown in the reactants and products are the same.

In considering an organic reaction, attention is usually focused on converting a particular starting material into a desired product. Because they aren't of interest, possible minor side products are often ignored in writing organic equations. Also, nonorganic products such as the HCl formed in the chlorination of methane, are often of little interest. Therefore, it's frequently not necessary to balance the equation for an organic reaction as long as the reactant, the major product, and any necessary **reagents** and conditions are shown. A chemist who plans to convert methane into methyl bromide might therefore write the equation as follows:

A reactant used to bring about a specific chemical reaction.

$$CH_4 \xrightarrow[\text{light, heat}]{Br_2} CH_3Br$$

Like many equations for organic reactions, this equation isn't balanced.

Practice Problem   **1.14**   Write the structures of all possible products with one or two chlorine atoms that could be formed in the substitution reaction of propane with chlorine.

1.9 CYCLOALKANES

The organic compounds described thus far have all been open-chain or **acyclic alkanes. Cycloalkanes,** which contain rings of carbon atoms, are also well known and are widespread throughout nature. Cholesterol, for example (Figure 1.2), has three rings of 6 carbon atoms and one of 5 carbon atoms. Compounds of all ring sizes from 3 through 30 carbon atoms and beyond have been prepared in the laboratory. The two simplest cycloalkanes contain 3 (cyclopropane) and 4 carbon atoms (cyclobutane) joined in rings:

An alkane that contains no rings.

An alkane that contains a ring of carbon atoms.

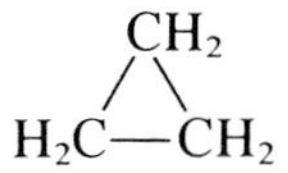

$$CH_2$$
$$H_2C{-}CH_2$$

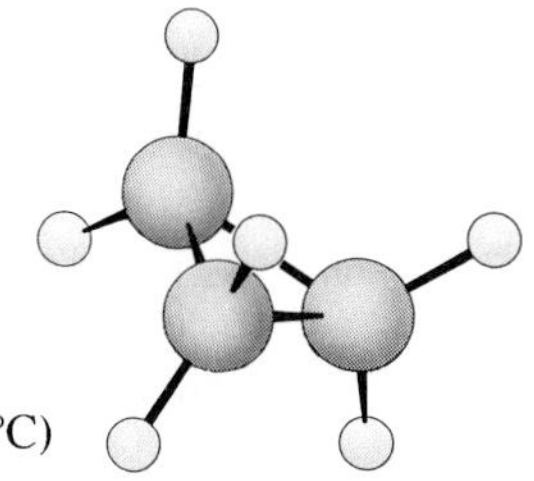

Cyclopropane

(mp − 128°C, bp − 33°C)

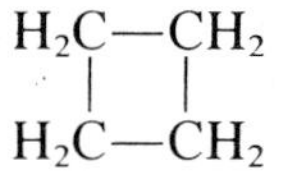

$$H_2C{-}CH_2$$
$$H_2C{-}CH_2$$

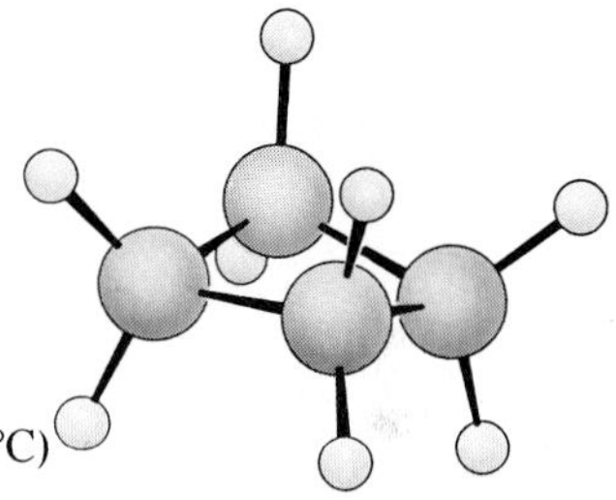

Cyclobutane

(mp − 50°C, bp − 12°C)

Cyclic and acyclic alkanes are similar in many of their properties, for the cycloalkanes are also nonpolar molecules. Cyclopropane and cyclobutane are gases, and those with more carbon atoms are liquids and solids. Like the alkanes, the cycloalkanes are insoluble in water and flammable.

Because of their cyclic structures, cycloalkane molecules are more rigid and less flexible than their open-chain counterparts. Rotation is not possible around the carbon–carbon bonds in cycloalkanes without breaking open the ring. To maintain closed rings, the carbon bond angles in the small cyclopropane and cyclobutane rings are compressed, causing these compounds to be less stable and more reactive than other cycloalkanes. The six-membered cyclohexane ring exists in a puckered, nonplanar shape known as the *chair conformation* (Figure 1.5), in which the carbon atoms have nearly tetrahedral bond angles. The cyclohexane ring is therefore very stable, and many naturally occurring and biochemically active molecules contain cyclohexane rings.

Figure 1.5
The chair conformation of cyclohexane. All bond angles are close to the 109.5° tetrahedral value.

1.10 DRAWING AND NAMING CYCLOALKANES

Even condensed structures become cluttered and awkward when working with large molecules that contain rings. Thus, a more streamlined way of drawing structures is often used in which the cycloalkanes are represented simply by polygons. A triangle represents cyclopropane, a square represents cyclobutane, a pentagon represents cyclopentane, and so on.

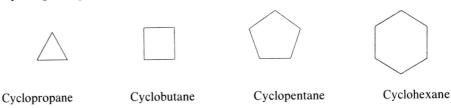

Cyclopropane   Cyclobutane   Cyclopentane   Cyclohexane

Line structure A shorthand way of representing ring structures as polygons without showing individual carbon atoms.

Notice that carbon and hydrogen atoms aren't even shown in these **line structures**. A carbon atom is simply "understood" to be at every intersection of two lines, and the proper number of hydrogen atoms necessary to give each carbon atom four covalent bonds is supplied mentally. Methylcyclohexane looks like this in a line structure:

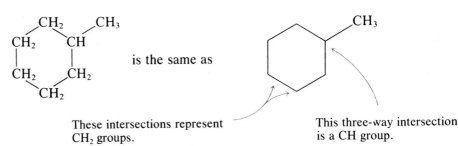

is the same as

These intersections represent CH_2 groups.

This three-way intersection is a CH group.

Cycloalkanes are named by a straightforward extension of the rules for open-chain alkanes. The carbon ring is equivalent to the main chain. In most cases, only two rules are needed:

1. *Use the cycloalkane name as the parent.* That is, compounds should be named as alkyl-substituted cycloalkanes rather than as cycloalkyl-substituted alkanes. If there is only one substituent on the ring, it's not even necessary to assign a number since all ring positions are equivalent.

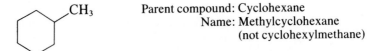

Parent compound: Cyclohexane
Name: Methylcyclohexane
(not cyclohexylmethane)

2. *Number the substituents*. Start numbering at the group that has alphabetical priority and proceed around the ring in the direction that gives the second substituent the lowest possible number.

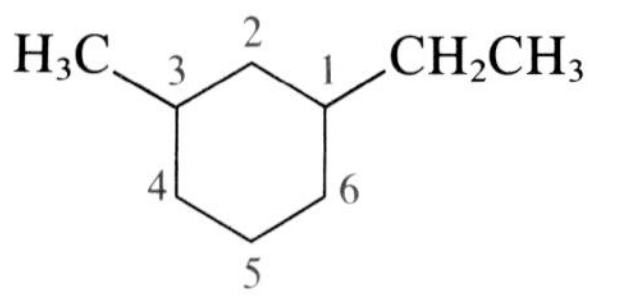

1-Ethyl-3-methylcyclohexane
(not 1-ethyl-5-methylcyclohexane or
1-methyl-3-ethylcyclohexane or
1-methyl-5-ethylcyclohexane)

Solved Problem 1.5 What is the IUPAC name of this cycloalkane?

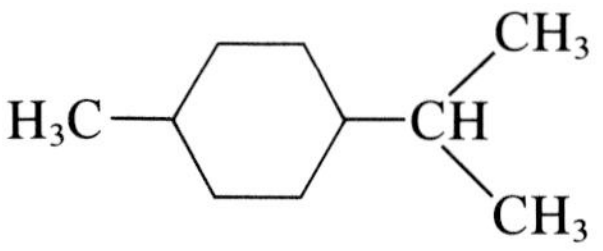

Solution First, identify the parent cycloalkane (cyclohexane in this case). Second, identify the two substituents (a methyl group and an isopropyl group). Third, number the compound beginning at the group having alphabetical priority (isopropyl rather than methyl) and proceed around the ring in a direction that gives the second group the lowest possible number.

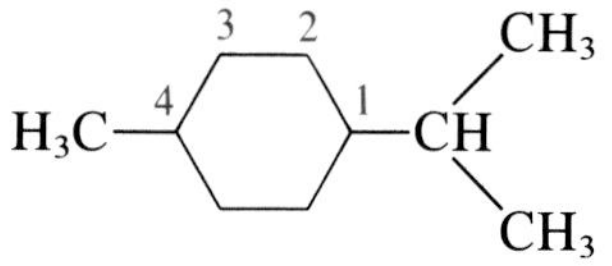

1-Isopropyl-4-methylcyclohexane

Solved Problem 1.6 Draw a line structure for 1,4-dimethylcyclohexane.

Solution First, draw a hexagon to represent a cyclohexane ring and then attach a —CH_3 (methyl) group at an arbitrary position that becomes C1. Then count around the ring to C4 and attach another —CH_3 group (written here as H_3C— since it is attached on the left side of the ring).

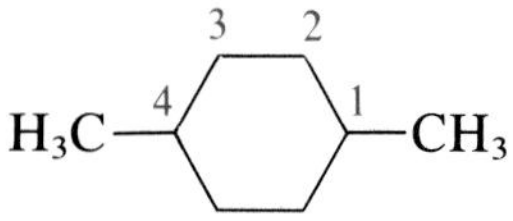

1,4-Dimethylcyclohexane

Practice Problems **1.15** What are the IUPAC names of these cycloalkanes?

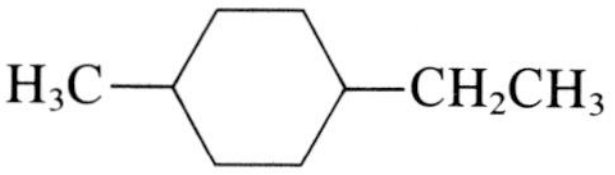

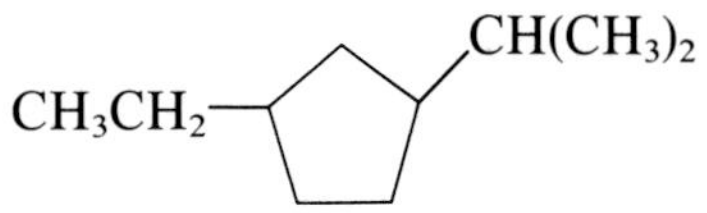

1.16 Draw structures representing these IUPAC names. Use simplified line structures rather than condensed structures.
(a) 1,1-diethylcyclohexane   (b) 1,3,5-trimethylcycloheptane

INTERLUDE: PETROLEUM

Although many alkanes occur naturally throughout the plant and animal world, natural gas and petroleum deposits provide the most abundant supply. Laid down eons ago, these deposits are largely derived from the decomposition of marine organic matter.

Natural gas consists chiefly of methane, with smaller amounts of ethane, propane, and butane also present. **Petroleum** is a complex mixture of hydrocarbons that must be separated, or *refined,* into different fractions before it can be used. Petroleum refining begins with **distillation** to separate the crude oil into three main fractions according to boiling points: straight-run gasoline (bp 30–200°C), kerosene (bp 175–300°C), and gas oil (bp 275–400°C). The residue is then further distilled under reduced pressure to recover lubricating oils, waxes, and asphalt, as shown in the diagram.

Distillation of petroleum is just the beginning of the process for making automobile fuel. It's long been known that straight-chain alkanes burn far less smoothly than branched-chain compounds, a quality measured by determining a compound's **octane number.** Heptane, a particularly bad fuel, is assigned an octane rating of 0, and 2,2,4-trimethylpentane (known as isooctane) is given a rating of 100. Straight-run gasoline,

with its high percentage of unbranched alkanes, is thus a poor fuel. Petroleum chemists, however, have devised sophisticated methods to remedy the problem. One of these methods, **catalytic cracking,** involves taking the kerosene cut (C_{11}–C_{14}) and "cracking" it into smaller C_3–C_5 molecules at high temperature. These small hydrocarbons are then catalytically recombined to yield C_7–C_{10} branched-chain molecules that are perfectly suited for use as automobile fuel.

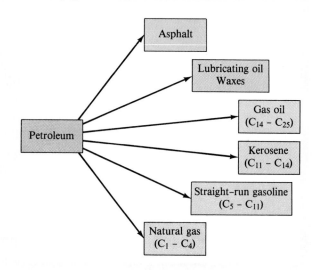

The principal products of petroleum refining.

SUMMARY

Carbon has the unique ability to form strong bonds to other carbon atoms, resulting in the formation of rings or long chains and giving rise to many millions of possible structures. Organic compounds can be classified into various families according to the functional groups they contain. A **functional group** is a part of a larger molecule and is composed of an atom or group of atoms that has characteristic chemical reactivity. A given functional group undergoes the same chemical reactions in every molecule where it occurs.

Compounds of carbon and hydrogen that contain only single bonds are called **alkanes. Straight-chain alkanes** have all their carbons connected in a row; **branched-chain alkanes** have a branching connection of atoms some-

where along their chains; and **cycloalkanes** have a ring of carbon atoms. **Isomerism** is possible in alkanes having four or more carbons. **Constitutional isomers** are compounds that have the same molecular formula but have different structures and properties because of different connections between atoms. Organic compounds may be represented by **structural formulas** that show all atoms and bonds, by condensed structures in which not all bonds are drawn, or by **line structures** in which the carbon skeleton is represented by lines and the locations of C and H atoms are understood.

Alkanes are named in the **IUPAC system** by applying a set of **nomenclature** rules. Straight-chain alkanes are named by adding the family ending -*ane* to a root that tells how many carbon atoms are present. Branched-chain alkanes are named by using the longest continuous chain of carbon atoms for the root and then identifying the alkyl groups present as branches off the main chain. An **alkyl group** is the part of an alkane that remains when one hydrogen is removed. Isomerism is possible in alkyl groups just as in alkanes themselves. Cycloalkanes are named in the same system by adding *cyclo-* as a prefix to the name of the alkane with the same number of carbon atoms.

Alkanes are generally soluble only in nonpolar organic solvents, have weak intermolecular forces, increase in melting and boiling points with formula weight, and are nontoxic. Their principal chemical reactions are **combustion** and **halogenation,** a **substitution reaction** in which halogen atoms are substituted for hydrogen atoms.

REVIEW PROBLEMS

Organic Molecules and Functional Groups

1.17 What special characteristic of carbon makes possible the existence of so many different organic compounds?

1.18 What are functional groups; and why are they important?

1.19 Why are most organic compounds insoluble in water and nonconducting?

1.20* If you were given two unlabeled bottles, one containing hexane and one containing water, how could you tell them apart?

1.21 What is meant by the term *polar covalent bond?* Give an example of such a bond.

1.22* Give examples of compounds that are members of the following families:
(a) alcohol   (b) amine   (c) carboxylic acid
(d) ether

1.23 Locate and identify the functional groups in these molecules:
(a)

Menthol

(b)

Aspirin (acetylsalicylic acid)

1.24 Propose structures for molecules that fit these descriptions:
(a) a ketone with the formula $C_5H_{10}O$
(b) an ester with the formula $C_6H_{12}O_2$

(c) a compound with the formula $C_2H_5NO_2$ that is both an amine and a carboxylic acid
(d) an amide with the formula C_4H_9NO

1.25 Identify the functional groups in these molecules:

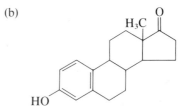

(a)

Vitamin A

(b)

Estrone, a female sex hormone

Alkanes and Isomers

1.26* What structural feature distinguishes a straight-chain alkane from a branched-chain alkane?

1.27 What requirement must be met in order for two compounds to be isomers?

1.28* If one compound has the formula C_5H_{10} and another has the formula C_4H_{10}, are the two compounds isomers? Explain.

1.29 What is the difference between a secondary carbon and a tertiary carbon? Between a primary carbon and a quaternary carbon?

1.30* Why can't a compound have a *quintary* carbon (five R groups attached to C)?

1.31 Give examples of compounds that meet these descriptions:
(a) an alkane that has two tertiary carbons
(b) a cycloalkane that has only secondary carbons

---

*Problems marked with an asterisk are answered at the back of the book.

1.32* There are three isomers with the formula C_3H_8O. Draw their structures.

1.33 Write condensed structures for each of the following molecular formulas. You may have to use rings and/or multiple bonds in some instances.
(a) C_2H_7N (b) C_4H_8 (c) C_2H_4O (d) CH_2O_2

1.34* If someone reported the preparation of a compound with the formula CH_{39}, most chemists would be skeptical. Why?

1.35 How many isomers can you write that fit these descriptions?
(a) alcohols with formula $C_4H_{10}O$
(b) amines with formula C_3H_9N
(c) ketones with formula $C_5H_{10}O$
(d) aldehydes with formula $C_5H_{10}O$
(e) esters with formula $C_4H_8O_2$
(f) carboxylic acids with formula $C_4H_8O_2$

1.36* Which of the following pairs of structures are identical, which are isomers, and which are unrelated?
(a) $CH_3CH_2CH_3$ and

$$\begin{array}{c} CH_3 \\ | \\ CH_2CH_3 \end{array}$$

(b) $CH_3\!-\!\underset{\underset{H}{|}}{N}\!-\!CH_3$ and $CH_3CH_2\!-\!\underset{\underset{H}{|}}{N}\!-\!H$

(c) $CH_3CH_2CH_2\!-\!O\!-\!CH_3$ and

$$CH_3CH_2CH_2\!-\!\overset{\overset{\displaystyle O}{\|}}{C}\!-\!CH_3$$

(d) $CH_3\!-\!\overset{\overset{\displaystyle O}{\|}}{C}\!-\!CH_2CH_2CH(CH_3)_2$ and

$$CH_3CH_2\!-\!\overset{\overset{\displaystyle O}{\|}}{C}\!-\!CH_2CH_2CH_2CH_3$$

(e) $CH_3CH\!=\!CHCH_2CH_2\!-\!O\!-\!H$ and

$$CH_3CH_2\underset{\underset{CH_3}{|}}{CH}\!-\!\overset{\overset{\displaystyle O}{\|}}{C}\!-\!H$$

1.37 What is wrong with each of the following structures?
(a) $CH_3\!=\!CHCH_2CH_2OH$

(b) $CH_3CH_2CH\!=\!\overset{\overset{\displaystyle O}{\|}}{C}\!-\!CH_3$

(c) $CH_2CH_2CH_2C\!\equiv\!\underset{\underset{CH_3}{|}}{C}CH_3$

1.38* Which of the sturctures in each group represent the same compound, and which represent isomers?
(a)

$$H\!-\!\overset{\overset{\displaystyle H}{|}}{\underset{\underset{H}{|}}{C}}\!-\!\overset{\overset{\displaystyle H}{|}}{\underset{\underset{H}{|}}{C}}\!-\!\overset{\overset{\displaystyle H}{|}}{\underset{\underset{H}{|}}{C}}\!-\!\overset{\overset{\displaystyle H}{|}}{\underset{\underset{H}{|}}{C}}\!-\!H$$

$$H\!-\!\overset{\overset{\displaystyle H}{|}}{\underset{\underset{H}{|}}{C}}\!-\!\overset{\overset{\overset{\displaystyle H}{|}}{\overset{\displaystyle C}{|}}\overset{\displaystyle H}{}}{\underset{\underset{\underset{\underset{H}{|}}{C}}{|}}{C}}\!-\!\overset{\overset{\displaystyle H}{|}}{\underset{\underset{H}{|}}{C}}\!-\!H$$

$$H\!-\!\overset{\overset{\displaystyle H}{|}}{\underset{\underset{H}{|}}{C}}\!-\!\overset{\overset{\displaystyle H}{|}}{\underset{\underset{\underset{\underset{H}{|}}{C}\,\,H}{|}}{C}}\!-\!\overset{\overset{\displaystyle H}{|}}{\underset{\underset{H}{|}}{C}}\!-\!H$$

(b)
$$\underset{\underset{Br}{|}}{CH_2CHCH_2CH_3} \overset{\overset{CH_3}{|}}{} \qquad \underset{\underset{Br}{|}}{CH_3CHCHCH_3}\overset{\overset{CH_3}{|}}{}$$

$$\underset{\underset{Br}{|}}{CH_2CHCH_2CH_3}\overset{\overset{CH_3}{|}}{}$$

Alkane Nomenclature

1.39 What are the IUPAC names of these alkanes?
(a)
$$CH_3CH_2CH_2CH_2\underset{\underset{CH_3}{|}}{CH}CHCH_2CH_3 \overset{\overset{CH_2CH_3}{|}}{}$$

(b)
$$CH_3CH_2CH_2\underset{}{CH}CH_2\underset{\underset{CH_2CH_3}{|}}{CH}CH_3 \overset{\overset{CH_3CHCH_3}{|}}{}$$

(c)
$$CH_3\underset{\underset{CH_3}{|}}{\overset{\overset{CH_3}{|}}{C}}CH_2CH_2CH_2\underset{}{\overset{\overset{CH_3}{|}}{C}}HCH_3$$

(d)
$$CH_3CH_2CH_2\underset{\underset{CH_3CHCH_3}{|}}{\overset{\overset{CH_2CH_2CH_2CH_3}{|}}{C}}CH_3$$

(e)
$$CH_3\underset{\underset{CH_3}{|}}{\overset{\overset{CH_3}{|}}{C}}CH_2\underset{\underset{CH_3}{|}}{\overset{\overset{CH_3}{|}}{C}}CH_3$$

(f)
$$CH_3CH_2\underset{\underset{CH_3CH_2}{|}}{\overset{\overset{CH_3CH_2}{|}}{C}}CH_2\underset{\underset{CH_3}{|}}{\overset{\overset{CH_3}{|}}{C}}H$$

(g) $CH_3(CH_2)_7\underset{\underset{CH_3}{|}}{\overset{\overset{CH_3}{|}}{C}}\!-\!CH_3$

1.40 Write condensed structures for each of these compounds:
(a) 3-ethylhexane   (b) 2,2,3-trimethylpentane
(c) 3-ethyl-3,4-dimethylheptane
(d) 5-isopropyl-2-methyloctane
(e) 2,2,6,6-tetramethyl-4-propylnonane

1.41 The following compound, known trivially as isooctane, is important as a reference substance for determining the "octane" rating of gasoline. What is the proper IUPAC name of isooctane?

$$\text{Isooctane} \qquad \underset{\overset{|}{CH_3}}{CH_3}\overset{\overset{CH_3}{|}}{\underset{}{C}}CH_2\overset{\overset{CH_3}{|}}{\underset{}{C}}HCH_3$$

1.42* Provide IUPAC names for each of the five isomers with the formula C_6H_{14}.

1.43 Draw structures corresponding to these IUPAC names:
(a) cyclooctane   (b) 1,1-dimethylcyclopentane
(c) 1,2,3,4-tetramethylcyclobutane
(d) 4-ethyl-1,1-dimethylcyclohexane
(e) ethylcycloheptane   (f) 1,3,5-triethylcyclohexane
(g) 2-isopropyl-1,1-dimethyl-3-propylcyclobutane
(h) 1,2,4,5-tetramethylcyclooctane

1.44* Provide IUPAC names for these cycloalkanes:
(a)

(b)

(c) $CH_3CH_2CH_2-$

(d) $CH_3CH_2CH_2CH_2-$

(e)

(f)

(g)

1.45 The following names are incorrect. Tell what is wrong with each and provide the correct names.

(a)
$$\underset{\overset{|}{CH_3}}{\overset{\overset{CH_3}{|}}{CH_3C}}CH_2CH_2CH_3$$
2,2-Methylpentane

(b)
$$\underset{}{CH_3}\overset{\overset{CH_2CH_3}{|}}{\underset{}{CH}}CH_2\overset{\overset{}{}}{\underset{\overset{|}{CH_3}}{CH}}CH_2CH_3$$
5-Ethyl-3-methylhexane

(c)
$$CH_3\overset{\overset{CH_3}{|}}{\underset{}{CH}}CH_2-◇$$
1-Cyclobutyl-2-methylpropane

1.46* Draw structures and give IUPAC names for the nine isomers of C_7H_{16}.

Reactions of Alkanes

1.47 Propane, more commonly known as LP gas, burns in air to yield CO_2 and H_2O. Write a balanced equation for the reaction.

1.48* Write the balanced equation for the combustion of isooctane (Problem 1.41).

1.49 Write the balanced equation for the combustion of a component of paraffin, $C_{25}H_{52}$.

1.50* In a substitution reaction in alkanes, what atom does the halogen replace?

1.51 Write the formulas of the three singly substituted isomers formed when 2,2-dimethylbutane reacts with chlorine in the presence of light.

1.52 Write the formulas of the seven doubly substituted isomers formed when 2,2-dimethylbutane reacts with bromine in the presence of light.

Applications

1.53 Why does a synthetically produced version of a compound have exactly the same properties as a "natural" compound? [App: Natural vs. Synthetic]

1.54* What does it mean when a compound is semisynthetic? [App: Natural vs. Synthetic]

1.55 Must "natural" products be beneficial? Try to find examples of five "natural" chemicals that are hazardous or toxic to human beings. [App: Natural vs. Synthetic]

1.56* Why is it important to know the shape of a molecule? [App: Displaying Molecular Shapes]

1.57 How does petroleum differ from natural gas? [Int: Petroleum]

1.58* What types of hydrocarbons burn most efficiently in an internal combustion engine? [Int: Petroleum]

1.59 Discuss how different alkanes can be separated by the process of distillation. [Int: Petroleum]

1.60* Methane and ethane have boiling points of $-164°C$ and $-89°C$, respectively. Hexadecane ($C_{16}H_{34}$) and heptadecane ($C_{17}H_{36}$) having boiling points of $287°C$ and $303°C$, respectively. Suggest a reason why the boiling points of the latter compounds are so much closer to each other and so much higher than those of the first members of the alkane family? [Int: Petroleum]

Additional Questions and Problems

1.61 Label the functional groups present in the following molecules:

(a) Testosterone, a male sex hormone

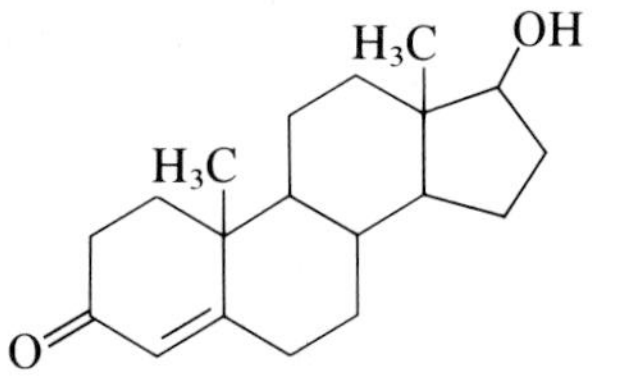

(b) Aspartame (an artificial sweetener)

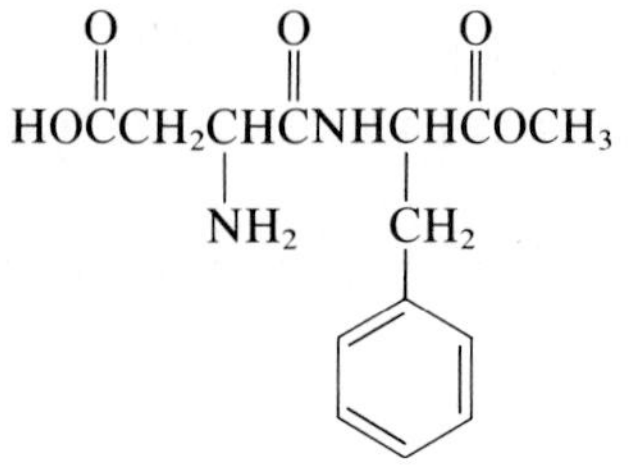

1.62* Label each carbon in Problem 1.61 as primary, secondary, tertiary, or quaternary.

1.63 An analysis proves that two different pure samples each have the molecular formula C_5H_{12}. At atmospheric pressure one boils at $36°C$, while the other boils at $30°C$. How can this be?

1.64* Most lipsticks are about 70% castor oil and wax. Why is lipstick more easily removed with petroleum jelly than with water?

1.65 Write the balanced combustion reaction for dodecane ($C_{12}H_{26}$), a component of kerosene.

1.66 When cyclohexane is exposed to bromine in the presence of light, substitution occurs. Write the formulas of (a) all possible monosubstituted products, and (b) all possible disubstituted products.

1.67 The following names are incorrect. Write the structural formula that agrees with the apparent name and then write the correct name of the compound.
(a) 2-ethylbutane
(b) 2-isopropyl-2-methylpentane
(c) 5-ethyl-1,1-methylcyclopentane
(d) 3-ethyl-3,5,5-trimethylhexane
(e) 1,2-dimethyl-4-ethylcyclohexane
(f) 2,4-diethylpentane
(g) 5,5,6,6-methyl-7,7-ethyldecane

1.68* Draw the structural formulas and name all the cyclic isomers with the formula C_5H_{10}.

1.69 Which should have a higher boiling point, pentane or neopentane? Why?

2

Alkenes, Alkynes, and Aromatic Compounds

A fennel plant is an aromatic herb; a phenyl group, pronounced exactly the same way, is the characteristic structural feature of "aromatic" organic compounds. And the connection is more than a pun because anethole ($CH_3OC_6H_4CH{=}CHCH_3$), the compound that gives fennel its characteristic licorice aroma and flavor, contains a phenyl group.

In this and the remaining four chapters on organic chemistry, we'll examine families of organic compounds whose functional groups give them characteristic properties. Compounds in the three families described in this chapter all contain carbon–carbon multiple bonds. Alkenes, such as ethylene, contain a carbon–carbon double bond; alkynes, such as acetylene, contain a carbon–carbon triple bond; and aromatic compounds, such as benzene, contain a six-membered ring of carbon atoms with three double bonds or other rings with similar arrangements of alternating double bonds. Note that the computer-generated representations of these compounds shown below indicate only the connection between atoms. The multiple bonds themselves are understood rather than specifically represented.

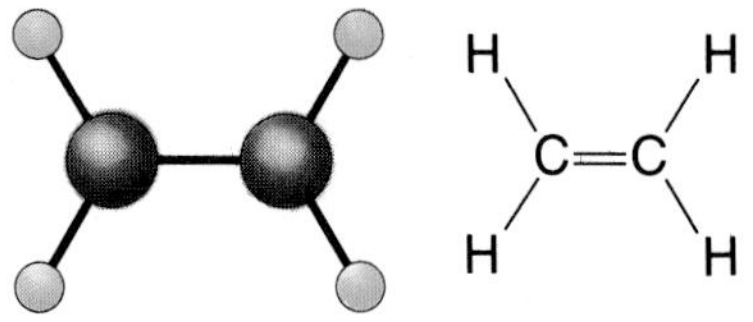

Ethylene
(an *alkene*—has
a C—C double bond)

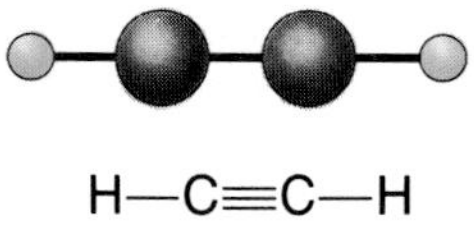

Acetylene
(an *alkyne*—has
a C—C triple bond)

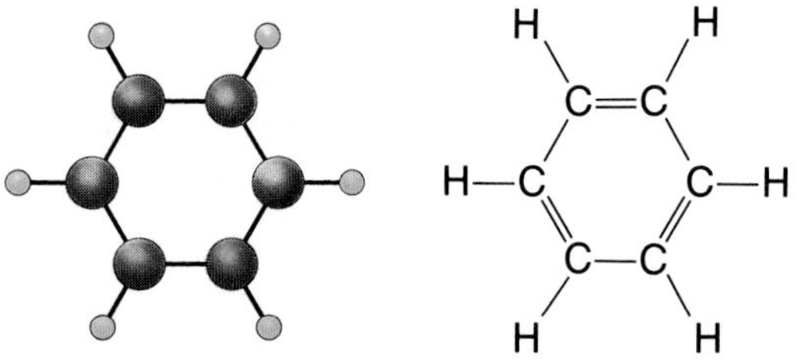

Benzene
(an *aromatic compound*—
has a six-membered ring
and three double bonds)

In this chapter, we'll look at answers to these questions:

1. **What are alkenes, alkynes, and aromatic compounds?** The goal: Be able to recognize the structures of members of these three families of unsaturated organic compounds and give examples of each.
2. **How are alkenes, alkynes, and aromatic compounds named?** The goal: Be able to name an alkene, alkyne, or simple aromatic compound from its structure or write the structure, given the name.
3. **What are cis–trans isomers?** The goal: Be able to identify cis–trans isomers and predict their occurrence.
4. **What are the general properties of alkenes, alkynes, and aromatic compounds?** The goal: Be able to describe and compare such properties as polarity, water solubility, flammability, bonding, and chemical reactivity.
5. **What are the common chemical reactions of alkenes, alkynes, and aromatic compounds?** The goal: Be able to predict the products of reactions of alkenes, alkynes, and aromatic compounds.
6. **How do organic reactions take place?** The goal: Be able to show how addition reactions occur and describe the difference between addition and substitution reactions.

2.1 SATURATED AND UNSATURATED HYDROCARBONS

Saturated Containing only single bonds between carbon atoms and thus unable to accommodate additional hydrogen atoms.

Unsaturated Containing one or more double or triple bonds between carbon atoms and thus able to add more hydrogen atoms.

Alkene A compound that contains only carbon and hydrogen and has a carbon–carbon double bond.

Alkyne A compound that contains only carbon and hydrogen and has a carbon–carbon triple bond.

The alkanes, introduced in Chapter 1, are often referred to as **saturated** hydrocarbons because each carbon atom is fully "saturated" by bonds to the maximum number of hydrogen atoms: No more hydrogen atoms can be added. Alkenes and alkynes, however, contain carbon–carbon multiple bonds to which more hydrogen atoms can be added and are therefore known as **unsaturated** hydrocarbons. **Alkenes** are hydrocarbons that contain carbon–carbon double bonds and **alkynes** are hydrocarbons that contain carbon–carbon triple bonds.

$$CH_3CH_2CH_3 \qquad CH_3CH{=}CH_2 \qquad CH_3C{\equiv}CH$$

An alkane   An alkene   An alkyne

2.2 ALKENES

Simple alkenes are made in huge quantities by *thermal cracking* of natural gas and petroleum and are the basic building blocks of the vast *petrochemical industry,* which produces organic chemicals from petroleum. Alkanes are rapidly heated to such high temperatures (750–900°C) that they "crack" apart, forming reactive fragments that then reunite or rearrange into lower-molecular-weight molecules such as ethylene ($CH_2{=}CH_2$) and propylene ($CH_2{=}CHCH_3$).

$$CH_3(CH_2)_nCH_3 \xrightarrow{750-900°C} H_2 + CH_4 + CH_2{=}CH_2 + CH_3CH{=}CH_2 + CH_3CH_2CH{=}CH_2$$

By far the majority of organic chemicals used in making drugs, explosives, paints, plastics, pesticides, and many other products are synthesized by routes that depend on alkene starting materials. More ethylene is produced in the United States each year than any other organic chemical: 18 million tons in 1992. Much of it goes into making polyethylene.

Curiously enough, ethylene is also a natural product, being formed in the leaves, flowers, and roots of plants. At each stage in the development of a plant, ethylene has a role to play as a hormone: It controls seedling growth, stimulates root formation, and regulates fruit ripening. During ripening, a fruit produces ethylene that then speeds up ripening in other nearby fruits.

The ability of ethylene to hasten ripening was discovered indirectly by citrus growers who at one time ripened green oranges in rooms heated with kerosene stoves. When the stoves were replaced, the fruit, to their surprise, no longer ripened. Ethylene from incomplete combustion of the kerosene had caused the ripening, not heat from the stoves. Today, fruit is intentionally exposed to ethylene to continue the ripening process during shipment to market.

2.3 NAMING ALKENES AND ALKYNES

Alkenes and alkynes are named systematically by a series of three rules similar to those used for alkanes. The roots indicating the number of carbon atoms in the main chain are the same as those for alkanes (Table 1.4). For an alkene, the

-ene suffix is used in place of *-ane*, and for an alkyne the *-yne* suffix is used to give the name of the parent compound.

Step 1. Name the parent compound. Find the longest chain *containing the double or triple bond* and name the parent compound by adding the suffix *-ene* or *-yne* to the root for this main chain.

$CH_3CH_2CH_2CH{=}CH_2$ Name as a *pentene*—a five-carbon chain containing a double bond.

$CH_3CH_2CH_2C{\equiv}CCH_3$ Name as a *hexyne*—a six-carbon chain containing a triple bond.

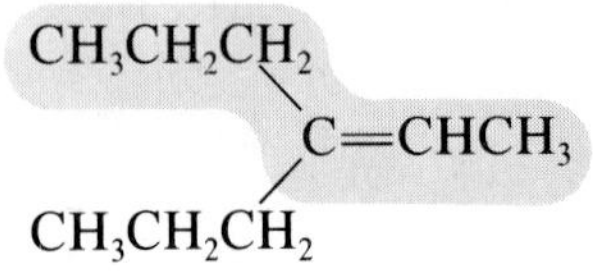

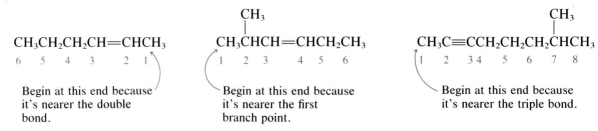

Name as a *hexene*—a six-carbon chain containing a double bond—

not

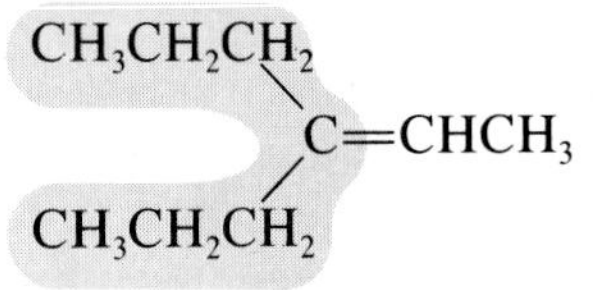

as a heptene, since the double bond is not contained in the seven-carbon chain.

Step 2. Number the carbon atoms in the main chain. Beginning at the end nearer the multiple bond, number each carbon atom in the chain. If the multiple bond is an equal distance from both ends, as in the middle structure shown here, begin numbering at the end nearer the first branch point.

$$CH_3CH_2CH_2CH{=}CHCH_3$$
$$\text{6} \quad \text{5} \quad \text{4} \quad \text{3} \quad \text{2} \quad \text{1}$$

Begin at this end because it's nearer the double bond.

$$\overset{\displaystyle CH_3}{|}$$
$$CH_3CHCH{=}CHCH_2CH_3$$
$$\text{1} \quad \text{2} \quad \text{3} \quad \text{4} \quad \text{5} \quad \text{6}$$

Begin at this end because it's nearer the first branch point.

$$\overset{\displaystyle CH_3}{|}$$
$$CH_3C{\equiv}CCH_2CH_2CHCH_3$$
$$\text{1} \quad \text{2} \quad \text{3} \quad \text{4} \quad \text{5} \quad \text{6} \quad \text{7} \quad \text{8}$$

Begin at this end because it's nearer the triple bond.

For cyclic alkenes, the *-ane* ending of the cycloalkane is replaced by *-ene*. No numbers are needed for an unsubstituted cycloalkene. In substituted cycloalkenes, the double-bond carbon atoms are numbered 1 and 2 so that the first substituent has the lowest number:

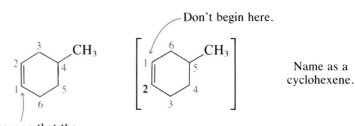

Don't begin here.

Name as a cyclohexene.

Begin here so that the substituent has the lowest number.

Step 3. Write out the full name. Identify and number branching substituents in the same way as for alkanes: Assign numbers to the branching substituents on the chain and list them alphabetically. Use commas to separate numbers and use hyphens to separate words from numbers. Indicate the position of the multiple bond in the chain by giving the number of the *first* multiple-bonded carbon. If more than one double bond is present, identify the position of each and use the name endings *-diene, -triene, -tetraene,* and so forth, to give the exact number of double bonds. In naming cycloalkenes, it isn't necessary to include the number 1 to locate the double bond. For example,

$$CH_3CH_2CH_2CH{=}CH_2$$
$$5 \quad 4 \quad 3 \quad 2 \quad \quad 1$$

1-Pentene

$$CH_3CH_2CH_2C{\equiv}CCH_3$$
$$6 \quad 5 \quad 4 \quad 3 \quad 2\,1$$

2-Hexyne

$$\begin{array}{c} 6 \quad 5 \quad 4 \\ CH_3CH_2CH_2 \\ \diagdown {}^3 \quad {}^2 \quad {}^1 \\ C{=}CHCH_3 \\ \diagup \\ CH_3CH_2CH_2 \end{array}$$

3-Propyl-2-hexene

$$\begin{array}{c} CH_3 \\ | \\ CH_3C{\equiv}CCH_2CH_2CH_2CHCH_3 \\ 1 \quad 2 \quad \;3\;4 \quad\; 5 \quad\; 6 \quad\;\; 7 \quad\; 8 \end{array}$$

7-Methyl-2-octyne

$$\begin{array}{c} CH_2CH_3 \\ | \\ H_2C{=}C{-}CH{=}CH_2 \\ 1 \quad\; 2 \quad\; 3 \quad\;\; 4 \end{array}$$

2-Ethyl-1,3-butadiene

4-Methylcyclohexene

For historical reasons, there are a small number of alkenes and alkynes whose names don't conform strictly to the rules. The two-carbon alkene $H_2C{=}CH_2$ should properly be called *ethene,* but the name *ethylene* has been used for so long that it's now accepted by IUPAC. Similarly, the three-carbon alkene *propene* ($CH_3CH{=}CH_2$) is often called *propylene.* The simplest alkyne, $HC{\equiv}CH$, should be known as *ethyne* but is more often called *acetylene.*

Solved Problem 2.1   What is the IUPAC name of the following alkene?

$$\begin{array}{c} H_3C \quad\;\; CH_2CH_3 \\ |\qquad\quad | \\ CH_3CH_2CH_2{-}C{=}C{-}CH_3 \end{array}$$

Solution   First, find and circle the longest chain containing the double bond. In this case, we have to turn a corner to find that it's a heptene:

$$\begin{array}{c} H_3C \quad\;\; CH_2CH_3 \\ |\qquad\quad | \\ CH_3CH_2CH_2{-}C{=}C{-}CH_3 \end{array}$$ Name as a *heptene.*

Next, number the chain from the end nearer the double bond. The first double-bond carbon is C4 starting from the left-hand end, but C3 starting from the right:

$$\begin{array}{c} \qquad\quad 2 \quad\; 1 \\ H_3C \quad\;\; CH_2CH_3 \\ 7 \;\; 6 \;\; 5 \;\;\; 4| \qquad |3 \\ CH_3CH_2CH_2{-}C{=}C{-}CH_3 \end{array}$$ Name as a substituted *3-heptene.*

Finally, identify the substituents and write the name.

$$\begin{array}{c} \qquad\quad 2 \quad\; 1 \\ H_3C \quad\;\; CH_2CH_3 \\ 7 \;\; 6 \;\; 5 \;\;\; 4| \qquad |3 \\ CH_3CH_2CH_2{-}C{=}C{-}CH_3 \end{array}$$

Substituents: 3-methyl, 4-methyl

Name: *3,4-dimethyl-3-heptene.*

Solved Problem 2.2   Draw the structure of 3-ethyl-4-methyl-2-pentene.

Solution   First, identify the root name (*pent*) and write the number of carbon atoms it indicates in a straight line. Then, counting from one end, put in the double bond between the number 2 and 3 carbon atoms:

$$\overset{3}{C} \quad \overset{}{C} \quad \overset{3}{C}=\overset{2}{C} \quad \overset{1}{C} \qquad \text{2 Pentene}$$

Next, add the ethyl and methyl substituents on carbons 3 and 4 and write in the additional hydrogen atoms so that each carbon atom has four bonds.

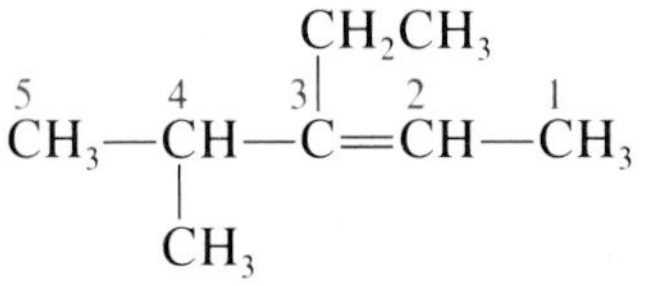

$$\overset{5}{CH_3}-\overset{4}{CH}-\overset{3}{\underset{|}{C}}=\overset{2}{CH}-\overset{1}{CH_3} \qquad \text{3-Ethyl-4-methyl-2-pentene}$$

with CH$_2$CH$_3$ above carbon 3 and CH$_3$ below carbon 4.

Practice Problems   **2.1**   What are the IUPAC names of the following compounds?
(a) $CH_3CH_2CH_2CH=CHCH(CH_3)_2$
(b) $H_2C=CHCH_2CH_2\underset{\underset{CH_3}{|}}{C}=CH_2$

2.2   Draw structures corresponding to these IUPAC names:
(a) 3-methyl-1-heptene   (b) 4,4-dimethyl-2-pentyne
(b) 2-methyl-2-hexene   (d) 3-ethyl-2,2-dimethyl-3-hexene

2.4 THE STRUCTURE OF ALKENES: CIS–TRANS ISOMERISM

Alkenes and alkynes differ from alkanes in shape because of their different kinds of bonds. Thus, whereas methane is tetrahedral, ethylene is flat (planar) and acetylene is linear (straight).

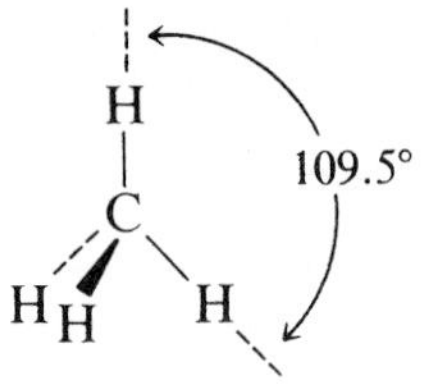

Methane—a tetrahedral molecule with bond angles of 109.5°.

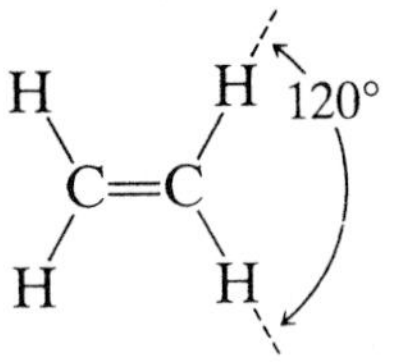

Ethylene—a flat molecule with bond angles of 120°.

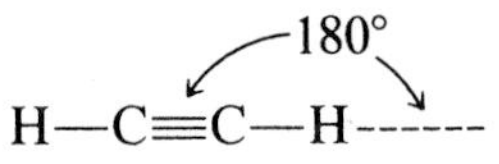

Acetylene—a straight molecule with bond angles of 180°.

The two carbons and four attached atoms that make up the double-bond functional group always lie in a plane. Unlike the situation in alkanes where free rotation around the C—C single bond occurs (Section 1.5), there is no free rotation around double bonds. As a consequence, a new kind of isomerism is possible for alkenes.

To see this new kind of isomerism, look at the following list of C_4H_8 compounds. When written as condensed structures, there appear to be three alkene isomers of formula C_4H_8: 1-butene, 2-butene, and 2-methylpropene. In fact, though, there are *four*. 1-Butene and 2-butene are isomers because their double bonds occur at different positions along the chain, while 2-methylpropene is isomeric with both because it has a different connection of carbon atoms. But because rotation can't occur around carbon–carbon double bonds, *there are two different 2-butenes*. In one isomer, the two methyl groups are close together on the same side of the double bond, but in the other isomer, the two methyl groups are far apart on opposite sides of the double bond.

1-Butene $H_2C=CHCH_2CH_3$ $=$

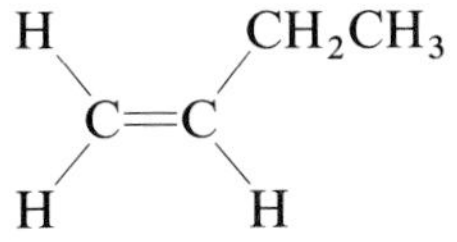

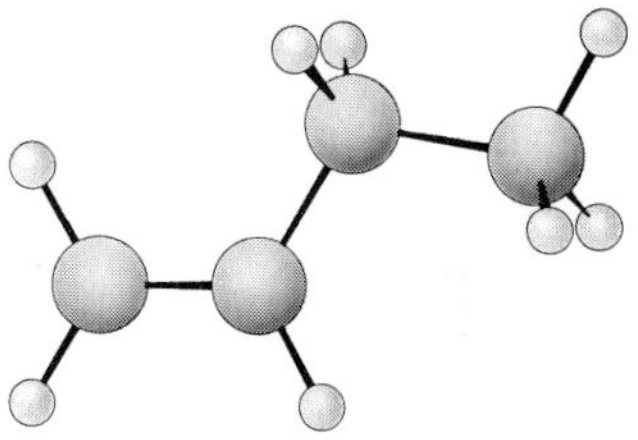

2-Butene $CH_3CH=CHCH_3$ $=$

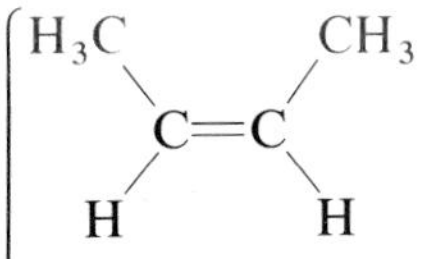

cis

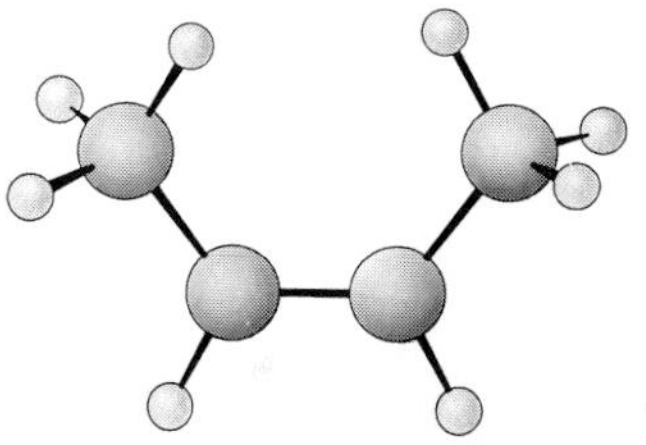

and

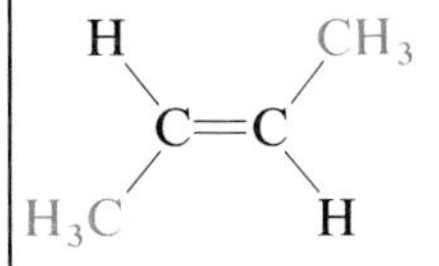

trans

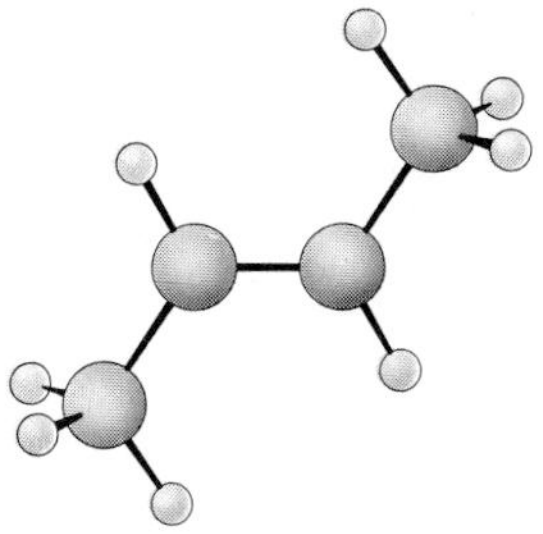

2-Methylpropene $H_2C=C(CH_3)_2$ $=$

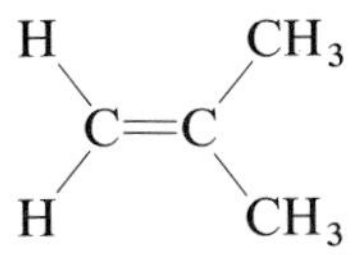

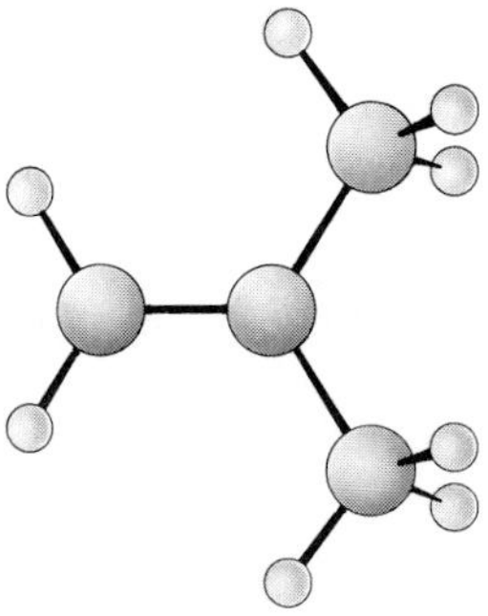

Isomers, cis–trans
Alkenes that have the same formula and connections between atoms but differ in having pairs of groups on opposite sides of the double bond.

Cis isomer Isomer with a specific pair of atoms or groups on the same side of the double bond.

Trans isomer Isomer with a specific pair of atoms or groups on opposite sides of the double bond.

The two 2-butenes are called **cis–trans isomers;** they have the same formula and connections between atoms but have different structures because of the way that groups are attached to different sides of the double bond. The **cis isomer,** the one with its methyl groups on the same side of the double bond, is named *cis*-2-butene. The **trans isomer,** the one with its methyl groups on opposite sides of the double bond, is named *trans*-2-butene. Notice above the considerable difference in shape between cis and trans isomers.

Side-view condensed formulas also illustrate the different geometries of cis and trans isomers. The following formulas are shown in a manner frequently used in organic chemistry: Bonds in the plane of the paper are solid lines; bonds that extend behind this plane are dashed lines; and bonds that come toward you from the paper are wedges.

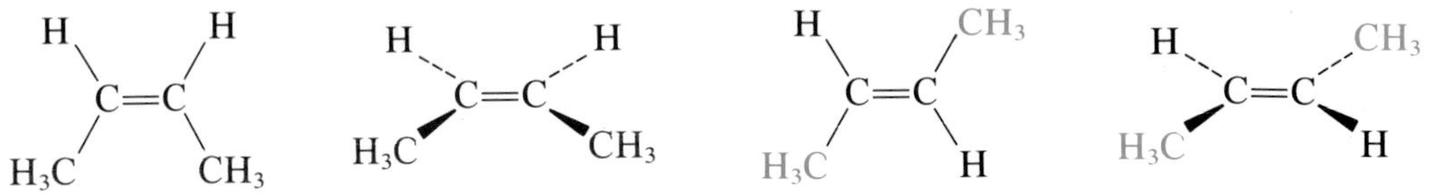

(Top view)	(Side view)	(Top view)	(Side view)
cis-2-Butene		*trans*-2-Butene	

Cis–trans isomerism occurs in an alkene whenever each double-bond carbon is bonded to two different substituent groups. If either double-bond carbon is attached to two identical groups, however, cis–trans isomerism is not possible. The two following compounds are identical, and cis–trans isomerism is impossible for them. You can convince yourself of this by mentally flipping one of the structures and seeing that it becomes identical to the other structure.

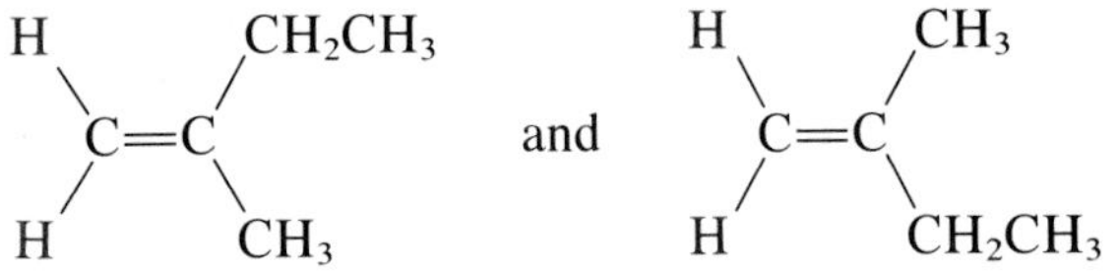

These compounds are identical. Because the left-hand carbon of the double bond has two —H's attached, cis–trans isomerism is impossible.

In the next two compounds, however, the structures do not become identical when flipped.

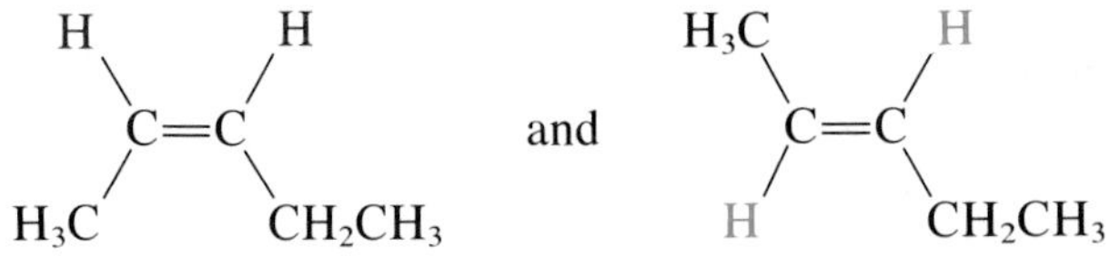

These compounds are not identical. Neither carbon of the double bond has two identical groups attached to it.

Solved Problem 2.3   Draw structures for the cis–trans isomers of 2-hexene.

Solution   First, draw a condensed structure of 2-hexene to see what groups are attached to the double-bond carbons.

$$CH_3CH{=}CHCH_2CH_2CH_3$$
$$123456$$

Attached to C2: —H and —CH$_3$
Attached to C3: —H and —CH$_2$CH$_2$CH$_3$

Next, draw two double bonds. Choose one end of each double bond and attach its groups in the same way to generate two identical part-structures:

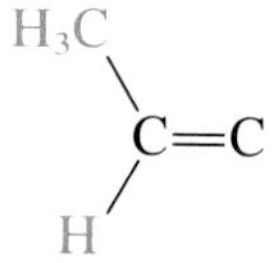

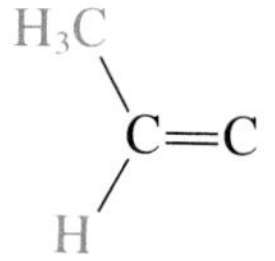

Finally, attach the proper groups to the other end in the two possible ways:

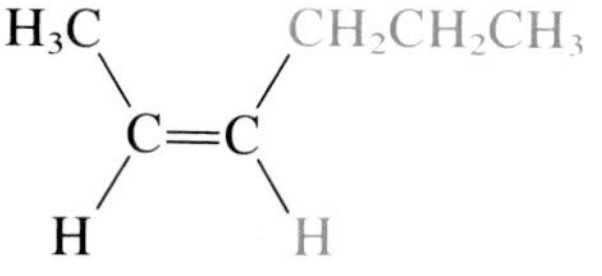

cis-2-Hexene and *trans*-2-Hexene

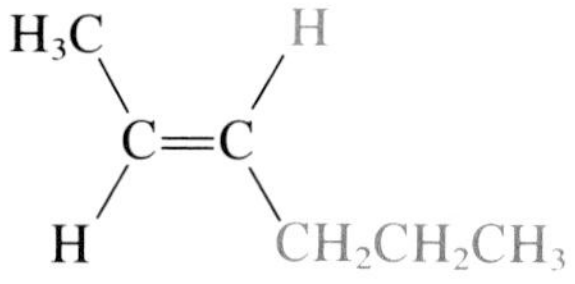

The compound with the two hydrogens on the same side of the double bond is named the cis isomer, and that with the two hydrogens on opposite sides is named the trans isomer.

Practice Problems   **2.3**   Which of these substances exist as cis–trans isomers?
(a) 3-heptene   (b) 2-methyl-2-hexene   (c) 5-methyl-2-hexene

2.4   Draw the cis–trans isomers of 3,4-dimethyl-3-hexene.

2.5 PROPERTIES OF ALKENES

Alkenes resemble alkanes in most general properties except for the chemical reactivity of their double bonds. The bonds in alkenes are essentially nonpolar, and the physical properties of alkenes, like those of alkanes, are influenced mainly by weak London forces. Alkenes with one to four carbon atoms are gases, and boiling points increase with the size of the molecules, as illustrated in Table 2.1.

Like alkanes, alkenes are insoluble in water, soluble in nonpolar solvents, and less dense than water. Notice in the table that the different geometrical arrangement in cis and trans isomers is reflected in different physical properties. The alkenes are flammable, and those that are gases present explosion hazards when mixed with air. Gaseous alkenes are not toxic but cause suffocation because of lack of oxygen. As you will see in the next section, alkenes react by addition across their double bonds.

Properties of Alkenes

● Nonpolar; insoluble in water; soluble in nonpolar organic solvents; less dense than water
● Alkane-like in melting points, boiling points, and other physical properties
● Flammable; gaseous alkenes not toxic
● Display cis–trans isomerism when each double-bond C atom has different substituents
● Chemically reactive at double bond, which undergoes addition reactions.

AN APPLICATION: THE CHEMISTRY OF VISION

Does eating carrots really improve your vision? Although carrots probably don't do much to help someone who's already on a proper diet, it's nevertheless true that the chemistry of carrots and the chemistry of vision are related.

Carrots, peaches, sweet potatoes, and other yellow vegetables are rich in β-carotene, an orange-colored alkene that provides our main dietary source of vitamin A. The conversion of β-carotene to vitamin A takes place in the liver, where enzymes first cut the molecule in half and then change the geometry of the C11-C12 double bond to produce a compound named 11-*cis*-retinal. After transport from the liver to the eye, 11-*cis*-retinal reacts with the protein *opsin* to produce the light-sensitive substance *rhodopsin*.

The human eye has two kinds of light-sensitive cells: *rod cells,* which are responsible for seeing in dim light, and *cone cells,* which are responsible for color vision. When light strikes the rod cells of the eye, cis–trans isomerization of the C11-C12 double bond occurs and 11-*trans*-rhodopsin, also called metarhodopsin II, is produced. This cis–trans isomerization is accompanied by a change in molecular geometry, which in turn causes a nerve impulse to be sent to the brain where it is perceived as vision. Metarhodopsin II is then changed back to 11-*cis*-retinal for use in another vision cycle.

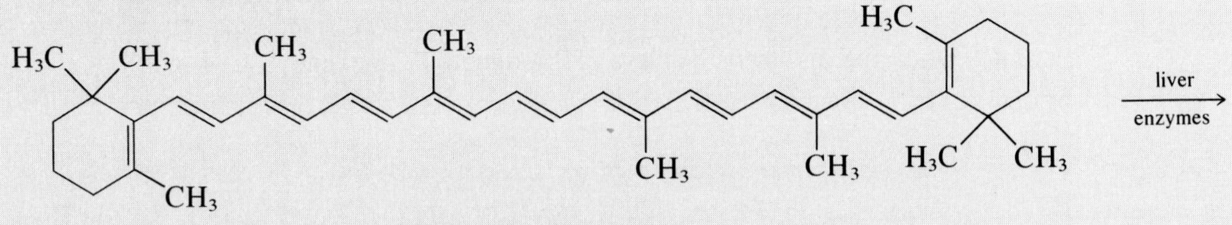

β-Carotene (a polyalkene)

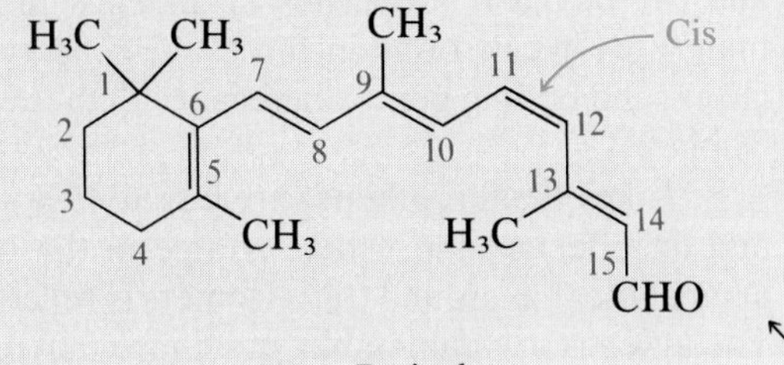

11-*cis*-Retinal

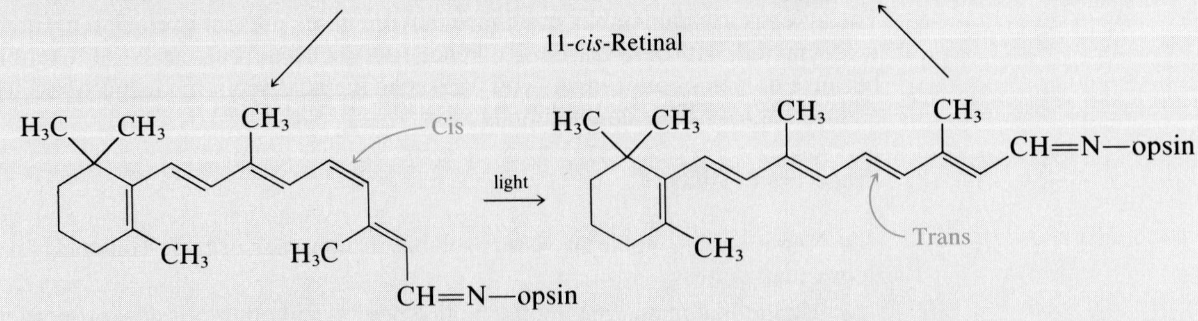

Rhodopsin

Metarhodopsin II

Table 2.1 Physical Properties of Simple Alkenes

Structure	Name	Boiling Point (°C)
$CH_2{=}CH_2$	Ethylene	−104
$CH_3CH{=}CH_2$	Propylene	−47
$CH_3CH_2CH{=}CH_2$	1-Butene	−6
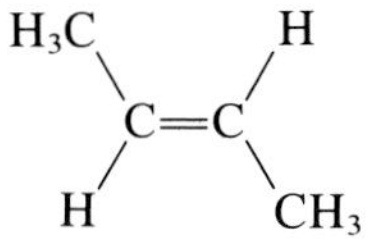	*trans*-2-Butene	1
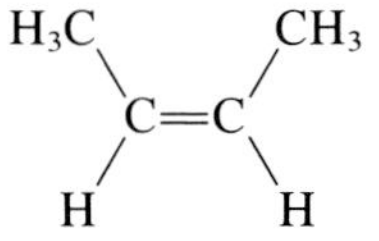	*cis*-2-Butene	4
$CH_3CH_2CH_2CH{=}CH_2$	1-Pentene	30
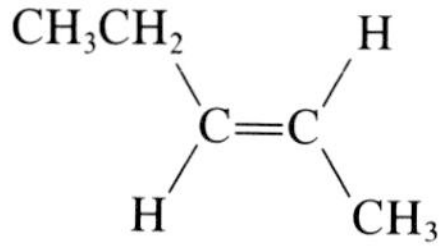	*trans*-2-Pentene	36
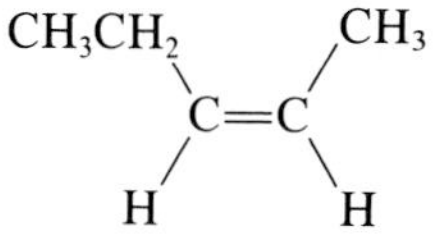	*cis*-2-Pentene	37

2.6 CHEMICAL REACTIONS OF ALKENES AND ALKYNES

Most of the reactions of carbon–carbon multiple bonds can be grouped under the category of **addition reactions,** which have the general pattern A + B → C. A reagent we might write as X—Y adds to the multiple bond in the unsaturated starting material to yield a product that has only single bonds. Alkenes and alkynes react similarly in many ways, but we'll look mainly at alkenes in this chapter because they're more common and more important.

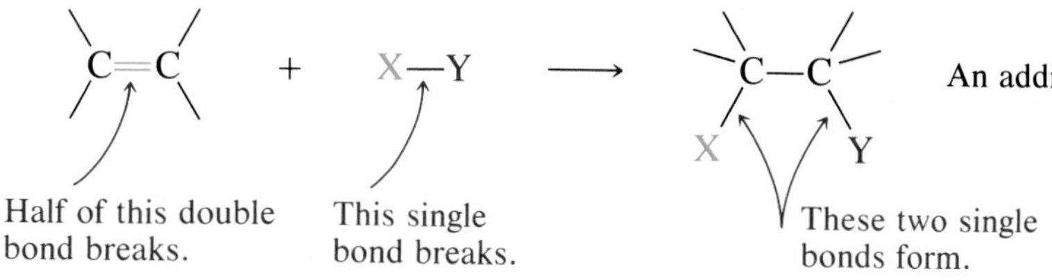

An addition reaction

Half of this double bond breaks. This single bond breaks. These two single bonds form.

Addition of Hydrogen to Alkenes and Alkynes: Hydrogenation   Alkenes and alkynes react with hydrogen in the presence of a metal catalyst to yield the corresponding alkane product:

REVIEW A catalyst speeds up a chemical reaction without itself undergoing any permanent chemical change.

(An alkene) $\text{C}=\text{C}$ + H_2 $\xrightarrow{\text{catalyst}}$ $\text{C}-\text{C}$ (An alkane)

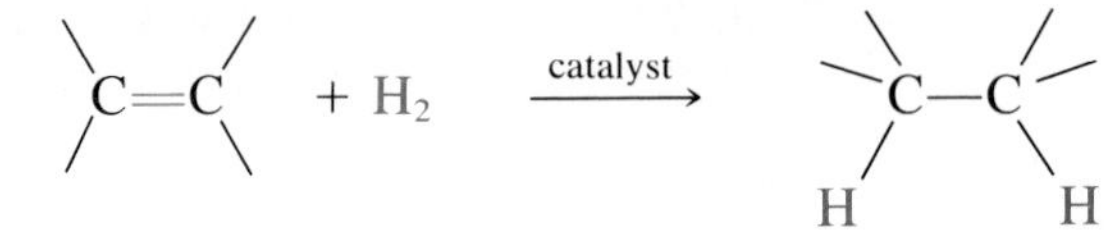

(An alkyne) $-\text{C}\equiv\text{C}-$ + 2 H_2 $\xrightarrow{\text{catalyst}}$ (An alkane)

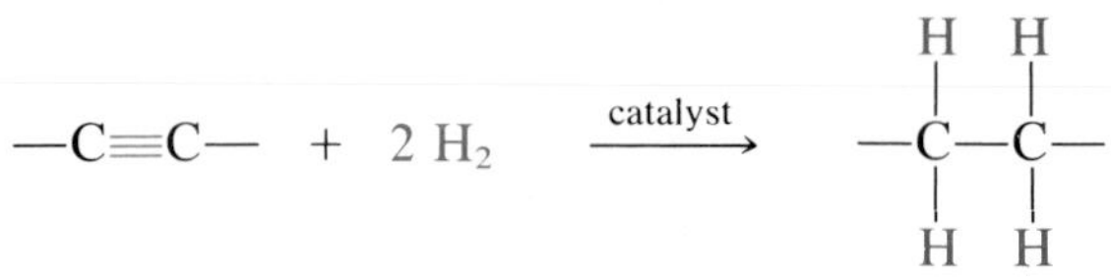

For example,

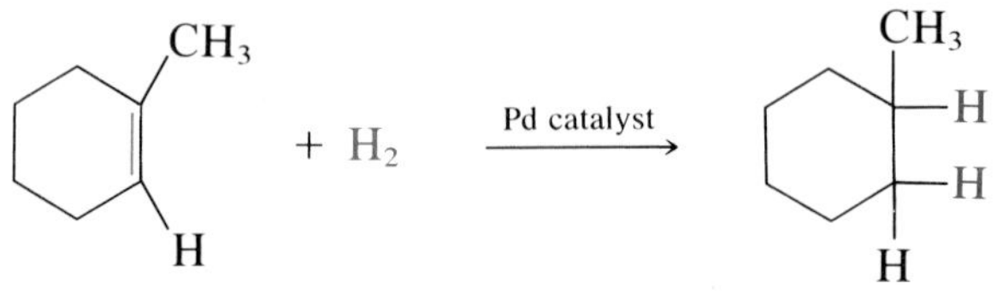

1-Methylcyclohexene + H_2 $\xrightarrow{\text{Pd catalyst}}$ Methylcyclohexane (85% yield)

Hydrogenation The reaction of an alkene (or alkyne) with H_2 to yield an alkane product.

The addition of hydrogen to an alkene, often called **hydrogenation,** is used commercially to convert unsaturated vegetable oils, which contain numerous double bonds, to the saturated fats used in margarine and cooking fats. We'll see exact structures for these fats and oils in Chapter 12.

Solved Problem 2.4 What product would you obtain from the following reaction?

$$\text{CH}_3\text{CH}_2\text{CH}_2\text{CH}=\text{CHCH}_3 + \text{H}_2 \xrightarrow{\text{Pd}} \text{?}$$

Solution First, rewrite the starting material, showing a single bond and two partial bonds in place of the double bond:

$$\text{CH}_3\text{CH}_2\text{CH}_2\text{CH}-\text{CHCH}_3$$

Next, add one hydrogen to each carbon atom of the double bond and rewrite the product in condensed form:

$$\text{CH}_3\text{CH}_2\text{CH}_2\text{CH}-\text{CHCH}_3 = \text{CH}_3\text{CH}_2\text{CH}_2\text{CH}_2\text{CH}_2\text{CH}_3 \quad \text{Hexane}$$

The reaction is

$$\text{CH}_3\text{CH}_2\text{CH}_2\text{CH}=\text{CHCH}_3 + \text{H}_2 \xrightarrow{\text{Pd}} \text{CH}_3\text{CH}_2\text{CH}_2\text{CH}_2\text{CH}_2\text{CH}_3$$

Practice Problem 2.5 Write the structures of products of these hydrogenation reactions:

(a) $\text{CH}_3\text{CH}_2\text{CH}=\text{CH}_2 + \text{H}_2 \xrightarrow{\text{Pd}} \text{?}$ (b) *cis*-2-butene + $\text{H}_2 \xrightarrow{\text{Pd}} \text{?}$

(c) *trans*-2-butene + $\text{H}_2 \xrightarrow{\text{Pd}} \text{?}$ (d) 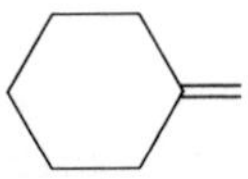 $\bigcirc=\text{CH}_2 + \text{H}_2 \xrightarrow{\text{Pd}} \text{?}$

Halogenation, alkene The reaction of an alkene with a halogen (Cl_2 or Br_2) to yield a 1,2-dihaloalkane product.

Addition of Cl_2 and Br_2 to Alkenes: Halogenation Alkenes react with the halogens Br_2 and Cl_2 to give 1,2-dihaloalkane addition products in a **halogenation** reaction:

$$\begin{array}{c}\diagdown\diagup\\ C=C\\ \diagup\diagdown\end{array} + X_2 \longrightarrow \begin{array}{c}\diagdown\diagup\\ -C-C-\\ \diagup||\diagdown\\ XX\end{array}$$

(A 1,2-dihaloalkane where X = Br or Cl)

For example,

$$\begin{array}{c}HH\\ \diagdown\diagup\\ C=C\\ \diagup\diagdown\\ HH\end{array} + Cl_2 \longrightarrow \begin{array}{c}HH\\ ||\\ H-C-C-H\\ ||\\ ClCl\end{array}$$

Ethylene   1,2-Dichloroethane

Approximately 7 million tons per year of 1,2-dichloroethane are manufactured by this reaction as the first step in making PVC [poly(vinyl chloride)] plastics.

Bromination provides a convenient test for the presence in a molecule of a carbon–carbon double or triple bond (Figure 2.1). A drop of the reddish-brown solution of bromine in carbon tetrachloride (CCl_4) is added to a sample of an unknown compound. Immediate disappearance of the color as the bromine reacts to form a colorless dibromide reveals the presence of the multiple bond.

Practice Problem   2.6 What products would you expect from the following halogen addition reactions?

(a) 2-methylpropene + Br_2 → ?   (b) 1-pentene + Cl_2 → ?

Figure 2.1
Testing for unsaturation with bromine. (a) No color change results when the bromine solution is added to hexane (C_6H_{14}). (b) Disappearance of the bromine color when it is added to 1-hexene (C_6H_{12}) indicates the presence of a double bond.

(a)   (b)

Hydrohalogenation The reaction of an alkane with HCl or HBr to yield an alkyl halide product.

Addition of HBr and HCl to Alkenes Alkenes react with hydrogen bromide (HBr) to yield *alkyl bromides* (R—Br) and with hydrogen chloride (HCl) to yield *alkyl chlorides* (R—Cl), in what are called **hydrohalogenation** reactions:

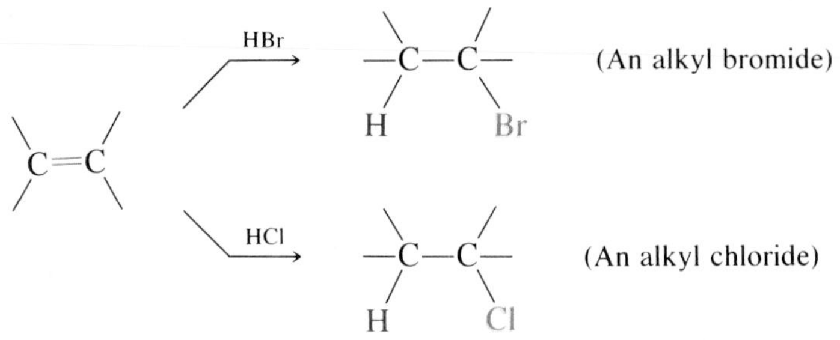

$$\begin{array}{c} \text{HBr} \\ \longrightarrow \end{array} \quad \overset{\displaystyle|}{\underset{\displaystyle H}{C}}\!-\!\overset{\displaystyle|}{\underset{\displaystyle Br}{C}} \qquad \text{(An alkyl bromide)}$$

$$C\!=\!C$$

$$\begin{array}{c} \text{HCl} \\ \longrightarrow \end{array} \quad \overset{\displaystyle|}{\underset{\displaystyle H}{C}}\!-\!\overset{\displaystyle|}{\underset{\displaystyle Cl}{C}} \qquad \text{(An alkyl chloride)}$$

The addition of HBr to 2-methylpropene is typical of such reactions:

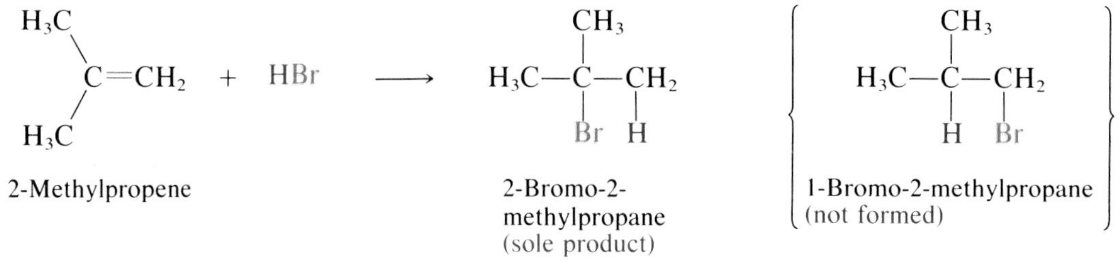

$$\underset{\text{2-Methylpropene}}{\overset{\displaystyle H_3C}{\underset{\displaystyle H_3C}{\Large>}}C\!=\!CH_2} \; + \; HBr \; \longrightarrow \; \underset{\substack{\text{2-Bromo-2-} \\ \text{methylpropane} \\ \text{(sole product)}}}{H_3C\!-\!\overset{\displaystyle CH_3}{\underset{\displaystyle Br}{C}}\!-\!\overset{}{\underset{\displaystyle H}{CH_2}}} \qquad \left\{\underset{\substack{\text{1-Bromo-2-methylpropane} \\ \text{(not formed)}}}{H_3C\!-\!\overset{\displaystyle CH_3}{\underset{\displaystyle H}{C}}\!-\!\overset{}{\underset{\displaystyle Br}{CH_2}}}\right\}$$

Look carefully at the previous example. Only one of the two possible addition products is obtained. 2-Methylpropene *could* add HBr to give 1-bromo-2-methylpropane, but it doesn't; it gives only 2-bromo-2-methylpropane. This is what usually happens when HBr and HCl (or other reagents) add to alkenes with unsymmetrically substituted double bonds, that is, double bonds in which one carbon is bonded to more hydrogens than the other. Such addition takes place according to a rule formulated in 1869 by the Russian chemist Vladimir Markovnikov:

Markovnikov's rule In the addition of HX to an alkene, the H becomes attached to the carbon that already has the most H's, and the X becomes attached to the carbon that has fewer H's.

Markovnikov's rule: In the addition of HX to an alkene, the H becomes attached to the carbon that already has the most H's, and the X becomes attached to the carbon that has fewer H's.

This rule allows prediction of the results of hydrohalogenation:

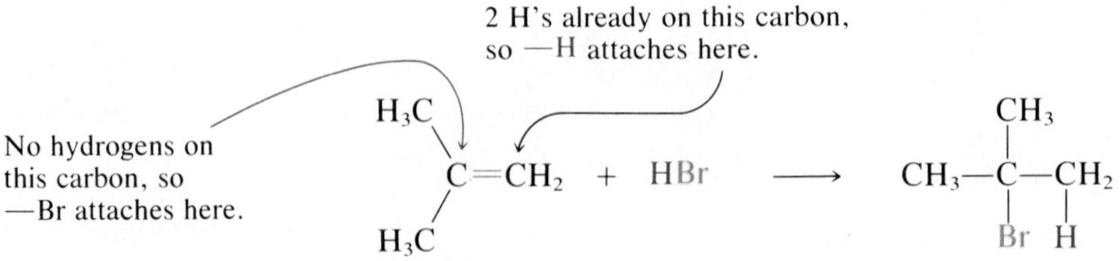

2 H's already on this carbon, so —H attaches here.

No hydrogens on this carbon, so —Br attaches here.

$$\underset{\displaystyle H_3C}{\overset{\displaystyle H_3C}{\Large>}}C\!=\!CH_2 \; + \; HBr \; \longrightarrow \; CH_3\!-\!\overset{\displaystyle CH_3}{\underset{\displaystyle Br}{C}}\!-\!\overset{}{\underset{\displaystyle H}{CH_2}}$$

Solved Problem 2.5   What product would you expect from the following reaction?

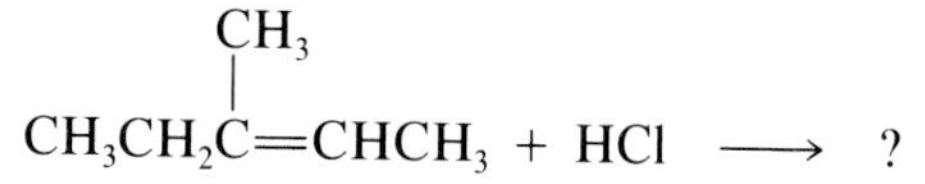

$$\underset{\displaystyle CH_3CH_2\overset{\displaystyle \overset{\textstyle CH_3}{|}}{C}=CHCH_3}{} + HCl \longrightarrow ?$$

Solution   We know that reaction of an alkene with HCl leads to formation of an alkyl chloride addition product according to Markovnikov's rule. To make a prediction, look at the starting alkene and count the number of hydrogens already attached to each double-bond carbon. Then write the product by attaching H to the carbon that already has more hydrogens and attaching Cl to the carbon that has fewer hydrogens.

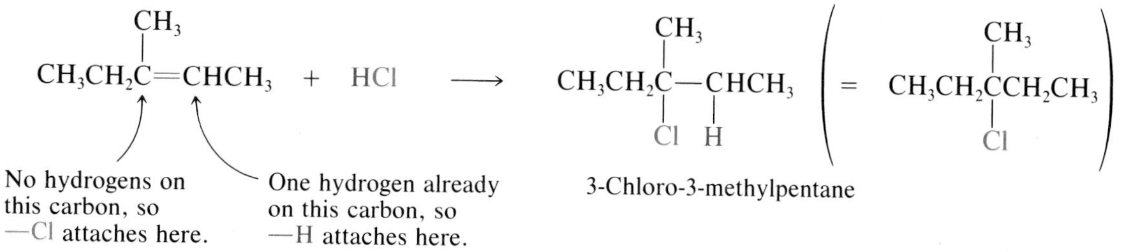

$$CH_3CH_2\overset{\overset{\textstyle CH_3}{|}}{C}=CHCH_3 \;+\; HCl \;\longrightarrow\; CH_3CH_2\overset{\overset{\textstyle CH_3}{|}}{\underset{\underset{\textstyle Cl}{|}}{C}}-\overset{}{\underset{\underset{\textstyle H}{|}}{C}}HCH_3 \quad \left(= CH_3CH_2\overset{\overset{\textstyle CH_3}{|}}{\underset{\underset{\textstyle Cl}{|}}{C}}CH_2CH_3 \right)$$

No hydrogens on this carbon, so —Cl attaches here.   One hydrogen already on this carbon, so —H attaches here.   3-Chloro-3-methylpentane

---

Practice Problems   **2.7**   What products would you expect from these reactions?
(a) cyclohexene + HBr → ?   (b) 1-hexene + HCl → ?
(c) $(CH_3)_2CHCH=CH_2$ + HI → ?

2.8   What alkenes are the following alkyl halides likely to have been made from? (Careful, there may be more than one answer.)
(a) 3-chloro-3-ethylpentane   (b) $(CH_3)_2CHCBr(CH_3)_2$

Addition of Water to Alkenes: Hydration   An alkene doesn't react with pure water if the two are just mixed together, but if the right experimental conditions are used, an addition reaction takes place to yield an *alcohol* (R—OH). This **hydration** reaction occurs on treatment of the alkene with water in the presence of a strong acid catalyst such as H_2SO_4. In fact, nearly 300 million gallons of ethyl alcohol (ethanol) are produced each year in the United States by this method.

Hydration, alkene   The reaction of an alkene with water to yield an alcohol.

REVIEW   Acids and bases are reviewed in Appendix B.

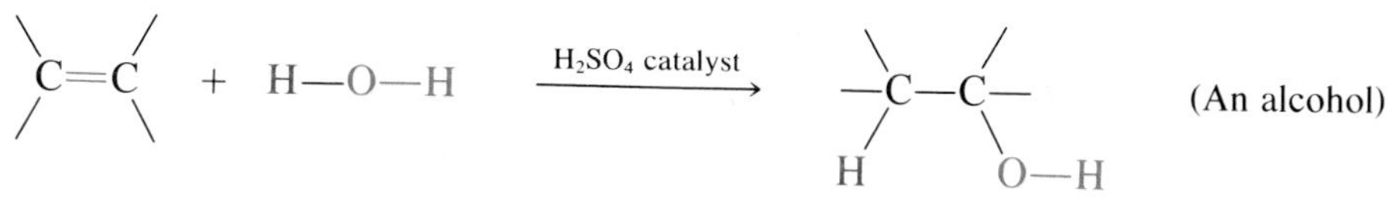

$$\underset{}{\overset{}{C}}{=}\underset{}{\overset{}{C}} \;+\; H{-}O{-}H \;\xrightarrow{\;H_2SO_4 \text{ catalyst}\;}\; -\overset{}{\underset{\underset{\textstyle H}{|}}{C}}-\overset{}{\underset{\underset{\textstyle O-H}{|}}{C}}- \qquad \text{(An alcohol)}$$

For example,

$$H_2C{=}CH_2 \;+\; H_2O \;\xrightarrow{\;H_2SO_4,\ 250°C\;}\; H_2\overset{}{\underset{\underset{\textstyle H}{|}}{C}}-\overset{}{\underset{\underset{\textstyle O-H}{|}}{C}}H_2 \qquad (= CH_3CH_2OH)$$

Ethylene   Ethyl alcohol

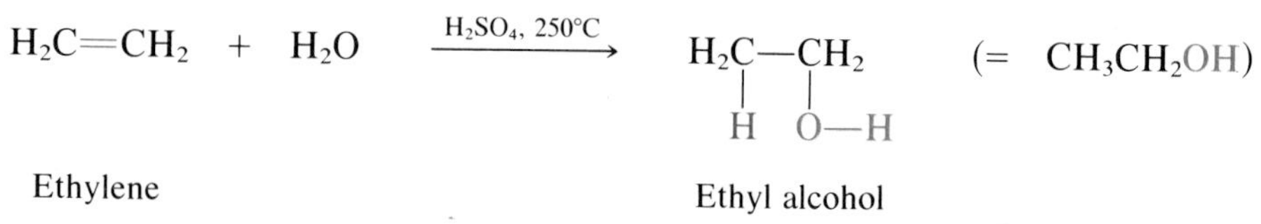

As with the addition of HBr and HCl, Markovnikov's rule can be used to predict the product when water adds to an unsymmetrically substituted alkene:

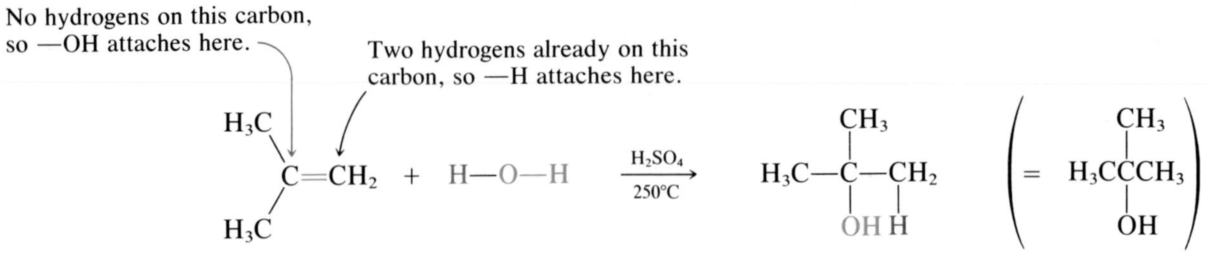

No hydrogens on this carbon, so —OH attaches here.

Two hydrogens already on this carbon, so —H attaches here.

$$H_3C \\ C{=}CH_2 + H{-}O{-}H \xrightarrow[250°C]{H_2SO_4} H_3C{-}\underset{\underset{OH}{|}}{\overset{\overset{CH_3}{|}}{C}}{-}\underset{\underset{H}{|}}{C}H_2 \quad \left(= H_3C\underset{\underset{OH}{|}}{\overset{\overset{CH_3}{|}}{C}}CH_3 \right)$$

2-Methyl-2-propanol

Practice Problems   **2.9**   What products would you expect from these hydration reactions?

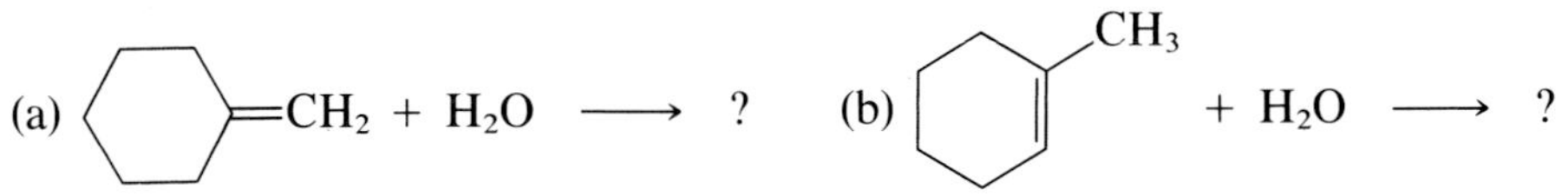

(a) [structure] $={CH_2} + H_2O \longrightarrow$?   (b) [structure with CH_3] $+ H_2O \longrightarrow$?

2.10   What alkene starting material might 3-methyl-3-pentanol be made from?

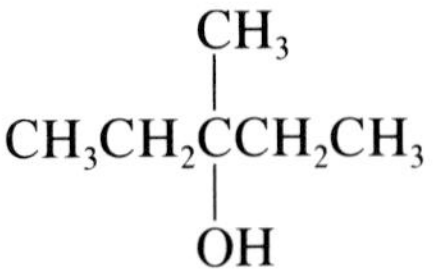

$$CH_3CH_2\underset{\underset{OH}{|}}{\overset{\overset{CH_3}{|}}{C}}CH_2CH_3 \qquad \text{3-Methyl-3-pentanol}$$

2.7   HOW AN ALKENE ADDITION REACTION OCCURS

How do addition reactions take place? Do two molecules, say ethylene and HBr, simply collide and immediately form a product molecule of bromoethane, or is the process more complex? Studies have shown that alkene addition reactions take place in two distinct steps, as illustrated in Figure 2.2 for the addition of HBr to ethylene.

In the first step, the alkene reacts with H^+ from the acid HBr. The carbon–carbon double bond partially breaks, and two electrons go to form a new covalent bond between one of the carbons and the incoming hydrogen. The other double-bond carbon, having had electrons removed from it, now has only six electrons in its outer shell and bears a positive charge. Unlike sodium ion (Na^+) and other metal cations, which are stable and easily isolated in salts like NaCl, such **carbocations** are quite unstable. As soon as it's formed by reaction of an alkene with H^+, the carbocation immediately reacts with bromide ion to form a neutral product.

An accounting like that in Figure 2.2 of the individual steps by which old bonds are broken and new bonds formed in a reaction is known as a **reaction mechanism.** Although we won't examine many reaction mechanisms in this book, understanding mechanisms is an important part of organic chemistry.

Carbocation   A polyatomic ion with a positively charged carbon atom.

Reaction mechanism   A complete description of how a reaction occurs, including the details of each individual step in the overall process.

$$H^+$$

$$\begin{array}{c} H \qquad\quad H \\ \diagdown \qquad \diagup \\ C::C \\ \diagup \qquad \diagdown \\ H \qquad\quad H \end{array}$$

Step 1
Two electrons from the C—C double bond are used to form a new single bond between the incoming hydrogen ion and one of the carbons. The other carbon now has only six electrons in its outer shell and has a positive charge.

$$\left[\begin{array}{c} \quad H \quad H \\ \quad | \quad\; | \\ H-C-C-H \\ \quad^+ \quad\; | \\ \qquad\qquad H \\ \;\; :\!Br\!:^- \end{array} \right] \quad a\ carbocation$$

Step 2
The positively charged *carbocation* then reacts with the negative bromide ion, using an electron pair from bromide to form a single bond between carbon and bromine. The second carbon thus regains an outer-shell octet.

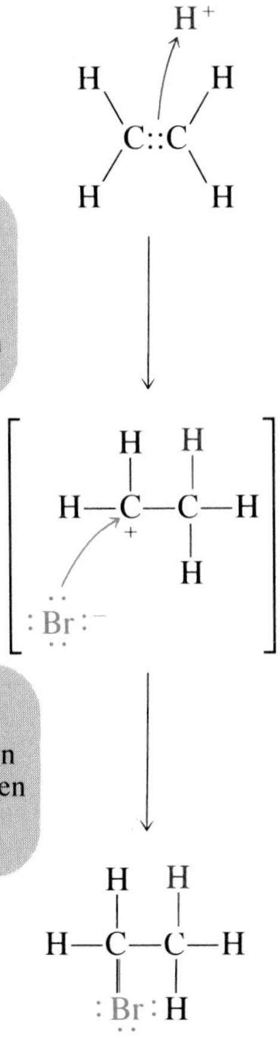

$$\begin{array}{c} H \quad H \\ | \quad\; | \\ H-C-C-H \\ | \quad\; | \\ :Br:\ H \end{array}$$

Figure 2.2
How the addition reaction of HBr to an alkene occurs: a reaction mechanism. The reaction takes place in two steps and involves a carbocation intermediate.

Their study is, for example, essential to our ever-expanding ability to understand biochemistry and the physiological effects of drugs. If you continue your study of chemistry, you'll see reaction mechanisms often.

Practice Problem   ***2.11***   Draw the structure of the carbocation intermediate formed during the reaction of 2-methylpropene with HCl.

2.8   ALKENE POLYMERS

Polymer   A very large molecule composed of identical repeating units and formed by the combination of small molecules.

Monomer   A small molecule combined to form a polymer.

Polymers are large molecules formed by the bonding together of many smaller molecules called **monomers.** As we'll see in later chapters, *biological polymers* occur throughout nature. Cellulose and starch, for example, are polymers built from sugars. Although the basic idea is the same, synthetic polymers are much simpler than biopolymers, since the starting monomer units are usually small and simple organic molecules.

AN APPLICATION: ISOPRENE, TERPENES, AND NATURAL RUBBER

Nature is a clever architect, constructing many complex biomolecules by combination of smaller, simpler units. The isoprene structure and the molecules that contain it provide a beautiful example of this principle.

A branched alkene with two double bonds

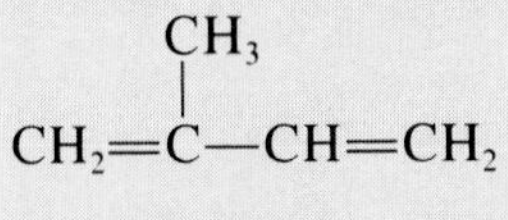

$$CH_2=C-CH=CH_2$$
$$\overset{\displaystyle CH_3}{|}$$

Isoprene

Isoprene itself is not a natural product. Nevertheless, an amazing array of molecules that contain the five-carbon isoprene skeleton are found in plants and animals, where they perform a wide variety of functions. Some variations on the isoprene unit are drawn here as line structures. In some, a double bond is preserved, and in others it is lost.

Isoprene   Some isoprene units in terpenes

Many essential oils owe their distinctive aromas to simple *terpenes*, compounds composed entirely of isoprene units and found in all plants, for example,

Myrcene—two isoprene units
(oil of bay)

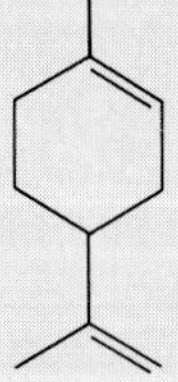

Limonene—two isoprene units
(oil of lemon, orange, caraway)

More complex terpenes and compounds with various functional groups on their isoprene units are also widely found in nature. The hydrocarbon squalene, for example, is an intermediate used by organisms in synthesizing steroids, an important class of biomolecules (Section 12.10). Retinal, a key substance in vision, is also a terpene.

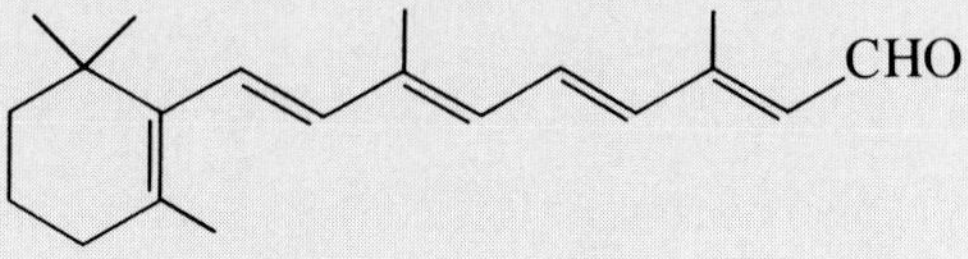

CHO

Retinal—four isoprene units

The raw material from rubber trees is a polyisoprene with an average of 5000 isoprene units per molecule. Interestingly, the double bonds that remain all have cis geometry.

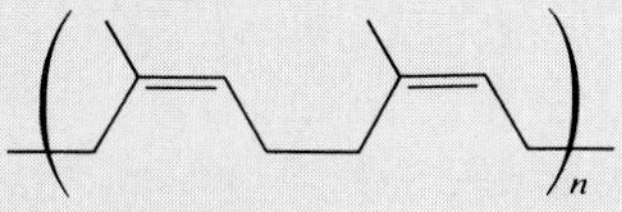

Natural rubber (*cis*-polyisoprene)

Rubber collected in the Amazon region in Brazil is being smoked to dry it. Natural rubber is sticky and not too useful until the polymer molecules are connected (*vulcanized*) by cross links that pull them back together when rubber is stretched.

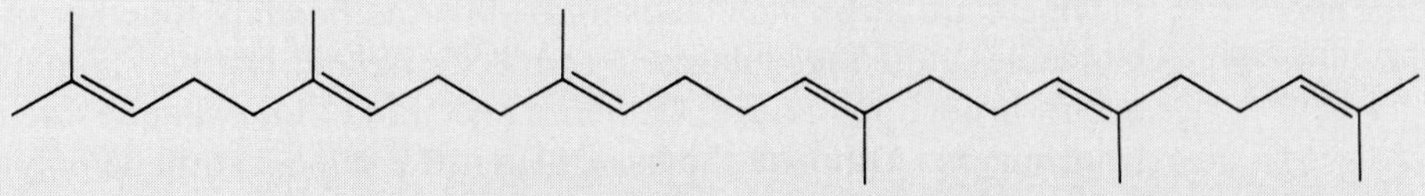

Squalene—six isoprene units

Vinyl monomer A compound with the structure CH_2=CHZ that undergoes polymerization to give a polymer with the repeating unit —CH_2—CHZ—.

Polymerization A reaction in which monomers combine to form a polymer.

Chain-growth polymer A polymer formed by the addition of monomer molecules one by one to the end of a growing chain.

Many simple alkenes (called **vinyl monomers** because H_2C=CH_2 is known as a *vinyl group*) undergo **polymerization** reactions when treated with the proper catalysts. Ethylene yields polyethylene on polymerization, propylene yields polypropylene (Figure 2.3), and styrene yields polystyrene. The polymer product might have anywhere from a few hundred to a few thousand monomer units incorporated into a long chain.

The fundamental reaction in the polymerization of a vinyl monomer resembles the additions to double bonds described in the previous section. The addition to an alkene of a species called an *initiator* yields a reactive intermediate, which in turn adds to a second alkene molecule to produce another reactive intermediate, which adds to a third alkene molecule, and so on. Because the result is continuous addition of one monomer after another to the end of the growing polymer chain, polymers made in this way are classified as **chain-growth polymers** (also known as addition polymers).

$$H_2C=\overset{\overset{\displaystyle Z}{|}}{CH} \quad \xrightarrow{\text{polymerization}} \quad -(-CH_2-\overset{\overset{\displaystyle Z}{|}}{CH}-CH_2-\overset{\overset{\displaystyle Z}{|}}{CH}-CH_2-\overset{\overset{\displaystyle Z}{|}}{CH}-)_{\overline{n}}$$

Vinyl monomer (where the colored Z represents a substituent group)

Polymer

REVIEW Density is mass per unit volume.

Variations in the Z group impart very different properties, as illustrated by the familiar alkene polymers listed in Table 2.2. Some vinyl polymers, like styrene–butadiene rubber shown in Table 2.2, are *copolymers:* polymers made from more than one kind of monomer.

Polymer properties also vary depending on the average size of the huge molecules (*macromolecules*) in a particular sample and on how much they are branched. The long molecules in straight-chain polyethylene pack close together, giving a rigid material (*high-density polyethylene*) mainly used in bottles for products such as milk and motor oil (Figure 2.4a). When polyethylene molecules contain many branches, they can't pack together so tightly and instead form a flexible material (*low-density polyethylene*) used mainly in packaging films (Figure 2.4b).

Figure 2.3
Plastics and creative endeavor. In 1983, the artist Christo surrounded 11 islands in Biscayne Bay with 6.5 million square feet of floating pink fabric to create a striking work of environmental art. The fabric used was woven polypropylene, an alkene polymer. (© Christo, 1983.)

Table 2.2   Some Alkene Polymers and Their Uses

Monomer Name	Monomer Structure	Polymer Name	Uses
Ethylene	$H_2C{=}CH_2$	Polyethylene	Packaging, bottles
Propylene	$H_2C{=}CH{-}CH_3$	Polypropylene	Bottles, rope, pails, medical tubing
Vinyl chloride	$H_2C{=}CH{-}Cl$	Poly(vinyl chloride)	Insulation, plastic pipe
Styrene	$H_2C{=}CH{-}$⬡	Polystyrene	Foams and molded plastics
Styrene and butadiene	$H_2C{=}CH{-}$⬡ and $H_2C{=}CHCH{=}CH_2$	Styrene–butadiene rubber (SBR)	Synthetic rubber for tires
Acrylonitrile	$H_2C{=}CH{-}C{\equiv}N$	Orlon, Acrilan	Fibers, outdoor carpeting
Methyl methacrylate	$H_2C{=}\overset{\displaystyle O}{\overset{\|}{C}}\underset{\underset{\displaystyle CH_3}{\|}}{C}OCH_3$	Plexiglas, Lucite	Windows, contact lenses, fiber optics
Tetrafluoroethylene	$F_2C{=}CF_2$	Teflon	Nonstick coatings, bearings, replacement heart valves and blood vessels

Figure 2.4
Polyethylene items. (a) Low-density polyethylene (LDPE) is moisture-proof but flexible and is used in plastic bags and packaging of all kinds. (b) High-density polyethylene (HDPE) is strong and rigid enough to be used in many kinds of bottles.

(a)   (b)

Polymer technology has come a long way since the development of synthetic rubber, nylon, Plexiglas, and Teflon, which was hastened by the need for materials during World War II. The use of polymers has changed the nature of activities from plumbing and carpentry to clothing and auto manufacture. In the health care fields one result has been the proliferation of inexpensive, disposable equipment (Figure 2.5).

Figure 2.5
Disposable syringes. These syringes have been removed from their packaging, used once, and discarded.

Practice Problem   ***2.12***   Write the structure of the repeating units in (a) polystyrene and (b) polyacrylonitrile (Acrilan).

2.9 ALKYNES

Alkynes are named systematically in the same manner as alkenes and are like alkenes in physical and chemical properties.

$$HC \equiv CH \qquad CH_3C \equiv CH \qquad \underset{\underset{CH_3}{|}}{CH_3CHC} \equiv CH$$

Acetylene   Propyne   3-Methyl-1-butyne
(ethyne)

     Acetylene (ethyne), the commercially most important alkyne, is a colorless gas with a light odor like that of ether. Industrial-grade acetylene used in cutting and welding, however, often has a bad odor because of impurities. When mixed with oxygen in an oxyacetylene welding torch, burning acetylene produces a clear, light-blue flame with temperatures up to 3300°C.

     Alkynes are not plentiful in nature, but they have been found in a number of plants and some animals. The compound known as *ichthyothereol* was identified in a natural poison used by Amazonian indians on their arrowheads. Not many alkynes play essential roles in human biochemistry, but one group of alkynes is indeed significant: oral contraceptives.

Some physiologically active alkynes

Ichthyothereol
(a poison of plant origin)

Norethynodrel
(a contraceptive, used in Enovid and others)

2.10 AROMATIC COMPOUNDS AND THE STRUCTURE OF BENZENE

In the early days of organic chemistry, the word *aromatic* was used to describe certain fragrant substances from fruits, trees, and other natural sources. It was soon realized, however, that substances grouped as aromatic behaved in a chemically different manner from most other organic compounds. Today, the term *aromatic* is used to refer to the class of compounds containing benzene-like rings. The association with fragrance has long been lost.

Aromatic compound   A compound that contains a six-membered ring of carbon atoms with three double bonds or a similarly stable ring.

Benzene, the simplest **aromatic compound,** is a flat, symmetrical molecule with the molecular formula C_6H_6. It's often represented as cyclohexatriene: a six-membered carbon ring with three double bonds. The problem with this representation is that it gives the wrong impression about benzene's chemical reactivity and bonding. Since benzene appears to have three double bonds, you might expect it to react with H_2, Br_2, HCl, and H_2O to give the same kinds of addition products that alkenes do. But this expectation is wrong. Benzene and other aromatic compounds are much less reactive than alkenes and don't normally undergo addition reactions.

Benzene's remarkable stability is a consequence of its structure. Forming a ring of six carbons joined by single bonds to each other and to six hydrogen atoms (Figure 2.6a) leaves six electrons left over. Placing these electrons in three alternating double bonds gives each carbon atom an electron octet. But where are these double bonds located in the ring? There are two equivalent possibilities, as shown in Figure 2.6b.

The problem with the cyclohexatriene representations for benzene is that neither of the two equivalent structures in Figure 2.6b is fully correct by itself. Experimental evidence shows that all six carbon–carbon bonds in benzene are identical, so a picture with three double bonds and three single bonds can't be right.

Resonance   The existence of a molecule in a single structure intermediate among two or more correct possible double-bond-containing structures that can be drawn.

The stability of benzene can best be explained by assuming it to be a completely symmetrical molecule with the six "double-bond" electrons circulating around the *entire* ring (Figure 2.6c). Such a structure is hard to represent by the standard conventions using lines for covalent bonds. Sometimes it's done by representing the six electrons as a circle inside the six-membered ring (Figure 2.6d). Usually, though, we draw the ring with three double bonds, with the understanding that it's an aromatic ring with equivalent bonding all around. Similar limitations arise in placing double bonds in the structures of many molecules. The name **resonance** is given to this phenomenon: the existence of a molecule in a single structure intermediate among two or more possible correct double-bond-containing structures that can be drawn.

Figure 2.6
Some representations of benzene. The planar molecular structure is shown in (a), and an electron density cloud of bonding electrons distributed around the ring is shown in (c). Benzene is usually represented by the two equivalent structures in (b) or by the symmetrical structure in (d).

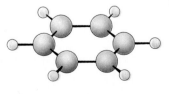

(a)

and

(b)

(c)

(d)

The simple aromatic hydrocarbons are nonpolar, insoluble in water, volatile, and flammable. Unlike the alkanes and alkenes, however, several are toxic. Benzene exposure must be limited because it is carcinogenic, and the dimethyl-substituted benzenes are central-nervous-system depressants.

Everything that we've said about the structure and stability of the benzene ring (C_6H_6) also applies to the ring when it has substituents, for example, in compounds such as hexachlorophene and vanillin:

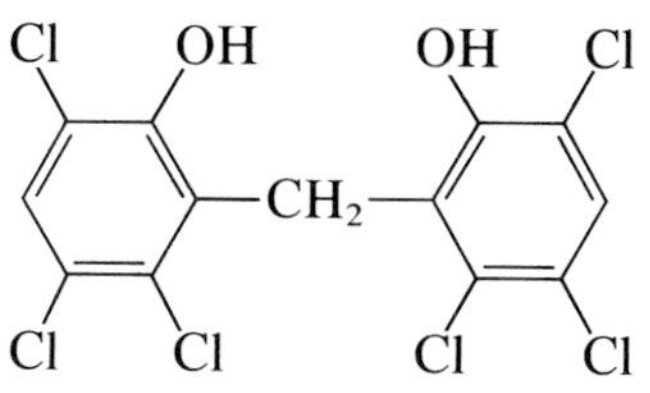

Hexachlorophene
(a germicide)

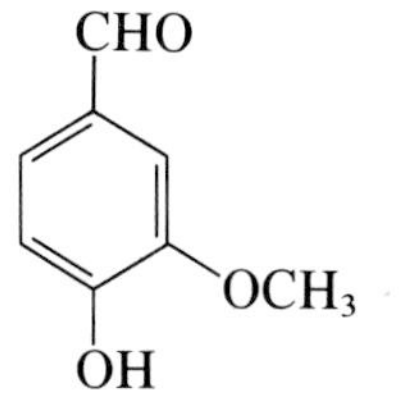

Vanillin
(vanilla flavoring)

The benzene ring is also present in many biomolecules and retains its characteristic properties in these compounds as well. "Aromaticity" is not limited to rings containing only carbon. Many aromatic compounds contain nitrogen atoms, as shown here for indole and adenine. These and all compounds that contain a substituted benzene ring, or a similar stable ring in which electrons are equally shared around the ring, are classified as aromatic compounds.

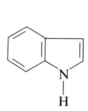

Indole
(from indigo plant and coal tar,
used in perfume)

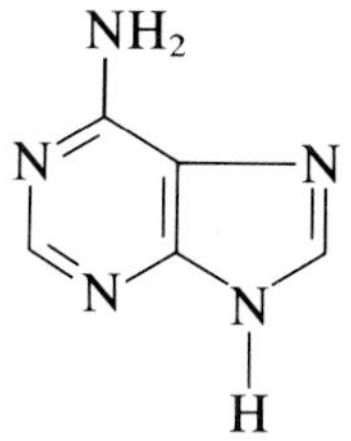

Adenine
(a nucleic acid component)

2.11 NAMING AROMATIC COMPOUNDS

Compounds in which one or more hydrogen atoms in the benzene ring are replaced by other groups use *-benzene* (or *-benz-*) as the root for their systematic, IUPAC names. Thus, C_6H_5Br is bromobenzene, $C_6H_5CH_2CH_3$ is ethylbenzene, and so on. No number is needed for monosubstituted benzenes because all the ring positions are equivalent.

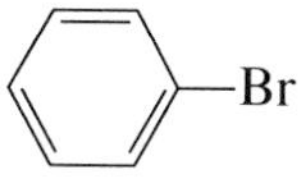

Ortho Indicates 1,2 substituents on a benzene ring.

Meta Indicates 1,3 substituents on a benzene ring.

Para Indicates 1,4 substituents on a benzene ring.

Bromobenzene

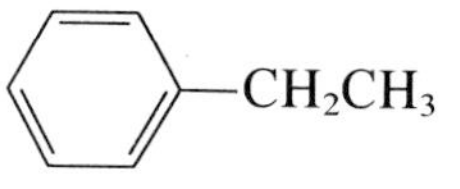

Ethylbenzene

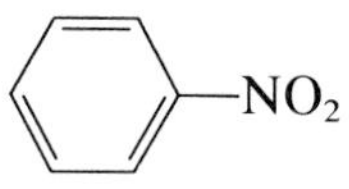

Nitrobenzene

Disubstituted aromatic compounds are named using one of the prefixes **ortho-, meta-,** or **para-.** An *ortho-* or *o*-disubstituted benzene has its two substituents in a 1,2 relationship on the ring; a *meta-* or *m*-disubstituted benzene has its two substituents in a 1,3 relationship; and a *para-* or *p*-disubstituted benzene has its substituents in a 1,4 relationship.

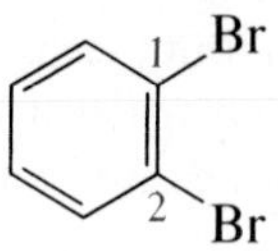

ortho-Dibromobenzene 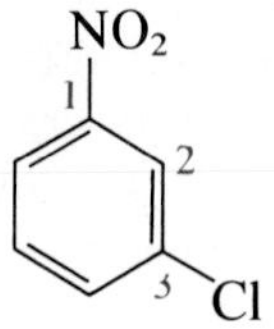   *meta*-Chloronitrobenzene 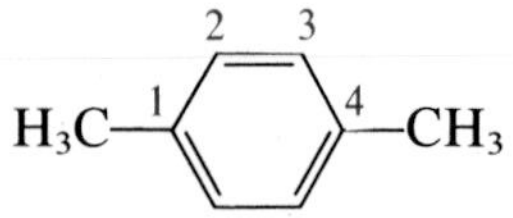   *para*-Dimethylbenzene

Although all substituted aromatic compounds can be named systematically, many have common, nonsystematic names as well. For example, methylbenzene is familiarly known as *toluene,* hydroxybenzene as *phenol,* aminobenzene as *aniline,* and so on, as shown in Table 2.3. Frequently, these common names ae used together with *ortho, para-,* or *meta-,* for example,

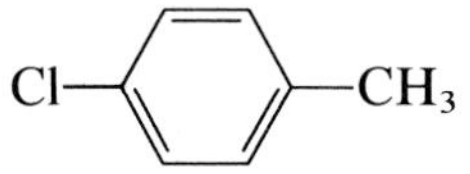

p-Chlorotoluene 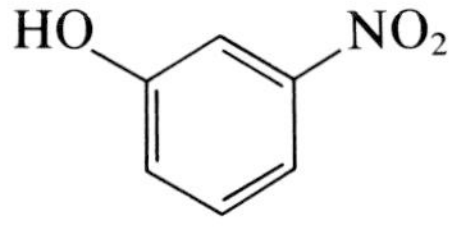   *m*-Nitrophenol 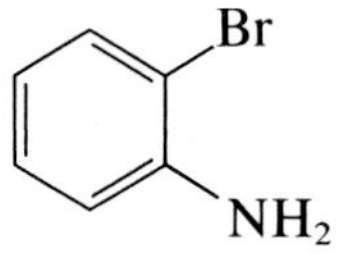   *o*-Bromoaniline

Numbers must be included in the name when a benzene ring has three or more substituents. In the same manner as for alkanes, the numbers are chosen so that substituents have the lowest possible numbers and the substituents are named in alphabetical order. Here too, the common names for monosubstituted benzenes are often used as the parent names:

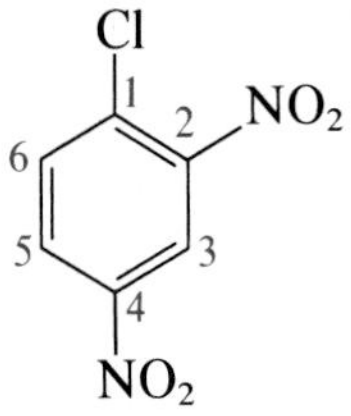

1-Chloro-2,4-dinitrobenzene 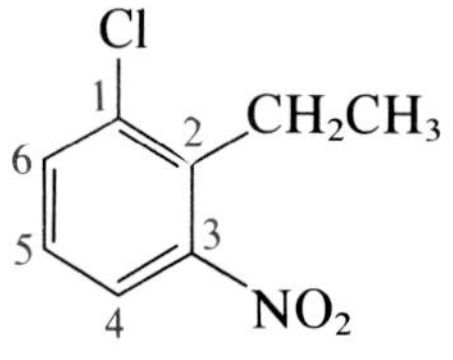   1-Chloro-2-ethyl-3-nitrobenzene 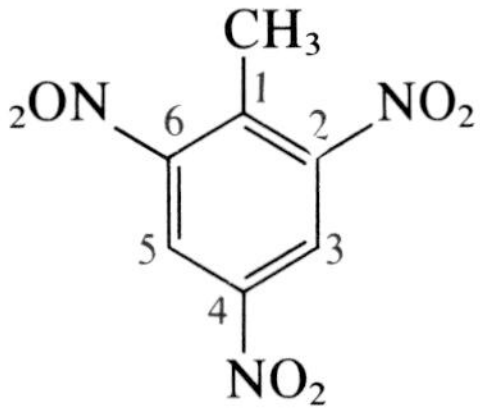   2,4,6-Trinitrotoluene (TNT)

Table 2.3 Common Names of Some Aromatic Compounds

Structure	Name	Structure	Name
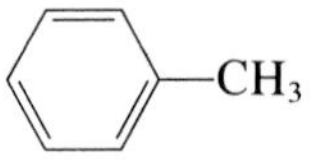 —CH$_3$	Toluene	 H$_3$C— —CH$_3$	*para*-Xylene
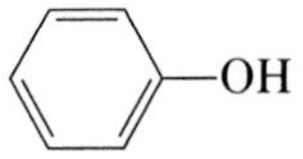 —OH	Phenol	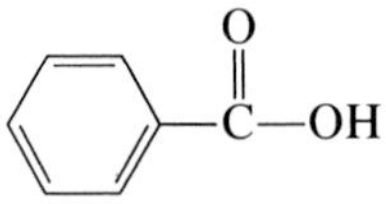 $\overset{O}{\underset{\parallel}{}}$ —C—OH	Benzoic acid
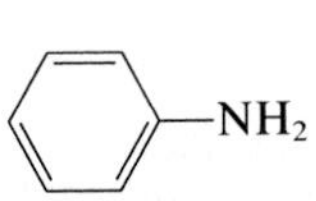 —NH$_2$	Aniline	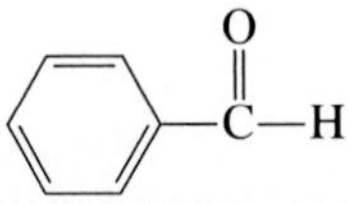 $\overset{O}{\underset{\parallel}{}}$ —C—H	Benzaldehyde

Occasionally, the benzene ring itself might be considered a substituent group attached to another parent compound. When this happens, the name **phenyl** is used for the C_6H_5— unit:

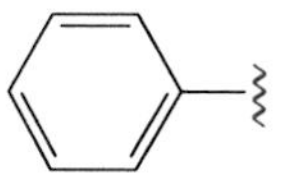

Phenyl group The name of the C_6H_5— unit when a benzene ring is considered a substituent group.

A phenyl group
(C_6H_5—§)

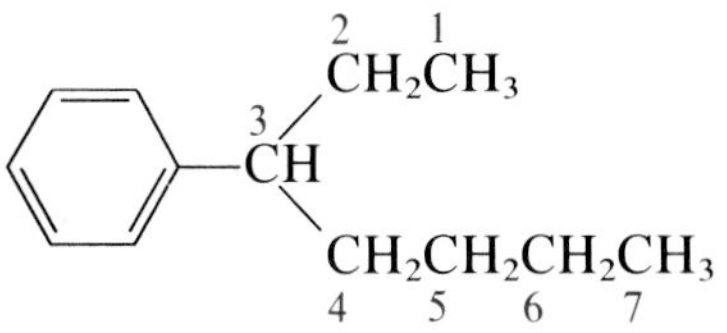

3-Phenylheptane

Aryl group The general name for any substituent derived from an aromatic compound.

Aryl is a general term for any substituent group that contains an aromatic ring.

Solved Problem 2.6 Draw the structure of *m*-chloroethylbenzene.

Solution First, look at the name and determine that the root is benzene. *m*-Chloroethylbenzene must have a benzene ring with two substituents, chloro and ethyl, in a meta relationship. Next, draw a benzene ring and attach one of the substituents, say chloro, to any position:

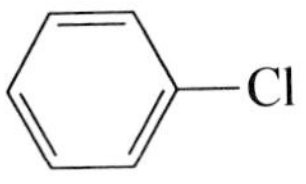

Now, go the meta position two carbons away from the chloro-substituted carbon and attach the second (ethyl) substituent:

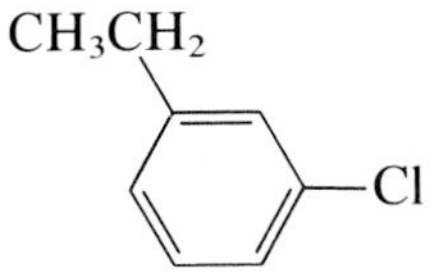

CH_3CH_2

—Cl   *m*-Chloroethylbenzene

Practice Problems **2.13** What are the correct IUPAC names for these compounds?

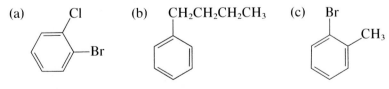

(a) Cl —Br   (b) $CH_2CH_2CH_2CH_3$ (c) Br CH_3

2.14 Draw structures corresponding to these names:
(a) *o*-dibromobenzene   (b) *p*-nitrotoluene   (c) *m*-diethylbenzene
(d) 2-chloro-4-isopropylphenol

2.12   CHEMICAL REACTIONS OF AROMATIC COMPOUNDS

Unlike alkenes, which undergo addition reactions, aromatic compounds usually undergo **aromatic substitution reactions.** That is, a group Y *substitutes* for one of the hydrogen atoms on the aromatic ring without changing the highly stable ring itself. Of course, it doesn't matter which of the six ring hydrogens is replaced since all six are equivalent.

Substitution reaction, aromatic   Substitution of an atom or group for one of the hydrogens on an aromatic ring.

REVIEW   Nitric acid, usually written HNO_3, is a strong acid.

Nitration   Substitution of a nitro group ($-NO_2$) for one of the ring hydrogens occurs when benzene reacts with nitric acid in the presence of sulfuric acid as catalyst:

Benzene   Nitric acid   Nitrobenzene

Nitration of aromatic rings is a key step in the synthesis both of explosives like TNT (trinitrotoluene) and of many important pharmaceutical agents. Nitrobenzene itself is the industrial starting material for preparation of many of the brightly colored dyes used in clothing (Figure 2.7).

Figure 2.7
A variety of aniline dyes. Many dyes are derivatives of aminobenzene, or aniline. Aniline itself is made by the reduction of nitrobenzene.

Halogenation   Substitution of a bromine or chlorine for one of the ring hydrogens occurs when benzene reacts with Br_2 or Cl_2 in the presence of iron as catalyst:

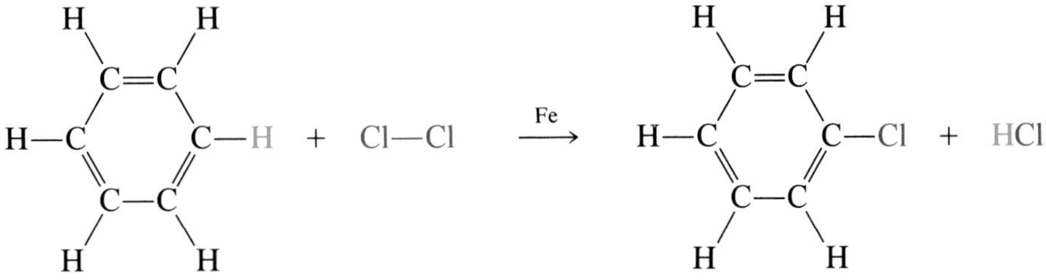

Benzene   Chlorine   Chlorobenzene

Sulfonation   Substitution of a sulfonic acid group ($-SO_3H$) for one of the ring hydrogens occurs when benzene reacts with concentrated sulfuric acid and sulfur trioxide:

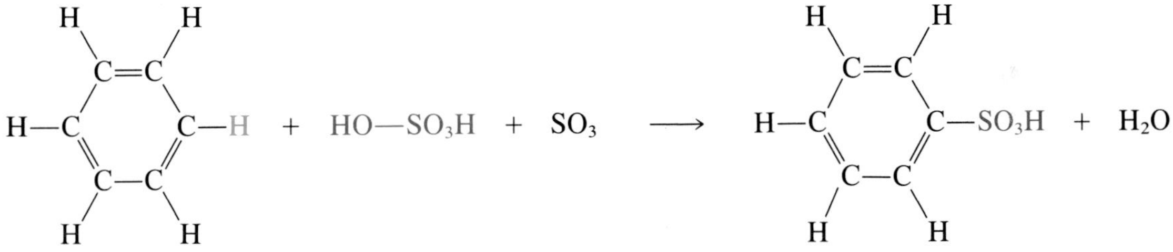

Benzene   Sulfuric acid   Benzenesulfonic acid

Aromatic-ring sulfonation is a key step in the synthesis of such compounds as the sulfa-drug family of antibiotics.

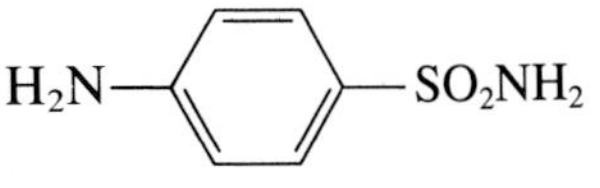

$$H_2N-\langle\ \rangle-SO_2NH_2$$

Sulfanilamide—a sulfa antibiotic

Practice Problems   **2.15**   Write the products from reaction of these reagents with *para*-xylene (*p*-dimethylbenzene).
(a) Br_2 and Fe   (b) HNO_3 and H_2SO_4 (c) SO_3 and H_2SO_4

2.16   Reaction of Br_2 and Fe with toluene (methylbenzene) can lead to a mixture of *three* substitution products. Show the structure of each.

2.13 POLYCYCLIC AROMATIC COMPOUNDS AND CANCER

Polycyclic aromatic compound A substance that has two or more benzene-like rings fused together along their edges.

The definition of the term *aromatic* can be extended beyond simple monocyclic (one-ring) compounds to include **polycyclic aromatic compounds:** substances that have two or more benzene-like rings joined together by sharing a common bond. Naphthalene, familiar for its use in mothballs, is the simplest and best known polycyclic aromatic compound.

INTERLUDE: COLOR IN UNSATURATED COMPOUNDS

The compounds below are brightly colored, as is β-carotene (look at its structure in the earlier Application in this chapter on vision). What do all these compounds have in common?

Have you noticed that each has numerous alternating double and single bonds?

In describing benzene, we said that the double-bond electrons are spread out over the whole

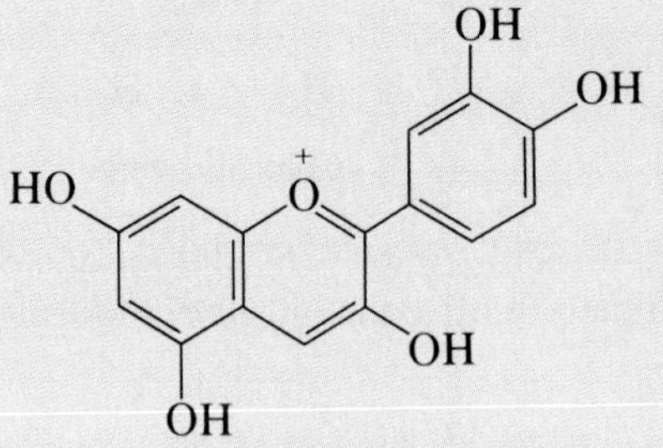

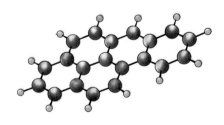

Cyanidin
(bluish-red color in flowers and cranberries)

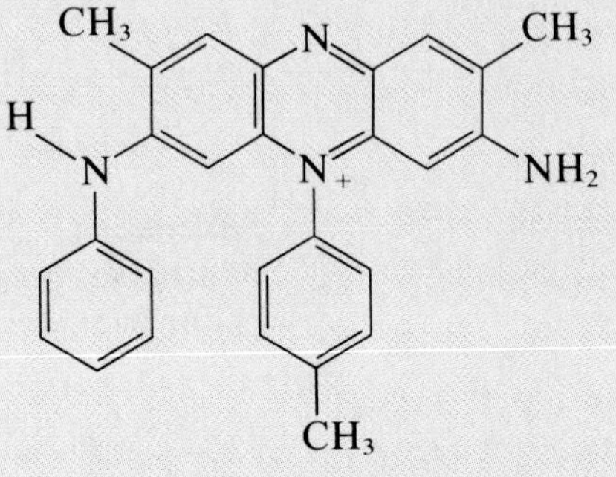

Mauve
(the first synthetic dye)

In addition to naphthalene, there are many more-complex polycyclic aromatic compounds. 1,2-Benzpyrene, for example, contains five benzene-like rings joined together; ordinary graphite (the "lead" in pencils) consists of enormous two-dimensional sheets of benzene rings.

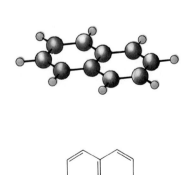

Naphthalene

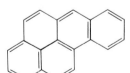

Benz[a]pyrene

A graphite segment

Carcinogenic Cancer-causing.

Benz[a]pyrene is particularly well known because it is one of the **carcinogenic** (cancer-causing) substances found in chimney soot and cigarette smoke. When taken into the body by eating or inhaling, benz[a]pyrene is converted by enzymes into an oxygen-containing metabolite. This metabolite is able to bind to cellular DNA, causing mutations and interfering with the normal flow of genetic information.

molecule. The same phenomenon occurs whenever there are many alternating double and single bonds. The double-bond electrons form a uniform region of electron density that is capable of absorbing light. Organic compounds such as benzene, with small numbers of delocalized (spread-out) electrons, absorb in the ultraviolet region of the electromagnetic spectrum which our eyes can't detect. Compounds with long stretches of alternating double and single bonds absorb in the visible region. The color that we see is complementary to the color that's absorbed; that is, we see what's left of the white light after certain colors have been absorbed. For example, β-carotene absorbs blue light and appears yellow-orange. The same principle applies to almost all colored organic compounds: They contain large regions of delocalized electrons and absorb some portion of the visible spectrum. When the delocalized electron system is broken up, for example by a bleach that breaks double bonds, the color is lost.

The sodium hypochlorite bleach has oxidized the dyes in the colored papers.

Benz[*a*]pyrene →(enzymes)→ An oxygenated metabolite

Even benzene itself can cause certain types of cancer on prolonged exposure. Breathing the fumes of benzene and other volatile aromatic compounds in the laboratory should therefore be strictly avoided.

SUMMARY

Alkenes contain a carbon–carbon double bond, and **alkynes** contain a carbon–carbon triple bond. Both families are said to be **unsaturated** since they have fewer hydrogens that corresponding alkanes. Alkenes are named using the family ending *-ene*, while alkynes use the family ending *-yne*.

The planar arrangement of doubly bonded carbons and their substituents and the lack of rotation around carbon–carbon double bonds lead to **cis–trans isomerism** for disubstituted alkenes. The cis isomer has a specific pair of atoms or groups on the same side of the double bond; the trans isomer has them on opposite sides.

Alkenes and alkynes resemble alkanes in their melting and boiling points and in being nonpolar, flammable, and generally nontoxic. They differ, however, in readily undergoing chemical reactions at their multiple bonds. Alkenes undergo **addition reactions** readily. Addition of hydrogen to an alkene (**hydrogenation**) yields an alkane product; addition of Cl_2 or Br_2 (**halogenation**) yields a 1,2-dihaloalkane product; addition of HBr and HCl (**hydrohalogenation**) yields an alkyl halide product; and addition of water (**hydration**) yields an alcohol product. **Markovnikov's rule** predicts that in the addition of HX or HOH to a double bond, the H becomes attached to the carbon that has the most H's, and the X or OH to the other carbon atom. **Polymers** such as polyethylene are formed when vinyl monomers undergo **chain-growth polymerization** (also known as addition polymerization).

    Aromatic compounds contain a six-membered ring of carbon atoms (or one like it). Benzene rings are usually written with three double bonds but actually have equal bonding between neighboring carbon atoms because the double-bond electrons are delocalized around the entire ring. Simple aromatic compounds are named using the root **-benzene**. Disubstituted benzenes are named with one of the prefixes **ortho-** (1,2 substitution), **meta-** (1,3 substitution), or **para-** (1,4 substitution). Aromatic compounds are unusually stable but can be made to undergo **substitution reactions,** in which one of the ring hydrogens is replaced by some other group ($C_6H_6 \rightarrow C_6H_5Y$). Among these substitutions are **nitration** (substitution of $-NO_2$ for $-H$), **halogenation** (substitution of $-Br$ or $-Cl$ for $-H$), and **sulfonation** (substitution of $-SO_3H$ for $-H$).

REVIEW PROBLEMS

Naming Alkenes, Alkynes, and Aromatic Compounds

2.17 Why are alkenes, alkynes, and aromatic compounds said to be unsaturated?

2.18* Not all compounds that smell nice are called "aromatic," and not all compounds called "aromatic" smell nice. Explain.

2.19 Circle the aromatic portions of these molecules:

(a)

Epinephrine

(b)

Penicillin V

2.20* What family-name endings are used for alkenes, alkynes, and substituted benzenes?

2.21 Write structural formulas for compounds that meet these descriptions:
(a) an alkene with five carbons
(b) an alkyne with four carbons
(c) a substituted aromatic hydrocarbon with eight carbons

2.22* What are the IUPAC names of these compounds?
(a) $CH_3CH_2CH_2CH{=}CH_2$

(b) $CH_3CHCH_2C{\equiv}CCH_3$ with CH_3 branch

(c) $(CH_3)_2C{=}C(CH_3)_2$

(d) $CH_3CH{=}C{-}C{=}CH_2$ with CH_3 and CH_2CH_3 substituents

(e)

(f)

2.23 Draw structures corresponding to these IUPAC names:
(a) *cis*-2-hexene   (b) 2-methyl-3-hexene
(c) 2-methyl-1,3-butadiene   (d) *trans*-3-heptene
(e) 3,3-diethyl-6-methyl-4-nonene
(f) 3-isopropyl-3,4-dimethyl-1,4-hexadiene

2.24* There are only three alkynes with the formula C_5H_8. Draw and name them.

2.25 Provide correct IUPAC names for these aromatic compounds:

(a) with $-CHCH_3$ group   (b) $Br{-}\!\!\!\!\!\!{-}NO_2$

(c) with two NO_2 groups

2.26 Draw structures corresponding to these names:
(a) aniline   (b) phenol   (c) *o*-xylene
(d) toluene   (e) benzoic acid

2.27 Draw structures corresponding to these names:
(a) *p*-nitrophenol   (b) *o*-chloroaniline
(c) *m*-bromotoluene   (d) 1,3,5-trimethylbenzene
(e) *o*-ethylphenol   (f) *m*-dipropylbenzene

2.28 Draw and name all aromatic compounds with the formula C_7H_7Br.

2.29 The following names are incorrect by IUPAC rules. Draw the structures represented by these names and write correct names.
(a) 2-methyl-4-hexene   (b) 5,5-dimethyl-3-hexyne
(c) 2-butyl-1-propene   (d) 1,5-dibromobenzene
(e) 1,2-dimethyl-3-cyclohexene
(f) 3-methyl-2,4-pentadiene
(g) 1,3,4-trichlorobenzene

2.30* How many dienes (compounds with two double bonds) are there with the formula C_5H_8? Draw and name structures of as many as you can.

Alkene Cis–Trans Isomers

2.31 What requirement must be met for an alkene to show cis–trans isomerism?

2.32* Why don't alkynes show cis–trans isomerism?

2.33 Excluding cis–trans isomers, there are five alkenes with the formula C_5H_{10}. Draw structures for as many as you can and give their IUPAC names.

2.34* Which of the alkenes in Problem 2.33 can exist as cis–trans isomers?

2.35 Draw structures of the double-bond isomers of these compounds:

(a)

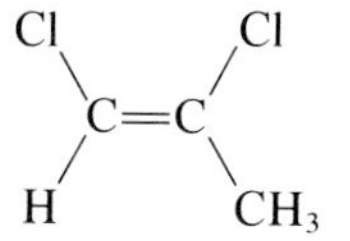

(b)

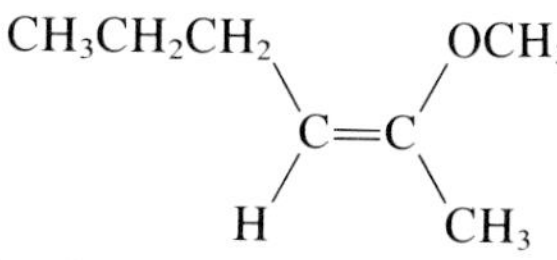

2.36* Which of the following pairs are isomers, and which are identical?

(a)

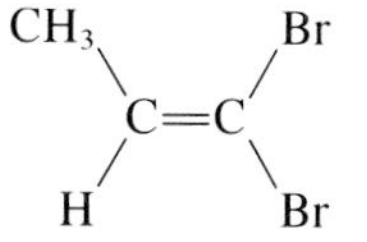

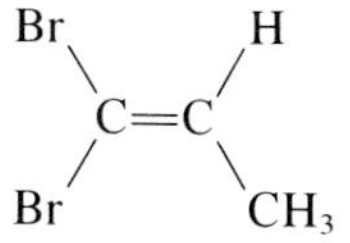

(b)

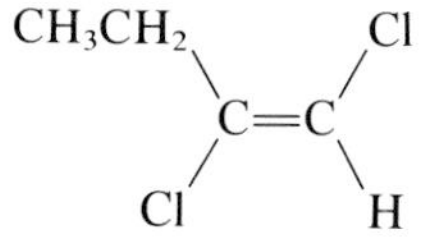

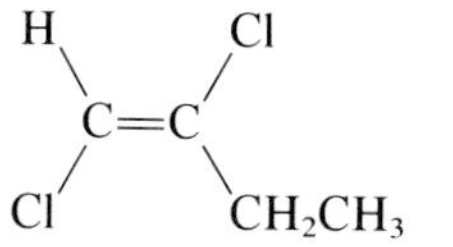

2.37 Why do you suppose small-ring cycloalkenes like cyclohexene don't exist as cis–trans isomers, whereas large-ring cycloalkenes like cyclodecene *do* show isomerism?

Reactions of Alkenes and Alkynes

2.38 What is meant by the term *addition reaction?*

2.39 Give a specific example of an alkene addition reaction.

2.40 Write equations for the reaction of 2,3-dimethyl-2-butene with these reagents:

(a) H_2 and Pd catalyst   (b) Br_2 (c) HBr
(d) H_2O and H_2SO_4 catalyst

2.41 Write equations for the reaction of 2-methyl-2-butene with the reagents shown in Problem 2.40.

2.42 Write the equations for the reaction of 3-methyl-1-butyne with
(a) excess H_2 with a Pd catalyst   (b) excess Cl_2
(c) excess HBr

2.43 What alkene could you use to make the following products. Draw the structural formula of the alkene and tell what inorganic reagent is also required for the reaction to occur.

(a) 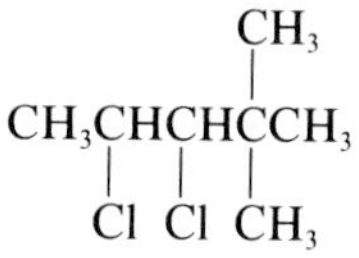
(b) $CH_3CH_2CH_3$

(c) 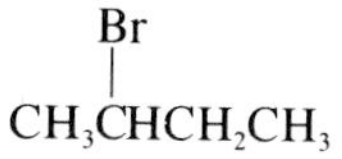
(d)

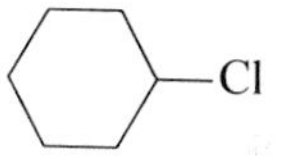

(e) 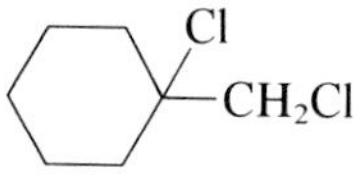
(f)

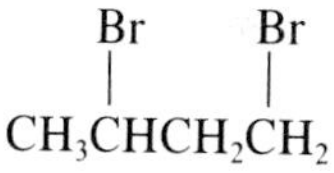

2.44* Write a balanced equation for the combustion of acetylene, used in high-temperature oxyacetylene torches.

2.45 Draw the carbocation formed as an intermediate when HCl adds to styrene (Table 2.2).

2.46 Polyvinylpyrrolidine (PVP) is often used in hair sprays as the substance that holds hair in place. Draw a few units of the PVP polymer. The vinylpyrrolidine monomer unit has the structure

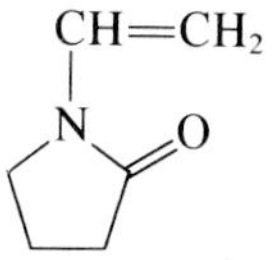

2.47 Saran, used as a plastic wrap for foods, is a polymer with the following structure. What is the monomer unit of Saran?

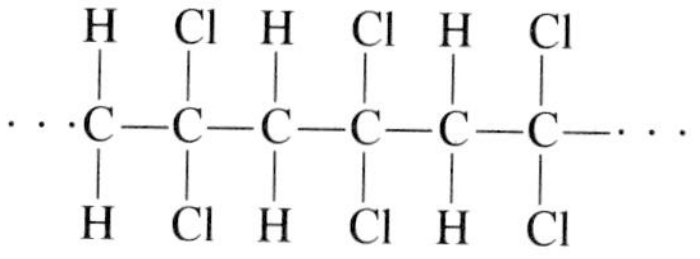

Reactions of Aromatic Compounds

2.48 What is meant by the term *substitution reaction?*

2.49 Give a specific example of a substitution reaction of an aromatic compound.

2.50* Benzene reacts with only one of the following four reagents. Which of the four is it, and what is the structure of the product?
(a) H_2 and Pd catalyst   (b) Br_2 and catalyst
(c) HBr   (d) H_2O and H_2SO_4 catalyst

2.51 Write equations for the reaction of *p*-dichlorobenzene with these reagents:
(a) Br_2 and Fe catalyst   (b) HNO_3 and H_2SO_4 catalyst
(c) H_2SO_4 and SO_3 (d) Cl_2 and Fe catalyst

2.52* Benzene and other aromatic compounds don't normally react with hydrogen in the presence of a palladium catalyst. If very high pressures (200 atm) and high temperatures are used, however, benzene will add three molecules of H_2 to give an addition product. What is a likely structure for the product?

2.53 How can you account for the fact that reaction of benzene with Br_2 and an iron catalyst yields a single substitution product, but the similar reaction of *o*-xylene (*o*-dimethylbenzene) with Br_2 and iron yields a mixture of two different substitution products? What are their structures?

2.54 The explosive trinitrotoluene, or TNT, is made by carrying out three successive nitration reactions on toluene. If these nitrations take place in the ortho and para positions relative to the methyl group, what is the structure of TNT?

Applications

2.55 What is the difference in the purpose of the rod and the cone cells in the eye? [App: Chemistry of Vision]

2.56* Describe the isomerization that occurs when light strikes the rhodopsin in the eye. [App: Chemistry of Vision]

2.57 Circle the isoprene units in the structure of β-carotene shown in Chemistry of Vision. [App: Terpenes]

2.58 Carvone is a component of spearmint oil. Circle the isoprene units in carvone. [App: Terpenes]

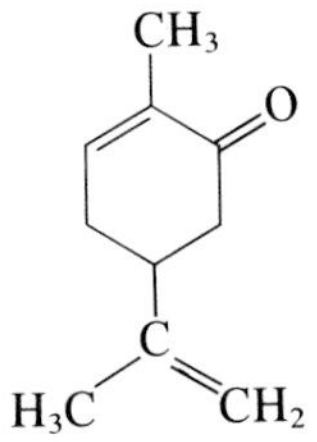

2.59 Isoprene, the monomer unit of rubber, is not capable of cis–trans isomerism, yet the natural rubber polymer exhibits cis–trans isomerism (primarily cis). What is responsible for this difference? [App: Terpenes]

2.60* Naphthalene is a white solid. Does it absorb light in the visible or in the ultraviolet range? [App: Color in Unsaturated Compounds]

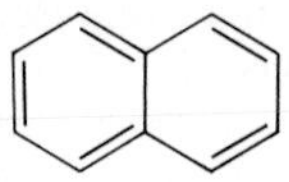

2.61 Tetrabromofluorescein, a purple dye often used in lipsticks, has the following structure. What structural feature makes it colored? If the dye is purple, what color must it absorb? [App: Color in Unsaturated Compounds]

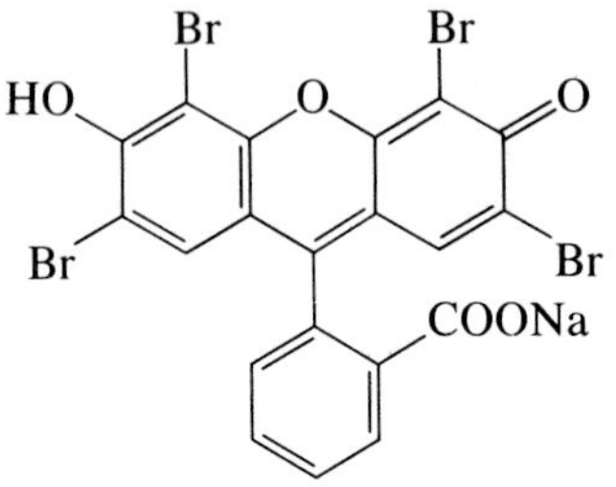

Additional Problems

2.62* Why can't alkanes form addition polymers?

2.63 Salicylic acid, or *o*-hydroxybenzoic acid, is used as starting material for the industrial preparation of aspirin. Draw the structure of salicylic acid.

2.64* Assume that you have two unlabeled bottles, one with cyclohexane and one with cyclohexene. How could you tell them apart by carrying out chemical reactions?

2.65 Assume you have two unlabeled bottles, one with cyclohexene and one with benzene. How could you tell them apart by carrying out chemical reactions?

2.66 The compound *p*-dichlorobenzene has been used as an insecticide. Draw its structure.

2.67 Menthene, a compound found in mint plants, has the formula $C_{10}H_{18}$ and the IUPAC name 1-isopropyl-4-methylcyclohexene. What is the structure of menthene?

2.68 Cinnamaldehyde, the pleasant-smelling substance found in cinnamon oil, has the following structure. What product would you expect to obtain from reaction of cinnamladehyde with hydrogen and a palladium catalyst?

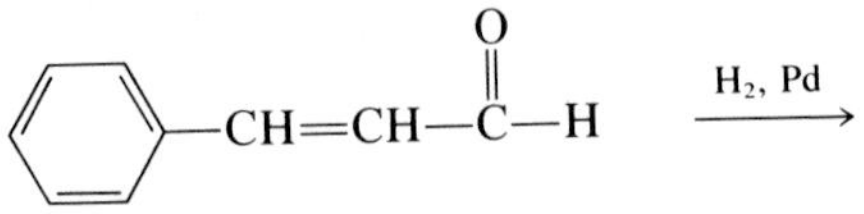

2.69 Write the products of these reactions:

(a)

(b)

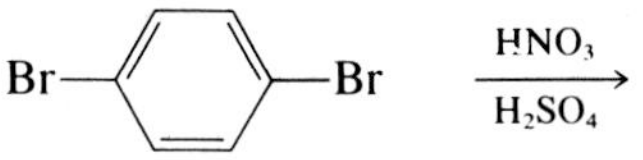

Br—⟨benzene ring⟩—Br $\xrightarrow[\text{H}_2\text{SO}_4]{\text{HNO}_3}$

(c)

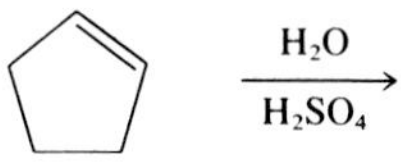

⟨cyclopentene⟩ $\xrightarrow[\text{H}_2\text{SO}_4]{\text{H}_2\text{O}}$

(d) CH$_3$CHCH$_3$
       |
     CH$_3$CHCH$_2$CH=CH$_2$ $\xrightarrow{\text{HBr}}$

(e) CH$_3$C≡CCH$_2$CH$_3$ $\xrightarrow{\text{H}_2, \text{ Pd}}$

(f) ⟨benzene ring⟩—CH=CH$_2$ + HCl $\longrightarrow$

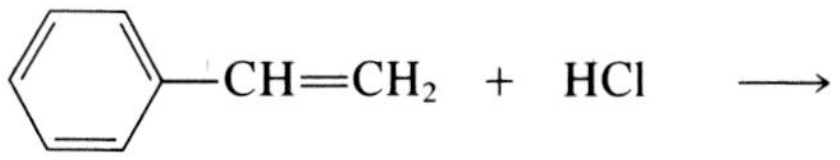

2.70* Two different products are possible when 2-pentene is treated with HBr. Write the structural formulas of the possible products and discuss why they are made in about equal amounts.

2.71 Benzene is a liquid at room temperature, while naphthalene, another aromatic compound (Problem 2.60),

is a solid. Account for this difference in physical properties.

2.72 The compound 4-chloro-3,5-dimethylphenol is used as an antiseptic, as a germicide, and for mildew prevention. Draw its structure.

2.73 Ocimene, a compound isolated from the herb basil, has the IUPAC name, 3-7-dimethyl-1,3-6-octatriene.
(a) Draw its structure
(b) Circle the isoprene units in the formula.
(c) Draw the structure of the compound formed if enough HBr is added to react with all the double bonds in ocimene.

2.74* Describe how you could prepare the following compound from an alkene. Draw the formula of the alkene, name it, and list the inorganic reactants needed for the conversion.

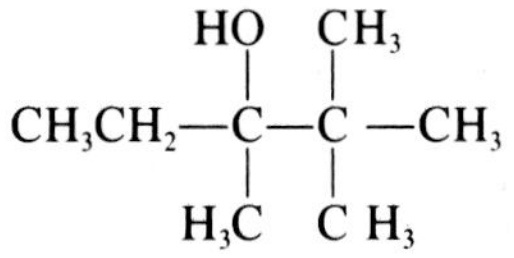

$$\begin{array}{c} \text{HO} \quad \text{CH}_3 \\ | \qquad | \\ \text{CH}_3\text{CH}_2-\text{C}-\text{C}-\text{CH}_3 \\ | \qquad | \\ \text{H}_3\text{C} \quad \text{CH}_3 \end{array}$$

CHAPTER

3

Some Compounds with Oxygen, Sulfur, or Halogens

Nighttime balloon launching in January in Sweden. The helium-filled balloon carries instruments that will measure ozone and pollutants in polar stratospheric clouds at 16–25 km altitude. The pollutants responsible for ozone depletion include chlorofluorocarbons, one of a wide variety of compounds introduced in this chapter.

Having introduced the hydrocarbons that form the skeletons of all organic compounds, we're ready to move on to compounds with functional groups containing other elements. In this chapter we'll concentrate on functional groups with single bonds to oxygen, sulfur, and the halogens. In future chapters you'll be introduced to groups with bonds to nitrogen and those containing the C=O group.

At this point in studying organic chemistry, students often begin to wonder what they're meant to "learn" about organic compounds. There are so many different kinds. First, it's important to recognize the structures of the functional groups, whether they're present in simple or complex compounds. Then, it's useful to know the general properties imparted by these groups and a few of their most important chemical reactions. It's also valuable to learn something about some representative members of each family of compounds, such as their properties, hazards, occurrences in nature, and uses.

Questions we'll answer in this chapter include the following:

1. **How do alcohols, phenols, and ethers compare with water and alkanes?** The goal: Be able to describe the major differences in properties among members of these families.

2. **What are the distinguishing features of ethers, sulfur-containing compounds, and halogen-containing compounds?** The goal: Be able to describe the structures and outstanding properties of compounds with these functional groups.

3. **How are alcohols, phenols, ethers, sulfur-containing compounds, and halogen-containing compounds named?** The goal: Be able to give systematic names for the simple members of these families and write their structures, given the names.

4. **What are the general properties of alcohols and phenols?** The goal: Be able to describe such properties as polarity and hydrogen bonding, water solubility, and boiling points.

5. **Why are alcohols and phenols weak acids?** The goal: Be able to show why alcohols and phenols are acids.

6. **What are the major chemical reactions of alcohols?** The goal: Be able to describe and predict the products of the dehydration and oxidation of alcohols.

7. **What are some of the significant applications of alcohols, phenols, ethers, sulfur-containing compounds, and halogen-containing compounds?** The goal: Be able to identify major applications of each family of compounds and describe some important members of each family.

3.1 ALCOHOLS, PHENOLS, AND ETHERS

Alcohol   A compound that contains an —OH functional group covalently bonded to an alkane-like carbon atom, R—OH.

Hydroxyl group   A name tor the —OH group in an organic compound.

Phenol   A compound that has an —OH functional group bonded directly to an aromatic, benzene-like ring.

Ether   A compound that has an oxygen atom bonded to two carbon atoms, R—O—R.

Alcohols are compounds that have an —OH group (a **hydroxyl group**) covalently bonded to a saturated, alkane-like carbon atom; **phenols** have an —OH group bonded directly to an aromatic, benzene-like ring (an aryl group); and **ethers** have an oxygen atom bonded to two organic groups. Compounds in all three families can be thought of as organic derivatives of water in which one or both of the water hydrogens have been replaced by an organic substituent. Using R for any organic group and Ar for one that is specifically aromatic, the general formulas for these families are

$$\underset{\text{Water}}{\text{H—O—H}} \qquad \underset{\text{Alcohol}}{\text{R—O—H}} \qquad \underset{\text{Phenol}}{\text{Ar—O—H}} \qquad \underset{\text{Ether}}{\text{R—O—R}}$$

An alcohol, a phenol, and an ether are shown here. Note that each functional group contains an oxygen atom with single bonds to its neighbors.

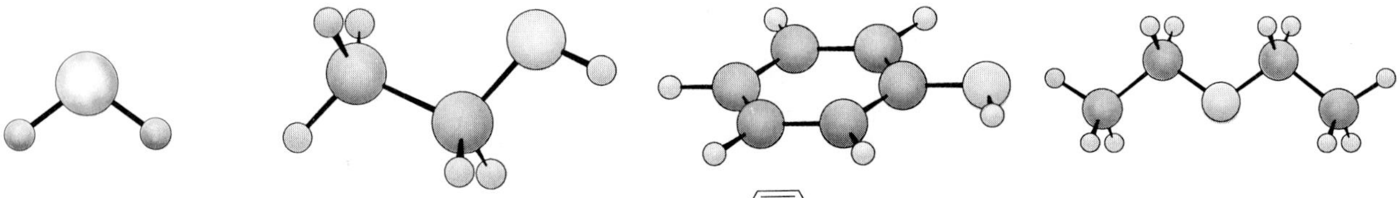

H—O—H	CH_3CH_2—O—H	 ⬡—O—H	CH_3CH_2—O—CH_2CH_3
Water	Ethyl *alcohol*	*Phenol*	Diethyl *ether*

The structural similarity between alcohols and water extends to many of their properties. For example, compare the boiling points of ethyl alcohol, dimethyl ether, propane, and water:

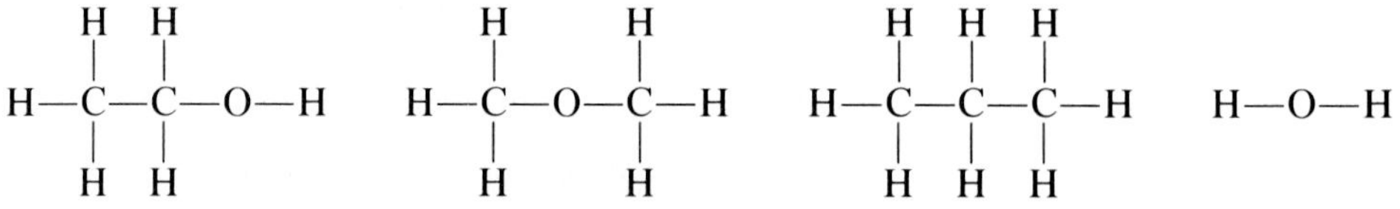

Ethyl alcohol (mol wt 46, bp 78.5°C)	Dimethyl ether (mol wt 46, bp −23°C)	Propane (mol wt 44, bp −42°C)	Water (mol wt 18, bp 100°C)

REVIEW   Hydrogen bonding (Appendix A.10) is a strong intermolecular force due to attraction between a highly electronegative atom (O, N, or F) and a hydrogen atom bonded to another highly electronegative atom.

REVIEW   The stronger the intermolecular forces, the higher the boiling point.

Ethyl alcohol, dimethyl ether, and propane have similar molecular weights, yet ethyl alcohol boils more than 100° higher than the other two. In fact, the boiling point of ethyl alcohol is close to that of water. Why should this be?

We know that water has a high boiling point because of the formation of hydrogen bonds. These hydrogen bonds cause an attraction between individual molecules that prevents their easy escape into the vapor phase. In a similar manner, hydrogen bonds form between alcohol (or phenol) molecules (Figure 3.1). Since alkanes and ethers don't have hydroxyl groups, they can't form hydrogen bonds and therefore have much lower boiling points. In many properties ethers resemble alkanes, while alcohols resemble water.

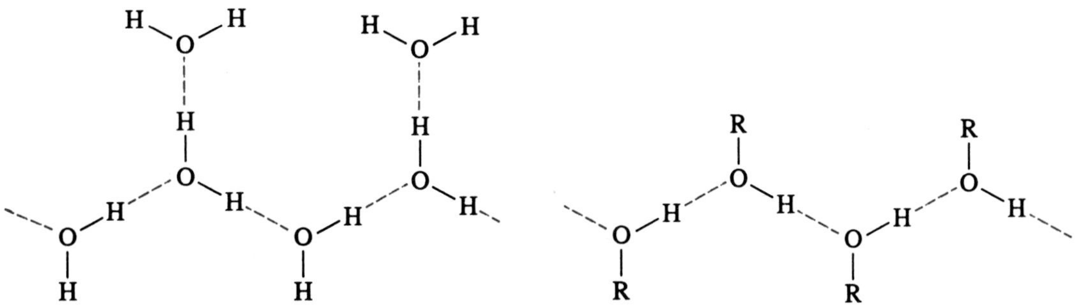

Figure 3.1
The formation of hydrogen bonds in water (a) and in alcohols (b). Due to the attractive forces of the hydrogen bonds, the easy escape of molecules into the vapor phase is prevented, resulting in high boiling points.

Practice Problems   ***3.1***   Identify each of these compounds as an alcohol, a phenol, or an ether.

(a) $CH_3CH_2CHCH_3$
          |
          OH

(b) 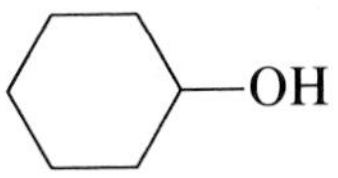 —OH

(c) 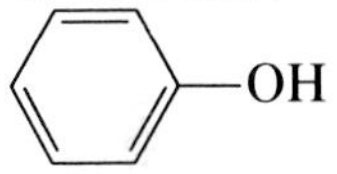 —OH

(d) 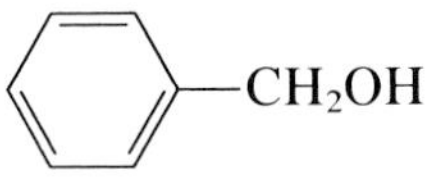 —CH_2OH

(e) 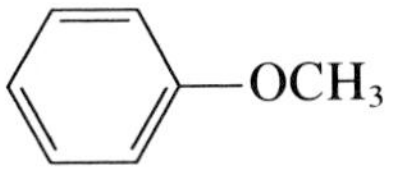 —OCH_3

(f) $CH_3CHOCH_2CH_3$
          |
          CH_3

3.2   Explain the difference between a hydroxyl group and a hydroxide ion.

3.2   SOME COMMON ALCOHOLS

Simple alcohols are among the most commonly encountered of all organic chemicals (Figure 3.2). They are useful as solvents, antifreeze agents, and disinfectants, and they are valuable in the laboratory and in industry because compounds containing many other functional groups can be synthesized from alcohols. In nature, lower-molecular-weight alcohols are formed during the fermentation of organic matter. "Higher alcohols," those with six or more carbon atoms, are obtained by chemical breakdown of natural fats and oils and are used in the production of plastics and detergents.

Wood alcohol   A common name for methyl alcohol, CH_3OH.

Methyl Alcohol (CH_3OH, Methanol)   Methyl alcohol, the simplest alcohol, is commonly known as **wood alcohol** because it was once prepared by heating wood in the absence of air. Today it is made in large quantities from a mixture of carbon monoxide and hydrogen,

$$CO(g) + 2\ H_2(g) \xrightarrow[\text{high pressure, 250°C}]{\text{Cu catalyst}} CH_3OH(l)$$

Coal and natural gas are sources for the mixture of CO and H_2, and there is great interest in methyl alcohol as a fuel and a raw material for organic chemicals now produced from petroleum. Methyl alcohol is colorless, miscible

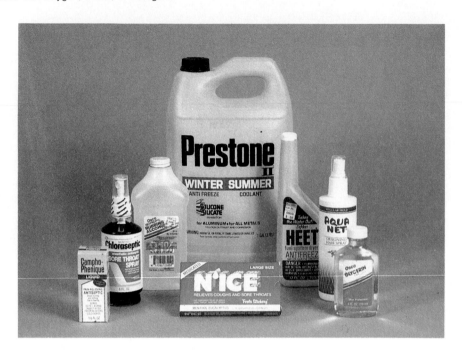

Figure 3.2
Some common household products made entirely or mostly of alcohols.

with water, and toxic to humans whether ingested or inhaled, causing blindness in low doses (about 15 mL) and death in larger amounts (about 30 mL).

Ethyl Alcohol (CH_3CH_2OH, Ethanol)   Ethyl alcohol is one of the oldest known pure organic chemicals, and its production by fermentation of grain and sugar goes back for many thousands of years. Sometimes called **grain alcohol,** ethyl alcohol is the "alcohol" present in all table wines (10–13%), beers (3–5%), and distilled liquors (35–90%). During fermentation, starches or complex sugars are converted to simple sugars ($C_6H_{12}O_6$) which are then converted to ethyl alcohol:

Grain alcohol   A common name for ethyl alcohol, CH_3CH_2OH.

$$C_6H_{12}O_6 \xrightarrow{\text{yeast enzymes}} 2\ CH_3CH_2OH + 2\ CO_2$$

About 14% (by volume) is the maximum alcohol concentration produced by fermentation. Higher alcohol concentrations in beverages are produced by distillation of the fermentation product or addition of distilled products such as brandy. The production of ethyl alcohol and alcoholic beverages in the United States is supervised by the government. To avoid the government tax on the sale of consumable alcohol, a toxic substance such as methyl alcohol or benzene is added to ethyl alcohol to produce *denatured alcohol* for nonbeverage uses.

Industrially, most ethyl alcohol is made by the hydration of ethylene (Section 2.6). Distillation yields a 95% ethyl alcohol–5% water mixture useful for any application with which water does not interfere. Reaction of the mixture with a dehydrating agent gives 100% ethyl alcohol, known as *absolute alcohol*. In some states, a blend of ethyl alcohol and gasoline known as *gasohol* is commercially available, and numerous experiments are under way with cars that burn methyl or ethyl alcohol. These fuels are desirable because they produce fewer air pollutants than gasoline; also, ethyl alcohol can be produced from corn rather than from dwindling fossil fuel reserves.

Isopropyl alcohol ((CH₃)₂CHOH, Isopropanol)   Isopropyl alcohol, or *rubbing alcohol,* is used as a 70% mixture with water for rubdowns because it cools the skin through evaporation and causes pores to close. It's also used as a solvent for medicines, as a sterilant for instruments, and to cleanse the skin before drawing blood or giving injections. The "medicinal" odor we associate with doctors' offices is often that of isopropyl alcohol. Although not as toxic as methyl alcohol, isopropyl alcohol is much more toxic than ethyl alcohol.

Ethylene Glycol (HOCH₂CH₂OH)   A dialcohol, ethylene glycol is a colorless liquid miscible with water and insoluble in nonpolar solvents. Its two major uses are in antifreeze and in making polymer films and fibers such as Dacron. In humans it's a central nervous system depressant with a lethal dose of about 100 mL (Figure 3.3).

Glycerol (HOCH₂CH(OH)CH₂OH)   Like ethylene glycol, the trialcohol glycerol is a colorless liquid that is miscible with water. Unlike ethylene glycol, it's not toxic but has a sweet taste that's put to use in candy and prepared foods. Glycerol, often called glycerin, is also used in cosmetics and tobacco as a moisturizer, in plastics manufacture, in antifreeze and shock absorber fluids, and as a solvent. The glycerol molecule provides the structural backbone of natural fats and oils (Section 12.1).

3.3 NAMING ALCOHOLS

Alcohols with one —OH group are often referred to by common names like those used in the preceding section, which consist of the name of the alkyl group followed by the word "alcohol." Thus, the two-carbon alcohol is ethyl alcohol, a four-carbon alcohol is a butyl alcohol, and so on. Both the common and the systematic names of some representative alcohols are included in Table 3.1. Notice that *dialcohols* are often called **glycols.** Ethylene glycol is the simplest glycol; propylene glycol is often used as a solvent for medicines that need to be inhaled or rubbed onto the skin.

Alcohols are classified according to the number of carbon substituents bonded to the hydroxyl-bearing carbon. Alcohols with one substituent are classified as **primary** (1°), those with two substituents as **secondary** (2°), and those with three substituents as **tertiary** (3°). Of course, the substituent groups don't have to be the same, so we'll use the representations R, R′, and R″ to indicate different groups.

Figure 3.3
Ethylene glycol, a poisonous dialcohol used in antifreeze, and glycerol, a sweet trialcohol used as a food additive. Can you tell by their appearance which is which?

Glycol A dialcohol; a compound that contains two —OH groups.

Primary alcohol An alcohol in which the OH-bearing carbon atom is bonded to one other carbon (and two hydrogens), RCH₂OH.

Secondary alcohol An alcohol in which the OH-bearing carbon atom is bonded to two other carbons (and one hydrogen), R₂CHOH.

Tertiary alcohol An alcohol in which the OH-bearing carbon atom is bonded to three other carbons (and no hydrogens), R₃COH.

$$
\begin{array}{ccc}
\text{H} & \text{R}' & \text{R}' \\
| & | & | \\
\text{R}-\text{C}-\text{OH} & \text{R}-\text{C}-\text{OH} & \text{R}-\text{C}-\text{OH} \\
| & | & | \\
\text{H} & \text{H} & \text{R}''
\end{array}
$$

A *primary* alcohol
(one R group on
OH-bearing carbon)

A *secondary* alcohol
(two R groups on
OH-bearing carbon)

A *tertiary* alcohol
(three R groups on
OH-bearing carbon)

Table 3.1 Physical Properties of Some Simple Alcohols[a]

Structure	Common Name	IUPAC Name	Boiling Point (°C)
CH_3OH	Methyl alcohol	Methanol	65
CH_3CH_2OH	Ethyl alcohol	Ethanol	78
$CH_3CH_2CH_2OH$	*n*-Propyl alcohol	1-Propanol	97
$CH_3\overset{\underset{\displaystyle |}{OH}}{C}HCH_3$	Isopropyl alcohol	2-Propanol	82
$CH_3CH_2CH_2CH_2OH$	*n*-Butyl alcohol	1-Butanol	118
$CH_3{-}\underset{\underset{\displaystyle CH_3}{|}}{\overset{\overset{\displaystyle CH_3}{|}}{C}}{-}OH$	*tert*-Butyl alcohol	2-Methyl-2-propanol	83
$\underset{\underset{\displaystyle OH}{|}}{CH_2}{-}\underset{\underset{\displaystyle OH}{|}}{CH_2}$	Ethylene glycol	1,2-Ethanediol	198
$\underset{\underset{\displaystyle OH}{|}}{CH_2}{-}\underset{\underset{\displaystyle OH}{|}}{CH}{-}CH_3$	Propylene glycol	1,2-Propanediol	187
$\underset{\underset{\displaystyle OH}{|}}{CH_2}{-}\underset{\underset{\displaystyle OH}{|}}{CH}{-}\underset{\underset{\displaystyle OH}{|}}{CH_2}$	Glycerol (glycerin)	1,2,3-Propanetriol	290

[a] All the alcohols listed are liquids at room temperature and completely miscible with water, except 1-butanol (7 g/100 mL water).

Alcohols are named in the IUPAC system as derivatives of alkanes, using the *-ol* ending for the parent compound, according to the following steps.

Step 1. Name the parent compound. Find the longest chain *containing the hydroxyl group* and name the chain by replacing the *-e* ending of the corresponding alkane (Table 1.4) with $-ol$:

$$CH_3CH_2CH_2CH_2OH$$

Name as a *butanol*—a four-carbon chain containing a hydroxyl group.

If the compound is a cycloalkane, its name provides the root name, for example, *cyclohexan*ol.

Step 2. Number the carbon atoms in the main chain. Begin at the end nearer the hydroxyl group, ignoring the location of other substituents:

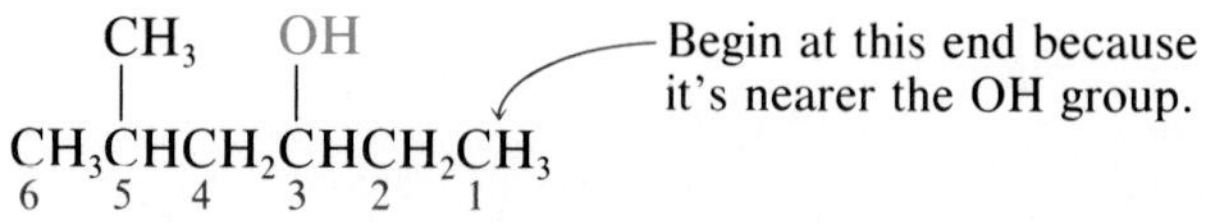

$$\underset{\displaystyle \underset{6\ \ \ \ 5\ \ \ 4\ \ \ \ 3\ \ \ 2\ \ \ 1}{CH_3CHCH_2CHCH_2CH_3}}{\overset{\displaystyle \overset{CH_3\ \ \ \ \ \ OH}{|\ \ \ \ \ \ \ \ |}}{}}$$

Begin at this end because it's nearer the OH group.

In an OH-substituted cycloalkane, begin with the —OH group.

Step 3. *Identify and number the hydroxyl group and the substituents.*

A 1 hydroxyl

$$CH_3CH_2CH_2CH_2OH$$
$$4321$$

A 5 methyl   A 3 hydroxyl

$$\overset{\displaystyle CH_3}{|}\quad\overset{\displaystyle OH}{|}$$
$$CH_3CHCH_2CHCH_2CH_3$$
$$654321$$

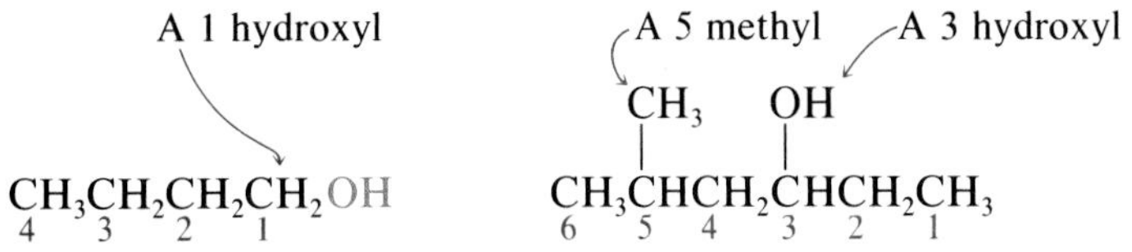

Step 4. *Write the full name.* Place the number that locates the hydroxyl group immediately before the parent compound name (with the *-ol* ending). Number all other substituents according to their positions and list them alphabetically:

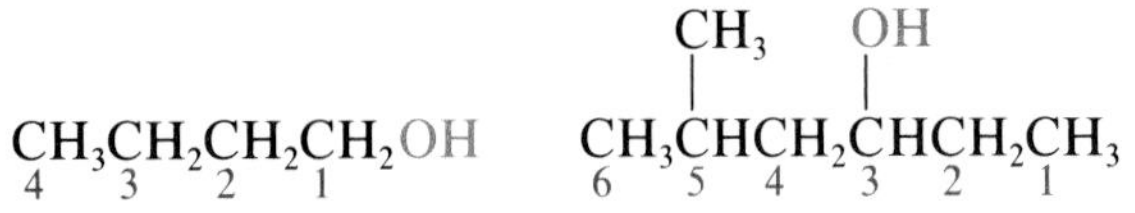

$$CH_3CH_2CH_2CH_2OH$$
$$4321$$

1-Butanol

$$\overset{\displaystyle CH_3}{|}\quad\overset{\displaystyle OH}{|}$$
$$CH_3CHCH_2CHCH_2CH_3$$
$$654321$$

5-Methyl-3-hexanol

$$\overset{\displaystyle OH}{|}\quad\overset{\displaystyle Cl}{|}\quad\overset{\displaystyle CH_3}{|}$$
$$CH_3CHCH_2CHCH_2CHCH_3$$
$$1234567$$

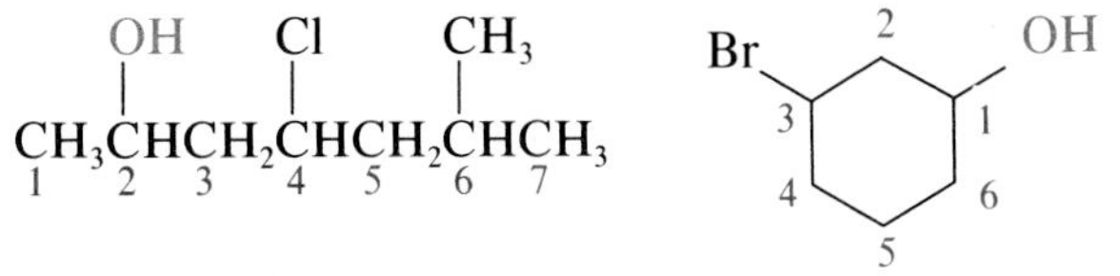

4-Chloro-6-methyl-2-heptanol

3-Bromocyclohexanol

Solved Problem 3.1   Give the systematic name of the following alcohol and classify it as primary, secondary, or tertiary.

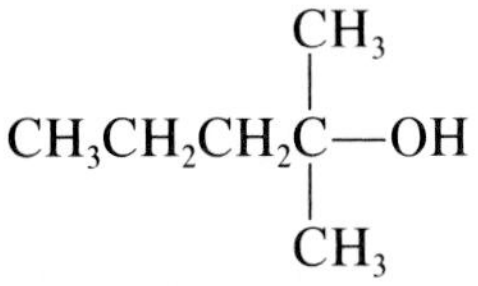

$$\overset{\displaystyle CH_3}{|}$$
$$CH_3CH_2CH_2C-OH$$
$$\underset{\displaystyle CH_3}{|}$$

Solution   First, identify the longest carbon chain and number the carbon atoms beginning at the end nearest —OH. The longest of the two chains attached to the —OH here has five carbon atoms.

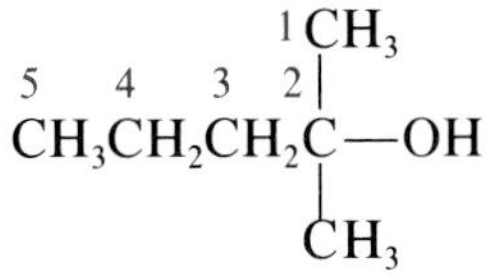

$$\overset{1\,CH_3}{|}$$
$$\underset{5432|}{CH_3CH_2CH_2C-OH}$$
$$\underset{\displaystyle CH_3}{|}$$

Name as a pentanol.

Next, identify and number the hydroxyl group and the substituents. Finally, write the full name of the compound.

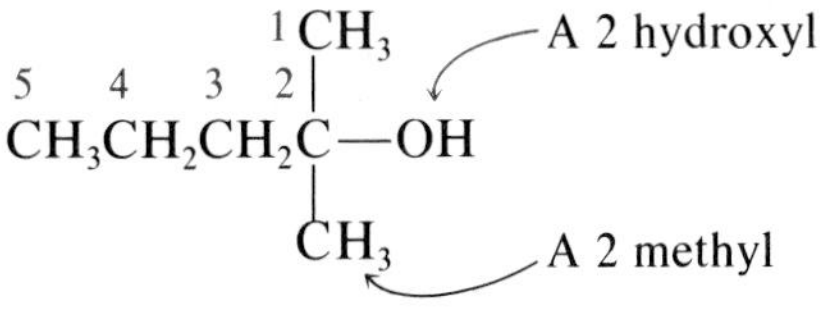

$$\overset{1\,CH_3}{|}$$
$$\underset{5432|}{CH_3CH_2CH_2C-OH}$$
$$\underset{\displaystyle CH_3}{|}$$

A 2 hydroxyl

A 2 methyl

2-Methyl-2-pentanol

The —OH group is bonded to a carbon atom that has three alkyl substituents, so this is a tertiary alcohol.

Practice Problems   **3.3**   Draw structures corresponding to these names:
(a) 3-methyl-3-pentanol   (b) cyclohexanol
(c) 2-methyl-4-heptanol   (d) 2-butanol   (e) 3-chloro-1-propanol

3.4   Give systematic names for these compounds:

(a) $CH_3CH_2\overset{\overset{\displaystyle OH}{|}}{C}HCH_2CH_3$ (b) $CH_3CH_2\overset{\overset{\displaystyle CH_2OH}{|}}{C}HCH_2CH_2CH_3$

(c) $CH_3CH_2\overset{\overset{\displaystyle CH_2OH}{|}}{C}HCH_2CH_2CH_2Br$ (d)

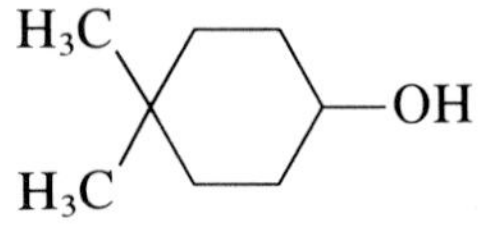

3.5   Identify each of the alcohols in Problems 3.3 and 3.4 as primary, secondary, or tertiary.

3.4   PROPERTIES OF ALCOHOLS

Hydrogen bonding is the major influence on the physical properties of alcohols. Straight-chain alcohols with up to 12 carbon atoms are liquids, and each boils at a higher temperature than similar alkanes. Most familiar alcohols, in fact, are liquids.

   Solubility behavior, like hydrogen bonding, is a property shared by water and alcohols. Water, because of its polar bonds and its ability to solvate ions, is an excellent solvent for ionic compounds and polar molecules but a poor solvent for nonpolar organic molecules. Low-molecular-weight alcohols such as methanol and ethanol are water-like in their solubility behavior. Methanol and ethanol are both infinitely soluble in water, with which they can form hydrogen bonds. Both can also dissolve small amounts of many salts, and both are miscible with many organic solvents. Higher-molecular-weight alcohols, such as 1-heptanol, however, are much more alkane-like and less water-like. 1-Heptanol is nearly insoluble in water and can't dissolve salts but does dissolve alkanes.

> **REVIEW**   In water solution, ions or polar molecules are "solvated," meaning they are surrounded and held in solution by attraction to polar water molecules.

> **REVIEW**   Miscible substances are mutually soluble in all proportions without limit.

$$CH_3{-}OH \qquad\qquad CH_3CH_2CH_2CH_2CH_2CH_2CH_2{-}OH$$

Methanol—has a small organic part and is therefore water-like.   1-Heptanol—has a large organic part and is therefore alkane-like.

   Alcohols with two or more —OH groups can form more than one hydrogen bond. They are therefore higher boiling and more water-soluble than similar alcohols with one —OH group. For example, compare 1-butanol and 1,4-butanediol,

$CH_3CH_2CH_2CH_2OH$ Bp 117°C, water solubility of 7 g/100 mL $HOCH_2CH_2CH_2CH_2OH$ Added —OH raises bp to 230°C and gives miscibility with water

1-Butanol   1,4-Butanediol

Properties of Alcohols

- Alcohols are hydrogen-bonded and higher boiling than similar alkanes.
- Common alcohols are liquids.
- Lower alcohols are miscible with water; higher dialcohols are more water-soluble than similar monoalcohols.
- Alcohols are very weak acids (proton donors); water solutions are neutral (Section 3.7).
- Common alcohols are flammable; can be toxic.
- Dehydration and oxidation are major alcohol reactions.

Practice Problems **3.6** Which of the following compounds has the highest boiling point?
(a) $CH_3CH_2CH_2OH$ (b) $CH_3CH_2OCH_3$ (c) $CH_3CH_2CH_3$

3.7 Which of the following compounds is most water-soluble?
(a) $CH_3(CH_2)_{10}CH_2OH$ (b) $CH_3CH_2CHCH_3$ (c) $CH_3CH_2OCH_3$
$\qquad\qquad\qquad\qquad\qquad\qquad\qquad\quad$ |
$\qquad\qquad\qquad\qquad\qquad\qquad\qquad$ OH

3.5 CHEMICAL REACTIONS OF ALCOHOLS

Dehydration Loss of water, as from an alcohol to yield an alkene.

Dehydration of Alcohols   Alcohols undergo loss of water (**dehydration**) on treatment with a strong acid catalyst. The —OH group is lost from one carbon, and an —H is lost from an adjacent carbon to yield an alkene product:

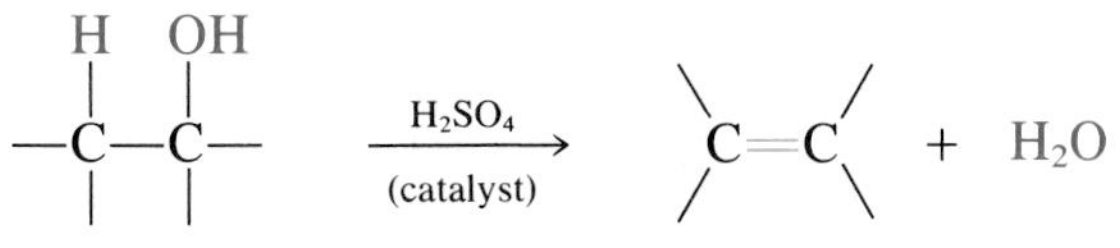

An alcohol   An alkene

For example,

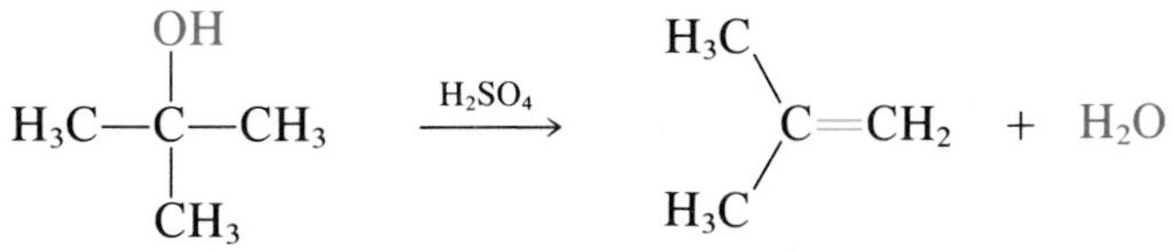

tert-Butyl alcohol   2-Methylpropene

In cases where more than one alkene can result from dehydration, a mixture of products is usually formed. A good rule of thumb is that the *major* product is the one that has the greater number of alkyl groups attached to the double-bond carbons. For example,

$$\underset{\underset{\text{Dehydration from}}{\text{this position?}}}{\text{CH}_3}\underset{}{\overset{\overset{\displaystyle\text{OH}}{|}}{\text{CH}_2\text{CH}}}\underset{\underset{\text{Or this}}{\text{position?}}}{\text{CH}_3} \xrightarrow{\text{H}_2\text{SO}_4} \underset{\text{2-Butene (80\%)}}{\text{CH}_3-\text{CH}=\text{CH}-\text{CH}_3} + \underset{\text{1-Butene (20\%)}}{\text{CH}_3\text{CH}_2-\text{CH}=\text{CH}_2}$$

Two alkyl groups on double-bond carbons

One alkyl group on double-bond carbons

Solved Problem 14.2   What products would you expect from the following dehydration reaction? Which product will be major, and which minor?

$$\underset{\underset{|}{\overset{|}{\text{CH}_3}}}{\text{CH}_3\overset{\overset{\displaystyle\text{OH}}{|}}{\text{CH}}\text{CHCH}_3} \xrightarrow{\text{H}_2\text{SO}_4} \quad ?$$

Solution   First, find the hydrogens on carbons next to the OH-bearing carbon and rewrite the structure to emphasize these hydrogens:

$$\underset{\underset{|}{\overset{|}{\text{CH}_3}}}{\text{CH}_3\overset{\overset{\displaystyle\text{OH}}{|}}{\text{CH}}\text{CHCH}_3} \;=\; \text{CH}_3-\overset{\overset{\displaystyle\text{H}}{|}}{\underset{\underset{|}{\text{CH}_3}}{\text{C}}}-\overset{\overset{\displaystyle\text{OH}}{|}}{\text{CH}}-\overset{\overset{\displaystyle\text{H}}{|}}{\text{CH}_2}$$

Next, circle and remove the possible combinations of H and OH, drawing a double bond where H and OH have come from:

$$\text{CH}_3-\overset{\overset{\displaystyle\text{H}}{|}}{\underset{\underset{|}{\text{CH}_3}}{\text{C}}}-\overset{\overset{\displaystyle\text{OH}}{|}}{\text{CH}}-\overset{\overset{\displaystyle\text{H}}{|}}{\text{CH}_2} \longrightarrow \underset{\underset{|}{\text{CH}_3}}{\text{CH}_3-\text{C}=\text{CH}-\text{CH}_3} \quad \text{and} \quad \underset{\underset{|}{\text{CH}_3}}{\text{CH}_3-\text{CH}-\text{CH}=\text{CH}_2}$$

Finally, determine which of the alkenes has the larger number of alkyl substituents on its double-bond carbons and is therefore the major product:

$$\underset{\text{H}_3\text{C}}{\overset{\text{H}_3\text{C}}{}}\text{C}=\text{C}\underset{\text{CH}_3}{\overset{\text{H}}{}} \qquad \text{and} \qquad \underset{\text{H}}{\overset{(\text{CH}_3)_2\text{CH}}{}}\text{C}=\text{C}\underset{\text{H}}{\overset{\text{H}}{}}$$

2-Methyl-2-butene
major product (three alkyl groups)

3-Methyl-1-butene
minor product (one alkyl group)

Practice Problems   **3.8**   What alkenes might be formed by dehydration of the following alcohols? If more than one product is possible, indicate which is major.

(a) $\text{CH}_3\text{CH}_2\text{CH}_2\text{OH}$ (b) ⬡—OH   (c) $\text{CH}_3\overset{\overset{\displaystyle\text{OH}}{|}}{\text{CH}}\text{CH}_2\overset{\overset{\displaystyle\text{CH}_3}{|}}{\text{CH}}\text{CH}_3$

3.9   Dehydration of what alcohols would yield these alkenes?
(a) $(\text{CH}_3)_2\text{C}=\text{C}(\text{CH}_3)_2$ (b) $\text{CH}_3\text{CH}_2\text{CH}=\text{CH}_2$

Carbonyl group A functional group that has a carbon atom joined to an oxygen atom by a double bond, C=O.

Oxidation of Alcohols   Primary and secondary alcohols are converted into carbonyl-containing compounds on treatment with an oxidizing agent. A **carbonyl group** (pronounced car-bo-**neel**) is a functional group that has a carbon atom joined to an oxygen atom by a double bond, C=O. Many different oxidizing agents can be used—potassium permanganate ($KMnO_4$), sodium dichromate ($Na_2Cr_2O_7$), or nitric acid (HNO_3), for example—and it usually doesn't matter which specific reagent is chosen. Thus, we'll simply use the symbol [O] to indicate a generalized reagent.

Oxidation In organic chemistry, the removal of hydrogen from a molecule or the addition of oxygen to a molecule.

The net effect of an alcohol **oxidation** is the removal of two hydrogen atoms. One hydrogen comes from the —OH group, and the other comes from the carbon atom bonded to the —OH group. These hydrogens are converted into water during the reaction by the oxidizing agent [O].

REVIEW An oxidizing agent is a reactant that causes oxidation and is itself reduced.

An alcohol   A carbonyl compound

Different kinds of carbonyl-containing products are formed depending on the structure of the starting alcohol and on the exact reaction conditions. Thus, primary alcohols (RCH_2OH) are converted either into **aldehydes** (RCH=O) if carefully controlled conditions are used, or into **carboxylic acids** (RCOOH) if an excess of oxidant is used.

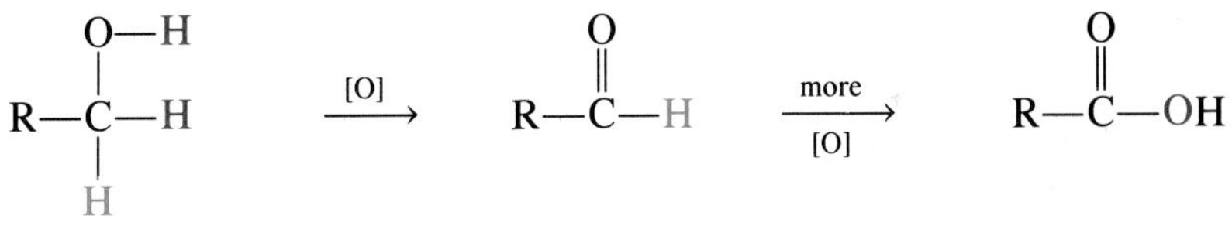

Aldehyde A compound with the —CH=O functional group.

Carboxylic acid A compound with the —COOH functional group.

A primary alcohol   An aldehyde   A carboxylic acid

For example,

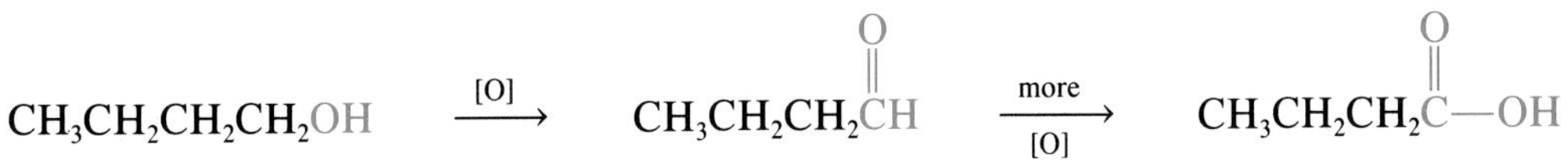

1-Butanol   Butanal   Butanoic acid

Ketone A compound that has a carbonyl group bonded to two carbon atoms, $R_2C=O$.

Secondary alcohols (R_2CHOH) are converted into **ketones** ($R_2C=O$) on treatment with oxidizing agents:

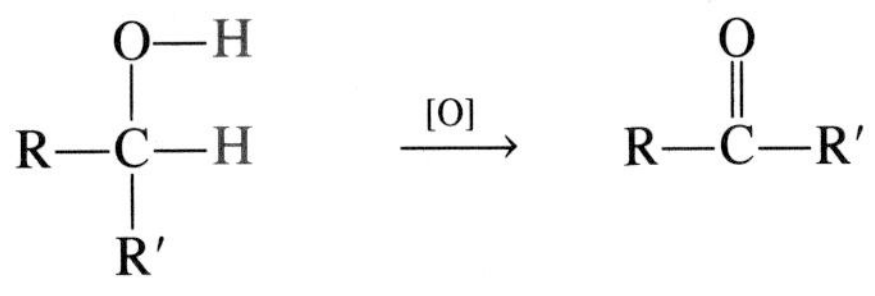

A secondary alcohol   A ketone

For example,

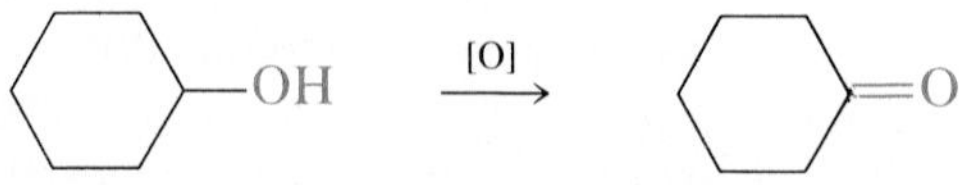

Cyclohexanol   Cyclohexanone

Tertiary alcohols don't normally react with oxidizing agents because they don't have a hydrogen on the carbon atom next to the —OH group:

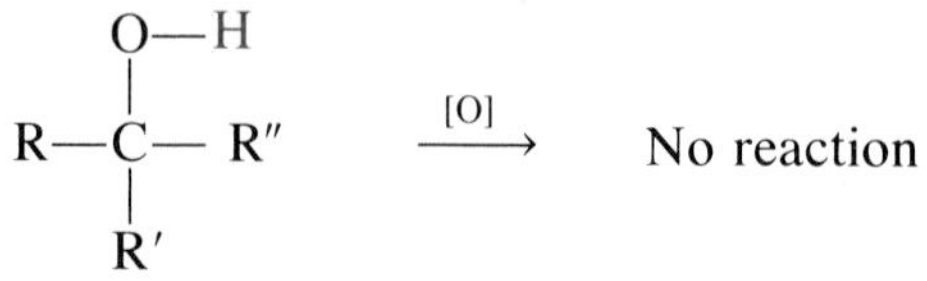

A tertiary alcohol

Alcohol oxidations are critically important steps in many biological processes. For example, when lactic acid builds up in tired, overworked muscles, the liver removes it by oxidizing it to pyruvic acid. Our bodies, of course, don't use $Na_2Cr_2O_7$ or $KMnO_4$ for the oxidation; they use specialized, highly selective enzymes to carry out their chemistry. Regardless of the details, though, the net chemical transformation is the same whether carried out in a laboratory flask or in a living cell. (In writing biochemical reactions, we'll write acids in the ionized forms that exist in body fluids.)

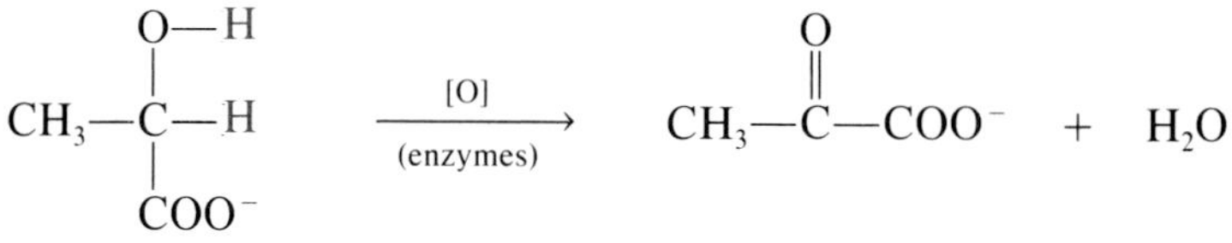

REVIEW   Ionization of a carboxylic acid:
$$RCOOH \rightarrow RCOO^- + H^+$$

Lactic acid anion   Pyruvic acid anion
(an alcohol-acid)

Solved Problem 14.3   What are the products of the following reaction?

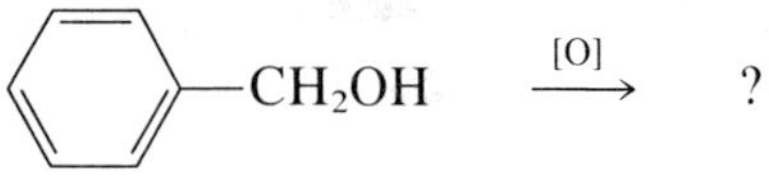

Benzyl alcohol

Solution   Since the starting material is a primary alcohol, it will be converted first to an aldehyde and then to a carboxylic acid. To find the structures of these products, first rewrite the starting alcohol to emphasize the hydrogen atoms on the OH-bearing carbon:

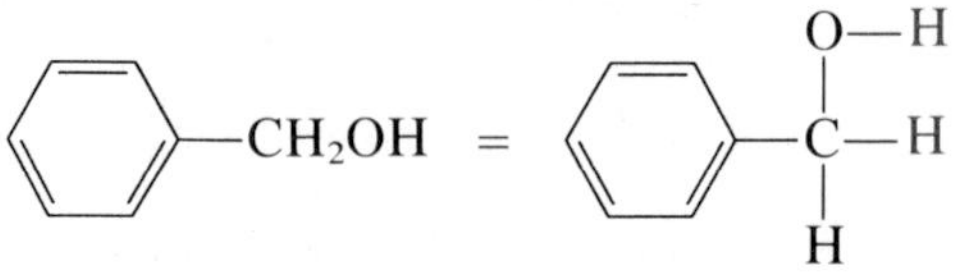

Next, circle and remove two hydrogens, one from the —OH group and one from the neighboring carbon. In their place, make a C—O double bond. This is the aldehyde product that will form initially:

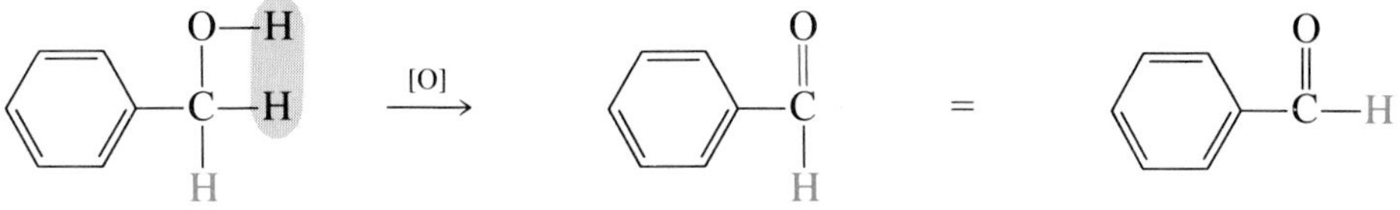

Finally, convert the aldehyde to a carboxylic acid by removing the —CH=O hydrogen and replacing it with an —OH group:

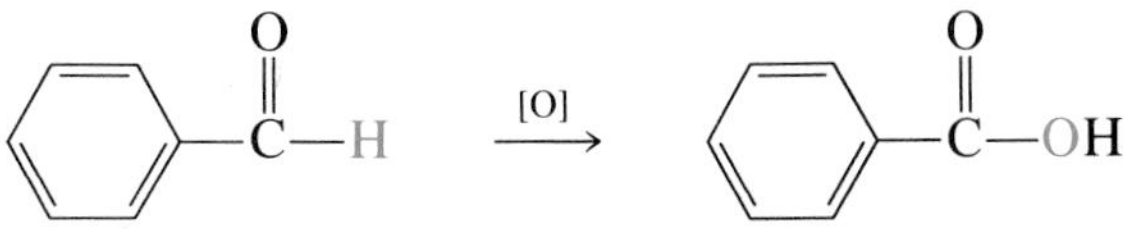

Practice Problems   **3.10**   What products would you expect from oxidation of these alcohols?

(a) $CH_3CH_2CH_2OH$ (b) $CH_3CHCH_2CH_2CH_3$ with OH

(c) 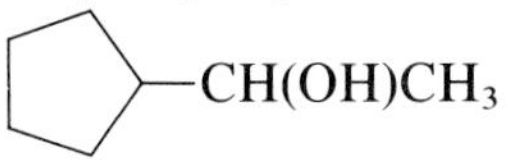 —CH(OH)CH₃

3.11   What alcohols might these carbonyl products have come from?

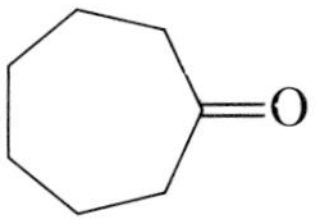

(a) CH_3CCH_3 (with O double bond)   (b) =O   (c) CH_3CHCH_2COOH with CH_3

3.6   PHENOLS

The word *phenol* is the name both of a specific compound (hydroxybenzene, C_6H_5OH) and of a family of compounds. Phenol itself, also called *carbolic acid,* is a medical **antiseptic** first used by Joseph Lister in 1867. Lister showed that the occurrence of postoperative infection dramatically decreased when phenol was used to cleanse the operating room and the patient's skin, and in dressings for surgical wounds. Because phenol numbs the skin, it also became popular in topical drugs for pain and itching, and in treating sore throats.

The medical use of phenol is now restricted because it can cause severe skin burns and has been found to be toxic, both by ingestion and by absorption through the skin and inhalation. The once-common use of phenol for treating diaper rash is especially hazardous because phenol is more readily absorbed through a rash. Only water or alcohol solutions containing a maximum of 1.5% phenol and lozenges containing a maximum of 50 mg of phenol are now allowed in nonprescription drugs. Many mouthwashes and throat lozenges contain alkyl-substituted phenols as active ingredients for pain relief, for example,

OH

CH₂CH₂CH₂CH₂CH₂CH₃

4-Hexylresorcinol
(a topical anesthetic)

CH₃

CH₃CHCH₃

Thymol
(a topical anesthetic; occurs
naturally in the herb thyme)

Disinfectant An agent that can be used to destroy or prevent the growth of harmful microorganisms on inanimate objects only.

Phenols and substituted phenols such as the cresols (methylphenols) are common as **disinfectants** in hospitals and elsewhere. By contrast with an anti-

AN APPLICATION: ETHYL ALCOHOL AS A DRUG AND POISON

Ethyl alcohol as a drug is classified as a central-nervous-system (CNS) depressant. Its direct effects (being "drunk") resemble the response to anesthetics and are quite predictable. At first, there is an appearance of stimulation—excitability and outgoing sociable behavior—but this results from depression of inhibition rather than stimulation. At a blood alcohol concentration of 100–300 mg/dL, motor coordination and pain perception are affected, accompanied by loss of balance, slurred speech, and amnesia. At the next stage (300–400 mg/dL), voluntary responses to stimuli are affected, and there may be nausea and loss of consciousness. Further increases in blood alcohol levels cause progressive loss of protective reflexes in stages like those of surgical anesthesia. Above 600 mg/dL of blood alcohol, spontaneous respiration and cardiovascular regulation are affected, and the ultimate result can be death.

The pathway of ethyl alcohol through the body begins with its ready absorption in the stomach and small intestine, followed by rapid distribution to all body fluids and organs. In the pituitary gland, alcohol inhibits the production of a hormone that regulates urine flow, causing increased urine production and dehydration. In the stomach, it stimulates production of acid. Throughout the body, it causes blood vessels to dilate, resulting in flushing of the skin and a sensation of warmth as blood moves into capillaries beneath the surface. The result, though, is not a warming of the body but an increased loss of heat at the surface, making alcoholic beverages a poor choice for help in enduring cold weather.

The metabolism of alcohol occurs mainly in the liver and proceeds by oxidation in two steps, first to acetaldehyde and then to acetic acid. One of the hydrogen atoms lost in the oxidation at each stage binds to the biochemical oxidizing agent NAD^+ (nicotinamide adenine dinucleotide), and the other leaves as a hydrogen ion. When continuously present in the bodies of chronic alcoholics, alcohol and acetaldehyde are toxic, leading to devastating physical and metabolic deterioration. The liver usually suffers the worst damage, as it is the major site of alcohol metabolism.

Other alcohols are oxidized in the same manner as ethyl alcohol. The toxicity of methyl

Alcohol metabolism

$$CH_3CH_2OH + NAD^+ \xrightarrow[\text{enzyme}]{\substack{\text{alcohol} \\ \text{dehydrogenase}}} CH_3\overset{\overset{\displaystyle O}{\|}}{C}-H + NADH/H^+ \quad CH_3\overset{\overset{\displaystyle O}{\|}}{C}-H + NAD^+ + H_2O \xrightarrow[\text{enzyme}]{\substack{\text{aldehyde} \\ \text{dehydrogenase}}}$$

septic, which safely kills microorganisms on living tissue, a disinfectant should be used only on inanimate objects. The germicidal properties of phenols can be partially explained by their ability to disrupt the permeability of cell walls of microorganisms.

Phenols are usually named with the ending *-phenol* rather than *-benzene*. For example,

o-Chlorophenol   *p*-Methylphenol

alcohol is due to the formation of formaldehyde (HCHO), a more toxic chemical than acetaldehyde. The danger in ingesting ethylene glycol is from its oxidation product, oxalic acid, which is not soluble in body fluids.

Because ethyl alcohol competes with other alcohols in binding to alcohol dehydrogenase, it slows their conversion to harmful oxidation products, giving the body a chance to eliminate the alcohols before damage is done. Therefore, ethyl alcohol is often administered as an antidote in methyl alcohol or ethylene glycol poisoning.

The quick and uniform distribution of ethyl alcohol in body fluids, the ease with which it crosses lung membranes, and its ready oxidizability provide the basis for tests for blood alcohol concentration. The Breathalyzer test measures alcohol concentration in expired air by the color change that occurs when the chromium in the bright yellow-orange oxidizing agent potassium dichromate ($K_2Cr_2O_7$) is reduced to blue-green chromium(III). The color change can be interpreted by instruments to give an accurate measure of alcohol concentration in the blood. In most states, a blood alcohol level above 0.10% (100 mg/dL) is evidence for legal charges of driving while intoxicated (DWI).

$$\overset{O}{\overset{\|}{CH_3C}}OH + NADH + H^+$$

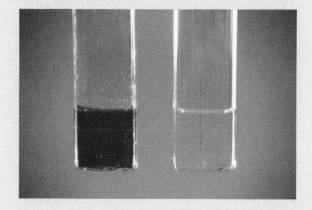 

The Breathalyzer test is based on the chemical reaction between orange potassium dichromate ($K_2Cr_2O_7$) and ethyl alcohol (in the vessels at the left). When they are mixed, the alcohol is oxidized to acetaldehyde and chromium is reduced to the green Cr^{3+} ion (on the right).

The properties of phenols, like those of alcohols, are influenced by hydrogen bonding. Most phenols are water-soluble to some degree and have higher melting and boiling points than similarly substituted alkylbenzenes.

Among biomolecules, the amino acid tyrosine contains a hydroxyl-substituted benzene ring, as do many compounds that give plants their characteristic properties, such as eugenol and urushiol (Figure 3.4).

Some naturally occurring phenols

Tyrosine
(an *amino acid*)

Eugenol
(in cloves, bananas, and
other fruits; used for
toothache pain)

A urushiol
(skin irritant in
poison ivy)

Practice Problems   **3.12**   Draw structures for (a) *m*-bromophenol, (b) 4-ethylphenol.

3.13   Name the following compounds:
(a)   (b)

Figure 3.4
Do you recognize this plant? If you don't know what poison ivy looks like, you may experience firsthand some chemical interactions between urushiol and your body.

AN APPLICATION: ANTIOXIDANTS

If you're prone to reading food ingredient labels, the names "butylated hydroxytoluene" and "butylated hydroxyanisole" or their abbreviations BHT and BHA are familiar to you. You can see them on most cereals, cookie, and cracker boxes. Both compounds are substituted phenols.

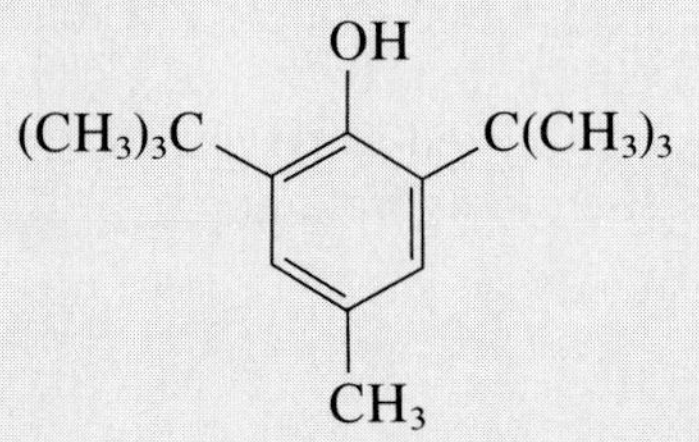

Butylated hydroxytoluene (BHT)

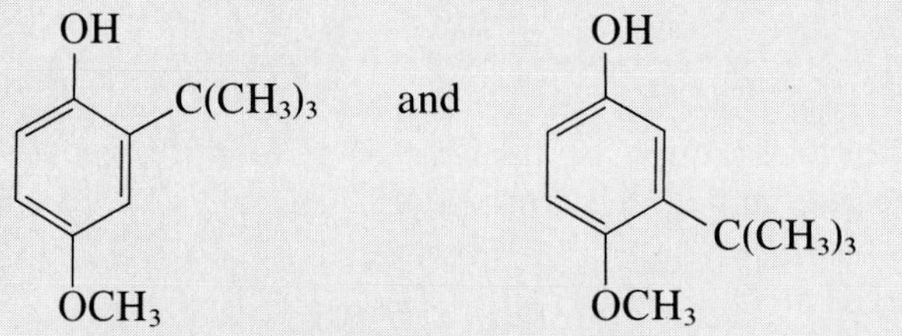

Butylated hydroxyanisole (BHA)
(a mixture of two isomers)

Foods that contain unsaturated fats (those with carbon–carbon double bonds) become rancid when oxygen from the air reacts with their double bonds, breaking them apart and producing bad-smelling and bad-tasting compounds. The oxidation reactions proceed by formation of extremely reactive intermediates that contain unpaired electrons and are known as **free radicals.** Each free radical reacts with a stable molecule to form another radical, which then reacts with another molecule, and so on, in long series of *chain reactions.*

BHT and BHA are called *antioxidants* because they interrupt a free radical chain reaction by combining with the free radicals to form stable compounds. The chain reaction is thus terminated rather than carried forward. Vitamin E is a natural antioxidant that traps free radicals within the body.

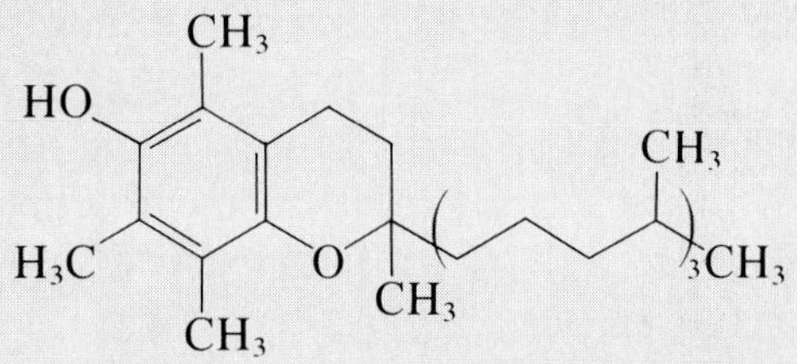

Free radical formation is suspected of playing a role in both cancer and normal aging of living tissue. Although there is no conclusive evidence, questions have been raised about the possible effectiveness of antioxidants in slowing the progress of these two conditions.

Free radical An atom or group that has an unpaired electron.

3.7 ACIDITY OF ALCOHOLS AND PHENOLS

Alcohols and phenols are weakly acidic. They dissociate only slightly in aqueous solution and therefore establish equilibria.

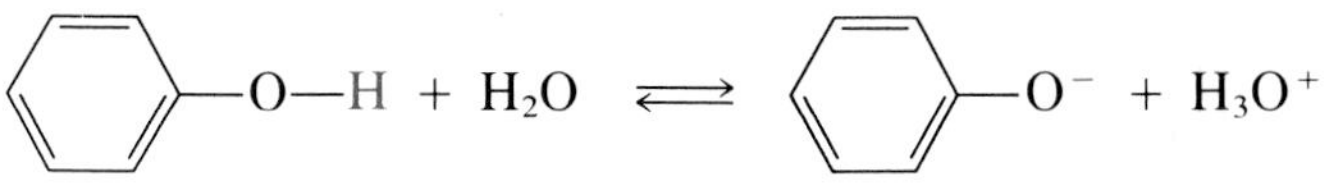

REVIEW It will be helpful here to review the properties of acids and bases described in Appendix B.

Water: $\quad H{-}O{-}H + H_2O \rightleftharpoons H{-}O^- + H_3O^+$

An alcohol: $\quad CH_3{-}O{-}H + H_2O \rightleftharpoons CH_3{-}O^- + H_3O^+$

A phenol:

Methanol and ethanol are no more acidic than water itself and are so slightly dissociated in water that their aqueous solutions are neutral (pH 7). The anion of an alcohol, RO^-, known as an *alkoxide ion,* is a strong base, as expected for the anion of a weak acid. Alkoxides are prepared by reaction of an alcohol with an alkali metal. For example,

$$\underset{\text{Methyl alcohol}}{2\,CH_3OH} + 2\,Na \longrightarrow \underset{\text{Sodium methoxide}}{2\,CH_3O^-Na^+} + H_2(g)$$

REVIEW The smaller the value of the acid ionization constant, K_a, the weaker an acid and the smaller the concentration of its ionization products at equilibrium (Appendix B.7).

Phenols are considerably more acidic than water. Phenol itself ($K_a = 1 \times 10^{-10}$, for example, is close to HCN ($K_a = 6 \times 10^{-10}$) and HCO_3^- ($K_a = 5 \times 10^{-11}$) in acidity. Phenols react as weak Brønsted-Lowry acids with hydroxide ion and are soluble in dilute aqueous sodium hydroxide.

A phenol: 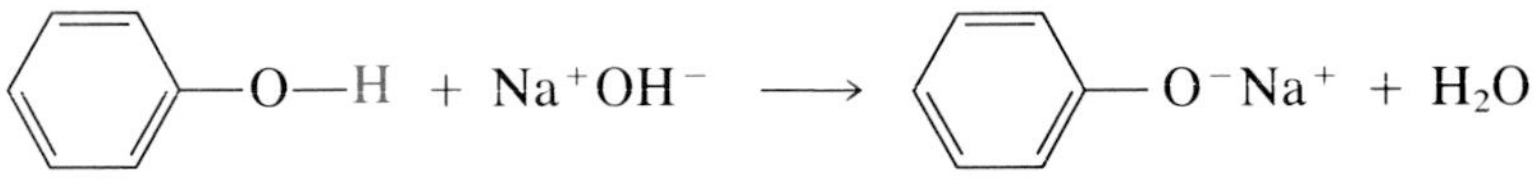

Sodium phenoxide

3.8 NAMES AND PROPERTIES OF ETHERS

Ethers with simple alkyl or aryl groups are named just by identifying the two groups bonded to oxygen and adding the word *ether:*

$CH_3—O—CH_3$	$CH_3—O—CH_2CH_3$	$CH_3CH_2—O—CH_2CH_3$
Dimethyl ether (bp −24.5°C)	Ethyl methyl ether (bp 10.8°C)	Diethyl ether (bp 34.5°C)

If both groups are the same, the "di-" is often left out and we refer to "methyl ether" or "ethyl ether." A reference to just "ether" usually means diethyl ether. Compounds containing the C—O—C group in a ring are classified as cyclic ethers but are not named as such. Many have common names.

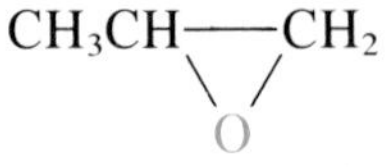

Reactive intermediate in polymer synthesis

Propylene oxide

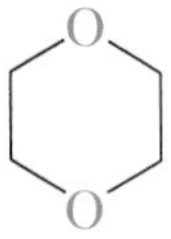

A solvent

Dioxane

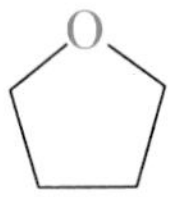

A solvent

Tetrahydrofuran

Alkoxy group An RO— group.

The RO— group is referred to as an **alkoxy group;** $CH_3O—$ is a *methoxy group,* $CH_3CH_2O—$ is an *ethoxy group,* and so on. These names are used when the ether functional group is present with other functional groups. For example,

$CH_3CH_2OCH_2CH_2OH$

2-Ethoxyethanol

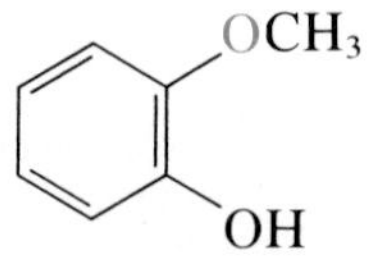

o-Methoxyphenol

Though polar, ethers lack the hydroxyl group of water and alcohols, and ether molecules do not hydrogen-bond to each other. Thus, the simple ethers are higher boiling than comparable alkanes but lower boiling than alcohols. The ether oxygen can hydrogen-bond with water, causing dimethyl ether to be water-soluble and diethyl ether to be partially miscible with water. Higher ethers are only slightly soluble or insoluble in water. Ethers are good solvents for most organic molecules.

Ethers are alkane-like in many chemicals properties and don't react with most acids, bases, or other chemical reagents. Ethers do, however, react readily with oxygen, and the simple ethers are highly flammable. On standing in air, many ethers form explosive *peroxides*, compounds which contain the unstable —O—O— group. Thus, ethers must be handled with care and stored in the absence of oxygen.

Properties of Ethers

● No hydrogen bonding occurs between ether molecules, but they are polar; lower boiling than alcohols, but higher boiling than alkanes.

● Lower ethers are volatile, flammable liquids.

● Ethers are slightly soluble or insoluble in water (except dimethyl ether, which is water-soluble).

● Are good solvents for organic compounds.

● Are not very reactive.

● Form explosive peroxides on standing in air.

3.9 SOME COMMON ETHERS

Diethyl ether, the most common ether, is best known as a solvent and anesthetic. Its value as an inhalation anesthetic was discovered in the 1840s, and until around 1930, diethyl ether, nitrous oxide (N_2O), and chloroform ($CHCl_3$) were the mainstays of the operating room. The ideal general anesthetic should act quickly, produce a deep sleep, allow a quick and smooth return to consciousness, produce few side effects, exit the body unchanged, and be safe to handle.

Ether acts quickly and is a very effective anesthetic, but it is far from ideal because recovery is not quick and it often induces nausea. Moreover, its effectiveness is strongly counterbalanced by the hazards of handling it. Diethyl ether is a highly volatile, flammable liquid whose vapor forms explosive mixtures with air that are ignited by the slightest spark. Ether has now been replaced by safer, less flammable anesthetics, two of which are halogenated ethers. It's interesting to note that these two compounds were products of an intensive effort during the 1960s in which more than 400 halogenated ethers were synthesized in a search for improved anesthetics (Figure 3.5).

Anesthetics

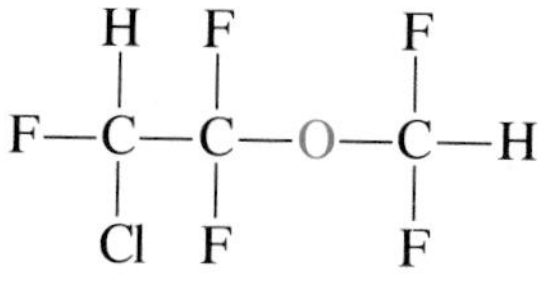

Enflurane

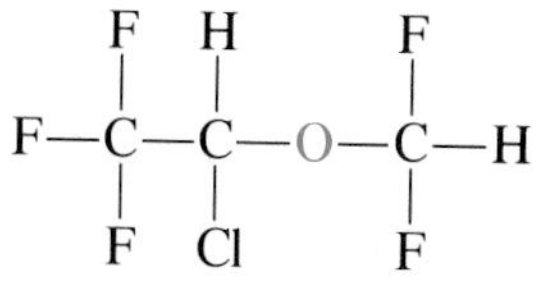

Isoflurane

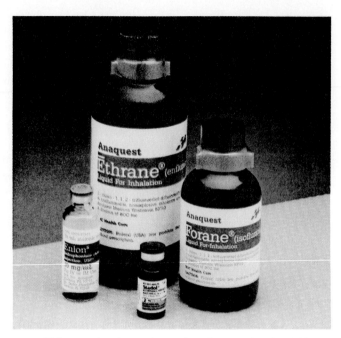

Figure 3.5
Inhalation anesthetics. Enflurane (trade name Ethrane) and isoflurane (trade name Forane) are widely used nonflammable, nonexplosive anesthetics that produce a rapid anesthesia from which recovery is also rapid. They are both low-boiling liquids administered as vapors (enflurane, bp 56.5°C; isoflurane, bp 48.5°C).

Ethers exist throughout the plant and animal kingdoms. Some are present in essential oils and are used in perfumes; others have a variety of biological roles. Juvenile hormone, for example, is a cyclic ether that helps govern the growth of insects and has attracted some interest as an insecticide. The three-membered *epoxide ring* in juvenile hormone is quite reactive because the oxygen and carbon bond angles are highly strained in the small ring.

Some biomolecules containing ether groups

From anise; a flavoring and perfume in soap and toothpaste

Anethole

Controls insect growth; as an insecticide prevents formation of the adult

juvenile hormone

3.10 SULFUR-CONTAINING COMPOUNDS: THIOLS AND DISULFIDES

Thiol   A compound that contains the —SH functional group, R—SH.

Mercaptan   An alternate name for a thiol, R—SH.

Sulfur is the element just below oxygen in the periodic table, and many oxygen-containing compounds have sulfur analogs. For example, **thiols (R—SH)**, also called **mercaptans,** are sulfur analogs of alcohols. The systematic parent name of a thiol is formed by adding *-thiol* to the parent hydrocarbon name. Otherwise, thiols are named in the same way as alcohols.

$$CH_3$$
$$|$$

$$CH_3CH_2SH \qquad CH_3CHCH_2CH_2SH \qquad CH_3CH{=}CHCH_2SH$$

Ethanethiol   3-Methyl-1-butanethiol   2-Butene-1-thiol

The most outstanding characteristic of thiols is their appalling odor. Skunk scent is caused by two of the simple thiols shown above, 3-methyl-1-butanethiol and 2-butene-1-thiol. Thiols are also in the air whenever garlic and onions are being sliced, or when there's a natural gas leak. Natural gas itself is odorless, so a low concentration of methanethiol (CH_3SH) is added as a safety measure, making it easy to detect the leak with the slightest sniff.

Disulfide A compound that contains a sulfur–sulfur single bond, R—S—S—R.

Thiols react with mild oxidizing agents to yield **disulfides**, R—S—S—R. Two thiols join together in this reaction, the hydrogen from each is lost, and the two sulfurs bond together.

REVIEW Inorganic sulfides such as PbS contain the sulfide ion, S^{2-}.

$$RSH + HSR \xrightarrow{\text{[O]}} RSSR$$

Two thiol molecules   A disulfide

For example,

$$H_3C{-}S{-}H + H{-}S{-}CH_3 \xrightarrow[\text{(oxidizing agent)}]{\text{[O]}} CH_3{-}S{-}S{-}CH_3 + H_2O$$

Methanethiol   Dimethyl disulfide

The reverse of this reaction occurs easily in the presence of many reducing agents, represented here by [H],

REVIEW A reducing agent is a reactant that causes reduction and is itself oxidized; oxidation and reduction always occur together in what are called "redox" reactions.

$$RSSR \xrightarrow{\text{[H]}} RSH + RSH$$

Thiols are important biologically because they occur as a functional group in the amino acid cysteine, which is part of many proteins.

$$O$$
$$\|$$
$$HSCH_2CHCOH \qquad \text{An amino acid}$$
$$|$$
$$NH_2$$

Cysteine

The easy formation of —S—S— bonds between cysteine groups helps to pull large protein molecules into the shapes they need to function. Hair protein, for example, is unusually rich in —SH groups. When hair is "permed," a mild oxidizing agent causes disulfide bonds to form between —SH groups, resulting in the introduction of bends and kinks into the hair (Figure 3.6).

Practice Problem   **3.14**   What disulfides would you obtain from oxidation of these thiols?
           (a) $CH_3CH_2CH_2SH$ (b) 3-methyl-1-butanethiol (skunk scent)

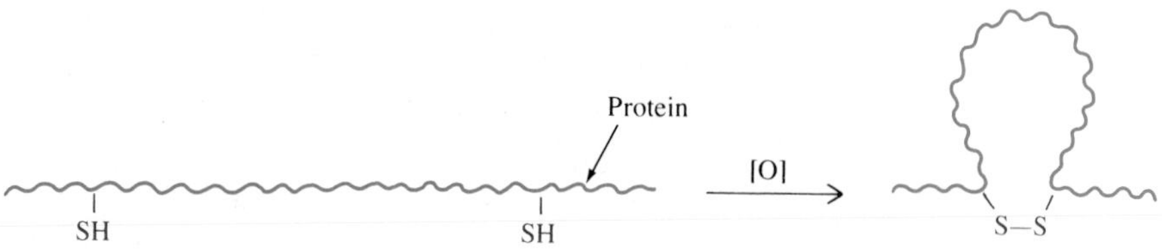

Figure 3.6
Sulfide bridges in hair. Chemistry can give you curly hair. A permanent wave results when disulfide bridges are formed between —SH groups in hair protein molecules.

3.11 HALOGEN-CONTAINING COMPOUNDS

Alkyl halide A compound that contains an alkyl group bonded to a halogen atom, RX.

Halogen-containing organic compounds result from the replacement of hydrogen atoms by halogen atoms. The simplest such compounds are the **alkyl halides,** RX, where R is an alkyl group and X is a halogen. Their common names are formed by giving the name of the alkyl group followed by the halogen name with an -*ide* ending. Systematic names for alkyl halides are derived like those of other families of compounds, by identifying the hydrocarbon root, using the alkane ending -*ane*, numbering the halogen substituents, and naming them as fluoro-, chloro-, bromo-, or iodo- groups. Some halogenated compounds, such as chloroform and halothane, are also known by common names.

CH_3CH_2Cl

Ethyl chloride, bp 12.5°C
(chloroethane)

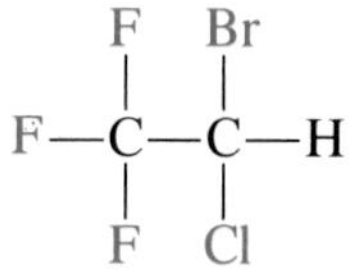

Halothane, bp 50°C
(1-bromo-1-chloro-2,2,2-
trifluoroethane)

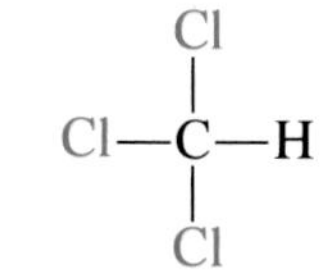

Chloroform, bp 61°C
(trichloromethane)

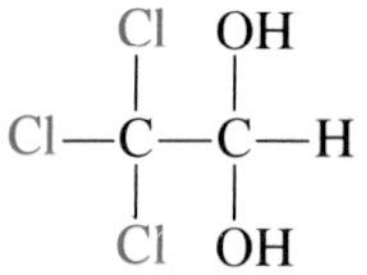

Chloral hydrate, bp 97.5°C
(1,1,1-trichloro-2,2-ethanediol)

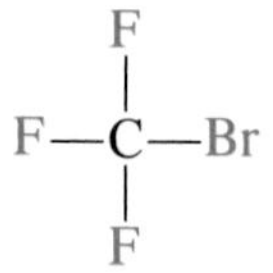

A Halon, bp −58°C
(bromotrifluoromethane)

Halogenated organic compounds have a variety of medical and industrial uses. Ethyl chloride is used as a topical anesthetic because it cools the skin through rapid evaporation; halothane is the most important non-ether anesthetic. Chloroform was once employed as an anesthetic (often by criminals in detective stories) and as a solvent for cough syrups and other medicines, but is now considered too toxic for such uses. Chloral hydrate is a sleep-inducing prescription drug (the "knockout" drops of detective stories). Bromotrifluoromethane is useful for extinguishing fires in aircraft and electronic equipment because it is not flammable, is nontoxic, and evaporates without a trace.

Although numerous halogen-containing organic compounds are found in nature, especially in marine organisms, few play significant roles in human biochemistry. One exception is thyroxine, an iodine-containing hormone secreted by the thyroid gland. A deficiency of iodine in the human diet leads to a low thyroxine level, which causes a condition called *goiter* and a resulting swelling of the thyroid gland. To ensure adequate iodine in the diet and prevent goiter, potassium iodide is added to table salt.

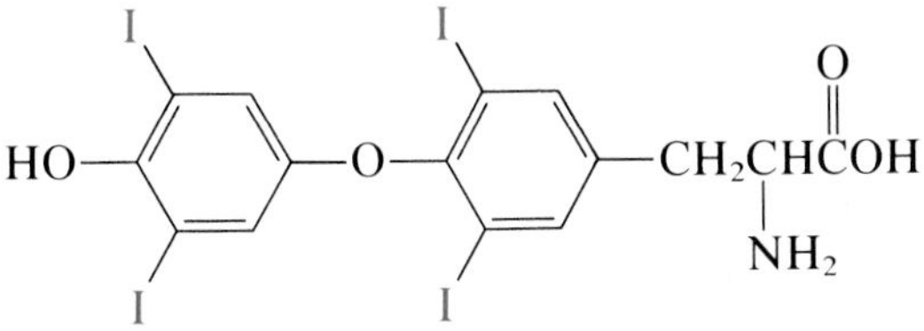

Thyroid gland hormone; deficiency causes goiter

Thyroxine

Halogenated compounds are also of great importance in industry and agriculture. Dichloromethane (CH_2Cl_2, methylene chloride) and trichloromethane ($CHCl_3$, chloroform) are common solvents, and trichloroethylene ($Cl_2C=CHCl$) is used to degrease machined metal parts. Because these substances are excellent solvents for the greases in skin, continued exposure often causes dermatitis.

Agricultural use of herbicides such as 2,4-D and fungicides such as Captan has resulted in vastly increased crop yields in recent decades, and the widespread application of chlorinated insecticides such as DDT is largely responsible for the progress made toward worldwide control of malaria and typhus. Despite their enormous benefits, chlorinated pesticides present problems because they are not broken down by natural processes and persist in the environment. They remain in the fatty tissues of organisms and accumulate up the food chain as larger organisms consume smaller ones. Eventually the concentration in some animals becomes high enough to cause harm. In an effort to maintain a balance between the value of halogenated pesticides and the harm they can do, the use of many has been restricted and some have been banned.

Some chlorinated pesticides

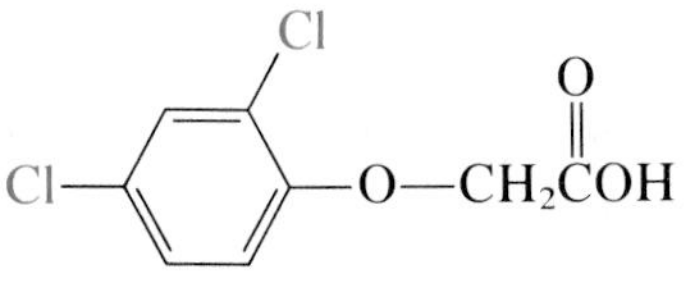

2,4-D

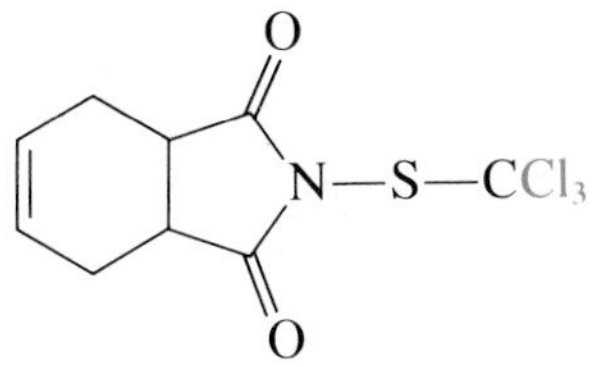

Captan

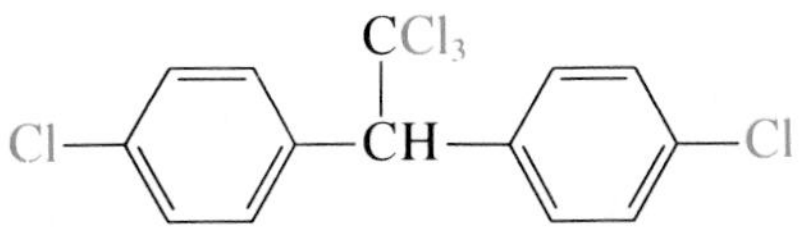

DDT

INTERLUDE: CHLOROFLUOROCARBONS AND THE OZONE HOLE

In recent years, newspaper stories about a "hole" in the ozone layer have appeared with regularity. What began as speculation about potential problems is now accepted as fact: Up to 75% of the ozone over the South Pole disappears in the fall when the so-called *ozone hole* develops. Furthermore, the overall ozone concentration in the region of the atmosphere extending from about 20 to 40 km above the earth's surface has decreased 2.5% in the last 10 years.

Although toxic to all life forms at high concentrations, ozone (O_3) is nevertheless critically important in the upper atmosphere because it acts as a shield to protect the earth from intense solar radiation. If the ozone layer were depleted, a great deal more solar radiation would reach the earth, causing an increase in the incidence of skin cancer and eye cataracts. The effects on microscopic plants in the oceans and crop plants are less predictable but could also be troublesome.

The causes of ozone depletion, although not fully understood, almost certainly involve a group of halogen-substituted alkanes called chlorofluorocarbons, familiar to most as *Freons* (DuPont). The chlorofluorocarbons (CFCs) are simple alkanes in which all the hydrogens have been replaced by either chlorine or fluorine. Fluorotrichloromethane (CCl_3F, Freon 11) and dichlorodifluoromethane (CCl_2F_2, Freon 12) are two of the most common CFCs in industrial use.

Because they are inexpensive and highly stable, yet not toxic, flammable, or corrosive, CFCs are ideal for use as propellants in aerosol cans, refrigerants, solvents, and fire extinguishers, and for blowing bubbles into foamed plastics such as those in insulation and mattresses. Unfortunately, the stability that makes CFCs so useful results in their persistence in the environment. The molecules slowly find their way into the upper atmosphere where they undergo a complex series of reactions that ultimately result in ozone destruction. When ultraviolet (UV) light strikes a CFC molecule, a carbon–chlorine bond breaks, producing a chlorine atom, which is highly reactive. The chlorine atom then reacts with ozone to yield oxygen and $ClO\cdot$, another reactive intermediate (a free radical):

$$CCl_2F_2 \xrightarrow{\text{UV light}} \cdot CClF_2 + Cl\cdot$$

$$Cl\cdot + O_3 \longrightarrow O_2 + ClO\cdot$$

Worldwide efforts to protect the ozone layer began in 1987 with an agreement, ultimately ratified by 36 countries and the European Economic Community, to gradually phase out CFC production. These efforts have since accelerated beyond the terms of the 1987 agreement as evidence for the serious nature of the problem has accumulated. Major producers of the chemicals are pooling their research efforts to find alternative, safe chemicals, and many nations have pledged to cease use of CFCs and other chemicals that release chlorine to the atmosphere by the year 2000. According to some predictions, even with these stringent efforts chlorine concentration will continue to rise due to molecules already in the atmosphere and will not return to current levels until the year 2045.

The atmosphere scientist is reviewing data on the ozone hole gathered by flights over the Antarctic.

SUMMARY

Alcohols have an —OH group (**hydroxyl**) bonded to a saturated alkane-like carbon atom and are given the family-name ending *-ol*. **Phenols** have an —OH group bonded directly to an aromatic ring and are named as phenols. **Ethers** have an oxygen atom bonded to two groups that may be either alkyl or aryl groups or may be in a ring, and their common names combine both group names with the word "ether." Both alcohols and phenols are "water-like" in their ability to form hydrogen bonds. The lower alcohols are miscible with water. Also like water, alcohols and phenols are weak acids that can donate their —OH protons to strong bases. Phenols are acidic enough to dissolve in aqueous NaOH. Ethers do not hydrogen-bond and are more alkane-like in their properties.

Alcohols undergo loss of water (**dehydration**) to yield alkenes when treated with a strong acid, and undergo **oxidation** to yield carbonyl-group-containing compounds. **Primary alcohols** (RCH_2OH) are oxidized to yield either aldehydes (RCH=O) or carboxylic acids (RCOOH); **secondary alcohols** (R_2CHOH) are oxidized to yield ketones (R_2C=O); and **tertiary** alcohols are not oxidized.

The —OH group is present in all carbohydrates and many other biochemically active molecules. Some common alcohols are methyl, ethyl, and isopropyl alcohol and ethylene glycol and glycerol. Except for glycerol, all are toxic in varying degrees. Phenols are notable for their use as disinfectants and antiseptics, and ethers for their solvent properties, flammability, and use as anesthetics.

Thiols, or **mercaptans** (RSH), are sulfur analogs of alcohols, many of which have unpleasant odors. They react with mild oxidizing agents to yield **disulfides** (RSSR), a reaction of importance in protein chemistry.

Alkyl halides, RX, contain a halogen atom bonded to an alkyl group. With increasing numbers of halogen atoms, compounds become higher boiling and less flammable. Halogenated compounds are rare in human biochemistry except for iodine-containing thyroxine but are widely used in industry as solvents and in agriculture as herbicides, fungicides, and insecticides. They are noted for their persistence in the environment.

REVIEW PROBLEMS

Alcohols, Ethers, and Phenols

3.15 How do alcohols, ethers, and phenols differ structurally?

3.16* What is the structural difference between primary, secondary, and tertiary alcohols?

3.17 Why do alcohols have higher boiling points than ethers of the same formula weight?

3.18* Which is the stronger acid, ethanol or phenol?

3.19 The steroidal compound prednisone is often used to treat poison ivy and poison oak inflammations. Identify the functional groups present in prednisone.

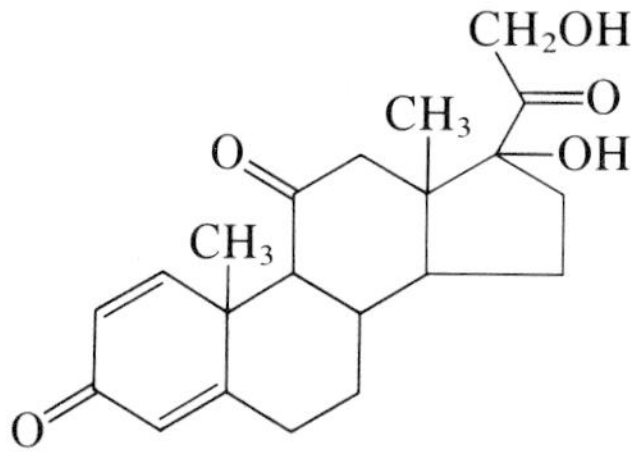

Prednisone

3.20* Vitamin E has the following structure. Identify the functional-group class to which each oxygen belongs.

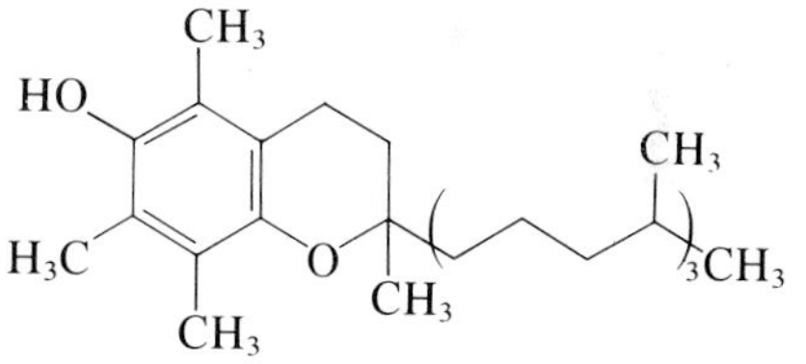

3.21 Give systematic names for these alcohols:

(a) 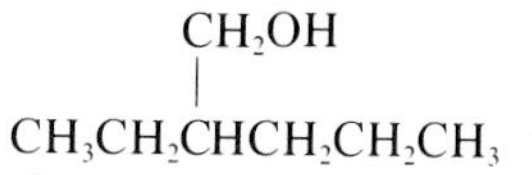 $CH_3CH_2\overset{\underset{\displaystyle |}{CH_2OH}}{CH}CH_2CH_2CH_3$

(b) $(CH_3)_2CHCH_2CH_2OH$

(c) 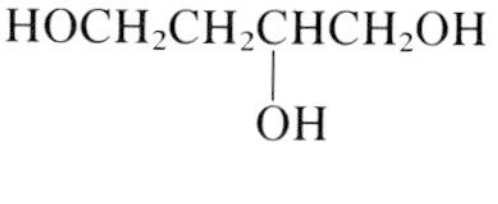 $HOCH_2CH_2\overset{\underset{\displaystyle |}{OH}}{CH}CH_2OH$

(d)

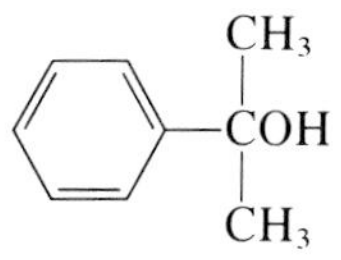

(e)

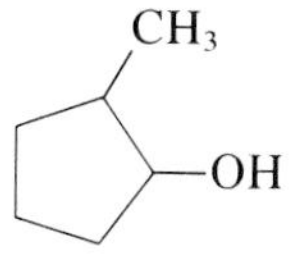

(f) 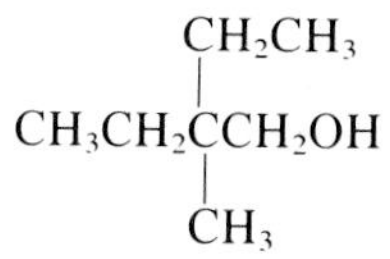 $CH_3CH_2\overset{\overset{\displaystyle CH_2CH_3}{|}}{\underset{\underset{\displaystyle CH_3}{|}}{C}}CH_2OH$

3.22 Draw structures corresponding to these names:
(a) 2,4-dimethyl-2-pentanol

(b) 2,2-dimethylcyclohexanol
(c) 5,5-diethyl-1-heptanol
(d) 3-ethyl-3-hexanol
(e) 2,3,7-trimethylcyclooctanol
(f) 3,3-diethyl-1,6-hexanediol

3.23 Identify each of the alcohols named in Problem 3.22 as primary, secondary, or tertiary.

3.24* Give systematic names for these compounds:

(a)

(b)

$$CH_3-\overset{\overset{\displaystyle CH_3}{|}}{CH}-O-CH_3$$

(c) O_2N-

$-O-CH_3$

(d)

(e)

(f) $CH_3CH_2CH_2OCH_2CH_2CH_3$

3.25 Draw structures corresponding to these names:
(a) ethyl phenyl ether
(b) *o*-dihydroxybenzene (catechol)
(c) *tert*-butyl *p*-bromophenyl ether
(d) *p*-nitrophenol
(e) 2,4-diethoxy-3-methylpentane
(f) 4-methoxy-3-methyl-1-pentene

3.26* Arrange the following six-carbon compounds in order of their expected boiling points and explain your ranking:
(a) hexane   (b) 1-hexanol   (c) dipropyl ether

Reactions of Alcohols

3.27 Give a specific example of an alcohol dehydration reaction.

3.28* What product is formed on oxidation of a secondary alcohol?

3.29 What structural feature prevents tertiary alcohols from undergoing oxidation reactions?

3.30* What product(s) can form on oxidation of a primary alcohol?

3.31 Which of these three compounds would you expect to be the most soluble in water, and which the least soluble? Explain.
(a) ethane   (b) 1-pentanol   (c) 1,2,3,-propanetriol

3.32* Assume that you have samples of the following two compounds, both with formula C_7H_8O. Both compounds dissolve in ether, but only one of the two dissolves in aqueous NaOH. How could you use this information to distinguish between them?

and

3.33 Assume that you have samples of the following two compounds, both with formula $C_6H_{12}O$. What simple chemical reaction will allow you to distinguish between them? Explain.

and

3.34 The following alkenes can be prepared by dehydration of an appropriate alcohol. Show the structure of the alcohol in each case. If the alkene can arise from dehydration of more than one alcohol, show all possibilities.

(a)

(b)

(c) 3-hexene

(d) $CH_3C=CHCH_2CH_3$
         $\overset{\overset{\displaystyle |}{CH_3}}{}$

(e) 1,3-butadiene

(f)

3.35 Phenols undergo the same kind of substitution reactions that other aromatic compounds do (Section 2.12). Formulate the reaction of *p*-methylphenol with bromine to give a mixture of two substitution products.

3.36 What carbonyl-containing products would you obtain from oxidation of these alcohols? If no reaction occurs, write "NR."

(a)

(b) CH_3CHCH_2OH
         $\overset{\overset{\displaystyle |}{CH_3}}{}$

(c) 3-methyl-3-pentanol

(d)

(e) $CH_3CH_2\overset{\overset{\displaystyle H_3C}{|}}{C}\overset{\overset{\displaystyle OH}{|}}{H}CH_3$
         $\overset{\overset{\displaystyle |}{CH_3}}{}$

(f)

3.37 What alkenes might be formed by dehydration of

these alcohols? If more than one product is possible, indicate which you expect to be major.

(a)

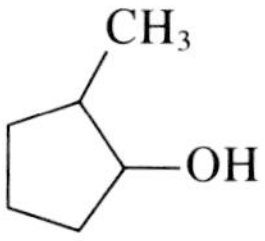

(b)

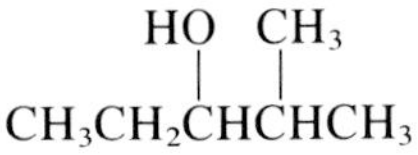

$$\underset{\underset{}{}}{HO}\ \underset{}{CH_3}$$
$$CH_3CH_2CHCHCH_3$$

(c)

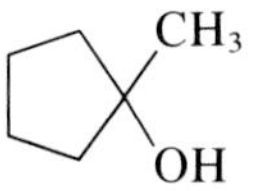

(d)

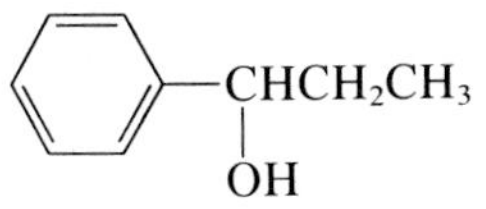

(e)

$$\underset{\underset{CH_2CH_3}{|}}{\overset{OH}{\underset{|}{CH_3CH_2CCH_2CH_3}}}$$

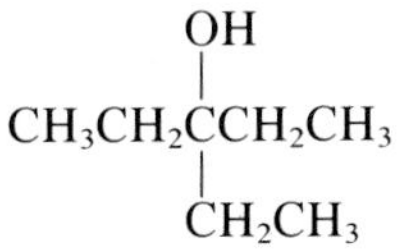

Thiols and Disulfides

3.38* What is the most noticeable characteristic of thiols?

3.39 What is the structural relationship between a thiol and an alcohol?

3.40 The amino acid cysteine forms a disulfide when oxidized. What is the structure of this disulfide?

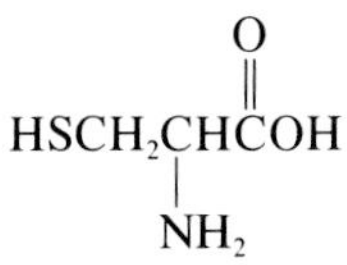

$$\underset{\underset{NH_2}{|}}{HSCH_2CH\overset{O}{\overset{||}{C}}OH}$$

3.41 Name these compounds.

(a) $\underset{\underset{SH}{|}}{CH_3CHCH_3}$ (b)

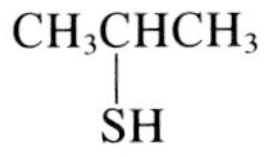

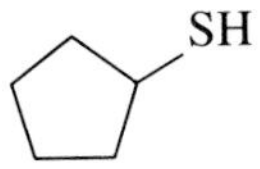

3.42* The boiling point of propanol is 97°C, while that of ethanethiol, with about the same molar mass, is 37°C. Chloroethane, with a similar molar mass, has a boiling point of 13°C. Explain.

3.43 Propanol is very soluble in water, while ethanethiol and chloroethane are only very slightly soluble. Explain.

Applications

3.44* Is ethanol a stimulant or a depressant? Justify your answer. [App: Ethyl Alcohol]

3.45 At what blood alcohol concentration does speech begin to be slurred? What is the approximate lethal concentration of ethyl alcohol in the blood? [App: Ethyl Alcohol]

3.46 Why does alcohol consumption cause dehydration? [App: Ethyl Alcohol]

3.47 Cirrhosis of the liver is a common disease of alcoholics. Why is the liver particularly affected by alcohol consumption? [App: Ethyl Alcohol]

3.48* Ethyl alcohol is toxic in high concentrations, yet it is administered to counteract the effects of both methanol and ethylene glycol poisoning. Name the enzyme responsible for alcohol metabolism and explain the rationale for this treatment. [App: Ethyl Alcohol]

3.49 Describe the basis of the Breathalyzer test for alcohol concentration. [App: Ethyl Alcohol]

3.50* Old westerns often show whiskey being consumed before bullets or arrows are removed from the body. Often the whiskey is poured into the wounds as well. What purpose does the whiskey serve? [App: Ethyl Alcohol]

3.51 What is a free radical? [App. Antioxidants]

3.52* What vitamin appears to be an antioxidant? [App: Antioxidants]

3.53 Ozone is considered to be an air pollutant at the earth's surface. Why is it of great benefit in the upper atmosphere? [Int: Chlorofluorocarbons]

3.54* Chlorofluorocarbons (CFCs) are still widely used as coolants in refrigerators and air conditioners, but states are beginning to legislate methods of transfer for CFCs and disposal of CFC-containing appliances. Why? [Int: Chlorofluorocarbons]

Additional Problems

3.55 Name the ether and alcohol isomers with formula $C_4H_{10}O$ and write their structural formulas.

3.56 What are the advantages and the disadvantages associated with the use of chlorinated pesticides?

3.57 Thyroxine (Section 3.11) is synthesized in the body by reaction of thyronine with iodine. Formulate the reaction and tell what kind of process is occurring.

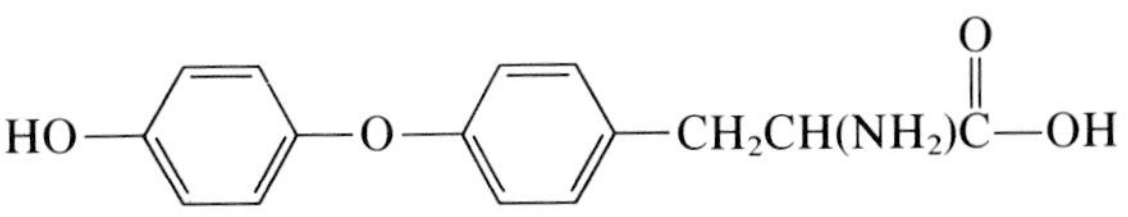

Thyronine

3.58* Neither 1-nonanol nor *n*-decane is water-soluble. Explain.

3.59 Thymol, mentioned in the chapter, has the following structure. Provide an IUPAC name for the compound.

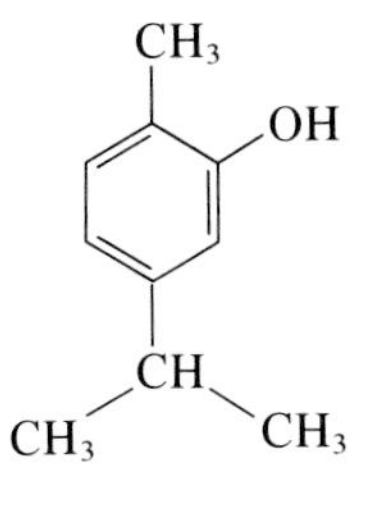

3.60* What is the difference between an antiseptic and a disinfectant?

3.61 Write the formulas and IUPAC names for the following common alcohols:
(a) rubbing alcohol (b) wood alcohol
(c) grain alcohol (d) diol used as antifreeze

3.62* Why is diethyl ether no longer ordinarily used as a general anesthetic?

3.63 Name the following compounds.

(a)

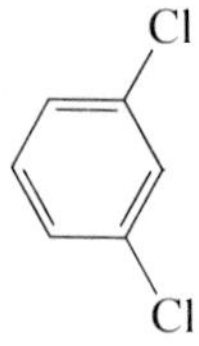

(b) 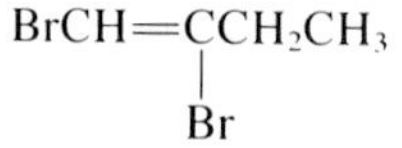 $BrCH\!\!=\!\!CCH_2CH_3$
      |
      Br

(c)

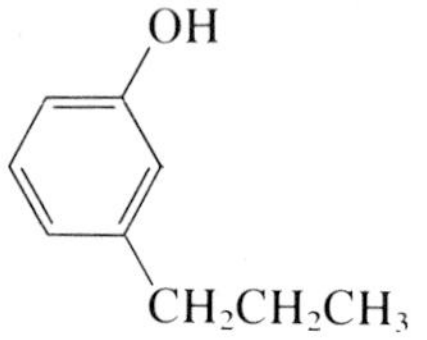

OH

(d)

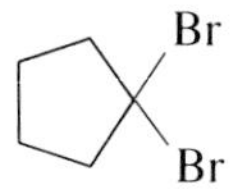

Br

Br

$CH_2CH_2CH_3$

(e) 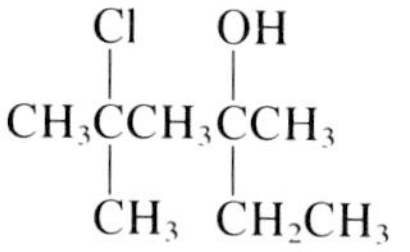 $\;\;$ Cl $\;$ OH
       | |
$CH_3CCH_2CCH_3$
       | |
      CH_3 CH_2CH_3

(f) 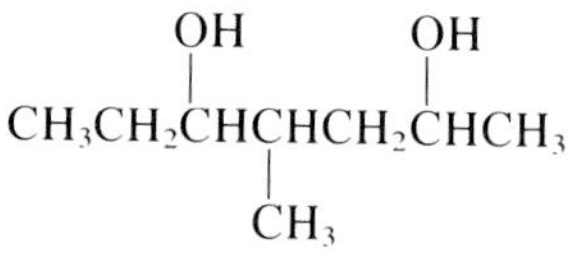 $\;\;$ OH $\;\;\;\;\;$ OH
      | |
$CH_3CH_2CHCHCH_2CHCH_3$
              |
             CH_3

(g) 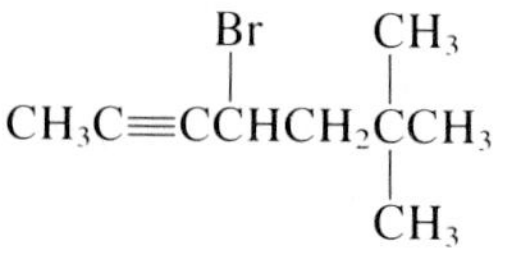 $\;\;\;\;$ Br $\;\;$ CH_3
         | |
$CH_3C\!\!\equiv\!\!CCHCH_2CCH_3$
                     |
                    CH_3

(h)

Br $\;\;\;\;$ Cl

3.64 Complete these reactions.
(a) 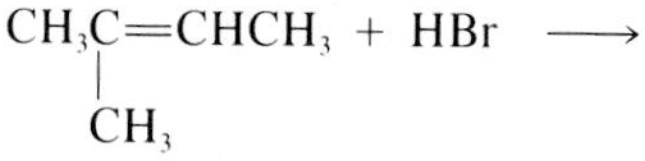 $CH_3C\!\!=\!\!CHCH_3$ + HBr $\longrightarrow$
       |
      CH_3

(b) 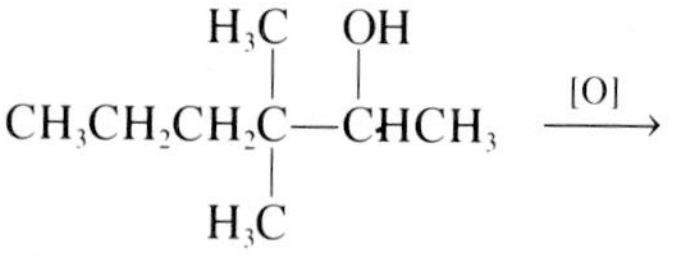 $\;\;\;\;\;\;\;\;$ H_3C $\;$ OH
             | |
$CH_3CH_2CH_2C\!-\!CHCH_3$ $\xrightarrow{[O]}$
             |
            H_3C

(c) 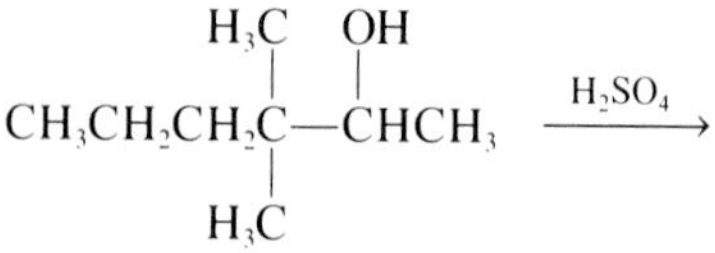 $\;\;\;\;\;\;\;\;$ H_3C $\;$ OH
             | |
$CH_3CH_2CH_2C\!-\!CHCH_3$ $\xrightarrow{H_2SO_4}$
             |
            H_3C

(d) 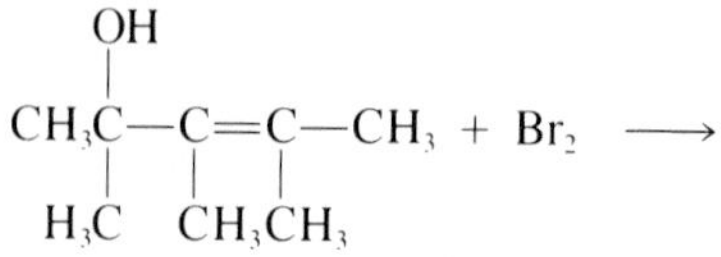 $\;\;\;\;$ OH
       |
$CH_3C\!-\!C\!\!=\!\!C\!-\!CH_3$ + Br_2 $\longrightarrow$
       | |
      H_3C CH_3CH_3

(e) $2\;(CH_3)_3CSH$ $\xrightarrow{[O]}$

(f) 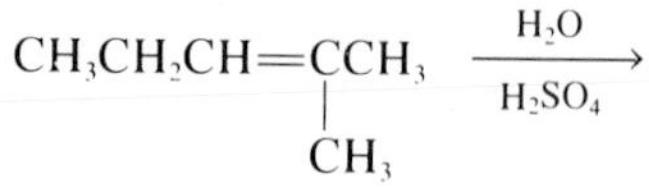 $CH_3CH_2CH\!\!=\!\!CCH_3$ $\xrightarrow[H_2SO_4]{H_2O}$
                       |
                      CH_3

(g)

$-CH_2CHCH_3$ $\xrightarrow{[O]}$
        |
       OH

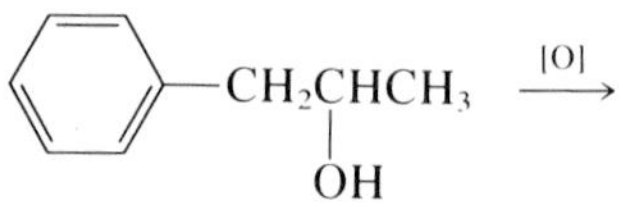

3.65 The odor of roses is due to geraniol.

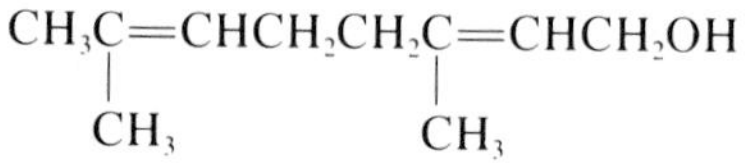

$CH_3C\!\!=\!\!CHCH_2CH_2C\!\!=\!\!CHCH_2OH$ $\;\;\;\;$ Geraniol
       | |
      CH_3 CH_3

(a) Name this compound by the IUPAC system.
(b) Circle the isoprene groups in this compound.
(c) When the alcohol group is oxidized to the aldehyde, citral, one of the compounds responsible for lemon scent is formed. Write the structure of citral.

3.66* Concentrated ethanol solutions can be used to kill microorganisms. However, at low concentrations, such as in some wines, the microorganisms can survive and cause oxidation of the alcohol. What is the structure of the acid formed?

3.67 "Flaming" desserts, such as cherries jubilee, use the ethanol in brandy or other distilled spirits as the flame carrier. Write the equation for the combustion of ethanol.

3.68* Simple sugars such as fructose, shown below, are very soluble in water in spite of having a long carbon chain. Why?

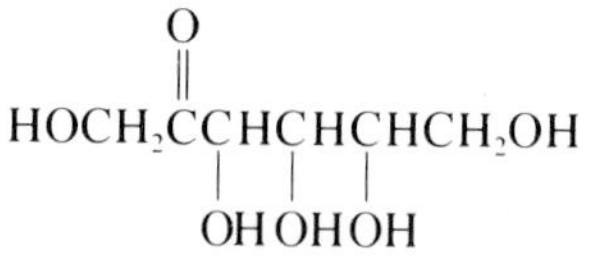

$\;\;\;\;\;\;\;\;\;\;$ O
$\;\;\;\;\;\;\;\;\;\;$ ||
$HOCH_2CCHCHCHCH_2OH$
$\;\;\;\;\;\;\;\;$ | | |
$\;\;\;\;\;\;\;\;$ OH OH OH

3.69 In Chapter 2, you saw that H_2SO_4 is a catalyst in the addition of water to alkenes to form alcohols. In this chapter, you found that H_2SO_4 is also used to dehydrate alcohols to make alkenes. Why do you think that sulfuric acid can serve two purposes—aiding in both hydration and dehydration?

CHAPTER

4 Amines

The active ingredients of many herbs in this early American apothecary shop in Shelburne, Vermont, were undoubtedly amines, the subject of this chapter.

In the preceding chapter we discussed organic compounds with single bonds between oxygen, sulfur, or halogen atoms and carbon atoms. Amines are organic compounds with single bonds between nitrogen and carbon atoms other than carbon atoms in C=O groups (amides, Section 6.9). There are many more kinds of amines in nature and in use in medicine than we can explore, but this chapter will introduce you to their diversity and numerous natural functions.

1. **What are the different types of amines?** The goal: Be able to recognize primary, secondary, tertiary, and heterocyclic amines.
2. **How are amines named?** The goal: Be able to name simple amines and write their structures, given the names.
3. **What are the general properties of amines?** The goal: Be able to describe amine properties such as hydrogen bonding, solubility, boiling point, and basicity.
4. **How do amines react with water and acids?** The goal: Be able to predict the structures of the ammonium ions and of the salts formed by reactions of amines with water and acids.
5. **What are some naturally occurring types of amines and some amine-containing drugs?** The goal: Be able to describe some types of amines found in biomolecules, plants, and drugs.

4.1 AMINES

Amine A compound that has one or more organic groups bonded to nitrogen, RNH_2, R_2NH, or R_3N.

Primary amine An amine that has one organic group bonded to nitrogen, RNH_2.

Secondary amine An amine that has two organic groups bonded to nitrogen, R_2NH.

Tertiary amine An amine that has three organic groups bonded to nitrogen, R_3N.

Amines are compounds that contain one or more organic groups bonded to nitrogen: RNH_2, R_2NH, or R_3N. Thus, they are organic derivatives of ammonia in the same way that alcohols and ethers are organic derivatives of water. Amines are classified as **primary, secondary,** or **tertiary,** depending on how many organic substituents are bonded to the nitrogen atom.

$$H—N—H$$
$$\vert$$
$$H$$

Ammonia

$$R—N—H \qquad R—N—H \qquad R—N—R''$$
$$\vert \qquad\qquad \vert \qquad\qquad \vert$$
$$H \qquad\qquad R' \qquad\qquad R'$$

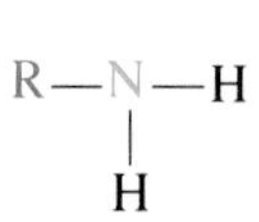

A *primary* amine (one R group on nitrogen)

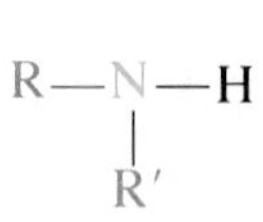

A *secondary* amine (two R groups on nitrogen)

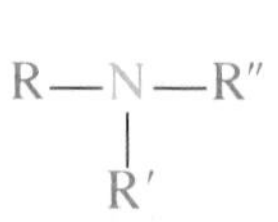

A *tertiary* amine (three R groups on nitrogen)

The groups bonded to the amine nitrogen atom may be alkyl or aryl groups. For example,

CH_3NH_2 [benzene ring]—NH_2 [naphthalene ring]—$NHCH_2CH_3$

Methylamine
(a primary alkyl amine)

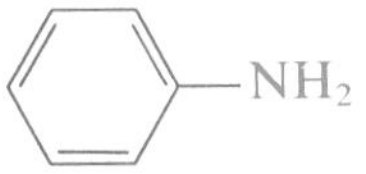

Aniline
(a primary aromatic amine)

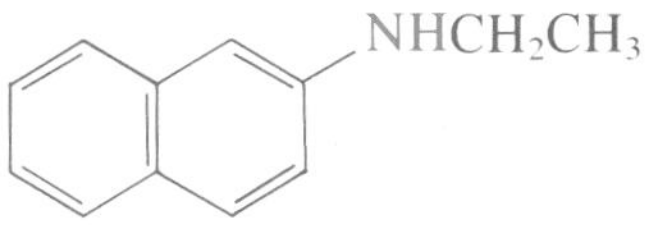

N-Ethylnaphthylamine
(a secondary aromatic amine)

4.2 NAMING AMINES

Primary alkyl amines, RNH_2, are named simply by identifying the alkyl group attached to nitrogen and adding the suffix *-amine* to the alkyl group name.

$$CH_3CH_2NH_2 \qquad CH_3\overset{\overset{\displaystyle CH_3}{|}}{C}HNH_2 \qquad \text{[cyclohexane ring]}—NH_2$$

Ethylamine   Isopropylamine   Cyclohexylamine

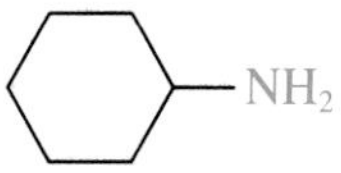

Secondary and tertiary amines with two or three identical groups are named by adding the appropriate prefix, *di-* or *tri-*, to the alkyl group name.

$$CH_3CH_2CH_2\underset{\underset{\displaystyle H}{|}}{N}CH_2CH_2CH_3 \qquad CH_3CH_2\underset{\underset{\displaystyle CH_2CH_3}{|}}{N}CH_2CH_3$$

Dipropylamine   Triethylamine

Secondary and tertiary amines with nonidentical R groups are named as *N*-substituted derivatives of a primary amine. The largest of the organic groups bonded to nitrogen is chosen as the parent compound, and the other groups are considered *N*-substituents (*N* because they're attached directly to nitrogen). The following compounds, for example, are named as propylamines because the propyl group in each is the largest alkyl group.

$$CH_3CH_2\underset{\underset{\displaystyle H}{|}}{N}CH_2CH_2CH_3 \qquad CH_3\underset{\underset{\displaystyle CH_3}{|}}{N}CH_2CH_2CH_3$$

N-Ethylpropylamine   *N,N*-Dimethylpropylamine

The simplest aromatic amine is known by the common name *aniline*. When the —NH_2 group must be named as a substituent, **amino-** is used as a prefix.

Amino group The —NH_2 functional group.

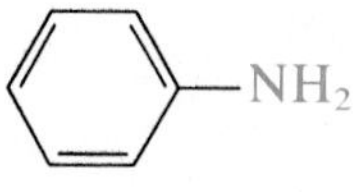

Aniline 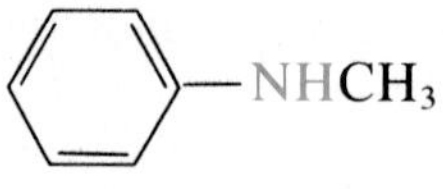   N-Methylaniline   3-Aminopropanoic acid

---

Practice Problems   **4.1**   Identify these compounds as primary, secondary, or tertiary amines.

(a) $CH_3CH_2CH_2NH_2$ (b) $CH_3CH_2NHCH_2CH_3$

(c)

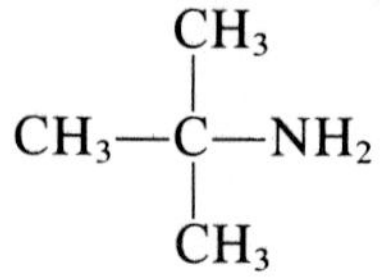

$$CH_3 - \underset{\underset{CH_3}{|}}{\overset{\overset{CH_3}{|}}{C}} - NH_2$$

(d) 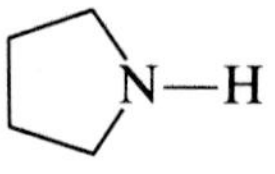 ⬡N—H   (e) 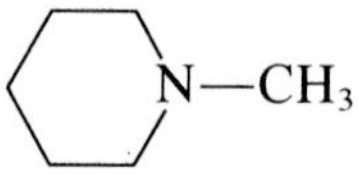 ⬡N—CH₃

4.2   What are the names of these amines?

(a) $CH_3CH_2CH_2NH_2$ (b)

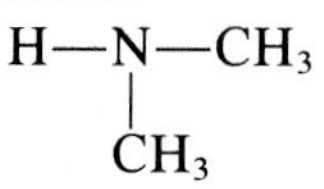

$$H - \underset{\underset{CH_3}{|}}{N} - CH_3$$

(c) 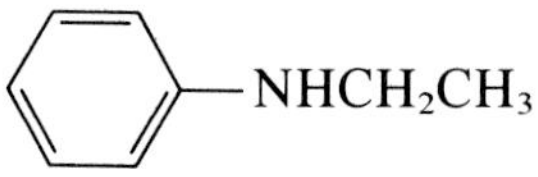 ⬡—NHCH₂CH₃

4.3   Draw structures corresponding to these names:

(a) butylamine   (b) N-methylethylamine

(c) N,N-dimethylaniline   (d) 2-aminobutanol

4.3 HETEROCYCLIC NITROGEN COMPOUNDS

Heterocycle A ring that contains nitrogen or some other atom in addition to carbon.

In many nitrogen-containing compounds, the nitrogen atom is in a ring with carbon atoms. Such **heterocyclic** nitrogen compounds may be nonaromatic or aromatic. Piperidine, for example, is a saturated heterocyclic amine with a six-membered ring, and pyridine is an aromatic heterocyclic amine which, like other aromatic compounds, is usually represented on paper by showing alternating double and single bonds in the ring.

REVIEW Nitrogen is in Group 5A. A nitrogen atom with three covalent bonds also has a lone pair—a pair of outer-shell electrons not used in a bond.

Piperidine   Pyridine

The names and structures of several heterocyclic nitrogen compounds are given in Table 4.1. You need not memorize all these names and structures but should realize that nitrogen heterocycles are very common in both plants and animals. Nicotine, found in tobacco leaves, contains two heterocyclic rings; quinine, an antimalarial drug isolated from the bark of the South American *Cinchona* tree, contains a quinoline ring system plus a nitrogen ring with a two-carbon bridge across it. The amino acid tryptophan contains an indole ring system.

Table 4.1   Some Heterocyclic Nitrogen Compounds

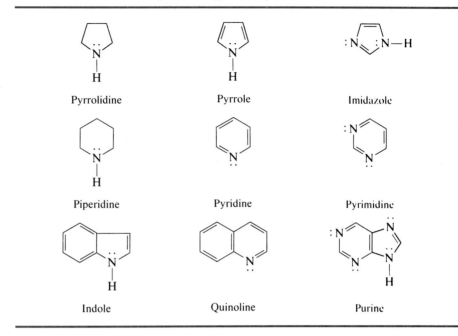

Pyrrolidine   Pyrrole   Imidazole

Piperidine   Pyridine   Pyrimidine

Indole   Quinoline   Purine

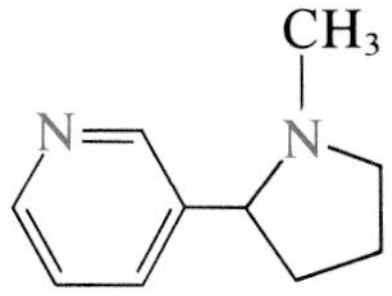

Nicotine
from tobacco
(an insecticide)

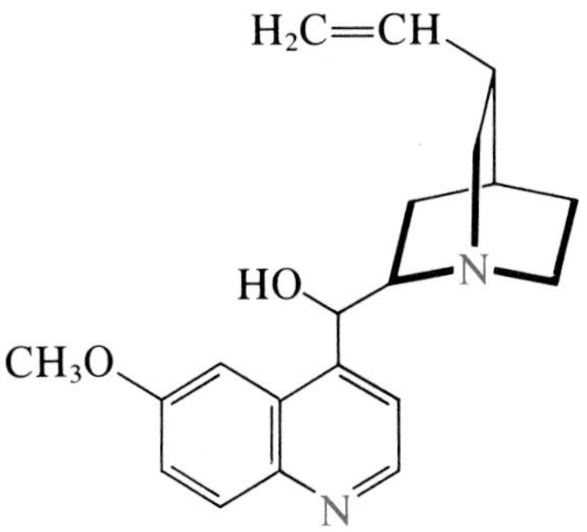

Quinine
from the *Cinchona* tree
(an antimalarial drug)

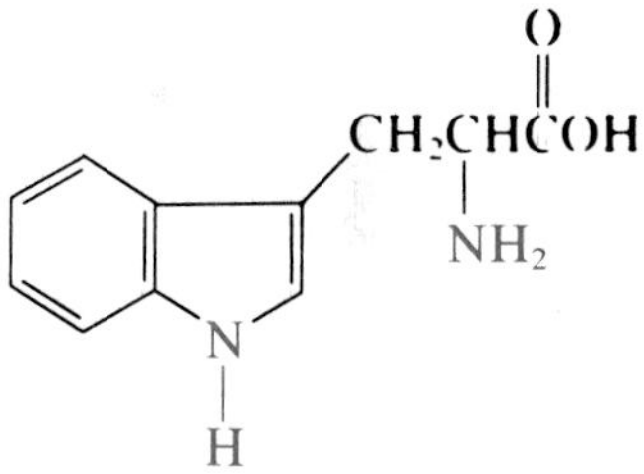

Tryptophan
(an amino acid)

4.4   PROPERTIES OF AMINES

REVIEW The most
electronegative elements:
F > O > Cl > N

Amines are analogous to ammonia in many of their physical properties, just as alcohols are analogous to water. Like ammonia, amines have an unshared electron pair on the nitrogen atom and have polar bonds because the nitrogen atom is electronegative. Thus, amines form hydrogen bonds in the same way that ammonia does (Figure 4.1) and are higher boiling than alkanes of similar size. Comparison of the following boiling points suggests, though, that intermolecular forces in amines are weaker than in alcohols:

$CH_3CH_2CH_2CH_3$ $CH_3CH_2CH_2NH_2$ $CH_3CH_2CH_2OH$

Butane, bp 0°C   Propylamine, bp 48°C   Propanol, bp 97°C

AN APPLICATION: CHEMICAL INFORMATION

Suppose you're reading an article in a magazine about caffeine in soft drinks and you decide you'd like to know something about the chemical called "caffeine." Where would you look for information?

The first place might be a chemical handbook, where you would find the structure and an entry like the one shown at the bottom of page 101 from *Lange's Handbook of Chemistry* which includes some physical properties and a reference to Beilstein's *Handbook of Organic Chemistry,* a publication that would give more chemical information (but it's written in German). A second standard chemical handbook is the *CRC Handbook of Chemistry and Physics* (CRC Press, Inc., Boca Raton, Fla.).

With your curiosity not yet satisfied, you could look next in another generally available and reliable source of chemical information, *The Merck Index: An Encyclopedia of Chemicals and*

Drugs, first published in 1889 as a list of products of the Merck pharmaceutical company. The index has since grown way beyond that to a standard reference work giving information on the preparation, properties, and uses of over 10,000 chemical compounds. It has a strong medical emphasis and, where appropriate, lists toxicity data, therapeutic uses in both human and veterinary medicine, and the physiological effects and cautions associated with hazardous chemicals.

The entry for caffeine in *The Merck Index,* reproduced below, begins with a list of alternate chemical names and a capitalized entry that's the name of a medication containing caffeine. One of the joys of using *The Merck Index* is that every one of the alternate names of every substance in the book, including the drug names, appears in the index, so no matter what name you come across, you can discover exactly what substance it refers to.

Caffeine information from *The Merck Index.* Reproduced from *The Merck Index,* 11th Ed. (1989), S. Budavari, M. J. O'Neil, A. Smith, P. E. Heckelman, Eds., by permission of the copyright owner, Merck & Co., Inc., Rahway, N.J., U.S.A., © Merck & Co., Inc., 1989.

1635. Caffeine. *3,7-Dihydro-1,3,7-trimethyl-1H-purine-2,6-dione;* 1,3,7-trimethylxanthine; 1,3,7-trimethyl-2,6-dioxopurine; coffeine; thein; guaranine; methyltheobromine; No-Doz. $C_8H_{10}N_4O_2$; mol wt 194.19. C 49.48%, H 5.19%, N 28.85%, O 16.48%. Occurs in tea, coffee, maté leaves; also in guarana paste and cola nuts: Shuman, U.S. pat. **2,508,-545** (1950 to General Foods). Obtained as a by-product from the manuf of caffeine-free coffee: Barch, U.S. pat. **2,817,588** (1957 to Standard Brands); Nutting, U.S. pat. **2,802,739** (1957 to Hill Bros. Coffee); Adler, Earle, U.S. pat. **2,933,395** (1960 to General Foods). Crystal structure: Sutor, *Acta Cryst.* **11,** 453 (1958). Synthesis: Fischer, Ach, *Ber.* **28,** 2473, 3135 (1895); . . . .

Hexagonal prisms by sublimation, mp 238°. Sublimes 178°. Fast sublimation is obtained at 160-165° under 1 mm press. at 5 mm distance. d_4^{18} 1.23. pH of 1% soln 6.9. Aq solns of caffeine salts dissociate quickly. Absorption spectrum: Hartley, *J. Chem. Soc.* **87,** 1802 (1905). One gram dissolves in 46 ml water, 5.5 ml water at 80°, 1.5 ml boiling water, 66 ml alcohol, 22 ml alcohol at 60°, 50 ml acetone, 5.5 ml chloroform, 530 ml ether, 100 ml benzene, 22 ml boiling benzene. Freely sol in pyrrole; in tetrahydrofuran contg about 4% water; also sol in ethyl acetate; slightly in petr ether. Soly in water is increased by alkali benzoates, cinnamates, citrates or salicylates. LD_{50} orally in mice, hamsters, rats, rabbits (mg/kg): 127, 230, 355, 246 (males); 137, 249, 247, 224 (females) (Palm).

Monohydrate, felted needles, contg 8.5% H_2O. Efflorescent in air; complete dehydration takes place at 80°. . . . .

THERAP CAT: CNS stimulant.

THERAP CAT (VET): Has been used as a cardiac and respiratory stimulant and as a diuretic.

Next comes information about the molecular formula and sources of the compound, followed by a list of patent and journal references to articles about caffeine, including an account of its first synthesis in 1895. After the structure is a paragraph of information about the physical properties of caffeine, followed by a listing of the properties of some important derivatives of caffeine. The final lines give the therapeutic uses of caffeine, which is a central-nervous-system (CNS) stimulant.

Having learned that caffeine is present in No-Doz, you can next turn to the *Physicians Desk Reference*, the *PDR*, to learn about the use of caffeine in drugs. While intended primarily for physicians, the *PDR* is readily available in bookstores and libraries. For each drug, it contains the full product labeling information provided by the manufacturers. You can go directly to the entry for No-Doz via the product name index and learn more about the effects and cautions accompanying the intended use of this medication. By using the Generic and Chemical Name Index in the *PDR* you can also discover that caffeine is present in 27 other medications ranging from nonprescription analgesics like Anacin to a potent prescription drug for migraine headaches that combines caffeine with belladonna and ergotamine (both alkaloids, Section 4.8), and sodium pentobarbital.

NO DOZ® Tablets
[nō'dōz]

COMPOSITION
Each tablet contains 100 mg. Caffeine. Other Ingredients: Cornstarch, Flavors, Mannitol, Microcrystalline Cellulose, Stearic Acid, Sucrose.

INDICATIONS
Helps restore mental alertness or wakefulness when experiencing fatigue or drowsiness.

DOSAGE AND ADMINISTRATION
Adults and children 12 years of age and over: One or two tablets not more often than every 3 to 4 hours.

CAUTION
Do not take without consulting physician if under medical care. No stimulant should be substituted for normal sleep in activities requiring physical alertness.

WARNING
For occasional use only. Not intended for use as a substitute for sleep. If fatigue or drowsiness persists or continues to recur, consult a doctor. The recommended dose of this product contains about as much caffeine as a cup of coffee. Limit the use of caffeine-containing medications, foods, or beverages while taking this product because too much caffeine may cause nervousness, irritability, sleeplessness and, occasionally, rapid heart beat. Do not give to children under 12 years of age. **KEEP THIS AND ALL MEDICINES OUT OF THE REACH OF CHILDREN. IN CASE OF ACCIDENTAL OVERDOSE, SEEK PROFESSIONAL ASSISTANCE OR CONTACT A POISON CONTROL CENTER IMMEDIATELY. As with any drug, if you are pregnant or nursing a baby, seek the advice of a health professional before using this product.**

OVERDOSE
Typical of caffeine.

HOW SUPPLIED
NO DOZ® is supplied as:
A circular white tablet with "NoDoz" debossed on one side.
. . . .

Entry on No-Doz® from *The Physicians Desk Reference*. Reproduced from *The Physicians Desk Reference* with permission of Medical Economics Company, Inc., Oradell, N.J.

Caffeine information from *Lange's Handbook of Chemistry*. Reproduced from *Dean/Lange's Handbook of Chemistry*, 13th Ed. (1987), Entry C1, p. 7-194, with permission of McGraw-Hill, Inc.

Name	Formula	Formula weight	Beilstein reference	Density	Refractive index	Melting point	Boiling point	Flash point	Solubility in 100 parts solvent
Caffeine		194.19	26, 461	1.23_4^{19}		238	subl 178		2.1 aq; 1.5 alc; 18 chl; 0.19 eth; 1 bz

As in so many other cases, the differences in properties between amines and alcohols are accounted for by hydrogen bonding. Because nitrogen atoms are less electronegative than oxygen atoms, they form weaker hydrogen bonds, and amines are therefore lower boiling than comparable alcohols. In fact, mono-, di-, and trimethylamine and ethylamine are gases at room temperature. Other common amines with molecules of moderate size are liquids (Table 4.2).

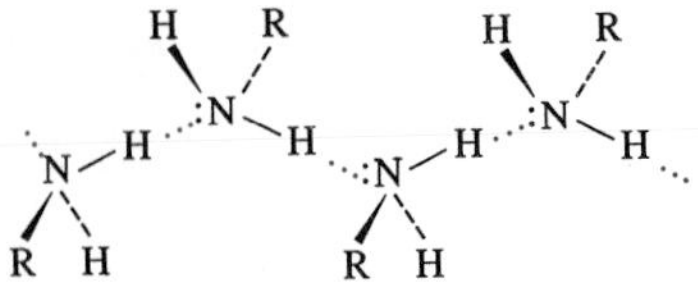

Tertiary amine molecules have no hydrogen atoms attached to nitrogen and, because they cannot hydrogen-bond with each other, are lower boiling than either amines or alcohols of similar molecular weight. Compare the boiling point of trimethylamine, which is 3°C, with those of the compounds shown above.

All amines, however, can hydrogen-bond to water molecules through the lone electron pair on their nitrogen atoms. As a result, amines with up to about six carbon atoms are quite soluble in water. Also like ammonia, amines are weak Brønsted-Lowry bases and raise the pH of aqueous solutions (Section 4.5).

One noticeable difference from alcohols is that many volatile amines have strong odors, some like ammonia and others like fish or decaying meat. The protein in flesh contains amine groups, and the smaller, volatile amines pro-

Table 4.2 Physical Properties of Some Simple Amines

Structure	Name	Melting Point (°C)	Boiling Point (°C)	Water Solubility (g/100 mL)	Base Ionization Constant K_b
NH_3	Ammonia	−77.7	−33.3	90	1.7×10^{-5}
Primary amines					
CH_3NH_2	Methylamine	−94	−6.3	Very soluble	1.8×10^{-5}
$CH_3CH_2NH_2$	Ethylamine	−81	16.6	Miscible	4.4×10^{-4}
$(CH_3)_3CNH_2$	tert-Butylamine	−67.5	44.4	Miscible	2.8×10^{-4}
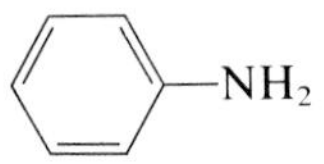 ⬡—NH_2	Aniline	−6.3	184.1	4	3.8×10^{-10}
Secondary amines					
$(CH_3)_2NH$	Dimethylamine	−93	7.4	Miscible	3.8×10^{-10}
$(CH_3CH_2)_2NH$	Diethylamine	−48	56.3	Miscible	9.6×10^{-4}
$[(CH_3)_2CH]_2NH$	Diisopropylamine	−61	84	11	—
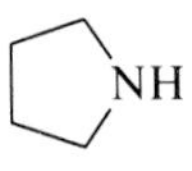 ⬠NH	Pyrrolidine	2	89	Miscible	—
Tertiary amines					
$(CH_3)_3N$	Trimethylamine	−117	3	41	5×10^{-5}
$(CH_3CH_2)_3N$	Triethylamine	−114	89.3	14	5.7×10^{-4}
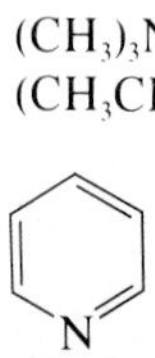 ⬡N	Pyridine	−42	115	Miscible	1.8×10^{-9}

duced during decay and protein breakdown are responsible for the odor of rotten meat. Two of the worst offenders have been given common names that are self-explanatory.

$$H_2NCH_2CH_2CH_2CH_2NH_2 \qquad H_2NCH_2CH_2CH_2CH_2CH_2NH_2$$

Putrescine   Cadaverine

Another significant property of amines is that many are physiologically active. The simpler amines are irritating to the skin, eyes, and mucous membranes and are toxic by ingestion. Some of the more complex amines from plants (Section 4.8) are among the most poisonous substances known, and some heterocyclic amines are believed to be carcinogens. On the other hand, all living things contain a wide variety of amines, and many useful drugs are amines.

Properties of Amines

- Primary and secondary amines are hydrogen-bonded and higher boiling than alkanes, but lower boiling than alcohols.
- Tertiary amines are lower boiling than secondary or primary amines because hydrogen bonding is not possible.
- The simplest amines are gases; other common amines are liquids.
- Volatile amines have unpleasant odors.
- Simple amines are water-soluble because of hydrogen bonding.
- Amines are weak Brønsted-Lowry bases (Section 4.5).
- Many amines are physiologically active, and many are toxic.

4.5 BASICITY OF AMINES

REVIEW In a reaction at equilibrium, forward and reverse reactions occur at equal rates and concentrations of reactants and products are constant.

Ammonia is a weak Brønsted-Lowry base. That is, ammonia can use its lone pair of electrons to accept a hydrogen ion (a proton) from an acid and form the ammonium ion, NH_4^+.

$$:NH_3 + H-O-H \rightleftharpoons NH_4^+ + OH^-$$

Ammonium ion   Hydroxide ion

$$:NH_3 + HCl(aq) \longrightarrow NH_4^+ Cl^-(aq)$$

Ammonium chloride

Amines behave similarly. The aqueous solutions of many amines are basic because of the following equilibrium:

$$RNH_2 + H_2O \rightleftharpoons RNH_3^+ + OH^-$$

Amine   Ammonium ion

Ammonium salt An ionic compound composed of an ammonium cation and an anion; an amine salt.

Also, like ammonia, amines react with strong acids such as hydrochloric acid to yield **ammonium salts** according to the general equation

$$RNH_2 + HX \longrightarrow RNH_3^+ \ X^-$$

Amine   Ammonium salt

For example,

$$CH_3-\overset{..}{\underset{\underset{H}{|}}{N}}-H \ + \ HCl(aq) \ \rightleftharpoons \ CH_3-\overset{\overset{H}{|}}{\underset{\underset{H}{|}}{\overset{+}{N}}}-HCl^-(aq)$$

Methylamine   Hydrochloric acid   Methylammonium chloride

$$CH_3CH_2-\overset{..}{\underset{\underset{CH_2CH_3}{|}}{N}}-H \ + \ HCl(aq) \ \rightleftharpoons \ CH_3CH_2-\overset{\overset{H}{|}}{\underset{\underset{CH_2CH_3}{|}}{\overset{+}{N}}}-HCl^-(aq)$$

Diethylamine   Hydrochloric acid   Diethylammonium chloride

The reaction between amines and acids is reversible (like many acid–base reactions) and can be made to take place in either direction depending on the reaction conditions. An amine is protonated to yield an ammonium ion when treated with acid, and an ammonium ion is deprotonated to yield a neutral amine when treated with base. Thus, the amount of pronation of the amine depends on the pH of the medium.

Acidic conditions:
(pH < 7)

$$RNH_2 \ + \ H_3O^+ \ \longrightarrow \ RNH_3^+ \ + \ H_2O$$

An amine   An acid   An ammonium ion

Basic conditions:
(pH > 7)

$$R-NH_3^+ \ + \ OH^- \ \longrightarrow \ R-NH_2(aq) \ + \ H_2O$$

An ammonium ion   A base   An amine

REVIEW Acid-base reactions are reviewed in Appendix B.4, and K_a is reviewed in Appendix B.7.

Alkylamines are weak bases, and aromatic and heterocyclic amines are weaker bases still. Base strengths are usually expressed in terms of base ionization constants (K_b) included in Table 4.2. The value of K_b is the equilibrium constant for the ionization of a base in water in the same way that K_a is the equilibrium constant for the ionization of an acid.

$$RNH_2 + H_2O \ \rightleftharpoons \ RNH_3^+ + OH^- \qquad\qquad K_b = \frac{[RNH_3^+][OH^-]}{[RNH_2]}$$

The smaller the value of K_b, the weaker the base (Table 4.2).

The positive ions formed by protonation of alkylamines are named by replacing the ending -*amine* by the ending -*ammonium*. To name the ions of heterocyclic amines, the amine name is modified by replacing the -*e* with -*ium*. For example,

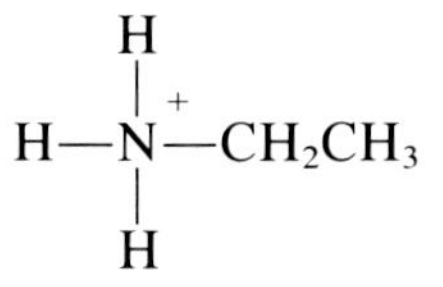

$$H-\overset{\overset{\displaystyle H}{|}}{\underset{\underset{\displaystyle H}{|}}{N}}{}^{+}-CH_2CH_3 \qquad CH_3CH_2CH_2\overset{\overset{\displaystyle H}{|}}{\underset{\underset{\displaystyle H}{|}}{N}}{}^{+}-CH_2CH_2CH_3$$

Ethylammonium ion
(from ethylamine)

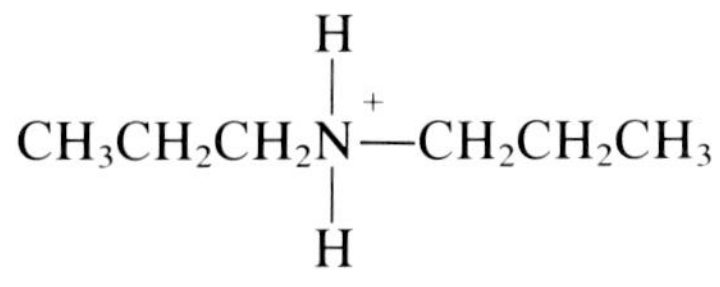

Dipropylammonium ion
(from dipropylamine)

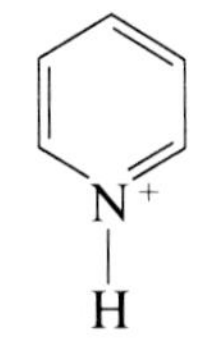

Pyridinium ion
(from pyridine)

Any nitrogen atom with four bonds has a positive charge and is an ammonium ion, even though the charge may not be written in the formula.

Practice Problems

4.4 Write an equation for the acid–base equilibrium of dimethylamine and water.

4.5 Complete the following equations:
(a) 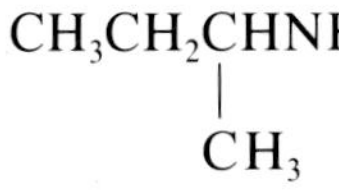 $CH_3CH_2\underset{\underset{\displaystyle CH_3}{|}}{C}HNH_2 + HBr(aq) \longrightarrow$

(b) 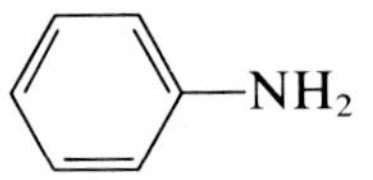 ⬡—NH_2 + HCl(aq) $\longrightarrow$

(c) $CH_3CH_2NH_2 + CH_3COOH(aq) \longrightarrow$
(d) $CH_3NH_3{}^{+}Cl^{-} + NaOH(aq) \longrightarrow$

4.6 Name the organic ions produced in reactions (a) through (c) in Problem 4.5.

4.7 Which is the stronger base?
(a) ammonia or ethylamine   (b) triethylamine or pyridine

4.6 AMMONIUM SALTS

Like any salt, an amine salt is composed of a cation and an anion and is usually named by combining the cation and anion names. In methylammonium chloride ($CH_3NH_3{}^{+}Cl^{-}$), for example, the methylammonium ion, $CH_3NH_3{}^{+}$, is the cation, and the chloride ion is the anion. In an older system, ammonium salts were written and named by combining the structures and names of the amine and the acid. Methylammonium chloride, in this system, is written $CH_3NH_2 \cdot HCl$ and named methylamine hydrochloride. You'll often see this system used with drugs that are ammonium salts. For example,

$$(C_6H_5)_2CHOCH_2CH_2N(CH_3)_2 \cdot HCl \qquad \text{or} \qquad (C_6H_5)_2CHOCH_2CH_2\overset{+}{N}H(CH_3)_2 \ Cl^{-}$$

Diphenhydramine hydrochloride (Benadryl), an antihistamine

Ammonium salts are generally odorless, white, crystalline solids that are much more water-soluble than neutral amines because they're ionic. Thus, ammonium salt formation, like carboxylate salt formation, provides a means for converting an insoluble compound into a water-soluble derivative.

$$CH_3CH_2{-}\underset{\underset{CH_2CH_3}{|}}{N}{-}CH_2CH_3 \quad + \quad HCl(aq) \quad \longrightarrow \quad CH_3CH_2{-}\underset{\underset{CH_2CH_3}{|}}{\overset{+}{N}H}{-}CH_2CH_3\ Cl^-(aq)$$

Triethylamine   Hydrochloric   Triethylammonium
(water-insoluble)   acid   chloride
(water-soluble)

Also, because many amines are unstable in air, salt formation allows the amine to be stored in a more stable form. When the free amine is needed, it is easily regenerated by treatment with a base:

$$CH_3NH_3{}^+\ Cl^-(aq) + NaOH(aq) \;\rightarrow\; CH_3NH_2(aq) + NaCl(aq) + H_2O(l)$$

Ammonium salts can be formed by reaction of an amine with an acid, as shown above, and also by reaction of an alkyl halide with an amine:

$$RNH_2 \quad + \quad R'X \quad \longrightarrow \quad RR'NH_2{}^+\ X^-$$

Amine   Alkyl halide   Alkyl ammonium halide

For example,

$$CH_3CH_2NH_2 \quad + \quad CH_3Cl \quad \longrightarrow \quad CH_3CH_2\underset{\underset{CH_3}{|}}{N}H_2{}^+\ Cl^-$$

Ethylamine   Methyl chloride   Ethylmethylammonium chloride

The reaction of a tertiary amine with an alkyl halide produces a **quaternary ammonium salt**—an ammonium salt with *four* R groups on the nitrogen atom:

Quaternary ammonium salt An ammonium salt with four organic groups bonded to the nitrogen atom.

$$R{-}\underset{\underset{R}{|}}{\overset{\overset{R}{|}}{N}} \quad + \quad R'X \quad \longrightarrow \quad R{-}\underset{\underset{R}{|}}{\overset{\overset{R}{|}}{\overset{+}{N}}}{-}R'\ X^-$$

A tertiary amine   An alkyl halide   A quaternary ammonium salt

For example,

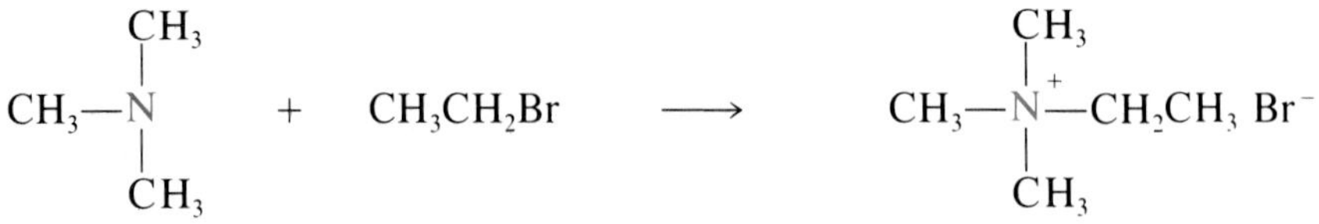

$$CH_3{-}\underset{\underset{CH_3}{|}}{\overset{\overset{CH_3}{|}}{N}} \quad + \quad CH_3CH_2Br \quad \longrightarrow \quad CH_3{-}\underset{\underset{CH_3}{|}}{\overset{\overset{CH_3}{|}}{\overset{+}{N}}}{-}CH_2CH_3\ Br^-$$

Trimethylamine   Ethyl bromide   Ethyltrimethylammonium bromide

Quaternary ammonium salts have no H atom that can be removed by a base and no lone pair that can be protonated, so their structure in solution is unaffected by changes in pH. One commonly encountered quaternary ammonium salt has the following structure, where R represents a range of C_8 to C_{18} alkyl groups.

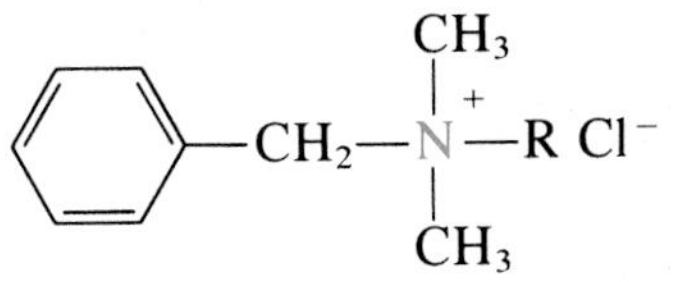

Benzalkonium chloride (an antiseptic and disinfectant)

$$R = -C_8H_{17} \text{ to } -C_{18}H_{37}$$

Practice Problems

4.8 Write the structures of the following compounds:
(a) hexyldimethylammonium chloride (b) ethylammonium bromide

4.9 Identify each compound in Problem 4.8 as the salt of a primary, secondary, or tertiary amine.

4.10 Complete the following equation and name the product:

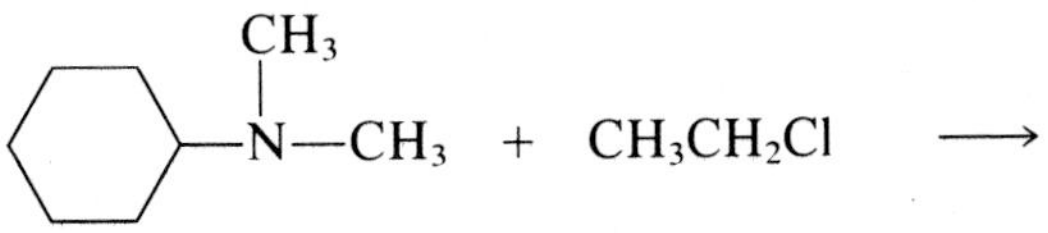

4.7 AMINES IN BIOMOLECULES

Proteins, we've noted, are polymers of amino acids, which all contain an amino group and a carboxylic acid group.

General formula of
α-amino acids found in proteins

Alanine
(an α-amino acid)

A second major class of biomolecules, the *nucleotides* (Chapter 15), all contain nitrogen heterocyclic rings derived from either purine or pyrimidine (see Table 4.1). Nitrogen heterocycles are also part of several vitamins, including the B vitamins:

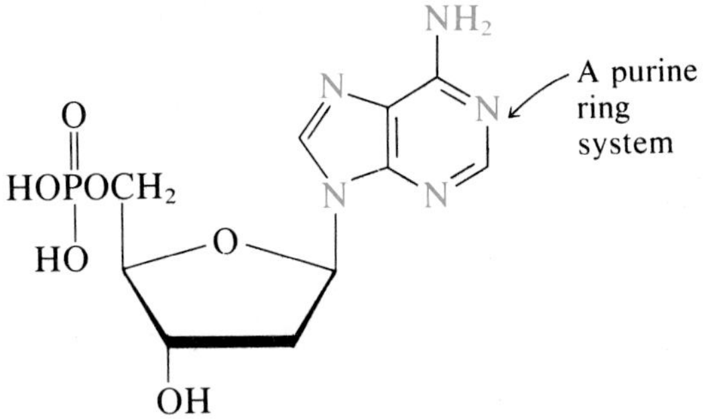

A DNA nucleotide containing
a purine heterocyclic system

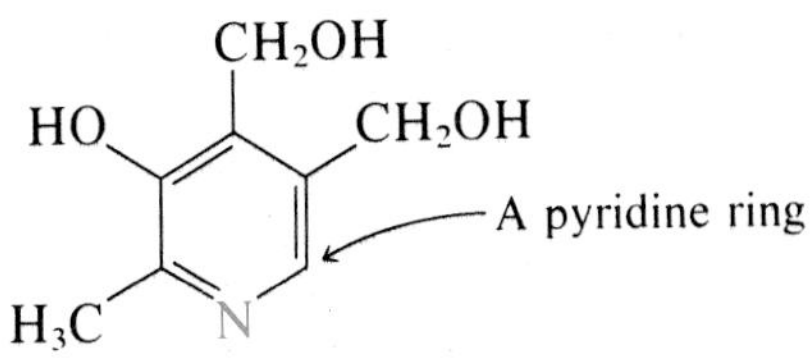

Pyridoxine
(one of the water-soluble B_6 vitamins)

AN APPLICATION: ORGANIC COMPOUNDS IN BODY FLUIDS

The chemical reactions that keep us alive and functioning occur in the aqueous solutions known as ''body fluids''—blood, digestive juices, urine, and the fluid inside cells. But you've seen that for organic compounds of all classes, water solubility decreases as the hydrocarbon-like portions of the molecules become larger and molecular weight increases. How does the body manage its reactions in water solution, especially when large and complex biomolecules are involved?

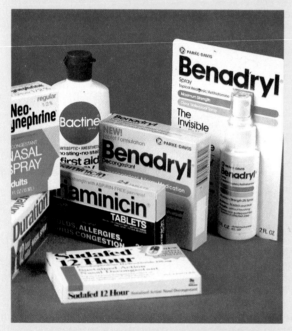

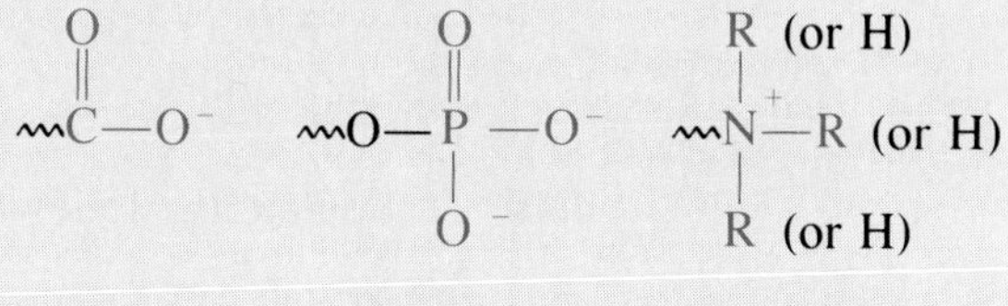

An ammonium chloride salt is the active ingredient in each of these over-the-counter medications.

Acidic and basic functional groups are part of many biomolecules, which then exist as soluble ions at the pH values of the various fluids. The most common ionized groups in biomolecules are carboxylate ions, phosphate ions, and ammonium ions:

Carboxylate ion | Phosphate ion | Ammonium ion

For example, NAD is an oxidizing agent that participates in a great many biochemical reactions:

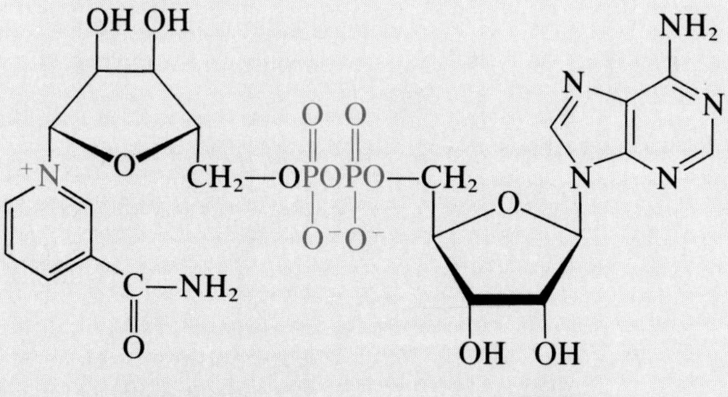

NAD (nicotinamide adenine dinucleotide)
(a coenzyme and biochemical oxidizing agent)

In addition, several members of the diverse class of compounds known as *neurotransmitters* (Section 8.15), which transmit nerve impulses throughout the body, are amines, including serotonin and dopamine. Neurotransmitters have a variety of physiological effects including the ability to dilate small airways in the lung (bronchodilation), to contract capillaries, and to increase blood pressure.

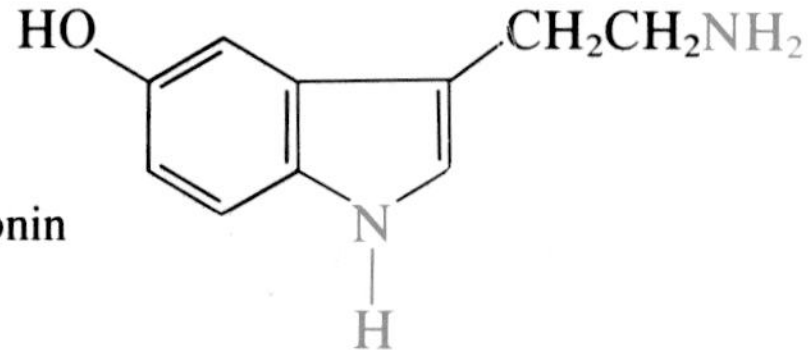

Serotonin

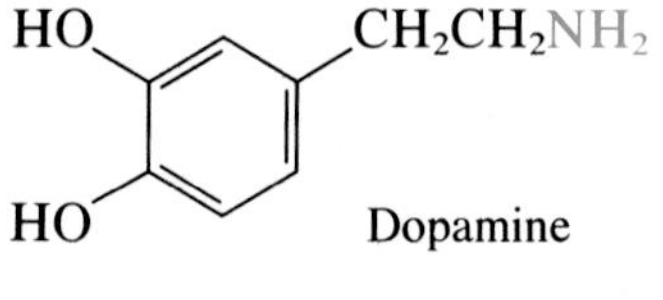

Dopamine

Because carboxylic acids (RCOOH) are ionized in body fluids, biochemists often refer to them by their carboxylate ion names rather than their acid names. You are just as likely, for example, to see pantothenic acid referred to as *pantothenate,* or pyruvic acid as *pyruvate,* or citric acid as *citrate.* Either way, the reference is to the same substance in solution. Also, you may see the same biomolecule structures written with —COOH or —COO$^-$, —PO$_3$H$_2$ or —PO$_3^{2-}$, and —NH$_2$ or —NH$_3^+$.

Drugs too must be soluble in body fluids and must be transported in fluids from their entry point in the body to their site of action. Most drugs are weak acids or bases and therefore are in equilibrium with their ions in body fluids. For example,

Aspirin
(an acid)

Amphetamine
(a base)

The extent of ionization of a drug helps determine how it's distributed in the body, because ions are less likely to cross cell membranes than uncharged molecules. Weak acids like aspirin are less dissociated in the acidic environment in the stomach and are absorbed readily there. Weak bases are not significantly absorbed in the stomach but are better absorbed in the more basic environment in the small intestine. It's even possible to control where a drug tablet dissolves by applying an *enteric coating,* which remains intact in the strongly acidic gastric juice but dissolves when it reaches the small intestine. This approach is taken for drugs that would be attacked and broken down by acid.

Many drugs must be delivered to the body in their more water-soluble forms as salts. Penicillin G, for example, is usually administered as its sodium or potassium salt to increase its water solubility.

Penicillin G potassium

Also, many amine-containing drugs are not very soluble in water. By converting such amines as phenylephrine, the decongestant in Neo-Synephrine, and phenylpropanolamine, the decongestant in Triaminic cough medicine, to ammonium hydrochlorides, however, their solubility increases to the point where delivery in solution is possible.

4.8 AMINES IN PLANTS

Alkaloid A naturally occurring nitrogen-containing compound isolated from a plant; usually basic, bitter, and poisonous.

The roots, leaves, and fruits of flowering plants (angiosperms) are a rich source of amines, including many that are physiologically active. These compounds, once called "vegetable alkali" because their water solutions are basic, are now referred to as **alkaloids.** The molecular structures of approximately 5500 alkaloids have been determined. Most are bitter-tasting and toxic to human beings and other animals in sufficiently high doses. Quinine, in fact, is used as a standard for bitterness: Even a 1×10^{-6} M solution tastes bitter.

The bitterness and poisonous nature of alkaloids probably evolved to protect plants from being devoured by animals. It's interesting to speculate on why, of the four basic taste sensations—sweet, bitter, salty, and sour—bitterness is the one we like least in foods. The three poisonous compounds described here—coniine, atropine, and solanine—illustrate some of the many types of alkaloid structures:

Coniine

Atropine

Solanine
(X = a group of three sugar molecules)

Figure 4.2
A potato plant. The potato grows in the absence of sunlight and should also be stored away from sunlight, which causes formation of poisonous solanine (along with green chlorophyll) under the skin.

● **Coniine** is extracted from poison hemlock (*Conium maculatum*). It was the poison with which Socrates ended his life after being convicted of corrupting Greek youth with philosophical discussions.

● **Atropine** is the toxic substance in the herb known as deadly nightshade or *belladonna* (*Atropa belladonna*). In Meyerbeer's opera, *L'Africaine*, the heroine sings of the peaceful death this plant can bring before committing suicide over her lost love. Like many other alkaloids, atropine acts on the central nervous system, a property sometimes applied (in appropriately low dosage!) in medications to reduce cramping of the digestive tract.

● **Solanine,** an even more potent poison than atropine (Figure 4.2), is found in potatoes and tomatoes, both of which belong to the same botanical family as the deadly nightshade (Solanaceae). If you've ever been warned that you must peel green potatoes and wondered why, herein lies the reason and it's a good one. The tiny amount of solanine in properly stored potatoes only contributes to their characteristic flavor. But when potatoes are exposed to sunlight, the production of solanine is increased to levels that can be dangerous. The alkaloids are formed under the skin and aren't destroyed by heating, but they can be removed by peeling. Sunlight fortunately also stimulates the formation of chlorophyll, and its green color provides a warning that potatoes must be peeled so that all the green flesh is removed. Potato sprouts also contain solanine and should be cut out before potatoes are cooked.

Some alkaloids are notable not as poisons but for their pain-relieving ability. The opium poppy, *Papaver somniferum* (Figure 4.3), has been used for this purpose at least since the seventeenth century. Morphine was the first pure compound to be isolated from the poppy, but several close relatives including codeine are also present in poppies. Heroin, another close relative of morphine, does not occur naturally but is easily synthesized in chemical laboratories. Within the body, hydrolysis of the $CH_3C{=}O$ groups converts heroin back to morphine.

Figure 4.3
The opium poppy, source of morphine and other addictive alkaloids.

Morphine   Codeine   Heroin

4.9   AMINES IN DRUGS

Given the variety of nitrogen-containing compounds that are physiologically active, it's not surprising that many drugs are amines or nitrogen heterocycles. The relationship between molecular structure and physiological activity is the key to understanding drug action and to discovering new drugs. We'll give just a few more examples of the many classes of drugs that are amines.

Histamine is the biomolecule responsible for the symptoms of the allergic reaction familiar to hay fever sufferers. It's also the chemical that causes an itchy bump when an insect bites you (Figure 4.4). The *antihistamines* are a

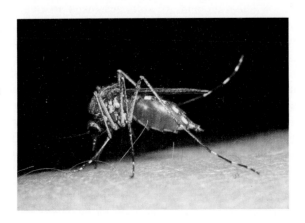

Figure 4.4
A mosquito in action. The mosquito injects antigens, substances that trigger a response by the body's immune system. The immune system in turn triggers release of histamine, which causes swelling, itchiness, and redness around the bite.

family of drugs that counteract histamine, and members of this family have in common a disubstituted ethylamine side chain, usually with two *N*-methyl groups.

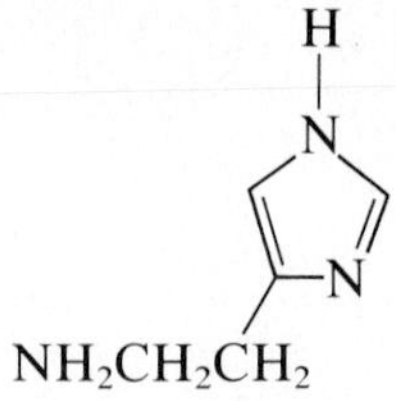

Histamine 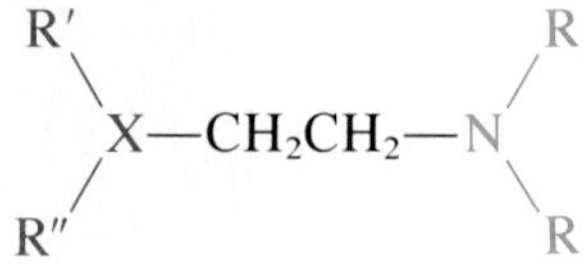   General antihistamine structure

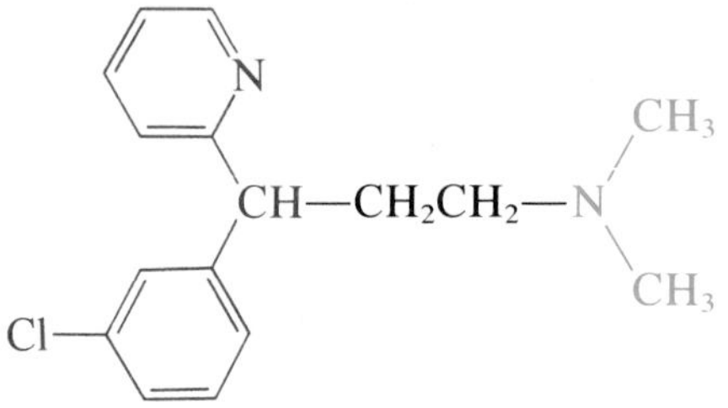

Chlorpheniramine
(an antihistamine)

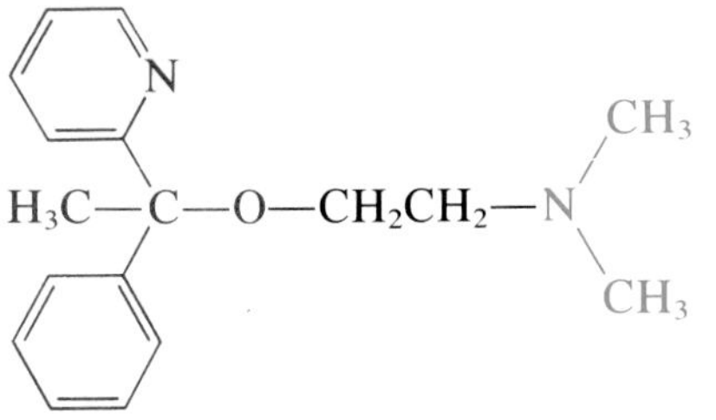

Doxylamine
(an antihistamine)

The R′ and R″ groups in the generalized structure tend to be bulky, and they may be bonded to carbon, nitrogen, or oxygen. Antihistamines act not by causing a physiological effect but by preventing one. They physically block the attachment of histamine to sites with which it must connect to create its unpleasant effects.

As another example of structure–activity relationships, compare the naturally occurring neurotransmitters on the left in Table 4.3 with the synthetic drugs on the right. Each is a variation on the same general structure, and each acts on the nervous system. The drugs are known as *sympathomimetics* because they mimic the action of the sympathetic nervous system. In the body, dopamine is a precursor of norepinephrine, a principal messenger in the nervous system. Epinephrine (adrenaline) is the compound specifically responsible for causing the surge of energy we feel in a frightening situation.

Ephedrine and its salts are used in over-the-counter medications for asthma and nasal congestion because they dilate bronchial and nasal passageways. Isoproterenol is a prescription drug for bronchial asthma. Amphetamine is the parent compound of a family of central-nervous-system stimulants that produce an array of effects like those of epinephrine, including excitement, increased blood pressure, quickened reflexes, and appetite suppression. The amphetamines are all hazardous because they are habit-forming.

Although enormously important as medicines, morphine and its alkaloid relatives pose a great problem because of their addictive properties. Much effort has therefore been devoted to understanding how morphine works and to

Table 4.3 Neurotransmitters and Drugs That Mimic Their Action (Sympathomimetics)

A general ephedrine-like structure
(groups in parentheses may or may not be present)

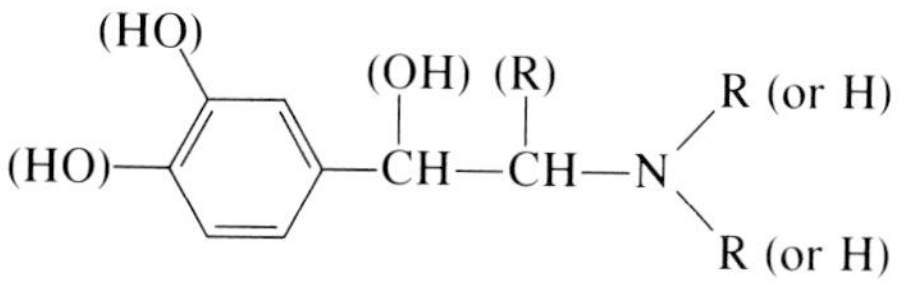

Natural Neurotransmitters

Dopamine

Norepinephrine

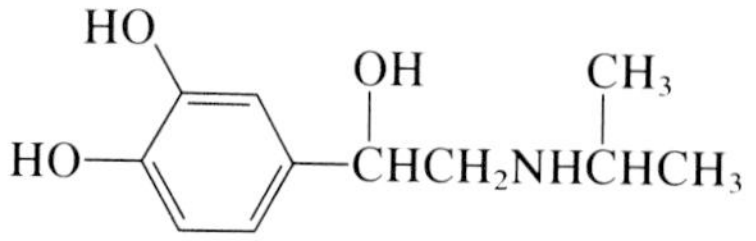

Epinephrine
(adrenaline)

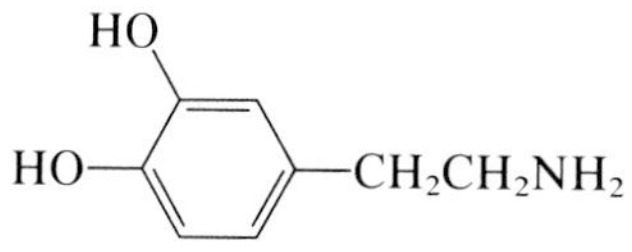

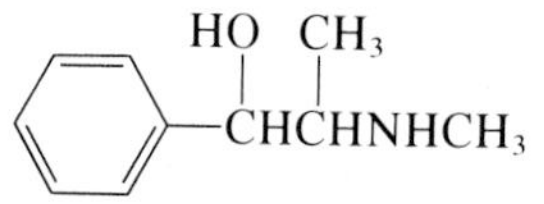

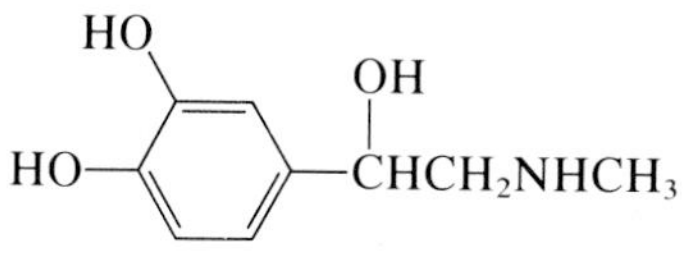

Sympathomimetic Drugs

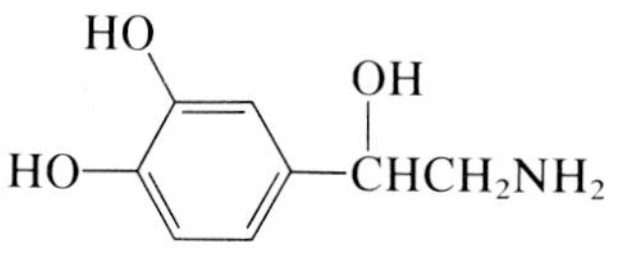

Ephedrine

Isoproterenol

Amphetamine

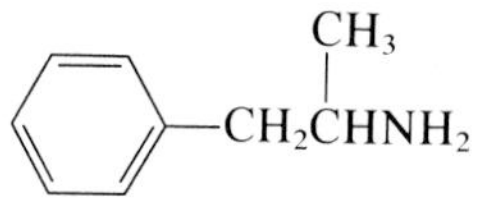

developing modified morphine derivatives that retain the desired painkilling activity but don't cause addiction. Many compounds with similar painkilling properties, including meperidine (Demerol) and methadone, have been synthesized, but no fully satisfactory morphine substitutes have yet been found. Demerol is widely used as a painkiller, and methadone is used in the treatment of heroin addiction.

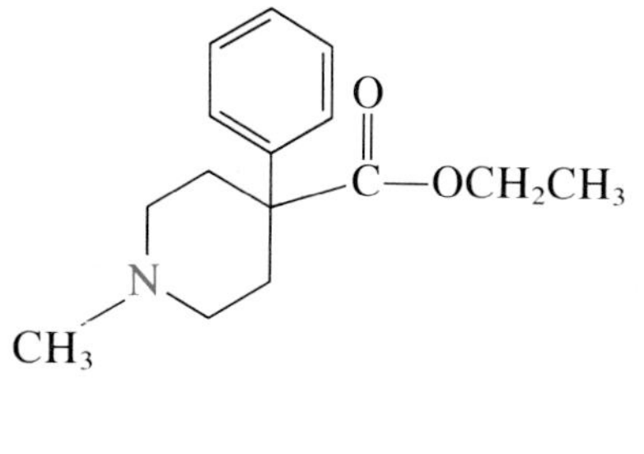

Demerol

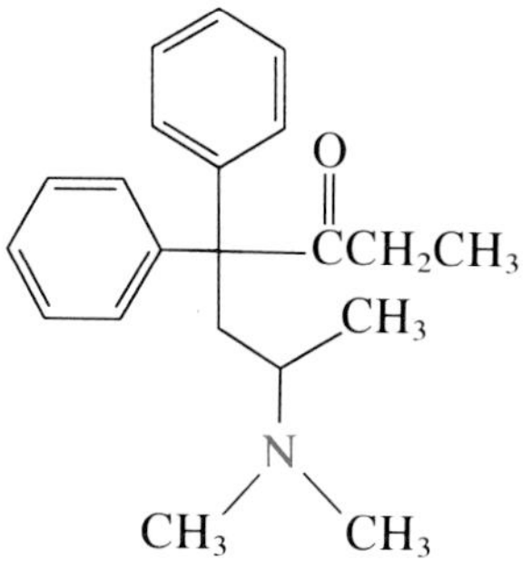

Methadone

Practice Problem 4.11 Compare the structures of Demerol and methadone with the structures in Section 4.8 of the morphine alkaloids and identify the structural unit that all have in common. (*Hint:* It includes the nitrogen atom.)

INTERLUDE: PRODRUGS

The modern approach to drug development has come to be called *drug design*, meaning the design of a drug molecule with a structure destined to meet a specific need. Problems arise in drug design when a molecule with the desired therapeutic effect can't be administered as a drug. This can happen for a variety of reasons: The compound might be unstable or insoluble in body fluids or taste terrible, it might have harmful side effects, or it might fail to reach the site where it must act.

A successful solution to such problems is the design of a *prodrug:* an inactive compound that's converted to the active drug *after* it's administered. Sulfasalazine, a sulfa antibacterial for treating ulcerative colitis, is an example of a site-specific prodrug. The active parent drug, sulfapyridine, does not reach the colon in effectively high concentrations because it is absorbed from the intestinal tract above the colon. Sulfasalazine remains unchanged until it is broken down by enzymes in the colon to generate the active drug (as shown below).

A prodrug based on epinephrine allows for safer and more effective use of this compound in relieving fluid pressure in the eye due to glaucoma. Relatively large concentrations of epinephrine must be used in eyedrops because the polar molecule does not easily cross the eye membrane. Reaction of epinephrine with a carboxylic acid yields dipivefrin, a prodrug that is less polar; it enters the eye more easily, is less

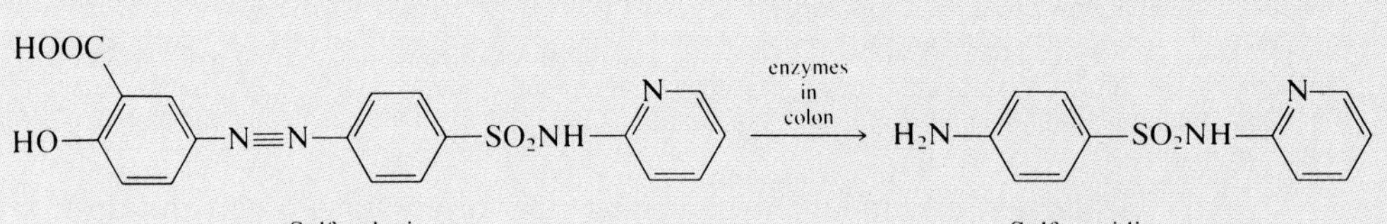

Sulfasalazine
(a prodrug)

Sulfapyridine
(an antibacterial drug)

SUMMARY

Amines are organic derivatives of ammonia in the same sense that alcohols and ethers are organic derivatives of water. Amines are classified according to the number of organic groups bonded to nitrogen as **primary amines,** RNH_2; **secondary amines,** R_2NH; and **tertiary amines,** R_3N. Nitrogen is also found in many types of **heterocyclic compounds,** both nonaromatic and aromatic.

Amines are polar compounds and can form hydrogen bonds with water through the lone pairs on nitrogen, making the simpler amines water-soluble. Primary and secondary amines are higher boiling than alkanes, but lower boiling than alcohols because they form weaker hydrogen bonds. Tertiary amines don't hydrogen-bond with each other and are lower boiling than other amines or alcohols of similar molecular weight. Volatile amines have bad odors, and many amines are physiologically active.

The most significant property of amines is their basicity. Like ammonia, amines can act as bases to accept protons from acids and are weak bases. Protonation of an amine by an acid yields an **ammonium salt** (RNH_2 + $HCl \rightarrow RNH_3^+ Cl^-$). The protonation reaction is reversible, and ammonium salts can be reconverted to amines by treatment with OH^-. Formation of an ammonium salt is often used to obtain a more stable, more water-soluble derivative of an amine. Reaction of tertiary amines with alkyl halides gives **quaternary ammonium salts** ($R_4N^+ X^-$), which are unaffected by changes in the pH of a solution.

Amino groups and nitrogen heterocycles are present in many biomolecules, including amino acids, nucleotides, and neurotransmitters. The **alkaloids,** which include many poisons and many drugs, are a large family of amines and nitrogen heterocycles found in plants. Many drugs are amines or nitrogen heterocycles.

irritating, and also reduces the danger of epinephrine entering the cardiovascular system where it can have undesirable side effects. Once dipivefrin has entered the eye tissue, enzymes release the epinephrine.

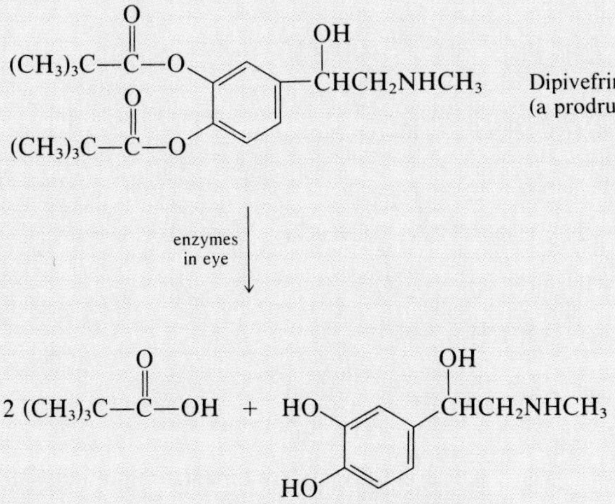

Dipivefrin
(a prodrug)

enzymes
in eye

2 (CH₃)₃C—C—OH +

Epinephrine
(an antiglaucoma
drug)

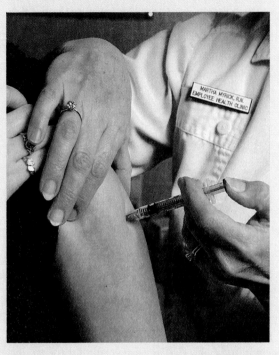

Some drugs can reach their sites of action through the bloodstream.

REVIEW PROBLEMS

Amines and Ammonium Salts

4.12* What family-name ending is used for amines?

4.13 What is the structural difference between primary, secondary, and tertiary amines?

4.14 Draw structures corresponding to these names:
(a) propylamine
(b) diethylamine
(c) *N*-methylpropylamine
(d) *N*-butyl-*N*-methylhexylamine
(e) *N*-ethylcyclopentylamine
(f) 2-methyl-3-propoxyaniline

4.15 Name these amines:

(a) CH₃CH₂CH₂CH₂NH₂

(b) CH₃CHCH₃ with NH₂

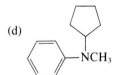

(c)

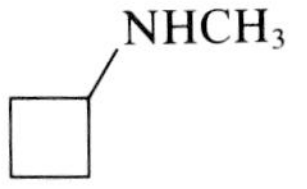

(d)

4.16* Identify each of the amines in Problems 4.14 and 4.15 as primary, secondary, or tertiary.

4.17 Propose structures for amines that fit these descriptions:
(a) a secondary amine with formula $C_5H_{13}N$
(b) a tertiary amine with formula $C_6H_{13}N$

4.18* Is pyridine a weaker or a stronger base than ammonia? (See Table 4.2).

4.19 Which solution has a higher pH, 0.20 M aniline or 0.20 M methylamine? (See Table 4.2).

4.20* Assume you have an aqueous solution of ethylamine.

(a) Write the equilibrium reaction for this weak base in aqueous solution. Circle the structure that is the predominant species.

(b) HNO_3 is added until the pH is 5.00. What is the predominant species now?

4.21 Name these ammonium salts. (See Table 4.2.)

(a) $CH_3NH_3^+ \ Cl^-$

(b) 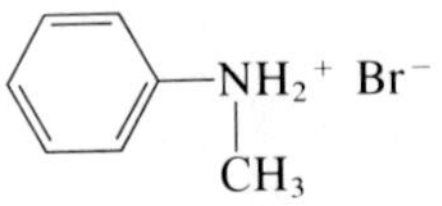 [structure: benzene ring]—$NH_2^+ \ Br^-$
     |
     CH_3

(c) 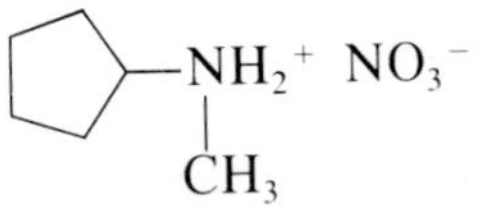 [structure: cyclopentane ring]—$NH_2^+ \ NO_3^-$
     |
     CH_3

4.22 Draw the structures of these ammonium salts. (See Table 4.2.)

(a) anilinium chloride
(b) N-propylbutylammonium bromide
(c) N-ethyl-N-isopropylhexylammonium chloride

4.23 Identify each salt in Problems 4.21 and 4.22 as the salt of a primary, secondary, or tertiary amine.

4.24* Which is the stronger base, diethyl ether or diethylamine?

4.25 The compound L-dopa (2-amino-3-(3,4-dihydroxyphenyl)propanoic acid) is used medically for its potent activity against Parkinson's disease, a chronic disease of the central nervous system. Identify the functional groups present in L-dopa.

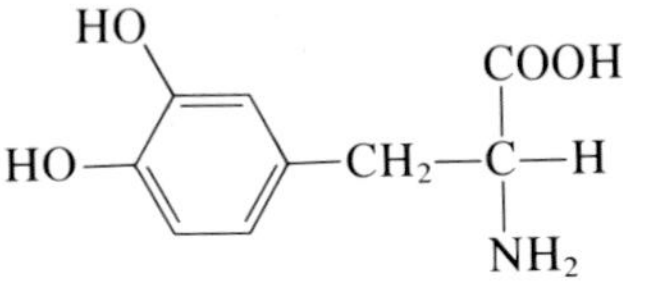

[structure of L-Dopa: HO and HO substituted benzene ring—CH_2—C(COOH)—H with NH_2] L-Dopa

4.26* Cocaine has the structure indicated. Is cocaine a primary, secondary, or tertiary amine?

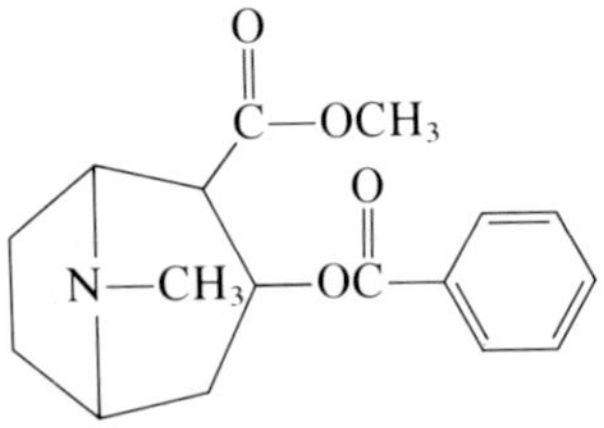

[structure of Cocaine] Cocaine

Reactions of Amines

4.27 Most illicit cocaine is actually cocaine hydrochloride—the product from the reaction of cocaine (Problem 4.26) with HCl. Show the structure of cocaine hydrochloride.

4.28* Assume that you have samples of quinine, an amine, and menthol, an alcohol. What simple chemical test could you do to distinguish between them?

4.29 Complete the following equations:

(a) 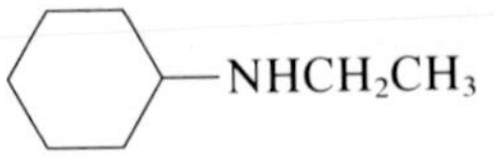 [cyclohexane ring]—$NHCH_2CH_3$ + HBr $\longrightarrow$

(b) 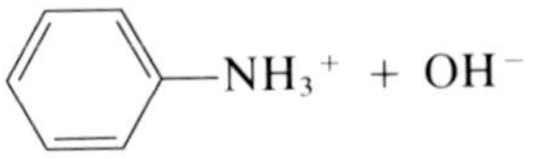 [benzene ring]—NH_3^+ + OH^- $\longrightarrow$

(c) 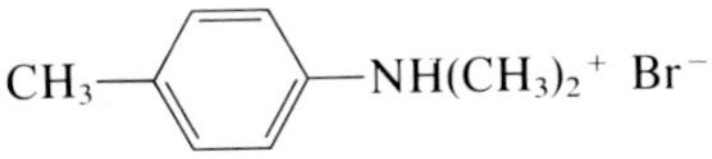 CH_3—[benzene ring]—$NH(CH_3)_2^+ \ Br^-$ + NaOH $\longrightarrow$

(d) CH_3CHNH_2 + H_2O $\rightleftharpoons$
    |
    CH_3

(e) $CH_3CH_2N(CH_3)_2$ + $CH_3CH_2CH_2CH_2Cl$ $\longrightarrow$

(f) CH_3CH_2NH + H_3O^+ $\longrightarrow$
    |
    CH_3

(g) 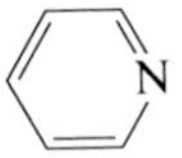 [pyridine ring] N + HCl $\longrightarrow$

4.30* Many hair conditioners contain an ammonium salt such as the following to help prevent "flyaway" hair. Will this salt react with acids or bases? Why or why not?

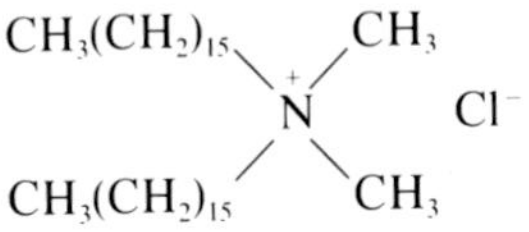

[structure: $CH_3(CH_2)_{15}$ and $CH_3(CH_2)_{15}$ on N with two CH_3 groups, Cl^-]

4.31 Zectran, a pesticide, has the following structure. Do you think that this substance reacts with aqueous hydrochloric acid? If so, what are the product(s)?

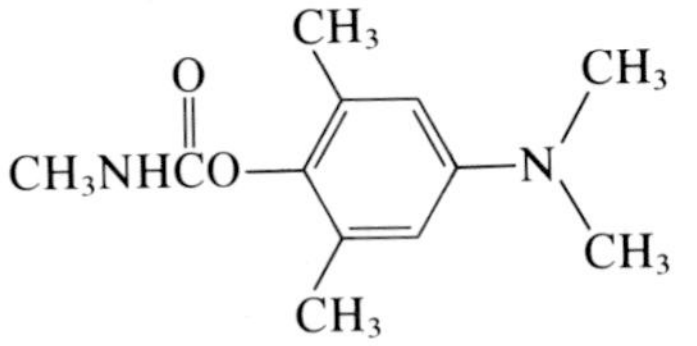

[structure of Zectran: CH_3NHCO—O—benzene ring with CH_3 groups—N with two CH_3 groups]

Applications

4.32 Acetaminophen is advertised as a mild pain reliever, superior to aspirin. Check the references listed in the Application on chemical information to determine (a) the structure of this compound; (b) its melting point, boiling point, and relative solubilities; (c) a few of the trade names under which it is sold; (d) uses for which it is prescribed; and (e) hazards associated with overdose. [App: Chemical Information]

4.33 The pH of most bodily fluids is about 7. Why do phosphoric, carbonic, and most organic acids exist in the ionized form at pH 7? [App: Organic Compounds in Body Fluids]

4.34* Why is it often necessary to use the salts of compounds rather than the neutral compound as drugs? [App: Organic Compounds in Body Fluids]

4.35 Salol, an ester of salicylic acid, is used as an enteric coating for other drugs. What is meant by the term "enteric coating"? [App: Organic Compounds in Body Fluids]

4.36* Promazine, a potent antipsychotic tranquilizer, is administered as the hydrochloride salt. Write the formula of this salt. [App: Organic Compounds in Body Fluids]

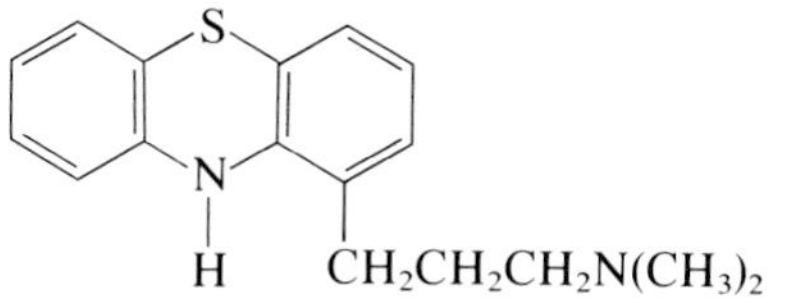

Promazine

4.37 What is a prodrug? [App: Prodrugs]

4.38 Explain why prodrugs are needed for certain applications. [App: Prodrugs]

Additional Questions and Problems

4.39 Propylamine, propanol, acetic acid, and *n*-butane have about the same molar masses. Which would you expect to have the (a) highest boiling point, (b) lowest boiling point, (c) least solubility in water, and (d) least chemical reactivity? Explain?

4.40* Explain why decylamine is much less soluble in water than ethylamine.

4.41 Each amino acid has a characteristic pH at which it exists in a form analogous to a salt where the acid group donates a proton to the amine group in the same molecule. Two naturally occurring amino acids are glycine and threonine. Show the salt form of each.

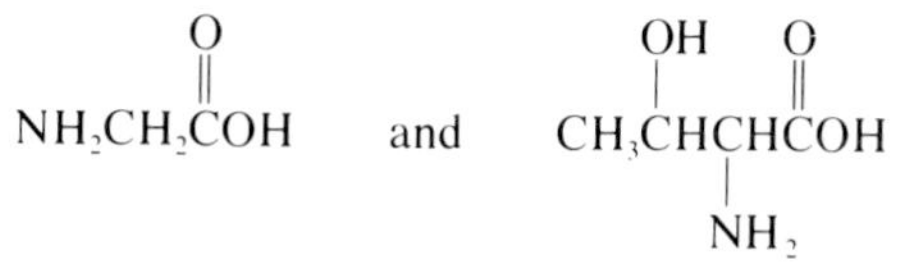

Glycine   Threonine

4.42 *para*-Aminobenzoic acid (PABA) is a common ingredient in sunscreens. Draw the structure of PABA.

4.43 PABA (Problem 4.42) is used by certain bacteria as a starting material from which folic acid (a necessary vitamin) is made. Sulfa drugs such as sodium sulfanilimide work because they resemble PABA. The bacteria try to metabolize the sulfa drug, fail to do so, and die due to lack of folic acid.

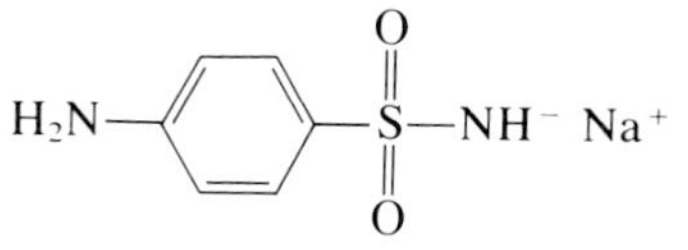

(a) Describe how this structure is similar to that of PABA.

(b) Why do you think the sodium salt, rather than the neutral compound, is used as the drug?

4.44 Acyclovir is an antiviral drug with the following structure:

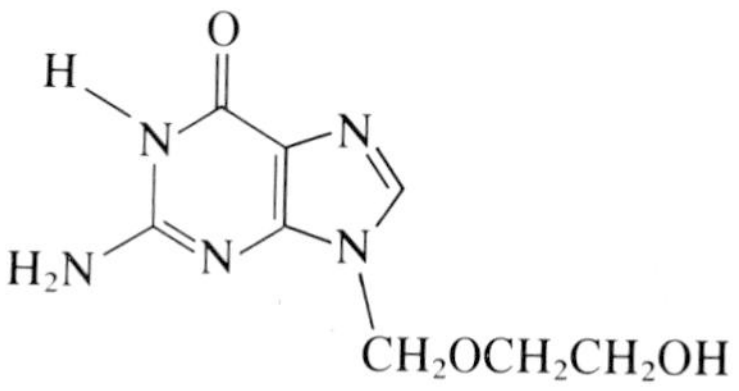

(a) What heterocyclic base (Table 4.1) is the parent of this compound?

(b) Label the other functional groups present.

(c) Label each nitrogen as primary, secondary, or tertiary.

4.45 Which is the stronger base, trimethylamine or pyridine? In which direction will the following reaction proceed?

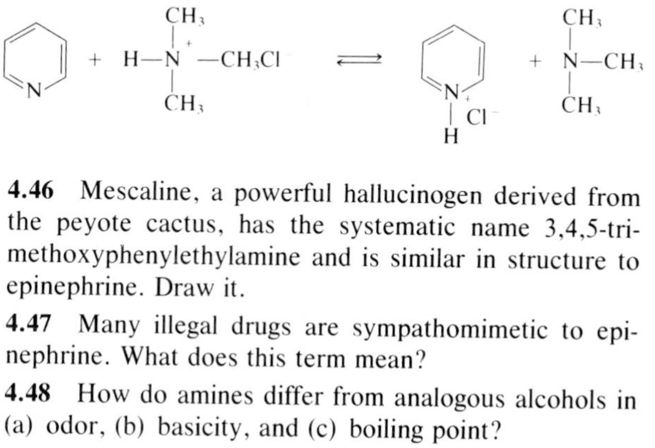

4.46 Mescaline, a powerful hallucinogen derived from the peyote cactus, has the systematic name 3,4,5-trimethoxyphenylethylamine and is similar in structure to epinephrine. Draw it.

4.47 Many illegal drugs are sympathomimetic to epinephrine. What does this term mean?

4.48 How do amines differ from analogous alcohols in (a) odor, (b) basicity, and (c) boiling point?

4.49 What two characteristics are often associated with alkaloids?

4.50 Describe how antihistamines function.

4.51 Name these compounds.

(a) 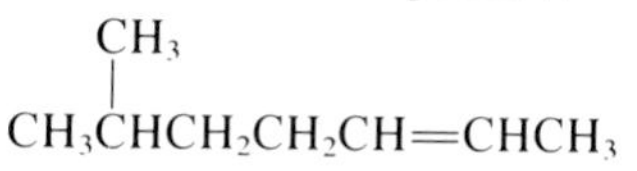 $CH_3CHCH_2CH_2CH=CHCH_3$

(b) 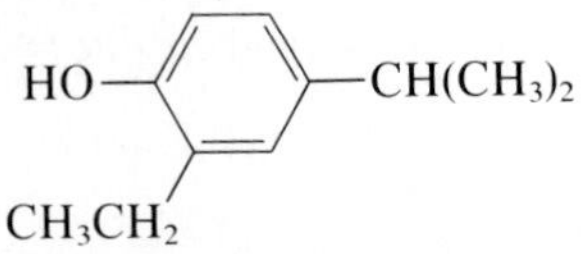 HO—⟨benzene ring⟩—CH(CH$_3$)$_2$

CH$_3$CH$_2$

(c) (CH$_3$CH$_2$CH$_2$CH$_2$)$_2$NH

4.52 Complete these equations.

(a)
$$CH_3CH_2\underset{\underset{CH_3}{|}}{\overset{\overset{CH_3}{|}}{C}}CH_2CH\!=\!\underset{\underset{CH_2CH_3}{|}}{C}CH_3 + HCl \longrightarrow$$

(b) CH$_3$CH$_2$$\overset{\overset{OH}{|}}{C}$HCH(CH$_3$)$_2$ + H$_2$SO$_4$ $\longrightarrow$

(c) 2 CH$_3$CH$_2$SH $\xrightarrow{[O]}$

(d) ⟨benzene ring⟩—CH$_2$$\overset{\overset{OH}{|}}{C}HCH_2CH_3$ $\xrightarrow{[O]}$

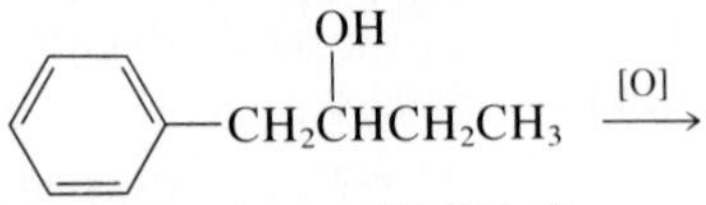

(e) CH$_3$CH$_2$NH$_2$ + CH$_3$CH$_2$Cl $\longrightarrow$

(f) (CH$_3$)$_3$N + H$_2$O $\longrightarrow$

(g) (CH$_3$)$_3$N + HCl $\longrightarrow$

(h) (CH$_3$)$_3$NH$^+$ + OH$^-$ $\longrightarrow$

5

Aldehydes
and Ketones

This bombardier beetle is spraying a potential predator with boiling-hot benzoquinones, produced in a fraction of a second by a chemical reaction that the insect carries out within its body. Benzoquinones are ketones, members of a class of compounds you'll learn about in this chapter. (For another example of insect chemical warfare, see the Interlude.)

In this and the next chapter, we'll discuss several families of compounds that are widespread in organic and biological chemistry: those that have a carbonyl group. Every carbonyl group contains a carbon–*oxygen* double bond in the same way that an alkene grouping contains a carbon–*carbon* double bond.

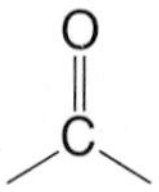

$$\underset{\diagdown \ \diagup}{\overset{\displaystyle \overset{O}{\underset{\|}{C}}}{}}$$ A carbonyl group

Here we'll look at the two most simple types of carbonyl compounds—aldehydes, in which the carbonyl group has one R group and one H atom on the carbonyl carbon atom, and ketones, in which there are two R groups on the carbonyl carbon.

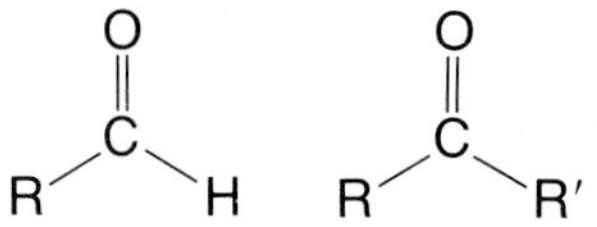

$$\underset{R \diagdown \diagup H}{\overset{\displaystyle \overset{O}{\underset{\|}{C}}}{}} \qquad \underset{R \diagdown \diagup R'}{\overset{\displaystyle \overset{O}{\underset{\|}{C}}}{}}$$

Aldehyde   Ketone

We'll answer the following questions in this chapter:

1. *What are the kinds of carbonyl groups, and how do they differ?* The goal: Be able to recognize and draw the structures of the important families of carbonyl compounds.
2. *What are the general properties of aldehydes and ketones?* The goal: Be able to describe such properties as polarity, hydrogen bonding, and water solubility.
3. *How are ketones and aldehydes named?* The goal: Be able to name the simple members of these families and write their structures, given the names.
4. *What are some of the significant applications and occurrences of aldehydes and ketones?* The goal: Be able to identify major applications and occurrences of each family of compounds and describe some important members of each family.
5. *What are the major chemical reactions of aldehydes and ketones?* The goal: Be able to describe and predict the products of the oxidation and reduction of aldehydes and ketones, and their addition of alcohols.
6. *What are hemiacetals and acetals?* The goal: Be able to recognize hemiacetals and acetals, describe the conditions under which they are formed, and predict the products of acetal hydrolysis.
7. *What is the aldol reaction and why is it important in biochemistry?* The goal: Be able to describe an aldol reaction, predict the products, and explain why it is important in biochemistry.

5.1 KINDS OF CARBONYL COMPOUNDS

There are many different kinds of carbonyl compounds, depending on what other substituents are bonded to the carbonyl-group carbon atom. As indicated in Table 5.1, *aldehydes* (often written as RCHO) have a carbon substituent and a hydrogen bonded to the carbonyl group; *ketones* (RCOR') have two carbon substituents; *carboxylic acids* (RCOOH) have a carbon and a hydroxyl (—OH); *anhydrides* (RCO₂COR') have a carbon and an oxygen bonded to another R—C=O group; *esters* (RCOOR') have a carbon and an alkoxyl (—OR); and *amides* (RCONH₂) have a carbon and a nitrogen group (—NH₂, NHR, or NR₂).

Since oxygen attracts electrons more strongly than carbon, carbonyl groups are strongly polarized, with a partial positive charge on carbon and a partial negative charge on oxygen. As we'll see in subsequent sections, this polarity of the carbonyl group helps to explain how many of its reactions take place.

REVIEW The VSEPR model explains the planarity of the carbonyl group and the bond angles by assuming that the three regions of electron density stay as far apart as possible (Appendix A.7).

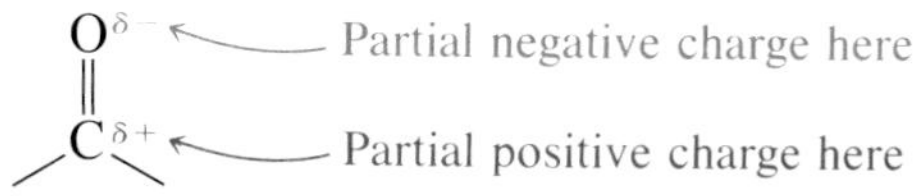

$O^{\delta-}$ ⟵ Partial negative charge here

$C^{\delta+}$ ⟵ Partial positive charge here

Another property common to all carbonyl groups is planarity. The carbonyl carbon atom is surrounded by three regions of electron density, and the bond angles between the three substituents on carbon are 120° or close to it.

120° angles in a planar triangle

Table 5.1   Some Kinds of Carbonyl Compounds

Name	Structure	Example	
Aldehyde	R—C(=O)—H	H₃C—C(=O)—H	Acetaldehyde
Ketone	R—C(=O)—R'	H₃C—C(=O)—CH₃	Acetone
Carboxylic acid	R—C(=O)—O—H	H₃C—C(=O)—O—H	Acetic acid
Anhydride	R—C(=O)—O—C(=O)—R	H₃C—C(=O)—O—C(=O)—CH₃	Acetic anhydride
Ester	R—C(=O)—O—R'	H₃C—C(=O)—O—CH₃	Methyl acetate
Amide	R—C(=O)—N	H₃C—C(=O)—NH₂	Acetamide

It's useful to classify carbonyl compounds into two groups based on their chemical properties. In one group are aldehydes and ketones. Since the carbonyl group in these compounds is bonded to atoms (H and C) that don't attract electrons strongly, the bonds to the carbonyl-group carbon of aldehydes and ketones are not strongly polar. Thus, the compounds behave similarly.

In the second group are carboxylic acids, esters, and amides. Since the carbonyl-group carbon in these compounds is bonded to an atom (O or N) that *does* attract electrons strongly, the bonds are strongly polar. Thus, carboxylic acids, esters, anhydrides, and amides behave similarly in many ways. This second group of carbonyl-containing compounds is discussed in Chapter 6.

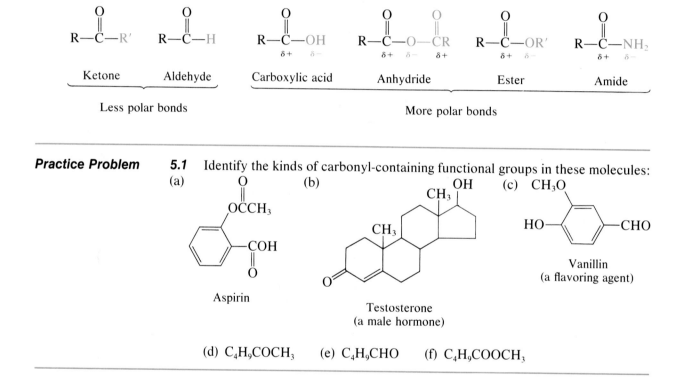

| Ketone | Aldehyde | Carboxylic acid | Anhydride | Ester | Amide |

Less polar bonds   More polar bonds

Practice Problem   **5.1**   Identify the kinds of carbonyl-containing functional groups in these molecules:

(a) Aspirin

(b) Testosterone (a male hormone)

(c) Vanillin (a flavoring agent)

(d) $C_4H_9COCH_3$ (e) C_4H_9CHO (f) $C_4H_9COOCH_3$

5.2 NAMING ALDEHYDES AND KETONES

Aldehyde A compound that has a carbonyl group bonded to one carbon atom and one hydrogen atom, RCHO.

Aldehydes are named systematically by replacing the final *-e* of the corresponding alkane name with *-al*. Thus, the three-carbon aldehyde derived from propane is propanal, the four-carbon aldehyde is butanal, and so on. When substituents are present, the chain is numbered beginning at the —CHO end. In addition to these systematic names, some simple aldehydes have common names that end in *-aldehyde*. The one-carbon aldehyde derived from methane (systematic name, *methanal*) is often called *formaldehyde*, the two-carbon aldehyde derived from ethane (systematic name, *ethanal*) is often called *acetaldehyde*, and the simplest aromatic aldehyde is *benzaldehyde*.

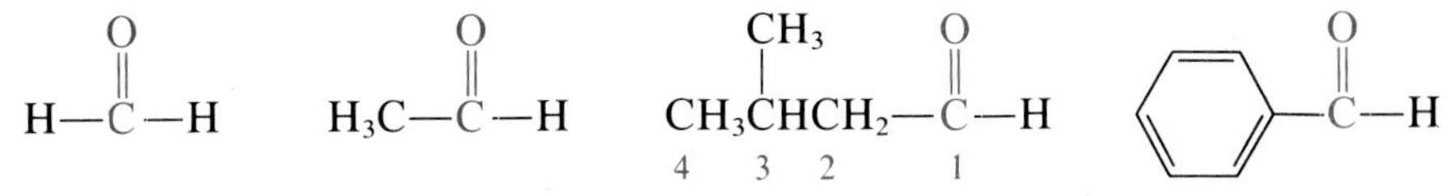

$$H-\overset{\overset{\displaystyle O}{\|}}{C}-H$$

Formaldehyde
(methanal)

$$H_3C-\overset{\overset{\displaystyle O}{\|}}{C}-H$$

Acetaldehyde
(ethanal)

$$\underset{4}{CH_3}\underset{3}{\overset{\overset{\displaystyle CH_3}{|}}{CH}}\underset{2}{CH_2}-\underset{1}{\overset{\overset{\displaystyle O}{\|}}{C}}-H$$

3-Methylbutanal

Benzaldehyde

Ketone A compound that has a carbonyl group bonded to two carbon atoms, $R_2C=O$.

Ketones are named systematically by replacing the final -*e* of the corresponding alkane name with -*one* (pronounced "own"). The numbering of the chain begins at the end nearer the carbonyl group, as shown for 2-pentanone, a five-carbon ketone derived from pentane. As with aldehydes, some simple ketones such as acetone and acetophenone also have common names. Often, the common name of a ketone gives the names of the two alkyl groups followed by the word *ketone*.

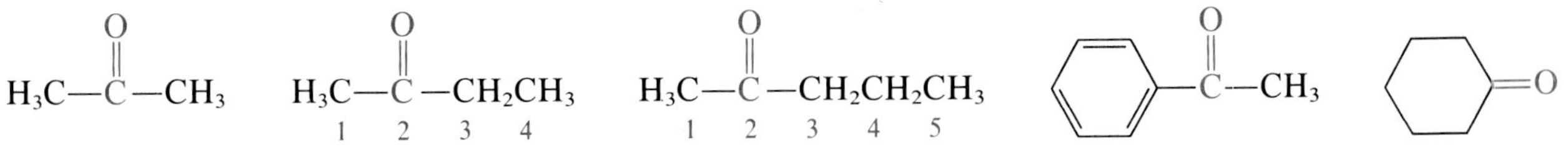

$$H_3C-\overset{\overset{\displaystyle O}{\|}}{C}-CH_3$$

Acetone
(propanone)

$$H_3C-\underset{2}{\overset{\overset{\displaystyle O}{\|}}{C}}-\underset{3}{CH_2}\underset{4}{CH_3}$$

Methyl ethyl ketone
(2-butanone)

$$H_3C-\underset{2}{\overset{\overset{\displaystyle O}{\|}}{C}}-\underset{3}{CH_2}\underset{4}{CH_2}\underset{5}{CH_3}$$

2-Pentanone

Acetophenone

Cyclohexanone

Practice Problems   **5.2**   Draw structures corresponding to these names:
(a) hexanal   (b) *p*-bromoacetophenone   (c) 4-methyl-2-hexanone
(d) methyl isopropyl ketone

5.3   Give systematic names for these compounds:

(a) $CH_3CH_2CH_2CH_2\overset{\overset{\displaystyle O}{\|}}{C}H$ (b) $CH_3CH_2\overset{\overset{\displaystyle O}{\|}}{C}CH_2CH_3$ (c) $CH_3CH_2\overset{\overset{\displaystyle CH_3}{|}}{C}HCH_2CH_2\overset{\overset{\displaystyle O}{\|}}{C}H$

5.3   PROPERTIES OF ALDEHYDES AND KETONES

The polarity of the carbonyl group makes aldehydes and ketones moderately polar compounds. As a result, they are higher boiling than alkanes with similar molecular weights. Because they have no hydrogen atoms bonded to oxygen or nitrogen, however, their molecules don't hydrogen-bond with each other, and this makes aldehydes and ketones lower boiling than alcohols. In a series of compounds with similar molecular weights, the alkane is lowest boiling, the alcohol is highest boiling, and the aldehyde and ketone fall in between.

$CH_3CH_2CH_2CH_3$

Butane, bp −43°C

$CH_3CH_2\overset{\overset{\displaystyle O}{\|}}{C}H$

Propanal, bp 50°C

$CH_3\overset{\overset{\displaystyle O}{\|}}{C}CH_3$

Acetone, bp 56°C

$CH_3CH_2CH_2OH$

Propanol, bp 97°C

Formaldehyde (HCHO), the simplest aldehyde, is a gas. The other simple aldehydes and ketones are liquids (Table 5.2), and those with more than 12 carbon atoms are solids.

Aldehydes and ketones are soluble in common organic solvents, and those with fewer than five carbon atoms are also soluble in water because they're able to hydrogen-bond with water molecules (Figure 5.1). These simple aldehydes and ketones are excellent solvents because they dissolve both polar and nonpolar compounds. With increasing numbers of carbon atoms, aldehydes and ketones become more alkane-like and less water-soluble. Many aldehydes and ketones, including several that are found naturally, have distinctive odors (Figure 5.2).

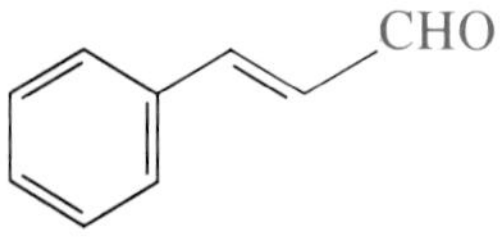

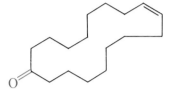

Cinnamaldehyde
(cinnamon flavor in
foods, drugs; from the
Chinese cinnamon plant)

Camphor
(cools skin in liniment and
itch remedies, but is not
safe at over 2.5% concentration;
from the camphor tree)

Civetone
(musky odor in perfumes;
from the scent gland
of the civet cat)

Camphor has been used in remedies for thousands of years but has been judged ineffective in most situations by the U.S. Food and Drug Administration. Although camphor does cool the skin, its major benefit may be that it has a "medicinal" smell.

The lower-boiling aldehydes and ketones are flammable and form explosive mixtures with air. The simple ketones have generally low toxicity, while the simple aldehydes, especially formaldehyde, are toxic.

Table 5.2   Physical Properties of Some Simple Aldehydes and Ketones[a]

Structure	Name	Boiling Point (°C)	Water Solubility (g/100 mL H$_2$O)
HCHO	Formaldehyde	−21	55
CH$_3$CHO	Acetaldehyde	21	Miscible
CH$_3$CH$_2$CHO	Propanal	49	16
CH$_3$CH$_2$CH$_2$CHO	Butanal	76	7
CH$_3$CH$_2$CH$_2$CH$_2$CHO	Pentanal	103	1
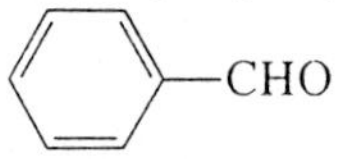 ⬡—CHO	Benzaldehyde	178	0.3
CH$_3$COCH$_3$	Acetone	56	Miscible
CH$_3$CH$_2$COCH$_3$	2-Butanone	80	26
CH$_3$CH$_2$CH$_2$COCH$_3$	2-Pentanone	102	6
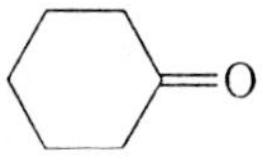 ⬡=O	Cyclohexanone	156	2

[a] All the compounds listed are liquids at room temperature.

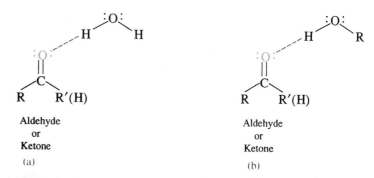

Figure 5.1
Hydrogen bonding between aldehyde or ketone molecules and (a) water and (b) alcohol molecules.

Figure 5.2
A civet, one of a large family of carnivores that live in Asia, Africa, and southern Europe. They mark their territory with musk, a potent perfume ingredient.

Properties of Aldehydes and Ketones

● No hydrogen bonding between aldehyde or ketone molecules, but they are polar and lower boiling than alcohols, but higher boiling than alkanes.

● Common aldehydes and ketones are liquids (except gaseous formaldehyde).

● Simple aldehydes and ketones are water-soluble due to hydrogen bonding with water molecules.

● Volatile aldehydes and ketones are flammable.

● Many have distinctive odors.

● Simple ketones are less toxic than simple aldehydes

5.4   SOME COMMON ALDEHYDES AND KETONES

Ketones are used in industry primarily as solvents, and aldehydes are valuable as reactants in organic synthesis and polymer formation. In living things, aldehyde and ketone functional groups are present in a great many compounds. Natural aldehydes and ketones range from simple compounds, such as those responsible for the odors and flavors of foods, spices, and flowers, to large molecules that contain numerous functional groups. Glucose and most other sugars contain —CHO groups in addition to their —OH groups. Take a good look at the glucose structure, for it plays a crucial role in the production of biochemical energy and we'll have much more to say about it. Testosterone, progesterone, and many other steroid hormones contain keto groups.

Some biomolecules containing —CHO or —C=O groups

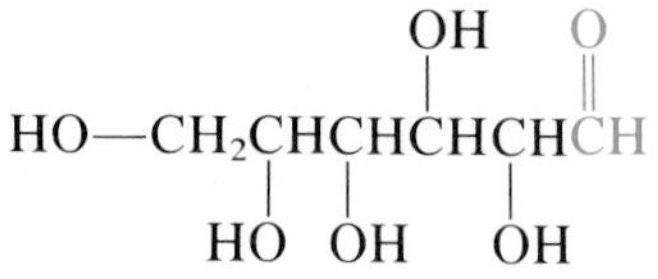

Glucose
(a pentahydroxyhexanal;
a sugar and a biochemical
energy source)

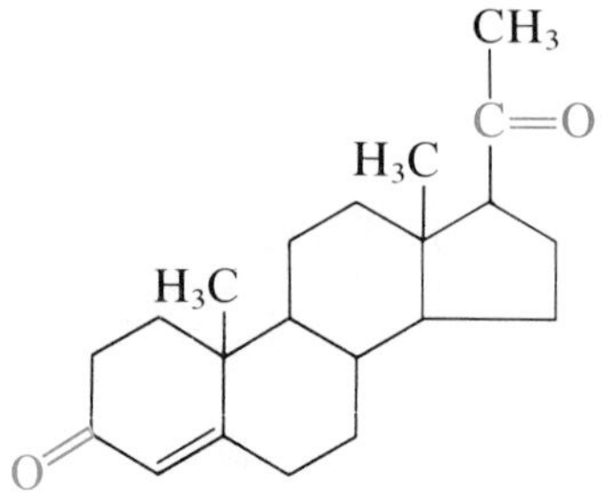

Progesterone
(a female sex hormone)

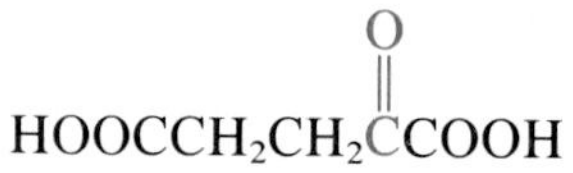

α-Ketoglutaric acid
(a key intermediate
in metabolism)

$$HOOCCH_2CH_2CCOOH$$

Formaldehyde and acetaldehyde, the two simplest aldehydes, and acetone, the simplest ketone, are described in the following paragraphs.

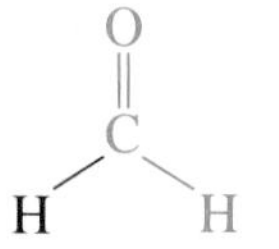

Formaldehyde 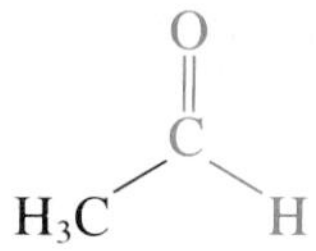   Acetaldehyde 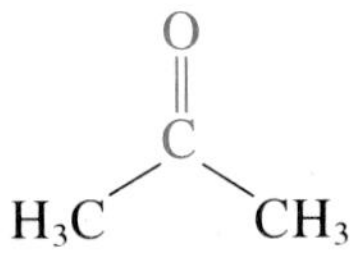   Acetone

Formaldehyde (HCHO)   At room temperature, pure formaldehyde is a colorless gas with a pungent, suffocating odor. Low concentrations in the air cause eye, throat, and bronchial irritation. Because formaldehyde is formed during incomplete combustion of coal and other organic substances, it's partly responsible for the irritation caused by smog-laden air. Higher concentrations of formaldehyde in the air cause bronchial pneumonia, and skin contact can produce dermatitis. Formaldehyde is highly toxic by ingestion, causing serious kidney damage, coma, and death. As noted in the Application on ethyl alcohol in Chapter 3, formaldehyde is responsible for the poisonous effects of methanol.

Formaldehyde is commonly sold as a 37% (w/w) aqueous solution under the name *formalin* for use as a disinfectant, antiseptic, and preservative for biological specimens. The activity of formaldehyde in these applications results from its reaction with the —NH₂ groups in proteins. Some other aldehydes are also good disinfectants, including the four-carbon dialdehyde succinaldehyde (OHCCH₂CH₂CHO), a chemical sterilant used for delicate surgical instruments that can't be sterilized by heat.

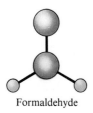

Formaldehyde

In the chemical industry, the major use of formaldehyde is in polymers with applications such as adhesives for binding plywood, foam insulation used in houses, textile finishes, and hard and durable manufactured objects such as telephone parts. *Bakelite*, one of the first commercial plastics, is a copolymer of phenol and formaldehyde. *Melamine*, familiar for its use in dinnerware and decorative laminated sheets for countertops, is a copolymer of formaldehyde and urea (H_2NCONH_2). Once the final polymerization of such a copolymer is carried out, no further melting and reshaping is possible because of the cross-linked, three-dimensional structure. In these general structures, wavy lines indicate bonds to the rest of the polymer; the CH_2 groups are from the formaldehyde molecules.

Phenol–formaldehyde polymer

Urea–formaldehyde polymer

Because concern has arisen over the toxicity and possible carcinogenicity of formaldehyde released from formaldehyde-based products, especially during fires, their use in some applications has been limited.

Acetaldehyde (CH_3CHO)   Acetaldehyde is a sweet-smelling, flammable liquid present in ripe fruits. It is less toxic than formaldehyde, and small amounts are produced in the normal breakdown of carbohydrates. Acetaldehyde is, however, a general narcotic, and large doses can cause respiratory failure. Chronic exposure produces symptoms like those of alcoholism. In industry, the major use of acetaldehyde is as an intermediate in the synthesis of other compounds.

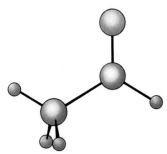

Acetaldehyde

Acetone (CH_3COCH_3)   Acetone, a liquid at room temperature, is perhaps the most widely used of all organic solvents. You may have seen cans of acetone sold in paint stores for general-purpose cleanup work, and it is the solvent in many varnishes, lacquers, and nail polish removers. Not only can acetone dissolve most organic compounds, it is also miscible with water.

Acetone is highly volatile and is a serious fire and explosion risk when allowed to evaporate in a closed space. No chronic health risk has been associated with acetone exposure, however, and it is not very toxic. When the biochemical breakdown of fats and carbohydrates to yield energy is out of balance, acetone is produced in the liver, a condition that can occur during starvation or diabetes and in severe cases leaves the odor of acetone on a patient's breath.

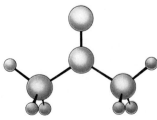

Acetone

AN APPLICATION: IS IT POISONOUS OR ISN'T IT?

In its broadest meaning, the term "drug" refers to any chemical agent other than food that affects living organisms. Mostly we refer either to substances used in preventing or treating disease, or to addictive substances obtained illegally, as "drugs." By contrast, a "poison" or "toxic substance" is any chemical agent that harms living organisms. How do we tell one from the other? Paracelsus, a Swiss physician who lived in the 1500s, understood the problem: "It is only the dose which makes a thing a poison."

The standard method for evaluating toxicity, in use since the 1920s, is the LD_{50}—lethal dose—test, and LD_{50} data are cited frequently. The LD_{50} is a measure of the toxicity of a single dose, known as *acute toxicity*. A substance is fed in varying doses to laboratory animals. The result of the test is reported as the LD_{50}, the dose that kills 50% of the animals in a uniform population. By comparing LD_{50} values, relative toxicities can be evaluated. The LD_{50} values in the accompanying table, reported as the dose in milligrams or grams per kilogram of body weight of the test

animal, show that in this group of compounds formaldehyde is the most toxic and acetone the least toxic by ingestion.

	LD_{50}
Formaldehyde, 37% aqueous solution	800 mg/kg
Acetaldehyde	1.9 g/kg
Butyraldehyde	5.8 g/kg
Acetone	8.4 g/kg

To put these values in further perspective, compare them with the LD_{50} of 17.2 mg/kg for *ouabain*, an arrow poison extracted from plants by natives in Africa, or the LD_{50} of 1.0 μg/kg for the even more toxic *batrachotoxin*, an arrow poison extracted from frogs in South America.

For therapeutic use in humans, a compound must show a comfortably wide margin between the dose that produces the desired effect and the dose that produces an acute toxic effect. Aspirin,

5.5 OXIDATION OF ALDEHYDES

Oxidation In organic chemistry, the removal of hydrogen from a molecule or addition of oxygen to a molecule.

We've already seen one important reaction of aldehydes: their **oxidation** to yield carboxylic acids (Section 3.5). In this reaction, the —CH=O hydrogen attached to the carbonyl carbon is removed and replaced by —OH. Ketones don't have this hydrogen, however, and thus don't react with most oxidizing agents.

$$\underset{\text{Aldehyde}}{R-\overset{\overset{\textstyle O}{\|}}{C}-H} \xrightarrow{[O]} \underset{\text{Carboxylic acid}}{R-\overset{\overset{\textstyle O}{\|}}{C}-OH} \qquad \underset{\text{Ketone}}{R-\overset{\overset{\textstyle O}{\|}}{C}-R'} \xrightarrow{[O]} \text{No reaction}$$

Tollens' reagent A reagent ($AgNO_3$ in aqueous NH_3) that converts an aldehyde into a carboxylic acid and deposits a silver mirror on the inside surface of the reaction flask.

Reaction with a mild oxidizing agent that doesn't affect other functional groups in a molecule can be used as a test for the presence of an aldehyde group. Among the many mild oxidizing agents that cause the aldehyde → carboxylic acid conversion, Tollens' reagent is the most visually appealing. Treatment of an aldehyde with **Tollens' reagent,** a dilute solution of silver nitrate in aqueous ammonia, rapidly yields the ammonium salt of the carboxylic acid and silver

for example, has an LD_{50} of 1.75 g/kg in rats. An average therapeutic dose of two aspirin tablets is only 650 mg, and the amount taken in 24 hr should not exceed 4000 mg.

The testing that precedes bringing a new drug to market includes many procedures in addition to the LD_{50} test. There must be *subacute toxicity* studies, in which the effects of less than lethal doses administered over several months are evaluated, and there must be *chronic toxicity* studies on the effects of low, medium, and high doses over the course of 6 months to 2 years. Additional studies include tests for effects on fertility and in pregnant animals, tests for behavioral effects, and tests for allergic reactions.

In the desire to decrease the enormous cost in time (over 8 years average) and money (millions of dollars) required to evaluate a new drug and to decrease the sacrifice of laboratory animals, the search for alternative methods of drug evaluation is under way. Significant progress is being made in procedures carried out *in vitro,* meaning literally "in glass," that is, not in ani-

mals. In the test pictured, for example, a value comparable to the LD_{50} is found by determining the concentration that kills 50% of the living cells in a cultured medium.

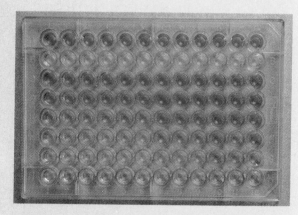

Each indentation in the test plate contains a red dye along with cells that were cultured in the presence of various concentrations of a chemical being tested for toxicity. Since only living cells take up the dye, the effect of the chemical is determined by the presence or absence of color.

REVIEW   Conversion of a metal ion like Ag^+ to the free metal is a reduction. In this redox reaction, the aldehyde is oxidized and Ag^+ is reduced.

metal as by-product. Addition of an acid such as HCl then yields the carboxylic acid. If the reaction is done in a clean flask or test tube, metallic silver deposits on the inner walls of the vessel, producing a beautiful, shiny mirror (Figure 5.3). In fact, a slight modification of this reaction is used in the industrial manufacture of glass mirrors.

$$\underset{\text{Aldehyde}}{R-\overset{\overset{\displaystyle O}{\|}}{C}-H} + 2\,AgNO_3(aq) + 3\,NH_3 + H_2O \longrightarrow \underset{\substack{\text{Ammonium salt} \\ \text{of acid}}}{R-\overset{\overset{\displaystyle O}{\|}}{C}-O^-\,NH_4^+} + \underset{\substack{\text{Silver} \\ \text{metal}}}{2\,Ag} + 2\,NH_4NO_3(aq)$$

For example,

Benzaldehyde $\xrightarrow[\text{2. } H_3O^+]{\text{1. Tollens' reagent}}$ Benzoic acid $+ 2\,Ag$

(a) (b) (c)

Figure 5.3
The Tollens' test for aldehydes: $RCHO + 2\ Ag(NH_3)_2^+ + 3\ OH^- \rightarrow RCOO^- + 2\ Ag + 4\ NH_3 + 2\ H_2O$. In a clean container, the silver deposits on the walls to form a mirror.

Benedict's reagent A reagent (Cu^{2+} in basic solution) that converts an aldehyde into a carboxylic acid and yields a brick-red precipitate of Cu_2O.

Another mild oxidizing agent, known as **Benedict's reagent,** also relies on reduction of a metal to produce a visible test for aldehydes. The reagent solution contains blue copper(II) ion that is reduced to give a precipitate of red copper(I) oxide in the reaction with an aldehyde (Figure 5.4):

$$R-\overset{\overset{\displaystyle O}{\|}}{C}-H \ + \ 2\ Cu^{2+} \ + \ 5\ OH^- \ \longrightarrow \ R-\overset{\overset{\displaystyle O}{\|}}{C}-O^- \ + \ Cu_2O(s) \ + \ 3\ H_2O$$

Deep-blue solution

Brick-red precipitate

Because many sugars contain aldehyde groups, Benedict's reagent has been used as a test for sugars in the urine (although more specific tests are now preferred; Chapter 10).

Figure 5.4
The Benedict's test for aldehydes: The Cu^{2+}-containing reagent solution is on the left. A few drops of glucose produce the greenish color in the center. A few milliliters produce the brick-red precipitate of Cu_2O on the right.

130

Practice Problem   5.4   Draw structures of the products you would obtain by treating the following compounds with Tollens' reagent. If no reaction occurs, write "NR."

$$\overset{\displaystyle CH_3}{\underset{\displaystyle |}{}}$$

(a) $CH_3CHCH_2CH_2CH_2CHO$ (b) 2,2-dimethylpentanal

(c) 2-methyl-3-pentanone

5.6 REDUCTION OF ALDEHYDES AND KETONES

Reduction In organic chemistry, the addition of hydrogen, for example, to a carbon–oxygen double bond or a carbon–carbon double bond, or the removal of oxygen.

Just as an alcohol can be converted into a carbonyl compound by *removing* hydrogen (an *oxidation*), a carbonyl compound can be converted back into an alcohol by *adding* hydrogen to the C=O double bond (a *reduction*). **Reduction** in organic chemistry usually refers to the addition of hydrogen to a molecule.

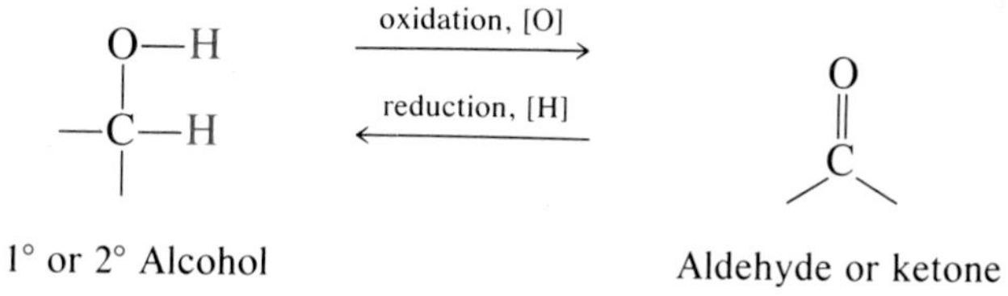

1° or 2° Alcohol   Aldehyde or ketone

The reduction of aldehydes and ketones is usually accomplished by treatment with sodium borohydride, $NaBH_4$. When mixed with a ketone or aldehyde, sodium borohydride transfers one of its hydrogens as a hydride ion ($:H^-$). The hydride ion contributes both electrons to a covalent bond to the carbonyl carbon, forming an anion. Aqueous acid is then added, H^+ adds to the oxygen, and a neutral alcohol product results. Aldehydes are converted into primary alcohols ($RCHO \rightarrow RCH_2OH$) by reaction with $NaBH_4$.

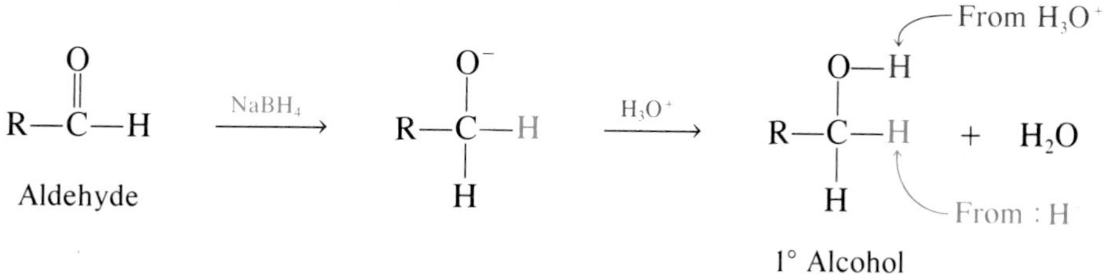

Note that the two new hydrogen atoms in the alcohol product come from different sources. The hydrogen atom attached to carbon comes from reaction with $:H^-$ (hydride ion) in the first step, whereas that attached to oxygen comes from reaction with H^+ (acid) in the second step.

Ketones are converted into secondary alcohols ($R_2CO \rightarrow R_2CHOH$) by reaction with $NaBH_4$.

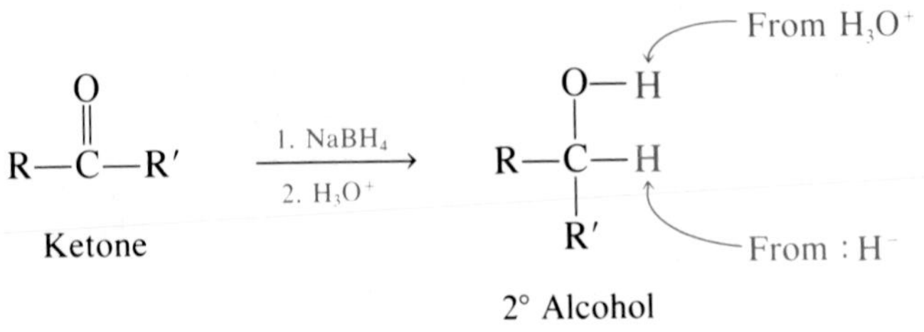

$$R-\overset{\overset{\displaystyle O}{\|}}{C}-R' \xrightarrow[\text{2. H}_3\text{O}^+]{\text{1. NaBH}_4} R-\overset{\overset{\displaystyle O-H}{|}}{\underset{\underset{\displaystyle R'}{|}}{C}}-H$$

From H$_3$O$^+$

From : H$^-$

Ketone

2° Alcohol

For example, the aldehyde propanal gives a primary alcohol, and the ketone cyclohexanone gives a secondary alcohol.

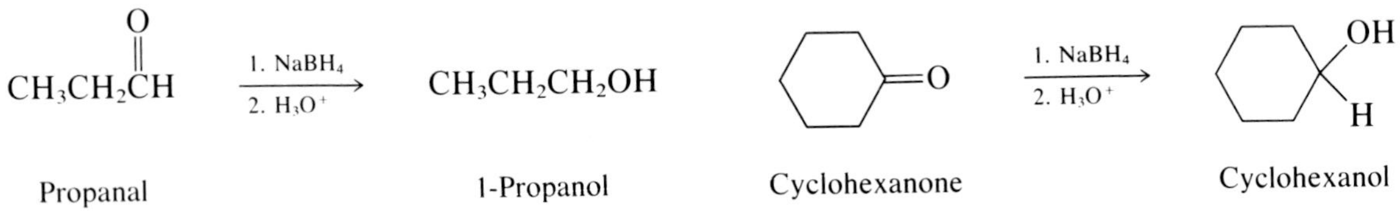

$$CH_3CH_2\overset{\overset{\displaystyle O}{\|}}{C}H \xrightarrow[\text{2. H}_3\text{O}^+]{\text{1. NaBH}_4} CH_3CH_2CH_2OH$$

Propanal   1-Propanol   Cyclohexanone   Cyclohexanol

The fact that the reduction of aldehydes and ketones takes two steps, addition of : H$^-$ and addition of H$^+$, makes good sense when you think about the polarity of the carbonyl group. Since the carbonyl-group carbon is polarized δ^+, the negatively charged hydride ion is naturally drawn to it. Similarly, a positively charged hydrogen ion is drawn to the negatively polarized carbonyl-group oxygen.

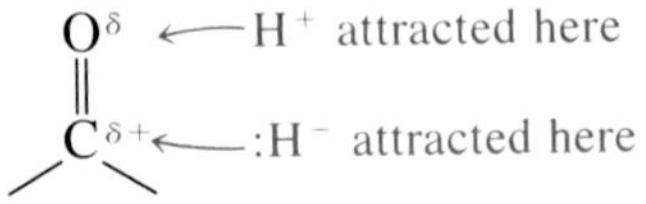

O$^\delta$ ←— H$^+$ attracted here

C$^{\delta+}$ ←— :H$^-$ attracted here

The reduction of aldehydes and ketones to yield alcohols occurs in living cells as well as in the laboratory. Although the body doesn't use NaBH$_4$ as its "reagent," there is nevertheless a remarkable similarity in the way the two kinds of reductions are done. Living organisms use *nicotinamide adenine dinucleotide*, abbreviated *NADH*, as a biological source of hydride ion. (For the structure of NAD$^+$, the oxidized form of NADH, see Application on organic compounds in body fluids, Chapter 4.) NADH serves exactly the same role in the cell that NaBH$_4$ serves in the laboratory: It donates : H$^-$ to an aldehyde or ketone to yield an anion that picks up H$^+$ from surrounding aqueous fluids. The net result is reduction.

An example of biochemical reduction occurs when pyruvic acid, an intermediate in glucose metabolism, is converted into lactic acid by active skeletal muscles. Vigorous exercise causes a buildup of lactic acid in muscles, leading to a tired, flat feeling.

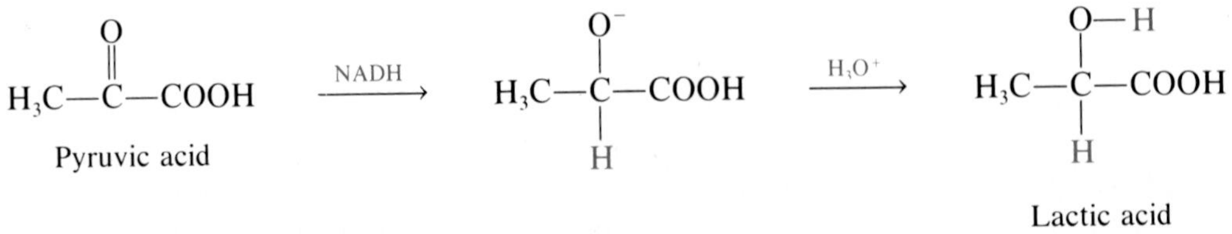

$$H_3C-\overset{\overset{\displaystyle O}{\|}}{C}-COOH \xrightarrow{\text{NADH}} H_3C-\overset{\overset{\displaystyle O^-}{|}}{\underset{\underset{\displaystyle H}{|}}{C}}-COOH \xrightarrow{\text{H}_3\text{O}^+} H_3C-\overset{\overset{\displaystyle O-H}{|}}{\underset{\underset{\displaystyle H}{|}}{C}}-COOH$$

Pyruvic acid

Lactic acid

Solved Problem 5.1 What product would you obtain by reduction of benzaldehyde?

Solution First, draw the structure of the starting material, showing the double bond in the carbonyl group. Then rewrite the structure showing only a *single* bond between C and O, along with *partial* bonds to both C and O:

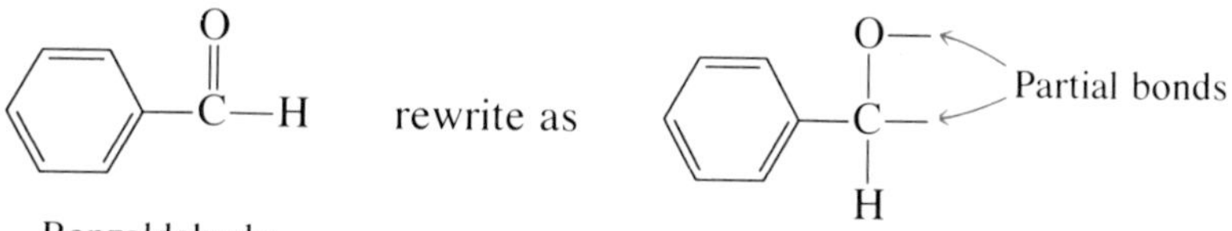

Benzaldehyde

rewrite as

Partial bonds

Finally, attach hydrogen atoms to the two part-bonds and rewrite the product:

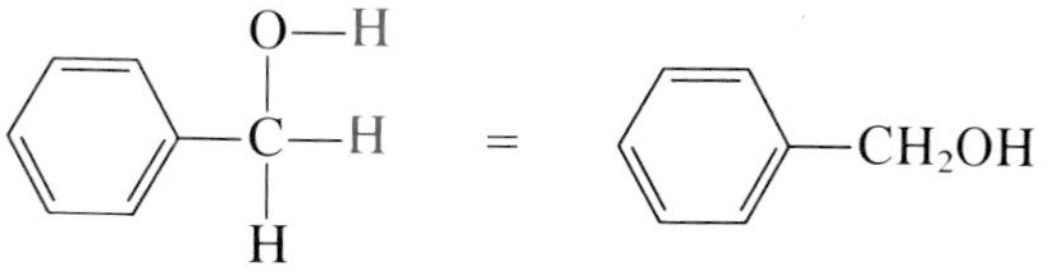

$=$ —CH$_2$OH   (Benzyl alcohol)

Practice Problems **5.5** What product would you obtain from reduction of the following ketones and aldehydes?

(a) $(CH_3)_2CHCHO$ (b) *m*-chlorobenzaldehyde   (c) cyclopentanone

5.6 What ketones or aldehydes might be reduced to yield the following alcohols?

(a)

(b) 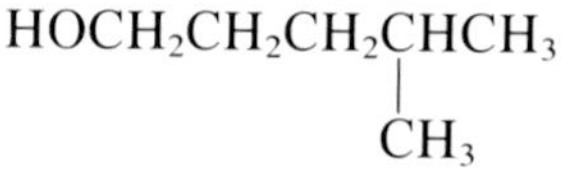 $HOCH_2CH_2CH_2CHCH_3$
        |
        CH_3

(c) 2-methyl-1-pentanol

5.7 REACTION WITH ALCOHOLS: HEMIACETALS AND ACETALS

Addition reaction, aldehydes and ketones Addition of an alcohol or other reagent to the carbon–oxygen double bond to give a carbon–oxygen single bond.

Hemiacetal A compound that has both an alcohol-like —OH group and an ether-like —OR group bonded to the same carbon atom.

Hemiacetal Formation Aldehydes and ketones undergo **addition reactions** with alcohols. Just as the C=C bond is converted to a single C—C bond when hydrogen adds to an alkene, the C=O bond is converted to a single C—O bond when an alcohol adds to an aldehyde or ketone. During the reaction, the H from the alcohol bonds to the carbonyl-group oxygen, and the OR from the alcohol bonds to the carbonyl-group carbon. The addition products, called **hemiacetals,** are compounds that have both an alcohol-like —OH group and an ether-like —OR group bonded to what was a carbonyl carbon atom.

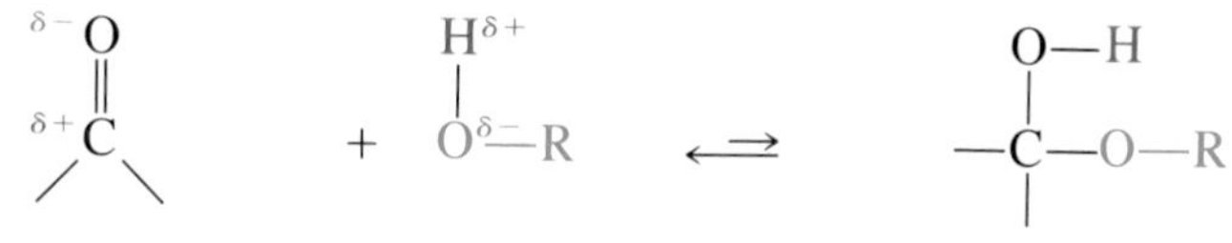

Aldehyde or ketone   Alcohol   Hemiacetal (unstable)

Two important features of this reaction need to be pointed out. It is the *negatively* polarized alcohol oxygen atom that adds to the *positively* polarized carbonyl carbon, just as negatively charged hydride ions adds to the positively polarized carbonyl oxygen during reduction (Section 5.6). Almost all carbonyl-group reactions follow this same polarity pattern. Also, note that the reaction is reversible. Hemiacetals rapidly revert back to aldehydes or ketones by loss of alcohol. Equilibrium is therefore always established between an aldehyde or ketone and its hemiacetal. For example, ethanol (CH_3CH_2OH) forms hemiacetals with acetaldehyde and acetone as follows:

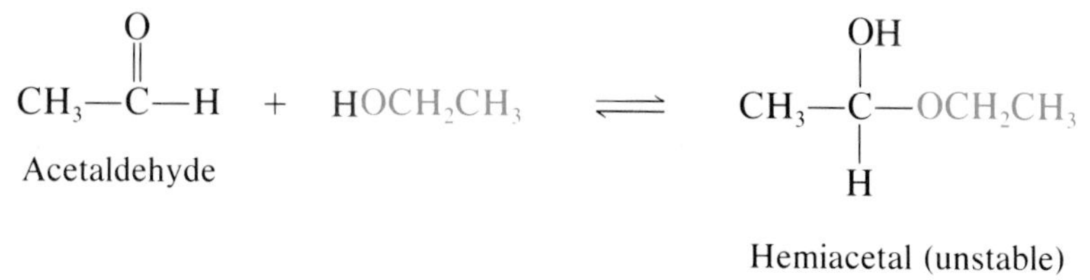

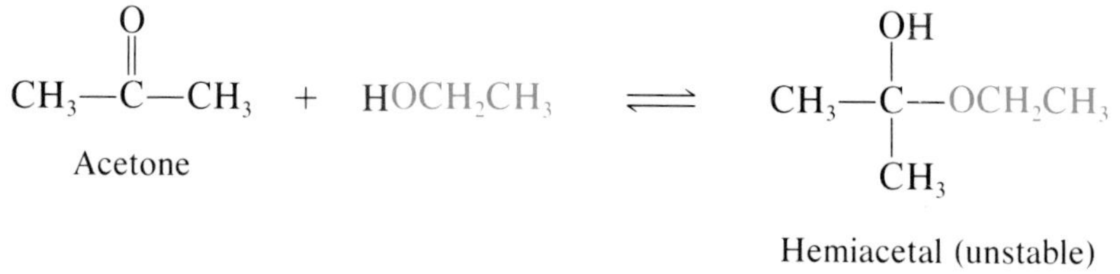

$$
\underset{\text{Acetaldehyde}}{CH_3-\overset{\overset{O}{\parallel}}{C}-H} \; + \; HOCH_2CH_3 \;\rightleftharpoons\; \underset{\underset{H}{|}}{CH_3-\overset{\overset{OH}{|}}{C}-OCH_2CH_3}
$$

Hemiacetal (unstable)

$$
\underset{\text{Acetone}}{CH_3-\overset{\overset{O}{\parallel}}{C}-CH_3} \; + \; HOCH_2CH_3 \;\rightleftharpoons\; \underset{\underset{CH_3}{|}}{CH_3-\overset{\overset{OH}{|}}{C}-OCH_2CH_3}
$$

Hemiacetal (unstable)

(In older nomenclature, a distinction was made by calling the compounds from aldehydes "hemiacetals," and those from ketones "hemiketals"; thus you may see "hemiketal" or "ketal" in some publications.)

In practice, hemiacetals are often too unstable to be isolated. When the equilibrium position for the reaction is reached, very little hemiacetal is present. The one major exception to this rule occurs in the case of sugars such as glucose in which the —OH and —CHO functional groups that react are part of the *same* molecule. The resulting cyclic hemiacetal is more stable than a noncyclic hemiacetal. As you'll see in Chapter 10, most simple sugars contain such a stable hemiacetal link.

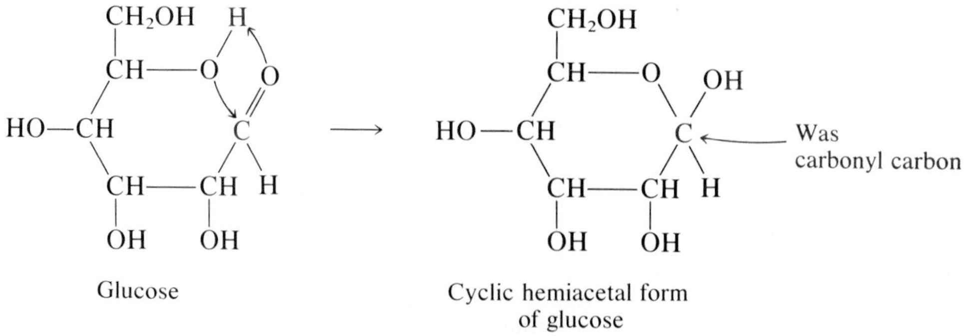

Glucose → Cyclic hemiacetal form of glucose (Was carbonyl carbon)

Solved Problem 5.2   Which of the following compounds is a hemiacetal?

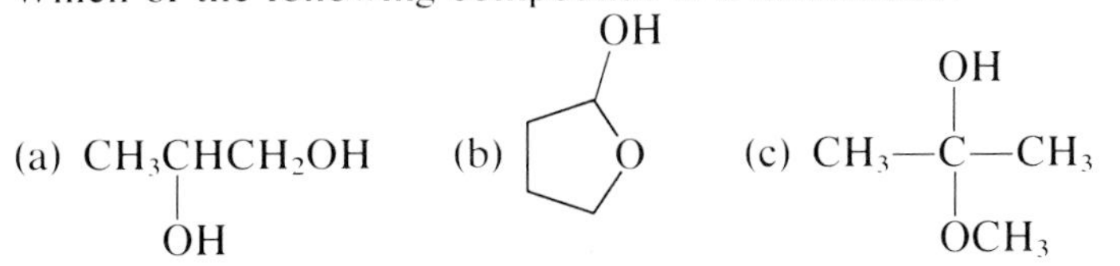

(a) CH_3CHCH_2OH (b)   (c) $CH_3—C—CH_3$
       | |
       OH   OCH_3

Solution   To identify a hemiacetal, look for a carbon atom attached to two oxygen atoms, one in an —OH group and one in an —OR group. Compound (a) contains two O atoms, but they are bonded to different C atoms; it is not a hemiacetal. Compound (b) has one ring C atom bonded to two oxygen atoms, one in the substituent —OH group and one bonded to the rest of the ring, the same as an R group; it is a cyclic hemiacetal. Compound (c) also contains a C atom bonded to one —OH group and one —OR group, so it too is a hemiacetal.

Acetal A compound that has two ether-like —OR groups bonded to the same carbon atom.

Acetal Formation   If a small amount of acid catalyst is added to the reaction of an alcohol and a carbonyl compound, the initially formed hemiacetal is converted into an acetal. An **acetal** is a compound that has *two* ether-like —OR groups bonded to what was a carbonyl carbon atom.

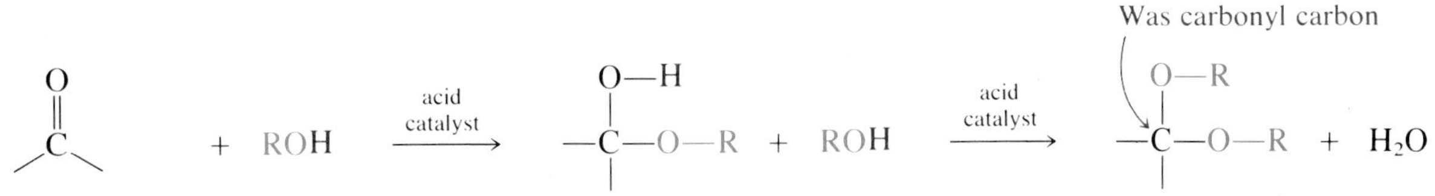

Aldehyde or ketone   + ROH →(acid catalyst)→ Hemiacetal   + ROH →(acid catalyst)→ Acetal (Was carbonyl carbon) + H_2O

For example,

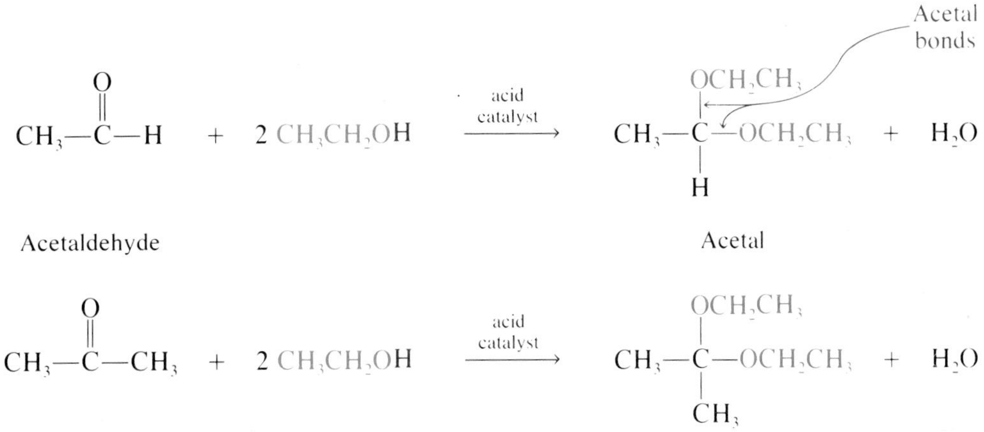

$CH_3—C—H$ (Acetaldehyde) + 2 CH_3CH_2OH →(acid catalyst)→ $CH_3—C—OCH_2CH_3$ (Acetal, Acetal bonds) + H_2O

$CH_3—C—CH_3$ (Acetone) + 2 CH_3CH_2OH →(acid catalyst)→ $CH_3—C—OCH_2CH_3$ (Acetal) + H_2O

Unlike hemiacetals, acetals are quite stable and are easily isolated. They are so stable, in fact, that some sugars and complex carbohydrates like cellulose and starch are held together simply by acetal groupings between individual sugar units. The sugar maltose, for example, is the product of acetal formation between the hemiacetal —OH group of one glucose molecule and an alcoholic —OH group of another glucose molecule. You'll learn more about carbohydrate acetals in Chapter 10.

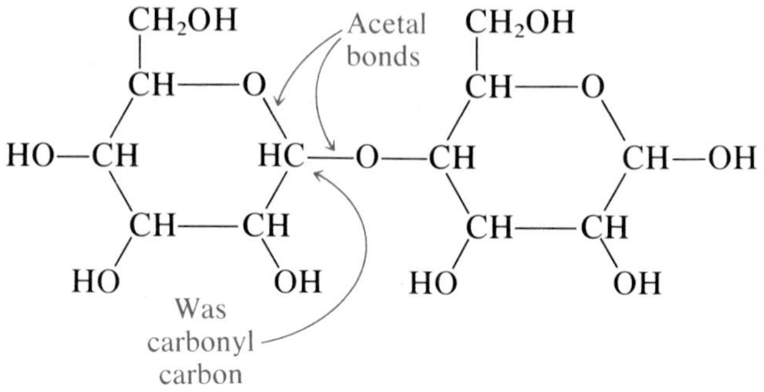

Maltose (a sugar)

Solved Problem 5.3 Write the structure of the intermediate hemiacetal and the acetal final product formed in the following reaction:

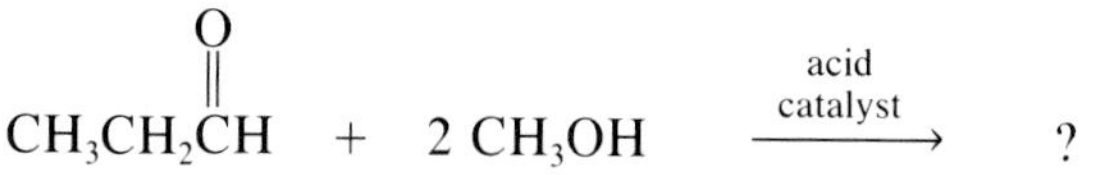

$$\underset{\substack{\\O}}{CH_3CH_2\overset{\displaystyle O}{\overset{\|}{C}}H} \ + \ 2\ CH_3OH \ \xrightarrow[\text{catalyst}]{\text{acid}} \ ?$$

Solution First, rewrite the structure showing only a single bond between C and O, along with partial bonds to both C and O:

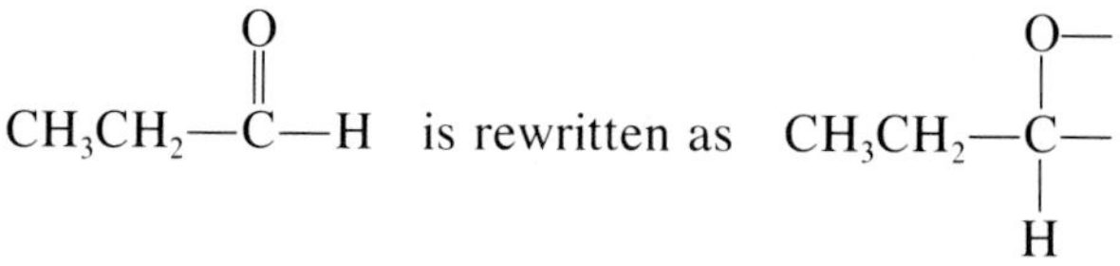

$$CH_3CH_2-\overset{\displaystyle O}{\overset{\|}{C}}-H \quad \text{is rewritten as} \quad CH_3CH_2-\overset{\displaystyle O-}{\underset{\displaystyle H}{C}}-$$

Next, add one molecule of the appropriate alcohol (CH_3OH in this case) by attaching —H to the oxygen part-bond and —OCH_3 to the carbon part-bond. This yields the hemiacetal intermediate:

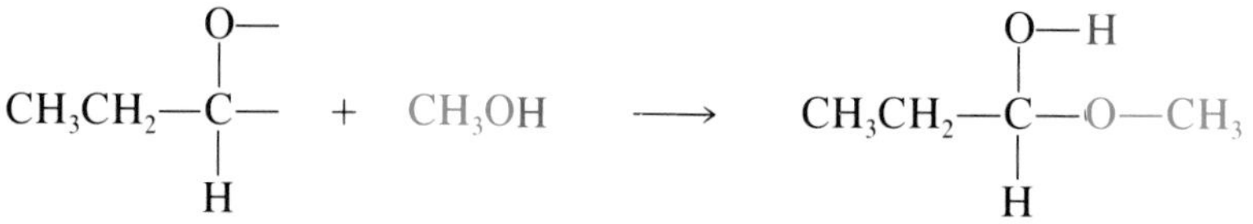

$$CH_3CH_2-\overset{\displaystyle O-}{\underset{\displaystyle H}{C}}- \ + \ CH_3OH \ \longrightarrow \ CH_3CH_2-\overset{\displaystyle O-H}{\underset{\displaystyle H}{C}}-O-CH_3 \qquad \text{Hemiacetal}$$

Finally, replace the —OH group of the hemiacetal with an —OCH_3 from a second molecule of alcohol. This yields the acetal product and water.

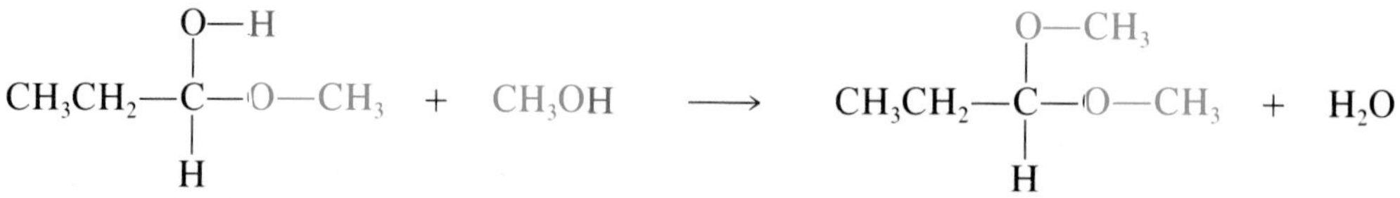

$$CH_3CH_2-\overset{\displaystyle O-H}{\underset{\displaystyle H}{C}}-O-CH_3 \ + \ CH_3OH \ \longrightarrow \ CH_3CH_2-\overset{\displaystyle O-CH_3}{\underset{\displaystyle H}{C}}-O-CH_3 \ + \ H_2O$$

Acetal product

Practice Problems   **5.7**   Draw the structures of the hemiacetals formed in these reactions:

(a) $CH_3CH_2CH_2CHO + CH_3CH_2OH \longrightarrow$?

(b) $CH_3CH_2\overset{\overset{\displaystyle O}{\displaystyle \|}}{C}CH_2CH(CH_3)_2 + CH_3OH \longrightarrow$?

5.8   Draw the structure of each acetal final product formed in the reactions shown in Problem 5.7.

5.9   Which of the following compounds are hemiacetals or acetals?

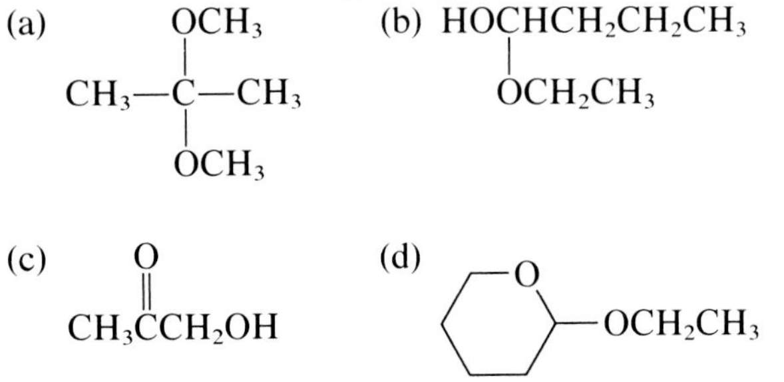

(a)

$$CH_3-\underset{\underset{\displaystyle OCH_3}{\displaystyle |}}{\overset{\overset{\displaystyle OCH_3}{\displaystyle |}}{C}}-CH_3$$

(b) $HOCHCH_2CH_2CH_3$
  $\quad\;\, OCH_2CH_3$

(c)

$$CH_3\overset{\overset{\displaystyle O}{\displaystyle \|}}{C}CH_2OH$$

(d)

[six-membered ring with O]—OCH_2CH_3

Hydrolysis   The breakdown of a compound, such as an acetal, by reaction with water; the H's and O of water add to the atoms of the broken bond.

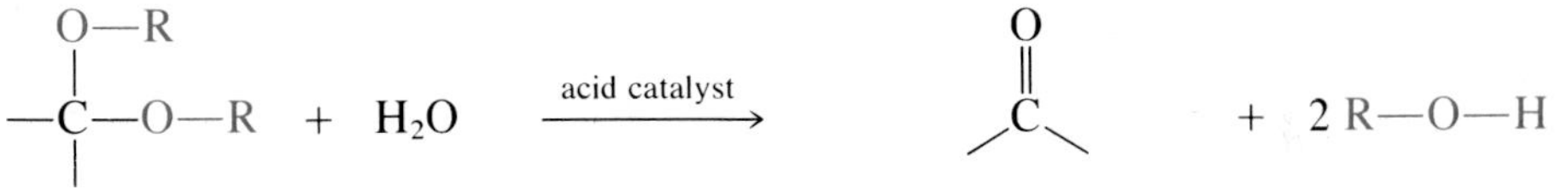

Acetal Hydrolysis   Although stable to most reagents, acetals react with aqueous acid to regenerate the original ketone or aldehyde plus two molecules of alcohol. This reaction, called a **hydrolysis** because water is used to break down the starting material, is exactly what happens in the mouth and small intestine when carbohydrates are digested.

$$-\underset{\underset{\displaystyle |}{\displaystyle |}}{\overset{\overset{\displaystyle O-R}{\displaystyle |}}{C}}-O-R + H_2O \xrightarrow{\text{acid catalyst}} \overset{\overset{\displaystyle O}{\displaystyle \|}}{\underset{\displaystyle C}{}} + 2\,R-O-H$$

Acetal   Aldehyde or ketone   Alcohol

For example,

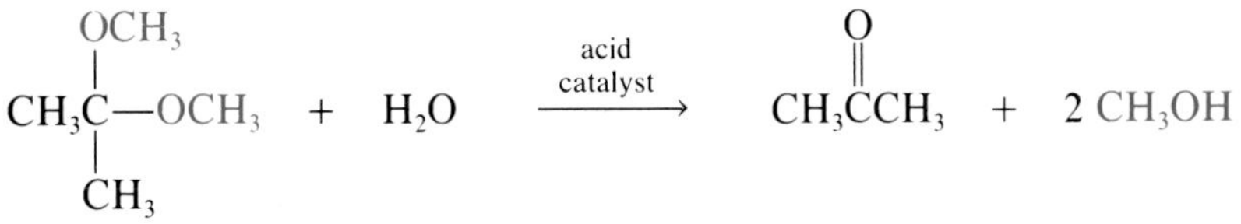

$$CH_3\underset{\underset{\displaystyle CH_3}{\displaystyle |}}{\overset{\overset{\displaystyle OCH_3}{\displaystyle |}}{C}}-OCH_3 + H_2O \xrightarrow[\text{catalyst}]{\text{acid}} CH_3\overset{\overset{\displaystyle O}{\displaystyle \|}}{C}CH_3 + 2\,CH_3OH$$

Solved Problem 5.4   Write the structure of the aldehyde or ketone that would be formed by hydrolysis of the following acetal:

$$(CH_3)_2CHCH_2CH(OCH_2CH_3)_2 \xrightarrow{H_3O^+} ?$$

Solution   First, rewrite the starting acetal so that the two ether-like acetal bonds are more evident:

$$(CH_3)_2CHCH_2CH(OCH_2CH_3)_2 \quad = \quad CH_3\overset{\overset{\displaystyle CH_3}{|}}{C}HCH_2 - \overset{\overset{\displaystyle O-CH_2CH_3}{|}}{\underset{\underset{\displaystyle H}{|}}{C}} \overset{}{-} O-CH_2CH_3$$

— Acetal bonds

Next, remove the two —OR groups from the acetal carbon, converting each into a molecule of alcohol and leaving two part-bonds on carbon:

$$CH_3\overset{\overset{\displaystyle CH_3}{|}}{C}HCH_2 - \overset{\overset{\displaystyle O-CH_2CH_3}{|}}{\underset{\underset{\displaystyle H}{|}}{C}} - O-CH_2CH_3 \quad = \quad CH_3\overset{\overset{\displaystyle CH_3}{|}}{C}HCH_2 - \overset{\overset{\displaystyle |}{|}}{\underset{\underset{\displaystyle H}{|}}{C}} - \quad + \quad 2\,CH_3CH_2-O-H$$

Ethanol

Finally, add an oxygen to the two part-bonds on the acetal carbon to form a carbonyl group and rewrite the structure of the product. In this example, the product is an aldehyde. Note that the carbonyl carbon is the one that was bonded to two oxygen atoms.

$$CH_3\overset{\overset{\displaystyle CH_3}{|}}{C}HCH_2 - \overset{\overset{\displaystyle |}{|}}{\underset{\underset{\displaystyle H}{|}}{C}} - \quad \longrightarrow \quad CH_3\overset{\overset{\displaystyle CH_3}{|}}{C}HCH_2 - \overset{\overset{\displaystyle O}{\|}}{C} - H \quad = \quad (CH_3)_2CHCH_2CHO$$

3-Methylbutanal

Practice Problem   5.10   What aldehydes or ketones result from the following acetal hydrolysis reactions? What alcohol is formed in each case?

(a)

$$\bigcirc\!\!\!\!\bigcirc - CH_2C(OCH_3)_2CH_2CH_3 \xrightarrow{H_3O^+} \; ?$$

(b) $CH_3CH_2CH_2OCH_2OCH_2CH_2CH_3 \xrightarrow{H_3O^+} \; ?$

5.8   ALDOL REACTION OF ALDEHYDES AND KETONES

All the organic reactions we've covered up to this point have been interconversions of one functional group with another. For example, alcohols can be oxidized to yield aldehydes and ketones, which in turn can be reduced to give back alcohols. Although important, these functional-group interconversion reactions are of limited use because they don't change the size or carbon framework of molecules.

In contrast to functional-group interconversions, certain other reactions form new carbon–carbon bonds. By using such reactions, it's possible to take two small pieces, join them together, and thereby make a larger molecule. Many of the biochemical processes in living cells do exactly this.

Aldol reaction   The reaction of a ketone or aldehyde to form a hydroxy ketone product on treatment with a base catalyst.

The **aldol reaction** occurs when an aldehyde or ketone is treated with a base. In the reaction, a bond forms between the carbon atom next to the carbonyl group of one molecule and the carbonyl-group carbon of the second molecule. The product, formed by joining together two molecules of starting material, is a hydroxy ketone or a hydroxy aldehyde.

AN APPLICATION: A BIOLOGICAL ALDOL REACTION

Aldol reactions are routinely used by living organisms as a key step in the biological synthesis of many different molecules. One particularly important example takes place in green leaves during the photosynthesis of carbohydrates when two three-carbon molecules are joined to form the six-carbon sugar, fructose, which is in turn converted into glucose, sucrose, and all other sugars.

As in most biochemical reactions, this transformation relies on enzymes as catalysts. As in many biochemical reactions involving sugars, some of the sugar —OH groups are replaced by phosphate ester groups. Although we haven't shown all the details, the reaction of glyceraldehyde 3-phosphate with dihydroxyacetone phosphate to yield the fructose diphosphate is clearly an aldol reaction. A hydrogen next to the dihydroxyacetone carbonyl group bonds to the carbonyl-group oxygen of glyceraldehyde, and a carbon–carbon bond forms between the two partners:

$$^{2-}O_3POCH_2\overset{\underset{\displaystyle |}{OH}}{CH}\overset{\underset{}{O}}{\overset{\|}{C}}-H \quad + \quad \overset{H}{\underset{\underset{\displaystyle |}{OH}}{CH}}-\overset{O}{\overset{\|}{C}}-CH_2OPO_3^{2-} \quad \longrightarrow \quad ^{2-}O_3POCH_2\overset{\underset{\displaystyle |}{OH}}{CH}\overset{\underset{\displaystyle |}{OH}}{CH}-\overset{OH}{\underset{\displaystyle |}{CH}}\overset{O}{\overset{\|}{C}}CH_2OPO_3^{2-}$$

| Glyceraldehyde 3-phosphate | Dihydroxyacetone phosphate | Fructose 1,6-diphosphate (a sugar phosphate) |

This oxygen and this hydrogen form a bond.

This new C—C bond is formed.

$$-\overset{|}{\underset{|}{C}}-\overset{O}{\overset{\|}{C}}- \quad + \quad -\overset{H}{\underset{|}{C}}-\overset{O}{\overset{\|}{C}}- \quad \underset{\text{catalyst}}{\overset{\text{NaOH}}{\rightleftharpoons}} \quad -\overset{|}{\underset{|}{C}}-\overset{OH}{\underset{|}{C}}-\overset{|}{\underset{|}{C}}-\overset{O}{\overset{\|}{C}}-$$

Two ketones or aldehydes

This carbon and this carbon form a bond.

A hydroxy ketone or aldehyde

For example,

$$H-\overset{H}{\underset{\displaystyle H}{C}}-\overset{O}{\overset{\|}{C}}-H \quad + \quad H-\overset{H}{\underset{\displaystyle H}{C}}-\overset{O}{\overset{\|}{C}}-H \quad \underset{\text{catalyst}}{\overset{\text{NaOH}}{\rightleftharpoons}} \quad H-\overset{H}{\underset{\displaystyle H}{C}}-\overset{OH}{\underset{\displaystyle H}{C}}-\overset{H}{\underset{\displaystyle H}{C}}-\overset{O}{\overset{\|}{C}}-H$$

Two acetaldehydes

3-Hydroxybutanal

One limitation of the aldol reaction is that the aldehyde or ketone starting material must have a hydrogen atom on the carbon next to the carbonyl group. If there is no such hydrogen present, an aldol reaction can't take place. For example, acetone can easily undergo an aldol reaction since it has six available hydrogens next to its carbonyl group, but benzaldehyde can't react in this way because it has no appropriately positioned hydrogen.

$$H_3C-\overset{\overset{\displaystyle O}{\|}}{C}-CH_3 \quad + \quad \overset{\overset{\displaystyle H}{|}}{H_2C}-\overset{\overset{\displaystyle O}{\|}}{C}-CH_3 \quad \underset{}{\overset{NaOH}{\rightleftharpoons}} \quad H_3C-\underset{\underset{\displaystyle CH_3}{|}}{\overset{\overset{\displaystyle OH}{|}}{C}}-CH_2-\overset{\overset{\displaystyle O}{\|}}{C}-CH_3$$

Two acetone molecules—
each with six hydrogens
next to the carbonyl group

4-Hydroxy-4-methyl-2-pentanone

$$\text{(benzaldehyde)}-\overset{\overset{\displaystyle O}{\|}}{C}-H \quad \xrightarrow{NaOH} \quad \text{No aldol reaction}$$

Benzaldehyde—
has no hydrogens on carbon
next to the carbonyl group

Solved Problem 5.5   What aldol product would you obtain from this reaction:

$$CH_3CH_2CHO \xrightarrow{NaOH} \quad ?$$

Solution   First, rewrite the reaction to emphasize the carbonyl group of one molecule and the C—H bond next to the carbonyl group of a second molecule:

$$CH_3CH_2-\overset{\overset{\displaystyle O}{\|}}{\underset{\underset{\displaystyle H}{|}}{C}} \quad + \quad \overset{\overset{\displaystyle H}{|}}{\underset{\underset{\displaystyle CH_3}{|}}{CH}}-\overset{\overset{\displaystyle O}{\|}}{\underset{\underset{\displaystyle H}{|}}{C}} \quad \xrightarrow{NaOH}$$

Next, draw the carbonyl group of the first molecule showing a C—O single bond and two part-bonds. Remove the appropriate hydrogen from the second molecule and connect it to the oxygen part-bond of the first molecule. Then connect the appropriate carbon from the second molecule to the carbon part-bond of the first. The structure that results is the final product.

Move this hydrogen to oxygen.

$$CH_3CH_2-\overset{\overset{\displaystyle O}{\cdot}}{\underset{\underset{\displaystyle H}{|}}{C}} \quad + \quad \overset{\overset{\displaystyle H}{}}{\underset{\underset{\displaystyle CH_3}{|}}{CH}}-\overset{\overset{\displaystyle O}{\|}}{\underset{\underset{\displaystyle H}{|}}{C}} \quad \rightleftharpoons \quad CH_3CH_2-\underset{\underset{\displaystyle H}{|}}{\overset{\overset{\displaystyle O-H}{|}}{C}}-\underset{\underset{\displaystyle CH_3}{|}}{CH}-\overset{\overset{\displaystyle O}{\|}}{\underset{\underset{\displaystyle H}{|}}{C}}$$

Connect these
carbons.

Practice Problems   **5.11**   Draw the aldol products from treatment of these aldehydes or ketones with base:

(a)

$$CH_3CH_2\overset{\overset{\displaystyle O}{\|}}{C}CH_2CH_3$$

(b)

$$\text{(phenyl)}-CH_2CHO$$

5.12   Which of the following compounds *can't* undergo aldol reactions?
(a) $CH_2{=}O$ (b) $(CH_3)_3CCHO$ (c) cyclopentanone

INTERLUDE: CHEMICAL WARFARE AMONG THE INSECTS

Life in the insect world is a jungle. Predators abound, just waiting to make a meal of any insect that happens along. Without missiles to protect themselves, many insects have evolved extraordinarily effective means of *chemical* protection. Take the humble millipede *Apheloria corrugata*, for example. When attacked by ants, the millipede protects itself by discharging a compound called *benzaldehyde cyanohydrin*.

Cyanohydrins [RCH(OH)C≡N] are compounds formed by addition of the toxic gas HCN (hydrogen cyanide) to ketones or aldehydes. Like the reaction of a ketone or aldehyde with an alcohol to yield a hemiacetal (Section 5.7), the reaction with HCN to yield a cyanohydrin is reversible. Thus, the benzaldehyde cyanohydrin secreted by the millipede can decompose to yield benzaldehyde and HCN. The millipede actually protects itself by discharging poisonous hydrogen cyanide at would-be attackers: a remarkably clever and very effective kind of chemical warfare.

Benzaldehyde
cyanohydrin

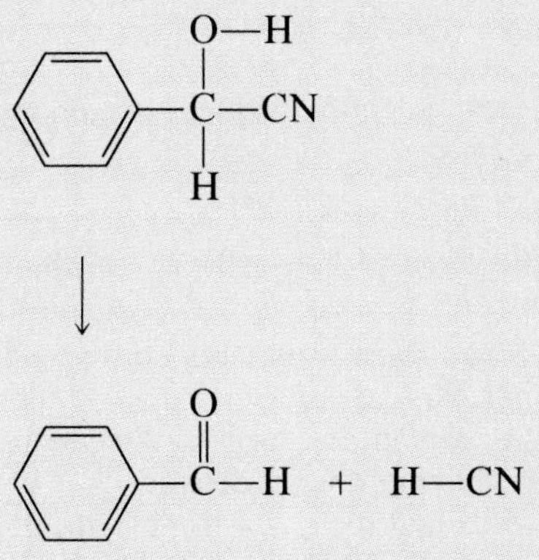

Benzaldehyde

Using a strategy similar to that of millipedes, apricots and peaches protect their seeds with a group of substances called *cyanogenic glycosides*. These compounds consist of benzaldehyde cyanohydrin bonded to glucose or another sugar. When eaten, the sugar unit is cleaved off by enzymes, and HCN is released. One of the cyanogenic glycosides, amygdalin, called Laetrile, was once well known because of its widely touted but never scientifically proven anticancer activity.

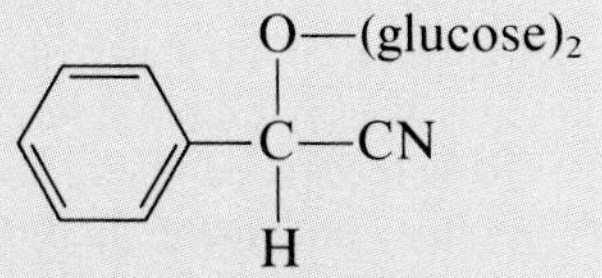

Amygdalin
(Laetrile)

The potent chemical weapon shown on the opening page of this chapter is benzoquinone, the simplest member of a class of compounds that are cyclohexadienediones (cyclohexene rings with two double bonds and two carbonyl groups). When threatened, the bombardier beetle initiates the enzyme-catalyzed oxidation of a dihydroxybenzene by hydrogen peroxide. A cloud of irritating benzoquinone vapor shoots out of the beetle's defensive organ at up to 100°C and with such force that it sounds like a pistol shot.

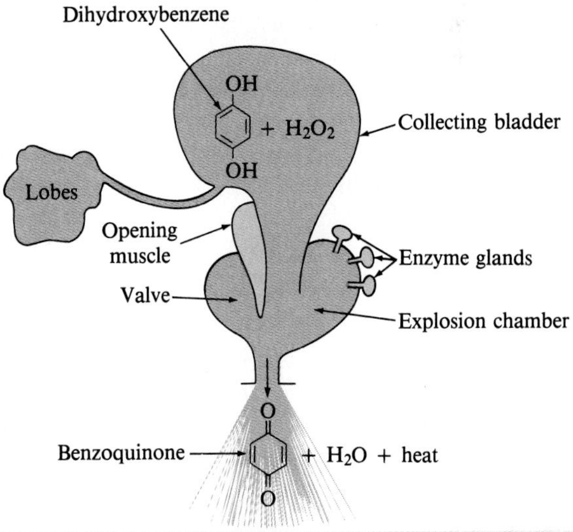

The defensive organ of the bombardier beetle.

SUMMARY

Carbonyl compounds contain the carbon–oxygen double bond, C=O. Different kinds of carbonyl compounds exist, depending on what other groups are bonded to the carbonyl carbon. **Aldehydes** (RCHO) and **ketones** (RCOR′) behave similarly because the atoms (—H and —C) bonded to the carbonyl-group carbon don't attract electrons strongly, and their bonds have similar polarity. They have no hydrogen bonding with each other, but hydrogen-bond with water or alcohols, making simple aldehydes and ketones water-soluble and excellent solvents. Many have characteristic odors and flavors; sugars contain aldehyde groups, and many hormones contain keto groups.

The major reaction of aldehydes and ketones is the **addition** of various reagents to the carbon–oxygen double bond. For example, aldehydes and ketones both add hydrogen to yield alcohol products. This **reduction** is carried out by treating the carbonyl compound first with NaBH₄ to add :H⁻, and then with aqueous acid to add H⁺. Aldehydes and ketones also add alcohols to yield **hemiacetals** or **acetals**, depending on the reaction conditions. A hemiacetal has an alcohol-like —OH group and an ether-like —OR group bonded to what was the carbonyl carbon; an acetal has two ether-like —OR groups bonded to what was the carbonyl carbon. Acetals react with aqueous acid (**hydrolysis**) to regenerate carbonyl compounds.

Aldehydes can be distinguished from ketones by their oxidation to carboxylic acids; the Tollens' and Benedict's reagents rely on metal reduction for this test.

Aldehydes and ketones that have a hydrogen atom at the position next to the carbonyl group can undergo the **aldol reaction.** Because this reaction results in formation of a carbon–carbon bond between two molecules of starting material, it is useful for building larger molecules from smaller pieces.

REVIEW PROBLEMS

Aldehydes and Ketones

5.13 What is the structural difference between an aldehyde and a ketone?

5.14* What family-name endings are used for aldehydes and ketones?

5.15 Use δ^+ and δ^- to show how the carbonyl group is polarized.

5.16* Draw structures for compounds that meet these descriptions:
(a) a ketone, C_5H_8O
(b) an aldehyde with eight carbons
(c) a keto aldehyde, $C_6H_{10}O_2$
(d) a hydroxy ketone, $C_5H_8O_2$

5.17 Which of these molecules contain carbonyl groups?

(a) $CH_3CH_2\overset{\underset{OH}{|}}{C}=O$ (b) $O=CCH_2CH_2CHCH_3$ ($\underset{NH_2}{|}$) ($\underset{CH_3}{|}$)

(c) $CH_3CH_2—O—CH=CH_2$ (d) $CH_3CH_2C(OCH_3)_3$
(e) $CH_3\underset{\underset{CH_3}{|}}{C}HCOOH$ (f) $CH_3COCH_2CH_2OH$

5.18* Which of these molecules contain carbonyl groups?
(a) CH_3CH_2CHO (b) $CH_3CH_2COCH_2CH_3$
(c) $(CH_3)_2C(OH)CH_2CH_2CH_3$ (d) CH_3COOCH_3

(e) $CH_3\text{—}\langle\text{benzene}\rangle\text{—}CONH_2$ (f) $CH_3CHCH_2CHCH_3$ with $\underset{OH}{|}$ and $\underset{OCH_3}{|}$

5.19 Identify the kinds of carbonyl groups in these molecules:

(a) $O=\langle\text{ring}\rangle—CHO$ (b) $\langle\text{ring}\rangle—\overset{\overset{O}{\|}}{C}OCH_3$

5.20 Draw structures corresponding to these aldehyde names:
(a) heptanal   (b) 4,4-dimethylpentanal
(c) *o*-chlorobenzaldehyde
(d) 3-hydroxy-2,2,4-trimethylpentanal
(e) 4-ethyl-2-isopropylhexanal
(f) *p*-bromobenzaldehyde

5.21 Draw structures corresponding to these ketone names:
(a) 3-heptanone   (b) 2,4-dimethyl-3-pentanone
(c) *m*-nitroacetophenone   (d) cyclohexanone
(e) 1,1,1-trichloro-2-butanone
(f) 2-methyl-3-hexanone

5.22* Give systematic names for these aldehydes:

(a) $CH_3CH_2\underset{\underset{CH_3}{|}}{C}HCHO$ (b) $CH_3CH_2CH_2\overset{\overset{CHO}{|}}{C}HCH_3$

(c) $(CH_3)_3CCHO$

(d) 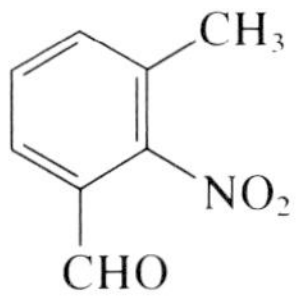

(e)
$$\begin{array}{c} CH_3\ \ O \\ |\ \ \ \ \ \| \\ CH_3CH_2CCH_2CH \\ | \\ CH_2CH_3 \end{array}$$

(f)
$$\begin{array}{c} Br\ \ \ O \\ |\ \ \ \ \| \\ CH_3CCH_2CH \\ | \\ Br \end{array}$$

5.23 Give systematic names for these ketones:

(a)
$$\begin{array}{c} O \\ \| \\ CH_3CCH_2CH_3 \end{array}$$

(b)
$$\begin{array}{c} O\ \ \ \ \ \ \ \ \ \ CH_3 \\ \|\ \ \ \ \ \ \ \ \ \ \ \ | \\ CH_3CCH_2CH_2CHCH_3 \end{array}$$

(c)
$$\begin{array}{c} O \\ \| \\ (CH_3)_3CCC(CH_3)_3 \end{array}$$

(d) 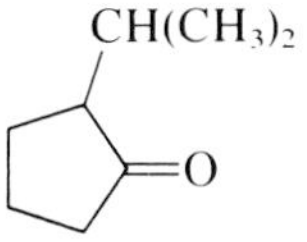

5.24* The following names are incorrect. What is wrong with each?
(a) 1-butanone (b) 4-butanone (c) 3-butanone
(d) cyclohexanal (e) 2-butanal

Reactions of Aldehydes and Ketones

5.25 What kind of compound is produced when an aldehyde reacts with an alcohol in a 1:1 ratio?

5.26* What kind of compound is produced when an aldehyde reacts with an alcohol in a 1:2 ratio in the presence of an acid catalyst?

5.27 Why does reduction of an aldehyde give a primary alcohol, whereas reduction of a ketone gives a secondary alcohol?

5.28* Give specific examples of these reactions:
(a) reduction of a ketone with $NaBH_4$
(b) oxidation of an aldehyde with Tollens' reagent
(c) formation of an acetal from an aldehyde and an alcohol

5.29 Which of the following compounds react with Tollens' reagent? Draw structures of the reaction products.
(a) cyclopentanone (b) hexanal

(c)
$$\begin{array}{c} CHO \\ | \\ CH_3CH_2CH_2CHCH_2CH_3 \end{array}$$

(d) 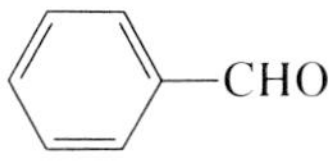

(e)
$$\begin{array}{c} O \\ \| \\ CH_3CH_2CCH_3 \end{array}$$

(f)
$$\begin{array}{c} O \\ \| \\ Cl_2CHCH \end{array}$$

5.30* Draw structures of the products obtained by reaction of the compounds in Problem 5.29 with $NaBH_4$ followed by acid treatment.

5.31 What is the difference between an acetal and a hemiacetal?

5.32* Assume that you are given two unlabeled bottles, one containing pentanal and the other containing 2-pentanone. What simple chemical test would allow you to distinguish between the contents of the two bottles?

5.33 Draw structures of the aldehydes that might be oxidized to yield these carboxylic acids:

(a) 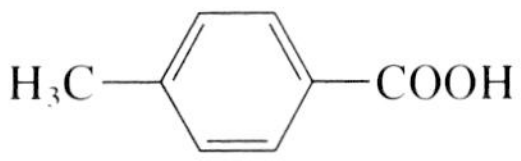

(b)
$$\begin{array}{c} COOH\ \ CH_3 \\ |\ \ \ \ \ \ \ \ \ \ \ | \\ CH_3CH_2CHCH_2CHCH_3 \end{array}$$

(c) $CH_3CH{=}CHCOOH$

(d) 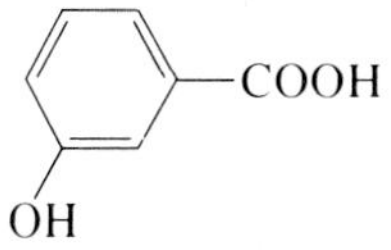

5.34* Draw structures of the primary alcohols that might be oxidized to yield the carboxylic acids shown in Problem 5.33.

5.35 What ketones or aldehydes might be reduced to yield these alcohols?

(a)
$$\begin{array}{c} CH_2OH \\ | \\ CH_3CH_2CH_2CHCH_3 \end{array}$$

(b) 2,2-dimethyl-1-hexanol

(c) 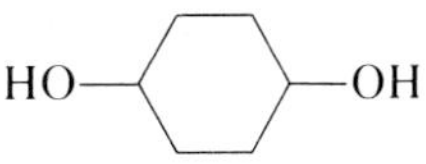

(d)
$$\begin{array}{c} OH \\ | \\ CH_3CH_2CHCHCH_3 \\ | \\ CH_3 \end{array}$$

5.36 Write the structures of the hemiacetals and acetals that result from the following reactions:
(a) 2-butanone + 1-propanol
(b) butanal + isopropyl alcohol
(c) acetone + ethanol
(d) hexanal + 2-butanol

5.37 Acetals are usually made by reaction of an aldehyde or ketone with two molecules of a monoalcohol. If an aldehyde or ketone reacts with *one* molecule of a dialcohol, however, a *cyclic* acetal results. Draw the structure of the cyclic acetal formed in the following reaction:

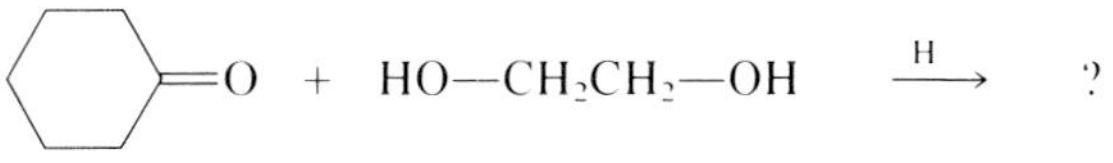

5.38 What products result from hydrolysis of these acetals?

(a)
$$\begin{array}{c} O{-}CH_2CH_3 \\ | \\ CH_3CH_2CH_2CH{-}O{-}CH_3 \end{array}$$

(b) 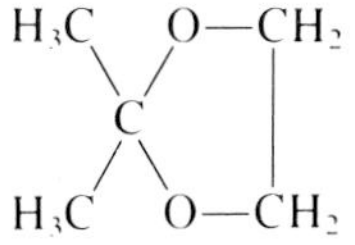

(c)

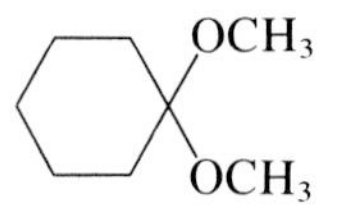

(d)

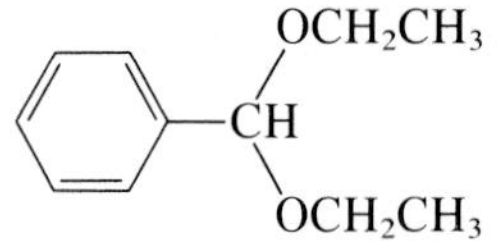

5.39 Cyclic hemiacetals sometimes form if an alcohol group in one part of a molecule adds to a carbonyl group elsewhere in the same molecule. What is the structure of the open-chain hydroxy aldehyde from which the following hemiacetal might form?

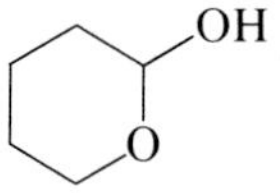

A cyclic hemiacetal

5.40* Like the cyclic hemiacetal in Problem 5.39, cyclic acetals are also known. What products would you expect on hydrolysis of the following cyclic acetal?

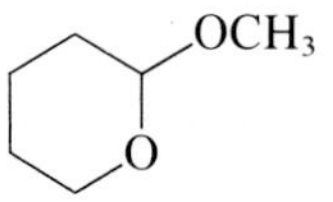

A cyclic acetal

5.41 Glucose exists largely in the cyclic hemiacetal form shown. Draw the structure of glucose in its open-chain hydroxy aldehyde form.

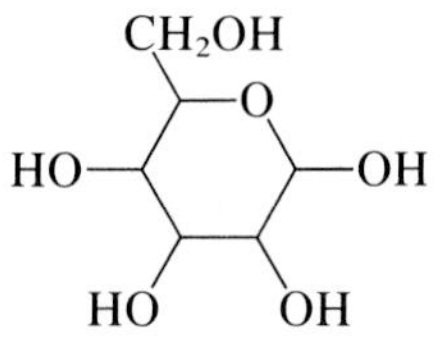

Glucose

5.42* In many respects, glucose acts as if it were an aldehyde, even though the structure shown in Problem 5.41 contains no —CHO group. For example, glucose reacts with Tollens' reagent to produce a silver mirror. Explain.

5.43 What products result from hydrolysis of the following cyclic acetal?

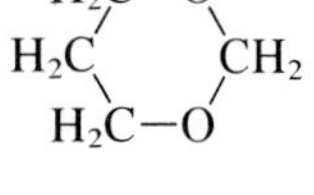

5.44 Aldosterone is a key steroid involved in controlling the sodium-potassium salt balance in the body. Identify the carbonyl groups in aldosterone.

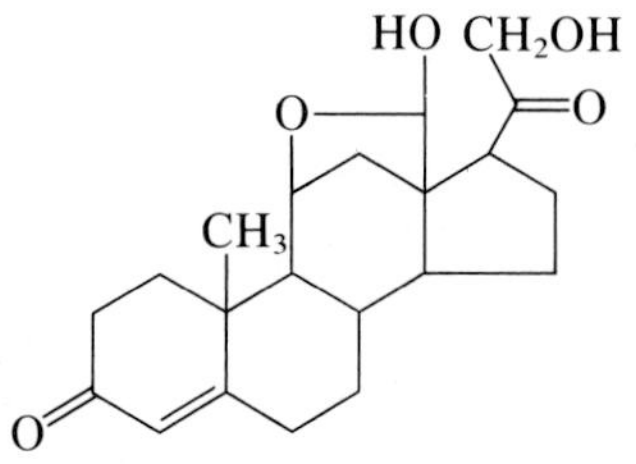

5.45 Aldosterone (Problem 5.44) also contains a cyclic hemiacetal linkage. Identify the linkage and tell whether it's derived from an aldehyde or a ketone.

5.46* 3-Cyclohexenone is an example of a *difunctional* molecule, a compound that has two different functional groups. What products would you expect from treatment of 3-cyclohexenone with the reagents listed?

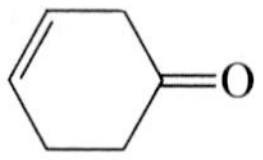

3-Cyclohexenone

(a) H_2 and a Pd catalyst (b) $NaBH_4$, then H_3O^+
(c) Br_2

5.47 What structural requirement must be met in order for an aldehyde or ketone to undergo an aldol reaction?

5.48 Write the structures of the aldol products that result from base treatment of these ketones or aldehydes.
(a) 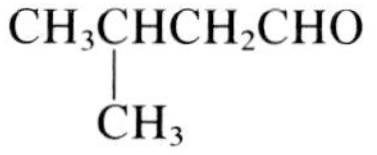 CH_3CHCH_2CHO (b) cyclohexanone
           |
          CH_3

(c) 3-pentanone   (d) 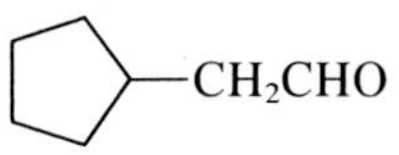 —CH_2CHO

5.49 When 2-butanone is treated with base, a mixture of two different aldol products results. What are their structures?

5.50* The aldol reaction is reversible; that is, aldol products can sometimes break apart to yield aldehydes or ketones. What products would result if the following hydroxy aldehyde broke apart in a reverse aldol reaction?

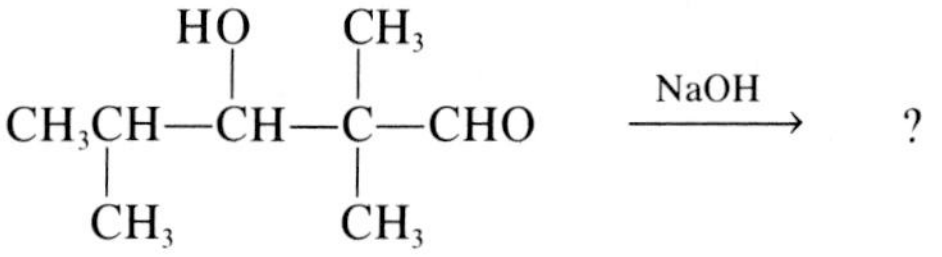

5.51 The following compound can be made by a mixed aldol reaction between two different carbonyl compounds. What two aldehyde or ketone starting materials would you use?

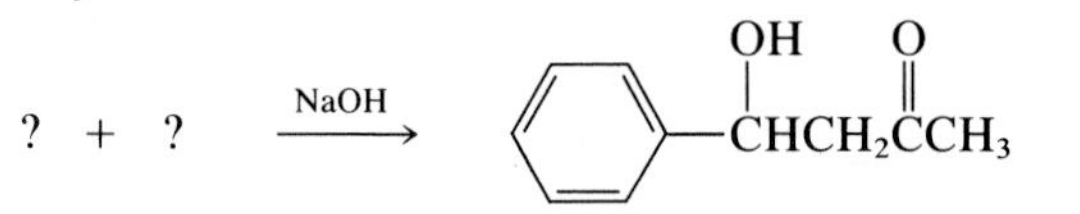

Applications

5.52 Discuss the differences among *acute, subacute,* and *chronic* toxicities of a drug. [App: Is It Poisonous?]

5.53 The LD_{50} value for the acute toxicity of botulism toxin, which is associated with improperly canned foods, is less than 0.01 mg/kg. This is less than 1 drop for an average adult. If 1 drop of botulism toxin is ingested, does this guarantee death? Discuss the significance of an LD_{50} rating. [App: Is It Poisonous?]

5.54 While many substances, such as ethanol, are only moderately toxic to adults, they can pose special dangers to embryos and fetuses. Why do you think this special susceptibility occurs? [App: Is It Poisonous?]

5.55 In a metabolism pathway in the body, oxaloacetic acid reacts with acetyl coenzyme A (acetyl SCoA) in the presence of an enzyme and water to form citric acid and free coenzyme A.

$$
\underset{\substack{\text{Oxaloacetic}\\\text{acid}}}{HOCCH_2-\overset{O}{\overset{\|}{C}}-\overset{O}{\overset{\|}{C}}-OH} \;+\; \underset{\text{Acetyl SCoA}}{CH_3CSCoA} \longrightarrow
$$

$$
\underset{\substack{\text{Citric acid}}}{HOCCH_2-\overset{OH}{\underset{\underset{O}{\overset{\|}{CH_2C-OH}}}{\overset{|}{C}}}-\overset{O}{\overset{\|}{C}}OH} \;+\; \underset{\text{Coenzyme A}}{CoASH}
$$

How is this reaction similar to an aldol reaction? [App: Biological Aldol Reaction]

5.56* Both HCN and benzaldehyde have the aroma of almonds. If a liquid had a faint almond odor, how would you test it to determine whether the aldehyde or the acid was present? [Int: Chemical Warfare]

5.57 Old recipes for peach jam suggested grinding up the peach pits and adding them to the jam for more flavor and a crunchy texture. Why is this not a good idea? [Int: Chemical Warfare]

5.58* HCN is quite toxic: How do you suppose that a millepede can use this weapon without killing himself? [Int: Chemical Warfare]

Additional Questions and Problems

5.59 Name vanillin (Practice Problem 5.1) systematically.

5.60* Can the following alcohol be formed by the reduction of an aldehyde or ketone? Why or why not?

$$(CH_3)_3COH$$

5.61 Many flavorings and perfumes are partially based on fragrant aldehydes and ketones. Why do you think the portion of the odor due to the ketone is more stable than that due to the aldehyde?

5.62 One problem with burning some plastics is the release of formaldehyde. What are some of the physiological effects of exposure to formaldehyde?

5.63 Chloral hydrate, a potent sedative and component in "knockout" drops, is formed by reacting 2,2,2-trichloroethanal with water in a reaction analogous to hemiacetal formation. Draw the formula of chloral hydrate.

5.64* Name these compounds.

(a) $CH_3CH_2\overset{O}{\overset{\|}{C}}CH(CH_3)_2$

(b) $CH_2=CHCHCH_2CH=CH_2$

(c)

(d) $(CH_3)_3C\overset{O}{\overset{\|}{C}}\underset{\underset{CH_3}{|}}{CH}CH_2CH_3$

(e)

(f) $CH_3CH_2C\equiv CC(CH_2CH_3)_3$

5.65 Draw the structural formulas of these compounds.
(a) *o*-nitrobenzaldehyde
(b) 2,3-dichlorocyclopentanone
(c) methyl *n*-butyl ether   (d) 2,4-dimethyl-3-pentanol
(e) 2,3,3-triiodobutanal   (f) 1,1,3-tribromoacetone

5.66 Complete these equations.

(a) $CH_3CH=C(CH_3)_2 + H_2 \xrightarrow{Pd}$

(b) $CH_3CH_2\underset{\underset{CH_2CH_3}{|}}{\overset{\overset{CH_2OH}{|}}{C}}CH_2CH_3 \xrightarrow{[O]}$

(c) $CH_3\overset{O}{\overset{\|}{C}}CH_2CH_2CH_3 \xrightarrow[H_3O^+]{NaBH_4}$

(d)
$$\text{(benzene)}CH_2\overset{O}{\overset{\|}{C}}H \;+$$
$$HOCH_2CH_2CH_3 \longrightarrow \quad (hemiacetal)$$

(e)
$$\text{(benzene)}CH_2\overset{O}{\overset{\|}{C}}H \;+$$
$$2\;HOCH_2CH_2CH_3 \longrightarrow \quad (acetal)$$

(f) $CH_3CH=\underset{\underset{CH_2CH_3}{|}}{\overset{\overset{CH_2CH_3}{|}}{C}}CH_2CH_2CH_3 \;+\; HCl \longrightarrow$

(g) $CH_3\text{—(benzene)—}CH_2CH_2OH \xrightarrow{H_2SO_4}$

5.67 How could you differentiate between 3-hexanol and hexanal using a simple chemical test?

CHAPTER

6

Carboxylic Acids and Their Derivatives

Every now and then, we should all take a moment to enjoy the aroma of some carboxylic acid esters, one of the kinds of compounds described in this chapter.

We said in the last chapter that carbonyl compounds can be classified into two groups, based on their structural and chemical similarities. In one group are aldehydes and ketones, and in the other group are carboxylic acids and their derivatives: esters, anhydrides, and amides.

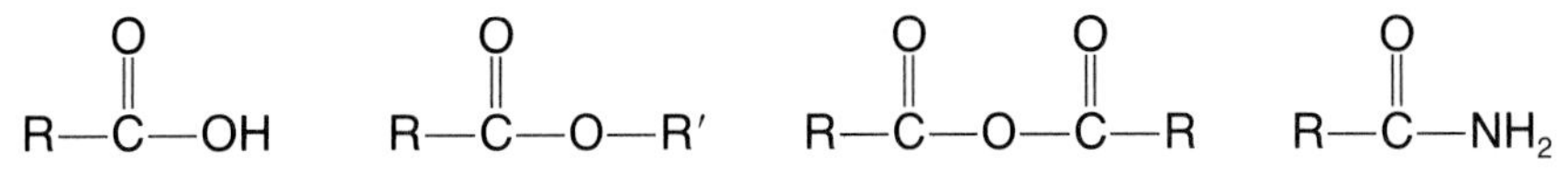

$$\underset{\text{Carboxylic acid}}{R-\overset{\overset{\displaystyle O}{\|}}{C}-OH} \qquad \underset{\text{Ester}}{R-\overset{\overset{\displaystyle O}{\|}}{C}-O-R'} \qquad \underset{\text{Acid anhydride}}{R-\overset{\overset{\displaystyle O}{\|}}{C}-O-\overset{\overset{\displaystyle O}{\|}}{C}-R} \qquad \underset{\text{Amide}}{R-\overset{\overset{\displaystyle O}{\|}}{C}-NH_2}$$

We'll look at the chemistry of this second group of carbonyl compounds in this chapter and answer the following questions:

1. **What are the general properties of carboxylic acids and their derivatives?** The goal: Be able to describe and compare such properties as hydrogen bonding, water solubility, boiling point, and acidity or basicity.

2. **How are carboxylic acids, anhydrides, esters, and amides named?** The goal: Be able to name the simple members of these families and write their structures, given the names.

3. **What are some of the significant applications and occurrences of carboxylic acids, anhydrides, esters, and amides?** The goal: Be able to identify major applications and occurrences of each family of compounds and describe some important members of each family.

4. **How are esters and amides synthesized from carboxylic acids and converted back to carboxylic acids?** The goal: Be able to describe and predict the products of the ester- and amide- forming reactions of carboxylic acids and the hydrolysis of esters and amides.

5. **What is the Claisen condensation reaction, and why is it important in biochemistry?** The goal: Be able to describe how the Claisen condensation reaction occurs, predict products of the reaction, and explain why it is important in biochemistry.

6. **What are phosphate esters and anhydrides, and why are they important?** The goal: Be able to recognize and write the structures of phosphate esters and anhydrides and explain why they are important in biochemistry.

6.1 PROPERTIES OF CARBOXYLIC ACIDS AND THEIR DERIVATIVES

Carboxylic acid A compound that has a carbonyl group bonded to one carbon substituent and one —OH group, RCOOH.

Unlike the aldehydes and ketones described in the last chapter, **carboxylic acids, acid anhydrides, esters,** and **amides** all have their carbonyl groups bonded to an atom (O or N) that strongly attracts electrons. Thus, the carbonyl groups are more strongly polarized in these families of compounds than in ketones and aldehydes.

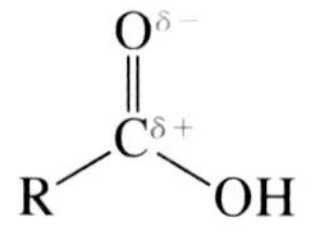

Acid anhydride A compound that has a carbonyl group bonded to a carbon substituent and an —OCOR group, RCO$_2$COR.

Ester A compound that has a carbonyl group bonded to one carbon atom and one —OR group, RCOOR′.

Amide A compound that has a carbonyl group bonded to one carbon atom and one nitrogen atom group, RCONR′$_2$, where R′ may be an H atom.

Carboxylic acid
(RCOOH or RCO$_2$H)

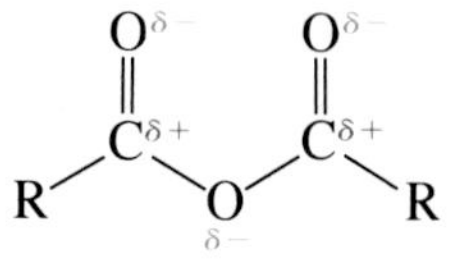

Acid anhydride
(RCO$_2$COR)

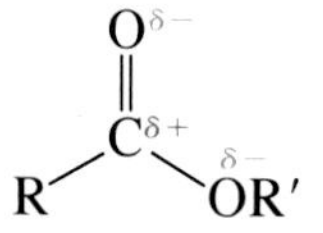

Ester
(RCOOR′ or RCO$_2$R′)

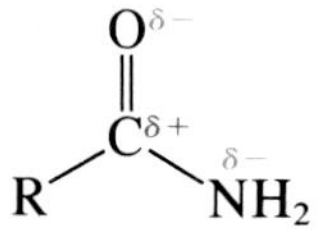

Amide
(RCONH$_2$)

The structural similarity of these families also leads to similarities in their chemistry. All undergo **carbonyl-group substitution reactions,** in which a group we can represent as —**X** replaces (substitutes for) the —OH, —OCOR, —OR, or —NH$_2$ group of the starting material.

A carbonyl-group substitution reaction

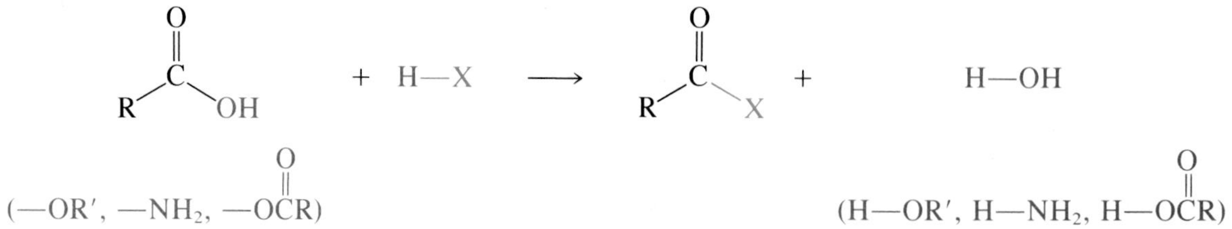

$$ \underset{R}{\underset{\text{(—OR′, —NH}_2\text{, —OCR)}}{\text{C—OH}}} \ + \ \text{H—X} \ \longrightarrow \ \underset{R}{\text{C—X}} \ + \ \underset{\text{(H—OR′, H—NH}_2\text{, H—OCR)}}{\text{H—OH}} $$

Carbonyl-group substitution reaction A reaction in which a new group X replaces (substitutes for) a group Y attached to a carbonyl-group carbon.

We'll give numerous examples of these carbonyl-group substitution reactions in the sections that follow. Also, we'll show that amides and esters can easily be either made from acids or converted back to acids. Acid anhydrides are the most reactive of these carboxylic acid derivatives, and you'll see that they cannot be used in aqueous solution because of their substitution reaction with water (Section 6.11).

Dimer A unit formed by the joining of two identical molecules.

Since carboxylic acids and their derivatives all contain polar functional groups, all are higher boiling than comparable alkanes. Like alcohols, carboxylic acids can form hydrogen bonds with each other. Molecules pair off to form **dimers** (Figure 6.1), and even formic acid (HCOOH), the lightest and simplest carboxylic acid, is a liquid at room temperature with a boiling point of 101°C.

Figure 6.1
Acetic acid hydrogen-bonded dimer. The ability to form hydrogen bonds makes simple carboxylic acids high boiling and soluble in water.

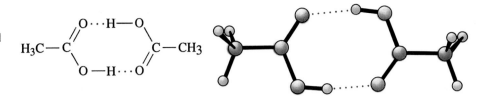

Acids with saturated straight-chain R groups containing up to nine carbon atoms are volatile liquids with strong, sharp odors, and those with up to four carbons are water-soluble. Higher saturated acids are waxy, odorless solids, and water solubility falls off as the size of the alkane-like portion increases relative to the size of the water-soluble —COOH portion.

Carboxyl group The —COOH functional group.

When the —OH of the **carboxyl group** (—COOH) is converted to the —OR of an ester, the ability of the molecules to hydrogen-bond with each other is lost. Simple esters are therefore lower boiling than the acids from which they are derived and are also lower boiling than comparable alcohols.

$$\underset{\text{Acetic acid, bp 118°C}}{CH_3\overset{\overset{\displaystyle O}{\|}}{C}OH} \qquad \underset{\text{Methyl ester, bp 57°C}}{CH_3\overset{\overset{\displaystyle O}{\|}}{C}OCH_3} \qquad \underset{\text{Ethyl ester, bp 77°C}}{CH_3\overset{\overset{\displaystyle O}{\|}}{C}OCH_2CH_3}$$

The simple esters are colorless, volatile liquids with pleasant odors, and many of them are responsible for the natural fragrance of flowers and ripe fruits. The lower esters have some solubility in water. Esters are neither acids nor bases in aqueous solution.

Replacement of the —OH group of an acid by an —NH₂ group produces an *unsubstituted amide (RCONH₂)*, that is, an amide with two hydrogen atoms on the nitrogen atom. Such amides can form three strong hydrogen bonds to other amide molecules and are thus higher melting and higher boiling than the acids from which they are derived. For example,

$$\underset{\text{Propanoic acid, mp}-21°C,\text{ bp 141°C}}{CH_3CH_2\overset{\overset{\displaystyle O}{\|}}{C}OH} \qquad \underset{\text{Propanamide, mp 81°C, bp 213°C}}{CH_3CH_2\overset{\overset{\displaystyle O}{\|}}{C}NH_2}$$

In fact, except for the simplest one (formamide, HCONH₂), all low-molecular-weight unsubstituted amides are relatively low-melting solids that are soluble in both water and organic solvents.

Note the distinction between amines, described in Chapter 4, and amides. The nitrogen atom is bonded to a carbonyl-group carbon atom in an amide, but not in an amine.

An amide
(RCONH₂)

An amine
(RNH₂)

As a result, amides are not basic like amines, mainly because the electron-withdrawing ability of the carbonyl group causes the unshared pair of electrons on nitrogen to be held tightly and prevents it from bonding to hydrogen.

The common carboxylic acids share the concentration-dependent corrosive properties of all acids but are not generally considered hazardous to human health. The vapors of volatile esters can be irritating and have a narcotic effect, while individual esters vary widely in their acute toxic effects. Amides can also be irritating to the skin, but the simple amides are not significant health hazards. Because anhydrides form acids on contact with water or water vapor, their corrosive and toxic properties are those of their parent acids.

Properties of Carboxylic Acid Derivatives

- All undergo carbonyl-group substitution reactions.
- Acid anhydrides react with water to form acids.
- Esters and amides are made from acids and easily converted back to acids.
- Hydrogen bonding is strong in pure acids and unsubstituted or monosubstituted amides; none in esters.
- Simpler acids and esters are liquids; all unsubstituted amides except formamide are solids.
- Acids give acidic aqueous solutions; esters and amides do not change pH.
- Volatile acids have strong odors; volatile esters have pleasant, fruity odors.

Practice Problems

6.1 Which of the following compounds has the highest boiling point and which has the lowest boiling point?
(a) CH_3OCH_3 (b) CH_3COOH (c) CH_3CH_2OH

6.2 Which of the following compounds has the highest melting and boiling points? Which has the lowest boiling point? Why?

$$\text{(a) } CH_3\overset{\overset{\textstyle O}{\|}}{C}NH_2 \qquad \text{(b) } CH_3\overset{\overset{\textstyle O}{\|}}{C}NHCH_2CH_3 \qquad \text{(c) } CH_3\overset{\overset{\textstyle O}{\|}}{C}N(CH_2CH_3)_2$$

6.3 In each of the following pairs of compounds, which would you expect to be more soluble in water?
(a) $CH_3CH_2CONH_2$ or $CH_3COOCH_2CH_3$
(b) $C_8H_{17}COOH$ or $CH_3CH_2CH_2COOH$
(c) $CH_3CHCOOH$ or $CH_3CH_2COOCHCH_3$
        | |
        CH_3 CH_3

6.2 NAMING CARBOXYLIC ACIDS AND THEIR DERIVATIVES

Carboxylic Acids   Carboxylic acids (RCOOH) are named systematically by replacing the final -*e* of the corresponding alkane name with -*oic acid*. The three-carbon acid is propanoic acid; the straight-chain four- carbon acid is butanoic

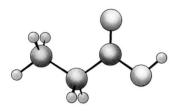

Propanoic acid

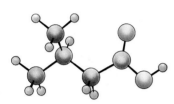

3-Methylbutanoic acid

acid; and so on. If alkyl substituents or other functional groups are present, the chain is numbered beginning at the —COOH end, as in 3-methylbutanoic acid.

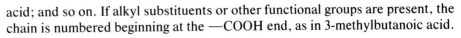

Propanoic acid   3-Methylbutanoic acid

Because many simple carboxylic acids were among the first organic compounds to be isolated and purified, a number of them also have common names. For example, the one-carbon compound (HCOOH) is always called *formic acid* rather than methanoic acid; the two-carbon compound (CH₃COOH) is always called *acetic acid* rather than ethanoic acid; and the three-carbon compound is sometimes called *propionic acid* rather than propanoic acid. The names and physical properties of some common carboxylic acids are given in Table 6.1.

Dicarboxylic acids, which contain two —COOH groups, are named systematically by adding the ending *-dioic* to the alkane name (the *-e* is retained), but many are usually referred to by their common names. Oxalic acid, for example, is found in plants of the genus *Oxalis,* which includes rhubarb and

Table 6.1 Physical Properties of Some Simple Carboxylic Acids

Structure	Common Name	IUPAC Name	Melting Point (°C)	Boiling Point (°C)	Water Solubility (g/100 mL)
HCOOH	Formic	Methanoic	8	101	Miscible
CH₃COOH	Acetic	Ethanoic	17	118	Miscible
CH₃CH₂COOH	Propionic	Propanoic	−22	141	Miscible
CH₃CH₂CH₂COOH	Butyric	Butanoic	−4	163	Miscible
CH₃CH₂CH₂CH₂COOH	Valeric	Pentanoic	−34	185	4
CH₃(CH₂)₁₆COOH	Stearic	Octadecanoic	70	383	Insoluble
HOOCCOOH	Oxalic	Ethanedioic	190	Decomposes	10
HOOCCH₂COOH	Malonic	Propanedioic	135	Decomposes	154
HOOCCH₂CH₂COOH	Succinic	Butanedioic	188	Decomposes	8
HOOCCH₂CH₂CH₂COOH	Glutaric	Pentanedioic	98	Decomposes	64
H₂C=CHCOOH	Acrylic	Propenoic	13	141	Miscible
CH₃CH=CHCOOH	Crotonic	2-Butenoic	72	185	55
(benzoic acid structure) COOH	Benzoic	Benzoic	122	249	0.3
(salicylic acid structure) COOH, OH	Salicylic	*o*-Hydroxybenzoic	159	Decomposes	0.2

spinach, and is rarely called by any other name (Figure 6.2). Malonic acid is an intermediate in the synthesis of pharmaceuticals.

$$
\underset{\substack{\text{Oxalic acid}\\ \text{(ethanedioic acid)}}}{\overset{\displaystyle \underset{\text{HOC}-\text{COH}}{\overset{\text{O} \quad \text{O}}{\|\quad\|}}}{}}
\qquad
\underset{\substack{\text{Malonic acid}\\ \text{(propanedioic acid)}}}{\overset{\displaystyle \underset{\text{HOC}-\text{CH}_2-\text{COH}}{\overset{\text{O} \qquad \text{O}}{\|\qquad\|}}}{}}
$$

Figure 6.2
Oxalic acid. The rhubarb leaves contain oxalic acid which, when pure, is a white crystalline solid. Oxalate ion is added to blood samples to prevent clotting because it forms a precipitate with Ca^{2+} needed for the clotting process.

Many other varieties of compounds containing —COOH groups are also common. There are unsaturated acids, which are named systematically with the ending *-enoic,* and there are acids containing other functional groups such as hydroxy acids and keto acids, several of which are important biomolecules.

$$
\underset{\substack{\text{Acrylic acid}\\ \text{(propenoic acid)}}}{\overset{\displaystyle \underset{\text{H}_2\text{C}=\text{CHCOH}}{\overset{\text{O}}{\|}}}{}}
\qquad
\underset{\substack{\text{Lactic acid}\\ \text{(2-hydroxypropanoic acid)}}}{\overset{\displaystyle \underset{\underset{\text{OH}}{|}}{\underset{\text{CH}_3\text{CHCOH}}{}}\overset{\text{O}}{\|}}{}}
\qquad
\underset{\substack{\text{Pyruvic acid}\\ \text{(2-oxopropanoic acid)}}}{\overset{\displaystyle \underset{\text{CH}_3-\text{C}-\text{COH}}{\overset{\text{O} \quad \text{O}}{\|\quad\|}}}{}}
$$

The carbon atoms *next to* the —COOH group are sometimes identified by Greek letters, beginning with α. In this system, pyruvic acid would be referred to as an α-keto acid.

$$
\underset{\substack{\text{Pyruvic acid}\\ \text{(an }\alpha\text{-keto acid)}}}{\overset{\displaystyle \underset{\underset{\beta\quad\ \alpha}{}}{\underset{\text{CH}_3-\text{C}-\text{COH}}{}}\overset{\text{O} \quad \text{O}}{\|\quad\|}}{}}
\qquad
\underset{\substack{\text{Alanine}\\ \text{(an }\alpha\text{-amino acid)}}}{\overset{\displaystyle \underset{\underset{\text{NH}_2}{|}}{\underset{\overset{\beta\qquad\ \alpha}{\text{CH}_3-\text{CH}-\text{COH}}}{}}\overset{\text{O}}{\|}}{}}
\qquad
\begin{array}{l}\text{In alanine,}\\ \text{as in all common amino acids,}\\ \text{the }-\text{NH}_2\text{ group is on the}\\ \text{alpha carbon atom}\end{array}
$$

Solved Problem 6.1   What is the IUPAC name for this acid?

$$
\underset{\substack{\ |\quad\ |\\ \text{HO}\ \ \text{CH}_3}}{\text{CH}_3\text{CHCHCOOH}}
$$

Solution   First, identify the longest chain containing the —COOH group and number it starting with the carboxyl-group carbon atom:

$$
\overset{\displaystyle 4\quad 3\ \ 2\ \ 1}{\underset{\substack{\ \ |\quad\ |\\ \ \ \text{HO}\ \ \text{CH}_3}}{\text{CH}_3\text{CHCHCOOH}}}
$$

The parent compound is the four-carbon acid, butanoic acid. It has a methyl group on position 2 and a hydroxy group on position 3, giving the name 3-hydroxy-2-methylbutanoic acid.

Acid Anhydrides   The word "anhydride" means "without water," and acid anhydrides have the structure that results from loss of a molecule of water by two acid molecules (whether or not the anhydride is made that way). Anhydride names are derived by replacing *acid* in the name of the original acid with *anhydride*.

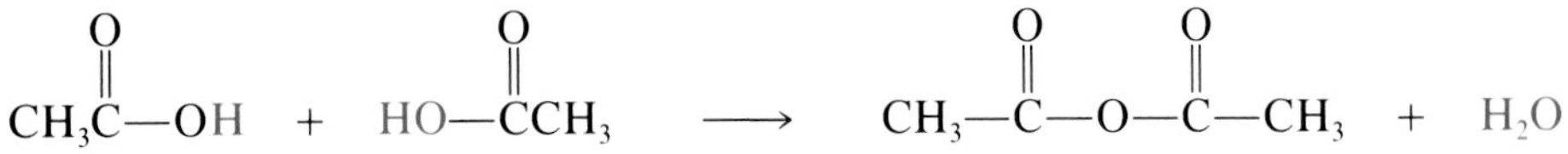

$$\underset{\text{2 Acetic acid}}{CH_3\overset{\overset{\displaystyle O}{\|}}{C}-OH \;+\; HO-\overset{\overset{\displaystyle O}{\|}}{C}CH_3} \longrightarrow \underset{\text{Acetic anhydride}}{CH_3-\overset{\overset{\displaystyle O}{\|}}{C}-O-\overset{\overset{\displaystyle O}{\|}}{C}-CH_3} \;+\; H_2O$$

Esters   Esters (RCOOR′) are named by first specifying the alkyl group attached to the ether-like oxygen (—O—R′) and then identifying the related carboxylic acid. The family-name ending *-ic acid* is replaced by *-ate*.

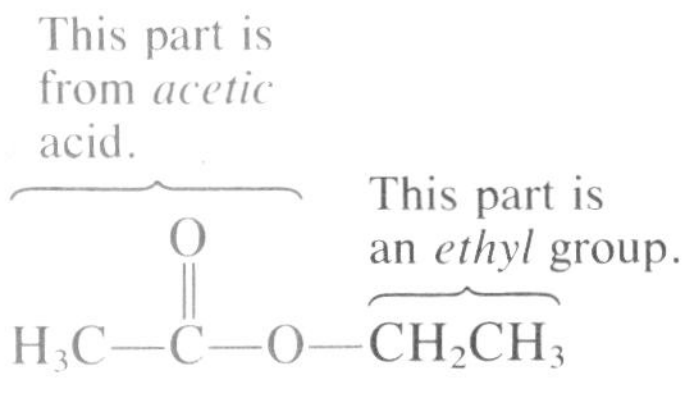

This part is from *acetic* acid. → $H_3C-\overset{\overset{\displaystyle O}{\|}}{C}-O-CH_2CH_3$ ← This part is an *ethyl* group.

Ethyl acetate

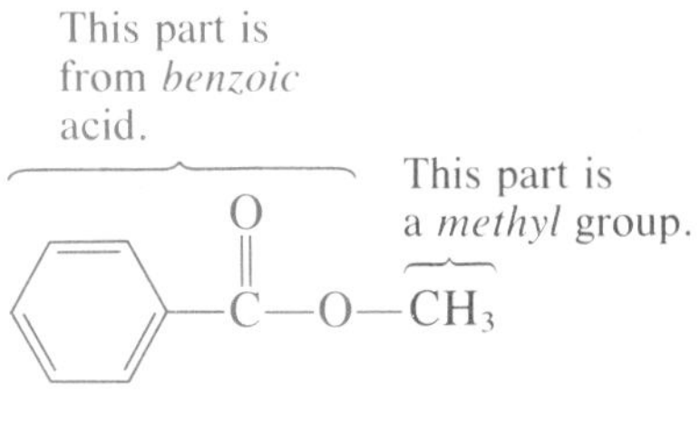

This part is from *benzoic* acid. → $\langle\!\!\!\bigcirc\!\!\!\rangle-\overset{\overset{\displaystyle O}{\|}}{C}-O-CH_3$ ← This part is a *methyl* group.

Methyl benzoate

Ester names always consist of two words—the name of the alkyl group in —COOR followed by the name of the acid with an *-ate* ending. Both common and systematic names are derived in this manner. An ester of the straight-chain, four-carbon carboxylic acid, for example, can be named as a butanoate or a butyrate:

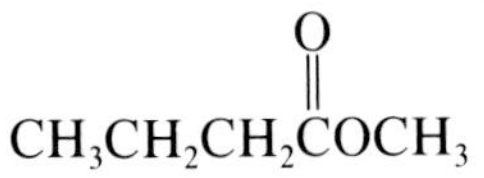

$$CH_3CH_2CH_2\overset{\overset{\displaystyle O}{\|}}{C}OCH_3$$

An ester used as a food flavoring to give the taste of apples

Methyl butyrate
(Methyl butanoate)

The names and properties of some common esters are given in Table 6.2.

Solved Problem 6.2   What is the structure of butyl acetate?

Solution   The two-word name consisting of an alkyl group name followed by an acid name with an *-ate* ending shows that the compound is an ester. The name "acetate" shows that the RCOO part of the molecule is from acetic acid (CH₃COOH). A butyl group is attached to the ether-like oxygen atom in the carboxyl group. Therefore the structure is

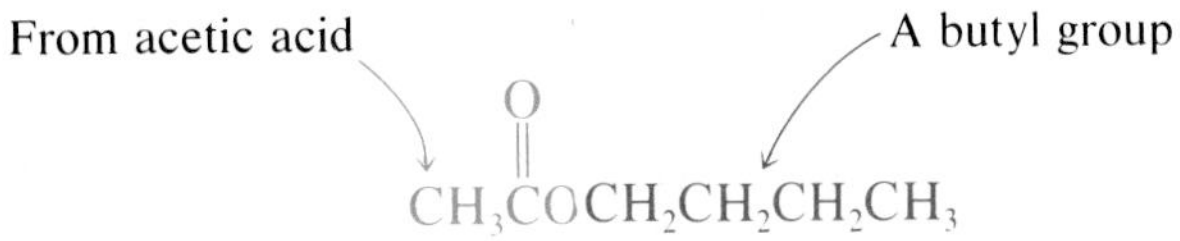

From acetic acid ↘   ↙ A butyl group

$$CH_3\overset{\overset{\displaystyle O}{\|}}{C}OCH_2CH_2CH_2CH_3$$

Table 6.2 Physical Properties of Some Simple Esters[a]

Structure	Name	Boiling Point (°C)	Water Solubility (20°C) (g/100 mL)
$HCOOCH_3$	Methyl formate	32	30
$CH_3COOCH_2CH_3$	Ethyl acetate	77	9
$CH_3COOCH_2(CH_2)_3CH_3$	Pentyl acetate	149	Slightly soluble
CH_3COO—⬡ 	Phenyl acetate	196	Insoluble
 ⬡—$COOCH_2CH_3$	Ethyl benzoate	212	Slightly soluble
$CH_3(CH_2)_{16}COOCH_3$	Methyl stearate	mp 39	Insoluble
$CH_3OOCCH_2CH_2COOCH_3$	Dimethyl succinate	mp 19.5	Slightly soluble

[a] All but methyl stearate and dimethyl succinate are liquids.

Solved Problem 17.3 What is the name of the compound shown?

$$\overset{\overset{\displaystyle O}{\|}}{H}COCH_2CH_2CH_3$$

Solution The compound has the general formula RCOOR′, so it is an ester in which R = H, showing that it is derived from formic acid, HCOOH. The R′ group has three carbon atoms and is therefore a propyl group. The compound is propyl formate.

Amides Amides ($RCONH_2$) with an unsubstituted —NH_2 group are named by replacing the *-oic acid* of the corresponding carboxylic acid name with *-amide*. For example, the two-carbon amide derived from acetic acid is called *acetamide*. If the nitrogen atom of the amide has alkyl substituents on it, the compound is named by first specifying the alkyl group and then identifying the amide name. The alkyl substituents are preceded by the letter *N* to identify them as being attached directly to nitrogen. Some common amides are illustrated in Table 6.3.

Table 6.3 Physical Properties of Some Simple Amides[a]

Structure	Name	Melting Point (°C)	Water Solubility (g/100 mL)
$HCONH_2$	Formamide	3	Miscible
CH_3CONH_2	Acetamide	81	70
$CH_3CONHCH_3$	N-Methylacetamide	31	Soluble
$CH_3CON(CH_3)_2$	N,N-Dimethylacetamide	−20	Miscible
$CH_3CH_2CONH_2$	Propanamide	81	Soluble
$CH_3CH_2CH_2CONH_2$	Butanamide	116	16
 ⬡—$CONH_2$	Benzamide	127	1.3

[a] All but formamide and N,N-dimethylacetamide are solids.

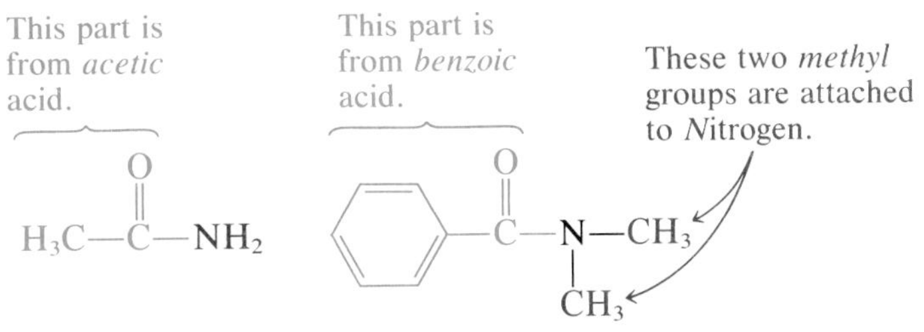

This part is from *acetic* acid.

This part is from *benzoic* acid.

These two *methyl* groups are attached to *Nitrogen.*

$$H_3C-\overset{\displaystyle O}{\overset{\displaystyle \|}{C}}-NH_2$$

Acetamide

N,N-Dimethylbenzamide

Practice Problems

6.4 What are the names of these compounds?

(a) $CH_3CHCH_2CH_2\overset{\displaystyle O}{\overset{\displaystyle \|}{C}}-OH$ with CH_3 branch

(b) $CH_3CH_2CH_2\overset{\displaystyle O}{\overset{\displaystyle \|}{C}}-O-CHCH_3$ with CH_3 branch

(c) 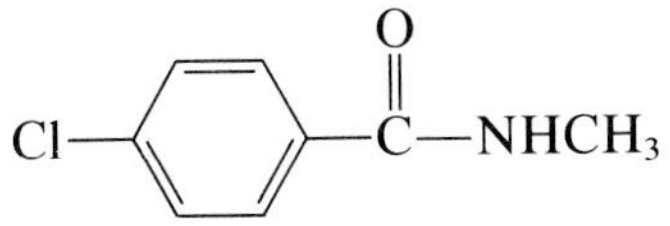 $Cl\!-\!\!\!\bigcirc\!\!\!-\overset{\displaystyle O}{\overset{\displaystyle \|}{C}}-NHCH_3$

6.5 Draw structures corresponding to these names:
(a) 3-methylhexanoic acid   (b) 4-methylpentanamide
(c) propyl benzoate   (d) *o*-nitrobenzoic acid
(e) *N*-methylbutanamide   (f) ethyl propanoate
(g) propanoic anhydride

6.3 SOME COMMON CARBOXYLIC ACIDS

Carboxylic acids occur widely throughout the plant and animal kingdoms. For example, butanoic acid (from the Latin, *butyrum,* "butter") is responsible for the odor of rancid butter, and cinnamic acid is partially responsible for the odor of cinnamon. Long-chain carboxylic acids such as stearic acid are components of all animal fats and vegetable oils, and many organic acids are formed in the body during the breakdown of foods.

Some naturally occurring carboxylic acids

$$CH_3CH_2CH_2-\overset{\displaystyle O}{\overset{\displaystyle \|}{C}}-OH$$

$$\bigcirc\!\!-CH\!=\!CH-\overset{\displaystyle O}{\overset{\displaystyle \|}{C}}-OH$$

Butanoic acid
(from butter)

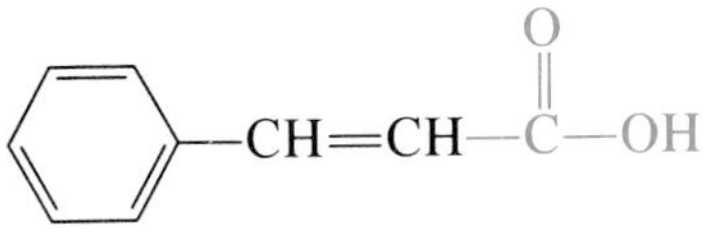

Cinnamic acid
(from oil of cinnamon)

$$CH_3CH_2CH_2CH_2CH_2CH_2CH_2CH_2CH_2CH_2CH_2CH_2CH_2CH_2CH_2CH_2CH_2-\overset{\displaystyle O}{\overset{\displaystyle \|}{C}}-OH$$

Stearic acid ($C_{18}H_{36}O_2$, from animal fat and vegetable oil)

$$H_3C-\overset{\displaystyle O}{\overset{\displaystyle \|}{C}}-OH \qquad\qquad HO\overset{\displaystyle O}{\overset{\displaystyle \|}{C}}CH_2\overset{\displaystyle OH}{\overset{\displaystyle |}{\underset{\displaystyle \underset{\displaystyle COOH}{|}}{C}}}CH_2\overset{\displaystyle O}{\overset{\displaystyle \|}{C}}OH$$

Acetic acid   Citric acid
(from vinegar)   (from citrus fruits)

Acetic Acid (CH₃COOH)   Everyone recognizes the sour taste of acetic acid, the most common carboxylic acid, because vinegar is a solution of 4–8% acetic acid in water (with various other flavoring agents). When fermentation of grapes, apples, and other fruits proceeds in the presence of ample oxygen, oxidation proceeds beyond the formation of ethanol to the formation of acetic acid (named from the Latin *acetum,* meaning "vinegar"). The acid in various concentrations in water solution is also a common laboratory reagent.

In concentrations over 50%, acetic acid is corrosive and can damage the skin, eyes, nose, and mouth. There is no pain when the concentrated acid is spilled on unbroken skin, but painful blisters form 30 min or more later. Pure acetic acid is known as *glacial* acetic acid because with just a slight amount of cooling the liquid forms icy-looking crystals at its melting point of 17°C.

Acetic acid is a reactant in many industrial processes and is sometimes used as a solvent, especially for other carboxylic acids. As a food additive, it is used to adjust pH. Earlier we noted that carboxylic acids often undergo carbonyl-group substitution reactions in which the —OH group is replaced by some other group, —X. The portion of acetic acid that bonds to X in any compound CH₃COX is known as an **acetyl group.**

Acetic acid

Acetyl group   The $CH_3\overset{\displaystyle O}{\overset{\displaystyle \|}{C}}-$ group.

$$CH_3\overset{\displaystyle O}{\overset{\displaystyle \|}{C}}-\xi \qquad\text{An acetyl group}$$

The acetyl group and its transfer from one molecule to another play a crucial role in biochemical reactions.

REVIEW   A buffer is a combination of substances, usually a weak acid and its anion, that act together to prevent a large change in the pH of a solution (Appendix B.10).

Citric Acid (A Tricarboxylic Hydroxy Acid)   Citric acid is produced by almost all plants and animals during metabolism, and its normal concentration in human blood is about 2 mg/100 mL. Citrus fruits owe their tartness to citric acid; for example, lemons contain 4–8% and oranges about 1% citric acid. Pure citric acid is a white, crystalline solid (mp 153°C) that is very soluble in water. Citric acid and its salts are extensively used in pharmaceuticals, foods, and cosmetics. It serves to buffer pH in shampoos and hair-setting lotions; it adds tartness to candies and soft drinks; and it is the acid that reacts with bicarbonate ion to produce the fizz in AlkaSeltzer.

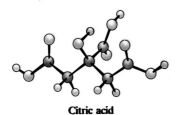

Citric acid

6.4   ACIDITY OF CARBOXYLIC ACIDS

Carboxylate anion   The anion that results from dissociation of a carboxylic acid, RCOO⁻.

Carboxylic acids are generally weak acids and establish equilibria in aqueous solution with **carboxylate anions,** RCOO⁻. Many carboxylic acids have about the same acid strength as acetic acid, as shown by the acid ionization constants in Table 6.4. There are some exceptions, though. Trichloracetic acid, used in preparing microscope slides and in precipitating protein in the analysis of body

Table 6.4   Carboxylic Acid Ionization Constants[a]

Structure	Name	K_a
CH_3COOH	Acetic acid	1.8×10^{-5}
$ClCH_2COOH$	Chloroacetic acid	1.4×10^{-3}
$HCOOH$	Formic acid	1.8×10^{-4}
CH_3CH_2COOH	Propanoic acid	1.3×10^{-5}
$HOOCCOOH$	Oxalic acid	5.4×10^{-2}
		5.2×10^{-5}
$HOOCCH_2CH_2COOH$	Succinic acid	6.4×10^{-5}
		2.3×10^{-6}
$H_2C{=}CHCOOH$	Acrylic acid	5.6×10^{-5}
⬡—COOH	Benzoic acid	6.5×10^{-5}

[a] The K_a is the equilibrium constant for the ionization of an acid; the smaller its value, the weaker the acid.

$$RCOOH + H_2O \; \rightleftharpoons \; RCOO^- + H_3O^+ \quad K_a = \frac{[RCOO^-]\,[H_3O^+]}{[RCOOH]}$$

REVIEW   A formula in square brackets represents concentration in moles per liter. For example, $[CH_3COOH] = 0.1$ indicates a solution with 0.1 mole of dissolved acetic acid per liter of solution. One mole is the mass in grams equal to the formula weight of the substance.

fluids, is a very strong acid that must be handled with the same respect as sulfuric acid.

Carboxylate ions are named by replacing the -*ic* ending in the carboxylic acid name with -*ate*. (The same names and endings are used in naming esters).

$$\underset{\text{Acetic acid}}{CH_3\overset{O}{\overset{\|}{C}}OH} \; + \; H_2O \; \rightleftharpoons \; \underset{\text{Acetate ion}}{CH_3\overset{O}{\overset{\|}{C}}O^-} \; + \; H_3O^+$$

$$\underset{\text{Propanoic acid}}{CH_3CH_2\overset{O}{\overset{\|}{C}}OH} \; + \; H_2O \; \rightleftharpoons \; \underset{\text{Propanoate ion}}{CH_3CH_2\overset{O}{\overset{\|}{C}}O^-} \; + \; H_3O^+$$

$$\underset{\text{Pyruvic acid}}{CH_3\overset{O}{\overset{\|}{C}}{-}\overset{O}{\overset{\|}{C}}OH} \; + \; H_2O \; \rightleftharpoons \; \underset{\text{Pyruvate ion}}{CH_3\overset{O}{\overset{\|}{C}}{-}\overset{O}{\overset{\|}{C}}O^-} \; + \; H_3O^+$$

Carboxylic acid salt   An ionic compound containing a carboxylate ion.

Carboxylic acids react completely with strong bases such as sodium hydroxide to give water and a **carboxylic acid salt.** Like any salt, a carboxylic acid salt is named with cation and anion names. For example,

$$\underset{\substack{\text{Acetic acid} \\ \text{(a weak acid)}}}{H_3C{-}\overset{O}{\overset{\|}{C}}{-}O{-}H(aq)} \; + \; \underset{\substack{\text{Sodium} \\ \text{hydroxide}}}{Na^+OH^-(aq)} \; \longrightarrow \; \underset{\text{Sodium acetate}}{H_3C{-}\overset{O}{\overset{\|}{C}}{-}O^- \, Na^+(aq)} \; + \; H_2O$$

Acetate ion, as the anion of a weak acid, is itself a base that can accept a proton from HCl or some other strong acid. Thus, carboxylate anions can be converted back into free carboxylic acids by reaction with acids.

$$H_3C-\overset{\overset{\displaystyle O}{\|}}{C}-O^- \ Na^+(aq) \ + \ HCl(aq) \ \longrightarrow \ H_3C-\overset{\overset{\displaystyle O}{\|}}{C}-O-H \ + \ Na^+ \ Cl^-(aq)$$

Sodium acetate   Acetic acid

The interconversion of a carboxylic acid and its carboxylate ion is easily reversible depending on the conditions. Which form is present depends on the pH of the medium and can be controlled by adjusting the pH. At low pH (acidic solution), carboxylic acid is present; at high pH (basic solution), carboxylate ion is present:

Acidic conditions: $RCOO^-$ + H_3O^+ $\longrightarrow$ $RCOOH$ + H_2O

(pH < 7)   Carboxylate ion   Acid   Carboxylic acid

Basic conditions: $RCOOH$ + OH^- $\longrightarrow$ $RCOO^-$ + H_2O

(pH > 7)   Carboxylic acid   Base   Carboxylate ion

AN APPLICATION: ACID SALTS AS FOOD ADDITIVES

Run your eye over the ingredients listed on the labels of soft drinks, cookies or cakes, dried sauce mixes, or preserved meats. Chances are you'll find the name of one or more acid salts.

Scanning the label on a package of strawberry jam-filled cookies, for example, turns up *sodium benzoate, sodium propionate, potassium sorbate,* and *sodium citrate.* What's the purpose of all these *food additives?* The first three are preservatives. Sodium benzoate prevents the growth of microorganisms, especially in acidic foods. It's present in many soft drinks, is one of the most common food additives, and has been used for over 70 years. Sodium propionate, also very common, prevents the growth of mold in baked goods. Potassium sorbate, the salt of an unsaturated acid that occurs naturally in many plants (sorbic acid, $CH_3CH=CHCH=CHCOOH$), is a good mold and fungus inhibitor. The fourth ingredient is the trisodium salt of citric acid. In the cookies, it's combined with citric acid to buffer the acidity of the strawberry jam.

A package of dehydrated cream sauce with bacon bits and noodles includes a group of less familiar acid salts. *Disodium guanylate, disodium inosinate,* and *monosodium glutamate* are salts of acids that occur naturally in meats. As food additives, these salts serve as flavor enhancers by imparting a "meaty" flavor and are sometimes found in packaged foods that should taste "meaty" but don't contain much meat.

Many foods also contain two other additives that are salts, not of carboxylic acids, but of compounds with acidic hydroxyl groups: *sodium ascorbate* and *sodium erythorbate.* Ascorbic acid is vitamin C, and erythorbic acid is an isomer of vitamin C. Both salts act as antioxidants and help to maintain the color of cured meat such as bacon, though only the ascorbate has value as a vitamin.

Sodium or potassium salts of carboxylic acids are ionic solids that are usually far more soluble in water than the carboxylic acids themselves and are used when acid solubility must be increased. For example, benzoic acid has a water solubility of only 3.4 g/L at 25°C, whereas sodium benzoate has a water solubility of 550 g/L.

Practice Problems   **6.6**   Write the products of these reactions:
(a) $CH_3CH_2CH_2COOH + KOH \longrightarrow$?
(b) 2-methylpentanoic acid + $Ba(OH)_2 \longrightarrow$?

6.7   Write the formulas of calcium formate and sodium acrylate. (See Table 6.1.)

6.5   REACTIONS OF CARBOXYLIC ACIDS: ESTER FORMATION

Esterification reaction The reaction between an alcohol and a carboxylic acid to yield an ester plus water.

Both in chemical laboratories and in living organisms, the conversion of carboxylic acids into esters is a commonly encountered reaction. In the laboratory, an **esterification reaction** is carried out by warming a carboxylic acid with an alcohol in the presence of a strong-acid catalyst such as H_2SO_4. In so doing, an —H is lost from the alcohol, an —OH is lost from the acid, and water is formed as a by-product. The net effect is substitution of —OR′ for —OH:

Two common food additives

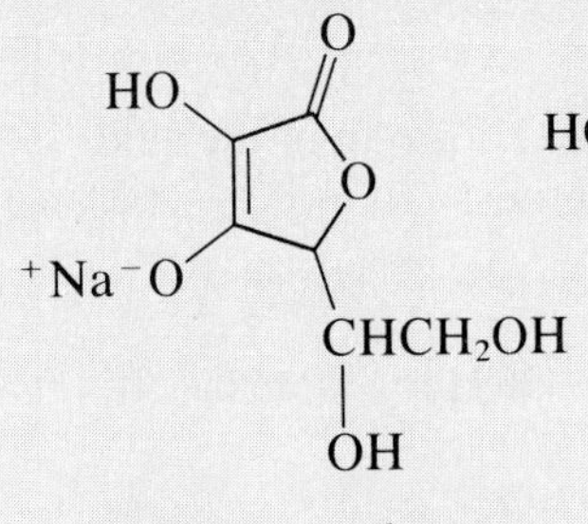

Sodium ascorbate
(an antioxidant)

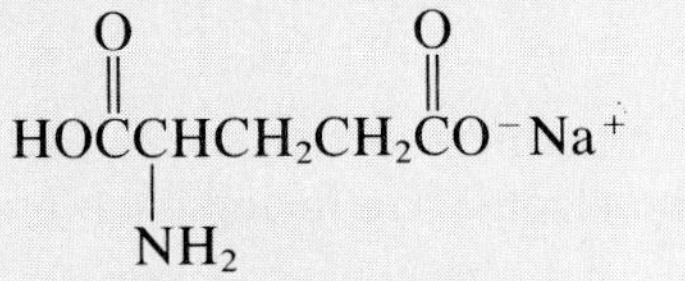

Monosodium glutamate
(flavor enhancer)

vor, color, or consistency. It could be argued that such additives are not essential, although that doesn't necessarily mean they are harmful. Do we really need dry powders that turn into cream sauce when water, butter, and milk are added? Each of us must decide this for ourselves.

All the additives mentioned are salts of acids that occur naturally in plants and animals, and all have been OK'd by the U.S. Food and Drug Administration (FDA). Some additives are essential—without preservatives certain foods would harbor disease-causing microorganisms. Other additives make convenience foods possible or make food more appealing by enhancing fla-

Without certain acid salts, this will happen to your bread more quickly than with them.

This —OH group is replaced by this —OR' group.

$$R\overset{\displaystyle O}{\underset{}{\parallel}}{C}-OH \ + \ H-OR' \ \xrightarrow{\text{H}^+ \text{ catalyst}} \ R\overset{\displaystyle O}{\underset{}{\parallel}}{C}-OR' \ + \ H_2O$$

A carboxylic acid   An alcohol   An ester

For example,

$$CH_3CH_2CH_2\overset{\displaystyle O}{\underset{}{\parallel}}{C}-OH \ + \ H-OCH_2CH_3 \ \rightleftharpoons \ CH_3CH_2CH_2\overset{\displaystyle O}{\underset{}{\parallel}}{C}-OCH_2CH_3 \ + \ H_2O$$

Butanoic acid   Ethanol   Ethyl butanoate
(in pineapple oil)

REVIEW LeChatelier's principle: When stress is applied to a system in equilibrium, the equilibrium shifts to relieve the stress.

Esterification reactions are reversible and often reach equilibrium with substantial amounts of both reactants and products present. In ester synthesis, good yields are obtained either by using a large excess of the alcohol or by continuously removing one of the products (for example, by distilling off a low-boiling ester). Both techniques are applications of LeChatelier's principle.

Solved Problem 17.4   The flavor ingredient in oil of wintergreen can be made by reaction of *o*-hydroxybenzoic acid with methanol. What is its structure?

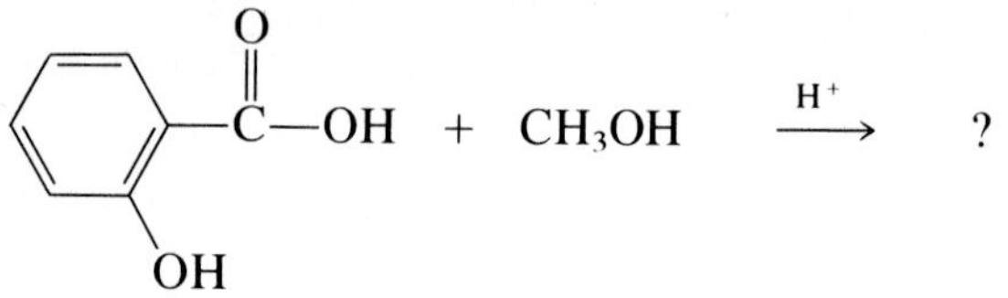

Solution   First, write the two reaction partners so that the —COOH group of the acid and the —OH group of the alcohol are facing each other:

(*o*-Hydroxybenzoic acid)

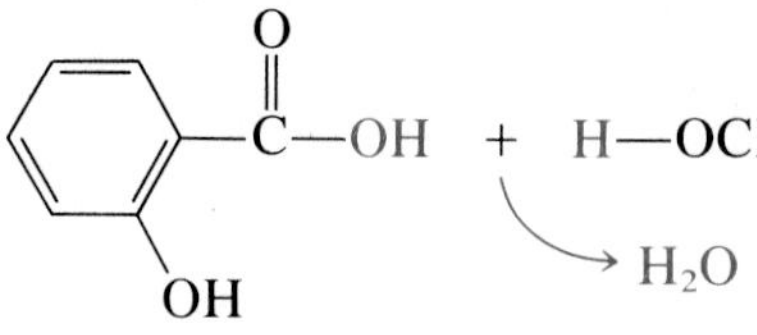

Next, remove —OH from the acid and —H from the alcohol to form water and then join the two resulting organic fragments. The product is the ester.

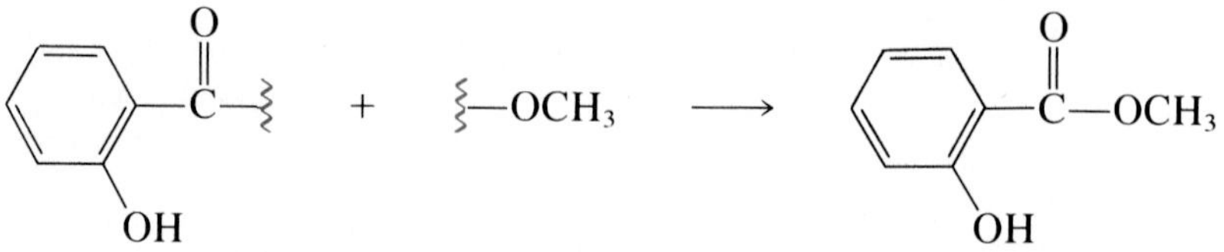

Methyl *o*-hydroxybenzoate
(in oil of wintergreen)

Practice Problems

6.8 Raspberry oil contains an ester that can be made by reaction of formic acid with 2-methyl-1-propanol. What is its structure?

$$HCOOH + (CH_3)_2CHCH_2OH \longrightarrow ?$$

6.9 What carboxylic acid and what alcohol are needed to make each of the following esters?

(a) 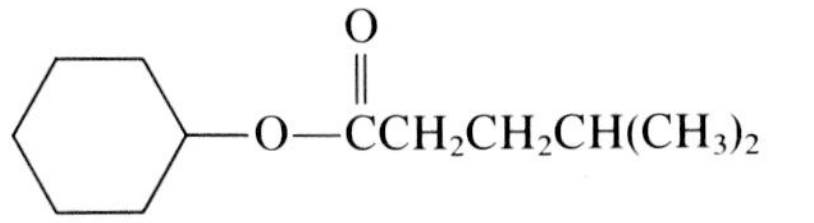
$$\begin{array}{c} O \\ \parallel \\ \text{—O—CCH}_2\text{CH}_2\text{CH(CH}_3)_2 \end{array}$$

(b) $CH_3CH_2CH_2CH_2\overset{\overset{\displaystyle O}{\parallel}}{C}—O—CH(CH_3)_2$

6.6 SOME COMMON ESTERS

Esters have many uses in medicine, in industry, and in living systems. In medicine, a number of important pharmaceutical agents, including aspirin and the local anesthetic benzocaine, are esters. In industry, the formation of polymers by esterification is an important reaction (see Interlude in this chapter). In nature, many simple esters are responsible for the fragrant odors of fruits and flowers, such as isopentyl acetate in bananas.

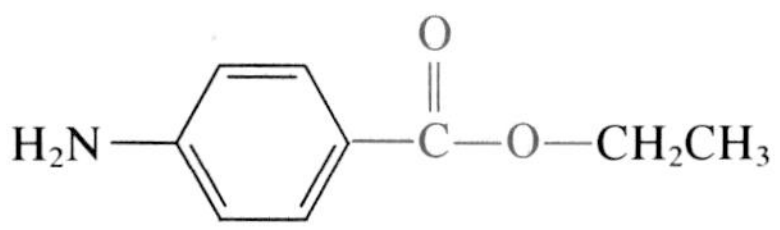

$$H_2N—\underset{}{\bigcirc}—\overset{\overset{\displaystyle O}{\parallel}}{C}—O—CH_2CH_3$$

Benzocaine
(a local anesthetic)

$$H_3C—\overset{\overset{\displaystyle O}{\parallel}}{C}—O—CH_2CH_2CH(CH_3)_2$$

Isopentyl acetate
(from bananas)

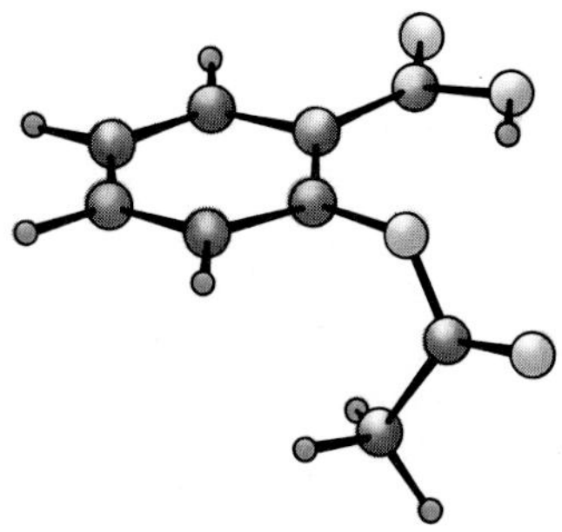

Acetylsalicylic acid

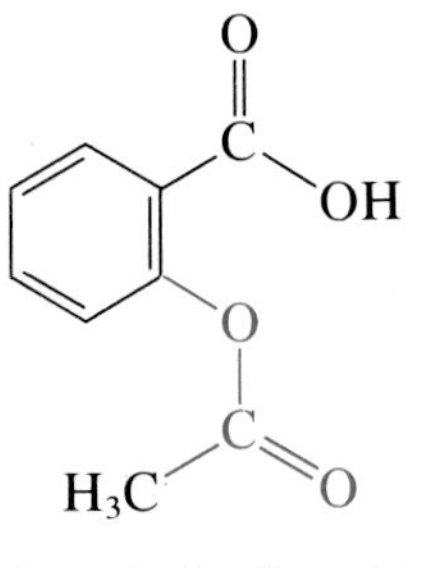

Acetylsalicylic acid
(aspirin)

Methyl salicylate
(methyl *o*-hydroxybenzoate)

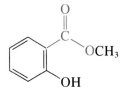

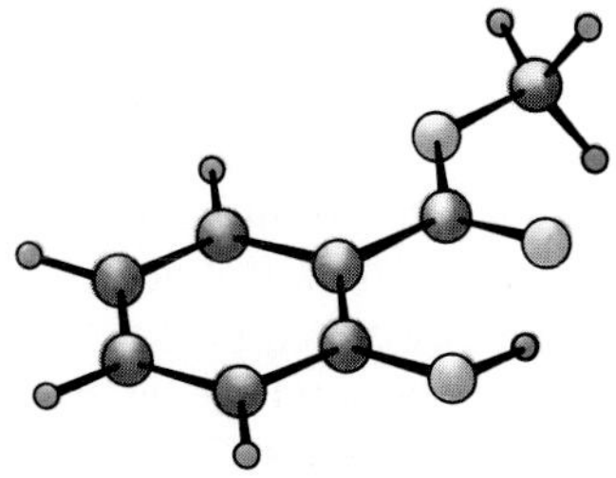

Methyl salicylate

Aspirin and Methyl Salicylate   Aspirin is a white, crystalline solid (mp 135°C) that's a member of the group of drugs known as *salicylates:* esters of salicylic acid. Methyl salicylate, the methyl ester formed at the carboxylic acid group, is a *counterirritant,* a substance that relieves internal pain by stimulating nerve endings in the skin for warmth, coolness, or pain. Poisonous if swallowed in significant amounts, methyl salicylate is one of the active ingredients in liniments such as Ben Gay and Heet. In low concentration it is the flavoring *oil of wintergreen.*

AN APPLICATION: THIOL ESTERS—BIOLOGICAL CARBOXYLIC ACID DERIVATIVES

Although the principles remain the same, many of the carbonyl-group substitution reactions that take place in living organisms use *thiol esters,* or *thioesters,* in place of carboxylic acid esters. A thiol ester, which is simply a sulfur-containing analog of a carboxylic acid ester, has its carbonyl group bonded to one —SR group (like that in a *thiol, RSH*) and has the general formula, RCOSR′. Many of the reactions used in the body to extract energy from food involve thiol esters.

Acetyl coenzyme A, usually abbreviated as acetyl SCoA, is the most common thiol ester in nature. Although it has a much more complex structure than esters such as ethyl acetate, it reacts in almost exactly the same way that such esters do. To give just one example of the use of acetyl coenzyme A by living organisms, *N*-acetylglucosamine, an important constituent of cell-surface membranes in mammals, is synthesized in nature by reaction of acetyl SCoA with glucosamine. Note how the result of the reaction is transfer of the acetyl group from one molecule to another.

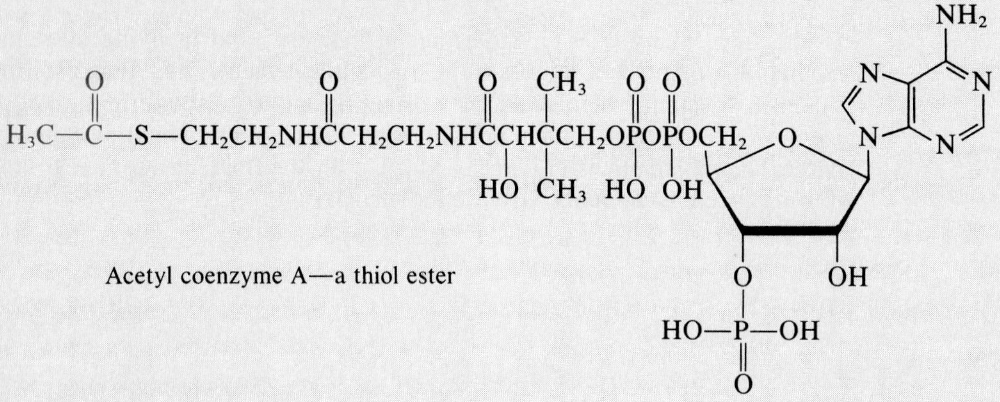

Acetyl coenzyme A—a thiol ester

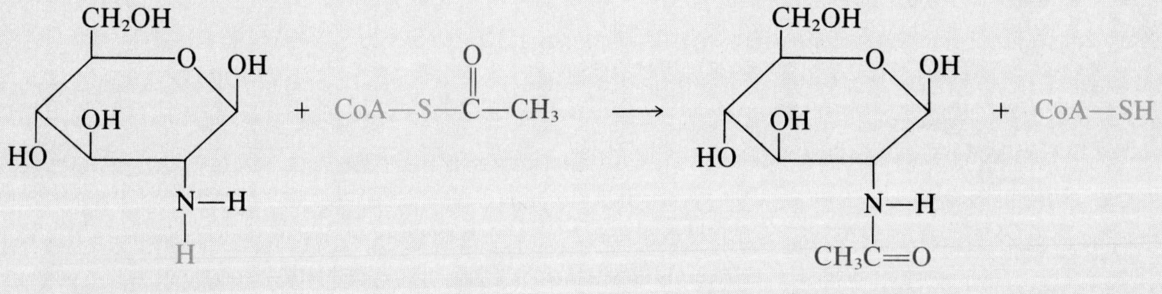

Glucosamine   Acetyl coenzyme A   *N*-Acetylglucosamine

Aspirin is only very slightly ionized in the acidic environment in the stomach but causes trouble once inside the stomach lining. There it ionizes to give acetylsalicylate ion and hydrogen ion, which do not easily exit across cell membranes. When sufficient ions have collected, which happens with even one aspirin tablet, the cell membranes are damaged and bleeding occurs. The usual blood loss of a few milliliters per tablet is not harmful, but in susceptible individuals more extensive bleeding can occur.

Aspirin was introduced as a medication in Europe in about 1900 by the Bayer Chemical Company after it was found to be as effective a pain reliever as other salicylates but with fewer unpleasant side effects (Figure 6.3). In addition to providing pain relief, aspirin lowers fever and acts to reduce inflammation. For a drug that has been in use so long and in such large quantity (about 100 aspirin tablets per U.S. resident per year), it's amazing that discoveries about the physiological effects of aspirin are still being made. Since 1982, aspirin has been known to cause a liver disorder called *Reye's syndrome* in children and is no longer recommended for use in childhood diseases such as chicken pox. Evidence is accumulating, however, that regular doses of aspirin may reduce the incidence of heart attacks in adults because of its ability to slow down blood coagulation.

Glycerol Esters   Glycerol is a trihydroxy alcohol that forms the backbone of natural fats and oils when it is esterified with carboxylic acids that have long hydrocarbon chains.

$$3\ RCOOH\ +\ \begin{array}{c} HO-CH_2 \\ | \\ HO-CH \\ | \\ HO-CH_2 \end{array} \longrightarrow \begin{array}{c} RCOOCH_2 \\ | \\ RCOOCH \\ | \\ RCOOCH_2 \end{array}$$

A fatty acid   Glycerol   A triglyceride
                                    (a fat or oil)

Such esters, generally known as *triglycerides,* are members of the lipid class of biomolecules. The R groups in a given fat or oil may be the same or different. The digestion of fats and oils, as you'll see in Chapter 13, is essentially a reverse of the esterification reaction.

Figure 6.3
Pharmaceutical ad from the 1900s. The ad points out that aspirin was a substitute for salicylates. Substitutes have since been found for some of the other pharmaceuticals listed, too.

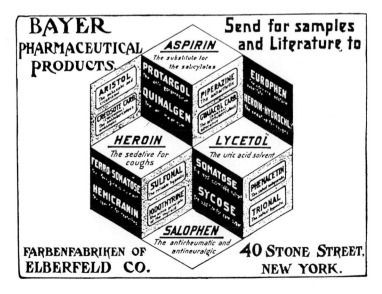

6.7 REACTIONS OF ESTERS: HYDROLYSIS

Hydrolysis The breakdown of a compound by reaction with water; the H's and O from water usually add to the atoms in the broken bond in one way or another.

Esters undergo a **hydrolysis** reaction with water that splits the ester molecule into the carboxylic acid and alcohol combined in the ester. The net effect of the hydrolysis is a substitution of —OH for —OR'.

This —OR' group is replaced by this —OH group.

$$
\underset{\text{An ester}}{R-\overset{\overset{\displaystyle O}{\|}}{C}-OR'} \;+\; H-OH \;\longrightarrow\; \underset{\text{Carboxylic acid}}{R-\overset{\overset{\displaystyle O}{\|}}{C}-OH} \;+\; \underset{\text{Alcohol}}{H-OR'}
$$

Ester hydrolysis is catalyzed both by acids and by bases. Acid-catalyzed hydrolysis is simply the reverse of the esterification reaction discussed in Section 6.5. An ester is treated with water in the presence of a strong acid such as H_2SO_4, and hydrolysis takes place. For example,

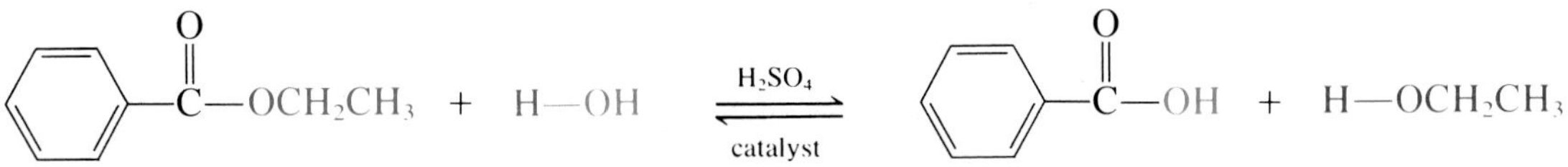

$$
\underset{\text{Ethyl benzoate}}{C_6H_5-\overset{\overset{\displaystyle O}{\|}}{C}-OCH_2CH_3} \;+\; H-OH \;\underset{\text{catalyst}}{\overset{H_2SO_4}{\rightleftharpoons}}\; \underset{\text{Benzoic acid}}{C_6H_5-\overset{\overset{\displaystyle O}{\|}}{C}-OH} \;+\; \underset{\text{Ethanol}}{H-OCH_2CH_3}
$$

Saponification reaction The reaction of an ester with aqueous hydroxide ion to yield an alcohol and the metal salt of a carboxylic acid.

Base-catalyzed hydrolysis, often called a **saponification reaction** (after the Latin word *sapo*, "soap"), takes place when an ester is treated with water in the presence of a base such as NaOH or KOH. The main difference between the acid- and base-catalyzed hydrolysis methods is that the base-catalyzed reaction yields the carboxylate anion as a product rather than a free carboxylic acid. In order to isolate the acid, the anion has to be protonated by treatment with HCl or H_2SO_4. (The use of saponification in making soap is discussed in Section 12.4.)

Saponification

$$
\underset{\text{Ester}}{R-\overset{\overset{\displaystyle O}{\|}}{C}-OR'} \;+\; NaOH(aq) \;\longrightarrow\; \underset{\text{Carboxylate salt}}{R-\overset{\overset{\displaystyle O}{\|}}{C}-O^-\,Na^+} \;+\; \underset{\text{Alcohol}}{R'O-H}
$$

For example,

$$
\underset{\text{Methyl butanoate}}{CH_3CH_2CH_2-\overset{\overset{\displaystyle O}{\|}}{C}-OCH_3} \;+\; NaOH(aq) \;\longrightarrow\; \underset{\text{Sodium butanoate}}{CH_3CH_2CH_2-\overset{\overset{\displaystyle O}{\|}}{C}-O^-\,Na^+} \;+\; \underset{\text{Methanol}}{CH_3OH}
$$

Solved Problem 17.5   What product would you obtain from acid-catalyzed hydrolysis of ethyl formate, a flavor constituent of rum?

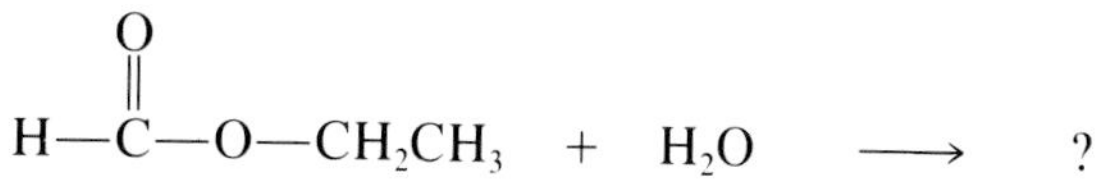

$$H-\underset{\underset{O}{\parallel}}{C}-O-CH_2CH_3 \ + \ H_2O \ \longrightarrow \ ?$$

Solution   First, look at the name of the starting ester. Usually, the name of the ester gives a good indication of the names of the two products. Thus, *ethyl formate* yields *ethyl* alcohol and *form*ic acid. To find the product structures in a more systematic way, write the structure of the ester and locate the bond between the carbonyl-group carbon and the —OR′ group.

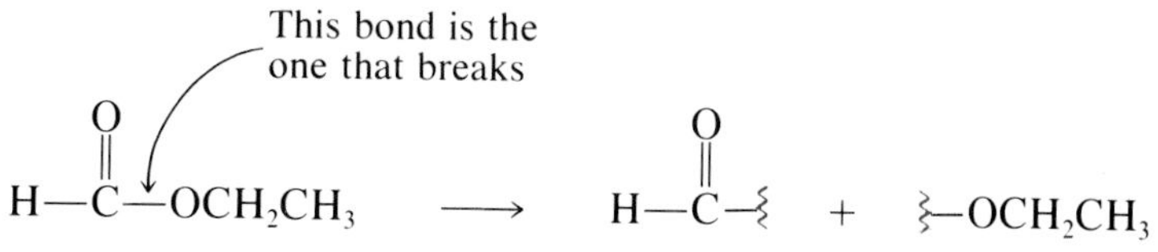

This bond is the
one that breaks

$$H-\underset{\underset{O}{\parallel}}{C}-OCH_2CH_3 \ \longrightarrow \ H-\underset{\underset{O}{\parallel}}{C}-\xi \ + \ \xi-OCH_2CH_3$$

Next, carry out a substitution reaction on paper. First form the carboxylic-acid product by connecting an —OH to the carbonyl-group carbon. Then add an —H to the —OCH₂CH₃ group to form the alcohol product.

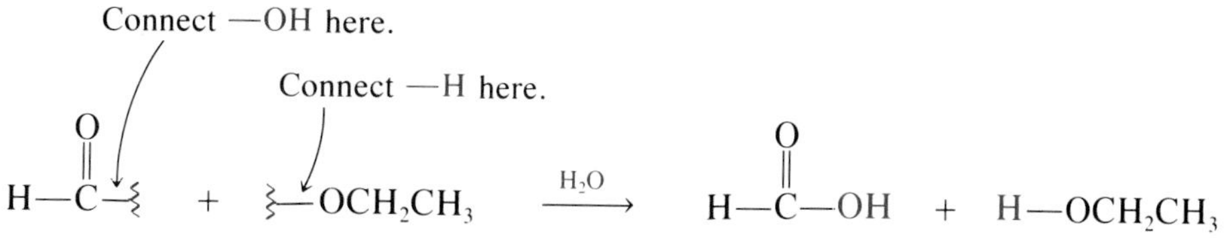

Connect —OH here.

Connect —H here.

$$H-\underset{\underset{O}{\parallel}}{C}-\xi \ + \ \xi-OCH_2CH_3 \ \xrightarrow{H_2O} \ H-\underset{\underset{O}{\parallel}}{C}-OH \ + \ H-OCH_2CH_3$$

Practice Problem   6.10   What products would you obtain from acid-catalyzed hydrolysis of these esters?

(a) 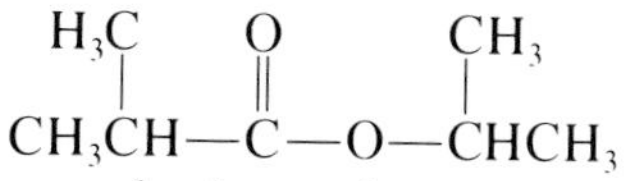
$$CH_3CH-\underset{\underset{O}{\parallel}}{\overset{\overset{H_3C}{|}}{C}}-O-\overset{\overset{CH_3}{|}}{CH}CH_3$$

(b) $CH_3CH=CHCOOCH_2CH_3$

(c) propyl *p*-bromobenzoate

6.8   REACTIONS OF ESTERS: CLAISEN CONDENSATION

Claisen condensation reaction   A reaction that joins two ester molecules to yield a keto ester product.

Just as the aldol reaction (Section 5.8) joins two aldehyde or ketone molecules together, the **Claisen condensation reaction** joins two ester molecules together. A Claisen condensation takes place when an ester is treated with a strong base such as sodium methoxide, $Na^+OCH_3^-$, the sodium salt of methanol. In the reaction, the —OR′ group is lost from the carbonyl-group carbon of one ester molecule, and a bond forms between that carbon and the carbon atom next to the carbonyl group of the second molecule. The product is a ketone-ester, or *keto* ester.

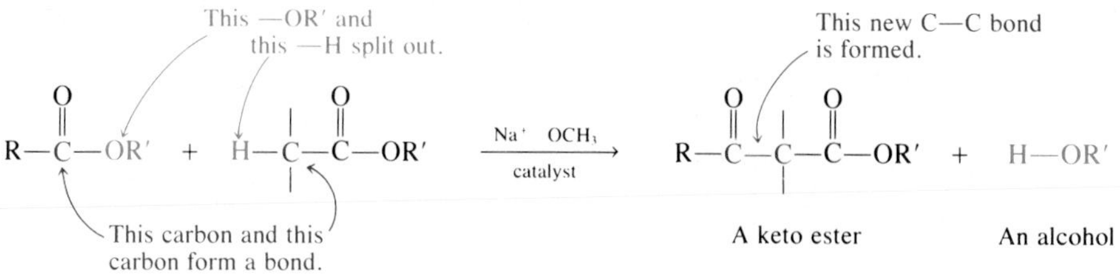

This —OR′ and
this —H split out.

This new C—C bond
is formed.

$$R-\overset{O}{\overset{\|}{C}}-OR' \;+\; H-\overset{\overset{\displaystyle |}{}}{C}-\overset{O}{\overset{\|}{C}}-OR' \;\xrightarrow[\text{catalyst}]{Na^+\ OCH_3}\; R-\overset{O}{\overset{\|}{C}}-\overset{\overset{\displaystyle |}{}}{C}-\overset{O}{\overset{\|}{C}}-OR' \;+\; H-OR'$$

This carbon and this
carbon form a bond.

A keto ester   An alcohol

For example,

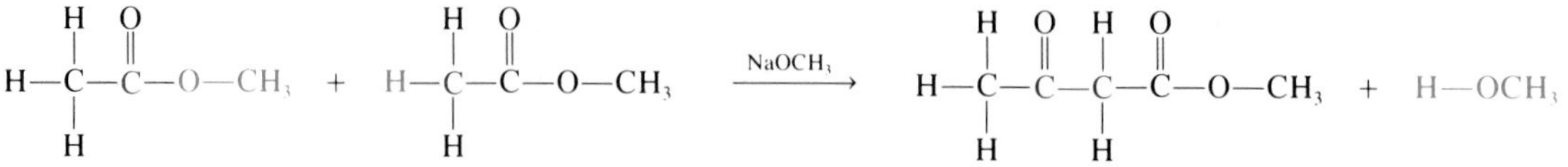

$$H-\overset{\overset{\displaystyle H}{|}}{\underset{\underset{\displaystyle H}{|}}{C}}-\overset{O}{\overset{\|}{C}}-O-CH_3 \;+\; H-\overset{\overset{\displaystyle H}{|}}{\underset{\underset{\displaystyle H}{|}}{C}}-\overset{O}{\overset{\|}{C}}-O-CH_3 \;\xrightarrow{NaOCH_3}\; H-\overset{\overset{\displaystyle H}{|}}{\underset{\underset{\displaystyle H}{|}}{C}}-\overset{O}{\overset{\|}{C}}-\overset{\overset{\displaystyle H}{|}}{\underset{\underset{\displaystyle H}{|}}{C}}-\overset{O}{\overset{\|}{C}}-O-CH_3 \;+\; H-OCH_3$$

2 Methyl acetate

Methyl 3-oxobutanoate
(methyl acetoacetate)   Methanol

Claisen condensation reactions are used by living organisms for the biological synthesis of many different kinds of molecules. Fats, sugars, steroid hormones, and many other classes of compounds are synthesized in the body by joining small ester molecules by Claisen condensations.

Solved Problem 17.6   What product would you obtain from the following Claisen condensation reaction?

$$2\ CH_3CH_2\overset{O}{\overset{\|}{C}}OCH_3 \;\xrightarrow{NaOCH_3}\; ?$$

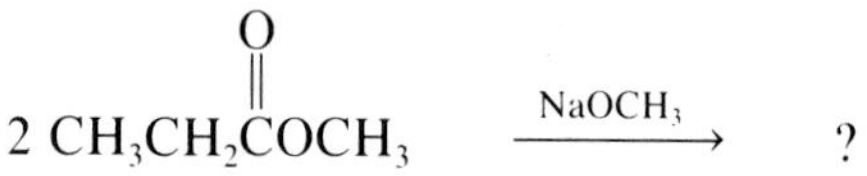

Solution   First, rewrite the reaction to emphasize the ester group of one molecule and a C—H bond next to the ester group of the second molecule:

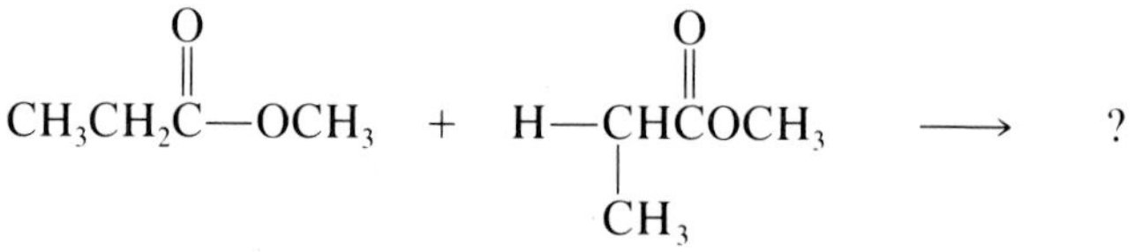

$$CH_3CH_2\overset{O}{\overset{\|}{C}}-OCH_3 \;+\; H-\underset{\underset{\displaystyle CH_3}{|}}{C}H\overset{O}{\overset{\|}{C}}OCH_3 \;\longrightarrow\; ?$$

Next, remove the —OCH₃ group from the first ester and the appropriate —H from the second ester to yield methanol. Then connect the remaining fragments to yield the keto ester product.

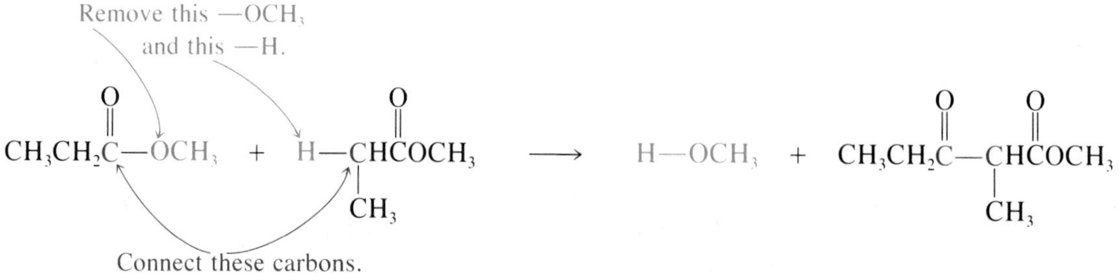

Remove this —OCH₃
and this —H.

$$CH_3CH_2\overset{O}{\overset{\|}{C}}-OCH_3 \;+\; H-\underset{\underset{\displaystyle CH_3}{|}}{C}H\overset{O}{\overset{\|}{C}}OCH_3 \;\longrightarrow\; H-OCH_3 \;+\; CH_3CH_2\overset{O}{\overset{\|}{C}}-\underset{\underset{\displaystyle CH_3}{|}}{C}H\overset{O}{\overset{\|}{C}}OCH_3$$

Connect these carbons.

Practice Problem   6.11 Draw the products from Claisen condensation of these esters:

$$\text{(a)}\quad \langle\;\rangle\!-\!CH_2\overset{\overset{\displaystyle O}{\|}}{C}OCH_3 \qquad \text{(b) methyl butanoate}$$

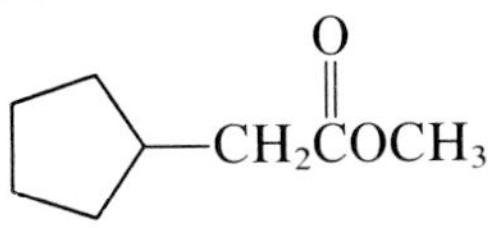

6.9 REACTIONS OF CARBOXYLIC ACIDS: AMIDE FORMATION

The reaction of a carboxylic acid with ammonia or an amine yields an amide, just as the reaction of a carboxylic acid with an alcohol yields an ester. In both cases, water is formed as a by-product, and the —OH part of the carboxylic acid is replaced.

$$\underset{\text{Acid}}{R\!-\!\overset{\overset{\displaystyle O}{\|}}{C}\!-\!OH} \;+\; \underset{\text{Amine}}{H\!-\!NR'_2} \;\longrightarrow\; \underset{\text{Amide}}{R\!-\!\overset{\overset{\displaystyle O}{\|}}{C}\!-\!NR'_2} \;+\; H_2O$$

The reaction of a carboxylic acid with ammonia or an amine to form an amide doesn't take place, however, unless the carboxyl —OH group is first replaced with a more reactive group. In the laboratory, a compound called *DCC* (dicyclohexylcarbodiimide) is often used.

$$\text{DCC} \;=\; \langle\;\rangle\!-\!N\!=\!C\!=\!N\!-\!\langle\;\rangle$$

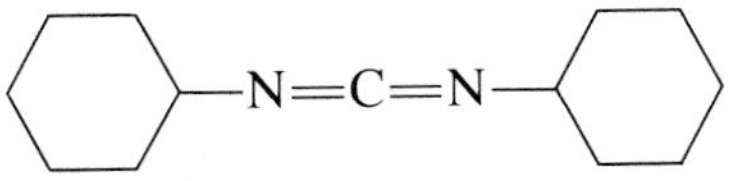

Although the exact way in which DCC works is a bit too complex to discuss in detail, its function is to *activate* the carboxylic acid by making it much more reactive. Once activated, the carboxylic acid then reacts rapidly with an amine to generate an amide.

$$R\!-\!\overset{\overset{\displaystyle O}{\|}}{C}\!-\!OH \;\xrightarrow{\text{DCC}}\; R\!-\!\overset{\overset{\displaystyle O}{\|}}{C}\!-\!\boxed{\text{activator}} \;\xrightarrow{NH_3}\; R\!-\!\overset{\overset{\displaystyle O}{\|}}{C}\!-\!NH_2$$

For example,

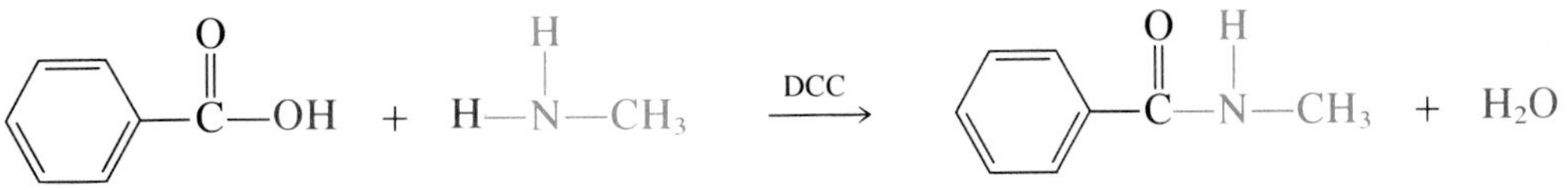

$$\underset{\text{Benzoic acid}}{\langle\;\rangle\!-\!\overset{\overset{\displaystyle O}{\|}}{C}\!-\!OH} \;+\; \underset{\text{Methylamine}}{H\!-\!\overset{\overset{\displaystyle H}{|}}{N}\!-\!CH_3} \;\xrightarrow{\text{DCC}}\; \underset{\text{N-Methylbenzamide}}{\langle\;\rangle\!-\!\overset{\overset{\displaystyle O}{\|}}{C}\!-\!\overset{\overset{\displaystyle H}{|}}{N}\!-\!CH_3} \;+\; H_2O$$

$$\underset{\text{Propanoic acid}}{CH_3CH_2\!-\!\overset{\overset{\displaystyle O}{\|}}{C}\!-\!OH} \;+\; \underset{\text{Dimethylamine}}{H\!-\!\overset{\overset{\displaystyle CH_3}{|}}{N}\!-\!CH_3} \;\xrightarrow{\text{DCC}}\; \underset{\text{N,N-Dimethylpropanamide}}{CH_3CH_2\!-\!\overset{\overset{\displaystyle O}{\|}}{C}\!-\!\overset{\overset{\displaystyle CH_3}{|}}{N}\!-\!CH_3} \;+\; H_2O$$

In living organisms, complex biomolecules function as activating reagents to allow amide formation from a carboxylic acid and an amine. Nevertheless, the principle behind a chemist's use of DCC in a laboratory and an organism's use of complex molecules in a cell to accomplish amide formation is the same.

The **amide bond** between nitrogen and a carbonyl-group carbon is the fundamental link in protein molecules (Figure 6.4). In addition, some synthetic polymers such as nylon contain amide groups (see the Interlude in this chapter), and important pharmaceutical agents such as acetaminophen, the aspirin substitute in Tylenol and Excedrin, are amides. Glutamic acid (glutamine), one of the natural amino acids, contains an unsubstituted amide group, and urea, which carries waste nitrogen from the body, is a diamide.

Amide bond The bond between a carbonyl group and the nitrogen atom in an amide.

Some common amides

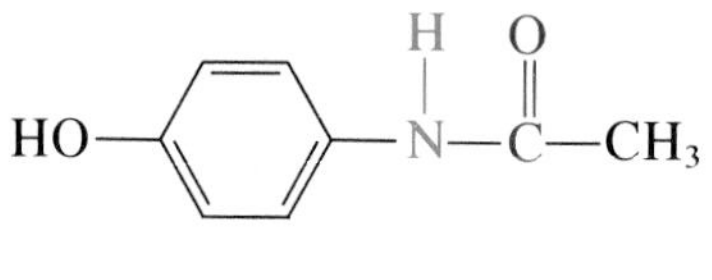

Acetaminophen 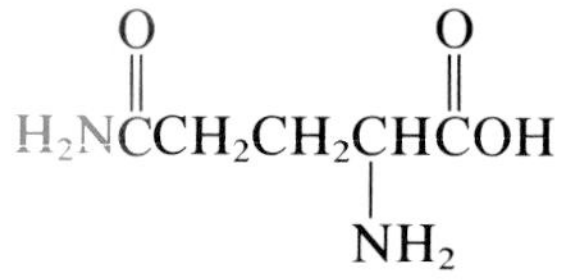   Glutamine 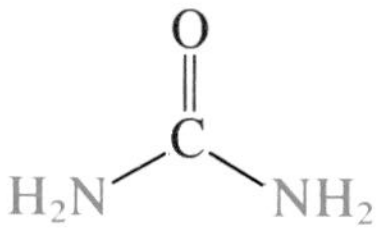   Urea

---

Solved Problem 17.7 The mosquito repellent DEET (diethyltoluamide) can be prepared by reaction of diethylamine with *p*-methylbenzoic acid (toluic acid) in the presence of DCC. What is the structure of DEET?

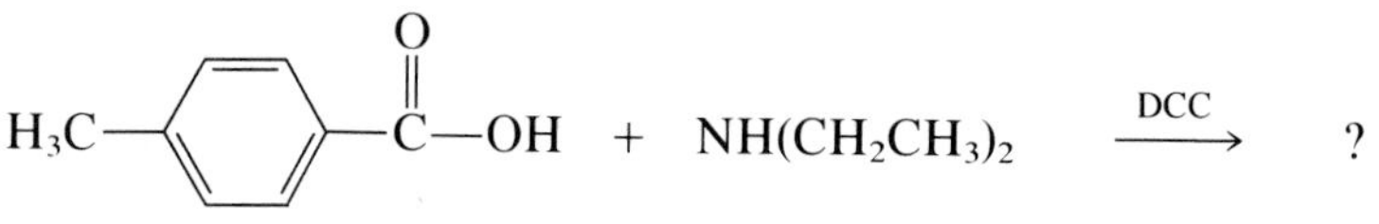

Solution First, rewrite the equation so that the —OH of the acid and the —H of the amine are facing each other:

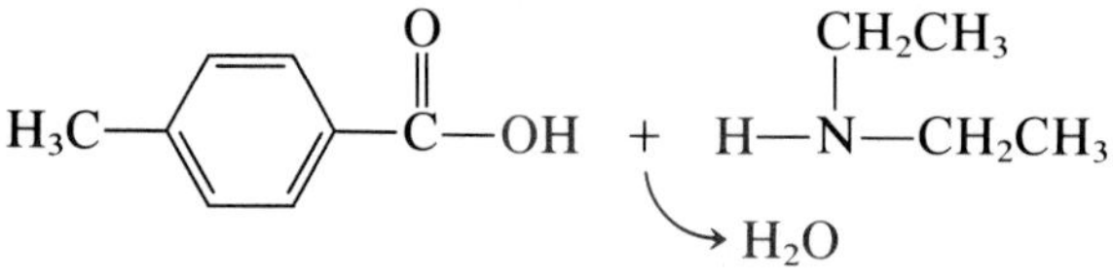

Next, remove the —OH from the acid and the —H from the nitrogen atom of the amine to form water. Then join the two resulting fragments together to form the amide product.

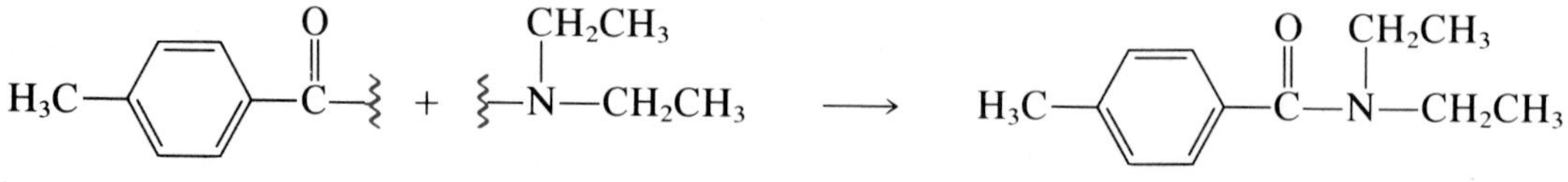

N, *N*-Diethyltoluamide (DEET)

Figure 6.4
Normal human striated muscle fibers. Protein molecules, which run lengthwise in each fiber, consist of amino acids held together by amide bonds.

Practice Problems   **6.12**   Draw structures of the amides formed in these reactions:

(a) CH_3NH_2 + $(CH_3)_2CHCOOH$ $\xrightarrow{\text{DCC}}$?

(b) [benzene ring]—NH_2 + [cyclopentane ring]—COOH $\xrightarrow{\text{DCC}}$?

6.13   What carboxylic acid and what amine would you use if you wanted to prepare the headache remedy phenacetin?

$$CH_3CH_2O-\text{[benzene ring]}-NHCCH_3$$

Phenacetin

6.10 REACTIONS OF AMIDES: HYDROLYSIS

Amides undergo a hydrolysis reaction with water in the same way that esters do. Just as an ester yields a carboxylic acid and an alcohol when hydrolyzed (Section 6.7), an amide yields a carboxylic acid and an amine. The net effect of amide hydrolysis is a substitution of —OH for NH_2 or the substituted amide nitrogen. As we'll see in Chapter 14 the cleavage of amide bonds by hydrolysis is the key process that occurs in the stomach during digestion of proteins.

This —NR'R″ group is replaced by this —OH group from water

$$R-\overset{O}{\overset{\|}{C}}-\underset{\underset{R''}{|}}{N}R' \ + \ H-OH \ \longrightarrow \ R-\overset{O}{\overset{\|}{C}}-OH \ + \ H-\underset{\underset{R''}{|}}{N}-R'$$

An amide   Carboxylic acid   Amine

Amide hydrolysis is catalyzed both by acids and by bases. If an acid catalyst such as HCl is used, the amine product is converted into its ammonium salt as soon as it's formed. If a basic catalyst such as NaOH is used, the carboxylic acid product is converted into its carboxylate ion. For simplicity, however, it's easiest to write the hydrolysis products as a free carboxylic acid and a free amine. For example,

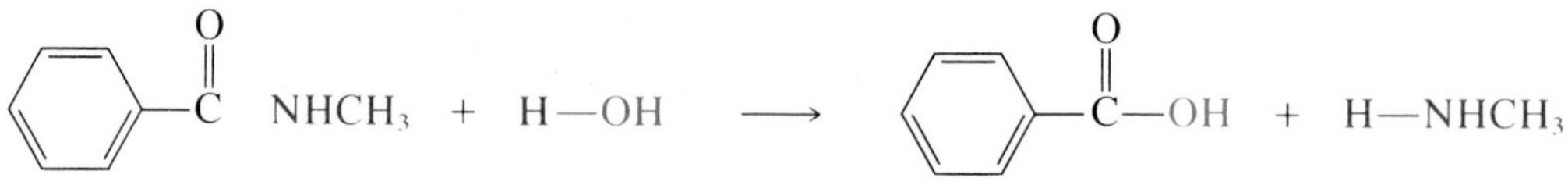

N-Methylbenzamide　　Benzoic acid　　Methylamine

Solved Problem 17.8 What products result from hydrolysis of *N*-ethylbutanamide?

$$CH_3CH_2CH_2—\overset{O}{\overset{\|}{C}}—NHCH_2CH_3 + H_2O \longrightarrow ?$$

Solution First, look at the name of the starting amide. Often, the name of the amide indicates the names of the two products. Thus, *N*-ethylbutanamide will yield ethylamine and butanoic acid. To be more systematic about finding the product structures, write the amide and locate the bond between the carbonyl-group carbon and the nitrogen. Then break this amide bond and write the two fragments:

This amide bond is the one that breaks.

$$CH_3CH_2CH_2\overset{O}{\overset{\|}{C}}—NHCH_2CH_3 \longrightarrow CH_3CH_2CH_2\overset{O}{\overset{\|}{C}}—\wr + \wr—NHCH_2CH_3$$

Next, carry out a hydrolysis reaction on paper and form the products by connecting an —OH to the carbonyl-group carbon and an —H to the nitrogen:

Connect —OH here.
Connect —H here.

$$CH_3CH_2CH_2\overset{O}{\overset{\|}{C}}—\wr + \wr—NHCH_2CH_3 \xrightarrow{H_2O} CH_3CH_2CH_2\overset{O}{\overset{\|}{C}}—OH + H—NHCH_2CH_3$$

Butanoic acid　　Ethylamine

Practice Problem 17.14 What products result from hydrolysis of these amides?

(a) $CH_3CH{=}CH\overset{O}{\overset{\|}{C}}NHCH_3$　　(b) *N,N*-diethyl-*p*-chlorobenzamide

6.11 ACID ANHYDRIDES

Acid anhydrides contain the R—C=O, or acyl group, portion of two carboxylic acids, both bonded to a central oxygen atom. The acyl groups may be from the same or from different acids, and the anhydride group may form between two carboxyl groups in the same molecule. For example,

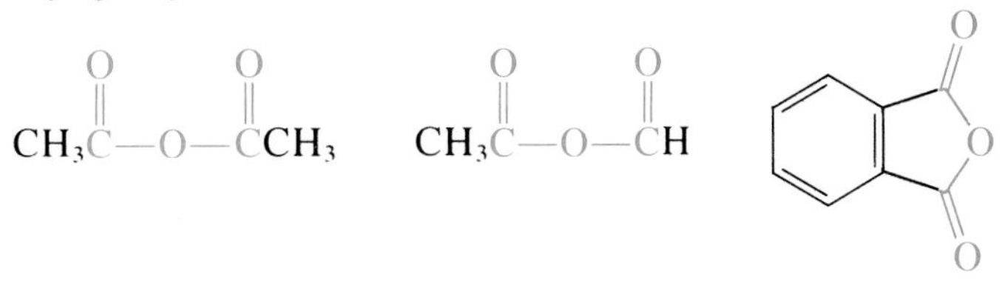

Acetic anhydride — Acetic formic anhydride — Phthalic anhydride

Because acid anhydrides (RCO_2OCR) react rapidly with water to regenerate acids, they are not found in plants and animals.

$$R{-}\overset{O}{\overset{\|}{C}}{-}O{-}\overset{O}{\overset{\|}{C}}{-}R + H_2O \longrightarrow 2\,R\overset{O}{\overset{\|}{C}}{-}OH$$

Their reactivity, however, makes anhydrides useful in the synthesis of other carboxylic acid derivatives. The R—C=O portion of an anhydride combines with an alcohol to give an ester or with an amine to give an amide, while the remainder of the anhydride adds hydrogen to give an acid:

$$R\overset{O}{\overset{\|}{C}}{-}O{-}\overset{O}{\overset{\|}{C}}R + R'OH \longrightarrow R\overset{O}{\overset{\|}{C}}OR' + R\overset{O}{\overset{\|}{C}}OH$$

$$R\overset{O}{\overset{\|}{C}}{-}O{-}\overset{O}{\overset{\|}{C}}R + R'NH_2 \longrightarrow R\overset{O}{\overset{\|}{C}}NHR' + R\overset{O}{\overset{\|}{C}}OH$$

Aspirin, an ester in which salicylic acid provides the —OH group, is synthesized by the reaction of salicylic acid and acetic anhydride:

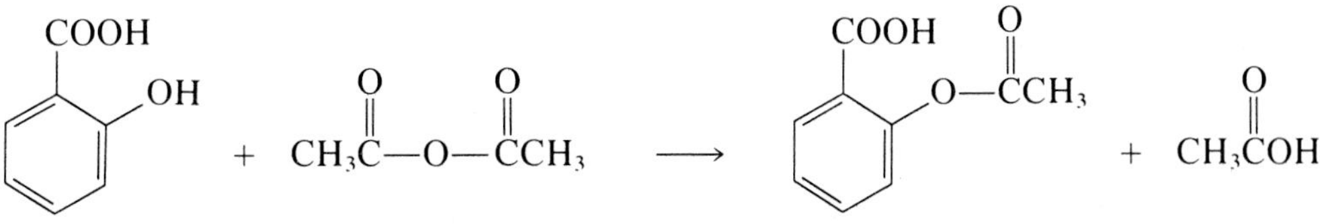

Salicylic acid — Acetic anhydride — Aspirin — Acetic acid

6.12 PHOSPHATE ESTERS AND ANHYDRIDES

Certain inorganic acids such as phosphoric acid, H_3PO_4, and nitric acid, HNO_3, have structures that are similar in many respects to the structures of carboxylic acids. All three contain an atom (C, P, or N) that is singly bonded to an —OH

Phosphate ester A compound formed by reaction of an alcohol with phosphoric acid.

Nitrate ester A compound formed by reaction of an alcohol with nitric acid.

group and doubly bonded to another oxygen. Thus, it's not surprising to find that both phosphoric acid and nitric acid react with alcohols to form **phosphate esters** and **nitrate esters.**

$$
\begin{array}{ccc}
\overset{\displaystyle O}{\underset{\displaystyle \parallel}{R-C-O-H}} & \overset{\displaystyle O}{\underset{\displaystyle \underset{\displaystyle OH}{\mid}}{HO-P-O-H}} & \overset{\displaystyle O}{\underset{\displaystyle \parallel}{^-O-\overset{+}{N}-O-H}} \\
\text{A carboxylic acid} & \text{Phosphoric acid} & \text{Nitric acid}
\end{array}
$$

$$\downarrow \text{R'OH} \qquad \downarrow \text{R'OH} \qquad \downarrow \text{R'OH}$$

$$
\begin{array}{ccc}
\overset{\displaystyle O}{\underset{\displaystyle \parallel}{R-C-O-R'}} & \overset{\displaystyle O}{\underset{\displaystyle \underset{\displaystyle OH}{\mid}}{HO-P-O-R'}} & \overset{\displaystyle O}{\underset{\displaystyle \parallel}{^-O-\overset{+}{N}-O-R'}} \\
\text{A carboxylic ester} & \text{A phosphate ester} & \text{A nitrate ester}
\end{array}
$$

REVIEW Inorganic phosphates and nitrates contain the phosphate ion, PO_4^{3-}, and the nitrate ion, NO_3^-, respectively.

Nitrate esters don't occur naturally, although nitroglycerin, a triester between glycerol (glycerin) and three molecules of nitric acid, is well known for its use in the treatment of heart disease.

$$
\begin{array}{ccc}
\begin{array}{l} CH_2-O-H \\ | \\ CH-O-H \\ | \\ CH_2-O-H \end{array} + 3\ HO-NO_2 & \longrightarrow & \begin{array}{l} CH_2-O-NO_2 \\ | \\ CH-O-NO_2 \\ | \\ CH_2-O-NO_2 \end{array} + 3\ H_2O \\
\text{Glycerin} \qquad \text{Nitric acid} & & \text{Nitroglycerin} \\
& & \text{(a nitrate triester)}
\end{array}
$$

A nitroglycerin tablet placed beneath the tongue provides relief within a few minutes for the pain (*angina pectoris*) caused by a brief interference in the flow of blood to the heart. By contrast, a mixture of nitroglycerin with oxygen-supplying salts and oxidizable materials is what we know as *dynamite.*

Phosphoric acid may be esterified at one, two, or all three of its —OH groups.

$$
\begin{array}{ccc}
\overset{\displaystyle O}{\underset{\displaystyle \underset{\displaystyle OH}{\mid}}{H-O-P-O-CH_3}} & \overset{\displaystyle O}{\underset{\displaystyle \underset{\displaystyle OH}{\mid}}{CH_3-O-P-O-CH_3}} & \overset{\displaystyle O}{\underset{\displaystyle \underset{\displaystyle O-CH_3}{\mid}}{CH_3-O-P-O-CH_3}} \\
\text{Methyl phosphate} & \text{Dimethyl phosphate} & \text{Trimethyl phosphate} \\
\text{(a phosphate monoester)} & \text{(a phosphate diester)} & \text{(a phosphate triester)}
\end{array}
$$

Because they still contain acidic hydrogen atoms, the monoester and diester are also acidic. Thus, in neutral or alkaline solutions, including body fluids that are slightly alkaline, they are present as ions:

$$\text{}^{-}O-\overset{\overset{\displaystyle O}{\|}}{\underset{\underset{\displaystyle O_{-}}{|}}{P}}-O-CH_3 \qquad CH_3-O-\overset{\overset{\displaystyle O}{\|}}{\underset{\underset{\displaystyle O^{-}}{|}}{P}}-O-CH_3$$

$$CH_3OPO_3^{2-} \qquad\qquad (CH_3O)_2PO_2^{-}$$

Phosphate esters are widespread throughout all living organisms and are key substances in nearly all metabolic pathways. Glyceraldehyde 3-phosphate, for example, is one of several phosphate esters that are intermediates in the breakdown of carbohydrates to yield energy.

$$\begin{array}{c}\text{CHO}\\|\\ \text{CHOH}\\|\\ \text{CH}_2-O-\overset{\overset{\displaystyle O}{\|}}{\underset{\underset{\displaystyle O^-}{|}}{P}}-O^-\end{array} \quad\text{or}\quad \begin{array}{c}\text{CHO}\\|\\ \text{CHOH}\\|\\ \text{CH}_2-O-PO_3^{2-}\end{array}$$

Glyceraldehyde 3-phosphate

Reaction of two molecules of phosphoric acid results in loss of water and produces a phosphorus anhydride that resembles carboxylic acid anhydrides but unlike them still contains four acidic hydrogens that will be ionized in solutions of appropriate pH. Reaction of this compound, known as *pyrophosphoric acid* (sometimes called diphosphoric acid), with yet another phosphoric acid molecule produces *triphosphoric acid,* also an acid anhydride with acidic hydrogens.

$$\text{HO}-\overset{\overset{\displaystyle O}{\|}}{\underset{\underset{\displaystyle OH}{|}}{P}}-O-\overset{\overset{\displaystyle O}{\|}}{\underset{\underset{\displaystyle OH}{|}}{P}}-OH \qquad \text{HO}-\overset{\overset{\displaystyle O}{\|}}{\underset{\underset{\displaystyle OH}{|}}{P}}-O-\overset{\overset{\displaystyle O}{\|}}{\underset{\underset{\displaystyle OH}{|}}{P}}-O-\overset{\overset{\displaystyle O}{\|}}{\underset{\underset{\displaystyle OH}{|}}{P}}-OH$$

Pyrophosphoric acid   Triphosphoric acid

Esters of these acids are referred to as diphosphates and triphosphates. In the following two methyl phosphate esters, note the difference between the P—O—C ester linkage and the P—O—P phosphorus anhydride linkages.

$$\text{HO}-\overset{\overset{\displaystyle O}{\|}}{\underset{\underset{\displaystyle OH}{|}}{P}}-O-\overset{\overset{\displaystyle O}{\|}}{\underset{\underset{\displaystyle OH}{|}}{P}}-O-CH_3$$

Methyl diphosphate

Anhydride

Ester

$$\text{HO}-\overset{\overset{\displaystyle O}{\|}}{\underset{\underset{\displaystyle OH}{|}}{P}}-O-\overset{\overset{\displaystyle O}{\|}}{\underset{\underset{\displaystyle OH}{|}}{P}}-O-\overset{\overset{\displaystyle O}{\|}}{\underset{\underset{\displaystyle OH}{|}}{P}}-O-CH_3$$

Methyl triphosphate

Phosphoric acid anhydride formation, like carboxylic acid anhydride formation, can be reversed by hydrolysis to give two acids. The diphosphate ADP and the triphosphate ATP, shown here ionized as they are in body fluids, are critical intermediates in the storage and use of energy in the body. Addition of the third phosphate group to ADP absorbs, or "stores," energy, and hydrolysis of ATP to give back ADP releases energy, a reaction we'll describe further in Chapter 9.

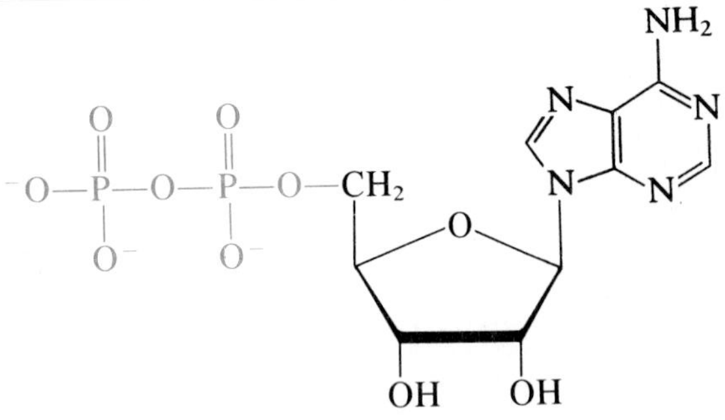

Adenosine diphosphate (ADP)

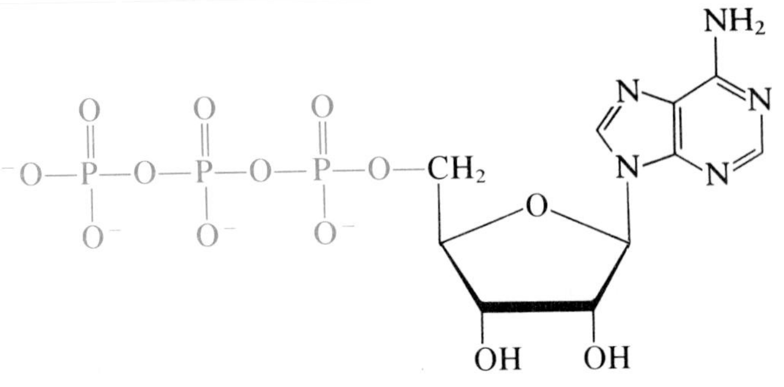

Adenosine triphosphate (ATP)

Practice Problems **6.15** Identify the functional group in each of the following compounds and give the structures of the products of hydrolysis of these compounds.

(a) CH_3CNH_2 (with $\overset{O}{\overset{||}{}}$ on carbonyl)   (b) $CH_3CH_2OPO_3^{2-}$ (c) $CH_3CH_2COCH_3$ (with $\overset{O}{\overset{||}{}}$ on carbonyl)

6.16 Draw the structure of ATP and identify its purine ring, ester group, and anhydride group.

6.13 ORGANIC REACTIONS

Equations for organic reactions tend to appear complicated. The bigger the molecules reacting or the longer the series of reactions, the more intimidating the equations. Remembering a few simple facts can help you to "see" what's going on, however. You've now been introduced to a variety of organic reactions. Let's examine what they have in common.

First, look for reaction patterns. For example, you've seen the *addition reactions* of alkenes, in which water (or halogens or hydrogen halides) add across the double bond:

$$A + B \longrightarrow C$$
$$H_2C = CH_2 + H_2O \longrightarrow CH_3CH_2OH$$

Addition reactions have the same pattern as "combination" reactions.

Elimination reactions, such as the elimination of water to give an alkene, are essentially the reverse of addition reactions.

$$\text{C} \longrightarrow \text{A} + \text{B}$$

$$CH_3CHCH_3 \longrightarrow CH_3CH{=}CH_2 + H_2O$$
$$\quad\quad\; |$$
$$\quad\quad OH$$

The "decomposition" reactions of inorganic compounds have the same pattern.

In this chapter you've seen several *substitution reactions*, which have the pattern called "exchange" when the reactants are ions.

$$A{-}B + X{-}Y \longrightarrow A{-}Y + X{-}B$$

The formation of esters and amides are substitution reactions, and so is the Claisen condensation.

$$\begin{array}{c} \; O \\ \| \\ RCOH \end{array} + R'OH \longrightarrow \begin{array}{c} O \\ \| \\ RCOR' \end{array} + HOH$$

$$\begin{array}{c} \; O \\ \| \\ RCOH \end{array} + R'NH_2 \longrightarrow \begin{array}{c} \; O \\ \| \\ RCNHR' \end{array} + HOH$$

$$\begin{array}{c} O \\ \| \\ RCH_2COR' \end{array} + \begin{array}{c} O \\ \| \\ RCH_2COR' \end{array} \longrightarrow \begin{array}{cc} O & O \\ \| & \| \\ RCH_2CCHCOR' \\ \quad\quad | \\ \quad\quad R \end{array} + R'OH$$

In organic substitution reactions, one product is often a small molecule, such as water or an alcohol, that splits out during the formation of a larger molecule. Sometimes, the small molecule isn't even written in the equation because it's unimportant for the result of greatest interest. In subsequent chapters you'll see many examples of substitution reactions in the buildup of biomolecules.

When looking at organic equations, in addition to the reaction patterns, remember that reactions take place at functional groups. We've shown here, as it might appear in a biochemistry textbook, the sequence of reactions for synthesis of the amino acid serine. You can ignore for now the other reactants and products above the arrows. Exactly what changes take place in these reactions?

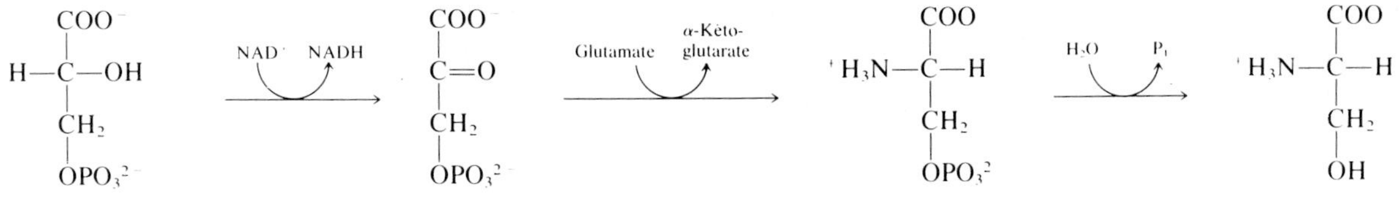

3-Phosphoglycerate   Serine

By comparing the functional groups in each product, you can immediately see that the first two reactions occur at the same carbon atom. One is the conversion of an —OH group to a C=O group, an oxidation reaction that is the elimination of two H atoms. The next reaction is the substitution of an —NH$_3^+$ group for the C=O. In the final step, the —NH$_3^+$ is unchanged, but an —OH group is substituted for the phosphate group. Whenever you come across a sequence of reactions, mentally highlight the changes as we have done here in color.

$$
\begin{array}{c}
COO^- \\
| \\
H-C-OH \\
| \\
CH_2 \\
| \\
OPO_3^{2-}
\end{array}
\xrightarrow[\;NAD^+\;NADH\;]{}
\begin{array}{c}
COO^- \\
| \\
C=O \\
| \\
CH_2 \\
| \\
OPO_3^{2-}
\end{array}
\xrightarrow[\;Glutamate\;\;glutarate\;]{\;\alpha\text{-Keto-}\;}
\begin{array}{c}
COO^- \\
| \\
{}^+H_3N-C-H \\
| \\
CH_2 \\
| \\
OPO_3^{2-}
\end{array}
\xrightarrow[\;H_2O\;\;P_i\;]{}
\begin{array}{c}
COO^- \\
| \\
{}^+H_3N-C-H \\
| \\
CH_2 \\
| \\
OH
\end{array}
$$

INTERLUDE: POLYAMIDES AND POLYESTERS

When a reaction takes place between a carboxylic acid and an amine, the two molecules link together to form an amide. Imagine what would happen, though, if a molecule with *two* carboxylic acid groups were to react with a molecule having *two* amino groups. An initial reaction would join two molecules together, but further reactions would then link more and more molecules together until a giant chain resulted. This is exactly what happens when certain kinds of synthetic polymers, known as *condensation polymers,* are made.

Nylons are *polyamides* that are prepared industrially by reaction of a diamine with a diacid. For example, nylon 66 is prepared by heating adipic acid (hexanedioic acid) with hexamethylenediamine (1,6-hexanediamine) at 280°C (see equation below).

Nylons have a great many uses, both in engineering applications and in fibers. High impact strength and abrasion resistance make nylon an excellent material for bearings and gears, and high tensile strength makes it suitable as fibers for a range of applications from clothing to mountaineering ropes to carpets.

Just as diacids and diamines react to yield polyamides, diacids and dialcohols react to yield *polyesters*.

The most industrially important polyester, made from reaction of terephthalic acid (1,4-benzenedicarboxylic acid) with ethylene glycol, is used under the trade name Dacron to make clothing fiber and under the name Mylar to make plastic film and recording tape. This same polymer, poly(ethylene terephthalate) or PET, is the material in clear, flexible soft-drink bottles.

$$
\begin{array}{c}
\quad\quad O \quad\quad\quad\quad O \\
\quad\quad \| \quad\quad\quad\quad \| \\
-\!\!\left[C-\!\!\left\langle\;\right\rangle\!\!-C-O-CH_2CH_2-O\right]_n\!\!-
\end{array}
$$

A polyester
(Dacron, PET)

Nylon 6,10 (like nylon 6,6, but with $(CH_2)_8$ replacing $(CH_2)_4$) is shown forming at the interface between layers of hexamethylenediamine (in NaOH) and a diacid chloride, $ClCO(CH_2)_8COCl$.

$$
\left.
\begin{array}{c}
HOOC-(CH_2)_4-COOH \\
\textit{Adipic acid} \\
+ \\
H_2N-(CH_2)_6-NH_2 \\
\textit{Hexamethylenediamine}
\end{array}
\right\}
\xrightarrow{280°C}
\begin{array}{c}
O \quad\quad\quad\quad O \\
\| \quad\quad\quad\quad \| \\
-\!\!\left[C-(CH_2)_4-C-NH-(CH_2)_6-NH\right]_n\!\!- \;+\; n\,H_2O
\end{array}
$$

Nylon 6,6, a polyamide

SUMMARY

Carboxylic acids (RCOOH), **esters** (RCOOR′), and **amides** (RCONH$_2$) occur widely throughout all living organisms. Structurally, these three families are related in that all have a carbonyl group bonded to a strongly electron-attracting atom (O or N). Compounds in all three families undergo **substitution reactions** in which a group we can represent by **—X** (for example, —OR, —NR$_2$′, or —OH) substitutes for (replaces) the —OH, —OR′, or —NH$_2$ group of the starting material.

A carbonyl-group substitution reaction

$$
\underset{\overset{\|}{O}}{R-C}-OH\;(-OR',\;-NH_2)\;+\;H-X\;\longrightarrow
$$

$$
\underset{\overset{\|}{O}}{R-C}-X\;+\;H-OH\;(H-OR',\;H-NH_2)
$$

Acid anhydrides (RCO$_2$COR) resemble other carboxylic acid derivatives in undergoing carbonyl-group substitution reactions but are not found in living things because they are hydrolyzed by water.

Carboxylic acids exist as hydrogen-bonded dimers and are higher boiling than comparable alcohols. The simpler acids are water-soluble liquids with strong odors. Diacids, unsaturated acids, and aromatic acids are all common. The simpler esters are volatile liquids with fruit-like odors and some water solubility, but esters are not hydrogen-bonded. Unsubstituted amides are hydrogen-

bonded, and all but formamide are solids and are water-soluble. Esters and amides are neither acidic nor basic.

Carboxylic acids are weak acids and therefore react with bases like NaOH to form water-soluble salts that contain **carboxylate anions** (RCOO$^-$). Carboxylic acids undergo conversion into esters by reaction with an alcohol in the presence of a strong-acid catalyst. The net effect of the **esterification reaction** is a substitution of —OR′ for —OH (RCOOH → RCOOR′).

Esters also undergo substitution reactions. For example, they undergo a **hydrolysis reaction** with water to yield a carboxylic acid and an alcohol. The reaction is catalyzed by both acids and, in what is known as **saponification,** by bases. Esters also undergo the **Claisen condensation reaction,** which joins two ester molecules together and forms a keto ester product.

Amides are usually prepared by reaction of a carboxylic acid with an amine in the presence of an activator. Like esters, amides undergo acid- and base-catalyzed hydrolysis, yielding carboxylic acid and amine products.

Certain inorganic acids such as nitric acid and phosphoric acid are analogous to organic carboxylic acids in that they react with alcohols to form esters. **Phosphate esters, diphosphate esters,** and **triphosphate esters** are particularly important in many biological processes.

Despite their complex appearance, many organic reactions fall into the categories known as **addition, elimination,** and **substitution.**

REVIEW PROBLEMS

Carboxylic Acids and Anhydrides

6.17 What are the structural differences among carboxylic acids, acid anhydrides, esters, and amides?

6.18* In what general way do carboxylic acids, acid anhydrides, esters, and amides differ from ketones and aldehydes?

6.19 Write the equation for the ionization of benzoic acid in water.

6.20 Show how two molecules of a carboxylic acid can hydrogen-bond to each other.

6.21 There are four carboxylic acids with the formula C$_5$H$_{10}$O$_2$. Draw and name each one.

6.22* Assume that you have a sample of propanoic acid dissolved in water.
(a) Draw the structure of the major species present in the water solution.
(b) Now assume that aqueous HCl is added to the propanoic acid solution until pH 2 is reached. Draw the structure of the major species present.

(c) Finally, assume that aqueous NaOH is added to the propanoic acid solution until pH 12 is reached. Draw the structure of the major species present.

6.23 Give IUPAC names for these carboxylic acids:

(a) CH$_3$CH$_2$CH$_2$CH$_2$CH$_2$COH (with =O)

(b) CH$_3$CH$_2$CH$_2$CHCH$_3$ with COOH

(c) CH$_3$CH$_2$CHCH$_2$CH$_3$ with COOH

(d) cyclopropyl—CH$_2$CH$_2$COH (with =O)

(e) BrCH$_2$CH$_2$CHCOH with CH$_3$ (and =O)

(f) o-CH$_3$ benzene—COOH

(g) (CH$_3$CH$_2$)$_3$CCOOH

(h) CH$_3$(CH$_2$)$_5$COOH

6.24*   Give IUPAC names for these carboxylic acid salts:

(a)

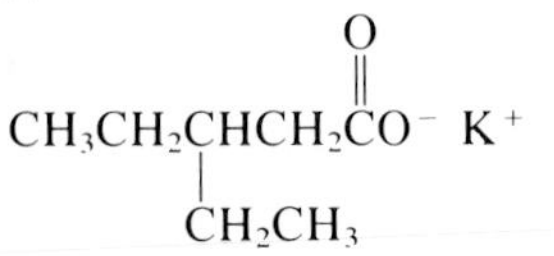

$$CH_3CH_2CHCH_2\overset{\overset{\displaystyle O}{\|}}{C}O^- \ K^+$$
$$|$$
$$CH_2CH_3$$

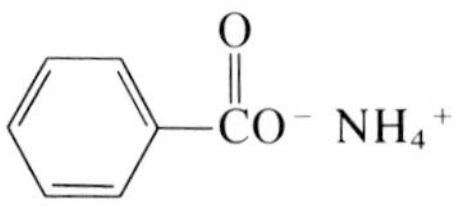

(b)

$$\text{(benzene ring)}\overset{\overset{\displaystyle O}{\|}}{C}O^- \ NH_4^+$$

(c)

$$[CH_3CH_2\overset{\overset{\displaystyle O}{\|}}{C}O^-]_2 \ Ca^{2+}$$

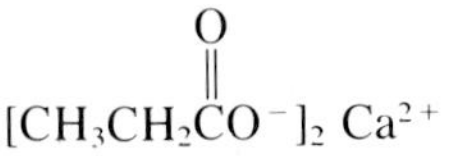

6.25   Draw structures corresponding to these names:
(a) 3,4-dimethylpentanoic acid
(b) triphenylacetic acid
(c) *m*-ethylbenzoic acid
(d) methylammonium butanoate
(e) 2,2-dichlorobutanoic acid
(f) 3-hydroxyhexanoic acid
(g) 3,3-dimethyl-4-phenylpentanoic acid
(h) benzoic anhydride

6.26*   Draw and name three different carboxylic acids with the formula $C_7H_{14}O_2$.

6.27   Malic acid, a dicarboxylic acid found in apples, has the IUPAC name *hydroxybutanedioic acid*. Draw its structure.

6.28   What is the formula of the disodium salt of malic acid (Problem 6.27)?

6.29   Aluminum acetate is used as an antiseptic ingredient in some skin-rash ointments. Draw its structure.

6.30*   How many grams of KOH does it take to neutralize these acids?
(a) 10.0 g of acetic acid
(b) 250 mL of 2.0 M propanoic acid

6.31   Write the reaction for the neutralization of citric acid (Section 6.3) by NaOH. What volume in milliliters of 0.040 M NaOH is required to neutralize 60.0 mL of 0.020 M citric acid?

6.32*   What volume of 0.20 M acetic acid is needed to react with 1.4 g of $Ca(OH)_2$?

6.33   What is the molarity of an oxalic acid solution if 15.0 mL is needed to react with (a) 0.100 g of NaOH (b) 20.0 mL of 0.300 M NaOH?

6.34   Look at Table 6.4 and tell which is the stronger acid, chloroacetic acid or acetic acid?

Esters and Amides

6.35   Draw and name compounds that meet these descriptions:
(a) three different amides with the formula $C_5H_{11}NO$
(b) three different esters with the formula $C_6H_{12}O_2$

6.36*   Give IUPAC names for these esters:

(a) 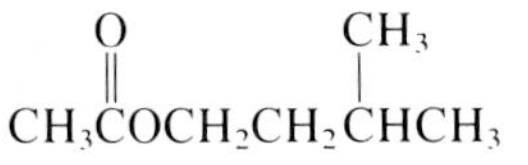
$$CH_3\overset{\overset{\displaystyle O}{\|}}{C}OCH_2CH_2\overset{\overset{\displaystyle CH_3}{|}}{C}HCH_3$$

(b)
$$CH_3\overset{\overset{\displaystyle CH_3}{|}}{C}HCH_2CH_2\overset{\overset{\displaystyle O}{\|}}{C}OCH_3$$

(c)
$$(CH_3)_3C\overset{\overset{\displaystyle O}{\|}}{C}OCH_2CH_3$$

(d) 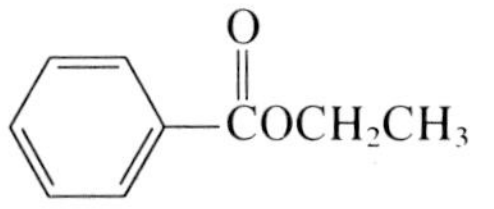
$$\text{(benzene ring)}\overset{\overset{\displaystyle O}{\|}}{C}OCH_2CH_3$$

(e) 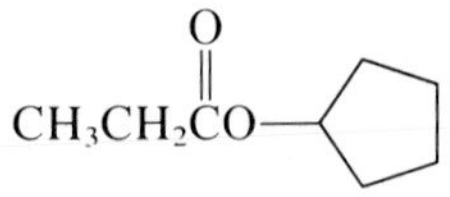
$$CH_3CH_2\overset{\overset{\displaystyle O}{\|}}{C}O\text{—(cyclopentane ring)}$$

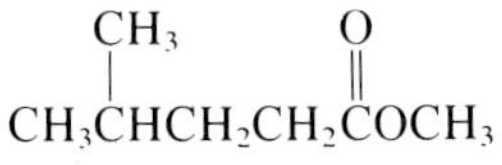

6.37   Draw structures corresponding to these IUPAC names:
(a) methyl pentanoate
(b) isopropyl 2-methylbutanoate
(c) cyclohexyl acetate   (d) phenyl *o*-hydroxybenzoate

6.38*   Show the structures of the carboxylic acids and alcohols you would use to prepare each of the esters in Problem 6.37.

6.39   Provide IUPAC names for these compounds:
(a) 
$$CH_3CH_2CHCH_2CH_3$$
$$|$$
$$CONH_2$$

(b) 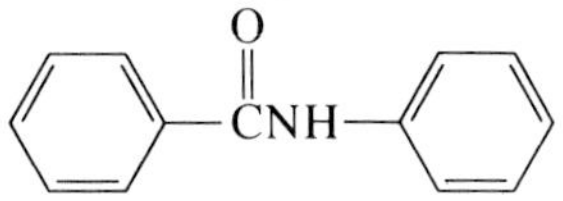
$$\text{(benzene ring)}\overset{\overset{\displaystyle O}{\|}}{C}NH\text{—(benzene ring)}$$

(c) $HCN(CH_3)_2$ with O above

(d) 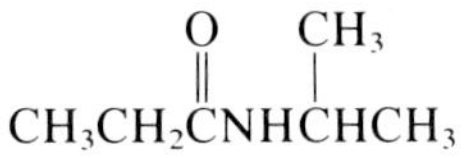
$$CH_3CH_2\overset{\overset{\displaystyle O}{\|}}{C}NH\overset{\overset{\displaystyle CH_3}{|}}{C}HCH_3$$

6.40*   Show how you could prepare each of the amides in Problem 6.39 from an appropriate carboxylic acid and amine.

6.41   Draw structures corresponding to these IUPAC names:
(a) 3-methylpentanamide   (b) *N*-phenylacetamide
(c) *N*-ethyl-*N*-methylbenzamide
(d) 2,3-dibromohexanamide

6.42*   What compounds would result from hydrolysis of each of the amides listed in Problem 6.41?

Reactions of Carboxylic Acids and Their Derivatives

6.43   What general kind of reaction do carboxylic acids and their derivatives undergo? Write the general equation for this type of reaction.

6.44*   Methyl *o*-aminobenzoate, commonly called *methyl anthranilate*, is used as a flavoring agent in grape drinks. Write an equation for the preparation of methyl anthranilate from the appropriate alcohol and carboxylic acid.

6.45   Procaine, a local anesthetic whose hydrochloride is Novocain, has the following structure. Identify the functional groups present and show the structures of the alcohol and carboxylic acids you would use to prepare it.

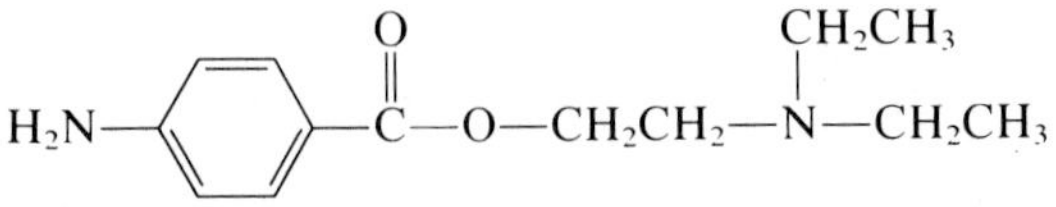

$$H_2N\text{—(benzene ring)—}\overset{\overset{\displaystyle O}{\|}}{C}\text{—O—}CH_2CH_2\text{—}\overset{\overset{\displaystyle CH_2CH_3}{|}}{N}\text{—}CH_2CH_3$$

Procaine

6.46* *Lactones* are cyclic esters in which the carboxylic acid part and the alcohol part are connected to form a ring. What product(s) would you expect to obtain from hydrolysis of butyrolactone?

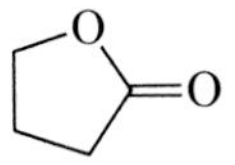

Butyrolactone (a cyclic ester)

6.47 Lidocaine (Xylocaine) is a local anesthetic closely related to procaine. Identify the functional groups present in lidocaine and show how you might prepare it from a carboxylic acid and an amine.

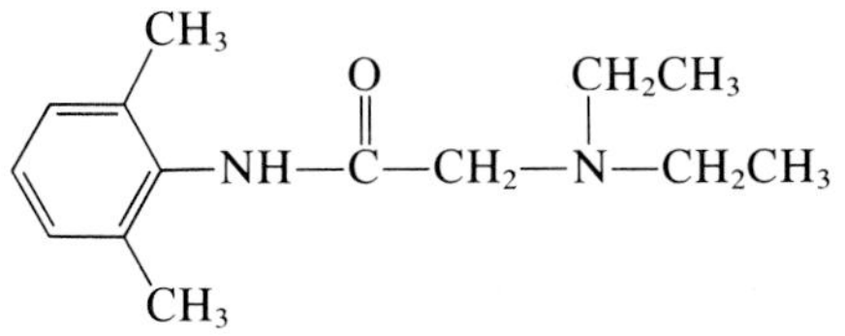

Lidocaine

6.48 Cocaine, an alkaloid isolated from the leaves of the South American coca plant, *Erythroxylon coca,* has the structure indicated. Identify the functional groups present and give the structures of the products you would obtain from hydrolysis of cocaine.

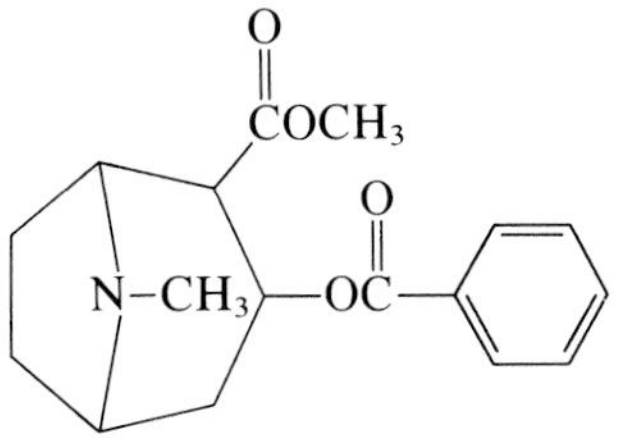

Cocaine

6.49 Household soap is a mixture of the sodium or potassium salts of long-chain carboxylic acids that arise from saponification of animal fat. Draw the structures of soap molecules produced in the following reaction:

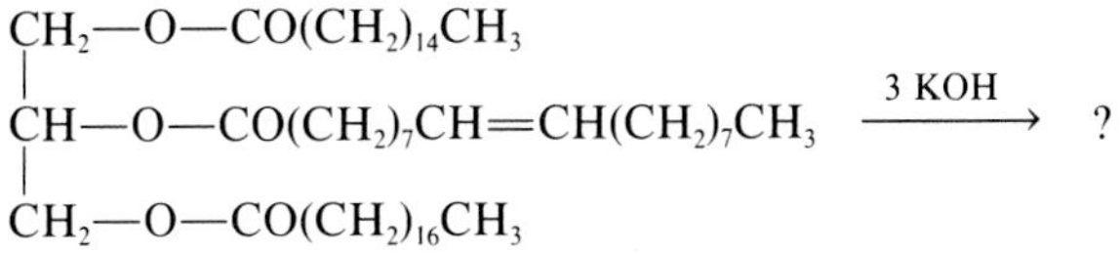

$$CH_2-O-CO(CH_2)_{14}CH_3$$
$$CH-O-CO(CH_2)_7CH=CH(CH_2)_7CH_3 \xrightarrow{\text{3 KOH}} ?$$
$$CH_2-O-CO(CH_2)_{16}CH_3$$

(A fat)

6.50* A *lactam* is a cyclic amide in which the carboxylic acid part and the amine part are connected. Draw the structure of the product(s) from hydrolysis of caprolactam, an industrial precursor of nylon.

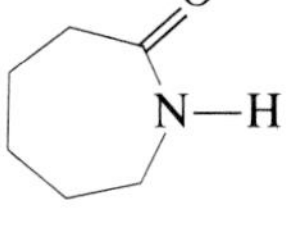

Caprolactam

6.51 Assume that you're given samples of pentanoic acid and methyl butanoate, both of which have the formula $C_5H_{10}O_2$. Describe how you can tell them apart.

6.52* Show how phenacetin, a drug once used in headache remedies, can be prepared from the reaction of an anhydride and an amine.

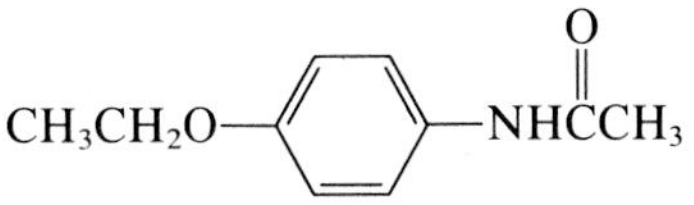

$$CH_3CH_2O-\!\!\!\bigcirc\!\!\!-NHCCH_3 \quad \text{Phenacetin}$$

6.53 The following phosphate ester is an important intermediate in carbohydrate metabolism. What products would result from hydrolysis of this ester?

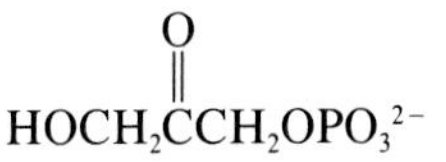

$$HOCH_2CCH_2OPO_3^{2-}$$

Claisen Condensation Reaction

6.54* What structural feature must an ester have in order to undergo a Claisen condensation reaction?

6.55 Which of the following esters can't undergo Claisen condensation reactions? Explain.

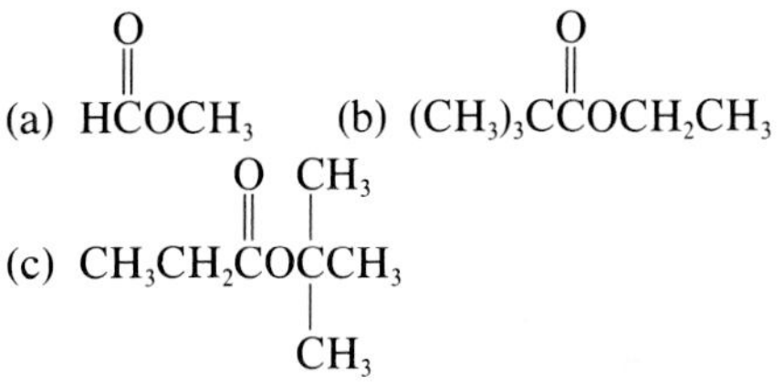

(a) $HCOCH_3$ (b) $(CH_3)_3CCOCH_2CH_3$

(c) $CH_3CH_2COCCH_3$
      |
      CH_3 ... CH_3

6.56 Draw the Claisen condensation products of the following esters:

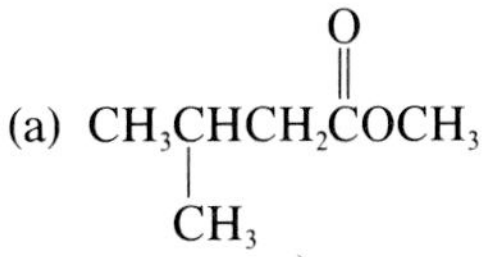

(a) $CH_3CHCH_2COCH_3$ (b) isopropyl acetate
      |
      CH_3

6.57 When a mixture of methyl acetate and methyl propanoate is treated with sodium methoxide in a Claisen condensation reaction, four keto ester products are formed. What are their structures?

Phosphate Esters

6.58* Why are pyrophosphoric acid and triphosphoric acid also acid anhydrides?

6.59 What is the result of complete acid hydrolysis of this diphosphate ester?

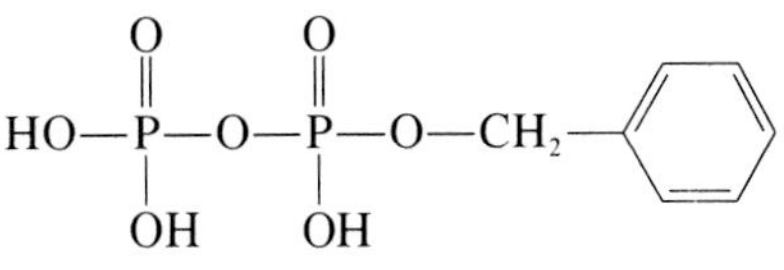

$$HO-P-O-P-O-CH_2-\!\!\!\bigcirc$$
$$\quad\; OH \quad\;\; OH$$

Organic Reactions

6.60* Categorize each of the following reactions as elimination, substitution, or addition.

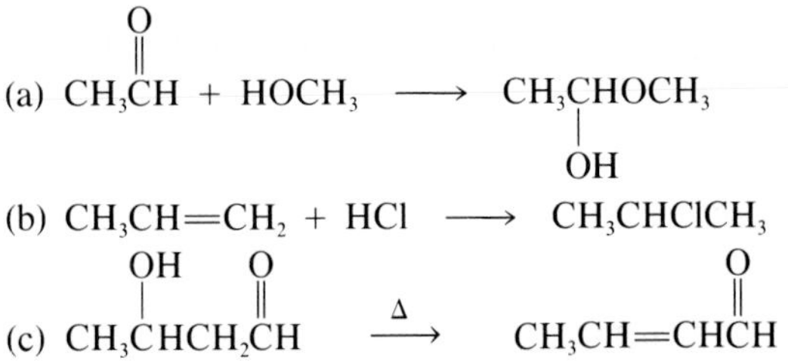

(a) $CH_3\overset{\overset{\displaystyle O}{\|}}{C}H + HOCH_3 \longrightarrow CH_3\underset{\underset{\displaystyle OH}{|}}{C}HOCH_3$

(b) $CH_3CH=CH_2 + HCl \longrightarrow CH_3CHClCH_3$

(c) $CH_3\underset{\underset{\displaystyle OH}{|}}{C}HCH_2\overset{\overset{\displaystyle O}{\|}}{C}H \overset{\Delta}{\longrightarrow} CH_3CH=CH\overset{\overset{\displaystyle O}{\|}}{C}H$

6.61 Describe in words the chemical changes that take place in the following series of biochemical reactions.

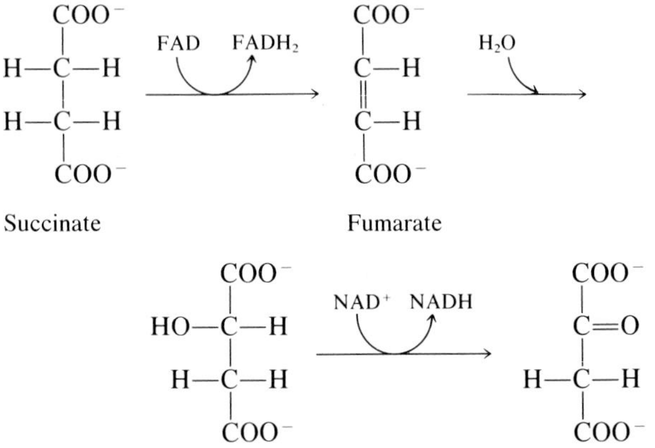

Succinate   Fumarate

Malate   Oxaloacetate

Applications

6.62* Against what type of microorganisms are sodium benzoate and potassium sorbate particularly effective? [App: Acid Salts as Food Additives]

6.63 What is the general formula of a thiol ester? [App: Thiol Esters]

6.64* Predict the structure of the thiol ester that could be formed from the reaction of acetic acid and methanethiol. [App: Thiol Esters]

6.65 Baked-on paints used for automobiles and many appliances are often based on alkyds, such as can be made from terephthalic acid and glycerol. Sketch a section of the resultant polyester polymer. Note that the glycerol can be esterified at any of the three alcohol groups, providing *cross linking* to form a very strong surface. [Int: Polyamides and Polyesters]

6.66 A simple polyamide can be made from ethylenediamine and oxalic acid. Draw a few units of the polymer formed. [Int: Polyamides and Polyesters]

$$H_2NCH_2CH_2NH_2 \quad \text{Ethylenediamine}$$

Additional Questions and Problems

6.67 Three amide isomers, *N,N*-dimethylformamide, *N*-methylacetamide, and propanamide, have respective boiling points of 153°C, 202°C, and 213°C. Explain these boiling points in light of their structural formulas.

6.68 Salol, the phenyl ester of salicylic acid, is used as an intestinal antiseptic. Draw the structure of phenyl salicylate.

6.69 Sketch the hydrogen-bonded dimer form of benzoic acid.

6.70* Propanamide and methyl acetate have about the same molar mass, both are quite soluble in water, and yet the boiling point of propanamide is 213°C while that of methyl acetate is 57°C. Explain.

6.71 Mention at least two simple chemical tests by which you could distinguish between benzaldehyde and benzoic acid.

6.72* What ester(s) would you need to produce this condensation product.

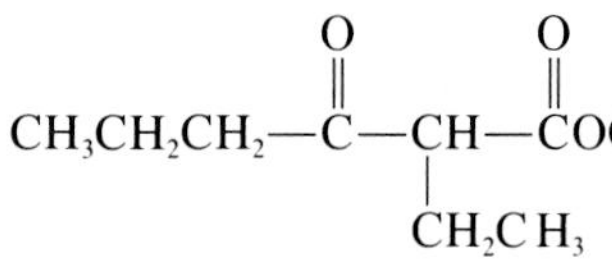

$$CH_3CH_2CH_2-\overset{\overset{\displaystyle O}{\|}}{C}-\underset{\underset{\displaystyle CH_2CH_3}{|}}{C}H-\overset{\overset{\displaystyle O}{\|}}{C}OCH_2CH_3$$

6.73 Write the formula of the triester formed from glycerol and stearic acid (Table 6.1).

6.74* Name these compounds.

(a) 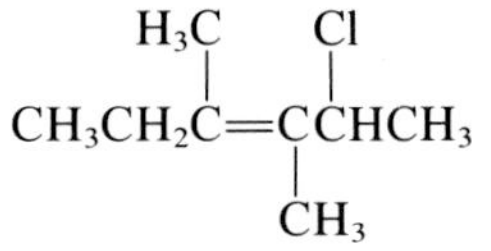 $CH_3CH_2\overset{\overset{\displaystyle H_3C}{|}}{C}=\overset{\overset{\displaystyle Cl}{|}}{\underset{\underset{\displaystyle CH_3}{|}}{C}}CHCH_3$

(b) $CH_3CH_2\overset{\overset{\displaystyle O}{\|}}{C}NCH_3$

(c) $(CH_3CH_2)_3C\overset{\overset{\displaystyle O}{\|}}{C}O-$

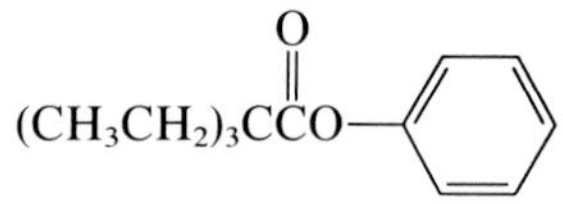

(d)

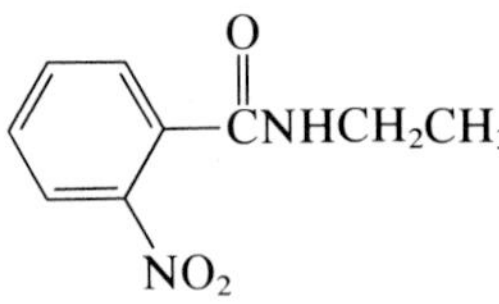

6.75 Complete these reactions.

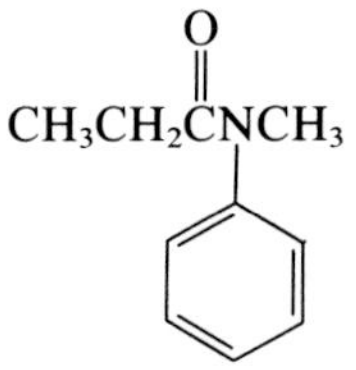

(a) 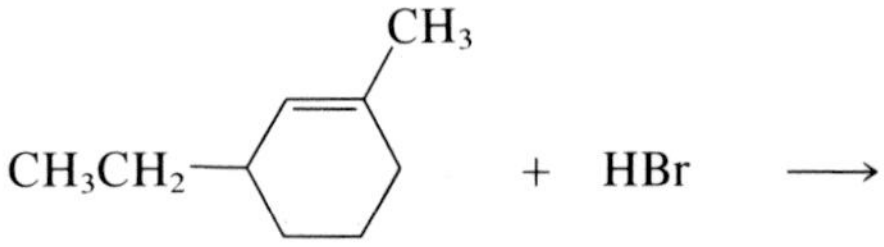 CH_3CH_2- + HBr $\longrightarrow$

(b) $CH_3\overset{O}{\overset{\|}{C}}H$ + $HOCH_2CH_2CH_3$ $\longrightarrow$ (hemiacetal)

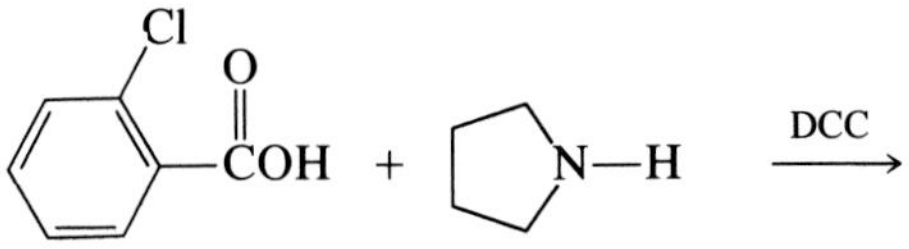

(e) $2\ CH_3CH_2\overset{O}{\overset{\|}{C}}CH_2CH_3$ $\xrightarrow{NaOH}$ (aldol)

(c) $CH_3CH_2\overset{O}{\overset{\|}{C}}O\overset{CH_3}{\overset{|}{C}}HCH_3$ + H_2O $\xrightarrow{H_2SO_4}$

(f) $CH_3CH_2CCl_2\overset{O}{\overset{\|}{C}}H$ $\xrightarrow{[O]}$

(d)

$\underset{\overset{}{\text{COH}}}{\overset{Cl}{\overset{O}{\overset{\|}{C}}}}$ + $\langle\ \rangle N{-}H$ $\xrightarrow{DCC}$

(g) $CH_3CH_2\overset{O}{\overset{\|}{C}}C(CH_3)_3$ $\xrightarrow[H_3O^+]{NaBH_4}$

CHAPTER

7

Amino Acids and Proteins

Muscle proteins are easy to see. Many other kinds of proteins that we'll be describing in this chapter are less visible, but equally important to the efforts of these athletes.

The word *protein* is familiar to everyone. Taken from the Greek *proteios,* meaning "primary," the name "protein" is an apt description for the group of biological molecules that are of primary importance to all living organisms. Approximately 50% of your body's dry weight is protein, and every reaction that occurs in your body is catalyzed by proteins. In fact, a human body contains well over *100,000* different kinds of proteins.

Proteins have many different biological functions. Some proteins, such as the *keratin* in skin, hair, and fingernails, and the *collagen* in connective tissue, serve a structural purpose. Other proteins, such as the *insulin* that controls glucose metabolism, serve as hormones to regulate specific body processes. And still other proteins, such as *DNA polymerase,* serve as enzymes, the biological catalysts that carry out all body chemistry.

In this chapter, we'll look at the following questions about proteins:

1. **What are the structures of amino acids?** The goal: Be able to recognize amino acid structures and give some representative structures of amino acids.

2. **What are the properties of amino acids?** The goal: Be able to describe how the properties of amino acids vary with their side chains and how their ionic charges vary with pH.

3. **Why do amino acids have "handedness"?** The goal: Be able to explain what is responsible for handedness in certain molecules and recognize simple molecules that display this property.

4. **What types of interactions occur between amino acids?** The goal: Be able to give examples of the disulfide bridge and the noncovalent interactions between backbone peptide links and between side chains.

5. **What are the primary structures of proteins?** The goal: Be able to use structural formulas or symbols to show the primary structures of proteins.

6. **What are the secondary, tertiary, and quaternary structures of proteins?** The goal: Be able to distinguish among these types of structures and explain the most common kinds of secondary protein structure.

7. **How are proteins classified?** The goal: Be able to describe three different ways of classifying proteins and give examples of each.

8. **What chemical properties do proteins have?** The goal: Be able to describe protein hydrolysis and denaturation and give some examples of agents that cause denaturation.

Protein A large biological molecule made of many amino acids linked together through amide bonds.

Amino acid A molecule that contains both an amino group and a carboxylic acid functional group; proteins are polymers of amino acids.

7.1 AN OVERVIEW OF PROTEIN STRUCTURE

Regardless of their differing biological functions, all proteins are chemically similar. **Proteins** are made up of many amino acids linked together to form a long chain. As their name implies, **amino acids** are molecules that contain two functional groups, an acidic carboxyl group (—COOH) and a basic amino group (—NH$_2$). Glycine is the simplest example of an amino acid.

Acidic carboxyl group

Basic amino group

$$H_2N-CH_2-\overset{\displaystyle O}{\overset{\displaystyle \|}{C}}-OH$$

Glycine—an amino acid

Peptide bond An amide bond that links two amino acids together.

Two or more amino acids can link together by forming amide bonds (Section 6.9), usually called **peptide bonds.** A *dipeptide* results from the linking together of two amino acids by formation of a peptide bond between the —NH$_2$ group of one and the —COOH group of the second; a *tripeptide* results from linkage of three amino acids via two peptide bonds; and so on. Any number of amino acids can link together to form a long chain. Chains with between about 10 and 100 amino acids are usually called **polypeptides,** while the term *protein* is usually reserved for larger chains.

Polypeptide A molecule composed of roughly 10–100 amino acids linked by peptide bonds.

Peptide bond

$$H_2N-\overset{\displaystyle }{\underset{\displaystyle R}{CH}}-\overset{\displaystyle O}{\overset{\displaystyle \|}{C}}-OH \; + \; H-NH-\overset{\displaystyle }{\underset{\displaystyle R'}{CH}}-\overset{\displaystyle O}{\overset{\displaystyle \|}{C}}-OH \; \longrightarrow \; H_2N-\overset{\displaystyle }{\underset{\displaystyle R}{CH}}-\overset{\displaystyle O}{\overset{\displaystyle \|}{C}}-NH-\overset{\displaystyle }{\underset{\displaystyle R'}{CH}}-\overset{\displaystyle O}{\overset{\displaystyle \|}{C}}-OH \; + \; H_2O$$

Two amino acids

A dipeptide

(R and R′ may be the same or different)

7.2 AMINO ACIDS

There are 20 different amino acids commonly found in proteins. As shown in Table 7.1, all 20 are **alpha (α) amino acids;** that is, the amino group in each is connected to the carbon atom *alpha to* (next to) the carboxylic acid group. The 20 amino acids differ only in the nature of the R group (called the *side chain*) attached to the α carbon).

Alpha (α) amino acid An amino acid in which the amino group is bonded to the carbon atom next to the —COOH group.

The —NH$_2$ group is on the carbon alpha to (next to) the —COOH.

$$H_2N-\overset{\displaystyle }{\underset{\displaystyle R}{\overset{\alpha}{CH}}}-\overset{\displaystyle O}{\overset{\displaystyle \|}{C}}-OH$$

This side-chain R group is different for each amino acid.

An α-amino acid

Nineteen of the 20 common amino acids are primary amines, RNH$_2$, and the remaining one (proline) is a secondary amine whose nitrogen and α-carbon atoms are joined together in a five-membered ring. Each amino acid is usually referred to by a three-letter shorthand code, such as Ala (alanine), Gly (glycine), Pro (proline), and so on. In addition, a space-saving one-letter code is often used. These codes are shown in Table 7.1.

The 20 common amino acids can be classified as *neutral, basic,* or *acidic,* depending on the structure of their side chains. Fifteen of the 20 have neutral side chains; 2 (aspartic acid and glutamic acid) have an additional carboxylic acid group in their side chains and are classified as acidic amino acids; and 3 (lysine, arginine, and histidine) have an additional amine nitrogen atom in their side chains and are classified as basic amino acids. The 15 neutral amino acids can be further divided into those with nonpolar hydrocarbon side chains and those with polar functional groups such as amide or hydroxyl groups. Nonpolar side chains are often described as **hydrophobic** (water-fearing) because they are repelled by water, while polar side chains are described as **hydrophilic** (water-loving) because they are attracted to water.

All 20 amino acids are needed to make proteins, but our bodies can synthesize only 10, known as **nonessential amino acids** because we don't have to get them from food. The remaining 10 are called **essential amino acids** (in red in Table 7.1) because they must be obtained from our diet. Failure to receive an adequate dietary supply of the essential amino acids leads to retarded growth and development in children and to disease and body deterioration in adults.

Hydrophobic Water-fearing; a hydrophobic substance does not dissolve in water.

Hydrophilic Water-loving; a hydrophilic substance dissolves in water.

Nonessential amino acid An amino acid that is synthesized by the body.

Essential amino acid An amino acid that can't be synthesized by the body and so must be obtained in the diet.

Practice Problems

7.1 Name the common amino acids that contain an aromatic ring. Name those that contain sulfur. Name those that are alcohols. Name those that have alkyl-group side chains.

7.2 Draw alanine showing the tetrahedral geometry of its α carbon.

7.3 DIPOLAR STRUCTURE OF AMINO ACIDS

Since amino acids contain both acidic and basic groups in the same molecule, they undergo an *internal* acid–base reaction and exist primarily as dipolar ions called **zwitterions** (German *zwitter*, "hybrid"). Because a zwitterion has one plus charge and one minus charge, it is electrically neutral overall. Although you'll often see amino acids written in the nonionized form for convenience, they're actually never in this form in either the solid state or aqueous solution.

Zwitterion A neutral dipolar compound that contains both + and – charges in its structure.

$$H_2N-\overset{\displaystyle |}{\underset{\displaystyle R}{CH}}-\overset{\displaystyle O}{\overset{\displaystyle \|}{C}}-OH \qquad\qquad \overset{+}{H_3N}-\overset{\displaystyle |}{\underset{\displaystyle R}{CH}}-\overset{\displaystyle O}{\overset{\displaystyle \|}{C}}-C$$

Amino acid—nonionized form   Amino acid—zwitterion form

Because they're zwitterions, amino acids have many of the physical properties we associate with salts. Thus, amino acids are crystalline, have high melting points, and are soluble in water but not in hydrocarbon solvents. In addition, amino acids can react either as acids or as bases depending on the circumstances. In acid solution (low pH), amino acid zwitterions *accept* a proton on their basic —COO⁻ group to yield a cation. In base solution (high pH), amino acid zwitterions *lose* a proton from their acidic —NH$_3^+$ group to yield an anion. The predominant structure and charge of an amino acid at any given time depend on the particular amino acid and on the pH of the medium.

Table 7.1   Structure of the 20 Common Amino Acids[a]

Name	Abbreviations	Molecular Weight	Structure	Isoelectric Point
Basic amino acids				
Arginine[a]	Arg (R)	174	 $H_2NC-NHCH_2CH_2CH_2-\overset{\overset{NH_2}{\mid}}{\underset{\underset{H}{\mid}}{C}}-COOH$ (with $\overset{NH}{\parallel}$ on H_2NC)	10.8
Histidine	His (H)	155	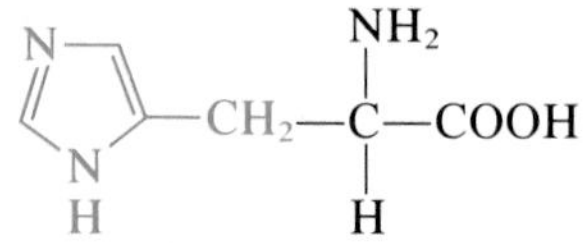 imidazole ring $-CH_2-\overset{\overset{NH_2}{\mid}}{\underset{\underset{H}{\mid}}{C}}-COOH$	7.6
Lysine	Lys (K)	146	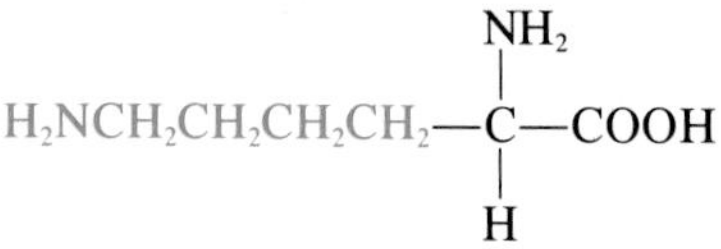 $H_2NCH_2CH_2CH_2CH_2-\overset{\overset{NH_2}{\mid}}{\underset{\underset{H}{\mid}}{C}}-COOH$	9.7
Neutral Amino Acids—Nonpolar Side Chains				
Alanine	Ala (A)	89	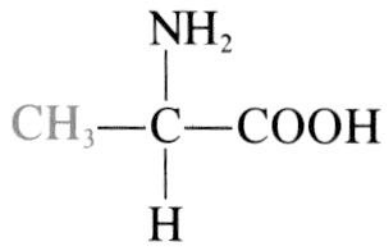 $CH_3-\overset{\overset{NH_2}{\mid}}{\underset{\underset{H}{\mid}}{C}}-COOH$	6.0
Glycine	Gly (G)	75	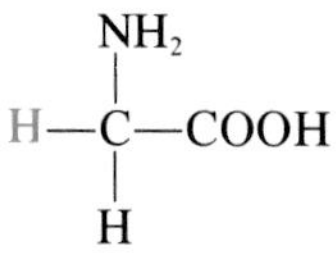 $H-\overset{\overset{NH_2}{\mid}}{\underset{\underset{H}{\mid}}{C}}-COOH$	6.0
Isoleucine	Ile (I)	131	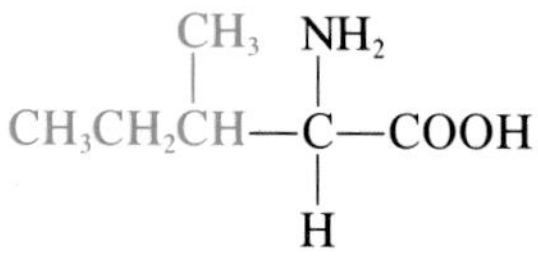 $CH_3CH_2\overset{\overset{CH_3}{\mid}}{CH}-\overset{\overset{NH_2}{\mid}}{\underset{\underset{H}{\mid}}{C}}-COOH$	6.0
Leucine	Leu (L)	131	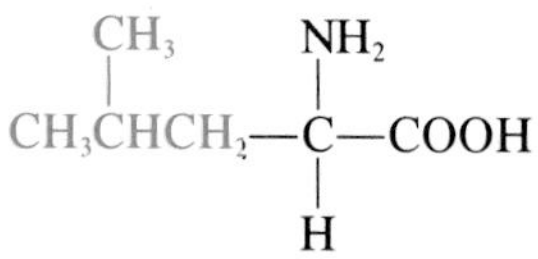 $CH_3\overset{\overset{CH_3}{\mid}}{CH}CH_2-\overset{\overset{NH_2}{\mid}}{\underset{\underset{H}{\mid}}{C}}-COOH$	6.0
Methionine	Met (M)	149	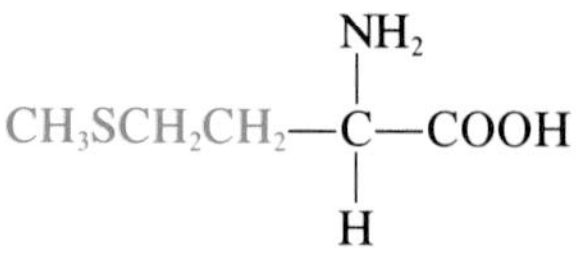 $CH_3SCH_2CH_2-\overset{\overset{NH_2}{\mid}}{\underset{\underset{H}{\mid}}{C}}-COOH$	5.7
Phenylalanine	Phe (F)	165	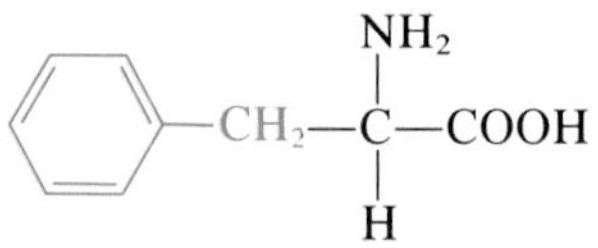 phenyl $-CH_2-\overset{\overset{NH_2}{\mid}}{\underset{\underset{H}{\mid}}{C}}-COOH$	5.5
Proline	Pro (P)	115	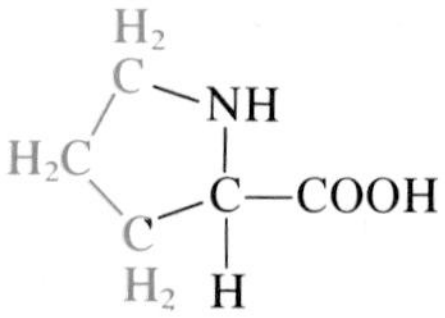 pyrrolidine ring $-\overset{\overset{}{}}{\underset{\underset{H}{\mid}}{C}}-COOH$	6.3
Valine	Val (V)	117	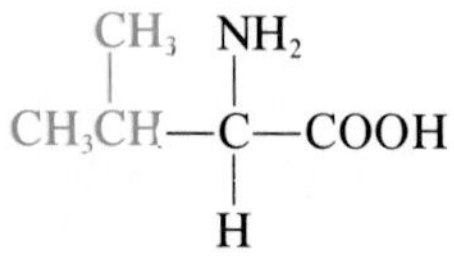 $CH_3\overset{\overset{CH_3}{\mid}}{CH}-\overset{\overset{NH_2}{\mid}}{\underset{\underset{H}{\mid}}{C}}-COOH$	6.0

Table 7.1   Structures of the 20 Common Amino Acids[a] (continued)

Name	Abbreviations	Molecular Weight	Structure	Isoelectric Point
Neutral Amino Acids—Polar Side Chains				
Asparagine	Asn (N)	132	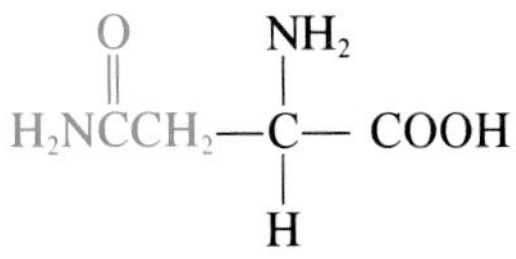 H₂NCCH₂—C—COOH (with O double bond and NH₂, H)	5.4
Cysteine	Cys (C)	121	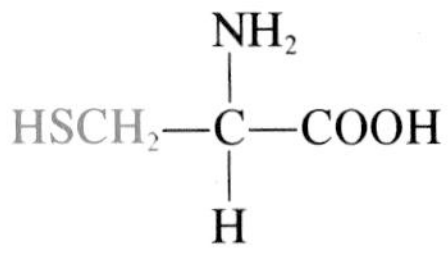 HSCH₂—C—COOH (with NH₂, H)	5.0
Glutamine	Gln (Q)	146	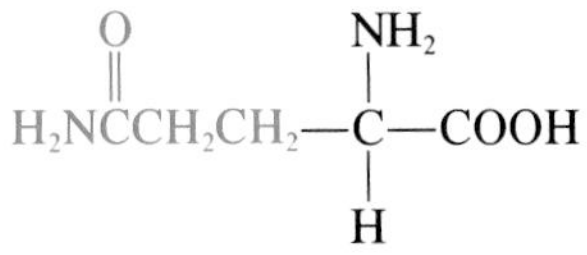 H₂NCCH₂CH₂—C—COOH (with O double bond and NH₂, H)	5.7
Serine	Ser (S)	105	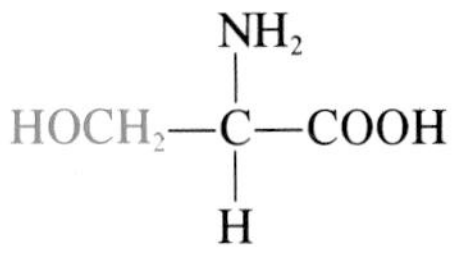 HOCH₂—C—COOH (with NH₂, H)	5.7
Threonine	Thr (T)	119	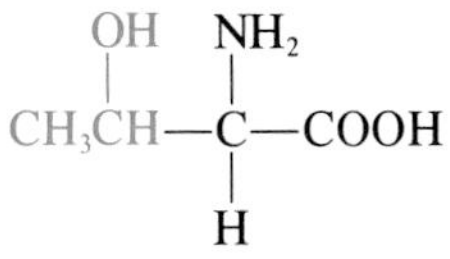 CH₃CH—C—COOH (with OH, NH₂, H)	5.6
Tryptophan	Trp (W)	204	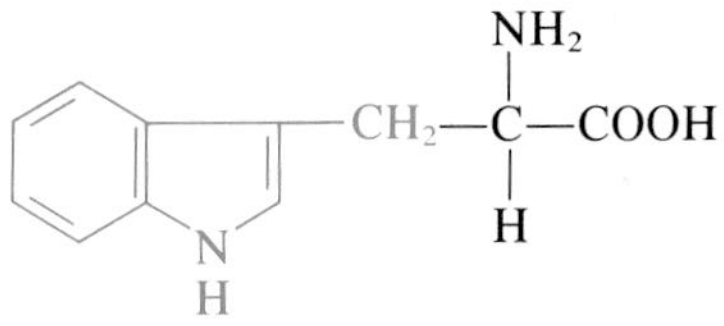 (indole ring)—CH₂—C—COOH (with NH₂, H)	5.9
Tyrosine	Tyr (Y)	181	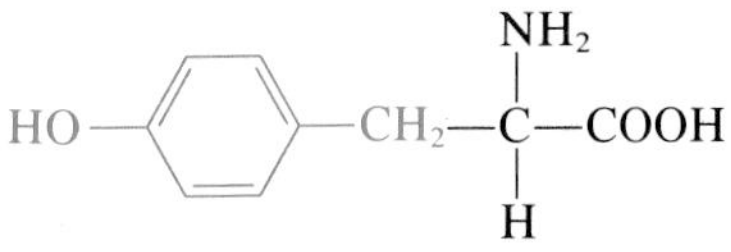 HO—(benzene ring)—CH₂—C—COOH (with NH₂, H)	5.7
Acidic Amino Acids				
Aspartic acid	Asp (D)	133	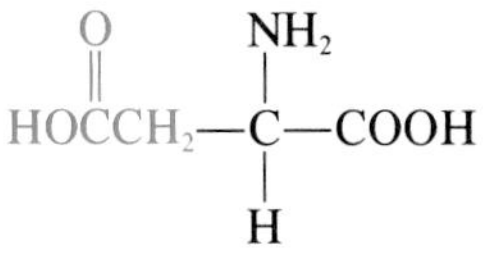 HOCCH₂—C—COOH (with O double bond and NH₂, H)	3.0
Glutamic acid	Glu (E)	147	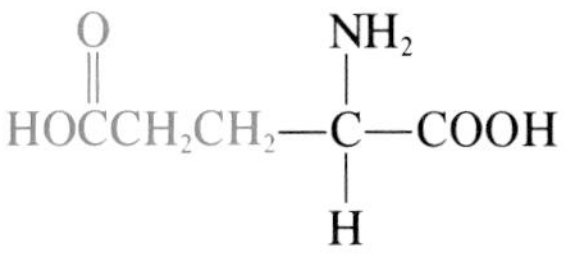 HOCCH₂CH₂—C—COOH (with O double bond and NH₂, H)	3.2

[a] Amino acids essential to the human diet are shown in red. Some lists, including that of Recommended Dietary Allowances, omit arginine as an essential amino acid because it is synthesized in the body; others include it as essential because, although synthesized, most is broken down and therefore not available for protein synthesis.

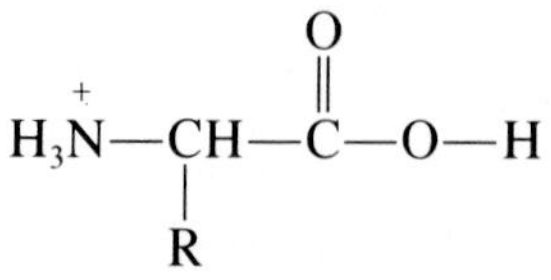

$$H_3\overset{+}{N}-CH-\overset{O}{\underset{R}{\overset{\|}{C}}}-O-H \qquad H_3\overset{+}{N}-CH-\overset{O}{\underset{R}{\overset{\|}{C}}}-O^- \qquad H_2N-CH-\overset{O}{\underset{R}{\overset{\|}{C}}}-O^-$$

| Predominant form at low pH (acidic) +1 charge | Predominant form at intermediate pH 0 charge | Predominant form at high pH (basic) −1 charge |

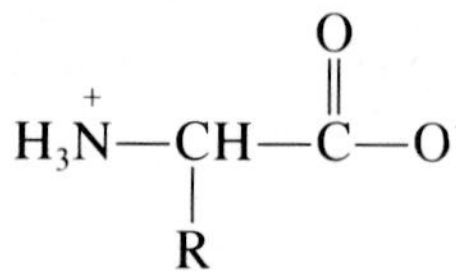

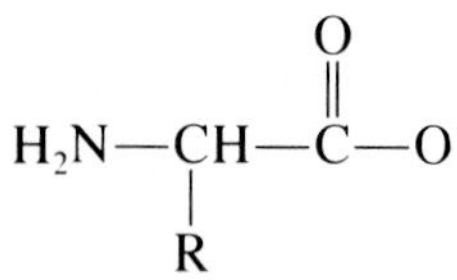

Isoelectric point The pH at which a large sample of amino acid molecules has equal numbers of + and − charges.

The pH at which the numbers of positive and negative charges in a large sample are equal and an amino acid exists mainly in its neutral dipolar form is called the amino acid's **isoelectric point (pI).** As indicated in Table 7.1, the 15 amino acids with neutral side chains have isoelectric points near neutrality in the pH range 5.0–6.5. Alanine, for example, has pI = 6.0. At the physiological pH of 7.4 a sample of an amino acid like alanine has a slight excess of negative charges because some acidic $-NH_3^+$ groups have lost their protons.

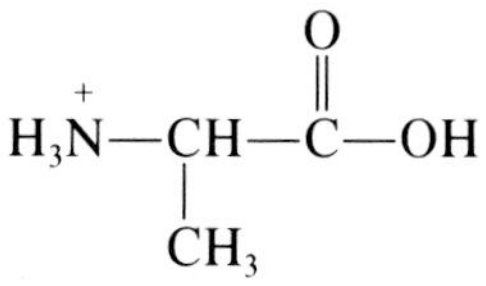

$$H_3\overset{+}{N}-CH-\overset{O}{\underset{CH_3}{\overset{\|}{C}}}-OH \qquad H_3\overset{+}{N}-CH-\overset{O}{\underset{CH_3}{\overset{\|}{C}}}-O^- \qquad H_2N-CH-\overset{O}{\underset{CH_3}{\overset{\|}{C}}}-O^-$$

| Alanine at pH 2.0; positively charged | Alanine at isoelectric point, pH 6; dipolar ion | Alanine at pH 12; negatively charged |

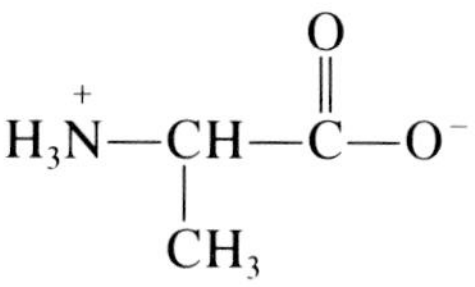

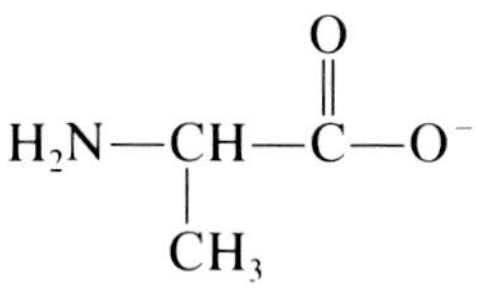

The two amino acids with acidic side chains, aspartic acid and glutamic acid, have isoelectric points at more acidic (lower) pH values than those with neutral side chains. Since the side chain —COOH groups of these compounds are substantially ionized at physiological pH of 7.4, they are often referred to as *aspartate* and *glutamate*.

The three amino acids with basic side chains have isoelectric points at more basic (higher) pH values than those with neutral side chains. In near-neutral solutions, lysine and arginine carry positive charges on their side chains, and the side chain in histidine is about 50% ionized.

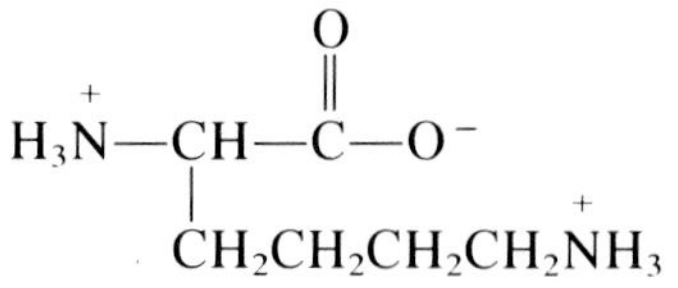

$$H_3\overset{+}{N}-CH-\overset{O}{\overset{\|}{C}}-O^- \\ \underset{CH_2CH_2CH_2CH_2\overset{+}{N}H_3}{}$$

Lysine

$$H_3\overset{+}{N}-CH-\overset{O}{\overset{\|}{C}}-O^- \qquad NH_2 \\ \underset{CH_2CH_2CH_2NH\overset{\|}{C}NH_2}{}$$

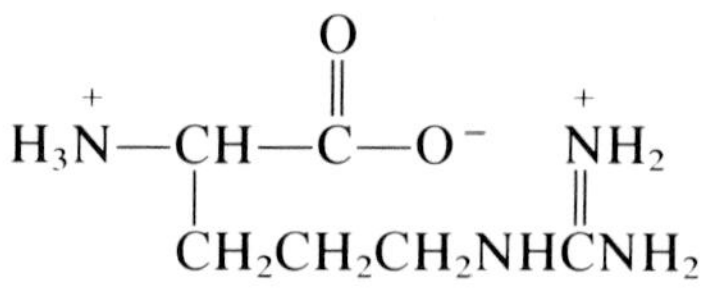

Arginine

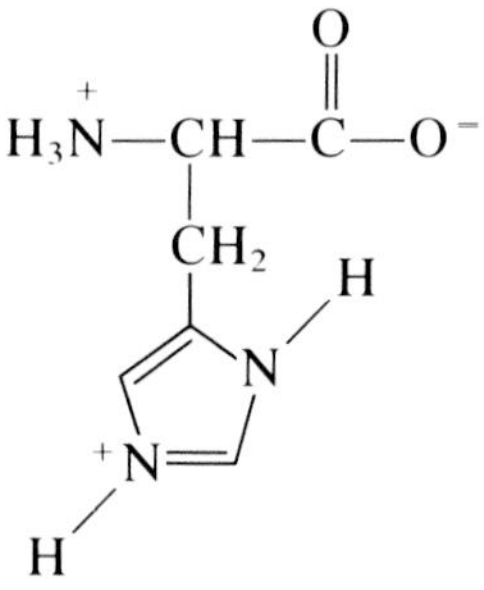

Histidine

Because most proteins contain large numbers of both acidic and basic side chains, they can react with both H_3O^+ and OH^- ions, allowing them to act as buffers and help maintain the constant pH of body fluids.

Solved Problem 7.1 Draw the zwitterion forms of phenylalanine and serine. Which of these two acids has a hydrophobic side chain and which has a hydrophilic side chain?

Solution The zwitterions are shown here. The side chain in phenylalanine ($C_6H_5CH_2$—) is a hydrocarbon and therefore is nonpolar and hydrophobic. The side chain in serine ($HOCH_2$—) contains a polar hydroxyl group and is therefore hydrophilic.

$$H_3\overset{+}{N}-CH-\overset{\overset{\displaystyle O}{\|}}{C}-O^-$$

$$\underset{|}{CH_2}$$

Phenylalanine

$$H_3\overset{+}{N}-CH-\overset{\overset{\displaystyle O}{\|}}{C}-O^-$$

$$\underset{|}{CH_2OH}$$

Serine

Practice Problems

7.3 Look up the structure of valine in Table 7.1 and draw it at low pH, at its isoelectric point, and at high pH.

7.4 Draw glutamic acid in its fully ionized form. Is this amino acid hydrophobic or hydrophilic?

7.4 HANDEDNESS

Are you right-handed or left-handed? Although you may not often think about it, handedness affects almost everything you do. Anyone who has played much softball knows that the last available glove always fits the wrong hand; any left-handed person taking notes in a lecture knows that it's awkward to do so in a right-handed chair. The reason for these difficulties is that your hands aren't identical. Rather, they're **mirror images.** When you hold your left hand up to a mirror, the image you see looks like your right hand (Figure 7.1). Try it.

Not all objects are handed, of course. There's no such thing as a right-

Mirror image The reverse image produced when an object is reflected in a mirror.

Figure 7.1
The meaning of *mirror image:* If you hold your left hand up to a mirror, the image you see looks like your right hand.

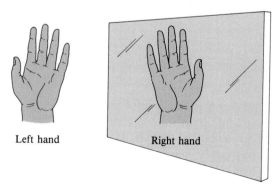

Left hand   Right hand

AN APPLICATION: THE HENDERSON-HASSELBALCH EQUATION AND AMINO ACIDS

The *Henderson-Hasselbalch* equation relates the pH of a weak-acid solution to the acid's pK_a (negative logarithm of the K_a) and to the concentrations of the acid and its anion.

The Henderson-Hasselbalch equation:

$$pH = pK_a + \log \frac{[A^-]}{[HA]}$$

The Henderson-Hasselbalch equation is particularly valuable for estimating $[A^-]/[HA]$ for amino acids at varying pHs. For example, when $pH = pK_a$, then $\log [A^-]/[HA] = 0$ and $[A^-]/[HA] = 1$. Thus, half of a weak acid is present as HA and half as A^-:

If $\qquad pH = pK_a$

then $\qquad \log [A^-]/[HA] = 0$

so $\quad [A^-]/[HA] = 1/1 \quad$ and $\quad [A^-] = [HA]$

At the point where $pH = pK_a - 2$, then $[A^-]/[HA] = 1/100$ and the acid is only about 1% ionized:

If $\qquad pH = pK_a - 2$

then $\qquad \log [A^-]/[HA] = -2$

so $\qquad [A^-]/[HA] = 1/100$

The Henderson-Hasselbalch equation thus gives the following relationships:

$pH = pK_a + 2$	$[A^-]/[HA] = 100/1$
$pH = pK_a + 1$	$[A^-]/[HA] = 10/1$
$pH = pK_a$	$[A^-]/[HA] = 1/1$
$pH = pK_a - 1$	$[A^-]/[HA] = 1/10$
$pH = pK_a - 2$	$[A^-]/[HA] = 1/100$

The pK_a of any acid can be experimentally determined by a titration in which a base is slowly added to the acid solution and the change in pH is recorded. The pH at which one-half the acid present has reacted is equal to the pK_a. If the acid has two acidic hydrogens, however, then two pK_a's are measured. Take alanine, for example. The accompanying *titration curve* shows the results of starting at low pH with fully protonated alanine (structure at the bottom) and titrating with 2 equivalents of NaOH until the alanine is fully deprotonated (structure at the top).

The two "legs" of the alanine titration curve—the first leg from pH 1 to 6 and the second leg from pH 6 to 11—show first the titration of the more acidic —COOH group and second the titration of the —NH_3^+ group. When 0.5 equivalent of NaOH is added, the first deprotonation is half complete and the pH equals the first pK_a; when 1.0 equivalent of NaOH is added, the first deprotonation is fully complete and the isoelectric point is reached; when 1.5 equivalents of NaOH is added, the second deprotonation is half complete and the pH equals the second pK_a; and when 2.0 equivalents of NaOH is added, the second deprotonation is complete. Of course, the procedure can also be run in reverse. That is,

Chiral Having (right or left) handedness; able to have two different mirror image forms.

Achiral The opposite of chiral; not having (right or left) handedness.

handed tennis ball or a left-handed coffee mug. When a tennis ball or a coffee mug is held up to a mirror, the image reflected is identical to the ball or mug itself. Objects that have handedness are said to be **chiral** (pronounced **ky**-ral, from the Greek *cheir,* meaning "hand"), and objects like the coffee mug that lack handedness are said to be nonchiral, or **achiral.**

Why is it that some objects are chiral (handed) but others aren't? In general, an object is not chiral if it is symmetrical. Conversely, an object *is*

once the pK_a's of an amino acid are known, a titration curve can be calculated with the Henderson-Hasselbalch equation.

The titration curve shows that the isoelectric point of an amino acid with two acidic groups is halfway between its two pK_a's. The value of pI, the pH at the isoelectric point, is therefore one-half the sum of the pK_a's, as illustrated here for alanine:

$$pI = \frac{pK_{a1} + pK_{a2}}{2} = \frac{2.3 + 9.8}{2} = 6.1$$

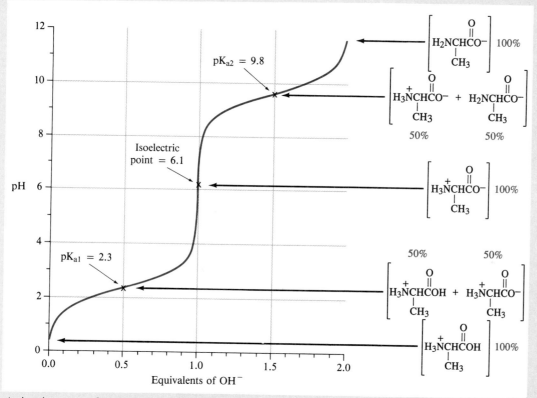

A titration curve for alanine. At pH < 1, alanine is entirely protonated; at pH 2.3, alanine is a 50:50 mix of protonated and neutral forms; at pH 6.1, alanine is entirely neutral; at pH 9.8, alanine is a 50:50 mix of neutral and deprotonated forms; at pH > 11, alanine is entirely deprotonated.

Symmetry plane An imaginary plane cutting through the middle of an object so that one half of the object is a mirror image of the other half.

chiral if it is *not* symmetrical. Symmetrical objects are those like the coffee mug that have an imaginary plane (a **symmetry plane**) cutting through their middle so that one half of the object is an exact mirror image of the other half. If you were to cut the mug in half, one half of the mug would be the mirror image of the other half. A hand, however, has no symmetry plane and is therefore chiral. If you were to cut a hand in two, one half of the hand would not be a mirror image of the other half (Figure 7.2).

Figure 7.2
The meaning of *symmetry plane:* An achiral object like the coffee mug has a symmetry plane passing through it, making the two halves mirror images. A chiral object like the hand has no symmetry plane because the two halves of the hand are not mirror images.

 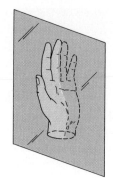

Practice Problems   **7.5**   Which of the following objects are handed?
(a) a glove   (b) a baseball   (c) a screw   (d) a nail

7.6   List three common objects that are handed and another three that aren't.

7.5   MOLECULAR HANDEDNESS AND AMINO ACIDS

Just as certain objects like a hand are chiral, certain *molecules* are also chiral. For example, compare propane and alanine (Figure 7.3). An alanine molecule has no symmetry plane because its two halves aren't mirror images. Like a hand, alanine is chiral and can exist in two forms: a "right-handed" form known as D-alanine and a "left-handed" form known as L-alanine. The two forms are not identical; they are related in the same way that your left and right hands are related. Propane, however, is achiral. It has a symmetry plane cutting through the three carbons such that one half of the molecule is a mirror image of the other half. Thus, propane exists in a single form.

Figure 7.3
Symmetry in molecules. Alanine (2-aminopropanoic acid) has no symmetry plane because the two halves of the molecule are not mirror images. Thus, alanine can exist in two forms—a "right-handed" form, referred to as D-alanine, and a "left-handed" form, referred to as L-alanine. Propane, however, has a symmetry plane and is achiral.

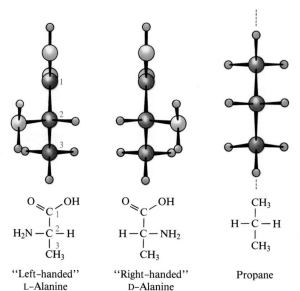

"Left-handed"   "Right-handed"   Propane
L-Alanine   D-Alanine

Why are some molecules chiral but others aren't? The answer has to do with the three-dimensional nature of organic compounds. Carbon forms four bonds that are oriented to the four corners of an imaginary tetrahedron. **Whenever a carbon atom is bonded to four *different* groups, chirality results.** If a carbon is bonded to two or three of the same groups, however, no chirality is present. In alanine, for example, carbon 2 is bonded to four different groups: a —COOH group, an —H atom, an —NH$_2$ group, and a —CH$_3$ group. Thus, alanine is chiral. In propane, however, each of the three carbons is bonded to at least two groups—the H atoms—that are identical. Thus, propane is achiral.

Groups attached to carbon 2			Groups attached to carbon 2	
1. —COOH			1. —CH$_3$	identical
2. —H		different	2. —CH$_3$	
3. —NH$_2$			3. —H	identical
4. —CH$_3$			4. —H	

COOH
|
H$_2$N—C—H
|
CH$_3$

Alanine
(chiral)

CH$_3$
|
H—C—H
|
CH$_3$

Propane
(achiral)

The easiest way to see how tetrahedral geometry leads to chirality is to make paper models of the sort shown in Figure 7.4. Cut two large equilateral triangles out of stiff paper, fold each one so that its three corners come together to form a tetrahedron, and then color each corner as indicated. When four different colors (groups) are used for the four corners of the tetrahedrons, the two models are not identical but have a right-hand–left-hand relationship.

The two mirror-image forms of a chiral molecule like alanine are called **optical isomers,** or **enantiomers:** "optical" because of their effect on polarized

Optical isomers, enantiomers The two mirror-image forms of a chiral molecule.

Figure 7.4
Paper molecular models. The two tetrahedrons are mirror images (that is, are chiral) when the four corners have four different colors.

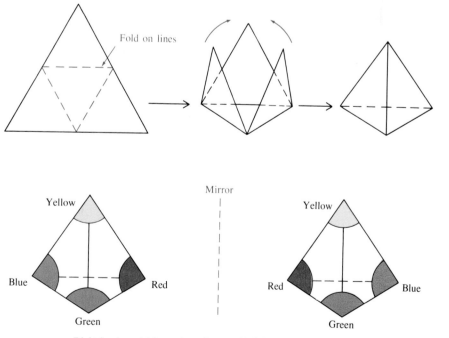

Fold on lines

Mirror

Yellow

Blue   Red

Green

Yellow

Red   Blue

Green

Right-hand model is a mirror image of left-hand model — *chiral.*

Stereoisomers Isomers that have the same molecular and structural formulas but different arrangements of their atoms in space.

light, as will be explained in Chapter 10. Like other isomers (Section 1.3), optical isomers have the same formula but have different arrangements of their atoms. More specifically, optical isomers are one kind of **stereoisomer,** compounds that have the same formula and the same connections of their atoms but have different arrangements of their atoms in space.

Although they are different compounds, optical isomers are closely related and have most of the same physical properties. Both optical isomers of alanine, for example, have the same melting point, the same solubility in water, the same density, and the same chemical reactivity. They differ only in their effect on polarized light. The mirror-image relationship of the optical isomers of a compound with four different groups on a **chiral carbon atom** is illustrated in Figure 7.5. Compare these structures with the tetrahedrons in Figure 7.4.

Chiral carbon atom   A carbon atom bonded to four different groups.

What about the amino acid structures in Table 7.1? Are any of them chiral? Of the 20 common amino acids, 19 are chiral because they have four different groups bonded to their α carbons, —H, —NH₂, —COOH, and —R (the side chain). Only glycine, H_2NCH_2COOH, is achiral. Even though the naturally occurring chiral α-amino acids can exist as pairs of optical isomers, nature uses only a single isomer of each for making proteins. For historical reasons that we'll discuss in Chapter 10, the naturally occurring isomers are all classified as left-handed or L-amino acids. The isomeric right-handed D-amino acids occur very rarely in nature.

Mirror

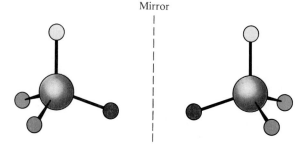

Figure 7.5
Mirror-image molecules. The central atom is chiral because it is bonded to four different groups.

Solved Problem 7.2   Lactic acid can be isolated from sour milk. Is lactic acid chiral?

$$\underset{3 \quad\quad 2 \quad\quad 1}{CH_3-\underset{\overset{|}{OH}}{CH}-\underset{\overset{\|}{O}}{C}OH} \qquad \text{Lactic acid}$$

Solution   To find out if lactic acid is chiral, list the groups attached to each carbon:

Groups on carbon 1	Groups on carbon 2	Groups on carbon 3
1. —OH	1. —COOH	1. —CH(OH)COOH
2. =O	2. —OH	2. —H
3. —CH(OH)CH₃	3. —H	3. —H
	4. —CH₃	4. —H

Next, look at the lists to see if any carbon is attached to four *different* groups. Of the three carbons, carbon 2 has four different groups, and lactic acid is therefore chiral.

Practice Problems   **7.7**   2-Aminopropane is an achiral molecule, but 2-aminobutane is chiral. Explain.

7.8   Which of these molecules are chiral?

$$CH_3 \quad CH_3$$

(a) 3-chloropentane   (b) 2-chloropentane   (c) 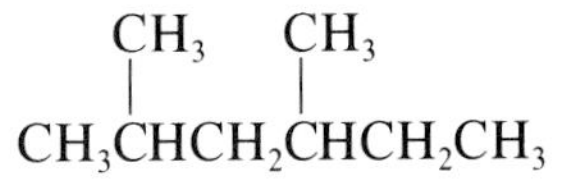 $CH_3CHCH_2CHCH_2CH_3$

7.9   Two of the 20 common amino acids have two chiral carbon atoms in their structures. Identify them.

7.6   THE PEPTIDE BOND AND PRIMARY STRUCTURE

Residue (amino acid) An alternative name for an amino acid unit in a polypeptide or protein.

Peptides and proteins are polymers in which individual amino acids, called **residues,** are linked together by peptide bonds. An amino group from one residue forms a peptide bond with the carboxylic acid group of a second residue; the amino group of the second forms a peptide bond with the carboxylic acid of a third; and so on. The repeating chain of amide linkages is called the *backbone* of the protein, and the amino acid side chains are substituents on the backbone.

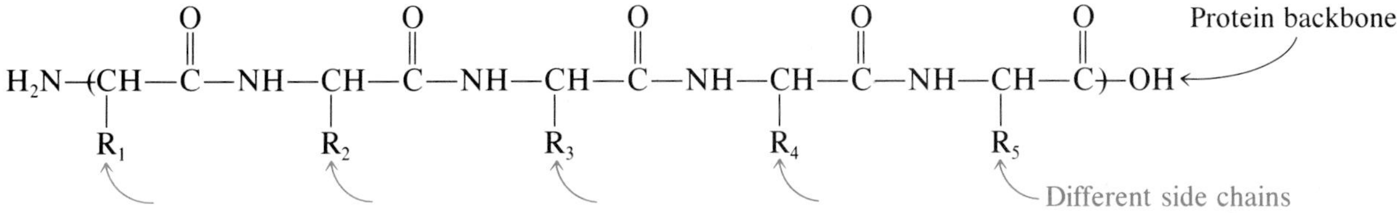

$$H_2N-(CH-\overset{\overset{\displaystyle O}{\|}}{C}-NH-CH-\overset{\overset{\displaystyle O}{\|}}{C}-NH-CH-\overset{\overset{\displaystyle O}{\|}}{C}-NH-CH-\overset{\overset{\displaystyle O}{\|}}{C}-NH-CH-\overset{\overset{\displaystyle O}{\|}}{C})-OH$$

Protein backbone

$$R_1 \qquad R_2 \qquad R_3 \qquad R_4 \qquad R_5$$

Different side chains

As a simple example, the dipeptide alanylserine results when a peptide bond is made between the alanine —COOH and the serine —NH$_2$:

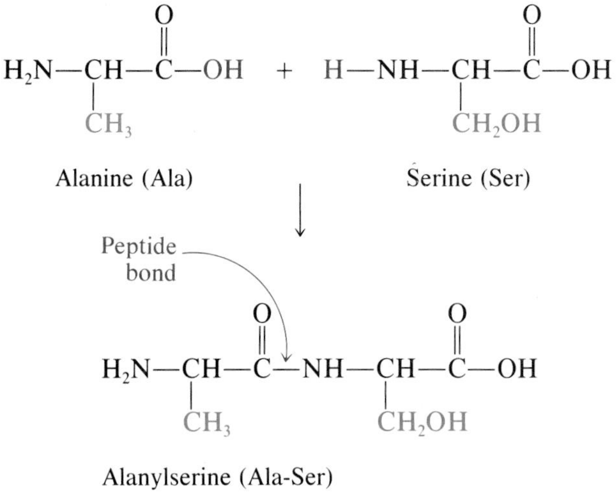

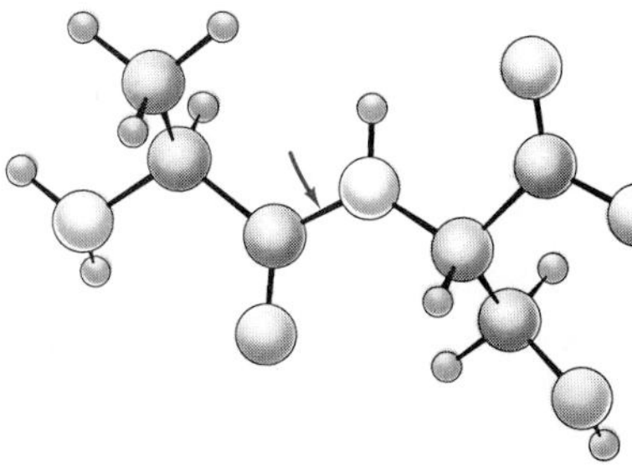

$$H_2N-CH-\overset{\overset{\displaystyle O}{\|}}{C}-OH \; + \; H-NH-CH-\overset{\overset{\displaystyle O}{\|}}{C}-OH$$

$$\overset{|}{CH_3} \qquad\qquad\qquad\qquad \overset{|}{CH_2OH}$$

Alanine (Ala)   Serine (Ser)

Peptide bond

$$H_2N-CH-\overset{\overset{\displaystyle O}{\|}}{C}-NH-CH-\overset{\overset{\displaystyle O}{\|}}{C}-OH$$

$$\overset{|}{CH_3} \qquad\qquad \overset{|}{CH_2OH}$$

Alanylserine (Ala-Ser)

Note that two *different* dipeptides can result from reaction of alanine with serine depending on which —COOH group reacts with which —NH$_2$ group. If the alanine —NH$_2$ reacts with the serine —COOH, serylalanine results.

Serine (Ser)   Alanine (Ala)

Serylalanine (Ser-Ala)

The peptide bond is marked with an arrow in the molecular model above, and the entire amide linkage between the two amino acids is circled. Notice that the C and N of the peptide bond and the four attached atoms all lie in the same plane with approximately 120° angles between them. If another amino acid is connected by a peptide bond to the nitrogen at the left end in the model above, that new linkage is also planar, as are all peptide bonds along the entire protein backbone.

N-terminal amino acid The amino acid with the free —NH$_2$ group at the end of a protein.

C-terminal amino acid The amino acid with the free —COOH group at the end of a protein.

By convention, peptides and proteins are always written with the **N-terminal amino acid** (the one with the free —NH$_2$ group) on the left, and the **C-terminal amino acid** (the one with the free —COOH group) on the right. A peptide is named by citing the amino acids in order starting at the N-terminal acid and ending with the C-terminal acid. All residues except the C-terminal one have the *-yl* ending instead of *-ine*, as in alanylserine (abbreviated Ala-Ser) or serylalanine (Ser-Ala). As a further example, the abbreviations listed at the bottom of Figure 7.6 are combined to show the structure of the blood-pressure-regulating hormone angiotensin II.

The number of possible isomeric peptides increases rapidly as the number of amino acid residues increases. There are 6 ways in which 3 different amino acids can be joined, more than 40,000 ways in which the 8 amino acids present in angiotensin II could be joined, and an inconceivably vast number of ways in which the *1800* amino acids in myosin, the major component of muscle filaments, could be arranged (approximately 10^{1800}—a far larger number than there are atoms in the universe).

Primary protein structure The sequence in which amino acids are linked together in a protein.

The size and complexity of protein chains built from up to 20 different amino acids is the basis for their ability to carry out so many different functions in living things. With molecular weights of up to 1/2 *million,* many proteins are so large that the word *structure* takes on a broader meaning when applied to these immense molecules than it does with simple organic molecules. In fact, chemists usually speak of four levels of structure when describing proteins. The **primary structure** of a protein is the sequence in which the various amino acids

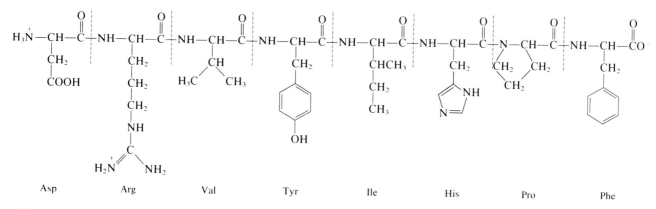

Figure 7.6

The structure of angiotensin II, a blood-pressure-regulating hormone present in blood plasma. Its primary structure is abbreviated Asp-Arg-Val-Tyr-Ile-His-Pro-Phe.

are linked together. The secondary, tertiary, and quaternary structures, as you'll see in later sections of this chapter, are various ways that proteins fold into the shapes essential for their functions.

Primary structure is the most important of the four structural levels because it is a protein's amino acid sequence that determines its overall shape, function, and properties. So crucial is primary structure to function that the change of only one amino acid out of several hundred can drastically alter a protein's biological properties. Sickle-cell anemia, for example, is caused by a genetic defect in blood hemoglobin whereby valine (with a hydrophobic side chain) is substituted for glutamic acid (with a hydrophilic side chain) at only one position, six amino acids from the N-terminal end in the chain of 146 amino acids.

Sickle-cell anemia is named for the sickle shape of affected red blood cells, which do not carry oxygen efficiently (Figure 7.7). Sickled cells are fragile and tend to collect and block capillaries, causing inflammation and pain. The percentage of individuals carrying the genetic trait for sickle-cell anemia is highest among ethnic groups with origins in tropical regions. The ancestors of these individuals survived because carriers of the sickle-cell trait are more resistant to malaria than those without this defect.

Figure 7.7
Sickled red blood cells of a patient with sickle-cell anemia.

Solved Problem 7.3   Draw the structure of the dipeptide Ala-Gly.

Solution   First look up the names and structures of the two amino acids, Ala (alanine) and Gly (glycine).

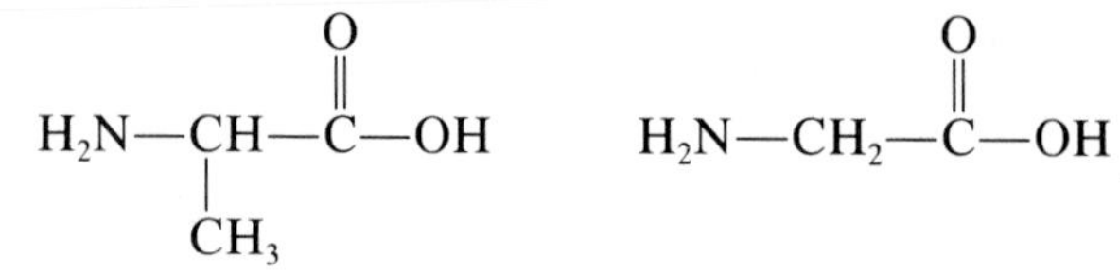

$$H_2N-\overset{\displaystyle CH}{\underset{\displaystyle CH_3}{|}}-\overset{\displaystyle O}{\overset{\displaystyle \|}{C}}-OH \qquad H_2N-CH_2-\overset{\displaystyle O}{\overset{\displaystyle \|}{C}}-OH$$

Alanine (Ala)   Glycine (Gly)

Since alanine is N-terminal and glycine is C-terminal, Ala-Gly must have a peptide bond between the alanine —COOH and the glycine —NH₂.

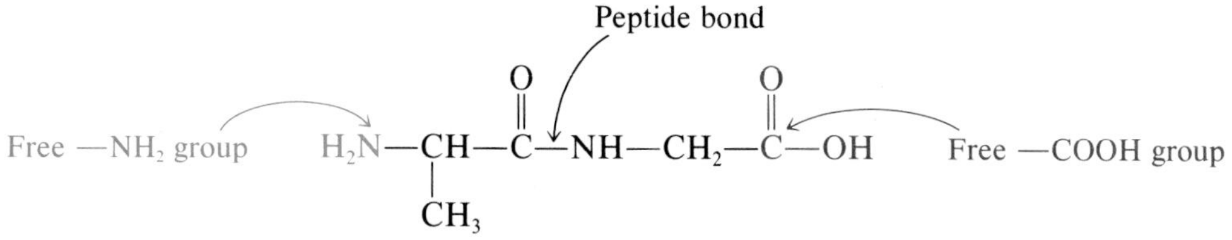

Peptide bond

Free —NH₂ group $H_2N-\overset{\displaystyle CH}{\underset{\displaystyle CH_3}{|}}-\overset{\displaystyle O}{\overset{\displaystyle \|}{C}}-NH-CH_2-\overset{\displaystyle O}{\overset{\displaystyle \|}{C}}-OH$ Free —COOH group

Alanine   Ala-Gly   Glycine

AN APPLICATION: PROTEIN ANALYSIS BY ELECTROPHORESIS

By taking advantage of their overall positive or negative charge, protein molecules can be separated from each other. When an electric field is created between two electrodes, a positively charged particle moves toward the negative electrode and a negatively charged particle moves toward the positive electrode. This movement is called *electrophoresis,* and its amount varies with the strength of the electric field, the charge of the particle, and the size and shape of the particle. Since the overall charge on a protein depends on pH, the electrophoresis of a protein also depends on the pH of the solution or other medium through which the particle moves.

Electrophoresis is routinely used in the clinical laboratory for determination of protein concentrations in blood serum. In the example of an electrophoresis apparatus shown here, a strip of buffer-soaked cellulose acetate is held between

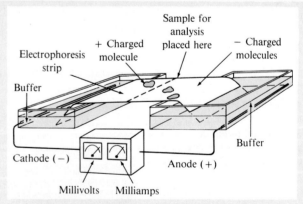

Electrophoresis apparatus for separation of blood serum proteins by overall charge.

buffer reservoirs in contact with the electrodes. The serum to be analyzed is placed in the center of the strip, and different proteins in the serum migrate different distances when the electric field

Practice Problems   **7.10**   Use the three-letter shorthand notations to name the two isomeric dipeptides that could be made from valine and cysteine. Draw the structure of each.

7.11   Name the six tripeptides that contain valine, tyrosine, and glycine.

7.12   Identify the amino acids in the following dipeptide and tripeptide and write the abbreviated forms of the peptide names.

(a)

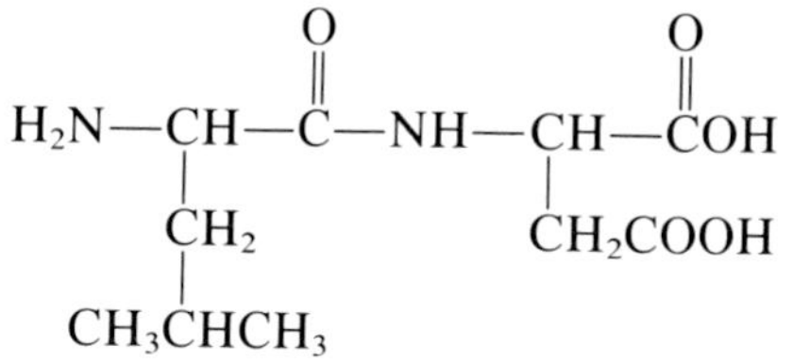

$$H_2N-CH-\overset{\displaystyle O}{\overset{\|}{C}}-NH-CH-\overset{\displaystyle O}{\overset{\|}{C}}OH$$

$$\underset{\underset{CH_3CHCH_3}{|}}{CH_2} \qquad \underset{}{CH_2COOH}$$

(b)

$$H_2N-CH-\overset{\displaystyle O}{\overset{\|}{C}}-NH-CH-\overset{\displaystyle O}{\overset{\|}{C}}-NH-CH-\overset{\displaystyle O}{\overset{\|}{C}}OH$$

$$\underset{}{CH_2} \qquad\qquad \underset{}{CH_2OH} \qquad\qquad CH_2(CH_2)_3NH_2$$

OH

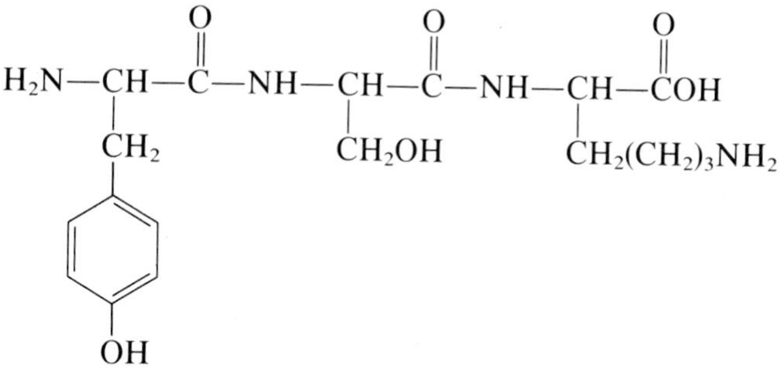

7.13   Draw the structures of the dipeptide and tripeptide shown in Practice Problem 7.12 with ions wherever ionization is possible.

is applied. After a standard time has passed, the electrophoresis strip is removed and a dye is added to make the pattern of the protein bands visible. Interpretation of the differences between bands is aided by an instrument that converts the intensity of the dye color into density plots like those shown here.

Blood serum contains over 125 proteins, but clinically useful electrophoresis need not separate them all. The distribution pattern of five or six different protein fractions (albumin, two α-globulins, one or two β-globulins, and the γ-globulins) is sufficient for diagnostic purposes, with abnormally high or low readings in each region indicative of different clinical conditions. More elaborate variations on electrophoresis are extensively used in biochemical research and allow separation of individual proteins according to their molecular weights and isoelectric points.

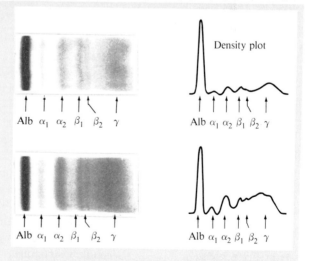

Normal electrophoresis pattern at top and abnormal pattern at bottom showing elevated γ-globulin, which indicates possibility of chronic liver disease, collagen disorder, or chronic infection.

7.7 DISULFIDE BRIDGES IN PROTEINS

Disulfide bridge An S—S bond formed between two cysteine residues that can join two peptide chains together or cause a loop in a peptide chain.

Although the peptide bond is the fundamental link between amino acid residues, a second kind of covalent bonding sometimes occurs when a **disulfide bridge** forms between two cysteine residues. Recall from Section 3.10 that thiols, RSH, react with mild oxidizing agents to yield disulfides, RS—SR. If the thiol groups are on the side chains of two cysteine amino acids, then disulfide bond formation links the two cysteine residues together. The linkage is sometimes indicated by writing CyS with a capital "S" (for sulfur) and then drawing a line from one CyS to the other: CyS—CyS.

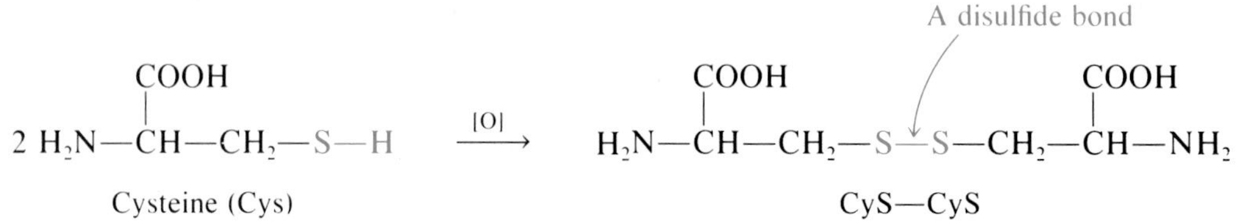

$$
\underset{\text{Cysteine (Cys)}}{2\ \text{H}_2\text{N}-\overset{\displaystyle\overset{\text{COOH}}{|}}{\text{CH}}-\text{CH}_2-\text{S}-\text{H}} \quad \xrightarrow{\text{[O]}} \quad \underset{\text{CyS}-\text{CyS}}{\text{H}_2\text{N}-\overset{\displaystyle\overset{\text{COOH}}{|}}{\text{CH}}-\text{CH}_2-\text{S}-\text{S}-\text{CH}_2-\overset{\displaystyle\overset{\text{COOH}}{|}}{\text{CH}}-\text{NH}_2}
$$

A disulfide bond

If a disulfide bond forms between two cysteine residues in different peptide chains, the otherwise separate chains are linked together. Alternatively, if a disulfide bond forms between two cysteines in the same chain, a loop is formed in the chain. Insulin, for example, consists of two polypeptide chains linked together by disulfide bridges in two places. One of the chains also has a loop in it caused by a third disulfide bridge (Figure 7.8).

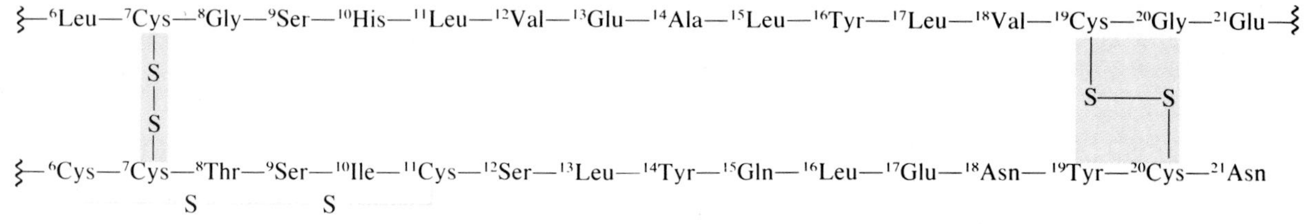

Figure 7.8
The structure of human insulin. There are two disulfide bridges linking the 21 amino acid A chain and the 30 amino acid B chain together and there is a disulfide loop in the A chain.

7.8 SECONDARY STRUCTURE OF PROTEINS

Secondary protein structure The way in which nearby segments of a protein chain are oriented into a regular pattern, for example, an α helix or a β-pleated sheet.

When looking at the primary structure of insulin in Figure 7.8, you might get the idea that the polypeptide chains are simply long threads, stretching from the N-terminal amino acids at one end to the C-terminal amino acids at the other end. The fact is, though, that proteins are not thread-like. Most proteins fold in such a way that nearby segments of the chain orient into a regular pattern called a **secondary structure.**

There are two common kinds of secondary structure patterns: the α helix and the β-pleated sheet. Many proteins have regions of α helix, regions of β-pleated sheet, and still other less organized regions where there is just a random coil. There are, in addition, other less common kinds of secondary structure such as the triple helix found in tropocollagen.

α Helix A common secondary protein structure in which a protein chain wraps into a coil stabilized by hydrogen bonds between peptide links in the backbone.

The *α* Helix: Secondary Structure of *α*-Keratin *α*-Keratin is a fibrous structural protein found in wool, hair, fingernails, and feathers. Studies have shown that *α*-keratin exists mostly as a helical coil, much like the coiled cord on a telephone (Figure 7.9a). Called an **α helix,** the coil is stabilized by the formation of hydrogen bonds between the N—H group of one peptide linkage and the carbonyl group of another peptide linkage four residues farther along the chain:

A hydrogen bond connecting two peptide linkages

The location of a hydrogen bond in the *α* helix is shown in red in Figure 7.9b. Although the strength of each individual hydrogen bond is low (about 5 kcal/mol), the large number present in the helix results in an extremely stable secondary structure. Each coil of the helix contains 3.6 amino acid residues, with a distance between coils of 0.54 nm (5.4 Å).

Figure 7.9
α-Helix secondary structure in *α*-keratin. (a) The amino acid backbone in the helical secondary structure of *α*-keratin winds in a spiral much like that in a telephone cord. (b) One example of the hydrogen bonds that hold the spiral in place is shown.

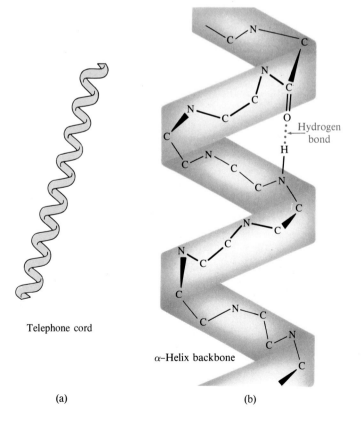

Telephone cord

α–Helix backbone

(a) (b)

AN APPLICATION: PROTEINS IN THE DIET

Protein is often thought of as "muscle food" because of its popularity with weightlifters and other athletes. Although it's true that protein is needed for growing muscles, it's also true that we *all* need substantial amounts of protein daily because our bodies don't store protein as they do carbohydrates and fats. Furthermore, our bodies can synthesize only 10 of the 20 common amino acids from simple precursor molecules; the remaining 10 amino acids must be obtained by digestion of edible proteins.

Children need large amounts of protein for proper growth, and adults need protein to replace what is lost each day by normal biochemical reactions in the body. The accompanying table shows the estimated daily requirements of essential amino acids for an infant and an adult. The *total* recommended daily amount of protein for an adult with ideal body weight is 0.8 g per kilogram of body weight, which converts to about 9% of daily calories from protein. For a 70 kg male, the recommended total protein is 56 g, and for a 55 kg female, it's 44 g. For reference, a McDonald's Big Mac contains 25 g of protein (along with 34 g of fat). Pregnant or lactating women and individuals recovering from illness or surgery need additional protein. Infants require 2.2 g of protein per kilogram of body weight.

Not all foods are equally good sources of protein. A high-quality protein source is one that provides the 10 essential amino acids in sufficient amount to meet our minimum daily needs. Most meat and dairy products meet this requirement, but many vegetable sources such as wheat and corn don't. The term *incomplete protein* refers to protein in which one or more of the 10 essential amino acids is present in too low a quantity to sustain the growth of laboratory animals. For example, wheat is low in lysine, and corn is low in both lysine and tryptophan. Eating only food with incomplete protein can cause nutritional deficiencies, particularly in growing children.

Some of the limiting amino acids found in various foods are listed below.

Limiting Amino Acids in Some Foods

Food Category	Limiting Amino Acid
Wheat, other grains	Lysine, threonine
Peas, beans, other legumes	Methionine, tryptophan
Nuts and seeds	Lysine
Leafy green vegetables	Methionine

Vegetarians must be certain to adopt a varied diet that provides proteins from several different sources. Legumes and nuts, for example, are particularly valuable in overcoming the deficiencies of wheat and grains. In some regions, the traditional cuisine includes combinations that provide complementary proteins, for example, rice and lentils in India, corn tortillas and beans in Mexico, and rice and black-eyed peas in the southern United States.

Estimated Amino Acid Requirements

Amino Acid	Daily Requirement (mg/kg body weight)	
	Infant	Adult
Arginine	?	?
Histidine	33	10
Isoleucine	83	12
Leucine	135	16
Lysine	99	12
Methionine[a]	49	10
Phenylalanine[b]	141	16
Threonine	68	8
Tryptophan	21	3
Valine	92	14

[a] Also supplied by synthesis from cysteine in diet.
[b] Also supplied by synthesis from tyrosine in diet.

β-Pleated sheet A
common secondary protein
structure in which segments
of a protein chain fold back
on themselves to form
parallel strands held together
by hydrogen bonds between
peptide links in the
backbone.

The β-Pleated Sheet: Secondary Structure of Fibroin Fibroin, the fibrous protein found in silk, contains mainly the **β-pleated sheet** type of secondary structure. In this structure, polypeptide chains line up next to each other and are held together by hydrogen bonds between peptide linkages (Figure 7.10a). The bond angles along the backbone are such that the sheet has a pleated contour, with side chains extending above and below the sheet (Figure 7.10b). Small β-pleated sheet regions are found in many proteins where sections of peptide chains double back on themselves.

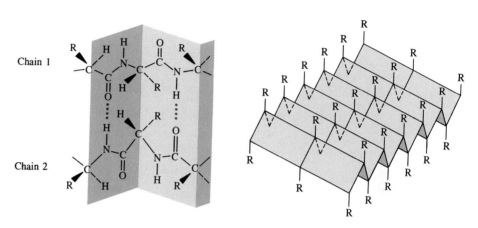

Figure 7.10
β-Pleated sheet secondary
structure. (a) Hydrogen
bonding between chains in
the β-pleated sheet. (b) The
β-pleated sheet secondary
structure of silk fibroin,
showing how the chains
line up.

REVIEW $1 \text{ nm} = 1 \times 10^{-9} \text{ m}$

The Triple Helix: Secondary Structure of Tropocollagen Collagen is the most abundant of all proteins in mammals, making up 30% or more of the total protein. A fibrous protein, collagen is the major constituent of skin, tendons, bones, blood vessels, and connective tissues. The basic structural unit of collagen is *tropocollagen,* a large protein that consists of three loosely coiled chains of about 1000 amino acids each, wrapped around each other to form a stiff, rod-like **triple helix** (Figure 7.11) held together by hydrogen bonds along the backbone. *Each* tropocollagen rod has a length of about 300 nm and a width of 1.5 nm. (For comparison, a simple organic molecule like propane is only about 0.5 nm long.)

Triple helix The secondary
protein structure of
tropocollagen in which three
protein chains coil around
each other to form a rod.

Figure 7.11
The triple helix in
tropocollagen. Three
individual protein strands
wrap around each other to
form a stiff helical cable.

7.9 NONCOVALENT INTERACTIONS IN PROTEINS

Noncovalent interactions such as the hydrogen bonds between backbone peptide bonds (which stabilize α helixes and β-pleated sheets) are essential to the structure and function of proteins. In addition, various other noncovalent interactions are possible between the amino acid side chains along a protein backbone. For instance, attractions between ionized acidic and basic side chains form what are known as *salt bridges*. In the imaginary protein shown in Figure 7.12, hydrogen bonds cause the chain to fold, and a salt bridge between an acidic glutamate and a basic lysine side chain pulls the chains together in the middle.

REVIEW Nonpolar substances, such as fats or oils, have only weak intermolecular forces (the London forces); they are hydrophobic because they cannot disrupt the strong hydrogen bonding forces among water molecules. When an oil-water mixture is shaken, the water layer always re-forms, squeezing out the oil molecules.

In addition to the salt bridges and hydrogen bonding interactions shown in Figure 7.12, the protein is also folded so that hydrophobic hydrocarbon side chains cluster together to exclude water. Here, the same intermolecular forces that hold an oil droplet together on the surface of a puddle are at work. Such a hydrophobic region within an enzyme can, for example, hold in place the hydrocarbon portion of a reactant while a chemical reaction occurs at another spot in the molecule.

To summarize the kinds of noncovalent interactions that occur between amino acids:

● **Hydrogen bonds** form between —OH or —NH$_2$ groups and oxygen or nitrogen atoms with unshared electron pairs.

● **Salt bridges,** or ionic attractions, occur between positively and negatively charged side chains.

● **Hydrophobic interactions** pull nonpolar side chains together and exclude water.

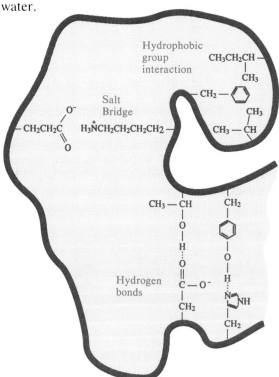

Figure 7.12
Noncovalent interactions of amino acid side chains in proteins.

Practice Problems **7.14** Look at Table 7.1 and identify the type of noncovalent interaction expected between each pair of amino acids:
(a) glutamine and tyrosine   (b) leucine and proline
(c) aspartate and arginine   (d) isoleucine and phenylalanine
(e) threonine and glutamine

7.10 TERTIARY AND QUATERNARY STRUCTURES OF PROTEINS

Tertiary protein structure
The way in which an entire protein chain is coiled and folded into its specific three-dimensional shape.

The overall three-dimensional shape that results from the coiling and folding of a protein chain is called the protein's **tertiary structure.** In contrast to secondary structure, which depends on relationships between amino acids close to each other in the chain, tertiary structure can be determined by interactions of amino acid side chains that are far apart.

Tertiary Protein Structure: Myoglobin   Myoglobin, a globular protein with a single chain of 153 amino acid residues, provides a good example of tertiary structure. A relative of hemoglobin, myoglobin is found in skeletal muscles, where it stores O_2. Sea mammals, for example, rely on myoglobin for oxygen to sustain them during long dives. Structurally, myoglobin consists of a number of straight segments, each of which adopts an α-helical secondary structure. These helical sections then fold up further to form a compact, nearly spherical, tertiary structure (Figure 7.13) in which a molecule of heme for binding the O_2 is embedded.

Although the bends appear to be irregular and the three-dimensional structure appears to be random, this is not the case. By bending and twisting in precisely this way, myoglobin can achieve maximum stability. All myoglobin molecules adopt this same shape because it's more stable than any other.

The most important forces stabilizing a protein's tertiary structure are the hydrophobic interactions of the nonpolar side chains, which congregate together in the hydrocarbon-like interior of a protein molecule, away from the aqueous medium. The acidic or basic amino acids with charged side chains and those with polar side chains that can form hydrogen bonds, by contrast, are usually found on the exterior of the protein where they can be solvated by water. Myoglobin has many such polar groups on its surface and is therefore water-soluble. Also important for stabilizing a protein's tertiary structure is the

Figure 7.13
Secondary and tertiary structure of myoglobin. The sausage-like shape is often used alone to represent the helical portions of a globular protein. The red structure embedded in the protein is a molecule of heme, to which O_2 binds.

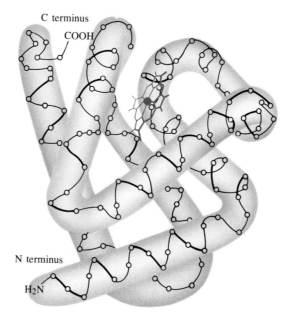

C terminus
COOH

N terminus

H₂N

Figure 7.14
Example of a globular protein, drawn with α helices as coils and β-pleated sheets as arrows pointing toward the C-terminal end. Ribonuclease is an enzyme that catalyzes hydrolysis of ribonucleic acid.

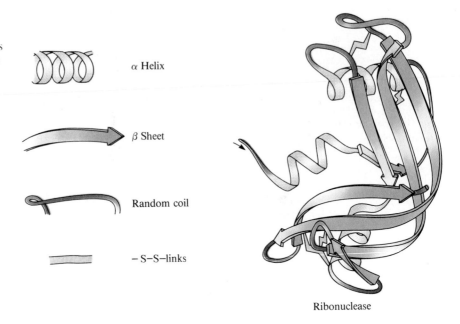

α Helix

β Sheet

Random coil

– S–S–links

Ribonuclease

formation of disulfide bridges between cysteine residues. Another example of tertiary protein structure is given in Figure 7.14. Note the combination of α-helix and β-pleated sheet regions as well as randomly coiled regions.

Quaternary Protein Structure: Collagen and Hemoglobin   When two or more protein chains combine in a single functional unit, the shape of the resulting protein is referred to as **quaternary structure.** We saw in the previous section, for example, that tropocollagen is a stiff, rod-like protein with a triple-helix secondary structure. Collagen itself, the actual protein present in skin, teeth, bones, and connective tissues, has a complex quaternary structure formed when a great many tropocollagen strands aggregate together by overlapping lengthwise in a *quarter-stagger arrangement,* as shown in Figure 7.15.

Depending on the exact purpose the collagen serves in the body, further structural modifications also occur. In connective tissue like tendons, chemical bonds form between strands to give collagen fibers with a rigid, cross-linked structure. In teeth and bones, a mineral called *calcium hydroxyapatite,* $Ca_5(PO_4)_3OH$, deposits in the gaps between chains to further harden the overall assembly.

Figure 7.16 shows another example of quaternary structure: that of hemoglobin, the oxygen carrier in blood, which consists of four polypeptide chains held together primarily by hydrophobic interactions.

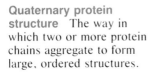

Quaternary protein structure   The way in which two or more protein chains aggregate to form large, ordered structures.

Figure 7.15
The quarter-stagger arrangement of tropocollagen triple helix units aggregating into a collagen fiber. Each tropocollagen chain is offset from its next neighbor by about one-fourth of its length, and the individual chains are connected by covalent bonds.

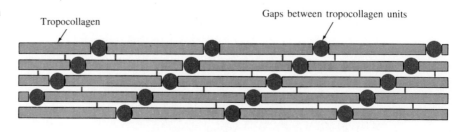

Tropocollagen

Gaps between tropocollagen units

Figure 7.16
Computer-generated model of the oxygen-carrying protein hemoglobin. The hemoglobin molecule consists of four polypeptide chains (yellow, blue, red, and green, with helical regions shown as cylinders). Within each polypeptide chain is embedded a planar heme molecule (gray carbon atoms and red oxygen atoms). Oxygen is carried by bonding to the single Fe^{2+} in the center of each of the four heme molecules.

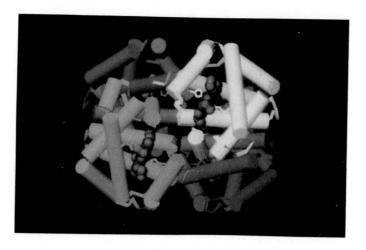

Protein structure can be summarized as follows:

- **Primary structure**—the amino acid sequence, for example,

Asp-Arg-Val-Tyr

- **Secondary structure**—the regular patterns of chain segments such as an α helix, a β-pleated sheet, or a triple helix, held together by hydrogen bonds between backbone peptide links near each other in the chain.

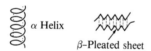

α Helix

β–Pleated sheet

- **Tertiary structure**—the folding of a protein molecule into a specific three-dimensional shape, held together mainly by noncovalent interactions between side-chain groups that can be quite far apart.

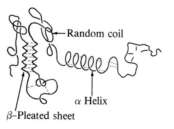

←Random coil

α Helix

β–Pleated sheet

- **Quaternary structure**—the combination of several protein chains into a larger three-dimensional structure held together by noncovalent interactions or covalent cross-links.

Simple protein A protein that yields only amino acids when hydrolyzed.

Conjugated protein A protein that yields one or more other substances in addition to amino acids when hydrolyzed.

Fibrous protein A tough, insoluble protein whose peptide chains are arranged in long filaments.

7.11 CLASSIFICATION OF PROTEINS

Proteins can be classified in several ways, the simplest of which is to group them according to composition. **Simple proteins,** such as blood-serum albumin, are those that yield only amino acids and no other compounds on hydrolysis. **Conjugated proteins,** which are far more common than simple proteins, yield nonprotein substances in addition to amino acids on hydrolysis. Table 7.2 lists some examples of conjugated proteins.

Another way to classify proteins is according to their three-dimensional shape as either *fibrous* or *globular*. **Fibrous proteins,** such as collagen and the keratins, consist of polypeptide chains arranged side by side in long filaments. Because these proteins are tough and insoluble in water, nature uses them for

Table 7.2   Some Different Kinds of Conjugated Proteins

Name	Nonprotein Part	Examples
Glycoproteins	Carbohydrates	Glycoproteins in cell membranes
Lipoproteins	Lipids	High- and low-density lipoproteins that transport cholesterol and other lipids through the body
Metalloproteins	Metal ions	The enzyme cytochrome oxidase, necessary for biological energy production, and many other enzymes
Phosphoproteins	Phosphate groups	Milk casein, which serves to store nutrients for a growing embryo
Hemoproteins	Heme	Hemoglobin and myoglobin, which transport and store oxygen, respectively
Nucleoproteins	RNA (ribonucleic acid)	Found in cell ribosomes, where they take part in protein synthesis

Table 7.3   Some Common Fibrous and Globular Proteins

Name	Occurrence and Function
Fibrous proteins (insoluble)	
Keratins	Found in skin, wool, feathers, hooves, silk, fingernails
Collagens	Found in animal hide, tendons, bone, eye cornea, and other connective tissue
Elastins	Found in blood vessels and ligaments, where ability of the tissue to stretch is important
Myosins	Found in muscle tissue
Fibrin	Found in blood clots
Globular proteins (soluble)	
Insulin	Regulatory hormone for controlling glucose metabolism
Ribonuclease	Enzyme controlling RNA hydrolysis
Immunoglobulins	Proteins involved in immune response
Hemoglobin	Protein involved in oxygen transport
Albumins	Proteins that perform many transport functions in blood; protein in egg white

Globular protein A water-soluble protein that adopts a compact, coiled-up shape.

structural materials like tendons, hair, ligaments, and muscle. **Globular proteins,** by contrast, are usually coiled into compact, nearly spherical shapes like myoglobin (Figure 7.13) and ribonuclease (Figure 7.14). Globular proteins, which might have anywhere from 100 to well over 1000 amino acids in their chains, include most of the 2000 or so known enzymes. Like myoglobin, they are mobile within cells and are generally soluble in water, with their hydrophobic groups folded in toward the center and their hydrophilic groups on the outside. Table 7.3 lists some common examples of both fibrous and globular proteins.

Yet a third way to classify proteins is to group them according to biological function. As indicated in Table 7.4, proteins have an extraordinary diversity of roles in the body.

Table 7.4   Some Biological Functions of Proteins

Type	Function and Example
Enzymes	Proteins such as alcohol dehydrogenase that act as biological catalysts
Hormones	Proteins such as insulin and growth hormone that regulate body processes
Storage proteins	Proteins such as ferritin that store nutrients
Transport proteins	Proteins such as hemoglobin that transport oxygen and other substances through the body
Structural proteins	Proteins such as keratin, elastin, and collagen that form an organism's structure
Protective proteins	Proteins such as the antibodies that help fight infection
Contractile proteins	Proteins such as actin and myosin found in muscles
Toxic proteins	Proteins such as the snake venoms that serve a defensive role for the plant or animal

7.12 CHEMICAL PROPERTIES OF PROTEINS

Protein Hydrolysis   Just as a simple amide can be hydrolyzed to yield an amine and a carboxylic acid (Section 6.10), a protein can be hydrolyzed to yield many amino acids. In fact, digestion of proteins involves nothing more than breaking the numerous peptide bonds that link amino acids together. For example,

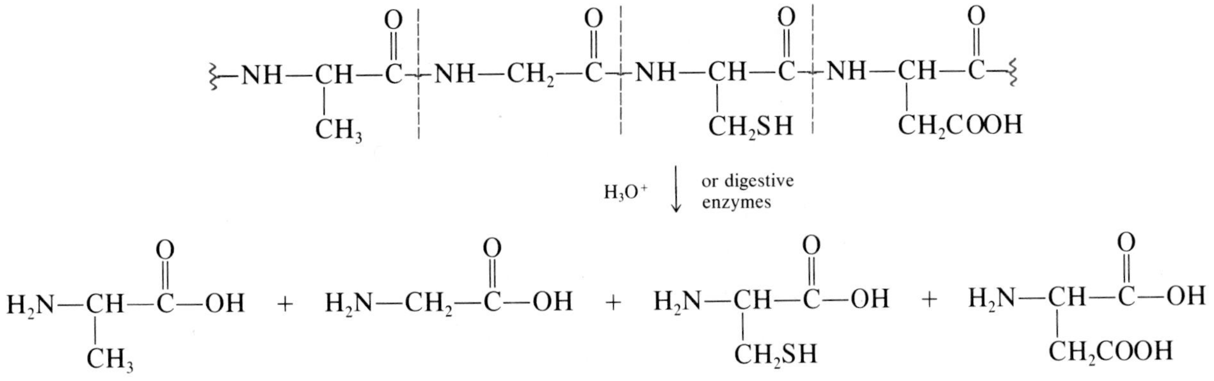

Alanine   Glycine   Cysteine   Aspartic Acid

Although a chemist in the laboratory might choose to hydrolyze a protein by heating it with a solution of hydrochloric acid, most digestion of proteins in the body takes place in the stomach and small intestine, where the process is catalyzed by enzymes. Once formed, individual amino acids are absorbed through the wall of the intestine and transported in the bloodstream to tissues.

Practice Problem   7.15   Look up the structure of angiotensin II in Figure 7.6 and draw the products that would be formed during digestion.

Denaturation The loss of secondary, tertiary, or quaternary protein structure due to disruption of noncovalent interactions that leaves peptide bonds intact.

Protein Denaturation   Since the overall shape of a protein is determined by a delicate balance of noncovalent forces, it's not surprising that a change in protein shape often results when the balance is disturbed. Such a disruption in shape without affecting the protein's primary structure is known as **denaturation.** When denaturation of a globular protein occurs, the structure unfolds from a well-defined globular shape to a randomly looped chain (Figure 7.17).

The agents that cause denaturation include heat, mechanical agitation, detergents, organic solvents, pH change, inorganic salts, and oxidizing agents:

● **Heat**   The weak side-chain attractions in globular proteins are easily disrupted by heating, in many cases only to temperatures above 50°C. Cooking meat, for example, converts some of the insoluble collagen into soluble gelatin, which can be used in glue and Jell-O, and for thickening sauces.

● **Mechanical agitation**   The most familiar example of denaturation by agitation is the foam produced by beating egg whites. Denaturation of proteins at the surface of the air bubbles stiffens the protein and causes the bubbles to be held in place.

● **Detergents**   Even very low concentrations of detergents can cause denaturation by disrupting the association of hydrophobic side chains.

● **Organic compounds**   Polar solvents such as acetone and ethanol interfere with hydrogen bonding by competing for hydrogen-bonding sites. The disinfectant action of ethanol, for example, results from its ability to denature bacterial protein.

● **pH change**   Excess H^+ or OH^- ions react with the basic or acidic side chains in amino acids and disrupt salt bridges. One familiar example of denaturation by pH change is the protein coagulation that occurs when milk turns sour.

● **Inorganic salts**   Ions can disturb salt bridges, and heavy metal ions such as those of lead, mercury, and silver react with —SH groups to form precipitates. Many heavy metals are poisons because they denature and inactivate crucial enzymes by such reactions.

Denaturation is accompanied by changes in both physical and biological properties. Solubility is often decreased by denaturation, as occurs when egg white is cooked and the albumins coagulate into an insoluble white mass. In addition, enzymes lose their catalytic activity and other proteins are no longer able to carry out their biological functions when their shapes are altered by denaturation.

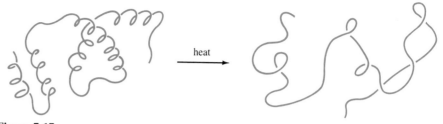

Figure 7.17

Denaturation. Disruption of its secondary, tertiary, or quaternary structure without breaking any peptide bonds results in denaturation of a protein.

Most denaturation is irreversible: Hard-boiled eggs don't soften when their temperature is lowered. Many cases are known, however, in which unfolded proteins spontaneously undergo *renaturation*—a return to their natural state. Renaturation is accompanied by full recovery of biological activity, indicating that the protein has completely returned to its stable secondary and tertiary structure. By refolding into their active shapes, proteins demonstrate that all the information needed to determine these shapes is present in the primary structure.

SUMMARY

Proteins are large biomolecules consisting of α-amino acid residues linked together by amide or **peptide bonds.** Twenty amino acids are commonly found in proteins. All exist as **zwitterions** containing both —NH_3^+ and —COO^- ions and in solution establish equilibria between neutral and charged forms. At the **isoelectric point** the numbers of positive and negative charges in an amino acid sample in solution are equal. Amino acids and proteins vary in structure and properties according to the amino acid side chains: Some side chains are nonpolar, some are polar, some are acidic, and some are basic.

Certain organic molecules have handedness and are said to be **chiral;** others have no handedness and are **achiral.** The difference is one of symmetry: Achiral molecules have an imaginary **symmetry plane** cutting through the middle so that one half of the molecule is an exact **mirror image** of the other half. Chiral molecules lack this symmetry plane. The two right- and left-handed forms of a chiral substance are called **optical isomers,** or **enantiomers.** Nineteen of the 20 common amino acids are chiral, and the naturally occurring optical isomers belong to the L **family.** Amino acids of the L **family** occur much more rarely in nature.

Large protein molecules are held in their characteristic shapes by **disulfide bridges, salt bridges,** hydrogen bonds, and **hydrophobic interactions.** Proteins are so large that the word *structure* has several meanings. A protein's

primary structure is its amino acid sequence. Its **secondary structure** is the way in which segments of the protein chain are oriented into a regular pattern, such as an α helix, a β-pleated sheet, or a triple helix. Its **tertiary structure** is the way in which the entire protein molecule is coiled or folded into a three-dimensional shape that may include α helix, β-pleated sheet, and randomly coiled regions. Its **quaternary structure** is the way in which several protein chains aggregate to form a larger structure. When the structure of a protein is disrupted by heating or other means, the protein is said to be **denatured.**

Proteins can be classified by composition, shape, or biological function. By composition, proteins are either simple or complex. **Simple proteins** yield only amino acids on hydrolysis; **conjugated proteins** yield other substances in addition to amino acids. By shape, proteins are either fibrous or globular. **Fibrous proteins** such as α-keratin are tough and water-insoluble; **globular proteins** such as myoglobin are water-soluble and mobile within cells. By biological function, proteins have an enormous diversity of roles: Some are enzymes, some are hormones, and some serve to store nutrients. Others act as structural, protective, or transport agents.

The chemistry of proteins is similar to that of simple amides. Hydrolysis, either by reaction with aqueous acid or by digestive enzymes, breaks a protein into its individual amino acid constituents.

INTERLUDE: DETERMINING PROTEIN STRUCTURE

Determining the primary structure of a peptide or protein requires answers to three questions: What amino acids are present? How many of each are present? In what order does each occur in the peptide chain? The answers to these questions are provided by two remarkable instruments, the amino acid analyzer and the protein sequenator.

An *amino acid analyzer* is an automated instrument for determining the identity and amount of each amino acid in a protein. The protein is first broken down into its constituent amino acids by reducing all disulfide bonds and hydrolyzing all amide bonds. The amino acid mixture that results is then separated by placing it at the top of a glass column filled with a special adsorbent material and pumping a series of aqueous buffer solutions through the column. Different amino acids pass down the column at different rates depending on their structures and are thus separated.

As each different amino acid passes from the end of the glass column, it is mixed with a solution of *ninhydrin*, a reagent that forms a deep purple color on reaction with α-amino acids, or with a solution of a reagent that forms a fluorescent compound on reaction with the amino acids. The purple color or the fluorescence is detected by an electronic sensor, and its intensity is measured. Since the amount of time required for a given amino acid to pass through a standard column is reproducible, the identity of all amino acids present in the sample can be determined simply by noting the time at which each comes off. The amount of each amino acid is measured by determining the intensity of the purple color or the fluorescence. The figure at the bottom of the page shows the results of amino acid analysis of a standard equimolar mixture of 17 amino acids.

With the identity and amount of each amino acid known, the next task is to *sequence* the peptide: to find out in what order the amino acids are linked together. The general idea of peptide sequencing is to cleave selectively one amino acid residue at a time from the end of the peptide chain, separate and identify that amino acid, and then repeat the process on the chain-shortened peptide until the entire structure is known.

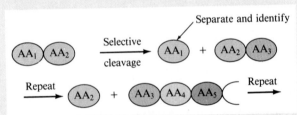

Although the chemistry of the cleavage process is complex, automated protein sequenators are available that allow a series of 30 or more repetitive sequencing steps to be carried out.

Amino acid analysis of an equimolar amino acid mixture. Each peak represents a different amino acid passing through the analysis column. The identities of the amino acids are determined by noting the times at which the peaks appear (horizontal axis), and the amounts are determined by measuring the area of each peak.

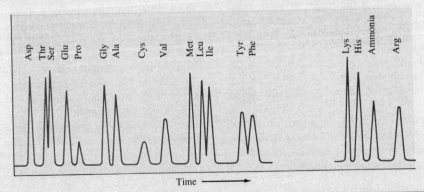

REVIEW PROBLEMS

Amino Acids

7.16* What is an amino acid?

7.17 What does the prefix "α-" mean when referring to α-amino acids?

7.18* What amino acids do these abbreviations stand for?
(a) Ser (b) Thr (c) Pro (d) Phe (e) Cys

7.19 Draw the structures of the amino acids listed in Problem 7.18.

7.20* Name and draw the structures of amino acids that fit these descriptions:
(a) contains an isopropyl group
(b) contains a secondary alcohol group
(c) contains a thiol group (d) contains a phenol group

7.21 What do the following terms mean as they apply to amino acids?
(a) zwitterion (b) essential amino acid
(c) isoelectric point

7.22* Classify these amino acids as neutral, basic, or acidic:
(a) lysine (b) phenylalanine (c) glutamic acid
(d) proline

7.23 At what pH would you expect aspartic acid to have the following structures, pH 3, pH 7, or pH 13?

(a) 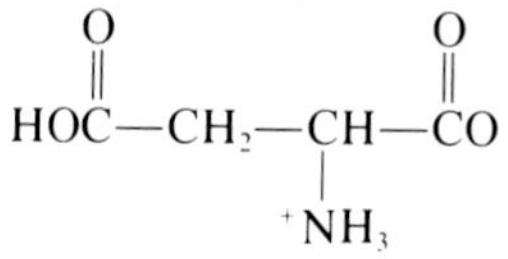 $\overset{\overset{\displaystyle O}{\|}}{HOC}-CH_2-\overset{\overset{\displaystyle}{|}}{\underset{\underset{\displaystyle {}^+NH_3}{|}}{CH}}-\overset{\overset{\displaystyle O}{\|}}{CO}$

(b) 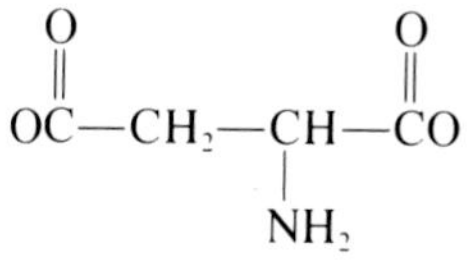 $\overset{\overset{\displaystyle O}{\|}}{OC}-CH_2-\overset{\overset{\displaystyle}{|}}{\underset{\underset{\displaystyle NH_2}{|}}{CH}}-\overset{\overset{\displaystyle O}{\|}}{CO}$

7.24 Draw the structures of proline at pH 1.0, pH 6.3, and pH 9.7.

7.25 Draw the structures of lysine at pH 1.0, pH 6.3, and pH 9.7.

Handedness in Molecules

7.26* What do the terms *chiral* and *achiral* mean?

7.27 Give two examples of chiral objects and two examples of achiral objects.

7.28* Which of the following objects are chiral?
(a) a shoe (b) a bed (c) a light bulb
(d) a flower pot (e) a house key
(f) a pair of scissors

7.29 2-Bromo-2-chloropropane is an achiral molecule, but 2-bromo-2-chlorobutane is chiral. Explain.

7.30* Draw the structures of the following compounds.

Which of them is chiral? Put an asterisk by each chiral carbon.
(a) 2-bromo-2-chloro-3-methylbutane
(b) 3-chloropentane (c) cyclopentanol
(d) 2-methylpropanol

7.31 Which of the carbon atoms marked with arrows in the following compounds are chiral?

(a) 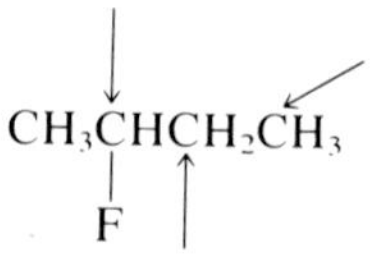 $CH_3CHCH_2CH_3$
   F

(b)

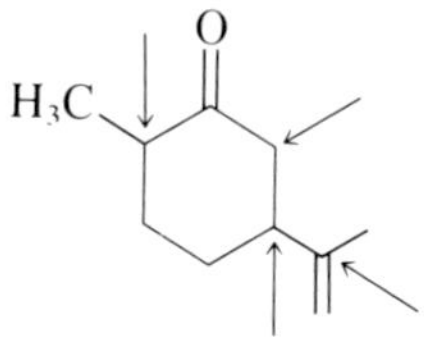

Peptides and Proteins

7.32* What is a protein?

7.33 What is the difference between a peptide and a protein?

7.34* What is the difference between a simple protein and a conjugated protein?

7.35 What kinds of molecules are found in these kinds of conjugated proteins in addition to the protein part?
(a) nucleoproteins (b) lipoproteins
(c) glycoproteins

7.36* What is the difference between fibrous and globular proteins?

7.37 Name three different biological functions that proteins have in the body.

7.38* What is meant by the following terms as they apply to proteins?
(a) primary structure (b) secondary structure
(c) tertiary structure (d) quaternary structure

7.39 What kind of bonding stabilizes helical and β-pleated sheet secondary protein structures?

7.40* Why is cysteine such an important amino acid for defining the tertiary structure of proteins?

7.41 How can cross-links form between cysteine residues in a protein?

7.42* How do the following noncovalent interactions help to stabilize the tertiary and quaternary structure of a protein?
(a) hydrophobic interactions (b) salt bridges
(c) hydrogen bonding

7.43 For each noncovalent interaction in Problem 7.42, list one pair of amino acids that could give rise to that interaction.

7.44* What kind of change takes place in a protein when it is denatured?

7.45 Explain how a protein is denatured by the following:
(a) heat (b) addition of a strong base (c) mercury ions

7.46* Look at the structure of angiotensin II in Figure 7.6 and identify both the N-terminal and C-terminal amino acids.

7.47 Use the three-letter abbreviations to name all tripeptides containing methionine, isoleucine, and lysine.

7.48 Write structural formulas for the two dipeptides containing phenylalanine and glutamic acid.

7.49 Which of the following amino acids are most likely to be found on the outside of a globular protein and which on the inside? Explain.
(a) valine   (b) leucine   (c) aspartic acid
(d) asparagine

7.50* Why do you suppose diabetics must receive insulin subcutaneously by injection rather than orally?

7.51 The *endorphins* are a group of naturally occurring neurotransmitters that act in a manner similar to morphine to control pain. Research has shown that the biologically active part of the endorphin molecule is a simple pentapeptide called an *enkephalin*, with the structure Tyr-Gly-Gly-Phe-Met. Draw the complete structure of this enkephalin.

7.52* Identify the N-terminal and C-terminal amino acids in enkephalin (Problem 7.51).

Properties and Reactions of Amino Acids and Proteins

7.53 Much of the chemistry of amino acids is the familiar chemistry of carboxylic acid and amine functional groups. What products would you expect to obtain from these reactions of glycine?

(a) $H_2N—CH_2—\overset{\overset{\displaystyle O}{\|}}{C}OH$ + CH_3OH $\xrightarrow{H^+ \text{ catalyst}}$

(b) $H_2N—CH_2—\overset{\overset{\displaystyle O}{\|}}{C}OH$ + HCl $\longrightarrow$

7.54* How can you account for the fact that glycine is a solid that decomposes without melting when heated to 262°C, whereas the ethyl ester of glycine is a liquid at room temperature?

7.55 We saw in Section 6.9 that amides can be prepared by reaction between a carboxylic acid and an amine in the presence of DCC. What problem would you expect to encounter if you tried to prepare the simple dipeptide glycylglycine by

$2 \; H_2N—CH_2—\overset{\overset{\displaystyle O}{\|}}{C}OH$ + $\xrightarrow{\text{DCC}}$

$H_2N—CH_2—\overset{\overset{\displaystyle O}{\|}}{C}—NH—CH_2—\overset{\overset{\displaystyle O}{\|}}{C}OH$ + many other products

7.56* (a) Identify the amino acids present in the hexapeptide shown at the bottom of the page.
(b) Identify the N-terminal and C-terminal amino acids of the hexapeptide.
(c) Show the structures of the products that would be obtained on digestion of the hexapeptide.

7.57 Which would you expect to be more soluble in water, a peptide rich in aspartic acid and lysine, or a peptide rich in valine and alanine? Explain.

7.58* Proteins are generally least soluble in water at their isoelectric points. Explain.

7.59 Proteins provide buffering action in cellular fluids. Explain or write equations to show they can do this.

7.60* *Aspartame*, marketed under the trade name Nutra-Sweet for use as a nonnutritive sweetener, is the methyl ester of a simple dipeptide.

$$H_2N—\underset{\underset{\underset{COOH}{|}}{\underset{CH_2}{|}}}{CH}—\overset{\overset{\displaystyle O}{\|}}{C}—NH—\underset{\underset{\text{(phenyl)}}{\underset{CH_2}{|}}}{CH}—\overset{\overset{\displaystyle O}{\|}}{C}—OCH_3$$

Aspartame

Identify the two amino acids present in aspartame and show all the products of digestion, assuming both amide and ester bonds are hydrolyzed in the stomach.

Applications

7.61 What is pK_a? [App: Henderson-Hasselbalch Equation]

7.62* Calculate what percent of the molecules in alanine are in the zwitterion form at pH 3.5. [App. Henderson-Hasselbalch Equation]

7.63 Why is it more important to have a daily source of protein than a daily source of fats or carbohydrates? [App: Proteins in the Diet]

7.64* What is an incomplete protein? [App: Proteins in the Diet]

For Problem 7.56

$$H_2N—\underset{\underset{CH_3CHCH_3}{|}}{CH}—\overset{\overset{\displaystyle O}{\|}}{C}—NH—\underset{\underset{H}{|}}{CH}—\overset{\overset{\displaystyle O}{\|}}{C}—NH—\underset{\underset{CH_2OH}{|}}{CH}—\overset{\overset{\displaystyle O}{\|}}{C}—NH—\underset{\underset{CH_2CH_2SCH_3}{|}}{CH}—\overset{\overset{\displaystyle O}{\|}}{C}—NH—\underset{\underset{CH_3}{|}}{CH}—\overset{\overset{\displaystyle O}{\|}}{C}—NH—\underset{\underset{CH_2COOH}{|}}{CH}—\overset{\overset{\displaystyle O}{\|}}{C}—OH$$

7.65 In general, which is more likely to contain complete protein—food from plant sources or from animal sources? [App: Proteins in the Diet]

7.66* What information can be obtained from an amino acid analyzer? [Int: Determining Protein Structure]

7.67 Ninhydrin is often used by forensic scientists to find fingerprints on paper and other surfaces. How do you think this technique works? [Int: Determining Protein Structure]

Additional Questions and Problems

7.68* Fresh pineapple can't be used in gelatin desserts because it contains an enzyme that hydrolyzes the proteins in gelatin, destroying the gelling action. Canned pineapple can be added to gelatin with no problem. Why?

7.69 In persons with sickle-cell anemia, a valine is substituted for a glutamic acid. What amino acid, if substituted for glutamic acid, might not cause a health problem?

7.70* Both α-keratin and tropocollagen have helical secondary structure. How do they differ?

7.71 Bradykinin, a peptide that helps to regulate blood pressure, has the primary structure Arg-Pro-Pro-Gly-Phe-Ser-Pro-Phe-Arg. Draw the complete structural formula of bradykinin.

7.72* Estimate the isoelectric pH of bradykinin (Problem 7.71).

7.73 For each amino acid listed, tell whether its influence on tertiary structure is largely through hydrophobic interactions, hydrogen bonding, formation of salt bridges, covalent bonding, or some combination of these effects.
(a) glutamic acid   (b) methionine   (c) glutamine
(d) threonine   (e) histidine   (f) phenylalanine
(g) cysteine   (h) valine

C H A P T E R

8

Enzymes, Vitamins, and Chemical Messengers

Messages cannot flow along nerve cells like these without the aid of neurotransmitters—one of several types of biomolecules that help to keep body chemistry under control.

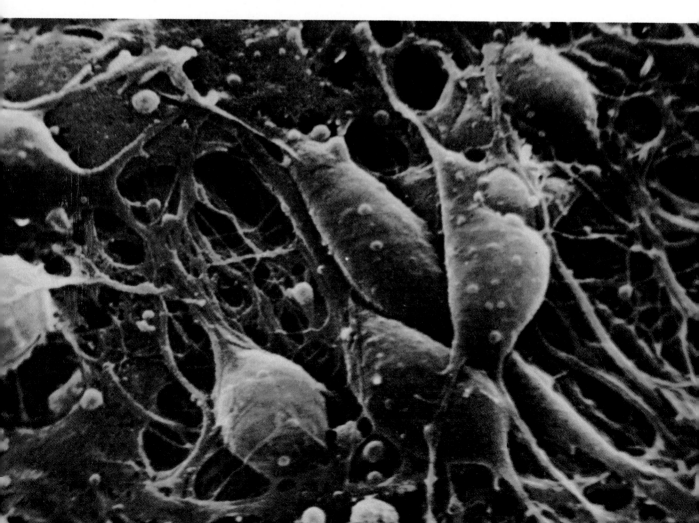

Think of your body as a walking chemical laboratory. Although the analogy isn't perfect, there's a good deal of truth to it. In a laboratory, chemical reactions are carried out one at a time by mixing pure chemicals in test tubes or flasks. In your body, chemical reactions take place in cells rather than in test tubes, and many thousands of reactions take place simultaneously.

The main difference between chemistry in a laboratory and chemistry in a living organism is *control.* In a laboratory, the speed of a reaction is controlled by adjusting experimental conditions such as temperature, solvent, reagent concentrations, and pH. In an organism, though, these conditions can't be adjusted. The human body must maintain a temperature of 37.0°C, the solvent must be water, and the pH must be 7.4.

How then does an organism control its thousands of different reactions so that all occur to the proper extent? The answer is that all reactions in living organisms are governed by biological catalysts called *enzymes.* It has been estimated that the body has at least 50,000 enzymes for regulating the multitude of reactions that take place inside us.

The activities of enzymes are controlled by several families of chemical messengers, including hormones and neurotransmitters. These messengers coordinate responses to changes in our environment and in conditions within our bodies. In this chapter, we'll explore how enzymes control biological reactions and how enzymes are in turn regulated by chemical messengers. Along the way we'll answer the following questions:

1. *What are enzymes?* The goal: Be able to describe the general structure of enzymes and their function in biological reactions.

2. *What kinds of enzymes are there and how are they classified?* The goal: Be able to name the classes of enzymes and describe the type of reaction catalyzed by each.

3. *How do enzymes work and why are they so specific?* The goal: Be able to use the lock-and-key and induced-fit models to explain why enzymes are specific and how they speed up reactions.

4. *What effect do temperature, pH, enzyme concentration, and substrate concentration have on enzyme activity?* The goal: Be able to describe what happens to enzyme activity when temperature, pH, enzyme concentration, and substrate concentration gradually change.

5. *How is enzyme activity regulated?* The goal: Be able to describe enzyme inhibition, feedback control, and allosteric control.

6. *What are vitamins, and how do they function?* The goal: Be able to describe the types of vitamins and their roles in biochemical reactions.

7. *What are hormones, and how do they function?* The goal: Be able to describe the origin, role in biochemical reactions, and mechanism of action of hormones.

8. *What are neurotransmitters, and how do they function?* The goal: Be able to describe the origin, role in biochemical reactions, and mechanism of action of neurotransmitters.

8.1 ENZYMES

Enzyme A protein or other molecule that acts as a catalyst for a biological reaction.

Enzymes are catalysts for biological reactions. Recall that a **catalyst** is a substance that speeds up the rate of a chemical reaction but that itself undergoes no permanent chemical change during the reaction. Sulfuric acid, for example, catalyzes the reaction of a carboxylic acid with an alcohol to yield an ester (Section 6.5). The reaction would occur very slowly if the catalyst were not present.

Reaction would occur very slowly without catalyst

$$CH_3-\overset{\overset{\displaystyle O}{\|}}{C}-OH \ + \ HOCH_3 \ \xrightarrow[\text{catalyst}]{H_2SO_4} \ CH_3-\overset{\overset{\displaystyle O}{\|}}{C}-OCH_3 \ + \ H_2O$$

Catalyst A substance that speeds up a chemical reaction without itself undergoing any permanent chemical change.

Enzymes are similar to sulfuric acid in that they catalyze reactions that might otherwise occur very slowly, but they differ in two important respects. First, enzymes are far larger, more complicated molecules than simple inorganic catalysts since, with just a few exceptions, enzymes are proteins. Second, enzymes are far more specific in their action. Whereas sulfuric acid catalyzes the reaction of nearly *every* carboxylic acid with nearly *every* alcohol, many enzymes catalyze only a single reaction of a single reactant.

Substrate The reactant in an enzyme-catalyzed reaction.

The reactant in an enzyme-catalyzed reaction is known as the enzyme's **substrate,** and the **specificity** of an enzyme is the extent to which it reacts with different substrates.

Specificity The extent to which an enzyme reacts with different substrates.

$$\text{Substrate} \ \xrightarrow{\text{enzyme catalyst}} \ \text{product}$$

REVIEW An energetically favorable reaction takes place spontaneously without any outside influence. An energetically unfavorable reaction occurs only with an outside influence—the addition of energy.

Enzymes differ greatly in their substrate specificity. For example, the enzyme *amylase* found in human digestive systems is able to catalyze the hydrolysis of starch to yield glucose but has no effect on cellulose, which is also composed of glucose. By contrast, the enzyme *papain,* a globular protein of 212 amino acids isolated from papaya fruit, catalyzes the hydrolysis of the peptide bonds in a variety of different polypeptides. It's this ability to hydrolyze peptides, in fact, that accounts for the use of papain in meat tenderizers and in contact-lens cleaners.

Bond that breaks

$$-\xi-NH-\underset{\underset{R}{|}}{CH}-\overset{\overset{\displaystyle O}{\|}}{C}-NH-\underset{\underset{R'}{|}}{CH}-\overset{\overset{\displaystyle O}{\|}}{C}-\xi- \ \xrightarrow[\text{papain}]{H_2O} \ -\xi-NH-\underset{\underset{R}{|}}{CH}-\overset{\overset{\displaystyle O}{\|}}{C}-OH \ + \ H_2N-\underset{\underset{R'}{|}}{CH}-\overset{\overset{\displaystyle O}{\|}}{C}-\xi-$$

REVIEW There is an energy barrier between reactants and products, and activation energy is the amount of energy reactants must have to get over that barrier.

Like all catalysts, an enzyme doesn't affect the equilibrium point of a reaction and can't bring about a reaction that is energetically unfavorable. What an enzyme *does* do is decrease the time it takes to reach equilibrium by lowering the activation energy. Catalase, for example, accelerates the rate of decomposition of hydrogen peroxide about 10^4 times more than a metal catalyst does (Figure 8.1).

Figure 8.1
The action of catalase on hydrogen peroxide (H_2O_2). When ground beef liver is added, the hydrogen peroxide foams up as it rapidly decomposes to water and oxygen. In the body, catalase catalyzes the same reaction, needed to destroy hydrogen peroxide produced by oxidase enzymes that remove hydrogen from substrates by combination with oxygen.

Turnover number The number of substrate molecules acted on by one molecule of enzyme per unit time.

The catalytic activity of an enzyme is measured by its **turnover number,** the number of substrate molecules acted on by one molecule of enzyme per unit time. As indicated in Table 8.1, enzymes vary greatly in their turnover number. Most enzymes have values in the 1–10,000 range, but some are much higher. Carbonic anhydrase in red blood cells, for example, is able to catalyze the reaction of *600,000* substrate molecules per second.

$$CO_2 + H_2O \xrightarrow{\text{carbonic anhydrase}} H_2CO_3 \longrightarrow HCO_3^- + H^+$$

8.2 ENZYME STRUCTURE

Cofactor A small, nonprotein part of an enzyme that is essential to the enzyme's catalytic activity.

Apoenzyme The protein portion of an enzyme.

Holoenzyme The combination of apoenzyme and cofactor that is active as a biological catalyst.

Many enzymes are globular proteins and thus have specific three-dimensional shapes determined by their secondary and tertiary structure. Many other enzymes consist of more than one protein chain and therefore have quaternary structure also.

In addition to their protein part, many enzymes are associated with small, nonprotein portions called **cofactors,** which may be held within the enzyme by either covalent bonds or noncovalent interactions. In such enzymes, the protein part is called an **apoenzyme,** while the entire assembly of apoenzyme plus cofactor is called a **holoenzyme.** Only holoenzymes are active as catalysts; neither apoenzyme nor cofactor alone can catalyze a reaction. An enzyme cofactor can be either an inorganic ion (usually a metal ion) or a small organic

Table 8.1 Turnover Numbers of Some Enzymes

Enzyme	Turnover Number (per second)
Carbonic anhydrase	600,000
Acetylcholinesterase	25,000
β-Amylase	18,000
Penicillinase	2,000
DNA polymerase I	15

Coenzyme A small, organic molecule that acts as an enzyme cofactor.

molecule called a **coenzyme.** Some enzymes need both a metal ion and a coenzyme to become active.

$$\text{Holoenzyme} = \text{apoenzyme} + \text{cofactor} \underset{\text{(coenzyme)}}{\overset{\text{Metal ion and/or small organic molecule}}{\diagdown}}$$

Why are cofactors necessary? The functional groups in proteins are limited to those of the amino acid side chains. By joining up with cofactors, however, enzymes can acquire chemically reactive groups not available from their amino acids, for example, groups needed to catalyze redox reactions.

The requirement that many enzymes have for metal ion cofactors is the main reason behind our dietary need for trace minerals. Iron, zinc, copper, manganese, molybdenum, cobalt, nickel, vanadium, and selenium are all known to function as enzyme cofactors. Every molecule of carboxypeptidase A, for example, contains one Zn^{2+} ion that serves as a catalyst in the hydrolysis of a peptide bond. Like the trace minerals, many vitamins are a necessary part of our diets because they are needed for enzyme activity (Section 8.11).

Practice Problem   8.1   Check the label on a bottle of vitamin-mineral supplements and look at the correlation between the metals listed and the metals known to be present in the body as enzyme cofactors.

8.3   ENZYME CLASSIFICATION

Enzymes are arranged into six main classes according to the general kind of reaction they catalyze, and each main class is further subdivided (Table 8.2). Most of the main-class names are self-explanatory; they're listed here with some examples:

● *Oxidoreductases* catalyze oxidation–reduction reactions of substrate molecules, which may involve addition or removal of oxygen or hydrogen:

$$\text{A(reduced)} + \text{B(oxidized)} \longrightarrow \text{A'(oxidized)} + \text{B'(reduced)}$$

$$\underset{\text{(Reduced)}}{CH_3-CH_2-OH} + \underset{\text{(Oxidized)}}{NAD^+} \xrightarrow[\text{dehydrogenase}]{\text{alcohol}} \underset{\text{(Oxidized)}}{CH_3-\overset{\displaystyle O}{\overset{\|}{C}}-H} + \underset{\text{(Reduced)}}{NADH} + H^+$$

● *Transferases* catalyze the transfer of a group from one substrate to another:

$$\text{A} + \text{B—C} \longrightarrow \text{A—B} + \text{C}$$

$$\text{Glucose} + \text{adenosine triphosphate (ATP)} \xrightarrow{\text{hexose kinase}}$$

$$\text{glucose 6-phosphate} + \text{adenosine diphosphate (ADP)}$$

Table 8.2   Classification of Enzymes

Main Class	Some Subclasses	Type of Reaction Catalyzed
Oxidoreductases	Oxidases	Oxidation of a substrate
	Reductases	Reduction of a substrate
	Dehydrogenases	Introduction of double bond (oxidation) by formal removal of H_2 from substrate
Transferases	Transaminases	Transfer of an amino group between substrates
	Kinases	Transfer of a phosphate group between substrates
Hydrolases	Lipases	Hydrolysis of ester groups in lipids
	Proteases	Hydrolysis of amide groups in proteins
	Nucleases	Hydrolysis of phosphate groups in nucleic acids
Lyases	Dehydrases	Loss of H_2O from substrate
	Decarboxylases	Loss of CO_2 from substrate
Isomerases	Epimerases	Isomerization of chiral center in substrate
Ligases	Synthetases	Formation of new bond between two substrates, with participation of ATP
	Carboxylases	Formation of new bond between substrate and CO_2, with participation of ATP

• *Hydrolases* catalyze the hydrolysis of substrates, that is, the breaking of bonds with addition of water:

$$A—B + H_2O \longrightarrow A—OH + B—H$$

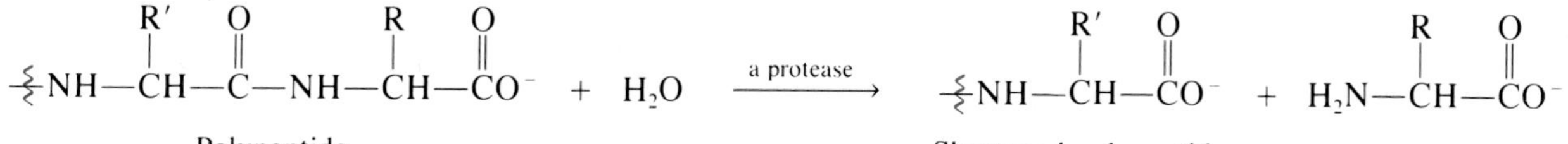

Polypeptide   Shortened polypeptide

• *Isomerases* catalyze the isomerization (rearrangement of atoms) of substrates:

$$A \longrightarrow B$$

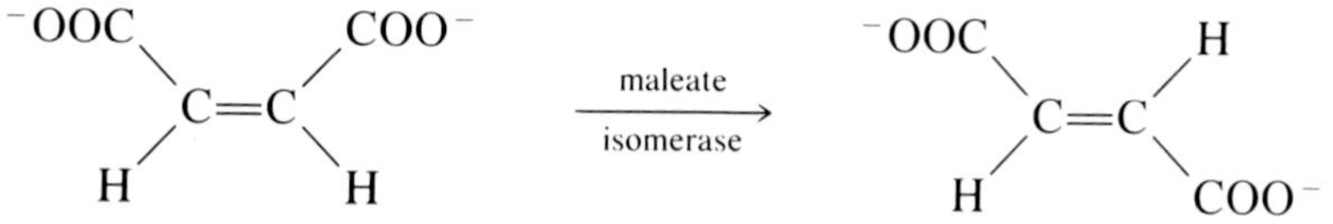

Maleate   Fumarate

• *Lyases* (from the Greek *lein*, meaning "to break") catalyze the addition of a small molecule such as H_2O or NH_3 (usually to a double bond) and the reverse reaction, the elimination of a small molecule (usually to leave behind a double bond).

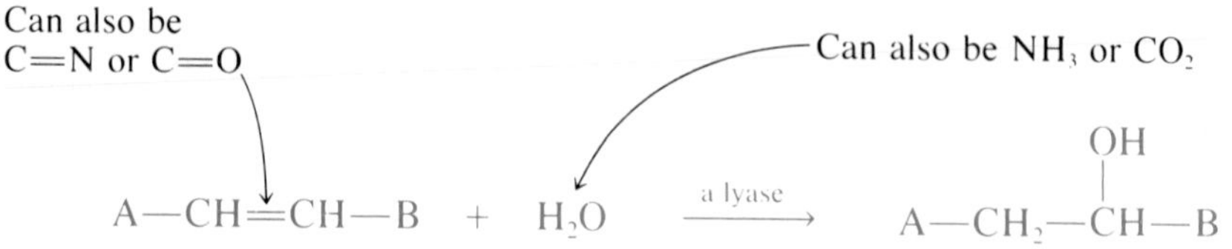

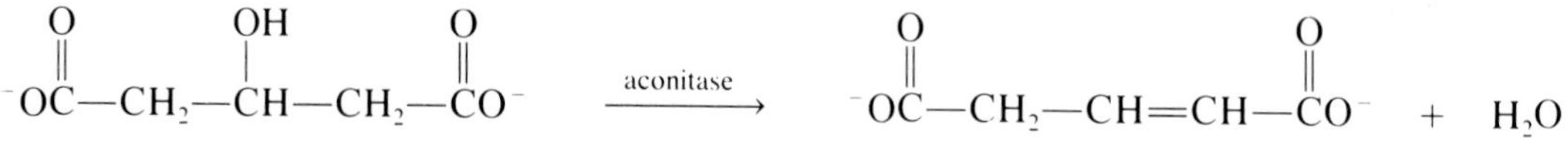

Can also be
C=N or C=O

Can also be NH$_3$ or CO$_2$

$$A-CH=CH-B \ + \ H_2O \xrightarrow{\text{a lyase}} A-CH_2-\overset{\overset{\displaystyle OH}{|}}{CH}-B$$

$$\overset{\overset{\displaystyle O}{\|}}{{}^-OC}-CH_2-\overset{\overset{\displaystyle OH}{|}}{CH}-CH_2-\overset{\overset{\displaystyle O}{\|}}{CO}{}^- \xrightarrow{\text{aconitase}} \overset{\overset{\displaystyle O}{\|}}{{}^-OC}-CH_2-CH=CH-\overset{\overset{\displaystyle O}{\|}}{CO}{}^- \ + \ H_2O$$

Citrate

cis-Aconitate

● *Ligases* (from the Latin *ligare*, meaning "to tie together") catalyze the bonding together of two substrate molecules, with the participation of ATP.

A + B + adenosine triphosphate (ATP) $\longrightarrow$ A—B + adenosine diphosphate (ADP) + PO$_4^{3-}$ (P$_i$)

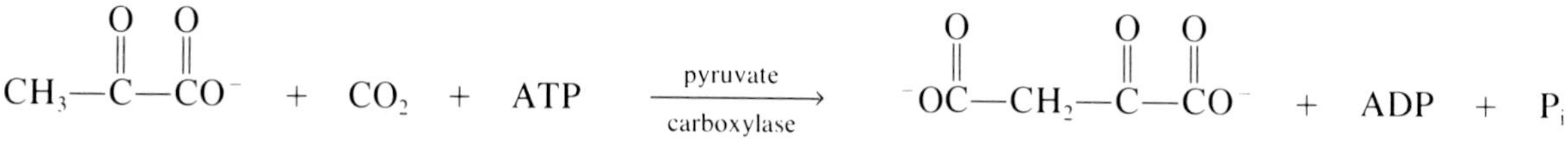

$$CH_3-\overset{\overset{\displaystyle O}{\|}}{C}-\overset{\overset{\displaystyle O}{\|}}{CO}{}^- \ + \ CO_2 \ + \ ATP \xrightarrow[\text{carboxylase}]{\text{pyruvate}} \overset{\overset{\displaystyle O}{\|}}{{}^-OC}-CH_2-\overset{\overset{\displaystyle O}{\|}}{C}-\overset{\overset{\displaystyle O}{\|}}{CO}{}^- \ + \ ADP \ + \ P_i$$

Pyruvate

Oxaloacetate

Note in the examples given above that all enzymes have the family-name ending *-ase*. Although many enzymes like papain and trypsin were given uninformative common names in the past, the modern systematic names have two parts: The first part identifies the substrate molecule on which the enzyme operates, and the second part is an enzyme subclass name like those shown in Table 8.2. For example, *alcohol dehydrogenase* is an oxidoreductase enzyme that oxidizes an alcohol to yield an aldehyde by removal of two hydrogens that are transferred to a coenzyme.

Solved Problem 8.1   To what class does the enzyme that catalyzes the following reaction belong?

$$CH_3\overset{\overset{\displaystyle O}{\|}}{\underset{\underset{\displaystyle NH_2}{|}}{CH}}CO^- \ + \ {}^-O\overset{\overset{\displaystyle O}{\|}}{C}CH_2CH_2\overset{\overset{\displaystyle O}{\|}}{C}-\overset{\overset{\displaystyle O}{\|}}{CO}{}^- \longrightarrow CH_3\overset{\overset{\displaystyle O}{\|}}{C}-\overset{\overset{\displaystyle O}{\|}}{CO}{}^- \ + \ {}^-O\overset{\overset{\displaystyle O}{\|}}{C}CH_2CH_2\overset{\underset{\underset{\displaystyle NH_2}{|}}{\displaystyle}}{CH}\overset{\overset{\displaystyle O}{\|}}{CO}{}^-$$

Solution   Comparing the functional groups in the reactants and products shows that an amino group and a carbonyl group have changed places. The reaction is therefore a functional-group transfer, and the enzyme is a transferase.

Practice Problems   **8.2**   What reactions would you expect these enzymes to catalyze?
(a) fumarate hydrase   (b) squalene oxidase   (c) glucose kinase
(d) cellulose hydrolase

8.3   To what class of enzymes does pyruvate decarboxylase, which acts with the participation of ATP, belong?

8.4 ENZYME SPECIFICITY

Lock-and-key model A model for enzyme specificity that pictures an enzyme as a large molecule with a cleft into which only specific substrate molecules can fit.

Active site A small, three-dimensional portion of an enzyme with the specific shape and structure necessary to bind a substrate.

Induced-fit model A model that pictures an enzyme with a conformationally flexible active site that can change shape to accommodate a range of different substrate molecules.

Any theory of how enzymes work must explain two facts: It must explain why enzymes are so specific, and it must explain how enzymes speed up reactions. Our current picture of enzymes relies on two complementary models to explain these facts. In the first, called the **lock-and-key model,** an enzyme is pictured as a large, irregularly shaped molecule with a cleft or crevice in its middle. Inside the crevice is an **active site,** a small, three-dimensional region of the enzyme with the specific shape, side chains, and cofactors necessary to bind the substrate, usually by noncovalent interactions, and catalyze the appropriate reaction. In other words, the shape of the active site is like a lock into which only a specific key, the substrate, can fit (Figure 8.2).

The lock-and-key model requires that enzyme and substrate have unchanging shapes, which means that only one substrate can fit a given enzyme. Although it's true that some enzymes are so specific they operate on only a single substrate molecule, other enzymes are less selective. For example, the lipase enzymes secreted by the pancreas hydrolyze all dietary fats and oils, regardless of their exact structure. To account for the broader specificity of many enzymes, the lock-and-key model has been expanded.

According to the **induced-fit model,** some enzymes are flexible enough to change the shapes and sizes of their active sites to fit the spatial requirements of different substrates. As the enzyme and substrate come together, their interaction *induces* exactly the right fit (Figure 8.3). Changes in the shape of the

Figure 8.2
The lock-and-key model of enzymes. Enzymes are large, three-dimensional molecules containing a crevice with a well-defined active site. Only a substrate whose shape and chemical nature is complementary to that of the active site can fit into the enzyme.

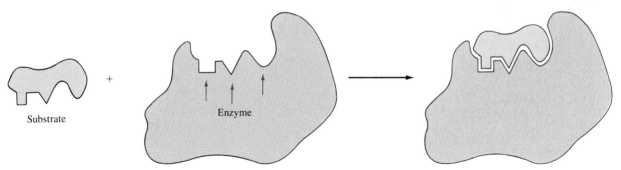

Substrate + Enzyme

Figure 8.3
The induced-fit model of enzyme action. Some enzymes have a conformation that is flexible enough to allow them to adapt the size and shape of their active site to the shapes of several different substrates.

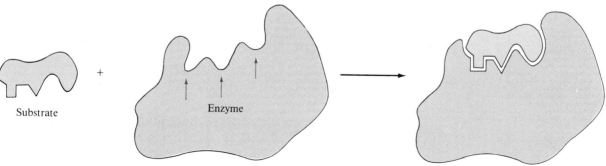

Substrate + Enzyme

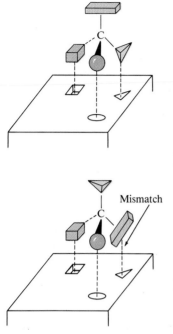

Figure 8.4
A chiral reactant and a chiral reaction site. The optical isomer on the left will fit the reaction site like a hand in a glove, but the optical isomer on the right can't fit and therefore can't react.

enzyme allow it to conform to the substrate, and changes in the shape of the substrate prepare it to react.

Just as an enzyme is often specific for a given substrate, it's also specific for one of a pair of optical isomers if the substrate is a chiral molecule (Section 7.5). For example, the enzyme lactate dehydrogenase catalyzes the removal of hydrogen from "left-handed" L-lactate but not from "right-handed" D-lactate.

$$\underset{\text{L-Lactate}}{\overset{\displaystyle O \!\! \diagup \!\! C \!\! \diagdown \!\! O^{-}}{HO-\overset{\displaystyle |}{\underset{\displaystyle |}{C}}-H}} \; + \; NAD^{+} \quad \xrightarrow[\text{dehydrogenase}]{\text{lactate}} \quad \underset{\text{Pyruvate}}{\overset{\displaystyle O \!\! \diagup \!\! C \!\! \diagdown \!\! O^{-}}{\overset{\displaystyle |}{\underset{\displaystyle CH_3}{C}}=O}} \; + \; NADH \; + \; H^{+}$$

$$\underset{\text{D-lactate}}{\overset{\displaystyle O \!\! \diagup \!\! C \!\! \diagdown \!\! O^{-}}{H-\overset{\displaystyle |}{\underset{\displaystyle CH_3}{C}}-OH}} \; + \; NAD^{+} \quad \xrightarrow[\text{dehydrogenase}]{\text{lactate}} \quad \text{no reaction}$$

As we saw in the previous chapter, amino acids and proteins have handedness. Thus, the specificity of an enzyme for one of two optical isomers is a matter of fit. As illustrated schematically in Figure 8.4, a left-handed enzyme can't fit with a right-handed substrate any more than a left-handed glove can fit on a right hand.

8.5 HOW ENZYMES WORK

Enzyme–substrate complex A complex of enzyme and substrate in which the two are bound together by noncovalent interactions with the substrate in position to react.

Now that you know why enzymes are so specific, we need to examine how they speed up reactions. Enzyme-catalyzed reactions begin with migration of the substrate into the active site to form an **enzyme–substrate complex.** Before complex formation, the substrate molecule is in its most stable, lowest-energy form. Within the complex, the molecule is forced into a higher-energy form in which certain bonds have been weakened. The result is to make the energy barrier between substrate and product smaller.

Within the enzyme–substrate complex, atoms that will form new bonds must connect with each other, and groups needed for catalysis must be close to the necessary locations in the substrate. Many organic reactions, for example, require acidic, basic, or metal ion catalysts. The active site can provide acidic or basic groups without disrupting the constant-pH environment in body fluids. Once the chemical reaction is completed, enzyme and product molecules separate from each other and the enzyme becomes available for another substrate.

$$\underset{\text{Enzyme}}{E} \; + \; \underset{\text{Substrate}}{S} \; \rightleftarrows \; \underset{\text{Complex}}{[E\!-\!S]} \; \longrightarrow \; \underset{\text{Enzyme}}{E} \; + \; \underset{\text{Product}}{P}$$

The hydrolysis of a peptide bond by chymotrypsin shown in Figure 8.5 illustrates how an enzyme functions. In the enzyme–substrate complex (Figure

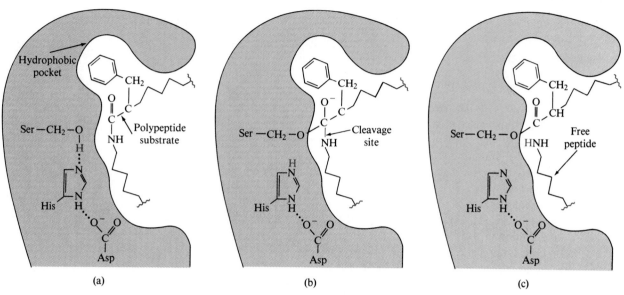

Figure 8.5

Hydrolysis of a peptide bond by chymotrypsin. (a) The polypeptide enters the active site with its hydrophobic side chain in the hydrophobic pocket and the peptide bond opposite serine and histidine residues. (b) Hydrogen transfer from serine to histidine allows bonding to the peptide. (c) The peptide bond is broken. Subsequent steps, not shown, release the other piece of the hydrolyzed polypeptide and restore the H to serine.

8.5a), attraction of a hydrophobic side chain into a hydrophobic pocket in the active site positions the substrate and places the bond to be broken next to the catalytic site. The enzyme has not only brought the two reactants together but has also oriented them correctly. Next (Figure 8.5b), the peptide bond carbon is temporarily bonded to serine in the active site, making it easier for the peptide bond to break (Figure 8.5c). In further steps not shown, the active site returns to its original condition and the rest of the polypeptide is set free.

In summary, enzymes act as catalysts because of their ability to

- Bring reactants together (*proximity effect*)
- Hold reactants at the exact distance and in the exact orientation necessary for reaction (*orientation effect*)
- Lower the energy barrier by inducing strain in the substrate (*energy effect*)
- Provide acidic, basic, or other types of sites required for catalysis (*catalytic effect*).

8.6 INFLUENCE OF TEMPERATURE AND pH ON ENZYMES

Enzymes have been finely tuned through evolution so that their maximum catalytic ability is highly dependent on pH and temperature. As you might expect, optimum conditions vary slightly for each enzyme but are generally near neutral pH and body temperature.

AN APPLICATION: MEDICAL USES OF ENZYMES AND ISOENZYMES

In a healthy person, most enzymes are found mainly within cells. When some diseases occur, however, enzymes are released from dying cells into the blood, where their increased levels can be measured by sensitive clinical instruments. The accompanying table lists some of the presently available enzyme tests, along with the medical conditions indicated by abnormal blood levels. Enzymes marked with an asterisk in the table are included in routine blood analysis.

Enzyme analysis relies on measuring the *activity* of an enzyme rather than its concentration. Because activity is influenced by pH, temperature, and substrate concentration, it is measured in international units at standard conditions. One unit (U) is defined as the amount of the enzyme that converts one micromole of substrate to product under defined standard conditions of pH, temperature, and substrate concentration. The analytical results are thus reported in units per liter (U/L).

Among the most useful enzyme assays are those done for diagnosis of heart disease. Three enzymes, creatine phosphokinase (CPK), aspartate transaminase (AST), and lactate dehydrogenase (LDH), are found in the muscle cells of a healthy heart, as well as in other tissues. When a heart attack, or *myocardial infarction* (MI), occurs, some cells are damaged and their enzymes leak into the bloodstream. As shown in the first figure, the blood levels of CPK, AST, and LDH all increase markedly in the hours immediately following a heart attack. The CPK level rises almost immediately following an MI, reaching a sixfold increase over normal values after about 30 hr; the AST level triples after about 40 hr; and the LDH level doubles after about 4 days.

Confirmation that a heart attack has occurred can be gained by a careful analysis of the individual enzyme levels. Recent work has shown that a number of enzymes, including CPK and LDH, are actually mixtures of several

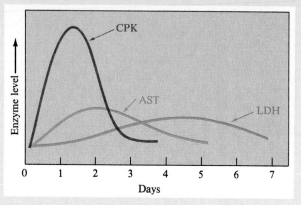

Blood levels of creatine phosphokinase (CPK), aspartate transaminase (AST), and lactate dehydrogenase (LDH) in the days following a heart attack (myocardial infarction).

closely related compounds called *isoenzymes* that have slightly different structures but that catalyze the same reaction. For example, creatine phosphokinase is a mixture of three isoenzymes, denoted CK(MM), CK(MB), and CK(BB). Brain tissue is rich in CK(BB), skeletal muscles are rich in CK(MM), and heart tissue is rich in CK(MB). Similarly, lactate dehydrogenase is a mixture of five isoenzymes, denoted LDH_1, LDH_2, LDH_3, LDH_4, and LDH_5. Of the five, heart tissue contains primarily LDH_1.

Since heart tissue contains primarily the CK(MB) and LDH_1 isoenzymes, their levels increase the most following a heart attack. Thus, separation and analysis of the five individual LDH isoenzymes shows that the LDH_2 level is higher than that of LDH_1 in a normal profile but that the two levels "flip" following a heart attack (facing page). Similarly, an analysis of the three CK isoenzymes shows that the CK(MB) level increases to a much greater extent than that of the other two following a heart attack.

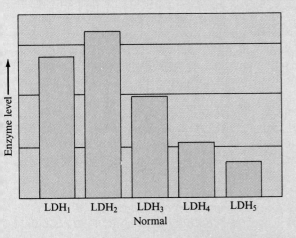 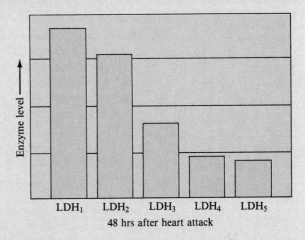

Blood levels of the five lactate dehydrogenase (LDH) isoenzymes in a normal person and in a heart attack victim after 48 hr. Note the flip of LDH_1 and LDH_2 levels following a heart attack.

Some Enzyme Assays in Body Fluids

Enzyme	Condition Indicated by Abnormal Level
Lactate dehydrogenase (LDH)[a]	Heart disease, liver diseases
Creatine phosphokinase (CPK)	Heart disease
Aspartate transaminase (AST)[a]	Heart disease, liver diseases, muscle damage
Alanine transaminase (ALT)[a]	Heart disease, liver diseases, muscle damage
Glutamyl transferase (GMT)	Liver diseases
Alkaline phosphatase (ALP)[a]	Bone disease, liver diseases
Amylase	Pancreatic diseases
Lipase (LPS)	Pancreatic diseases
Acid phosphatase (ACP)	Prostate cancer
Renin	Hypertension
Glucose-6-phosphate dehydrogenase (GPD)	Hemolytic anemia
γ-Glutamyl transferase (GGT)[a]	Liver disease, alcoholism

[a] Included in routine blood analysis.

Effect of Temperature   An increase in temperature leads to an increase in rate for most chemical reactions, and enzyme-catalyzed reactions are no exception. Unlike many simple reactions, however, enzyme-catalyzed processes always reach an optimum temperature, after which their rate again falls as shown in Figure 8.6. The falloff in reaction rate with heating beyond the optimum temperature occurs because enzymes begin to denature when heated too strongly (Section 7.12). The noncovalent attractions between protein side chains are disrupted, the delicately maintained three-dimensional shape of the enzyme begins to come apart, and as a result the active site needed for catalytic activity disappears.

Most enzymes denature and lose their catalytic activity above 50–60°C, a fact that explains why medical instruments and laboratory glassware can be sterilized by heating with steam in an autoclave. The high temperature of the autoclave denatures the enzymes of any bacteria present, thereby killing the organisms.

Effect of pH   The catalytic activity of many enzymes depends on the pH of the surroundings and often has a well-defined optimum point. For example, trypsin, a protease enzyme that aids digestion of proteins in the small intestine, has optimum activity at pH 8.0, an activity that falls dramatically at a pH either higher or lower than 8.0 (Figure 8.7). Changes in pH can change enzyme structure because of denaturation and can also change the ability of the active site to bind a substrate because of changes in the charges of acidic and basic side chains.

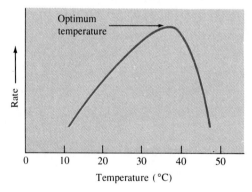

Figure 8.6
The effect of temperature on the rate of an enzyme-catalyzed reaction. There is always an optimum temperature at which the reaction rate reaches its maximum.

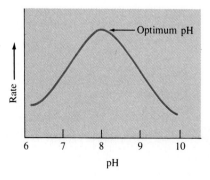

Figure 8.7
The effect of pH on the catalytic activity of trypsin, a protease digestive enzyme of the small intestine.

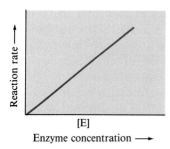

Figure 8.8
Change in reaction rate with enzyme concentration when [E] < [S] at constant temperature, pH, and substrate concentration. The reaction rate in the presence of excess substrate is directly proportional to the enzyme concentration. Doubling the enzyme concentration, for example, doubles the reaction rate.

8.7   EFFECT OF ENZYME AND SUBSTRATE CONCENTRATION ON ENZYME ACTIVITY

For a reaction to occur, the enzyme and substrate molecules must come together and form the enzyme–substrate complex. Therefore, some variation in the reaction rate can be expected if enzyme or substrate concentrations change.

First, consider what happens when the substrate concentration is high relative to the enzyme concentration. Under these conditions, all the enzymes are working to capacity. With *increasing enzyme concentration*, the additional enzyme molecules immediately start taking up substrate and the reaction rate increases. If the enzyme concentration is doubled, for example, the rate also **doubles, a directly proportional relationship as shown in Figure 8.8.**

Next, consider what happens when the substrate concentration is low relative to the enzyme concentration. Not all the enzymes are being utilized, and the rate is dependent on the concentration of substrate. With *increasing substrate concentration*, the rate increases because more of the enzyme molecules are put to work. Eventually, however, the substrate concentration reaches a point where all the available active sites are occupied. Since the reaction rate is now determined by how fast the enzyme–substrate complex is converted to product, the reaction rate levels off and becomes constant, and the enzyme is said to be *saturated*. Beyond this point, increasing substrate concentration has no effect on the rate (Figure 8.9).

Under most conditions, an enzyme is not likely to be saturated and the reaction rate is controlled by the overall efficiency of the enzyme. If the enzyme–substrate complex is rapidly converted to product, the rate at which enzyme and substrate combine to form the complex becomes the limiting factor. Calculations show there's an upper limit to this rate: Enzyme and substrate molecules moving at random in solution can collide with each other no more often than about 10^8 collisions per mole per second. Thus, no reaction can take place faster than this limit. Remarkably, a few enzymes actually operate with this maximum efficiency—every one of those 10^8 collisions results in the formation of product! One example is *triose phosphate isomerase*, which catalyzes a step in the breakdown of glucose.

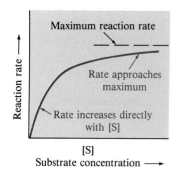

Figure 8.9
Change in reaction rate with substrate concentration at constant temperature, pH, and enzyme concentration. At low substrate concentration, the reaction rate is directly proportional to the concentration. At high substrate concentration, the rate approaches a maximum.

$$
\begin{array}{ccc}
CH_2OH & & H\!-\!\underset{\displaystyle\ \ }{\overset{\displaystyle O}{C}} \\
| & & | \\
C\!=\!O & \xrightarrow[\text{isomerase}]{\text{triose phosphate}} & H\!-\!C\!-\!OH \\
| & & | \\
CH_2OPO_3{}^{2-} & & CH_2OPO_3{}^{2-}
\end{array}
$$

Dihydroxyacetone phosphate Glyceraldehyde 3-phosphate

8.8   ENZYME INHIBITION

The control of biological reactions by enzymes is only half the overall picture. It's equally important that the enzymes themselves be regulated. After all, if an enzyme were continually functioning at full speed, it would soon run out of substrate and the body would soon have a huge oversupply of the enzyme's product.

Living things control their enzymes by a variety of strategies. Although we'll describe them separately, you should keep in mind that a number of control strategies operate simultaneously in most series of related biochemical reactions. Considering that a cell contains thousands of protein molecules and hundreds of other kinds of biomolecules, all in the concentrations required to maintain **homeostasis,** the achievement of enzyme control is awe-inspiring.

Any process that starts up or increases the action of an enzyme is called **activation.** Any process that slows down or stops the action of an enzyme is called **inhibition.** Inhibition is important for natural enzyme control and can also be put to work in medications that modify enzyme activity. Some inhibitors, however, are poisons because they prevent an enzyme from carrying out a necessary function. In this and the next two sections, we'll describe three types of inhibition and some of the major strategies for regulating enzyme activity.

Competitive Inhibition   What would happen if an enzyme were to encounter a molecule with a shape and size similar to that of its normal substrate? The imposter molecule could bind to the enzyme's active site and thereby prevent a normal substrate molecule from binding to the same site. As a result, the enzyme would be inactivated (Figure 8.10) and the overall effect would be equivalent to decreasing the enzyme concentration. Inhibition of this sort is called **competitive inhibition** because the inhibitor competes with substrate for binding to the active site. Of course, if the inhibitor later happens to migrate *out* of the active site, the enzyme will once again be able to bind with substrate and be fully active. Thus, competitive inhibition is reversible. As shown in Figure 8.11 (middle curve), the maximum reaction rate is unchanged by competitive inhibition, but a higher substrate concentration is required to reach that rate.

Competitive inhibition is used to good advantage in the treatment of methanol poisoning. Although not harmful itself, methanol (wood alcohol) is oxidized in the body to formaldehyde, which is highly toxic ($CH_3OH \rightarrow H_2C=O$). Because of its similarity to methanol, ethanol is able to act as a competitive inhibitor of the methanol oxidase enzyme, thereby blocking oxidation and allowing methanol to be excreted harmlessly. Thus, the medical treatment of methanol poisoning includes administering high levels of ethanol.

Homeostasis Maintenance of unchanging internal conditions by living things.

Activation (of an enzyme) Any process that initiates or increases the action of an enzyme.

Inhibition (of an enzyme) Any process that slows down or stops the action of an enzyme.

Competitive enzyme inhibition Enzyme regulation in which an inhibitor competes with a similarly shaped substrate for binding to the enzyme active site.

Figure 8.10
Competitive inhibition. A competitive inhibitor blocks an enzyme's active site by mimicking the shape and size of the normal substrate, thereby preventing the enzyme from functioning.

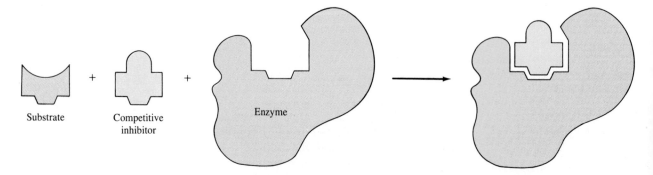

Substrate   +   Competitive inhibitor   +   Enzyme   →

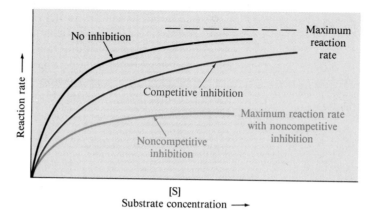

Figure 8.11
Change in reaction rate with inhibitors. The rate increases more slowly with a competitive inhibitor (red) but at high enough [S] reaches the maximum rate for the enzyme. A noncompetitive inhibitor (blue) lowers the maximum rate.

Noncompetitive enzyme inhibition Enzyme regulation in which an inhibitor binds to an enzyme elsewhere than at the active site, thereby changing the shape of the enzyme's active site.

Noncompetitive Inhibition   A second kind of reversible inhibition occurs when an inhibitor binds to an enzyme at some place other than the active site. Such binding can lead to a change in the shape of the enzyme, with a resultant change in the shape and binding ability of the active site. Binding of substrate thus becomes more difficult, enzyme activity diminishes, and the maximum reaction rate decreases (Figure 8.11, bottom curve). Called **noncompetitive inhibition** because the inhibitor does not compete with substrate for the active site, this kind of effect is illustrated in Figure 8.12.

Lead, mercury, and other heavy metals are toxic because of their ability to act as noncompetitive inhibitors. These metals bind strongly to thiol groups (—SH) on cysteine units in enzymes, thus altering the enzymes' shapes.

Irreversible Inhibition   Yet a third type of inhibition occurs when a molecule enters an enzyme's active site and forms a covalent bond to the enzyme. Since

Figure 8.12
Noncompetitive inhibition. By binding elsewhere on the enzyme, a noncompetitive inhibitor changes the shape of the active site and decreases the ability of the enzyme to bind with substrate.

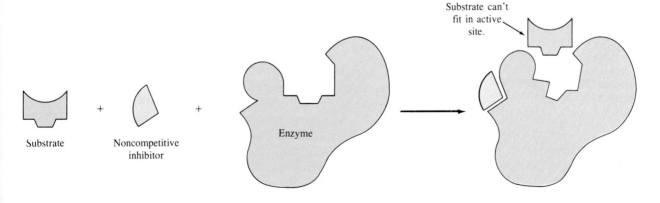

the inhibitor is firmly bonded in place, rather than just loosely held by noncovalent attractions, the active site is permanently blocked and the enzyme is irreversibly inactivated. **Irreversible inhibition** accounts for the toxicity of nerve gases like diisopropylphosphofluoridate, which reacts with the enzyme acetylcholinesterase to block nerve transmission. Acetylcholinesterase has at its active site a serine residue that covalently bonds to the inhibitor. Most irreversible inhibitors are poisons.

Irreversible enzyme inhibition A mechanism of enzyme deactivation in which an inhibitor forms covalent bonds to the active site.

Serine amino acid at active site of acetylcholinesterase

Diisopropylphosphofluoridate

Covalent bond that irreversibly binds the inhibitor to the enzyme

8.9 ENZYME REGULATION: FEEDBACK AND ALLOSTERIC CONTROL

"Feedback" is a general term applied whenever the result of a process feeds information back to affect the beginning of the process. For example, any device such as an oven that maintains a constant temperature has a sensor that detects temperature and feeds back that information to turn the heating elements on or off.

Consider a series of biochemical reactions in which A is converted to B, then B is converted to C, and so on.

$$A \xrightarrow{\text{enzyme 1}} B \xrightarrow{\text{enzyme 2}} C \xrightarrow{\text{enzyme 3}} D \xrightarrow{\text{enzyme 4}} E$$

Feedback control
Activation or inhibition of the first reaction in a sequence by a product of the sequence.

Allosteric control
Cooperative interaction by which binding a substrate or a regulator at one site in an enzyme influences shape and therefore binding at other sites.

Allosteric enzyme Enzyme with two or more protein chains (quaternary structure) and two or more interactive binding sites for substrate and regulators.

What will happen if product E is an inhibitor for enzyme 1? Since the amount of A converted to B will decrease, the amounts of B, C, and D will also decrease. The effect of this mechanism, known as **feedback control,** is to keep the concentration of E in a cell constant. When more E than needed is present, its synthesis is slowed down or stopped and no energy is wasted making the unneeded intermediates B, C, and D. When eventually there's not enough E, it leaves the binding sites in enzyme 1, the enzyme is no longer inhibited, and the production of E starts up again.

Generally, control strategies are more elaborate than simple feedback control because feedback by molecules with structures totally different than that of the substrate is necessary. Most biochemical pathways are regulated by **allosteric control**—a cooperative interaction in which binding a molecule at one site in an enzyme influences binding at other sites. All known **allosteric enzymes** have more than one protein chain and have two kinds of binding sites: those for substrate and those for regulators. Binding a *positive regulator* changes the shapes of active sites so that they accept substrate more readily and the rate accelerates. Binding a *negative regulator* (a noncompetitive inhibitor) changes the shapes of active sites so they accept substrate less readily, and the rate accelerates more slowly with increasing substrate concentration. The changes in enzyme shape that accompany binding either substrate or regulator are likely to be changes in quaternary structure. Because allosteric enzymes usually have several substrate and several regulator binding sites, and because there is cooperative interaction among them all, very fine control is achieved.

8.10 ENZYME REGULATION: ZYMOGENS AND GENETIC CONTROL

So far, we've discussed enzyme control by changes in enzyme shape. There are also two other important types of regulation.

Zymogen (proenzyme) A compound that becomes an active enzyme after undergoing a chemical change.

Some enzymes are synthesized in inactive forms that differ from the active forms in composition. Activation of such enzymes, known as **zymogens** or **proenzymes,** requires a chemical reaction that either adds or splits off some part of the molecule (Figure 8.13). Some of the enzymes that digest proteins, for example, are produced in the pancreas as the zymogens *trypsinogen, chymotrypsinogen,* and *proelastase.* These enzymes must not be active when they're synthesized, because they would attack the pancreas. One of the dangers of traumatic injury to the pancreas, in fact, is premature activation of these zymogens, resulting in potentially fatal *pancreatitis.* Each protease zymogen has a short polypeptide segment not present in the active enzymes, *trypsin, chymotrypsin,* and *elastase.* The unnecessary polypeptide segments are snipped off when the zymogens reach the small intestine, where protein digestion occurs.

Figure 8.13
Zymogens. Some enzymes are produced initially in inactive forms called zymogens, which are then activated at the proper time by snipping off a small segment.

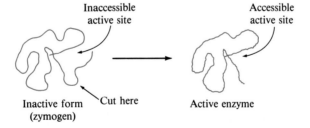

Inaccessible active site

Accessible active site

Inactive form (zymogen) Cut here

Active enzyme

Yet another enzyme control strategy goes back to the supply of the enzyme itself. The synthesis of enzymes, like that of all proteins, is regulated by genes (Chapter 15), and mechanisms exist that turn enzyme synthesis on and off. The **genetic control** strategy is especially useful for enzymes needed only at certain stages in an organism's growth.

Genetic enzyme control Regulation of enzyme activity by control of the synthesis of enzymes.

8.11 VITAMINS

Vitamin A small, organic molecule that must be obtained in the diet and that is essential in trace amounts for proper biological functioning.

It has been known since antiquity that there is a critical relationship between diet and health. Lime and other citrus juices cure scurvy, meat and milk cure pellagra, and cod-liver oil cures rickets. The active substances in all three cases are **vitamins,** small, organic molecules that must be obtained through the diet and that are required in trace amounts.

Precursor A compound necessary for the synthesis of another compound.

Vitamins are grouped by solubility into two classes: water-soluble and fat-soluble. The water-soluble vitamins (Table 8.3) are found in the aqueous environment inside cells where they function either as coenzymes or as **precursors** of coenzymes. Vitamin C, for example, is biologically active without any change in structure, but thiamine must be converted to the coenzyme thiamine pyrophosphate.

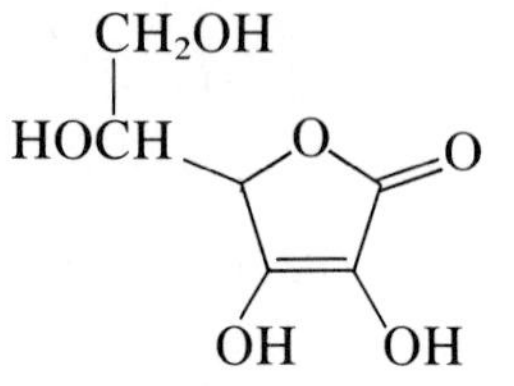

Vitamin C
(ascorbic acid)

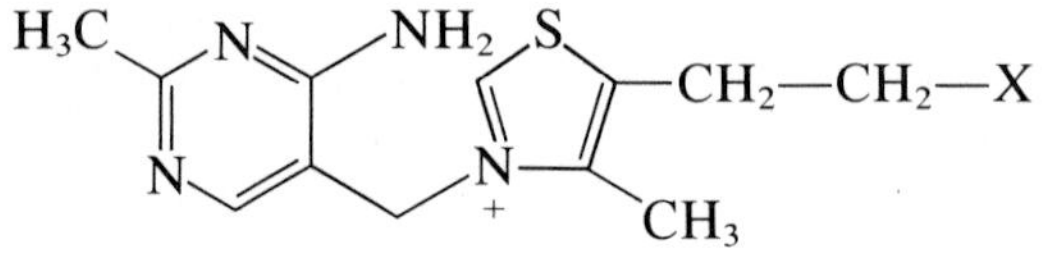

Thiamine, $X = {-}OH$

Thiamine pyrophosphate, $X = {-}OP_2O_6{}^{3-}$
(a coenzyme)

The fat-soluble vitamins A, D, E, and K are stored in the body's fat deposits, but their diverse functions are less well understood than those of the water-soluble vitamins.

Vitamin A, which is essential for normal development of epithelial tissue, night vision, and healthy eyes, has three active forms: retinol, retinal, and retinoic acid. The all-cis form of retinoic acid, the drug known as Retin-A, is

Table 8.3   Vitamins and Their Functions

Vitamin	Function (Coenzyme Example)	Deficiency Syndrome
Water-soluble vitamins		
Ascorbic acid (vitamin C)	Hydroxylases	Scurvy: bleeding gums, bruising
Thiamin (vitamin B_1)	Oxidoreductases (thiamine pyrophosphate)	Beriberi: fatigue, depression, heart disease
Riboflavin (vitamin B_2)	Oxidoreductases (flavin adenine dinucleotide, FAD)	Cracked lips, scaly skin, sore tongue
Pyridoxine (vitamin B_6)	Aminotransferases (pyridoxal phosphate)	Anemia, irritability, skin lesions, retarded growth
Niacin	Oxidoreductases (nicotinamide adenine dinucleotide, NAD)	Pellagra: dermatitis, dementia, breakdown of central nervous system, gastrointestinal function
Folic acid (vitamin M)	Methyltransferases	Megaloblastic anemia, retarded growth
Vitamin B_{12}	Isomerases (cobamide coenzymes)	Megaloblastic anemia, neurodegeneration
Pantothenic acid	Acyltransferases (coenzyme A)	Weight loss, irritability, retarded growth
Biotin (vitamin H)	Carboxylases	Dermatitis, anorexia, depression
Fat-soluble vitamins		
Vitamin A	Visual system	Night blindness, dry skin
Vitamin D	Calcium metabolism	Rickets, osteomalacia
Vitamin E	Antioxidant	Hemolysis of red blood cells
Vitamin K	Blood clotting	Hemorrhage, delayed blood clotting

used for treating skin diseases and has been reported to counteract the wrinkles of aging.

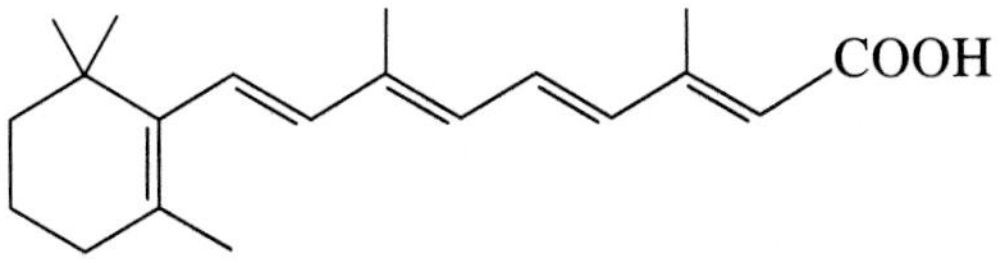

Retinoic acid

Vitamin D is related in structure to cholesterol and is synthesized when ultraviolet light from the sun strikes a cholesterol derivative in the skin. In the kidney, vitamin D is converted to a hormone that regulates calcium absorption and bone formation. Vitamin D deficiencies are most likely to occur in malnourished individuals living where there is little sunlight.

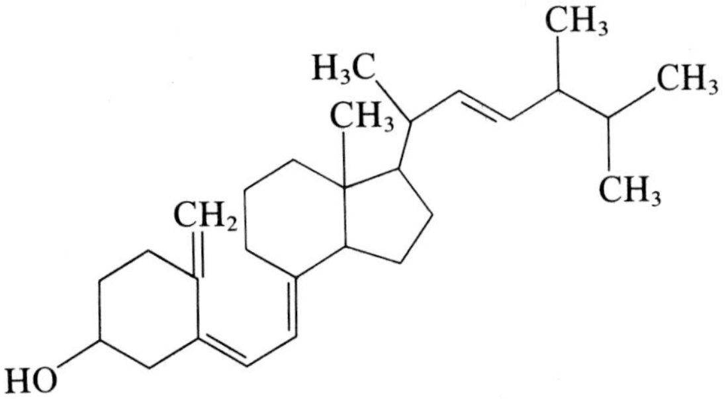

Vitamin D

Vitamin E refers to a group of structurally similar compounds called *tocopherols,* the most active of which is α-tocopherol. The best-understood function of vitamin E is as an antioxidant: It prevents the breakdown by oxidation of vitamin A and fats. The need for vitamin E apparently increases with the proportion of polyunsaturated fats in the diet.

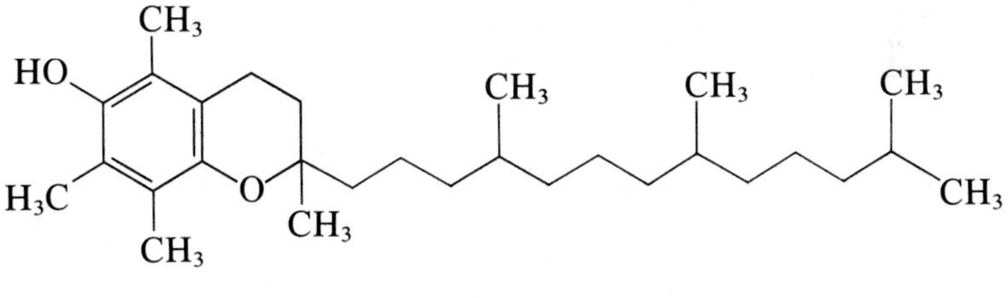

Vitamin E

Vitamin K includes the following structure and a number of related compounds with side chains of varying length. The letter K comes from the name *koagulations vitamin* bestowed by its Danish discoverer for the role of vitamin K in blood clotting. The vitamin is essential to the synthesis of several blood-clotting factors.

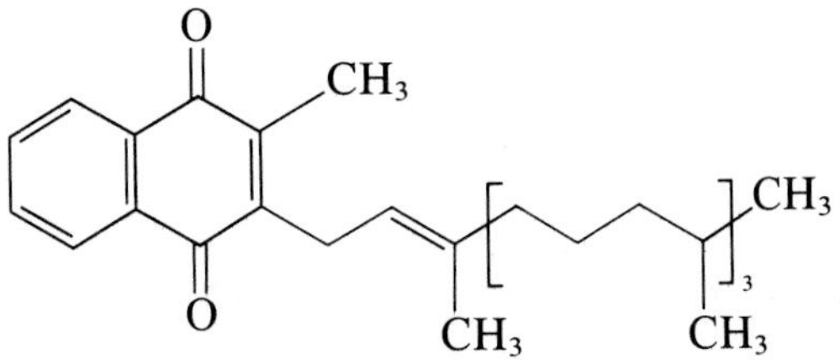

Vitamin K

AN APPLICATION: VITAMINS AND MINERALS IN THE DIET

It's not uncommon to encounter incomplete or incorrect information about vitamins and minerals in the diet. For instance, we've been frightened by the possibility that aluminum will cause Alzheimer's disease and tantalized by the possibilities that vitamin E will cure cancer and that vitamin C will defeat the common cold. Sorting out fact from fiction or distinguishing preliminary research results from scientifically proven relationships is especially difficult in this area of nutrition. Much is yet to be learned about the various functions of vitamins and minerals in the body, and new research results are continuously being reported. It's tempting for health-conscious individuals to look for guaranteed ways to better health by taking vitamins, and it's tempting to profit-making organizations to take advantage of this motivation.

One consistent source of information on nutrition is the Food and Nutrition Board of the National Academy of Sciences–National Research Council. Every 5 years they survey the latest nutritional information and publish Recommended Daily Dietary Allowances that are "designed for the maintenance of good nutrition of the majority of healthy persons in the United States." They also publish a list of estimated "safe and adequate" daily dietary ranges for selected vitamins and minerals for which there's not enough information to establish recommended allowances. Currently, biotin, pantothenic acid, copper, manganese, fluoride, chromium, molybdenum, sodium, potassium, and chloride are on this second list. The "percent of RDA" values given on food labels refers to U.S. RDAs derived from the National Academy of Sciences data by the Food and Drug Administration.

Daily vitamin intake way above the RDAs has become popular in recent years. The benefits of megadoses of any vitamins are doubted by many nutritionists, but there's no doubt about the potential for harm in megadoses of fat-soluble vitamins. Because they dissolve in fatty tissue, cell membranes, and the liver, excesses can build up in the body. Excess vitamin A, for example, can cause liver damage, skin peeling, nausea, and even death. Continuous overdoses of vitamin D lead to buildup of excess calcium in bones and also in soft tissues such as the kidneys (kidney stones), lungs, and inner ear. While overdoses of water-soluble vitamins can also disrupt body functions, this happens infrequently because water-soluble vitamins circulate in body fluids and excesses are eventually excreted in the urine.

Recommended Daily Dietary Allowances (Revised 1989)[a]

Fat-Soluble Vitamins	Male	Female	Water-Soluble Vitamins	Male	Female	Minerals	Male	Female
Vitamin A (μg RE)[b]	1000	800	Vitamin C (mg)	60	60	Ca (mg)	1200	1200
Vitamin D (μg)	10	10	Thiamin (mg)	1.5	1.1	P (mg)	1200	1200
Vitamin E (mg TE)[c]	10	8	Riboflavin (mg)	1.7	1.3	Mg (mg)	350	280
Vitamin K (μg)	70	60	Niacin (mg NE)[d]	19	15	Fe (mg)	10	15
			Vitamin B$_6$ (mg)	2.0	1.6	Zn (mg)	15	12
			Folate (μg)	200	180	I (μg)	150	150
			Vitamin B$_{12}$ (μg)	2.0	2.0	Se (μg)	70	55

[a] From *Recommended Dietary Allowances*, 10th ed., Food and Nutrition Board, National Research Council–National Academy of Sciences, 1989.

[b] 1 RE = 1 retinol equivalent = 1 μg retinol.

[c] 1 TE = 1 α-tocopherol equivalent = 1 mg α-tocopherol.

[d] 1 NE = 1 niacin equivalent = 1 mg of niacin or 60 mg of dietary tryptophan.

Practice Problem   *8.4*   Compare the structures of vitamin A and vitamin C. What structural features does each have that make one water-soluble but the other fat-soluble?

8.12 CHEMICAL MESSENGERS

At this point, you've seen some of the many kinds of enzyme-catalyzed reactions that take place in cells: oxidations, reductions, bond formations, and so forth. What hasn't been discussed yet is how the individual reactions are tied together. Clearly, the many thousands of separate reactions that take place in cells throughout the body don't occur randomly; there must be an overall control mechanism that coordinates them and keeps the entire organism in chemical balance.

Strategies much like those that inhibit or activate individual enzymes are also put to work in coordinating biochemical reactions. Messenger molecules unite with receptor molecules in induced-fit-type interactions, often in response to feedback control mechanisms, and are held together by noncovalent interactions long enough for messages to be delivered.

Hormone A chemical messenger, usually secreted by an endocrine gland and transported through the bloodstream to elicit response from a specific target tissue.

One group of chemical messengers is composed of the **hormones** produced in the endocrine glands. Another group is made up of **neurotransmitters,** compounds that help to transmit chemical messages in the nervous system. Together the endocrine and nervous systems bear the major responsibility for keeping track of internal and external conditions and for making changes when necessary. Chemical messengers also include such diverse compounds as the *pheromones,* which carry messages between different organisms, and *endorphins* and *enkephalins,* which act in the brain to relieve pain.

Neurotransmitter A chemical messenger that transmits a nerve impulse between neighboring nerve cells.

With increased understanding of biochemical interactions, the traditional definition of hormones as substances produced by the endocrine glands and acting on cells distant from the glands is being expanded. The prostaglandins and leukotrienes (Section 12.11), for example, don't meet the strict definition of hormones because they neither are synthesized by endocrine glands nor act on distant cells. They do, however, exert hormone-like influence on nearby cells. Similarly, the distinction between hormones and neurotransmitters is not absolute. It's been found that epinephrine (the fight-or-flight hormone discussed in Section 8.14) and norepinephrine meet both definitions—as hormones because they're secreted by endocrine glands and as neurotransmitters because they're also produced at nerve endings in the brain.

Endocrine system A system of specialized cells, tissues, and ductless glands that excretes hormones and shares with the nervous system the responsibility for maintaining constant internal body conditions and responding to changes in the environment.

8.13 HORMONES

The glands of the **endocrine system** are located in various parts of the body and secrete hormones that are transported through the bloodstream to their target tissues. Once at the target tissue, hormones interact with specific receptors to elicit a biological response. In no case, however, do hormones themselves carry out any chemical reactions; they act solely by turning existing biological mechanisms on or off.

The endocrine system is organized into successive stages under the overall control of the *hypothalamus,* a section of the brain just above the pituitary gland. The hypothalamus secretes tiny amounts of peptide hormones called *releasing factors* that in turn stimulate the release of other hormones by the pituitary gland. Several of these hormones then activate other endocrine glands or organs to produce yet a third round of hormones, which ultimately act on their target tissues, as diagrammed in Figure 8.14.

Endocrine hormones vary greatly in structure. Some, such as insulin and the releasing factors secreted by the hypothalamus, are polypeptides. Others, such as thyroxine and epinephrine, are small, organic molecules derived from amino acids. And still others, such as estrone and the other sex hormones, are steroids (Section 12.10).

Hormone receptors may be on the outer surface of the cell, in the cytoplasm inside the cell, or in the nucleus of the cell. Receptors on the surface act in several ways: Some release a substance within the cell that regulates an enzyme; some open up channels that quickly let needed ions or molecules into the cell; some extend through the cell membrane to catalytic sites within the cell that are activated when the hormone is bound to the receptor.

Figure 8.14
The primary organization of the endocrine system, together with the names and functions of some major hormones. The *hypothalamus* is a region in the brain that monitors information from the nervous system and coordinates the activities of the endocrine and nervous systems. Releasing factors from the hypothalamus act on the pituitary gland, which then produces hormones that act on either secondary targets (blue) or final target tissues (green).

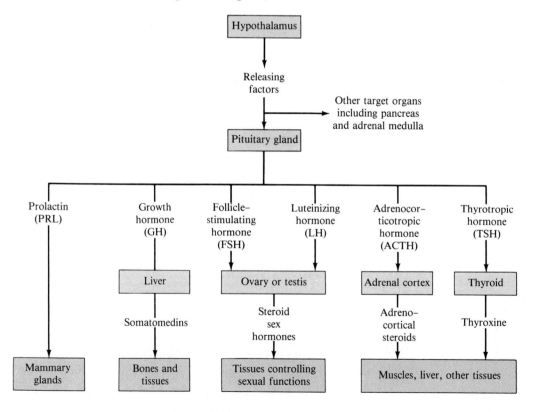

8.14 HOW HORMONES WORK: THE ADRENALINE RESPONSE

Let's look at a specific example of hormone action to see how the overall process works. *Epinephrine,* better known as *adrenaline,* is often called the fight-or-flight hormone because its release prepares the body for instant response to danger (Figure 8.15).

Figure 8.15
A grizzly bear exhibits the fight-or-flight reaction to a threatening situation. The hormone epinephrine produces a variety of physiological responses, such as increased heart rate and blood sugar level, and behavioral changes, such as these bared fangs.

We've all felt the rush of adrenaline that accompanies a near-miss accident or a sudden loud noise. Chemically, epinephrine is a simple aromatic amine that is secreted into the bloodstream by the adrenal medulla within the adrenal glands located just above each kidney. Although it produces a variety of responses from numerous target tissues, epinephrine's main function is to raise the body's blood sugar level by increasing the rate of its release from storage in muscles and the liver.

Epinephrine, like many of the body's chemical messengers, doesn't cross the target cell's membrane. Instead, epinephrine binds to a receptor site on the outside of a liver cell and stimulates an enzyme (adenylate cyclase) in the cell membrane to begin production of what is called a *second messenger.* The second messenger, cyclic adenosine monophosphate (cyclic AMP or cAMP) in the present case, is released into the cell's interior. There it interacts in a series of steps to activate *glycogen phosphorylase enzymes,* the enzymes needed to release glucose from storage. These enzymes convert glucose molecules from the glucose storage polymer, glycogen, to glucose phosphate. The phosphate is then converted to glucose and released to the bloodstream (Figure 8.16).

Figure 8.16
The series of events by which release of the hormone epinephrine raises blood glucose level. Interaction of the hormone with a cell membrane receptor causes the second messenger cyclic AMP to be produced inside the target cell from ATP. The cyclic AMP in turn activates a series of steps that initiates the phosphorylase enzymes needed to break down glycogen to glucose phosphate, which is then converted to glucose for transport in the bloodstream. Epinephrine also acts to inhibit the enzymes that synthesize glycogen from glucose.

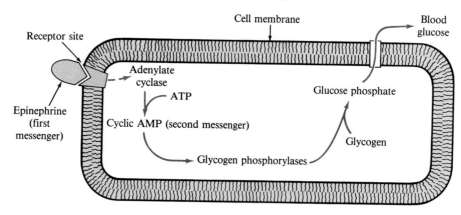

Muscle cells make energy directly from the phosphate by glycolysis (Section 11.3). When enough glucose has been produced, an enzyme called *phosphodiesterase* catalyzes the hydrolysis of cAMP, and glucose production ends. The entire series of events from initial stimulus to blood glucose production takes only seconds.

A number of hormones and neurotransmitters act via cyclic AMP as the second messenger, with the responses of different cells dependent on the different enzymes present within them. Ca^{2+} ion is also often a second messenger.

8.15 NEUROTRANSMITTERS

Neurotransmitters are substances that mediate the flow of nerve impulses by transmitting a signal between neighboring nerve cells, or **neurons** (Figure 8.17). Structurally, nerve cells have a bulb-like body connected to a long, thin stem called an *axon*. Short, tentacle-like appendages called *dendrites* protrude from the bulbous end of the neuron, while numerous feathery filaments protrude from the axon at the opposite end. When bundled together in a nerve fiber, the axon of one neuron butts up against the dendrites of the next neuron, separated only by a narrow gap called a **synapse.**

Transmission of a nerve impulse between cells occurs when neurotransmitter molecules are released from the axon of a *presynaptic* neuron, cross the synaptic cleft, and fit into appropriately shaped receptor sites on the dendrites of the next, *postsynaptic* neuron (Figure 8.18). Once the message has been

Synapse Narrow gap between nerve cells across which a signal is carried by a neurotransmitter.

Figure 8.17
A typical nerve cell.

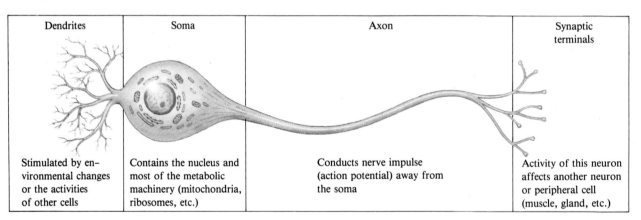

Dendrites	Soma	Axon	Synaptic terminals
Stimulated by environmental changes or the activities of other cells	Contains the nucleus and most of the metabolic machinery (mitochondria, ribosomes, etc.)	Conducts nerve impulse (action potential) away from the soma	Activity of this neuron affects another neuron or peripheral cell (muscle, gland, etc.)

Figure 8.18
Transmission of a nerve signal. Transmission occurs between neurons when a neurotransmitter molecule is released by the presynaptic neuron, crosses the synaptic cleft, and fits into a receptor site on the postsynaptic neuron.

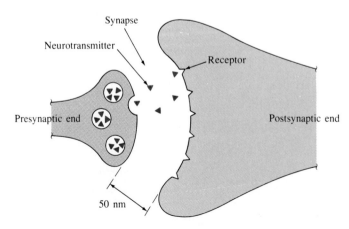

received, the cell transmits an electrical signal down its own axon and further passes along the nerve impulse.

The human nervous system has two main parts: the *central nervous system* (CNS), which consists of nerves in the brain and spinal cord, and the *peripheral nervous system* (PNS), which consists of sensory and motor nerves. The PNS, in turn, is divided into *autonomic* and *somatic* parts. The autonomic part deals with unconsciously controlled functions like blood circulation and digestion, while the somatic part deals with consciously controlled functions like movement.

Neurons in the PNS can be divided into *cholinergic* and *adrenergic* types, depending on what kind of neurotransmitter molecules they release. Cholinergic neurons, which include all those in the somatic part and many in the autonomic part, act by releasing *acetylcholine*. Adrenergic neurons, which are found only in the autonomic part, act by releasing *norepinephrine*.

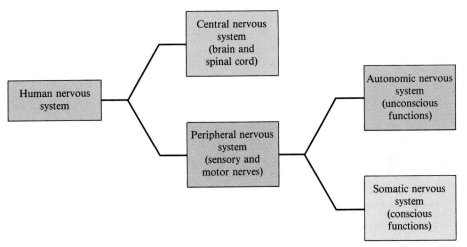

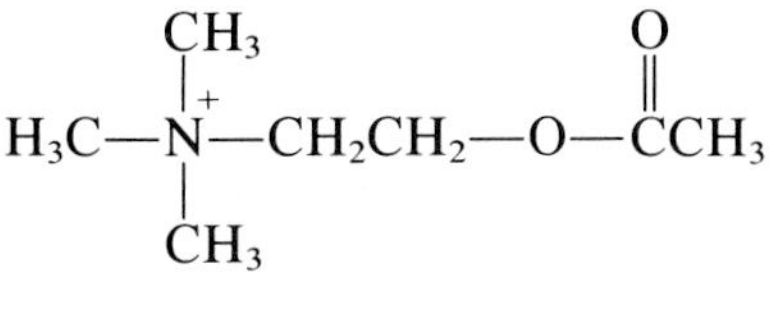

Acetylcholine

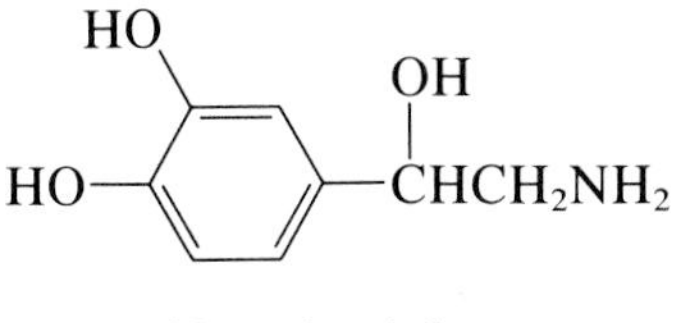

Norepinephrine

Unlike the PNS, the CNS uses a surprising variety of chemicals as neurotransmitters. Among the most interesting are *dopamine* and *serotonin,* which appear to be vital to such higher brain functions as perception and emotion. In addition, several amino acids and small peptides have been found to act as neurotransmitters in certain regions of the brain.

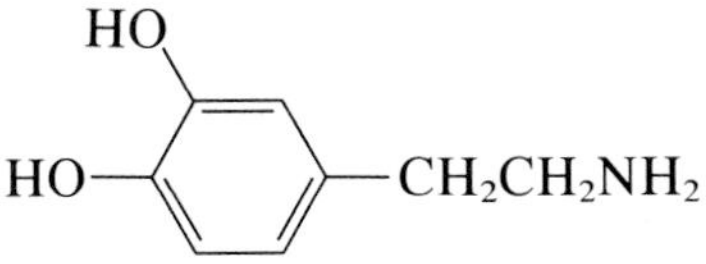

Dopamine

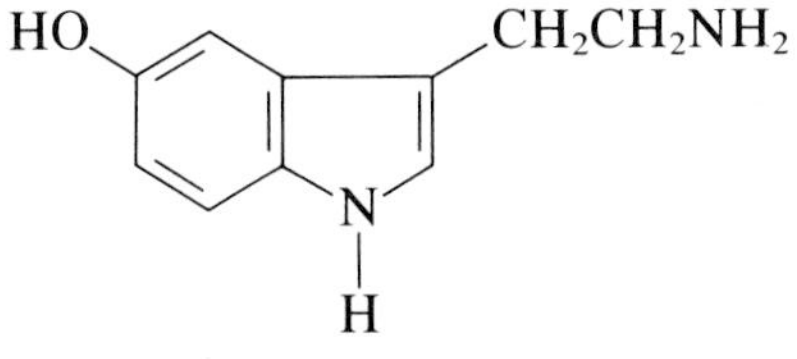

Serotonin

INTERLUDE: PENICILLIN, THE FIRST ANTIBIOTIC

Scarlet fever, diphtheria, bubonic plague, and rheumatic fever are all easily controlled diseases that are now almost unknown in modern, industrialized countries. But if you had the bad fortune to contract one of these diseases before 1940, you might well not have survived. The difference between then and now is due to the discovery in 1928 and the subsequent isolation in 1940 of penicillin, the first successful antibiotic.

Antibiotics are substances that are produced by one microorganism and are toxic to other microorganisms. Penicillin was discovered by the British bacteriologist Alexander Fleming as the result of a chance observation made when several culture plates of staphylococcus bacteria he was growing became contaminated. Fleming noticed that the contaminant, a mold subsequently identified as *Penicillium notatum,* produced a substance that appeared to kill the staphylococcus.

Chemically, the penicillins are known as *beta-lactam* antibiotics because they contain a four-membered amide ring. More than 30 naturally occurring penicillins with closely related structures are known, and several thousand synthetic derivatives have been prepared in the laboratories of pharmaceutical companies.

Different penicillins have different acyl side chains (in blue) attached to nitrogen. Penicillin G, the most common one, is shown.

The penicillins kill bacteria by acting as irreversible inhibitors of a critical transpeptidase enzyme that is involved in the synthesis of bacterial cell walls. Bacterial cells, unlike those of higher organisms, are enveloped by a protective coating called a *cell wall*. In the final stages of cell wall synthesis, strands of glycoprotein known as

---

Once a neurotransmitter has done its job, it must be rapidly removed from the receptor site so that the postsynaptic neuron is ready to receive another impulse. In cholinergic neurons, for example, this removal is accomplished by the enzyme *acetylcholinesterase*. Nerve gases (as illustrated in Section 8.8) and some insecticides act as cholinesterase inhibitors to block the activity of the enzyme, thereby allowing acetylcholine to accumulate at the receptor sites. Convulsions, paralysis of respiratory muscles, and death can result.

Malfunctions in the nervous system are involved as either symptoms or causes in a number of serious diseases. Parkinson's disease and schizophrenia, for example, have both been linked to defects in the brain's use of dopamine. Alzheimer's disease, a progressive disorder involving memory loss and deterioration of mental function, appears to involve a deficiency of the enzyme that synthesizes acetylcholine.

SUMMARY

Enzymes are large proteins that function as biological catalysts in converting **substrate** to product. Unlike typical inorganic catalysts, enzymes are large, complex molecules that are quite specific in their action. They are classified into six groups according to the kind of reaction they catalyze: **oxidoreductases, transferases, hydrolases, isomerases, lyases,** and **ligases.**

In addition to their protein part, many enzymes contain **cofactors** which can be either metal ions or small, organic molecules known as **coenzymes.** Often, the

peptidoglycans are cross-linked together to form the final three-dimensional web. The crucial cross-linking step involves enzyme-catalyzed reaction of an Ala-Ala terminus on one strand with a Gly terminus on another strand.

Peptidoglycan—Gly—NH$_2$ D-Ala— Peptidoglycan
                                    |
                              D-Ala

                    transpeptidase
                         ↓

Peptidoglycan—Gly—D-Ala—Peptidoglycan + D-Ala

Penicillin's three-dimensional shape is evidently similar enough to that of the Ala-Ala end of the peptidoglycan side chain that it is able to fit into the active site of the transpeptidase enzyme. Once in the active site, penicillin binds irreversibly by covalent bond formation between the enzyme and the carbonyl group of the β-lactam ring. With bacterial cell wall synthesis thus halted, the cell contents leak out through the weakened wall and the cell dies. Since the cells of higher organisms have no cell walls, penicillin is completely specific for bacteria and is nontoxic to all other organisms.

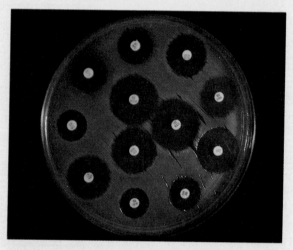

Testing the sensitivity of bacteria to different antibiotics. The clear circular areas on the culture dish are regions free of bacterial growth. The bacteria have been killed or prevented from reproducing by antibiotics diffusing outward from the smaller spots at the center of each circle.

coenzyme comes from a water-soluble **vitamin.** There are 13 known human vitamins, 9 of which are water-soluble and 4 of which are fat-soluble.

Enzymes contain an **active site,** a small, three-dimensional region with the specific shape and structure needed to bind a substrate in the **enzyme–substrate complex,** often by noncovalent interactions. Enzyme specificity is partly explained by the **lock-and-key model,** in which the substrate is thought to fit the enzyme like a key in a lock. Variations in specificity and the catalytic action of enzymes are further explained by the **induced-fit model,** in which both enzyme and substrate change shape in forming the complex. Overall, enzymes work by exercising effects on proximity, orientation, energy, and catalysis.

Enzymes function best at an optimum pH and temperature. Their action can be suppressed by **competitive inhibitors,** which compete with substrate for binding to the active site; by **noncompetitive inhibitors,** which act by changing the shape of the active site by binding elsewhere in the enzyme; or by **irreversible inhibitors,** which act by covalently bonding to and blocking the active site.

At low substrate concentration, reaction rate increases with the substrate concentration; at high substrate concentration, the rate increases more slowly and eventually reaches a maximum. Competitive inhibitors slow the approach to the maximum rate, but noncompetitive inhibitors decrease the maximum rate. Enzyme activity is controlled by **feedback,** in which a reaction product is an inhibitor or activator for the first enzyme in a series of reactions, and by **allosteric control,** in which an enzyme's quaternary structure changes shape in response to binding of activators or inhibitors. Control is also achieved by **zymogen** synthesis and **genetic control.**

Enzyme-catalyzed reactions throughout the body are coordinated by chemical messengers. **Hormones** from the **endocrine system** are produced under the overall control of the hypothalamus and pituitary gland and are carried in the bloodstream to target tissues. **Neurotransmitters** are secreted by nerve cells and transmit nerve impulses from cell to cell. Many hormones and neurotransmitters connect to receptors on cell surfaces, which then produce second messengers that carry the message to the cell's interior.

REVIEW PROBLEMS

Structure and Classification of Enzymes

8.5   What is the family-name ending for an enzyme?

8.6*   What is an enzyme, and what general kind of structure does it have?

8.7   There are approximately 50,000 different enzymes in the body. Why are so many needed?

8.8*   How do catalysts speed up reactions?

8.9   What are the two primary differences between an inorganic catalyst and an enzyme?

8.10*   What general kinds of reactions do these classes of enzymes catalyze?
(a) hydrolases   (b) isomerases   (c) lyases

8.11   What is a holoenzyme, and what are its two parts called?

8.12*   What is the difference between a cofactor and a coenzyme?

8.13   What feature of enzymes makes them so specific in their action?

8.14*   Describe in general terms how enzymes act as catalysts.

8.15   What classes of enzymes would you expect to catalyze these reactions:

(a)
$$H_2N-CH-C-NH-CH-COH + H_2O \longrightarrow$$
with R and O groups as shown

$$H_2N-CH-COH + H_2N-CH-COH$$
with R and R′ groups

(b) $HOCOH \longrightarrow H_2O + CO_2$

(c)
$$HOC-CH_2-CH_2-COH \longrightarrow$$
$$HOC-CH=CH-COH$$

8.16*   What kind of reaction does each of the following enzymes catalyze?
(a) aprotease   (b) a DNA ligase
(c) a transmethylase

8.17   The following reaction is catalyzed by an enzyme called *urease*. To what class of enzyme does urease belong?

$$H_2N-C-NH_2 + 2 H_2O \xrightarrow{\text{urease}} 2 NH_3 + H_2CO_3$$

8.18*   The catalytic activity of urease (Problem 8.17) can be inhibited by adding dimethylurea to the reaction. What kind of inhibition is probably occurring?

$$CH_3NH-C-NHCH_3 \qquad \text{Dimethylurea}$$

Enzyme Function and Regulation

8.19   What is the difference between the lock-and-key model of enzyme action and the induced-fit model?

8.20*   Fats are esters that are broken down during digestion to glycerol (1,2,3-propanetriol) and long-chain carboxylic acids. What class of enzyme would accomplish this?

8.21   Which enzyme is more specific in its action, a lock-and-key enzyme or an induced-fit enzyme?

8.22   Draw an energy diagram for the exothermic enzyme-catalyzed hydrolysis of urea (Problem 8.17). Label the energy levels of reactants and products, the activation energy, and the overall energy difference in the reaction.

8.23   Which part of a holoenzyme would you expect to be affected by denaturation, the coenzyme or the apoenzyme? Explain.

8.24*   Prior to the development of modern antibiotics, a drop of dilute $AgNO_3$ was often placed in the eyes of newborn infants to prevent gonorrheal eye infections. How might $AgNO_3$ kill bacteria?

8.25   Discuss how the rate of reaction changes as enzyme concentration is increased.

8.26*   Discuss how the rate of reaction changes as substrate concentration is increased.

8.27   Why don't increases in concentration of substrate or enzyme change reaction rates in exactly the same way?

8.28*   What general effects would you expect the following changes to have on the speed of an enzyme-catalyzed reaction?
(a) lowering the reaction temperature from 37°C to 27°C
(b) raising the pH from 7.5 to 10.5
(c) adding a heavy-metal salt like $Hg(NO_3)_2$
(d) raising the reaction temperature from 37°C to 40°C

8.29   How can you explain the observation that pepsin, a digestive enzyme found in the stomach, has a high catalytic activity at pH 1.5 while trypsin, a digestive enzyme of the small intestine, has no activity at that pH?

8.30*   What are the three kinds of enzyme inhibitors?

8.31   How does each of the three kinds of enzyme inhibitors work?

8.32*   What kinds of bonds are formed between an enzyme and each of the three kinds of enzyme inhibitors?

8.33 Poisoning by which of the three types of enzyme inhibitors is probably the most difficult to treat medically? Why?

8.34* The meat tenderizer used in cooking is primarily *papain*, a protease enzyme isolated from the papaya tree. Why do you suppose papain is so effective at tenderizing meat?

8.35 Papain (Problem 8.34) is also used to help relieve the pain from bee stings. Why do you suppose that works?

8.36* What is feedback inhibition?

8.37 Why do allosteric enzymes have two types of binding sites?

8.38* Discuss the purpose of positive and negative regulators.

8.39 What is a zymogen? Why must some enzymes be secreted as zymogens?

8.40* A substantial number of calories in regular beer are due to the presence of complex carbohydrates called *amylopectins*. By adding an amyloglucosidase enzyme during brewing, the amylopectins can be hydrolyzed into glucose and then fermented. What two characteristics does the resultant light beer have? (*Hint:* The first one is that it tastes great.)

Vitamins, Hormones, and Neurotransmitters

8.41 What is the difference between a hormone and a vitamin?

8.42* What is the difference between a hormone and a neurotransmitter?

8.43 What is the relationship between vitamins and enzymes?

8.44* Which of the following three are required in the diet: hormones, enzymes, or vitamins?

8.45 What are the four fat-soluble vitamins?

8.46* Why is daily ingestion of vitamin C more critical than daily ingestion of vitamin A?

8.47 What is the general purpose of the body's endocrine system?

8.48* What gland is primarily responsible for controlling the endocrine system?

8.49 How is a hormone transported from its secretory gland to its target tissue?

8.50 Name as many endocrine glands as you can.

8.51 Describe in general terms how a hormone works.

8.52* What is the structural difference between an enzyme and a hormone?

8.53 What is the relationship between enzyme specificity and tissue specificity of hormones?

8.54* What is the cellular role of cyclic AMP?

8.55 What biological effect does the hormone epinephrine have?

8.56* What are the two parts of the human nervous system?

8.57 What is a synapse, and what role does it play in nerve transmission?

8.58* Name two classes of neurotransmitters.

8.59 Describe in general terms how a nerve impulse is passed from one neuron to another.

8.60* How do the autonomic and somatic parts of the nervous system differ?

Applications

8.61 Why must enzyme activity be monitored under standard conditions? [App: Medical Uses of Enzymes and Isoenzymes]

8.62* What three enzymes show marked concentration increases in the blood after a heart attack? Why? [App: Medical Uses of Enzymes and Isoenzymes]

8.63 What are isoenzymes? [App: Medical Uses of Enzymes and Isoenzymes]

8.64* Why are overdoses less common with water-soluble than with fat-soluble vitamins? [App: Vitamins and Minerals in the Diet]

8.65 Read the labels of food that you eat for a day or look the foods up in a nutrition table and determine what percent of your daily dosage of vitamins and minerals you get from each. Are you getting the recommended amounts from the food you eat, or should you be taking a vitamin or mineral supplement? [App: Vitamins and Minerals in the Diet]

8.66* How does penicillin kill bacteria? [Int: Penicillin, the First Antibiotic]

8.67 Why is penicillin toxic to bacteria but not to higher organisms? [Int: Penicillin, the First Antibiotic]

Additional Questions and Problems

8.68 Look up the structures of vitamin C and vitamin E in Section 8.11 and identify the functional groups in these vitamins.

8.69 Thyrotropin-releasing factor, one of the primary hormones of the hypothalamus, is the C-terminal amide of the simple tripeptide PyroGlu-His-Pro, where PyroGlu is pyroglutamic acid. Draw the full structure of thyrotropin-releasing factor.

PyroGlu

8.70* What is the chemical relationship of pyroglutamic acid (Problem 8.69) to glutamic acid, one of the 20 common amino acids?

8.71   The adult recommended daily allowance (RDA) of riboflavin is 1.6 mg. If one glass (100 mL) of red wine contains 0.014 mg of riboflavin, how much wine would an adult have to consume to obtain the RDA?

8.72*   Some bacteria require *p*-aminobenzoic acid (PABA) to make folic acid, which they then use to produce a necessary enzyme. Sulfa drugs, which have the general structure shown here, interfere with folic acid synthesis. Draw the structure of PABA and propose a reason for the success of these drugs.

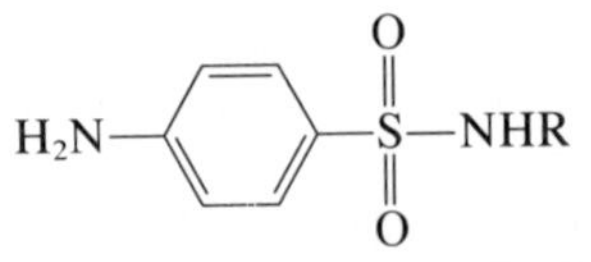

$$H_2N - \bigcirc - \overset{\overset{\displaystyle O}{\|}}{\underset{\underset{\displaystyle O}{\|}}{S}} - NHR \quad \text{A sulfa drug}$$

8.73   The active site of an enzyme is a very small portion of the molecule. What is the function of the rest of the huge molecule?

9 The Generation of Biochemical Energy

These athletes need a well-practiced strategy to pass the baton successfully. The energy they're using is produced with the aid of the strategies of metabolism discussed in this chapter.

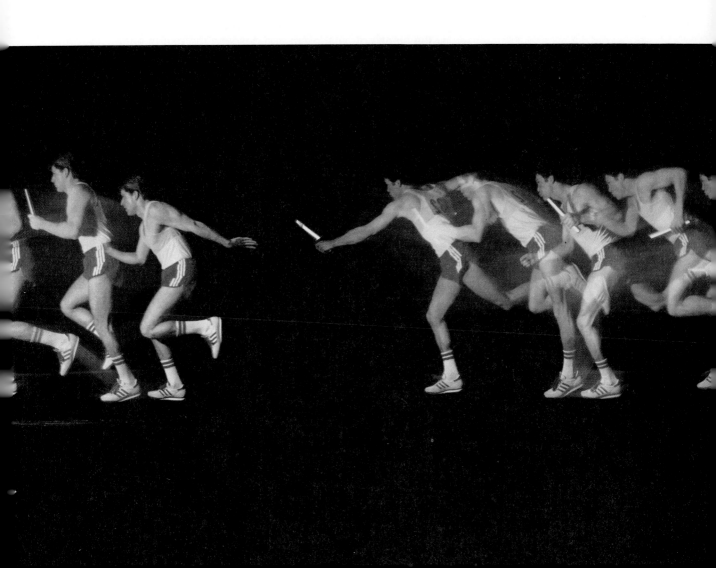

All organisms need to draw energy from their surroundings to stay alive. In animals, the energy comes from food and is released in the exquisitely interconnected reactions of **metabolism.** We are powered by the oxidation in cells of biomolecules containing mainly carbon, hydrogen, and oxygen. The end products are carbon dioxide, water, and energy.

$$\text{C, H, O (food molecules)} + O_2 \longrightarrow CO_2 + H_2O + \text{energy}$$

The principal food molecules—fats, proteins, and carbohydrates—differ in structure and are broken down by distinctive pathways that are examined in later chapters. For the present, though, we're going to concentrate on the final, common pathways by which energy is released from all types of food molecules. The questions to be answered include the following:

1. *What is the general structure of a eukaryotic cell?* The goal: Be able to identify the major parts of a eukaryotic cell, describe the structure of a mitochondrion, and tell how eukaryotic and prokaryotic cells differ.
2. *How are the reactions of catabolism organized?* The goal: Be able to list the stages of catabolism and describe the role of each.
3. *What are the major strategies of metabolism?* The goal: Be able to explain and give examples of the roles of ATP, coupled reactions, and oxidized and reduced coenzymes in metabolic pathways.
4. *What is the citric acid cycle?* The goal: Be able to describe what happens in the citric acid cycle and what its role is in energy production.
5. *How is ATP generated in the final stage of catabolism?* The goal: Be able to describe the respiratory chain, oxidative phosphorylation, and their coupling according to the chemiosmotic theory.

9.1 ENERGY AND LIFE

Metabolism The overall sum of the many reactions taking place in an organism.

Living things must do mechanical work—microorganisms engulf food, plants bend toward the sun, humans walk about. All organisms must also do the chemical work of synthesizing biomolecules needed for energy storage, growth, repair, and replacement (Figure 9.1). In addition, cells need energy for the work of moving certain molecules and ions across cell membranes. It is the energy released from food that allows these various kinds of work to be done.

Energy, which makes work possible, can be converted from one form to another but can be neither created nor destroyed. Ultimately, the energy used by all but a few living things on earth comes from the sun (Figure 9.2). Plants convert sunlight to potential energy stored mainly in the chemical bonds of carbohydrates. Plant-eating animals then convert some of this energy according to their own needs for staying alive and store the rest of it, mainly in the chemical bonds of fats. Other animals, including humans, are able to eat plants or animals and make use of the chemical energy these organisms have stored.

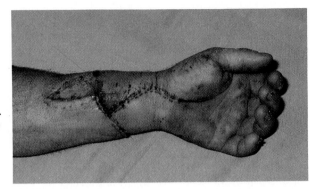

Figure 9.1
Energy at work in the body. The biomolecules needed to heal this wound will be synthesized using energy from the catabolism of food molecules.

Our bodies can't simply produce energy by burning up a large steak all at once because the large quantity of heat released would be harmful to us. Furthermore, it's difficult to capture energy for storage once it has been converted to heat. What we need is energy that can be stored and then released in the right amounts, whether we're running away from an angry dog or sleeping peacefully.

We therefore have some specific requirements for energy:

● Energy must be released from food gradually.

● Energy must be stored in a readily accessible form.

● The rate of energy release must be finely controlled.

● Just enough energy must be released as heat to maintain constant body temperature.

● Energy in a form other than heat must be available to drive reactions that aren't spontaneous at body temperatures.

In this chapter we'll look at some of the ways these requirements for energy management are met. We'll begin by reviewing some basic concepts about energy. Then we'll take an overview of metabolism and the strategies it

Figure 9.2
Flow of energy through the biosphere. Energy comes from the sun and is ultimately stored in chemical bonds, used for work, or lost as heat.

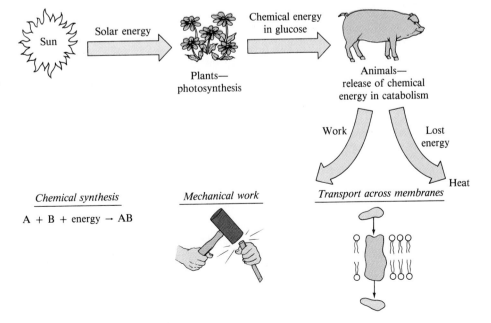

relies on, and in the rest of the chapter we'll look more closely at the generation of *adenosine triphosphate* (ATP), the molecule that carries stored energy to where it's needed.

9.2 FREE ENERGY AND BIOCHEMICAL REACTIONS

A chemical reaction is favorable, or *spontaneous*, if free energy is released during the process. Conversely, a reaction is not favorable if free energy is absorbed. The change in free energy (ΔG) depends on two factors: the release or absorption of heat (ΔH) and the increase or decrease in disorder that result from the reaction (ΔS). The release of heat (negative ΔH) and the increase in disorder (positive ΔS) both favor spontaneity, and the combined effect on spontaneity of ΔH and ΔS is given by

$$\Delta G = \Delta H - T\Delta S$$

Reactions in living organisms are no different from laboratory reactions in test tubes. Both follow the same laws, and both have the same kinds of energy requirements. The free energy change therefore applies equally well to biochemical reactions and laboratory reactions.

For a chemical reaction to occur spontaneously, energy must be given off, which is indicated by a negative value of ΔG. In other words, the products of a reaction must have less energy and be more stable than the reactants. Such a reaction is described as **exergonic.** As shown by the energy diagram in Figure 9.3, the favorable, exergonic reaction whose curve is on the left has its products farther "downhill" on the energy scale than the reactants.

Oxidation reactions are often downhill reactions that give off large amounts of energy. For example, oxidation of glucose, the principal source of energy for animals, releases 686 kcal per mole of glucose.

$$C_6H_{12}O_6 + 6\ O_2 \longrightarrow 6\ CO_2 + 6\ H_2O \qquad \Delta G = -686\ \text{kcal}$$

Glucose

The more negative the free energy change, the greater the amount of energy released and the further a reaction proceeds toward products before reaching equilibrium. In the combustion of glucose, virtually no reactant remains at equilibrium.

Figure 9.3
Energy diagrams for two reactions. In the favorable reaction, diagrammed on the left, the products have less energy than the reactants. In the unfavorable reaction, diagrammed on the right, the products have more energy than the reactants.

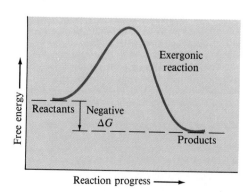
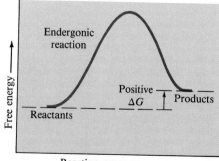

Endergonic Describes a process that absorbs free energy from the surroundings (positive ΔG).

Reactions in which the products are *higher* in energy than the reactants can also take place under the right circumstances, but such reactions can't be spontaneous. In other words, energy has to be added to the reactants for an energetically "uphill" change to occur (on the right in Figure 9.3). Such reactions have positive values of ΔG and are described as **endergonic.** An example is the conversion of glucose to glucose 6-phosphate, an important step in the breakdown of glucose from carbohydrates in the diet:

$\Delta G = +3.3$ kcal

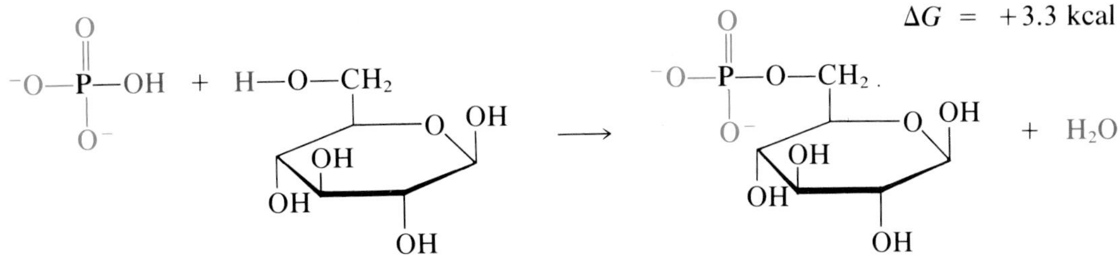

Hydrogen phosphate ion   Glucose   Glucose 6-phosphate

REVIEW Once a chemical reaction reaches equilibrium, the rates of the forward and reverse reactions are equal and the concentrations of reactants and products do not change (so long as conditions do not change).

The more positive the free energy change, the greater the amount of energy required and the less product present when a reaction reaches equilibrium.

Like the value of ΔH, the value of ΔG changes sign for the reverse of a given reaction. For photosynthesis, the process whereby plants store solar energy by converting CO_2 and H_2O to glucose plus O_2, ΔG is therefore positive and of the same numerical value as for the oxidation of glucose.

$$6\ CO_2\ +\ 6\ H_2O \qquad C_6H_{12}O_6\ +\ 6\ O_2$$

solar energy / photosynthesis $\Delta G = +686$ kcal

oxidation $\Delta G = -686$ kcal

Photosynthesis and glucose oxidation show how energy stored in the products of an *endergonic* reaction is released in the *exergonic* reverse of that reaction. Living systems make constant use of this principle.

Practice Problem   **9.1**   Each of the following reactions occurs in the citric acid cycle, an energy-producing sequence of reactions that we'll discuss later in this chapter. (Recall that organic acids are usually referred to in biochemistry with the -*ate* ending because they exist as ions in body fluids.) Which of the reactions listed is exergonic? Which is endergonic? Which will proceed farthest toward products at equilibrium?

(a) acetyl coenzyme A + oxaloacetate + H_2O $\longrightarrow$
$\qquad\qquad\qquad\qquad\qquad\qquad$ citrate + coenzyme A $\Delta G = -9$ kcal

(b) citrate $\longrightarrow$ isocitrate $\Delta G = +3$ kcal

(c) fumarate + H_2O $\longrightarrow$ L-malate $\Delta G = -0.9$ kcal

Prokaryotic cell Cell that has no nucleus; found in bacteria and algae.

Eukaryotic cell Cell with a membrane-enclosed nucleus; found in all higher organisms.

Cell membrane (plasma membrane) The membrane surrounding a cell.

Cytoplasm The region between the cell membrane and the nuclear membrane in a eukaryotic cell.

Organelle A small, organized unit in the cell that performs a specific function.

Cytosol The contents of the cytoplasm surrounding the organelles.

Mitochondrion An egg-shaped organelle where small molecules are broken down to provide the energy to power an organism.

Cristae The inner folds of a mitochondrion.

9.3 CELLS AND THEIR STRUCTURE

Before we take an overview of metabolism, it's important to see where the energy-generating reactions take place. There are two main categories of cells: **prokaryotic** cells, found in bacteria and algae, and **eukaryotic** cells, found in higher organisms, both plant and animal.

Prokaryotic cells are simple in structure: They have no nucleus, they are quite small, and the DNA that governs reproduction is dispersed throughout their contents. Eukaryotic cells, by contrast, have a membrane-enclosed nucleus containing their DNA and are about 1000 times larger than bacterial cells. These cells are separated from their environment by a **cell membrane,** sometimes called a *plasma membrane*. Everything between the cell membrane and the nuclear membrane in a eukaryotic cell is referred to as the **cytoplasm.**

A further important difference between cell types is that eukaryotic cells contain in their cytoplasm a large number of subcellular structures called **organelles**—small, functional units that perform specialized tasks. In addition to the nucleus, other organelles include ribosomes, endoplasmic reticulum, lysosomes, Golgi complex, and mitochondria. All these organelles are surrounded by the **cytosol,** the material that fills the interior of the cell and contains electrolytes, nutrients, and enzymes. A diagram of a typical eukaryotic cell is shown in Figure 9.4.

Among the organelles shown in Figure 9.4, the **mitochondria,** often called the cell's "power plants," are the most important for energy production. It is in the mitochondria that 90% of the body's energy-carrying molecule, ATP, is produced. A typical liver cell contains about 800 mitochondria, totaling about 20% of the cellular volume.

A mitochondrion is a roughly egg-shaped structure composed of a smooth outer membrane and a folded inner membrane (Figure 9.5). The folds in the inner membrane are called **cristae,** and the space surrounded by the inner

Figure 9.4
A typical eukaryotic cell, showing some of its organelles.

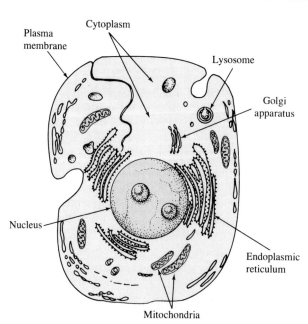

Cytoplasm

Plasma membrane

Lysosome

Golgi apparatus

Nucleus

Endoplasmic reticulum

Mitochondria

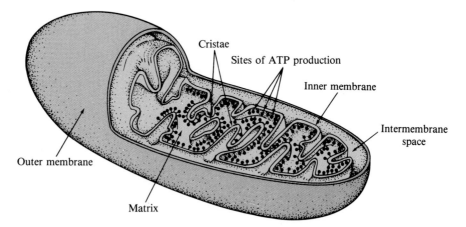

Figure 9.5
A mitochondrion. The citric acid cycle takes place in the matrix. The respiratory chain and oxidative phosphorylation take place at the inner membrane.

Mitochondrial matrix The space surrounded by the inner membrane.

membrane is called the mitochondrial **matrix.** The intermembrane space lies between the inner and outer membranes. You'll see in the following sections that some energy-producing reactions take place in the matrix, and others on the small knob-like protuberances on the cristae.

9.4 AN OVERVIEW OF METABOLISM AND ENERGY PRODUCTION

Most of the chemical reactions of metabolism occur in sequences called *metabolic pathways*. Such a pathway may be linear, in which the product of one reaction serves as the starting material for the next, or it may be cyclic, in which a series of reactions regenerates the first reactant.

Catabolism Metabolic reactions that break down molecules into smaller pieces.

Anabolism Metabolic reactions that build larger biological molecules from smaller pieces.

A linear sequence $\quad A \longrightarrow B \longrightarrow C \longrightarrow D \longrightarrow \cdots \quad$ A cyclic sequence $\quad \overset{\displaystyle A}{\underset{\displaystyle C}{D \quad B}}$

Those pathways that break molecules apart are known collectively as **catabolism,** while those that put building blocks back together to assemble larger molecules are known collectively as **anabolism.** The purpose of catabolism is to release energy from food, and the purpose of anabolism is to synthesize new biomolecules, including those that store energy.

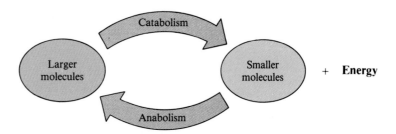

The overall picture of catabolism and energy production is simple: Eating provides fuel, breathing provides oxygen, and our bodies oxidize the fuel to extract energy. The process can be roughly divided into the four stages shown in Figure 9.6.

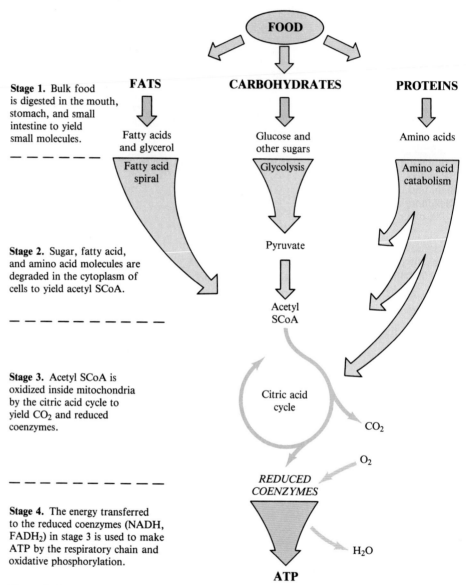

Stage 1. Bulk food is digested in the mouth, stomach, and small intestine to yield small molecules.

Stage 2. Sugar, fatty acid, and amino acid molecules are degraded in the cytoplasm of cells to yield acetyl SCoA.

Stage 3. Acetyl SCoA is oxidized inside mitochondria by the citric acid cycle to yield CO_2 and reduced coenzymes.

Stage 4. The energy transferred to the reduced coenzymes (NADH, $FADH_2$) in stage 3 is used to make ATP by the respiratory chain and oxidative phosphorylation.

Figure 9.6
An overview of catabolic pathways for the degradation of food and the production of biochemical energy stored in ATP.

The first catabolic stage, *digestion,* takes place in the mouth, stomach, and small intestine when bulk food is broken down into individual small molecules. Carbohydrates are broken down to glucose and other sugars; proteins are broken down into amino acids; lipids (commonly called "fats") are broken down into long-chain carboxylic acids called *fatty acids.* In stage 2, these small molecules are further degraded to yield two-carbon **acetyl groups** ($CH_3C=O$) that are attached to a large carrier molecule called **coenzyme A.** The resultant compound, *acetyl coenzyme A,* is an intermediate in the breakdown of all main classes of food molecules.

Acetyl group The $CH_3C=O$ group.

Coenzyme A Coenzyme that functions as a carrier of acetyl groups and other acyl ($R-C=O$) groups.

AN APPLICATION: BASAL METABOLISM

The minimum amount of energy expenditure required per unit of time to stay alive—to breathe, maintain body temperature, circulate blood, and keep all body systems functioning—is referred to as the **basal metabolic rate.** Ideally, it is measured in a person who is awake, is lying down at a comfortable temperature, has fasted and avoided strenuous exercise for 12 hr, and is not under the influence of any medications. The rate can be measured by monitoring respiration and finding the rate of oxygen consumption, which is proportional to the energy used.

An average basal metabolic rate is 70 kcal/hr, or about 1700 kcal/day. The rate varies with many factors, including sex, age, weight, and physical condition. A rule of thumb used by nutritionists to estimate basal energy needs per day is the requirement for 1 kcal/hr per kilogram of body weight by a male and 0.95 kcal/hr per kilogram of body weight by a female. For example, a 50. kg (110 lb) female has an estimated basal metabolic rate of (50. kg)/(0.95 kcal/kg hr) = 48 kcal/hr, giving a daily basal requirement of 1200 kcal.

The total calories a person needs each day is determined by his or her basal requirements plus the energy used in additional physical activities. The caloric consumption rates associated with some activities are listed in the following table. A relatively inactive person requires about 30% above basal requirements per day, a lightly active person requires about 50% above basal, and a very active person such as an athlete or construction worker can use 100% above basal requirements in a day. Each day that you consume food with more calories than you use, the excess calories are stored as potential energy in the chemical bonds of substances in your body and your weight rises. Each day that you consume food with fewer calories than you burn, some chemical energy in your body is taken out of storage and your weight drops.

The cola drink contains 160 Calories (kcal) and the hamburger contains 500 Calories. How long would you have to play tennis to burn off these calories?

Calories Used in Various Activities

Activity	Kilocalories used per minute
Sleeping	1.2
Sitting reading	1.3
Listening to lecture	1.7
Weeding garden	5.6
Walking, 3.5 mph	5.6
Pick-and-shovel work	6.7
Recreational tennis	7.0
Soccer, basketball	9.0
Walking up stairs	10.0–18.0
Running, 12 min/mi (5 mph)	10.0
Running, 5 min/mi (12 mph)	25.0

Acetyl groups are oxidized in the third stage, the *citric acid cycle,* to yield carbon dioxide, water, and a great deal of energy. Some of this energy is lost as heat, and some is carried by reduced coenzymes to the fourth stage, the *respiratory chain* and *oxidative phosphorylation,* in which molecules of ATP, the primary energy carrier, are produced.

In this chapter, we'll discuss stages 3 and 4, because that's where energy production mainly occurs. We'll discuss stages 1 and 2 along with other aspects of the metabolism of carbohydrates, lipids, and proteins in later chapters.

9.5 STRATEGIES OF METABOLISM: ATP AND ENERGY TRANSFER

Adenosine triphosphate (ATP) The triphosphate of adenosine; the "energy currency" molecule of metabolism.

The storage of energy in a readily accessible form is achieved in most organisms by the synthesis of **adenosine triphosphate (ATP)** (Figure 9.7). Recall from Section 6.12 that two molecules of phosphoric acid (H_3PO_4), which we'll write as the hydrogen phosphate ions present in body fluids (HPO_4^{2-}) and symbolize as P_i, can condense to form an anhydride. Pyrophosphoric acid, the resulting anhydride, exists in body fluids as pyrophosphate ion ($P_2O_7^{4-}$ or PP_i).

Hydrogen phosphate ions   Pyrophosphate ion

Phosphoric anhydride A —P—O—P— group, like those in ADP and ATP.

Note that the **phosphoric anhydride** link lies between two negatively charged groups. For this and other reasons, it's a relatively strained link that's often formed in an endergonic reaction and broken in an exergonic reaction.

Phosphate group A —PO_3^{2-} group in an organic molecule.

ATP, which has two phosphoric anhydride links, can be produced by addition of a **phosphate group** to ADP, which has one phosphoric anhydride link (Figure 9.7). The **phosphorylation** of ADP requires 7.3 kcal of energy per mole of ATP, and the hydrolysis of ATP releases this much energy.

Phosphorylation A reaction that results in addition of a phosphate group (—PO_3^{2-}).

$$ADP + HPO_4^{2-} + H^+ \longrightarrow ATP + H_2O \qquad \Delta G = + 7.3 \text{ kcal}$$

$$ATP + H_2O \longrightarrow ADP + HPO_4^{2-} + H^+ \qquad \Delta G = - 7.3 \text{ kcal}$$

REVIEW A ΔG value written as part of a chemical equation is the free energy change for amounts of reactants and products in moles, as represented by coefficients in the equation.

As the molecule that stores the energy needed to power all life processes, ATP has been called the "energy currency of the living cell." Catabolic reactions pay off in the endergonic synthesis of ATP, and anabolic reactions spend ATP by transferring a phosphate group to other molecules in exergonic reactions. The entire process of energy production and use thus revolves around the ATP $\rightleftarrows$ ADP interconversion.

Since the primary metabolic function of ATP is to store energy, it is often also referred to as being a "high-energy" molecule or as containing "high-energy" P—O bonds. These terms, while useful, are misleading because they promote the idea that ATP is somehow different from other compounds. The

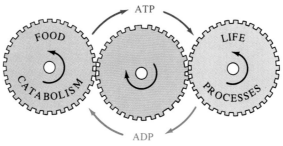

Figure 9.7
ATP and ADP. The interconversion of ATP and ADP serves as the central cog in the overall process of producing energy from food.

terms mean only that ATP is more reactive than many other biomolecules and releases a useful amount of energy when it transfers a phosphate group. In fact, if the hydrolysis of ATP released *unusually* large amounts of energy, the body wouldn't be able to provide enough energy to convert ADP back to ATP. ATP is a convenient energy carrier in metabolism *because* its hydrolysis is of intermediate energy, as illustrated in Table 9.1.

Table 9.1   Free Energies of Hydrolysis of Some Phosphates

Compound Name	ΔG (kcal/mol)
Phosphoenol pyruvate	−14.8
1,3-Bisphosphoglycerate	−11.8
Creatine phosphate	−10.3
ATP (→ ADP)	−7.3
Glucose 1-phosphate	−5.0
Fructose 6-phosphate	−3.3
Glucose 6-phosphate	−3.3

9.6 STRATEGIES OF METABOLISM: METABOLIC PATHWAYS AND COUPLED REACTIONS

Now that you're acquainted with ATP, we'll explore how stored chemical energy is gradually released and how it can be used to drive endergonic, or uphill, reactions. We've noted before that our bodies can't burn up a steak all at once. As shown by the diagram on the left in Figure 9.3, however, the energy difference between a reactant (the steak) and the ultimate products of its catabolism (mainly carbon dioxide and water) is a fixed quantity. *The same amount of energy is released no matter what pathway is taken between reactants and products.* The metabolic pathways of catabolism take advantage of this fact by releasing energy bit by bit in series of reactions, somewhat like the stepwise release of potential energy as water flows down an elaborate waterfall (Figure 9.8).

The overall reaction and the overall free energy change for any series of reactions can be found by summing up the equations and the free energy changes for each individual step. For example, the total free energy change (ΔG) for the 10 reactions in the *glycolysis* pathway by which one mole of glucose is converted to pyruvate (part of stage 2, Figure 9.6) is about -8 kcal, showing that the pathway is downhill and favorable. The reactions of all metabolic pathways *add up* to downhill processes with negative free energy changes.

Figure 9.8
Stepwise release of potential energy. No matter what the pathway from the top to the bottom of this waterfall, the amount of potential energy released as the water falls is the same.

Coupled reactions
Combination of an energy-requiring and energy-releasing reaction that occur together to give an overall energy-releasing reaction.

Unlike the waterfall, however, not every step in a metabolic pathway is downhill. The metabolic strategy for dealing with an energetically unfavorable, uphill reaction is to *couple* it with an energetically favorable, downhill reaction so that the overall energy change for the two reactions is favorable. For example, take the reaction of glucose with hydrogen phosphate ion (HPO_4^{2-}) to yield glucose 6-phosphate plus water, for which $\Delta G = +3.3$ kcal. The reaction doesn't take place spontaneously because the two products are about 3.3 kcal higher in energy than the starting materials.

But now let's see what the energy picture looks like when this reaction is coupled with the favorable hydrolysis of ATP to give ADP:

(*Unfavorable*)	Glucose + HPO_4^{2-}	$\longrightarrow$	glucose 6-phosphate + H_2O	$\Delta G =$	3.3 kcal
(*Favorable*)	ATP + H_2O	$\longrightarrow$	ADP + HPO_4^{2-}	$\Delta G =$	-7.3 kcal
(*Favorable*)	Glucose + ATP	$\longrightarrow$	glucose 6-phosphate + ADP	$\Delta G =$	-4.0 kcal

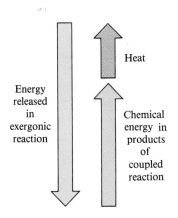

Figure 9.9
Energy exchange in coupled reactions.

The net energy change for these two **coupled reactions** for the synthesis of glucose 6-phosphate is favorable by about 4 kcal. Enough energy is transferred directly from the hydrolysis of the phosphate group in ATP to drive the addition of a phosphate group to glucose. The excess energy is released as heat and contributes to maintaining body temperature (Figure 9.9).

Although we've written the reactions separately to show how their energies combine, coupled reactions don't necessarily take place separately. The net change occurs all at once and, in fact, *it's only by such coupling that the energy stored in one chemical compound can be transferred to the reaction of other compounds.* ATP is used in this manner to drive many different kinds of unfavorable biochemical reactions.

What about the synthesis of ATP from ADP, which has $\Delta G = +7.3$ kcal and is an unfavorable reaction? The same principle of coupling is put to use. For this endergonic reaction to occur, it must be coupled with a reaction that releases *more* than 7.3 kcal. In a different reaction of glycolysis, the formation of ATP is coupled with the hydrolysis of phosphoenolpyruvate, a high-energy phosphate of higher energy than ATP (Table 9.1):

$$\underset{\text{Phosphoenolpyruvate}}{\overset{\overset{\textstyle O-PO_3^{2-}}{|}}{H_2C=C-COO^-}} + H_2O \longrightarrow \underset{\text{Pyruvate}}{\overset{\overset{\textstyle O}{\|}}{CH_3-C-COO^-}} + HPO_4^{2-} \qquad \Delta G = -14.8 \text{ kcal}$$

$$ADP + HPO_4^{2-} \longrightarrow ATP + H_2O \qquad \Delta G = +7.3 \text{ kcal}$$

$$\overset{\overset{\textstyle OPO_3^{2-}}{|}}{H_2C=C-COO^-} + ADP \longrightarrow \overset{\overset{\textstyle O}{\|}}{CH_3C-COO^-} + ATP \qquad \Delta G = -7.5 \text{ kcal}$$

To simplify matters, such coupled biochemical reactions are often written in a way that highlights the important transformation. The curved arrow intersecting the normal straight arrow shows that ADP is a reactant and ATP is a product:

$$\overset{\overset{\textstyle O-PO_3^{2-}}{|}}{H_2C=C-COO^-} \quad \overset{ADP \quad ATP}{\underset{\curvearrowright}{\longrightarrow}} \quad \overset{\overset{\textstyle O}{\|}}{CH_3-C-COO^-}$$

Practice Problems   **9.2**   One of the steps in lipid metabolism is the reaction of glycerol [1,2,3-propanetriol; $HOCH_2CH(OH)CH_2OH$] with ATP to yield glycerol 1-phosphate. Write the equation for this reaction using the curved arrow symbolism.

9.3   Why must a metabolic pathway that synthesizes a given molecule occur by a different series of reactions than a pathway that breaks down the same molecule?

9.4   The hydrolysis of acetyl phosphate to give acetate has $\Delta G = -10.3$ kcal. Combine the equations and ΔG values to show whether or not coupling of this reaction with phosphorylation of ADP is favorable. (You need only give compound names or abbreviations in the equations.)

9.7   STRATEGIES OF METABOLISM: OXIDIZED AND REDUCED COENZYMES

Many metabolic reactions are oxidation–reduction (redox) reactions, which means that a steady supply of oxidizing and reducing agents must be available. To deal with this requirement, a few coenzymes continuously cycle between their oxidized and reduced forms, just as adenosine continuously cycles between ATP and ADP.

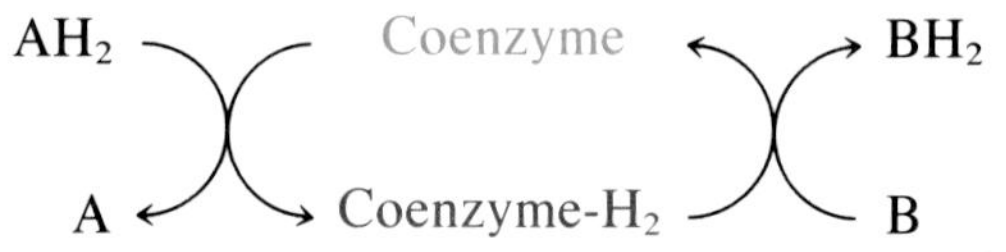

$$AH_2 \quad \text{Coenzyme} \quad BH_2$$
$$A \quad \text{Coenzyme-}H_2 \quad B$$

To review briefly the basics of oxidation and reduction: In every redox reaction, an oxidizing agent is reduced and a reducing agent is oxidized. In inorganic chemistry oxidations are the loss of electrons, reductions are the gain of electrons, and oxidation and reduction always occur together.

Oxidation:		$Zn \longrightarrow$	$Zn^{2+} + 2\,e^-$
Reduction:	$2\,Cu^+ + 2\,e^- \longrightarrow$		$2\,Cu$
Redox reaction:	$2\,Cu^+ + Zn \longrightarrow$		$2\,Cu + Zn^{2+}$

In organic chemistry, oxidation is often the addition of oxygen or the loss of hydrogen, and reduction is often the removal of oxygen or the addition of hydrogen.

Reduced form	Oxidized form
$R-\overset{\overset{\displaystyle O}{\|\|}}{C}-H$	$R-\overset{\overset{\displaystyle O}{\|\|}}{C}-OH$
RCH_2CH_2R	$RCH{=}CHR$

With these points in mind, let's use a reaction from the citric acid cycle to illustrate the action of a coenzyme as an oxidizing agent (Table 9.2). The entire cycle is described in Section 9.8.

Table 9.2 Coenzymes as Oxidizing and Reducing Agents

Coenzyme	As Oxidizing Agent	As Reducing Agent
Nicotinamide adenine dinucleotide	NAD^+	NADH (or $NADH/H^+$)
Nicotinamide adenine dinucleotide phosphate	$NADP^+$	NADPH (or $NADPH/H^+$)
Flavin adenine dinucleotide	FAD	$FADH_2$

Oxidation of 2° alcohol by NAD^+

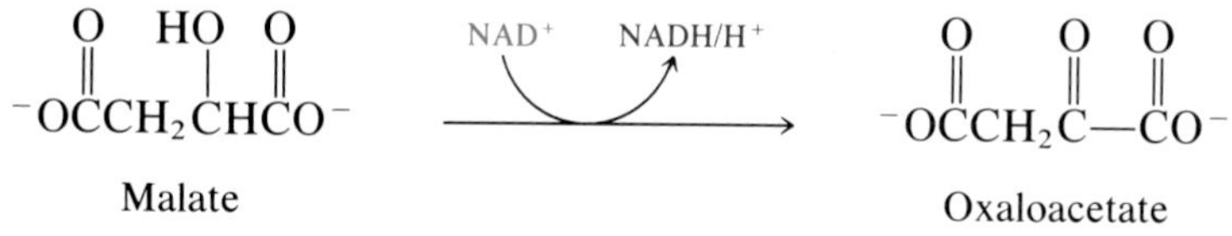

Nicotinamide adenine dinucleotide (NAD⁺)
Coenzyme that functions as an oxidizing agent and forms NADH/H⁺.

The oxidation of malate to oxaloacetate amounts to the removal of two hydrogen atoms. The oxidizing agent is **NAD⁺**, the oxidized form of **nicotinamide adenine dinucleotide,** whose structure is shown in Figure 9.10. NAD⁺ is a coenzyme for the dehydrogenase enzymes that often catalyze removal of hydrogen atoms from secondary alcohols to produce ketones. The structure of NADP⁺, a related coenzyme that plays a role in biosynthesis, is indicated in red in Figure 9.10.

It's important in considering enzyme-catalyzed redox reactions to recognize that a hydrogen *atom* is equivalent to a hydrogen *ion*, H⁺, plus an electron, e⁻. Thus, for the two hydrogen atoms removed in the oxidation of malate,

$$2 \text{ H atoms} = 2 \text{ H}^+ + 2 \text{ e}^-$$

When NAD⁺ is reduced, both electrons accompany one H⁺, in effect adding a hydride ion, H:⁻, to the molecule to give NADH. The second hydrogen re-

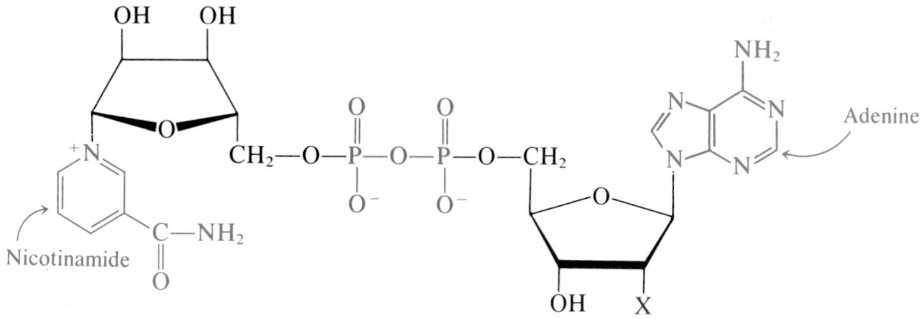

Figure 9.10
Oxidizing agents common in catabolism. NAD⁺ is reduced to NADH plus H⁺; NADP⁺ is reduced to NADPH plus H⁺; FAD is reduced to FADH₂.

Nicotinamide adenine dinucleotide (NAD⁺), X = —OH
Nicotinamide adenine dinucleotide phosphate (NADP⁺), X = —OPO₃²⁻

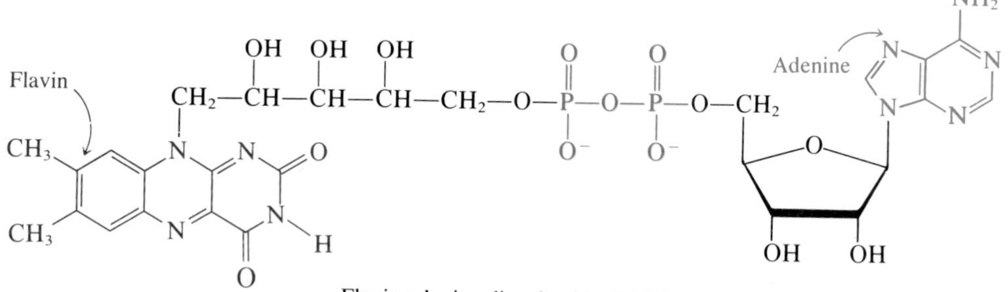

Flavin adenine dinucleotide (FAD)

moved from the oxidized substrate enters the solution as a hydrogen ion, H^+. The product of NAD^+ reduction is therefore often represented as $NADH/H^+$. $NADP^+$ is reduced in a similar manner to $NADPH/H^+$.

Reduction of NAD^+

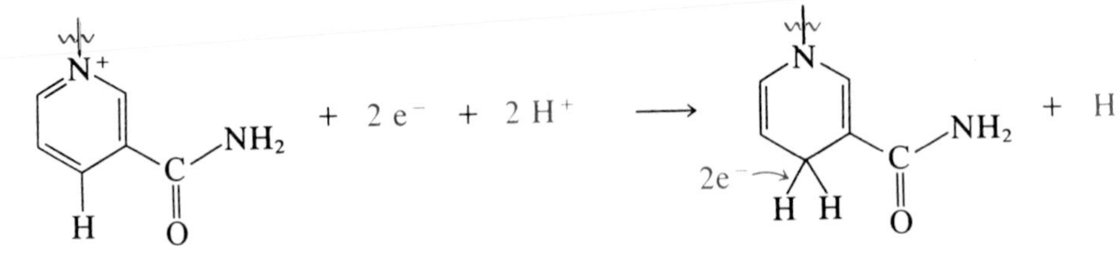

NAD$^+$ NADH/H$^+$

Flavin adenine dinucleotide (FAD) Coenzyme that functions as an oxidizing agent and forms $FADH_2$.

NAD^+ and **flavin adenine dinucleotide (FAD)**, also shown in Figure 9.10, are the most common oxidizing agents in the reactions of catabolism. FAD, which usually removes two hydrogen atoms from a carbon chain to form an alkene double bond, adds two hydrogen ions and two electrons to give $FADH_2$ when it is reduced. Because NADH and $FADH_2$ have picked up electrons, they are often referred to as "electon carriers."

Reduction of FAD

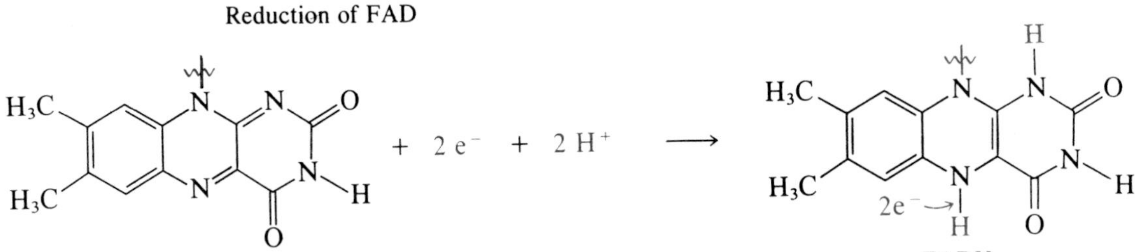

FAD   FADH$_2$

Oxidation by FAD to give alkene

During the reduction of NAD^+ or FAD, some of the chemical energy released from the oxidized substrate is captured in NADH and $FADH_2$. As these coenzymes cycle through their oxidized and reduced forms, they are therefore able to carry energy along from reaction to reaction. Ultimately, this energy is passed on to ATP for redistribution throughout an organism.

Practice Problems

9.5 Which of these reactions probably requires NAD^+ as a coenzyme and which requires FAD?

(a) $\underset{\qquad\quad}{HOOC-CH_2-\overset{\overset{\textstyle OH}{|}}{CH}-COOH} \longrightarrow HOOC-CH_2-\overset{\overset{\textstyle O}{\|}}{C}-COOH$

(b) $CH_3-\overset{\overset{\textstyle O}{\|}}{C}-CH_2-CH_2-COOH \longrightarrow CH_3-\overset{\overset{\textstyle O}{\|}}{C}-CH=CH-COOH$

9.6 Look ahead to Figure 9.11 for the citric acid cycle. In steps 3, 6, and 8, draw the structure of the reactant and indicate which hydrogen atoms are removed.

9.8 THE CITRIC ACID CYCLE

The first two stages of catabolism, shown earlier in Figure 9.6, result in the conversion of lipids, carbohydrates, and proteins into acetyl groups. These acetyl groups then enter the third stage, the citric acid cycle. The cycle takes place within cell mitochondria, and its reactions are coupled to the fourth stage of catabolism. Stages 1 and 2 of catabolism converge on **acetyl coenzyme A (acetyl SCoA)** when acetyl groups from the breakdown of food molecules become bonded to the thiol group of coenzyme A.

Acetyl coenzyme A (acetyl SCoA) Acetyl-substituted coenzyme A that is the common intermediate for carrying acetyl groups into the citric acid cycle.

Pantothenate, a B complex vitamin

$$X-S-CH_2-CH_2-\underset{\underset{H}{|}}{N}-\overset{\overset{O}{||}}{C}-CH_2-CH_2-\underset{\underset{H}{|}}{N}-\overset{\overset{O}{||}}{C}-\underset{\underset{CH_3}{|}}{\overset{\overset{HO}{|}}{C}}-\underset{\underset{CH_3}{|}}{\overset{\overset{CH_3}{|}}{C}}-CH_2-O-\overset{\overset{O}{||}}{\underset{\underset{O^-}{|}}{P}}-O-\overset{\overset{O}{||}}{\underset{\underset{O^-}{|}}{P}}OCH_2$$

ADP

NH_2

Coenzyme A, X = H—   Acetyl coenzyme A, X = $CH_3\overset{\overset{O}{||}}{C}-$

Like the phosphate groups in ATP, the acetyl group in acetyl SCoA is a high-energy group because it is readily removed in an exergonic hydrolysis reaction.

$$CH_3-\overset{\overset{O}{||}}{C}-S-CoA + H_2O \longrightarrow CH_3-\overset{\overset{O}{||}}{C}-O^- + H-S-CoA + H^+ \quad \Delta G = -7.5 \text{ kcal}$$

Acetyl coenzyme A   Coenzyme A

Citric acid cycle The series of biochemical reactions that breaks down acetyl groups to carbon dioxide and energy stored in reduced coenzymes.

Breakdown of the two-carbon acetyl groups to give CO_2 occurs in the **citric acid cycle,** also called the *tricarboxylic acid cycle (TCA)* or the *Krebs cycle* after Hans Krebs who unraveled its complexities in 1937. The eight steps of the citric acid cycle and a brief description of each reaction are given in Figure 9.11. The names of the enzymes for each of the steps are listed in Table 9.3.

As its name implies, the citric acid *cycle* is a closed loop of reactions in which the *product* of the final step (step 8) is a *reactant* in the first step. The intermediates are constantly regenerated and flow continuously through the cycle, which takes place in the mitochondrial matrix where the enzymes are either free or embedded in the inner mitochondrial membrane. Some of the intermediates also serve as precursors for the synthesis of other biomolecules.

The cycle operates as long as (1) acetyl groups are available from acetyl SCoA and (2) the oxidizing agents NAD^+ and FAD are available. To meet

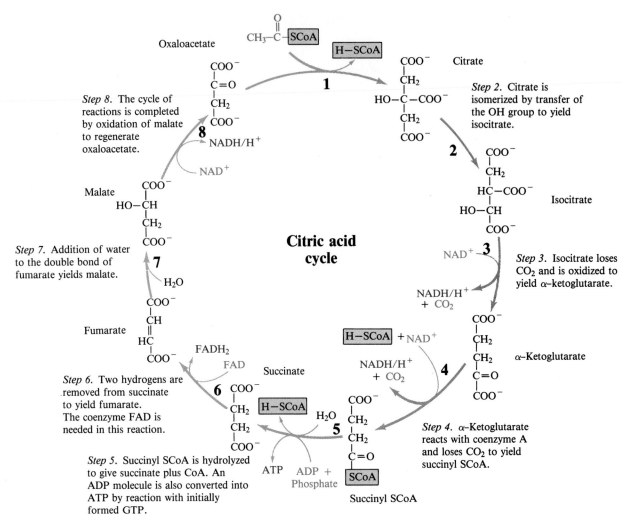

Figure 9.11

The citric acid cycle. The net effect of this eight-step series of reactions is the metabolic breakdown of acetyl groups (from acetyl SCoA) into two molecules of carbon dioxide plus reduced coenzymes. Here and throughout this and the following chapters, energy-rich forms (ATP, reduced coenzymes) are shown in red and their lower energy counterparts (ADP, oxidized coenzymes) are shown in blue.

Table 9.3   Enzymes of the Citric Acid Cycle

Step no. (Fig. 9.11)	Enzyme Name	Reaction Product
1	Citrate synthetase	Citrate
2	Aconitase	Isocitrate
3	Isocitrate dehydrogenase	α-Ketoglutarate
4	α-Ketoglutarate dehydrogenase complex	Succinyl SCoA
5	Succinyl CoA synthetase	Succinate
6	Succinate dehydrogenase	Fumarate
7	Fumarase	Malate
8	Malate dehydrogenase	Oxaloacetate

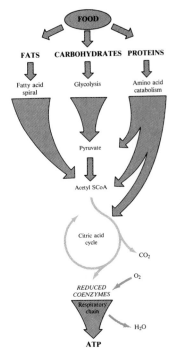

condition 2, the reduced coenzymes NADH and $FADH_2$ must be reoxidized via the respiratory chain, which in turn relies on oxygen as the final electron acceptor. Thus, the cycle is also dependent on the availability of oxygen and on the operation of the respiratory chain.

Steps 1 and 2 of the Citric Acid Cycle: Preparation   The first two steps set the stage for oxidation. Acetyl groups from acetyl SCoA enter the cycle at step 1 by addition to four-carbon oxaloacetate to give citrate, a six-carbon intermediate. Citrate is a tertiary alcohol and is converted in step 2 to its isomer, isocitrate, an easier-to-oxidize secondary alcohol. The isomerization occurs in two steps catalyzed by the same enzyme, aconitase; water is removed and then added back so that the —OH is on a different carbon atom.

$$
\begin{array}{ccccc}
\text{COO}^- & & \text{COO}^- & & \text{COO}^- \\
| & & | & & | \\
\text{CH}_2 & & \text{CH}_2 & & \text{CH}_2 \\
| & \xrightarrow[\text{aconitase}]{-\text{H}_2\text{O}} & | & \xrightarrow[\text{aconitase}]{+\text{H}_2\text{O}} & | \\
\text{HO--C--COO}^- & & \text{C--COO}^- & & \text{H--C--COO}^- \\
| & & \| & & | \\
\text{CH}_2 & & \text{CH} & & \text{HO--CH} \\
| & & | & & | \\
\text{COO}^- & & \text{COO}^- & & \text{COO}^- \\
\text{Citrate} & & \text{Aconitate} & & \text{Isocitrate}
\end{array}
$$

Steps 3 and 4 of the Citric Acid Cycle: Oxidative Decarboxylation   One molecule of CO_2 leaves at step 3 and one at step 4 as energy is released to the reduced coenzyme NADH. The reactions are oxidations in which first one and then another of the three carboxyl groups of isocitrate are removed and are thus known as *oxidative decarboxylations*. Step 4 requires coenzyme A and produces succinyl SCoA, which carries four carbon atoms along to the remaining reactions of the cycle.

Steps 5–8 of the Citric Acid Cycle: Oxidation of Succinate and Regeneration of Oxaloacetate   In step 5, the exergonic hydrolysis of succinyl SCoA is coupled with phosphorylation of guanosine diphosphate (GDP) to give guanosine triphosphate (GTP), a high-energy compound that transfers its phosphate directly to ADP. Such a reaction is known as a **substrate-level phosphorylation** to distinguish it from ATP formation in stage 4 of catabolism. Step 5 is the only step in the cycle that directly generates ATP instead of passing energy along to reduced coenzymes. In steps 6–8, two more reduced coenzyme molecules are produced as the four-carbon succinate is converted back to oxaloacetate, ready for the cycle to turn again.

Substrate-level phosphorylation
Formation of ATP by transfer of a phosphate to ADP from another phosphate-containing compound.

Net result of citric acid cycle

Acetyl SCoA $+$ 3 NAD$^+$ $+$ FAD $+$ ADP $+$ P$_i$ $+$ 2 H$_2$O $\longrightarrow$

$\qquad\qquad\qquad$ HSCoA $+$ 3 NADH $+$ 3 H$^+$ $+$ FADH$_2$ $+$ ATP $+$ 2 CO$_2$

- Production of four reduced coenzyme molecules
- Conversion of an acetyl group to two CO_2 molecules
- Production of one energy-rich ATP molecule.

The citric acid cycle as a whole is exergonic but includes several endergonic reactions, notably step 8 which regenerates oxaloacetate. Each endergonic reaction, however, is pulled along by subsequent exergonic reactions that consume their products.

The rate at which the citric acid cycle turns is controlled by the body's need for reduced coenzymes and for the energy produced from them. When the body's supply of NADH is high, for example, NADH inhibits the enzyme for step 3. When energy is being used at a high rate, ADP accumulates and activates the enzyme for step 3. When energy is in good supply, ATP accumulates and inhibits the enzyme for step 1. In other words, the cycle is activated when energy is needed and inhibited when energy is in good supply.

Practice Problems

9.7 Which of the substances in the citric acid cycle are tricarboxylic acids (thus giving the cycle its alternate name)?

9.8 Balance the equation for the conversion of acetyl SCoA into CO_2 and coenzyme A.

$$CH_3-\overset{\overset{\text{O}}{\|}}{C}-SCoA \; + \; O_2 \; \longrightarrow \; CO_2 \; + \; H-SCoA \; + \; H_2O$$

9.9 Why would you expect step 5 of the citric acid cycle to be sufficiently energy-releasing to be coupled with the phosphorylation of ADP?

9.9 THE RESPIRATORY CHAIN

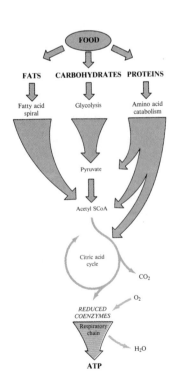

With carbon dioxide already released, reduced coenzymes from the citric acid cycle and other metabolic pathways are now ready to participate in forming the remaining two products of cellular respiration: water and energy stored in ATP. The end product of the **respiratory chain,** or *electron transport chain,* is water assembled from oxygen that we breathe, hydrogen ions, and electrons passed along from reduced coenzymes:

$$\tfrac{1}{2} O_2 + 2 e^- + 2 H^+ \longrightarrow H_2O$$

The reactions of the respiratory chain are coupled to *oxidative phosphorylation,* the conversion of ADP to ATP discussed in the next section.

The enzymes of the respiratory chain are embedded in the inner membrane of the mitochondrion and are arranged so that electrons pass directly from one to the next. Several types of electron carriers participate in the successive reactions. Some are located in fixed enzyme complexes and two, coenzyme Q and cytochrome c, are free to move as indicated in the schematic diagram in Figure 9.12.

The order of electron transfer, described in the following paragraphs, is from weaker to increasingly stronger electron acceptors, with energy released at each transfer, much like the waterfall shown in Figure 9.8. The coupled reactions of the pathway add up to the strongly exergonic oxidations of NADH (Figure 9.13) and $FADH_2$:

Figure 9.12
The respiratory chain. Symbols for the major electron carriers show their locations in the mitochondrial inner membrane. Coenzyme Q and cytochrome c are mobile, and the others are fixed within enzyme complexes. The red line shows the path of electrons along the chain. Electrons from $FADH_2$ join the chain at CoQ via an enzyme complex not shown in the figure.

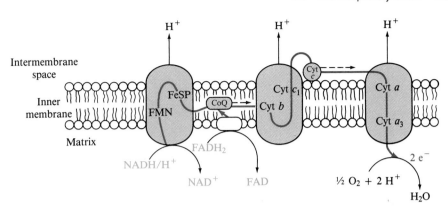

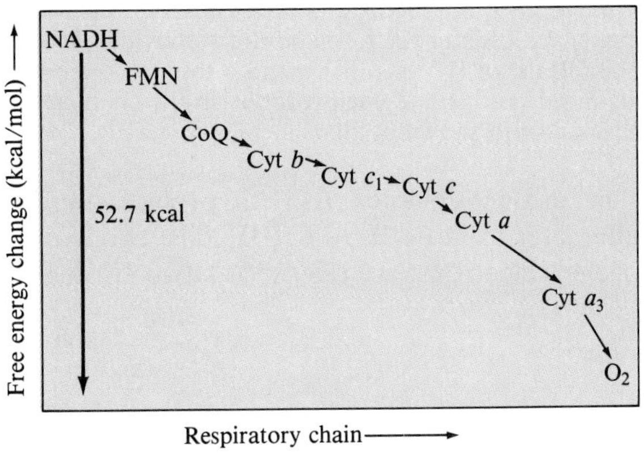

Figure 9.13
Energy released in the respiratory chain. The energy is used to pump protons across the mitochondrial inner membrane and into the intermembrane space.

$$NADH + H^+ + \tfrac{1}{2}\,O_2 \longrightarrow NAD^+ + H_2O + 52.7\text{ kcal}$$

$$FADH_2 + \tfrac{1}{2}\,O_2 \longrightarrow FAD + H_2O + 43.4\text{ kcal}$$

Transfer of 2 H Atoms to Flavin Mononucleotide (FMN) The chain begins when $NADH/H^+$ from the matrix is oxidized to NAD^+ as it passes two hydrogen atoms to *flavin mononucleotide (FMN)* (Figure 9.14). This coenzyme contains the same heterocycle as FAD and is reduced in the same way to $FMNH_2$ (Section 9.7).

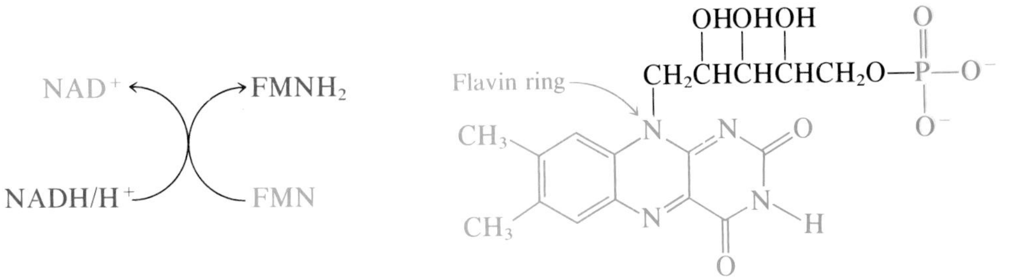

Flavin mononucleotide (FMN)

Respiratory chain (electron transport chain) The series of biochemical reactions that passes electrons from reduced coenzymes to oxygen and is coupled to ATP formation.

Transfer of 2 e⁻ to Iron/Sulfur Protein (FeSP) Following the formation of $FMNH_2$, electrons only are passed along to iron/sulfur protein (FeSP) within the

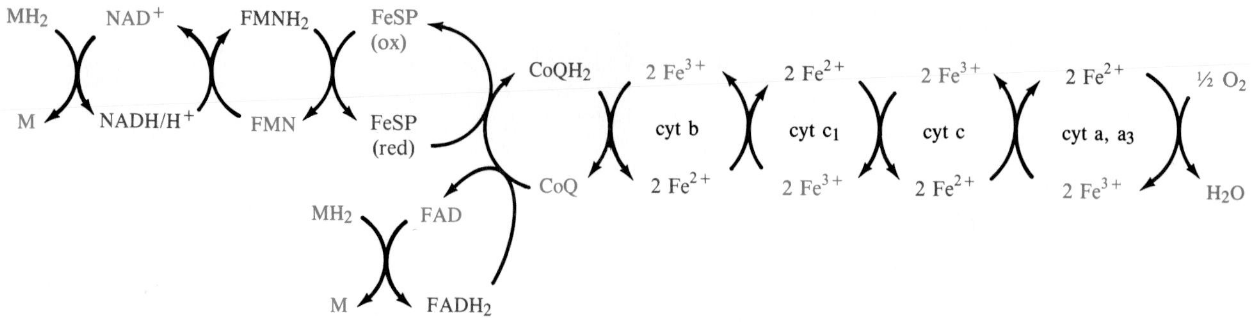

Figure 9.14
An overview of reactions in the respiratory chain. A complex series of coupled reactions leads to the formation of water, the generation of 52.7 kcal/mol energy, and the release of H^+ ions that migrate through the mitochondrial membrane. The oxidized products of each reaction in the chain are shown in red, and the reduced products are shown in blue.

REVIEW Roman numerals such as those in Fe (II) and Fe (III) represent oxidation states (in this case the ionic charges). In oxidation, the oxidation state increases; in reduction, the oxidation state decreases.

same enzyme complex. As FeSP cycles from reduced to oxidized forms, its iron alternates from Fe(II) to Fe(III). The two hydrogens of $FMNH_2$, now stripped of their electrons, are released as two H^+ ions into the intermembrane space.

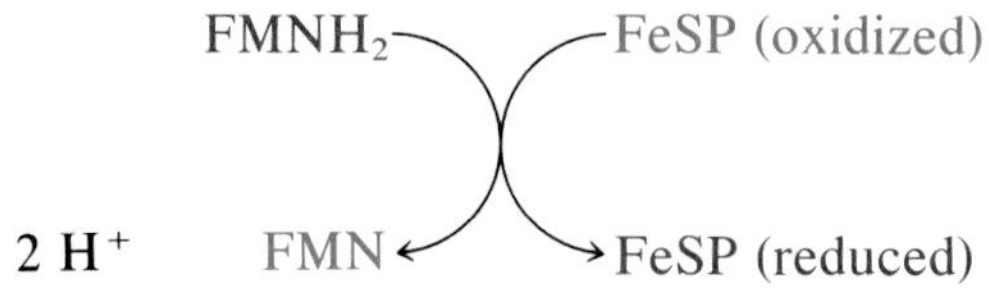

Transfer of Electrons from FeSP and FADH₂ to Coenzyme Q   The iron/sulfur protein next passes its electron pair to *coenzyme Q (CoQ),* sometimes known as *ubiquinone* because of its ubiquitous occurrence in most organisms. Electrons from $FADH_2$ also enter the chain here by transfer to CoQ. The CoQ picks up two H^+ ions to give reduced coenzyme Q ($CoQH_2$), and $FADH_2$ and FeSP are returned to their oxidized forms, ready for further reactions.

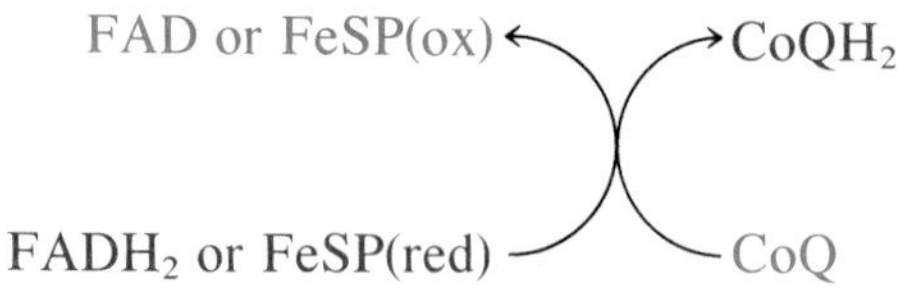

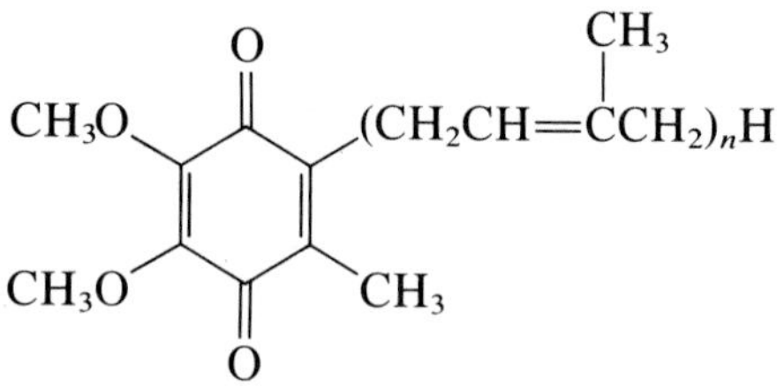

Coenzyme Q (CoQ)
($CoQH_2$ has two —OH's on a benzene ring)

Cytochromes Heme-containing coenzymes that function as electron carriers.

Cytochromes   Following the formation of $CoQH_2$, hydrogen ions no longer participate directly in the reductions of the respiratory chain. Instead, electrons are transferred to a series of enzymes called the **cytochromes.** There are a number of different cytochromes, abbreviated cyt a, cyt b, and so forth. All are closely related in structure, and each contains a heme coenzyme (Figure 9.15) with an iron ion that cycles between the $+2$ and $+3$ oxidation states.

Electrons are shuttled along one by one as an iron atom in one cytochrome gives up an electron to an iron atom in the next cytochrome and is oxidized from Fe(II) to Fe(III) in the process. At the same time, the iron atom in the cytochrome that accepts the electron is reduced from Fe(III) to Fe(II). Ultimately, cytochrome a_3, also called *cytochrome oxidase,* passes electrons along to an oxygen atom, which combines with two H^+ to yield water.

$$\tfrac{1}{2}\,O_2 + 2\,e^- + 2\,H^+ \longrightarrow H_2O$$

Now that we've described the most important parts of the respiratory chain, you should look back at the entire sequence shown in Figure 9.12. As the figure indicates, H^+ ions are released for transport through the inner mitochondrial membrane at each fixed enzyme complex. Evidence indicates that the total number of H^+ ions transported for each NADH oxidized may be 10, but the numbers transported at each location, exactly how they are transported, and the total are still questions for active investigation. Whatever the details, the transfer of hydrogen ions across the mitochondrial inner membrane as electrons move along is essential to the synthesis of ATP via oxidative phosphorylation, as described in the next section.

Practice Problems   **9.10**   Without looking at Figure 9.12 or 9.13, identify the missing substances in the following series of coupled reactions:

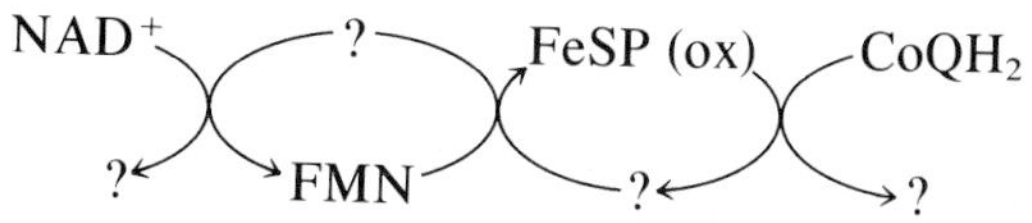

NAD$^+$?   FeSP (ox)   CoQH$_2$

?   FMN   ?   ?

9.11   Look at the structure of coenzyme Q and explain whether you would expect it to be soluble in water or in organic solvents.

Figure 9.15

(a) The structure of heme, the iron-containing coenzyme present in the cytochromes. (b) Computer-generated structure of two molecules of cytochrome c. The heme groups are shown in light green.

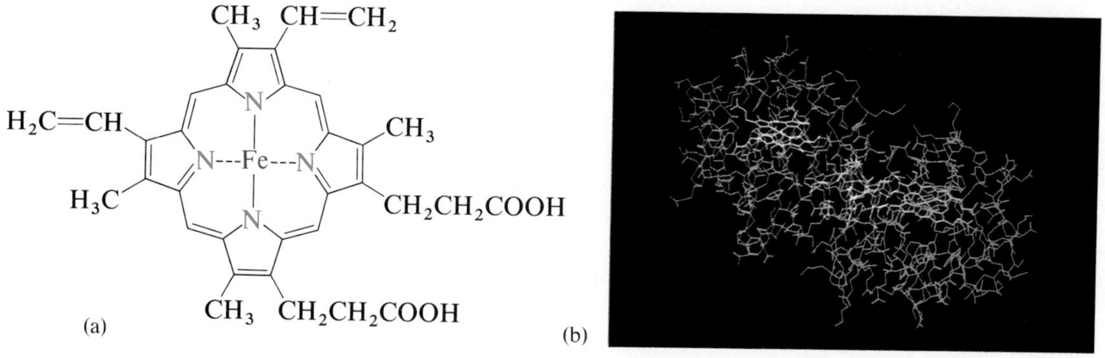

(a)

(b)

AN APPLICATION: BARBITURATES

The barbiturates, discovered in 1864 by the German chemist Adolph von Baeyer, were among the first pharmaceutical agents to come from the chemical laboratory rather than from nature. Different barbiturates are extremely easy to synthesize, and more than 2500 analogs have been prepared. All are cyclic amides that differ only in the nature of the two groups attached to the ring. Among the more common ones are barbital (sold as the drug Veronal, named after the Italian city of Verona), phenobarbital (Luminal), secobarbital (Seconal), and amobarbital (Amytal).

Barbiturates are usually classified into two groups: short-acting ones like secobarbital and long-acting ones like phenobarbital. All act as tranquilizers at low doses and as sleep inducers at somewhat higher doses. At one point several years ago it was estimated that enough barbiturates were being produced in the United States to put 10 million people to sleep every night.

Although medically useful at doses in the 50–100 mg range, barbiturates are toxic in higher amounts. At doses above 1.5 g, barbiturates depress the respiratory system, leading to severe oxygen deficiency in the brain and ultimately to death. These toxic effects appear to arise from an enzyme inhibition that blocks the participation of $NADH/H^+$ in the respiratory chain. Barbiturates somehow prevent electron transfer in the FMN/FeS/CoQ coenzymes, thereby preventing the major part of the chain from functioning.

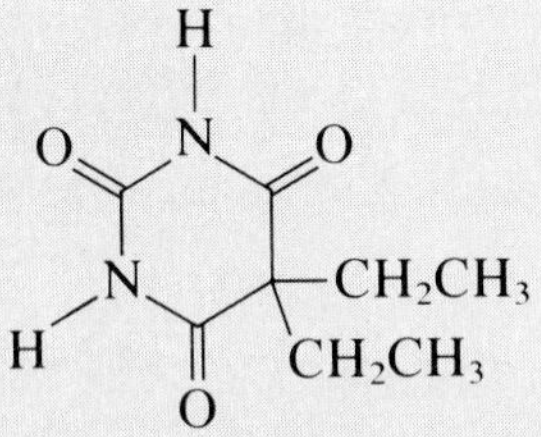

Veronal

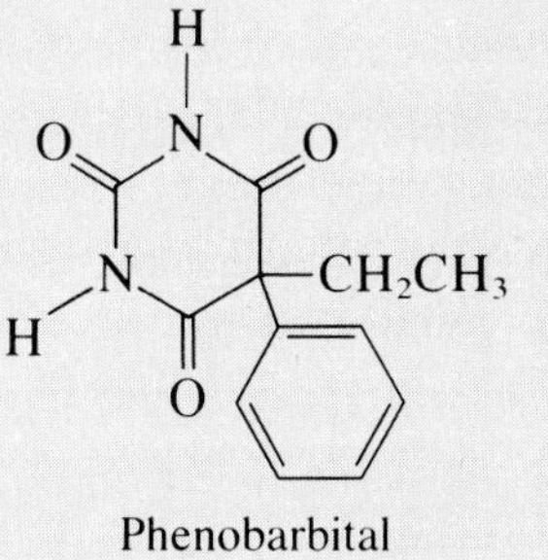

Phenobarbital

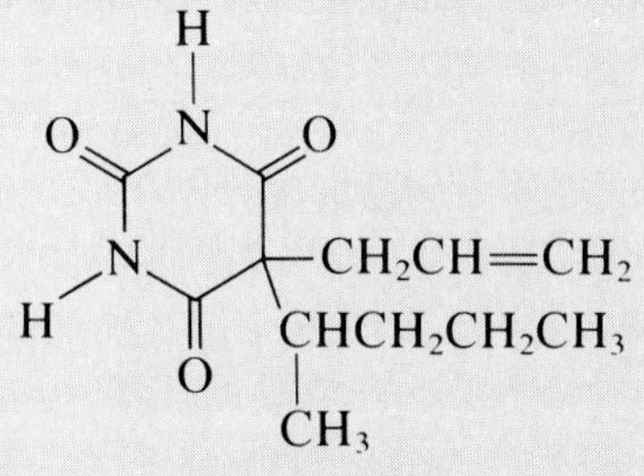

Secobarbital

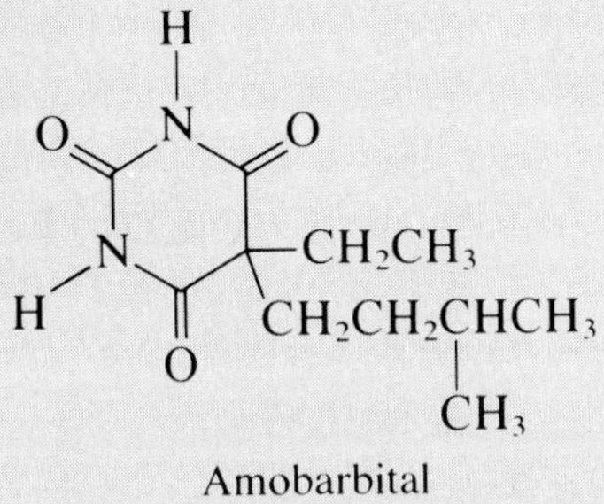

Amobarbital

9.10 OXIDATIVE PHOSPHORYLATION: THE SYNTHESIS OF ATP

Chemiosmotic theory An explanation of how the establishment of a pH difference across a mitochondrial membrane by the respiratory chain is coupled to ATP synthesis.

How does the complex sequence of reactions that make up the respiratory chain lead to the synthesis of ATP? Despite many years of work, the full answer to this question is still not known. It seems clear, however, that the key point is the establishment of a difference in hydrogen ion concentration between the inner and outer sides of the mitochondrial membrane.

According to the **chemiosmotic theory** introduced in 1961 by Peter Mitchell, a British biochemist, hydrogen ions are "pumped" across the inner mito-

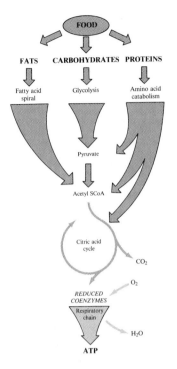

chondrial membrane during electron transport and are unable to diffuse back inside spontaneously. In other words, membrane passage of H^+ is easier in one direction than in the other, and a concentration difference is therefore maintained. The H^+ ion concentration difference between the inner and outer membranes of the mitochondrion is about 1.4 pH units, with the outside being more acidic. It's this pH difference that drives ATP formation.

By a mechanism whose details are unknown, the protons are able to cross back into the mitochondrial matrix only at certain sites. These sites, which are visible at very high magnification as knobby protrusions on the inner membrane, evidently house the ATP synthase enzyme necessary for combining ADP with phosphate to yield ATP. As the protons return to the matrix, the energy expended to maintain the concentration difference is released, and this proton flow activates the enzyme. Chemiosmotic coupling has now been demonstrated in numerous experiments, and Mitchell's achievement in proposing it was recognized in 1973 by award of a Nobel Prize.

Figure 9.16 summarizes schematically the relationships between the respiratory chain and **oxidative phosphorylation**: the synthesis of ATP from ADP. The passage of electrons from one mole of NADH down the respiratory chain releases 52.7 kcal and is coupled to the production of three molecules of ATP:

$$NADH + H^+ + \tfrac{1}{2}\,O_2 + 3\ ADP + 3\ P_i \longrightarrow NAD^+ + 3\ ATP + H_2O$$

Oxidative phosphorylation
The synthesis of ATP from ADP using energy released in the respiratory chain.

Since 7.3 kcal is required for each ADP → ATP conversion, about 22 kcal, or about 40% of the energy stored in NADH, is captured by oxidative phosphorylation. The oxidation of $FADH_2$, which enters the chain farther along than NADH, is coupled to the production of only two molecules of ATP.

Figure 9.16
Synthesis of ATP. The respiratory chain pumps H^+ ions from the matrix to the intermembrane space, creating a concentration imbalance. The ions can return only through channels in knobs that house the ATP synthetase enzyme complex. As the H^+ ions return, the energy expended to maintain the concentration imbalance is released and harnessed to synthesize ATP.

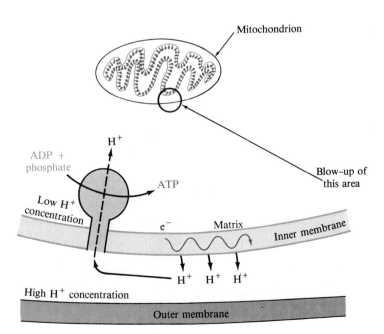

INTERLUDE: DIETS, BABIES, AND HIBERNATING BEARS

Most warm-blooded animals generate enough heat to maintain body temperature through normal metabolic reactions. In certain cases, however, normal metabolism is not able to satisfy the body's need for warmth, and a supplemental method of heat generation is needed. This occurs, for example, in newborn infants and in hibernating animals that must survive for long periods without food. Nature has therefore provided babies, bears, and many other creatures with *brown fat tissue*. Brown fat tissue, so called because its color contrasts with that of normal white fatty tissue, contains high concentrations of blood vessels and mitochondria. In addition, it contains a special protein that acts as an *uncoupler* of ATP synthesis.

Normally, the electron transport reactions of the respiratory chain are coupled to the synthesis of ATP. In other words, the purpose of the respiratory chain is to harness the energy released during catabolism to synthesize ATP. In the presence of an uncoupling agent like the protein in brown fat, however, the electron-transport reactions continue to take place, but no ATP is synthesized. As a result, the energy that would otherwise be used for ATP synthesis is released simply as heat, and the temperature of the organism rises. In essence, brown fat acts as a furnace to produce heat energy rather than stored energy.

The uncoupling of ATP synthesis from the respiratory chain can also be caused by chemical agents. It has been found, for example, that 2,4-dinitrophenol can cause uncoupling even in normal white fatty tissue of adults.

2,4-Dinitrophenol

Imagine the possibilities for a simple chemical agent that causes food to be converted directly into heat rather than ATP. The body, in critical need of ATP to drive its many reactions, would continue to burn fat at a frantic rate. Pound after pound of fat would literally melt away, giving every previously unsuccessful dieter the figure of a movie star. In fact, the idea works, and weight control pills containing 2,4-dinitrophenol were marketed for a brief period in the 1930s. Unfortunately, the dose needed to *slow* ATP synthesis is very close to the lethal dose that *stops* ATP synthesis, a fact that was discovered only after several fatalities had occurred. Thus, there's still no magic pill for dieters.

Brown fat and the uncoupling of the respiratory chain and ATP synthesis keep this harp seal warm.

SUMMARY

Metabolism is the sum total of all chemical reactions going on in the body. Those reactions that break down large molecules into smaller fragments are called **catabolism,** while those that build up large molecules from small pieces are called **anabolism.** Lipids, carbohydrates, and proteins are all catabolized to yield acetyl SCoA, which is then further degraded in the citric acid cycle. In the final stage of catabolism, energy is stored in ATP.

There are two broad categories of cells: **prokaryotic** cells, found in algae and bacteria, and **eukaryotic** cells, found in higher organisms. Eukaryotic cells contain a large number of small subcellular structures called **organelles** that perform specialized tasks in the cell. Among the most important organelles are the **mitochondria,** which house the enzymes necessary for producing energy from the breakdown of food-derived molecules.

Energy transfer is managed in metabolism by (1) storage in **adenosine triphosphate (ATP),** a molecule suitable for energy storage because it contains reactive P—O—P (**phosphoric anhydride**) bonds that allow it to transfer a phosphate group; (2) coupled reactions, in which the energy from an exergonic reaction is used to make possible an endergonic reaction; and (3) the cycling between their oxidized and reduced forms of enzymes, notably **nicotinamide adenine dinucleotide (NAD)** and **flavin adenine dinucleotide (FAD),** that act as electron and hydrogen carriers.

The **citric acid cycle** is a series of eight enzyme-catalyzed steps whose primary role is releasing energy from **acetyl SCoA,** which carries acetyl groups from earlier stages of catabolism. The acetyl groups are converted into two molecules of carbon dioxide plus a large amount of energy. The energy output of the citric acid cycle is carried by reduced coenzymes to the **respiratory chain,** in which electrons from the enzymes are carried in stepwise fashion through a series of enzymes ending with the **cytochromes** and with the reaction with oxygen to form water. According to the **chemiosmotic theory,** the protons released in the respiratory chain migrate back into the matrix at certain knob-like sites on the membrane, thereby activating the enzymes that carry out ATP synthesis (**oxidative phosphorylation**).

REVIEW PROBLEMS

Free Energy and Biochemical Reactions

9.12* What is the difference between an endergonic and an exergonic process?

9.13 What energy requirement must be met in order for a reaction to be spontaneous?

9.14* Why is ΔG a useful quantity for predicting the spontaneity of biochemical reactions?

9.15 Many biochemical reactions are catalyzed by enzymes. Do enzymes have an influence on the magnitude or sign of ΔG? Why or why not?

9.16* Each of the following reactions occurs during the catabolism of glucose (Chapter 11). Which is exergonic? Which is endergonic? Which proceeds farthest toward products at equilibrium?
(a) phosphoenol pyruvate + H_2O → pyruvate + phosphate $\Delta G = -14.8 \, \text{kcal/mol}$
(b) glucose + phosphate → glucose 6-phosphate + H_2O $\Delta G = 3.3 \, \text{kcal/mol}$
(c) 1,3-bisphosphoglycerate + H_2O → 3-phosphoglycerate + phosphate $\Delta G = -11.8 \, \text{kcal/mole}$

Cells and Their Structure

9.17 List several differences between prokaryotic and eukaryotic cells.

9.18* What kinds of organisms have prokaryotic cells, and what kinds have eukaryotic cells?

9.19 What is an organelle, and what is its general function?

9.20* What is the difference between the cytoplasm and the cytosol?

9.21 Describe in general terms the structural makeup of a mitochondrion.

9.22* Why are mitochondria called the body's "power plant"?

9.23 What are cristae, and why are they important?

Metabolism

9.24* What is the difference between digestion and metabolism?

9.25 What is the difference between catabolism and anabolism?

9.26* What key metabolic substance is formed from the catabolism of all three major classes of foods: carbohydrates, fats, and proteins?

9.27 Put the following events in the correct order of their occurrence: respiratory chain, digestion, oxidative phosphorylation, citric acid cycle.

Strategies of Metabolism

9.28* What is the full name of the substance formed during catabolism to store chemical energy?

9.29 What is the chemical difference between ATP and ADP?

9.30* Why is ATP called a high-energy molecule?

9.31 What general kind of chemical reaction does ATP carry out?

9.32 Adenosine *mono*phosphate is an important intermediate in certain biochemical pathways. Draw its structure.

9.33 What does it mean when we say that two reactions are coupled?

9.34* Show why coupling the reaction for the hydrolysis of 1,3-bisphosphoglycerate (Problem 9.16) to the phosphorylation of ADP is energetically favorable. Combine the equations and calculate ΔG for the coupled process. You need only give names or abbreviations, not chemical structures.

9.35 Write the reaction in Problem 9.34 with the curved arrow symbolism.

9.36* Would the hydrolysis of fructose 6-phosphate (Table 9.1) be favorable for phosphorylating ADP? Why or why not?

9.37 Both NAD^+ and FAD are coenzymes for dehydrogenation.
(a) When a molecule is dehydrogenated is it oxidized or reduced?
(b) Are NAD^+ and FAD oxidizing agents or reducing agents?
(c) What type of substrate is each coenzyme associated with, and what is the type of product molecule after dehydrogenation?
(d) What is the form of each coenzyme after the dehydrogenation process?
(e) Write a generalized equation for the operation of each coenzyme with the curved arrow symbolism.

The Citric Acid Cycle

9.38* Where in the cell does the citric acid cycle take place?

9.39 By what other names is the citric acid cycle known?

9.40* What substance acts as the starting point of the citric acid cycle, reacting with acetyl SCoA in the first step and being regenerated in the last step? Draw its structure.

9.41 Look at the eight steps of the citric acid cycle (Figure 9.11) and answer these questions:
(a) Which steps involve oxidation reactions?
(b) Which steps involve decarboxylations (loss of CO_2)?
(c) Which step involve a hydration reaction?

9.42* Consider step 5 of the citric acid cycle. What is substrate-level phosphorylation and what is the function of GTP in this process?

9.43 How many ATPs are directly formed as a result of the citric acid cycle?

9.44* How many molecules of NADH and $FADH_2$ are formed in the citric acid cycle?

9.45 What is the final fate of the carbons in the acetyl CoA after several turns of the citric acid cycle?

The Respiratory Chain

9.46* By what other name is the respiratory chain known?

9.47 What does the term "oxidative phosphorylation" mean? How does it differ from substrate-level phosphorylation?

9.48* In oxidative phosphorylation, what is oxidized and what is phosphorylated?

9.49 What two coenzymes initiate the events of the respiratory chain?

9.50* What are the ultimate products of the respiratory chain?

9.51 What do the following abbreviations stand for?
(a) FAD (b) CoQ (c) $NADH/H^+$ (d) cyt

9.52* What atom in the cytochromes undergoes oxidation and reduction in the respiratory chain?

9.53 Put the following substances in the correct order of their action in the respiratory chain: iron/sulfur protein, cytochrome a, coenzyme Q, NAD^+.

9.54* Fill in the missing substances in these coupled reactions:

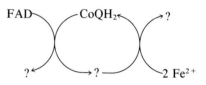

9.55 In the respiratory chain, how many ATPs are released by reoxidation of each NADH? Each $FADH_2$?

9.56* What would happen to the citric acid cycle if NADH and $FADH_2$ were not reoxidized?

9.57 According to the chemiosmotic theory, across what membrane is there a pH differential caused by the release of H^+ ions? On which side of the membrane are there more H^+ ions?

Applications

9.58* How is basal metabolic rate defined? [App: Basal Metabolism]

9.59 Estimate your basal metabolic rate using the guidelines in the application. [App: Basal Metabolism]

9.60* Why do activities such as walking raise a body's needs above the basal metabolic rate? [App: Basal Metabolism]

9.61 What is the medical function of a barbiturate? [App: Barbiturates]

9.62* By what mechanism do barbiturates become lethal at high doses? [App: Barbiturates]

9.63 Other than color, what is the difference between brown and white fat? [Int: Diets, Babies, and Hibernating Bears]

9.64* What compound has been used as an uncoupler of ATP synthesis? How can the uncoupling be used as a diet aid and why can it be dangerous? [Int: Diets, Babies, and Hibernating Bears]

Additional Questions and Problems

9.65 Why must the breakdown of molecules for energy in the body occur in several steps, rather than in one step?

9.66 The first step of the citric acid cycle involves an aldol condensation of acetyl CoA and oxaloacetic acid. Show the product of the aldol condensation before the hydrolysis to yield citrate.

9.67 The fumaric acid produced in step 6 of the citric acid cycle must have a trans double bond in order to continue on in the cycle. Suggest a reason why the corresponding cis double-bond isomer can't continue in the cycle.

9.68* With what class of enzymes (Section 8.3) are the coenzymes NAD^+ and FAD associated?

9.69 Considering both substrate-level phosphorylation and oxidative phosphorylation, how many molecules of ATP are generated by one turn of the citric acid cycle?

9.70* We talk of burning food in a combustion process, producing CO_2 and H_2O from food and oxygen. Explain how oxygen is involved in the process since there is no oxygen directly involved in the citric acid cycle. What enzyme is associated with the use of O_2?

C H A P T E R

10

Carbohydrates

Carbohydrates are stored in the liver and muscles in the polymer known as glycogen. This falsely colored micrograph of liver tissue shows a liver cell (at top), its large nucleus (green), packages of stored glycogen (pink-red), and vacuoles containing triacylglycerols (yellow), the stored lipids described in Chapter 12.

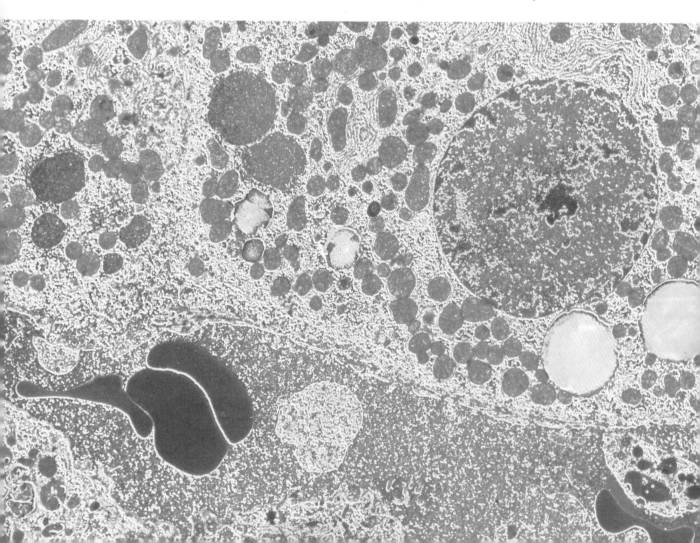

Carbohydrates occur in every living organism. The starch in food and the cellulose in grass are pure carbohydrate; modified carbohydrates form part of the coating around all living cells; other carbohydrates are found in the DNA that carries genetic information from one generation to the next; and still other carbohydrates such as streptomycin are valuable as medicines.

The word *carbohydrate* was used originally to describe glucose, the simplest and most readily available sugar. Because glucose has the formula $C_6H_{12}O_6$, it was once thought to be a "hydrate of carbon"—$C_6(H_2O)_6$. Although this view was soon abandoned, the name "carbohydrate" persisted until it is now used to refer to a large class of polyhydroxylated aldehydes and ketones.

$$\underset{\text{H}}{\overset{\text{H}}{\text{HO}-\text{C}}}-\underset{\text{OH}}{\overset{\text{H}}{\text{C}}}-\underset{\text{OH}}{\overset{\text{H}}{\text{C}}}-\underset{\text{H}}{\overset{\text{OH}}{\text{C}}}-\underset{\text{OH}}{\overset{\text{H}}{\text{C}}}-\overset{\text{O}}{\text{C}}-\text{H}$$

Glucose (a pentahydroxyhexanal)

Carbohydrates are synthesized in green leaves by the conversion of carbon dioxide into glucose during photosynthesis. Many molecules of glucose are then linked together to form either cellulose or starch. When starch is eaten and digested, the freed glucose provides the major source of energy required by living organisms. Thus, carbohydrates act as the intermediaries by which energy from the sun is converted into chemical energy.

In this chapter, we'll look for answers to these questions about carbohydrates:

1. **What are the different kinds of carbohydrates?** The goal: Be able to define the main classifications of carbohydrates and classify specific examples.

2. **Why do carbohydrates have handedness?** The goal: Be able to explain why carbohydrates have handedness and be able to show that handedness using Fischer projections.

3. **What are the structures of glucose and other important simple sugars?** The goal: Be able to describe the structural features of monosaccharides and distinguish between D and L sugars.

4. **How are monosaccharide molecules drawn?** The goal: Be able to draw the Fischer projection and cyclic formulas of D and L sugars.

5. **What is a reducing sugar?** The goal: Be able to recognize a reducing sugar and describe its oxidation.

6. **What is a glycoside?** The goal: Be able to predict the product of a given glycoside-forming reaction and classify the glycosidic bond formed.

7. **What are the structures of some important disaccharides?** The goal: Be able to draw the structures of the disaccharides maltose, sucrose, and lactose.

8. **What are the structures of some important polysaccharides?** The goal: Be able to describe how cellulose and starch are constructed and how they differ.

10.1 CLASSIFICATION OF CARBOHYDRATES

Carbohydrate A member of a large class of naturally occurring polyhydroxy ketones and aldehydes.

Monosaccharide (simple sugar) A carbohydrate that can't be chemically broken down into a smaller sugar by hydrolysis with aqueous acid.

Disaccharide A carbohydrate that yields two monosaccharides on hydrolysis.

Acetal A compound that has two —OR groups bonded to the same carbon atom.

Polysaccharide (complex carbohydrate) A carbohydrate composed of many monosaccharides bonded together.

Carbohydrates, a large class of naturally occurring polyhydroxylated aldehydes and ketones, are classified according to whether or not they can be broken down into smaller units. **Monosaccharides,** sometimes called **simple sugars,** are carbohydrates that can't be broken down into smaller molecules by hydrolysis with aqueous acid. Glucose, the pentahydroxyhexanal shown in the introduction to this chapter, is the most important monosaccharide. Naturally occurring monosaccharides have from three to seven carbon atoms, with an —OH group on each carbon atom except that with the aldehyde or ketone carbonyl group. The family-name ending *-ose* indicates a sugar, and specific sugars are known by common rather than systematic names.

Disaccharides are carbohydrates composed of two monosaccharides linked by C—O—C **acetal** bonds, while **polysaccharides** are compounds made of many simple sugar molecules bonded together. On hydrolysis, polysaccharides such as cellulose and starch are cleaved to yield many molecules of simple sugars. (Recall that in hydrolysis, larger molecules split apart with addition of the H— and —OH from water to the atoms in the broken bond.)

A hydrolysis reaction

Hydrolysis of a polysaccharide

$$\text{Cellulose or starch} + \text{H}_2\text{O} \xrightarrow[\text{catalyst}]{\text{H}^+} \text{thousands of glucose molecules}$$

Aldose A monosaccharide that contains an aldehyde carbonyl group.

Ketose A monosaccharide that contains a ketone carbonyl group.

Monosaccharides are classified as either aldoses or ketoses according to the kind of carbonyl group they have. An **aldose** contains an aldehyde carbonyl group; a **ketose** contains a ketone carbonyl group. The number of carbon atoms in an aldose or ketose is specified by using one of the prefixes *tri-, tetr-, pent-,* or *hex-.* Thus, glucose is an aldo-*hex*-ose (aldo- = aldehyde, *-hex* = six carbon, -ose = sugar); fructose is a keto*hex*ose (a six-carbon ketone sugar); and ribose is an aldo*pent*ose (a five-carbon aldehyde sugar). Most naturally occurring simple sugars are either aldopentoses or aldohexoses.

Glucose
(an aldohexose)

Fructose
(a ketohexose)

Ribose
(an aldopentose)

Practice Problems *10.1* Classify each of these monosaccharides:

(a) $HOCH_2$—$\overset{\overset{\displaystyle OH}{|}}{CH}$—$\overset{\overset{\displaystyle OH}{|}}{CH}$—$\overset{\overset{\displaystyle OH}{|}}{CH}$—$\overset{\overset{\displaystyle O}{\|}}{C}$—$H$ (b) $HOCH_2$—$\overset{\overset{\displaystyle O}{\|}}{C}$—$CH_2OH$

(c) $HOCH_2$—$\overset{\overset{\displaystyle OH}{|}}{CH}$—$\overset{\overset{\displaystyle OH}{|}}{CH}$—$\overset{\overset{\displaystyle O}{\|}}{C}$—$H$

10.2 Draw the structures of an aldohexose and a ketotetrose.

10.2 HANDEDNESS OF CARBOHYDRATES

Glyceraldehyde, the simplest naturally occurring carbohydrate, has the structure shown in Figure 10.1. Is glyceraldehyde chiral? As explained in Section 7.5, chiral compounds have a carbon atom bonded to four different atoms or groups of atoms. Such compounds lack a plane of symmetry and can exist as a pair of optical isomers (enantiomers) in either a "right-handed" D form or a "left-handed" L form. Glyceraldehyde is therefore chiral because it has four different groups bonded to C2: —CHO, —H, —OH, and —CH₂OH. Like all optical isomers, the two forms of glyceraldehyde have the same physical properties except for the way in which they interact with polarized light. This interaction is discussed in the Application on polarized light.

Compounds like glyceraldehyde have only one chiral carbon atom and can exist as two optical isomers. But what about compounds with more than one chiral carbon atom? How many isomers are there of compounds that have two, three, four, or more chiral carbons? Aldotetroses, for example, have two chiral carbon atoms and can exist in the four isomeric forms shown in Figure 10.2.

Figure 10.1
Glyceraldehyde (2,3-dihydroxypropanal), a chiral molecule. Glyceraldehyde is chiral because it has four different groups attached to carbon 2. Thus, glyceraldehyde can exist in two forms: a "right-handed" form referred to as D-glyceraldehyde and a "left-handed" form referred to as L-glyceraldehyde.

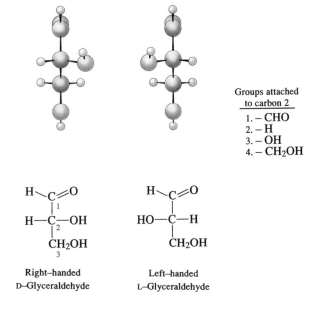

Groups attached
to carbon 2
1. – CHO
2. – H
3. – OH
4. – CH₂OH

Right–handed
D–Glyceraldehyde

Left–handed
L–Glyceraldehyde

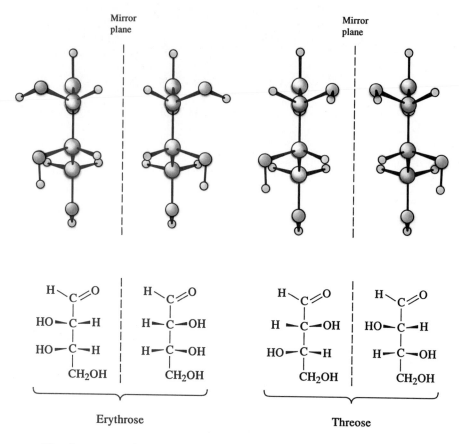

Figure 10.2
The four isomeric aldotetroses (2,3,4-trihydroxybutanals). Erythrose and threose differ in arrangement at C2.

Erythrose   Threose

Diastereomers
Stereoisomers that are not mirror images of each other.

The four stereoisomeric aldotetroses can be classifed into two mirror-image pairs of optical isomers, one pair named *erythrose* and one pair named *threose*. Erythrose and threose, however, are not mirror images of each other but are instead a different kind of stereoisomer. Non-mirror-image stereoisomers like erythrose and threose are called **diastereomers.**

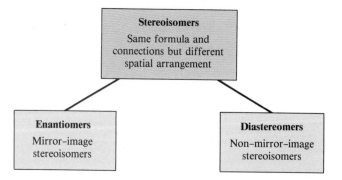

In general, a compound with n chiral carbon atoms has a maximum of 2^n possible stereoisomers and half that number of optical isomer pairs. The aldotetroses, for example, have $n = 2$ so that $2^n = 2^2 = 4$, showing that 4 stereoisomers are possible. Glucose, an aldohexose, has four chiral carbon atoms and a total of $2^4 = 16$ possible stereoisomers (8 pairs of optical isomers) that differ in the spatial arrangements of the substituents around chiral carbon atoms. All 16 are known.

AN APPLICATION: POLARIZED LIGHT AND OPTICAL ACTIVITY

Louis Pasteur, the French chemist best known for discovering the value of "pasteurizing" milk, was the first to propose that some compounds could exist as mirror-image pairs of molecules. In 1849, many years before molecular structure was understood, he was working with the crystalline salts of tartaric acid when he discovered two visibly different kinds of asymmetric crystals in a sample of sodium ammonium tartrate. By using tweezers, he patiently separated them and found that crystals of one type were *mirror images* of crystals of the second type. To explore the properties of these two tartrates, Pasteur placed their solutions in the path of *plane-polarized light* beams.

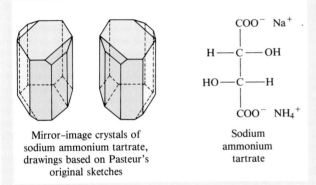

Mirror–image crystals of sodium ammonium tartrate, drawings based on Pasteur's original sketches

Sodium ammonium tartrate

Light as we usually see it consists of electromagnetic waves oscillating in all planes at right angles to the direction of travel of the light beam. When ordinary light is passed through a *polarizer,* only the waves in one plane get through, producing what is known as *plane-polarized light.* From earlier work by another French scientist, Jean Baptiste Biot, Pasteur knew that solutions of certain organic compounds change the plane in which the light is polarized. Because of their interaction with light, such compounds are described as *optically active.* The angle of rotation of plane-polarized light by an optically active compound is measured in an instrument known as a *polarimeter,* illustrated schematically below.

Pasteur discovered that solutions of his two types of crystals rotated plane-polarized light by equal amounts, but in *opposite directions.* One kind of crystal rotated the light in a right-handed direction, denoted $(+)$, and the other kind of crystal rotated the light in a left-handed direction, denoted $(-)$. To explain this phenomenon he proposed the existence of two types of tartaric acid molecules with opposite asymmetric arrangements that are, like the visible crystals, mirror images of each other. Although it could not be proven until many years later, Pasteur had indeed discovered what we now know as *optical isomers.*

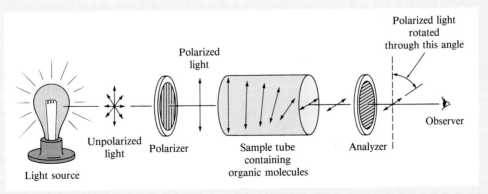

Diagram of a polarimeter, which measures the optical activity of a molecule

Note that we've waffled a bit in the above discussion by saying that 2^n represents the maximum number of "possible" stereoisomers. In some cases, variations around chiral atoms produce molecules that have symmetry planes and are thus identical to their mirror images. In such cases, there are fewer than the maximum possible number of stereoisomers.

Practice Problems

10.3 Draw tetrahedral representations of the two glyceraldehyde optical isomers using the standard method of wedged, dashed, and normal lines to show three-dimensionality.

10.4 Aldopentoses have three chiral carbon atoms. What is the maximum possible number of aldopentose stereoisomers?

Fischer projection
Structure that represents a chiral carbon atom as the intersection of two lines. The horizontal line represents bonds pointing out of the page, and the vertical line represents bonds pointing behind the page. For sugars, the aldehyde or ketone is at the top.

10.3 DRAWING SUGAR MOLECULES: THE D AND L FAMILIES OF SUGARS

A standard method of representation called a **Fischer projection** has been adopted for drawing stereoisomers on a flat page so that we can tell one from another. A chiral carbon atom is represented in Fischer projections as the intersection of two crossed lines. Bonds that point out of the page are shown as horizontal lines, and bonds that point behind the page are shown as vertical lines.

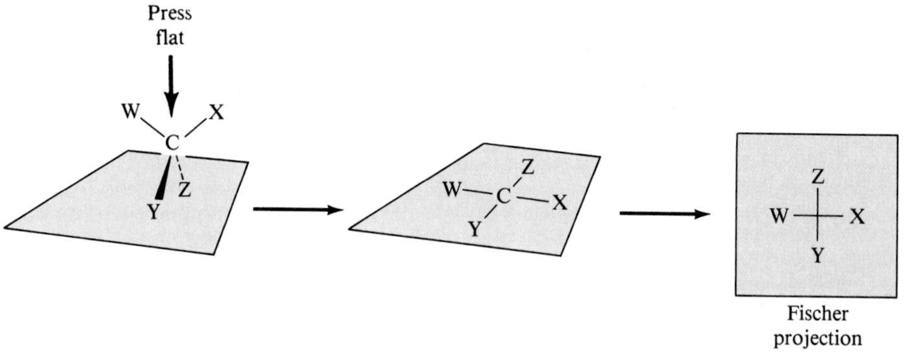

Press flat

Fischer projection

For consistency, the aldehyde or ketone carbonyl group is shown at or near the top when drawing a carbohydrate. Thus, one optical isomer of glyceraldehyde can be drawn as follows:

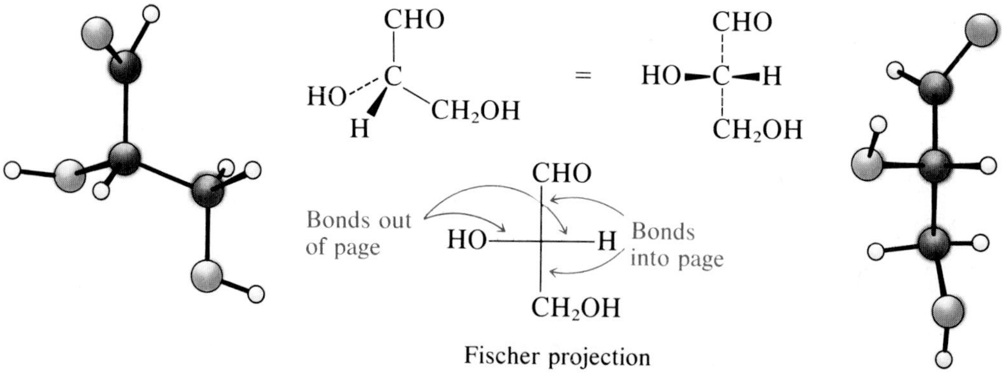

Bonds out of page
Bonds into page

Fischer projection

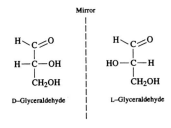

D–Glyceraldehyde | L–Glyceraldehyde

D Sugar Monosaccharide with —OH group on chiral carbon atom farthest from the carbonyl group pointing to the right in the Fischer projection.

L Sugar Monosaccharide with —OH group on chiral carbon atom farthest from the carbonyl group pointing to the left in the Fischer projection.

Monosaccharides are divided into two families—the D **sugars** and the L **sugars**—based on their structural relationships to glyceraldehyde. Note how the D and L forms of glyceraldehyde were shown back in Figure 10.1: In the D form, the —OH group on carbon 2 comes out of the plane of the paper and points to the right when the —CHO group is at the top; in the L form, the —OH group at carbon 2 comes out of the plane of the paper and points to the left when the —CHO group is at the top. If you mentally place a mirror plane between the Fischer projections of these molecules, you can see that they're mirror images.

Nature has a strong preference for one type of handedness in carbohydrates, just as it does in amino acids (Section 7.5) and also in snail shells (Figure 10.3). It happens, however, that carbohydrates and amino acids have opposite handedness. Most naturally occurring α-amino acids belong to the L family, but most carbohydrates belong to the D family. L-Carbohydrates occur only rarely in nature.

The designations D and L derive from the Latin *dextro* for "right" and *levo* for "left." In Fischer projections, the D form of a monosaccharide has *the hydroxyl group on the chiral carbon atom farthest from the carbonyl group* pointing toward the right, while the mirror-image L form has the hydroxyl group on this same carbon pointing toward the left. The D and L designations don't, however, have any relationship to the direction in which an isomer rotates plane-polarized light.

Fischer projections of molecules with more than one chiral carbon atom are written by stacking the chiral atoms one above the other. The following structures show two of the eight pairs of optical isomers of the aldohexoses written in this manner. Given the Fischer projection of one isomer, you can draw the other isomer by reversing the substituents on each chiral atom. Note that each pair of optical isomers has a different name.

Two pairs of aldohexose optical isomers

Chiral C atom farthest from C=O

D-Allose   L-Allose   D-Glucose   L-Glucose

Figure 10.3
Nature's preference. Not only molecules, but also snail shells have a preferred handedness. Most snail shells, like the one shown, are right-handed.

Practice Problems   **10.5**   Identify the following monosaccharides as (a) D-ribose or L-ribose, (b) D-mannose or L-mannose.

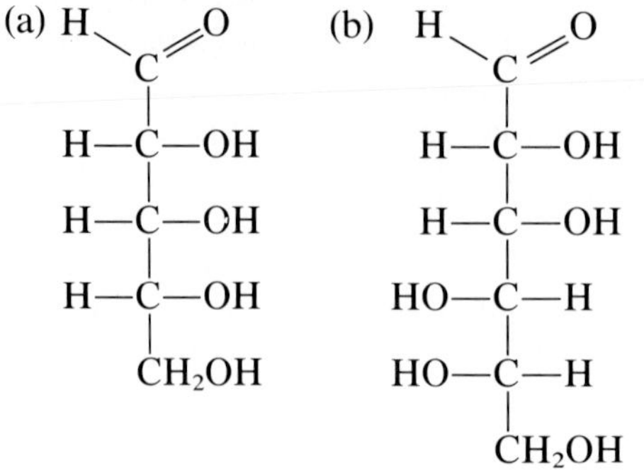

(a)
$$H\diagdown C \diagup\!\!\diagup O$$
H—C—OH
H—C—OH
H—C—OH
CH$_2$OH

(b)
$$H\diagdown C \diagup\!\!\diagup O$$
H—C—OH
H—C—OH
HO—C—H
HO—C—H
CH$_2$OH

10.6   Draw the optical isomers of the following monosaccharides and identify which is in the D family of sugars and which is in the L family in each pair.

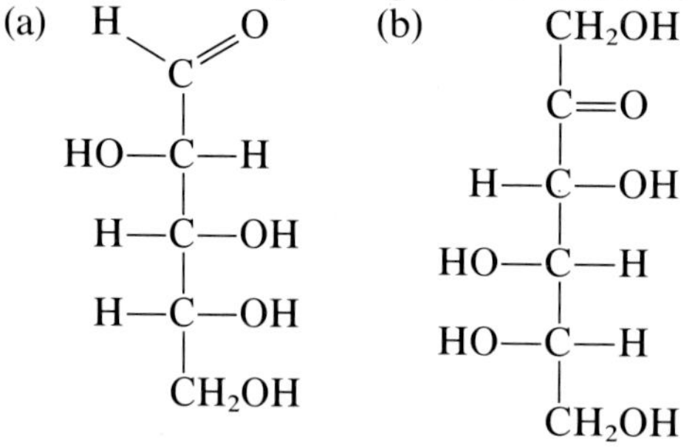

(a)
$$H\diagdown C \diagup\!\!\diagup O$$
HO—C—H
H—C—OH
H—C—OH
CH$_2$OH

(b)
CH$_2$OH
C=O
H—C—OH
HO—C—H
HO—C—H
CH$_2$OH

10.4   STRUCTURE OF GLUCOSE AND OTHER MONOSACCHARIDES

D-Glucose, sometimes called *dextrose* or *blood sugar,* is the most widely occurring of all monosaccharides. It is found in nearly all foods and in all living organisms, where it serves as a source of energy to fuel biochemical reactions. Before looking further at the chemical structure of glucose, glance back briefly at Section 5.7. We said there that aldehydes and ketones react reversibly with alcohols to yield **hemiacetal** addition products in which a carbon atom is bonded to both an —OH and an —OR group.

Hemiacetal A compound that has an —OH and —OR group bonded to the same carbon atom.

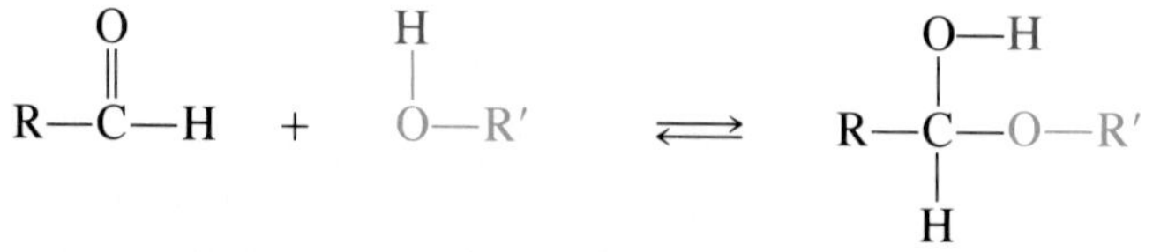

$$R-\overset{\overset{\textstyle O}{\|}}{C}-H \quad + \quad \overset{\overset{\textstyle H}{|}}{O}-R' \quad \rightleftharpoons \quad R-\overset{\overset{\textstyle O-H}{|}}{\underset{\underset{\textstyle H}{|}}{C}}-O-R'$$

An aldehyde   An alcohol   A hemiacetal

Now look at the Fischer projection structure of D-glucose shown in Figure 10.4. Since glucose is an aldohexose—a pentahydroxy aldehyde—it has alcohol

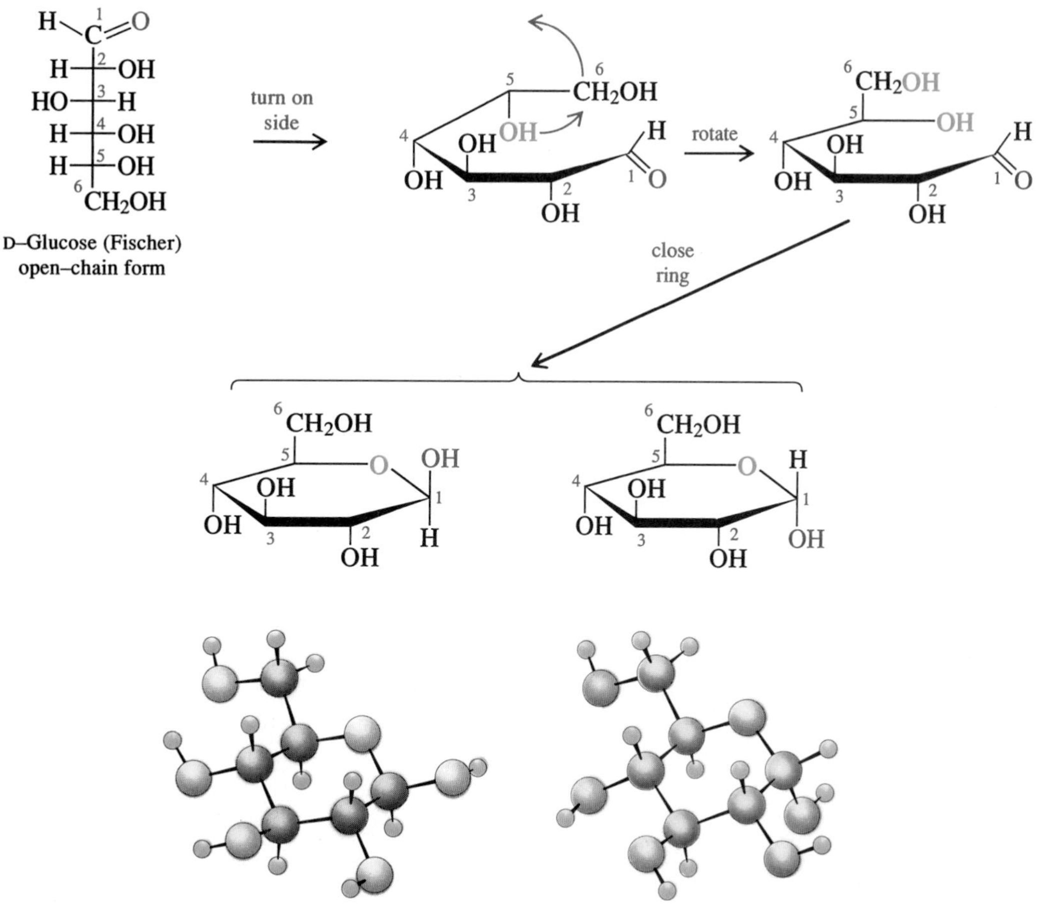

Figure 10.4
The structure of D-glucose. A glucose molecule can exist either in an open-chain hydroxy aldehyde form or in a cyclic hemiacetal form. There are two cyclic hemiacetal forms, called α-glucose and β-glucose, that differ in whether the hemiacetal hydroxyl group at C1 is on the opposite side of the six-membered ring from the CH_2OH (α) or on the same side (β). To convert the Fischer formula into the six-membered ring formula, the Fischer formula is laid down with C1 to the right in front and the other end curled around at the back. Then the single bond between C4 and C5 is rotated so that the —CH_2OH group is vertical. Finally, the hemiacetal O—R bond is formed by connecting oxygen from the —OH group on C5 to C1 and the hemiacetal O—H group is placed on C1.

hydroxyl groups and an aldehyde carbonyl group *in the same molecule.* Thus, an *internal* addition reaction can take place between the aldehyde carbonyl group at carbon 1 and one of the hydroxyl groups in the chain to yield a *cyclic* hemiacetal. In fact, it's the hydroxyl group at carbon 5 that reacts in glucose, leading to a six-membered-ring hemiacetal. The three structures at the top in Figure 10.4 show schematically how the 5-hydroxyl and the aldehyde group approach for acetal formation.

Actually, there are *three* forms of D-glucose—an open-chain form, a cyclic α form, and a cyclic β form—all shown in Figure 10.4. To see the difference between the α and β forms, compare the locations of the hemiacetal —OH groups on carbon 1, which is now a chiral carbon atom. In the β form, the hydroxyl at carbon 1 points *up* on the same side of the ring as the —CH₂OH group connected to carbon 5; in the α form, the hydroxyl at carbon 1 points *down* on the opposite side of the ring from the —CH₂OH group at carbon 5.

Cyclic sugars that differ only in the positions of substituents at carbon 1 are known as **anomers,** and carbon 1 is said to be an **anomeric carbon atom.** (Remember—such a carbon atom is bonded to two oxygen atoms. In the hemiacetal, they're in an —OH group and an —OR group in which R is the ring.) Note that anomers are not optical isomers because they aren't mirror images. Although the structural difference between anomers might seem small, it has enormous biological consequences. You'll see in Section 10.8, for instance, that this one small change in structure accounts for the vast difference between starch and cellulose.

The ordinary crystalline glucose you might take from a bottle is usually entirely in the cyclic α form. In solution, however, equilibria are established by both anomers with the open-chain form. As a result, the optical effect of the solution on polarized light changes, a phenomenon known as **mutarotation.** The equilibria favor the hemiacetal forms, and either anomer when placed in solution produces the identical equilibrium mixture containing mainly the two hemiacetals. For glucose at room temperature in water, at equilibrium the mixture contains 0.02% open chain, 36% α form, and 64% β form.

$$\alpha\text{-D-glucose} \quad \rightleftharpoons \quad \text{open-chain D-glucose} \quad \rightleftharpoons \quad \beta\text{-D-glucose}$$
$$(36\%) \qquad\qquad (0.02\%) \qquad\qquad (64\%)$$

To convert Fischer projections into cyclic structures so that the same relative arrangements are maintained at the chiral carbon atoms requires following the procedure illustrated in Figure 10.4. When cyclic structures, called *Haworth projections,* are drawn in this manner, the —CH₂OH group in D sugars is always above the plane of the ring. Thus, in β-D-glucose, the hydroxyl at carbon 1 points up, on the same side of the ring as the —CH₂OH on carbon 5. In α-D-glucose, the hydroxyl at carbon 1 points down, on the opposite side of the ring from the —CH₂OH group at carbon 5.

To summarize—monosaccharide structures have the following characteristics:

● Monosaccharides are polyhydroxy aldehydes or ketones.

● Monosaccharides have three to seven carbon atoms, and 2^n possible stereoisomers, where *n* is the number of chiral carbon atoms.

● D and L Optical isomers differ in the location of the —OH group on the chiral carbon atom farthest from C1; in Fischer projections, D sugars have —OH on the right, and L sugars have —OH on the left.

● α and β Anomers differ in the location of the —OH on the hemiacetal carbon atom in cyclic forms; α has the —OH on the opposite side from the —CH₂OH, and β has the —OH on the same side as —CH₂OH.

Anomers Cyclic sugars that differ only in positions of substituents at the hemiacetal carbon (the anomeric carbon); the α form has the —OH on the opposite side from the —CH₂OH; the β form has the —OH on the same side as the —CH₂OH.

Anomeric carbon atom The hemiacetal C atom in a cyclic sugar; the C atom bonded to an —OH group and an —OR group (R is the ring).

Mutarotation Change in rotation of plane-polarized light resulting from the equilibrium between cyclic anomers and the open-chain form of a sugar.

Solved Problem 10.1   The open-chain form of D-altrose, an aldohexose isomer of glucose, has the following structure. Draw D-altrose in its cyclic hemiacetal form.

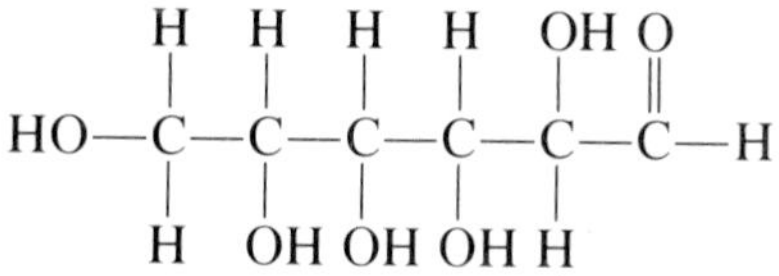

$$HO-C-C-C-C-C-C-H \qquad \text{D-Altrose}$$

with substituents H H H H OH O (top) and H OH OH OH H (bottom)

Solution   First, coil D-altrose into a circular shape by mentally grasping the end farthest from the carbonyl group and bending it backward into the plane of the paper:

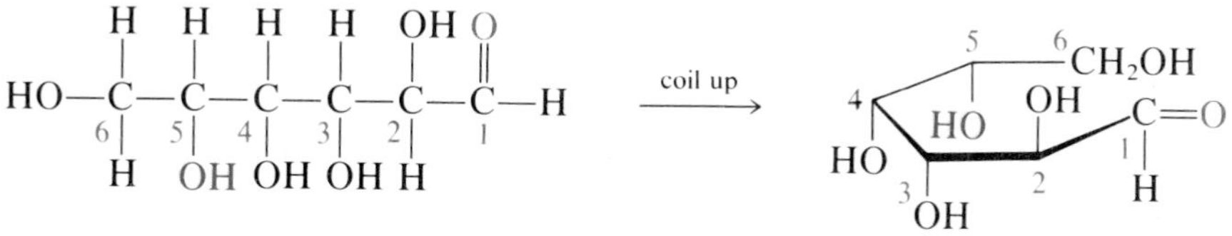

Next, rotate around the single bond between C4 and C5 so that the —CH$_2$OH group at the end of the chain is pointing up and the —OH group on C5 is pointing toward the aldehyde carbonyl group on the right:

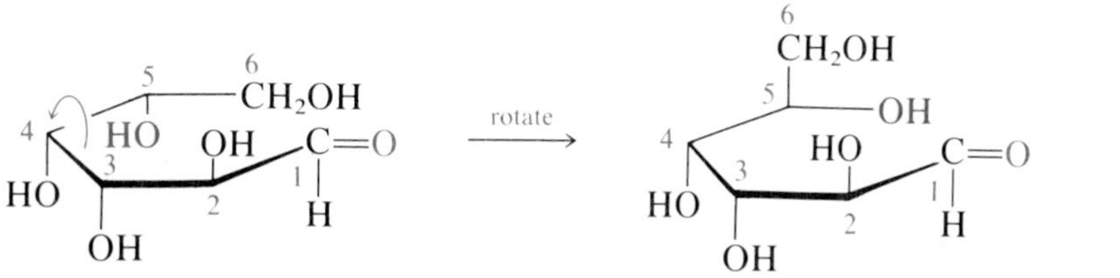

Finally, add the —OH group at C5 to the carbonyl C=O to form a hemiacetal ring. The new —OH group formed on C1 can be either up (β) or down (α).

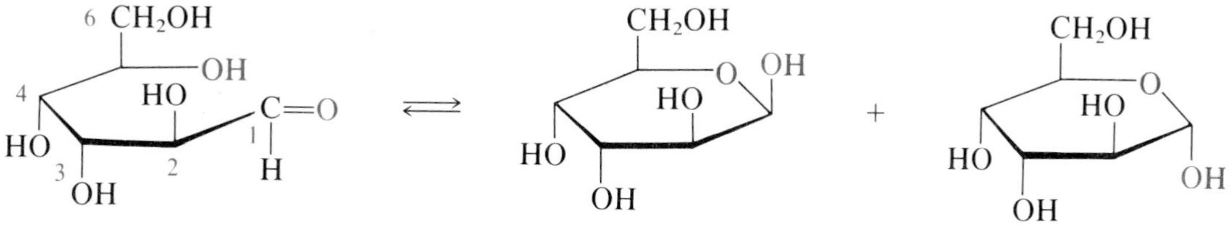

Practice Problem   **10.7**   D-Talose, a constituent of certain antibiotics, has the following open-chain structure. Draw D-talose in its cyclic hemiacetal form.

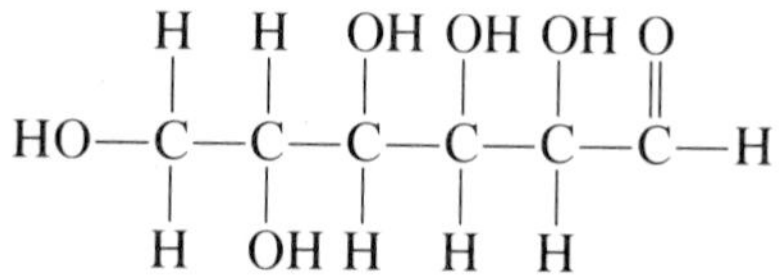

$$HO-C-C-C-C-C-C-H \qquad \text{D-Talose}$$

with substituents H H OH OH OH O (top) and H OH H H H (bottom)

AN APPLICATION: CARBOHYDRATES IN THE DIET

The major monosaccharides in our diets are fructose and glucose from fruits and honey. Sucrose (common table sugar) and lactose from milk are the major disaccharides. In addition, our diets contain large amounts of the digestible polysaccharide starch, present in grains such as wheat and rice, root vegetables such as potatoes, and legumes such as beans and peas. Nutritionists, it should be noted, often refer to polysaccharides as *complex carbohydrates*.

The body's major use of digestible carbohydrates is to provide energy, about 4 kcal per gram of carbohydrate. It's necessary to have some carbohydrate in our daily diet, even though it's not our only source of energy. A small amount of any excess carbohydrate is converted to glycogen for storage, but most dietary carbohydrate in excess of our immediate needs for energy is converted into fat.

As Americans have become increasingly health- and diet-conscious in recent years, many have modified their carbohydrate intake. On the one hand, the desire for weight loss has fostered the development of low-calorie sweeteners to replace sucrose. On the other hand, replacement of meats that have a high fat and high cholesterol content by polysaccharides is viewed as desirable, and a flurry of studies suggesting health benefits has created a desire for "high-fiber" foods.

Cereal boxes now list percentages of "dietary fiber" and even distinguish between "soluble" and "insoluble" fiber. Dietary fiber is the type of polysaccharide that can't be hydrolyzed to monosaccharides and absorbed into the bloodstream. Thus, "fiber" includes not only cellulose, the indigestible polysaccharide in vegetable stalks and leaves, but a variety of noncellulose polysaccharides. For example, pectins, which are present in fruits and provide the "gel" in jelly, contain polymers derived from a polyhydroxy compound with an aldehyde group at one end and a carboxylic acid group at the other [HOOC(CHOH)$_4$CHO]. Pectin and vegetable gums are either soluble or dispersible in water and make up the soluble portion of dietary fiber.

They are often added to prepared foods to retain moisture, to thicken sauces, or to give a creamier texture. Foods high in soluble fiber include fruits, carrots, barley, and oats. Foods high in insoluble fiber include wheat, bran cereals, and brown rice. Beans and peas have both types of fiber.

Fiber functions in the body to soften and add bulk to solid waste. Studies have shown that increased fiber in the diet may reduce the risk of colon and rectal cancer, hemorrhoids, diverticulosis, and cardiovascular disease. Cancer reduction may occur because potential carcinogenic substances are absorbed on fiber surfaces and eliminated before doing any harm. Pectin may also absorb and carry away bile acids, causing an increase in their synthesis from cholesterol in the liver and a resulting decrease in blood cholesterol levels. The mechanisms of action and the interrelated effects of different fibers are far from being understood.

Beans are an excellent source of dietary fiber, both soluble and insoluble.

10.5 SOME IMPORTANT MONOSACCHARIDES

The monosaccharides, with their many opportunities for hydrogen bonding through hydroxyl groups, are generally high-melting, white, crystalline solids that are soluble in water and insoluble in nonpolar solvents. Most monosaccharides (and also disaccharides) are sweet-tasting. The natural monosaccharides of interest in human biochemistry, except for glyceraldehyde (a triose) and fructose (a ketohexose), are aldopentoses and aldohexoses. Most are of the D family and in solution are in equilibrium with both cyclic hemiacetal forms. Glucose, which we've already discussed extensively, is the most important monosaccharide in human metabolism.

Galactose D-Galactose is widely distributed as a constituent of many plant gums and pectins. It's also a component of lactose (milk sugar) and is produced by lactose hydrolysis. Like glucose, galactose is an aldohexose. It's described as an *epimer* of glucose at carbon 4, meaning that it differs from glucose only in having the —OH on this one carbon atom on the opposite side of the ring from that in glucose. Also like glucose, galactose exists in solution as a mixture of cyclic α and β forms along with a small amount of open-chain form. In the body, galactose is converted to glucose to provide energy and is synthesized from glucose for use in lactose for milk and compounds needed in brain tissue.

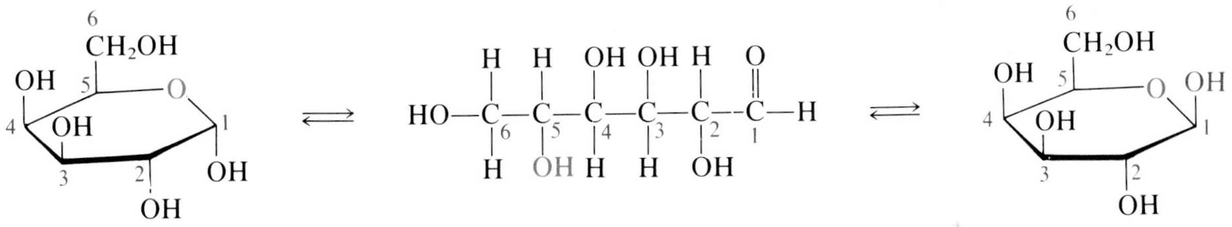

α-D-Galactose   Open-chain D-galactose   β-D-Galactose

Fructose D-Fructose, often called *levulose* or *fruit sugar*, occurs in honey and in a large number of fruits and is one of the monosaccharides in the disaccharide sucrose. Unlike glucose and galactose, fructose is a *keto*hexose: a six-carbon ketone sugar. Like glucose and galactose, however, fructose can exist in solution both in open-chain form and in cyclic forms. The cyclic hemiacetal form of fructose is a five-membered ring rather than a six-membered ring.

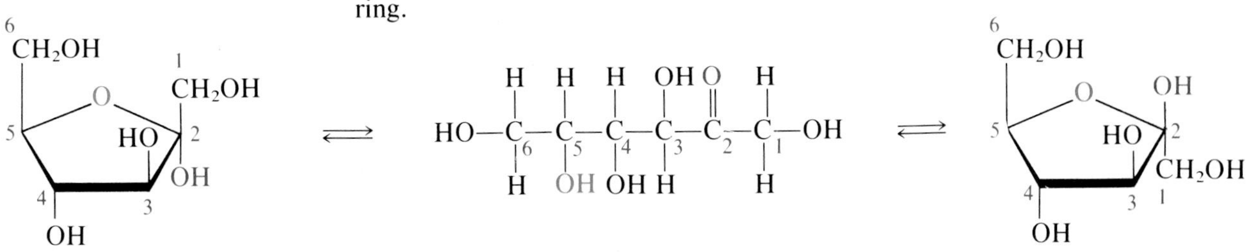

α-D-Fructose   Open-chain D-Fructose   β-D-Fructose

Ribose and Deoxyribose Ribose and its relative deoxyribose are both aldopentoses: five-carbon aldehyde sugars. Both occur in the cells of all living organisms as constituents of the nucleic acids RNA (ribonucleic acid) and DNA (deoxyribonucleic acid), substances that direct the synthesis of proteins and maintain genetic information.

As its name implies, *deoxy*ribose differs from ribose in that it is missing one oxygen atom. Since it's the hydroxyl at C2 that is missing, it's more accurate to name the compound 2-deoxyribose. Both ribose and 2-deoxyribose exist in the usual mixture of open-chain and cyclic hemiacetal forms.

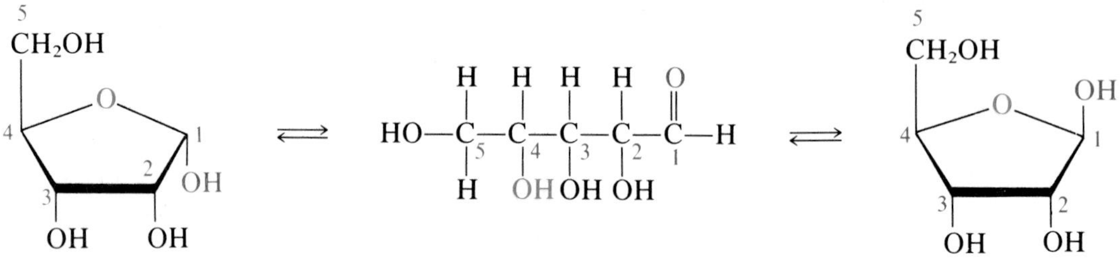

α-D-Ribose   Open-chain D-ribose   β-D-Ribose

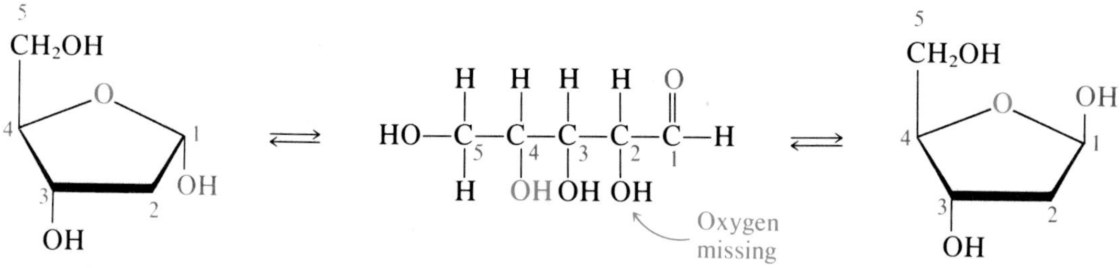

α-D-2-Deoxyribose   Open-chain D-2-deoxyribose   β-D-2-Deoxyribose

Practice Problems   **10.8**   In the following monosaccharide hemiacetal, identify the anomeric carbon atom, number all the carbon atoms, and identify it as the α or β anomer.

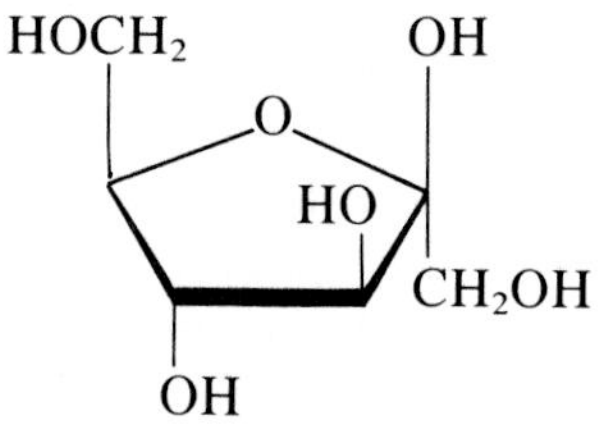

       10.9   Identify the chiral carbons in α-D-fructose, α-D-ribose, and β-D-2-deoxyribose.

10.6 REACTIONS OF MONOSACCHARIDES

Reaction With Oxidizing Agents: Reducing Sugars   You saw in Section 5.5 that aldehydes can be oxidized to yield carboxylic acids (RCHO ⟶ RCOOH). Because they too are aldehydes, aldoses like glucose undergo exactly the same oxidation reaction. Even though only a small amount of open-chain aldehyde form is present at any one time, the hemiacetal ring opening takes place so rapidly that the entire sample is eventually oxidized. Carbohydrates that react with oxidizing agents are called **reducing sugars** because they reduce the oxidizing agent.

Reducing sugar A carbohydrate that reacts with an oxidizing agent such as Benedict's reagent.

In basic solution, ketoses are also reducing sugars because equilibrium is established between the ketose and an *enol* (alk*ene* + alco*hol*), which then further equilibrates with an aldose. The keto form, the enol form, and the aldose form are isomers that differ only in the position of a hydrogen atom. Oxidation of the aldose to an acid drives the equilibrium toward the right, and complete oxidation of the ketose occurs. Thus, *in basic solution all monosaccharides, whether aldoses or ketoses, are reducing sugars.*

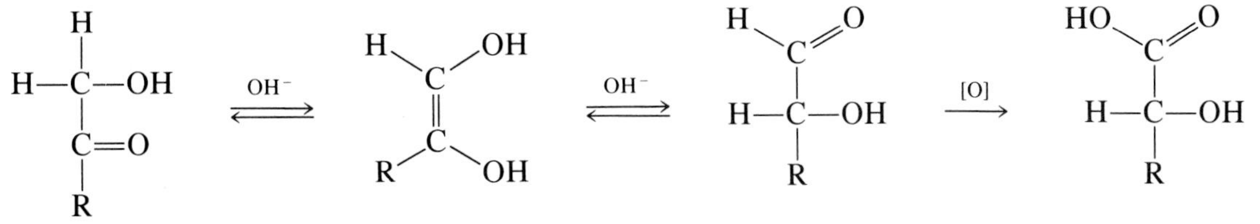

A ketose   An enol   An aldose   A carboxylic acid

Reaction With Alcohols: Glycoside and Disaccharide Formation Hemiacetals, as shown in Section 5.7, react with alcohols with the loss of H_2O to yield acetals, compounds that have two —OR groups bonded to the same carbon:

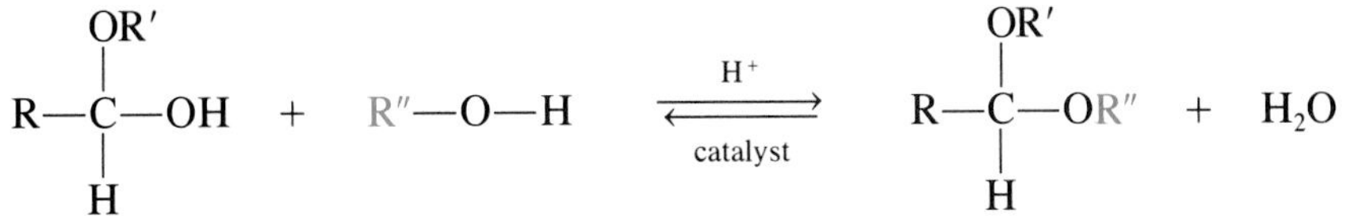

A hemiacetal   An alcohol   An acetal

Glycoside A cyclic acetal formed by reaction of a monosaccharide with an alcohol, accompanied by loss of H_2O.

Because glucose and other monosaccharides are cyclic hemiacetals, they react with alcohols to yield cyclic acetals called **glycosides.** In a glycoside, the —OH group on the anomeric carbon atom is replaced by an —OR group. For example, glucose reacts with methanol to produce methyl glucoside, which has an —OCH_3 group on carbon 1. (Note that a *glucoside* is a cyclic acetal formed by glucose. A *glycoside* is a cyclic acetal derived from *any* sugar.)

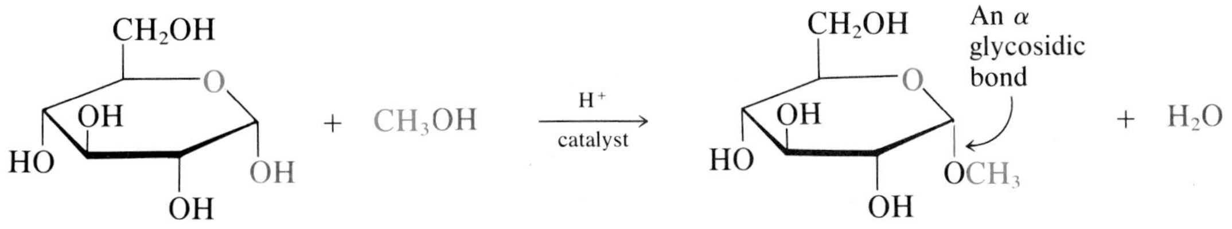

α-D-Glucose   Methyl α-D-glucoside, an acetal

Glycosidic bond Bond between the anomeric carbon atom of a monosaccharide and an —OR group.

The bond between the anomeric carbon atom of the monosaccharide and the oxygen atom of the —OR group is called a **glycosidic bond.** Such bonds may be either α or β according to the same definitions for monosaccharide anomers: α points below the ring and β points above the ring when the ring is drawn as described in Section 10.4. Since acetals are stable and not in equilibrium with an open-chain form, acetals like methyl α-D-glucoside are not reducing sugars.

When a monosaccharide forms a glycosidic bond to one of the —OH groups of a second monosaccharide, a disaccharide is produced. To describe the link between the two monosaccharides, the α or β orientation of the glycosidic bond (shown in red) and the numbers of the connected carbon atoms are specified. For example, the following two general structures show a **1,4 link,** a link between C1 of one monosaccharide and C4 of the second monosaccharide.

1,4 Link   An acetal link between the hydroxyl group at C1 of one sugar and the hydroxyl group at C4 of another sugar.

An α-1,4 disaccharide   A β-1,4 disaccharide

In some biomolecules, such as the glycoproteins (Figure 10.5) found in cell membranes, a carbohydrate and a protein are joined by a glycosidic bond to nitrogen:

Glycosidic bond in a glycoprotein

Figure 10.5
Glycoproteins at cell surfaces. At the top in the photo is the innermost layer of the small intestine (known as the *glycocalyx*). The layer consists of polysaccharide chains extending out from cell membrane proteins to which they are bonded. Such polysaccharide chains are associated with enzymes that aid in the digestion of food macromolecules.

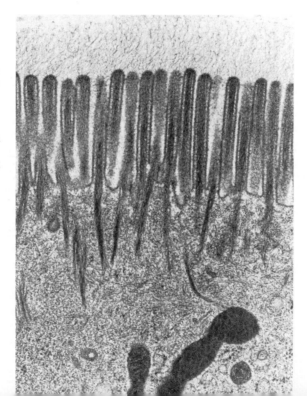

AN APPLICATION: GLUCOSE IN BLOOD AND URINE

Glucose measurements are essential in the diagnosis of *diabetes mellitus* and in the management of diabetic patients, either in a clinical setting or on a day-to-day basis by the patients themselves. One form of diabetes results when an individual has an insufficient supply of the hormone insulin. Without insulin, glucose does not move from the bloodstream into cells, where it is needed as a source of energy. As a result, glucose levels rise in blood and urine.

Most tests for glucose in urine or blood rely on detecting a color change that accompanies the oxidation of glucose. Because glucose and its oxidation product, gluconic acid, are colorless, the oxidation must be tied chemically to the color change of a suitable indicator.

H C=O		HO C=O
H—C—OH		H—C—OH
HO—C—H	$\xrightarrow{\text{oxidizing agent}}$	HO—C—H
H—C—OH		H—C—OH
H—C—OH		H—C—OH
CH$_2$OH		CH$_2$OH
D-Glucose		D-Gluconic acid

Benedict's reagent is one of the oxidizing agents mentioned in Section 5.5 as a test for aldehydes. The detectable color change with Benedict's reagent is the reduction of blue copper(II) ion to a brick-red precipitate of copper(I) oxide. Tablets containing the necessary Benedict's test chemicals in solid form (Clinitest tablets) are dropped into a diluted urine sample, and the color of the resulting mixture is compared to a standard chart to estimate glucose concentration. Unfortunately, Benedict's test is flawed because it is positive for any reducing sugar, for example, lactose in the urine of a pregnant woman. Although Clinitest tablets are still available in drugstores for home use by diabetics, there are now better methods.

Modern methods for glucose detection rely on the action of an enzyme specific to glucose, *glucose oxidase*. The reagent mixture includes a second enzyme called a *peroxidase* that catalyzes the reaction of hydrogen peroxide (H_2O_2) with a dye that gives a detectable color change.

$$\text{Glucose} + O_2 \xrightarrow{\text{glucose oxidase}} \text{gluconic acid} + H_2O_2$$

$$H_2O_2 + \begin{array}{c}\text{reduced dye}\\\text{(colorless)}\end{array} \xrightarrow{\text{peroxidase}}$$

$$H_2O + \begin{array}{c}\text{oxidized dye}\\\text{(colored)}\end{array}$$

The glucose oxidase test is available for urine and blood. Increasingly, diabetic individuals are monitoring their glucose levels in blood, often with a modestly priced instrument that reads the color change electronically. The blood test is desirable because it is more specific and it detects rising glucose levels earlier than the urine test.

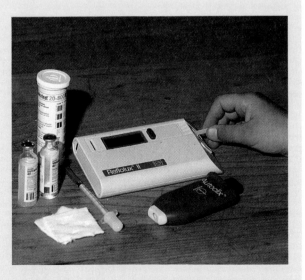

Equipment used by a diabetic individual for monitoring blood glucose and administering insulin.

Practice Problem 10.10 Look at the structure of α-D-galactose (Section 10.5) and write the two glyco-side products you would expect to obtain by reaction of galactose with methanol. Are these compounds reducing sugars?

10.7 SOME IMPORTANT DISACCHARIDES

Three of the most important naturally occurring disaccharides, maltose, lactose, and sucrose, illustrate three different ways monosaccharides can be linked together.

Maltose Maltose, often called *malt sugar*, is present in fermenting grains and can be prepared in about 80% yield by enzyme-catalyzed degradation of starch. In the body, it is produced in starch digestion. Two α-D-glucose molecules are joined in maltose by an α-1,4 link. A careful look at maltose shows that it is both an acetal and a hemiacetal. The α-D-glucose on the left contains an acetal grouping (C1 is bonded to two —OR groups), while that on the right contains a hemiacetal (C1 is bonded to an —OH and an —OR). Since the acetal ring on the left doesn't open and close spontaneously, the bond linking the two glucose units in maltose is not easily cleaved. The hemiacetal group of the glucose unit on the right, however, establishes equilibrium with the aldehyde, making maltose a reducing sugar.

Maltose

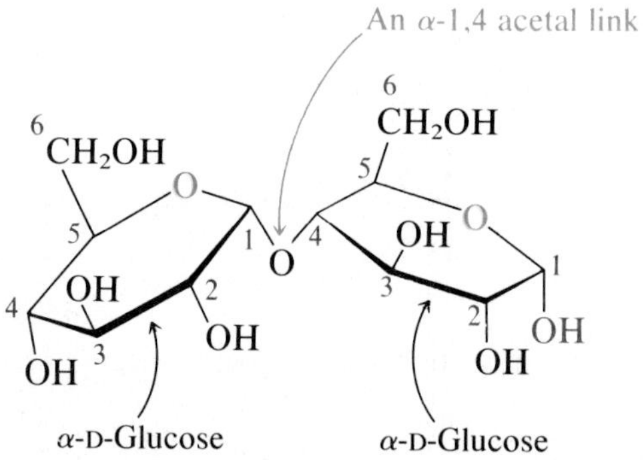

Practice Problem 10.11 Draw maltose in the form in which the glucose unit on the right is an open-chain aldehyde.

Lactose Lactose, or *milk sugar*, is the major carbohydrate present in mammalian milk. Human milk, for example, is about 7% lactose. Structurally, lactose is a disaccharide composed of galactose and glucose. The two sugars are joined in lactose by a β-1,4 acetal link between C1 of β-D-galactose and C4 of β-D-glucose. Like maltose, lactose is a reducing sugar because one glucose ring (on the right in the following structure) is a hemiacetal that is in equilibrium with an oxidizable aldehyde.

A β-1,4 acetal link

6
CH₂OH

6
CH₂OH

4

5

OH O

OH

Lactose

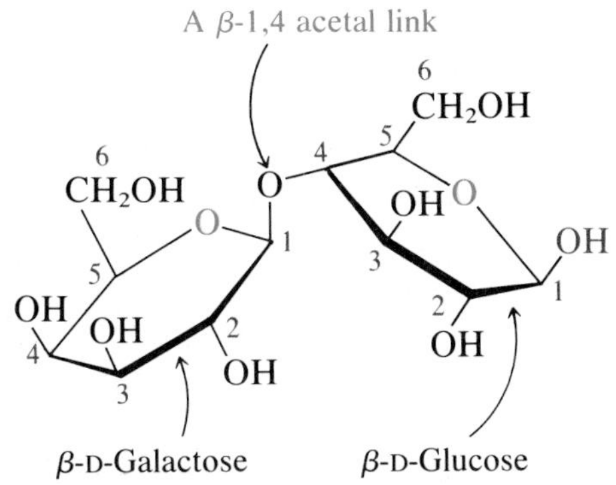

5

O

1

3

2

2

OH

1

OH

OH

4

OH

3

OH

β-D-Galactose

β-D-Glucose

Sucrose   Sucrose—plain table sugar—is probably the most common pure organic chemical in the world. Although it's found in many plants, sugar beets (20% by weight) and sugarcane (15% by weight) are the most common sources of sucrose. Hydrolysis of sucrose yields one molecule of glucose and one molecule of fructose. The 50:50 mixture of glucose and fructose that results, often referred to as *invert sugar*, is commonly used as a food additive because it's sweeter than sucrose.

Sucrose differs from maltose and lactose in that it has no hemiacetal group because a 1,2 link joins *both* anomeric carbon atoms. Since it has no hemiacetal group that opens to expose a free aldehyde, sucrose is not a reducing sugar.

6
CH₂OH

5

O

α-D-Glucose

4

OH

1

OH₃

2

Sucrose

OH O

HOCH₂

O

β-D-Fructose

5

HO

4

3

2

CH₂OH

OH

1

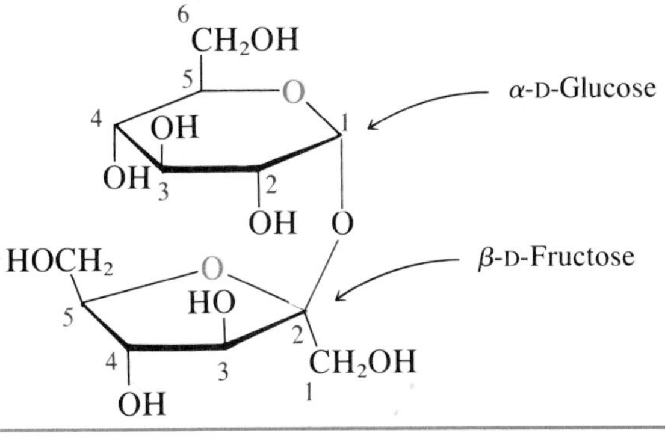

Practice Problems **10.12** The disaccharide cellobiose can be obtained by enzyme-catalyzed hydrolysis of cellulose. Would you expect cellobiose to be a reducing or a nonreducing sugar? Explain.

CH₂OH

CH₂OH

O

OH O

Cellobiose

O

OH

OH

OH

OH

OH

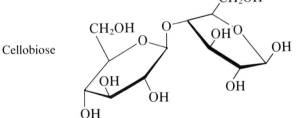

10.13 How would you classify the link between the monosaccharides in cellobiose?

10.14 Show the structures of the two monosaccharides that are formed on hydrolysis of cellobiose (Practice Problem 10.12). What are their names?

AN APPLICATION: CELL SURFACE CARBOHYDRATES AND BLOOD TYPE

It was discovered more than 80 years ago that human blood can be classified into four blood group types, called A, B, AB, and O. If a transfusion becomes necessary, blood from a donor of one type can't be given to a recipient with blood of another type unless the two types are compatible. If an incompatible mix is made, the red blood cells clump together, or *agglutinate*, and death can result. Agglutination indicates that the recipient's immune system has recognized foreign cells in the body and has formed antibodies to them.

Cells of types A, B, and O each have on their surfaces characteristic structural features called *antigenic determinants*, which can provoke an immune response that results in the production of antibodies. Cells of type AB have both A and B markers. As shown at the bottom of this box, the structures of the three blood group determinants are known. The marker for blood group O is a trisaccharide whose constituent sugars are common to all three types, whereas the markers for blood groups A and B have one additional sugar unit.

Blood to be used for transfusion must be labeled by type so that compatibility with the recipient's blood type can be assured.

Human Blood Group Compatibilities

Donor Blood Type	Acceptor Blood Type			
	A	B	AB	O
A	o	x	o	x
B	x	o	o	x
AB	x	x	o	o
O	o	o	o	o

o = compatible; x = incompatible

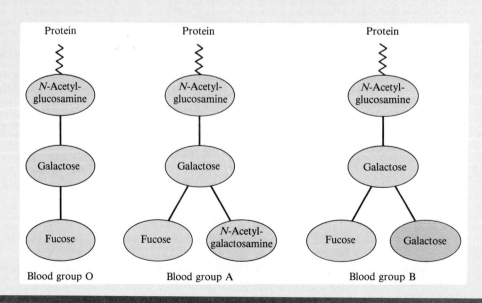

Blood group O   Blood group A   Blood group B

10.8 SOME IMPORTANT POLYSACCHARIDES

Polysaccharides have tens, hundreds, or even thousands of monosaccharides linked together through glycosidic bonds of the same sort as in maltose and lactose. Three of the most important polysaccharides are cellulose, starch, and glycogen.

Cellulose   Cellulose (Figure 10.6), the fibrous substance used by plants as a structural material in the cell walls of leaves, stems, and tree trunks, consists entirely of several thousand β-D-glucose units joined by β-1,4 links to form one large molecule. Cows and termites are able to digest cellulose because microorganisms living in their digestive tracts produce enzymes that hydrolyze the β acetal bonds. Humans, however, can't hydrolyze cellulose.

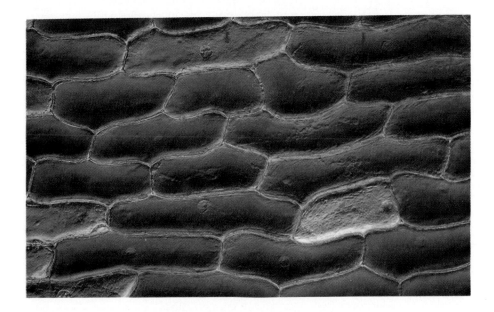

Starch   Starch, like cellulose, is a polymer of glucose. There is, however, a big difference between the two because starch, unlike cellulose, can be digested by human beings. Indeed, the starch in such vegetables as beans, wheat, rice, and potatoes is an essential part of the human diet. Structurally, starch differs from cellulose in that its individual glucose units are joined by α-1,4 links, as in maltose, rather than by the ß-1,4 links of cellulose (Figure 10.7).

Figure 10.6
Cellulose structure in the skin of a red onion.

Figure 10.7
Cellulose and the amylose form of starch. These structures represent the molecular shapes more correctly than the flat rings by showing the six-membered glucose rings in their chair forms. Note the difference between the α-1,4 links in amylose and the β-1,4 links in cellulose. In cellulose several thousand β-D-glucose units are joined by 1,4 acetal links between C1 of one sugar and C4 of its neighbor.

(a) Amylose
(in starch)

(b) Cellulose

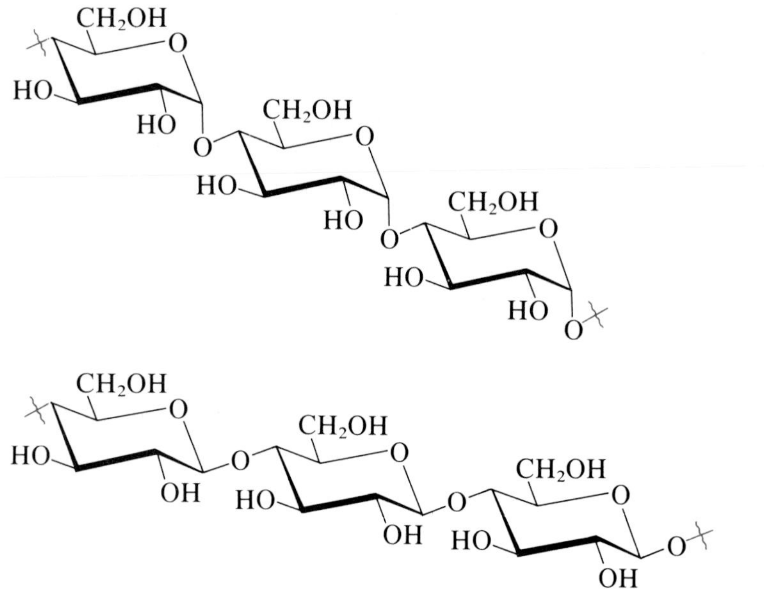

Unlike cellulose, which has only one form, there are two kinds of starch, called *amylose* and *amylopectin*. Amylose, which accounts for about 20% of starch, consists of several hundred to a thousand α-D-glucose units linked together in a long chain by α-1,4 acetal bonds (Figure 10.8a). Amylopectin, which accounts for about 80% of starch, is similar to amylose but is much larger (up to 100,000 glucose units per molecule) and has α-1,6 branches approximately every 25 units along its chain. A glucose molecule at one of these branch points uses *two* of its hydroxyl groups (those at C4 and C6) to form acetal links to two other sugars (Figure 10.8b).

When eaten, starch molecules are digested mainly in the small intestine by α-amylase, which catalyzes hydrolysis of the α-1,4 links so that the starch chain is broken down. As is usually the case in enzyme-catalyzed reactions, α-amylase is highly specific in its action. It hydrolyzes only α acetal links between monosaccharides (as in starch) while leaving β acetal links (as in cellulose) untouched. Thus, starch is easily digested, but cellulose is unaffected by these digestive enzymes.

Figure 10.8
The glucose polymers in starch. Amylose consists only of linear chains of α-D-glucose units linked by 1,4 acetal bonds, whereas amylopectin has branch points about every 25 sugars in the chain. A glucose unit at a branch point uses two of its hydroxyls (at C4 and C6) to form 1,4 and 1,6 acetal links to two other sugars.

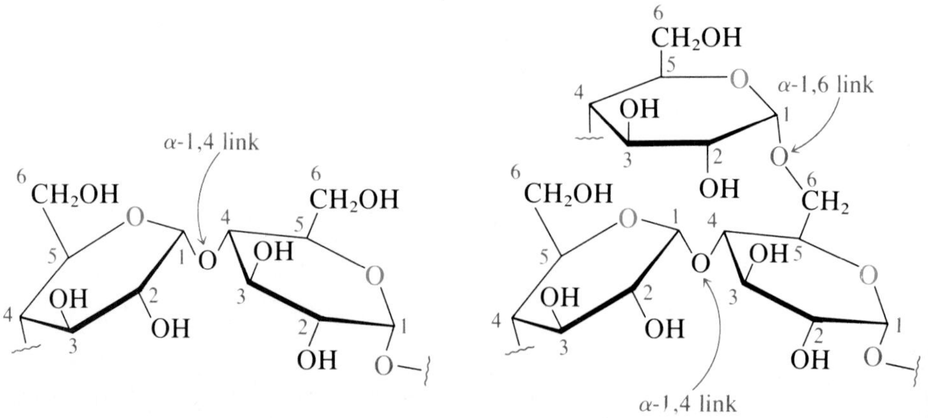

(a) Amylose

(b) Amylopectin

Practice Problem 10.15 An individual starch molecule contains thousands of glucose units but has only a single hemiacetal group at the end of the long polymer chain. Would you expect starch to be a reducing carbohydrate? Explain.

Glycogen Glycogen, sometimes called *animal starch,* serves the same food storage role in animals that starch serves in plants. After we eat starch and the body breaks it down into simple glucose units, some of the glucose is used immediately as fuel and some is stored in the body as glycogen for later use. Any additional excess is stored as fat.

Structurally, glycogen is similar to amylopectin in being a long polymer of α-D-glucose with branch points in its chain. Glycogen has many more branches than amylopectin, however, and is much larger: up to 1 million glucose units per molecule (Figure 10.9).

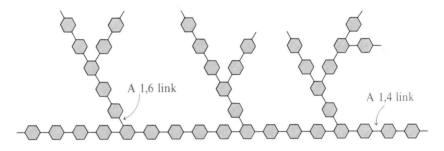

Figure 10.9
The structure of glycogen. The hexagons represent α-D-glucose units linked by 1,4 acetal bonds and (at branch points) 1,6 acetal bonds.

SUMMARY

Carbohydrates are polyhydroxy aldehydes and ketones. They are classified according to the number of carbon atoms and the kind of carbonyl group they contain. Glucose, for example, is an *aldohexose*—a six-carbon aldehyde sugar. **Monosaccharides** such as glucose can't be hydrolyzed to smaller molecules. All monosaccharides are chiral, and the naturally occurring optical isomers belong to the D **family.** Monosaccharides of the L **family** occur much more rarely in nature. **Polysaccharides** such as starch and cellulose contain many monosaccharides linked together.

Monosaccharides such as glucose, galactose, fructose, ribose, and 2-deoxyribose exist as a mixture of open-chain and cyclic hemiacetal forms in equilibrium with each other. There are two cyclic hemiacetal forms, called the α **form** and the β **form,** which differ in the orientation of the hemiacetal hydroxyl group, that is, the hydroxyl group on the **anomeric carbon.** Monosaccharides undergo many of the same reactions that other aldehydes do. Thus, they react with alcohols to yield cyclic acetals called **glycosides,** and they react with oxidizing agents to yield carboxylic acids. Sugars that can be oxidized in this way are called **reducing sugars.** All monosaccharides are reducing sugars because they establish equilibria with aldehyde forms that can be oxidized.

Disaccharides such as maltose, lactose, and sucrose contain two simple sugars joined by an acetal link classified according to whether the **glycosidic bond** is α or β and according to the carbon atoms it connects. Important polysaccharides include cellulose, starch, and glycogen. Cellulose is a linear polymer of up to 1000 glucose units bonded together by β**-1,4 acetal links.** Starch and glycogen are glucose polymers in which the individual sugar units are connected by α acetal links. The fraction of starch called amylose consists of linear chains, and the fraction called amylopectin has branched chains.

INTERLUDE: SWEETNESS

Mention the word *sugar* to most people and they'll automatically think of sweet-tasting foods or candies. In fact, most simple mono- and disaccharides *do* taste sweet, although the degree of sweetness varies from one compound to another. Using sucrose (table sugar) as a reference point, fructose is nearly twice as sweet, whereas glucose, galactose, lactose, and others are much less so. Exact comparisons between sugars are impossible, because sweetness is not a physical property and can't be accurately measured. Sweetness is simply a taste perception, and the ranking of different sugars is a matter of personal opinion. Nevertheless, the ordering shown in the accompanying table is generally agreed on.

Dietary concerns and the desire of many people to cut their caloric intake have led to widespread use of the artificial sweeteners aspartame, saccharin, and cyclamate. All are far sweeter than natural sugars, but doubts have been raised as to the long-term safety of all three. As their structures indicate, none of the three bear the slightest chemical resemblance to carbohydrates.

Relative Sweetness of Some Sugars and Sugar Substitutes

Name	Type	Sweetness
Lactose	Disaccharide	16
Galactose	Monosaccharide	30
Maltose	Disaccharide	33
Glucose	Monosaccharide	75
Sucrose	Disaccharide	100
Fructose	Monosaccharide	175
Cyclamate	Artificial	3,000
Aspartame	Artificial	15,000
Saccharin	Artificial	35,000

Aspartame

Saccharin

Sodium cyclamate

Sugarcane contains a high concentration of the disaccharide sucrose. The stems of the plant derive much of their strength and stiffness from another carbohydrate that is not digestible by humans: the polysaccharide cellulose.

REVIEW PROBLEMS

Classification and Structure of Carbohydrates

10.16* What is a carbohydrate?

10.17 What is the family-name ending for a sugar?

10.18* What is the structural difference between an aldose and a ketose?

10.19 Classify each of the following carbohydrates by indicating the nature of its carbonyl group and the number of carbon atoms present. For example, glucose is an aldohexose.

(a) 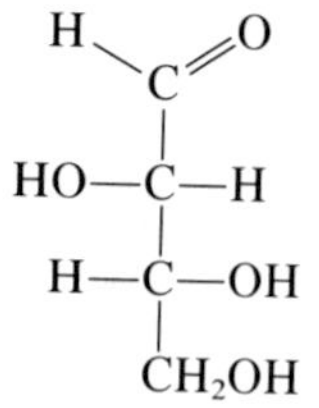
```
      H    O
       \  //
        C
   HO—C—H
    H—C—OH
       CH₂OH
```

(b) 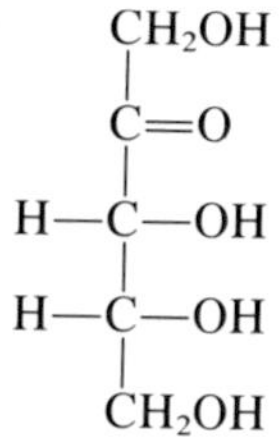
```
      CH₂OH
       |
      C=O
    H—C—OH
    H—C—OH
       CH₂OH
```

Threose   Ribulose

(c) 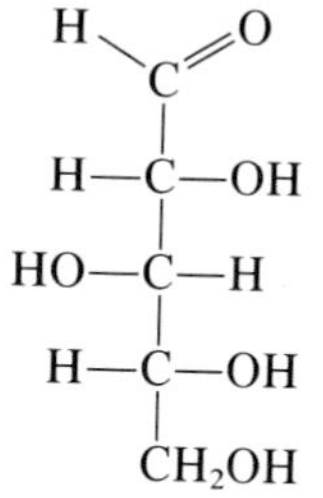
```
      H    O
       \  //
        C
    H—C—OH
   HO—C—H
    H—C—OH
       CH₂OH
```

(d) 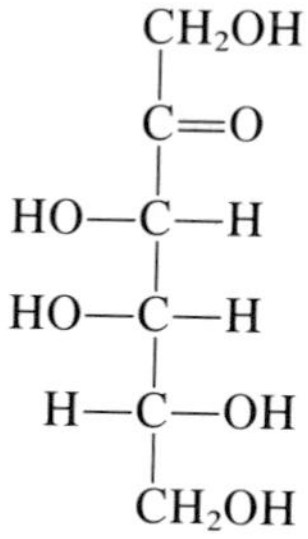
```
      CH₂OH
       |
      C=O
   HO—C—H
   HO—C—H
    H—C—OH
       CH₂OH
```

Xylose   Tagatose

10.20 Write the open-chain structure of a ketotetrose.

10.21 Write the open-chain structure of a four-carbon deoxy sugar.

10.22* Give the names of three important monosaccharides and tell where each occurs in nature.

10.23 "Dextrose" is a alternative name for what sugar?

Handedness in Carbohydrates

10.24* What is the difference between enantiomers and diastereomers?

10.25 What is the structural relationship of L-glucose to D-glucose?

10.26* Would you expect L-glucose to be a good food source in the human diet in the same way that D-glucose is?

10.27 Only three stereoisomers are possible for 2,3-dibromo-2,3-dichlorobutane. Draw them, indicating which pair are enantiomers (optical isomers). Why does the other isomer not have an enantiomer?

10.28* There are two D-aldotetroses, whose structures are shown. One of the two reacts with NaBH₄ to yield a chiral product, but the other yields an achiral product. Explain.

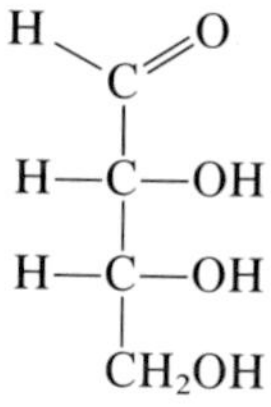

```
      H    O              H    O
       \  //               \  //
        C                   C
    H—C—OH             HO—C—H
    H—C—OH              H—C—OH
       CH₂OH               CH₂OH

   D-Erythrose          D-Threose
```

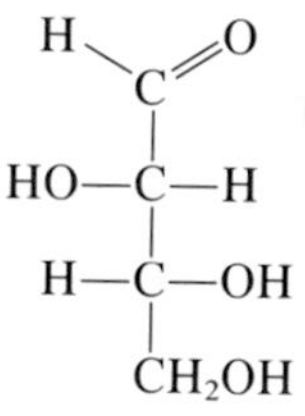

10.29 Draw the enantiomer of each molecule in Problem 10.19. Label each as a D or an L sugar.

Reactions of Carbohydrates

10.30* What does the term *reducing sugar* mean?

10.31 What is the structural difference between the α hemiacetal form of a carbohydrate and the β form?

10.32* D-Gulose, an aldohexose isomer of glucose, has the following cyclic structure. Which is shown, the α form or the β form?

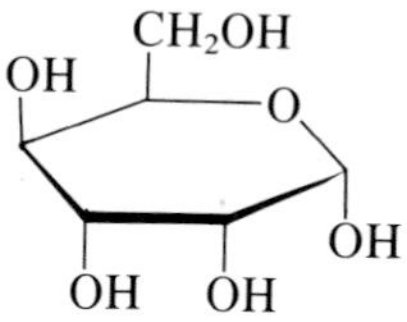

D-Gulose

10.33 Draw D-gulose (Problem 10.32) in its open-chain aldehyde form, both coiled and uncoiled.

10.34 D-Mannose, an aldohexose found in orange peels, has the following structure in open-chain form. Coil mannose around and draw it in cyclic hemiacetal α and β forms.

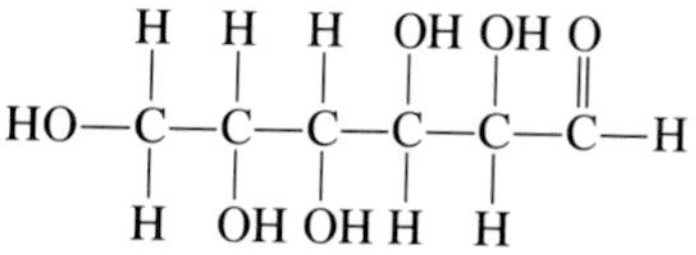

```
        H  H  H  OH OH  O
        |  |  |  |  |   ‖
  HO—C—C—C—C—C—C—H      D-Mannose
        |  |  |  |  |
        H  OH OH H   H
```

10.35 D-Ribulose, a ketopentose related to ribose, has the following structure in open-chain form. Coil ribulose around and draw it in its five-membered cyclic β hemiacetal form.

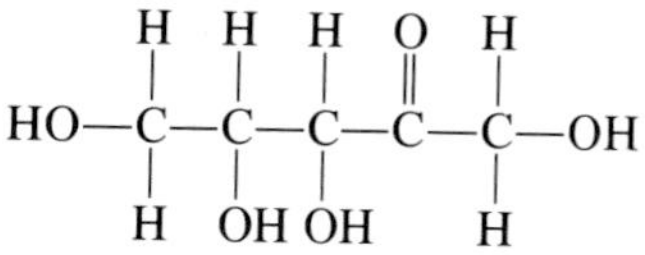

```
        H  H  H  O   H
        |  |  |  ‖   |
  HO—C—C—C—C—C—OH     D-Ribulose
        |  |  |       |
        H  OH OH      H
```

10.36 D-Allose, an aldohexose, is identical with D-glucose except that the hydroxyl group at C3 points down rather than up in the cyclic hemiacetal form. Draw the β form of this cyclic form of D-allose.

10.37 Draw D-allose (Problem 10.36) in its open-chain form.

10.38 We saw in Section 5.5 that aldehydes react with reducing agents like NaBH₄ to yield primary alcohols (RCH=O → RCH₂OH). Treatment of D-glucose with NaBH₄ yields *sorbitol*, a substance used as a sugar substitute by diabetics. Draw the structure of sorbitol.

10.39 Reduction of D-fructose with NaBH₄ (Problem 10.38) yields a mixture of D-sorbitol along with a second, isomeric product. What is the structure of the second product?

10.40* Refer to Problems 10.38 and 10.39 and explain why two products result from the reduction of D-fructose, while only one results from reduction of D-glucose.

10.41 Treatment of an aldose with an oxidizing agent like Tollens' reagent (Section 5.5) yields a carboxylic acid. Gluconic acid, the product of glucose oxidation, is used as its magnesium salt for the treatment of magnesium deficiency. Draw the structure of gluconic acid.

10.42* What is the structural difference between a hemiacetal and an acetal?

10.43 What are glycosides, and how can they be formed?

10.44 Look at the structure of D-mannose (Problem 10.34) and draw the two glycosidic products that you would expect to obtain by reacting D-mannose with methanol.

10.45 Show two D-mannose molecules (Problem 10.34) attached by an α-1,4 glycosidic linkage.

Disaccharides and Polysaccharides

10.46* Give the names of three important disaccharides. Tell where each occurs in nature. From which two monosaccharides is each made?

10.47 Starch and cellulose are both polymers of glucose. What is the main structural difference between them, and what different roles do they serve in nature?

10.48* How are amylose and amylopectin similar and how are they different?

10.49 Starch and glycogen are both α-linked polymers of glucose. What is the structural difference between them, and what different roles do they serve in nature?

10.50* Lactose and maltose are reducing disaccharides, but sucrose is a nonreducing disaccharide. Explain.

10.51 Gentiobiose, a rare disaccharide found in saffron, has the following structure. What simple sugars would you obtain on hydrolysis of gentiobiose?

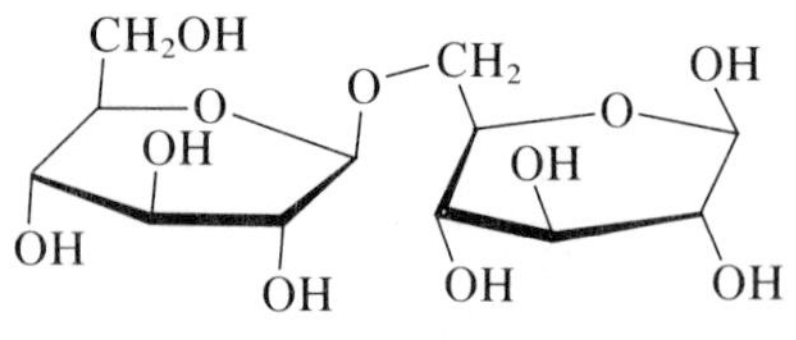

Gentiobiose

10.52* Look carefully at the structure of gentiobiose (Problem 10.51). Does gentiobiose have an acetal grouping? A hemiacetal grouping? Would you expect gentiobiose to be a reducing or a nonreducing sugar? How would you classify the linkage (α or β and carbon numbers) between the two monosaccharides?

10.53 Trehalose, a disaccharide found in the blood of insects, has the following structure. What simple sugars would you obtain on hydrolysis of trehalose?

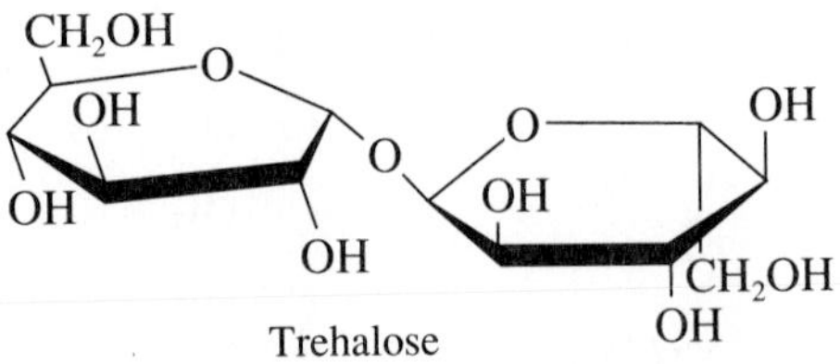

Trehalose

10.54* Does trehalose (Problem 10.53) have an acetal grouping? A hemiacetal grouping? Would you expect trehalose to be a reducing or a nonreducing sugar?

10.55 Amygdalin, or Laetrile, is a glycoside isolated in 1830 from almond and apricot seeds. It is called a *cyanogenic glycoside* because hydrolysis with aqueous acid liberates hydrogen cyanide (HCN) along with benzaldehyde and two molecules of glucose. Structurally, amygdalin is a glycoside between the hemiacetal, gentiobiose (Problem 10.51), and the alcohol, mandelonitrile. Draw the structure of amygdalin.

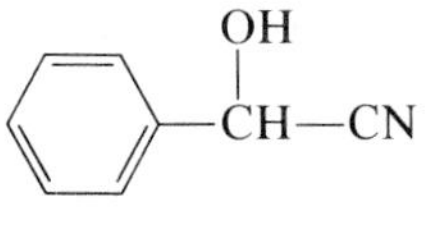

Mandelonitrile

Applications

10.56* What is an optically active compound? [App: Polarized Light and Optical Activity]

10.57 What does a polarimeter measure? [App: Polarized Light and Optical Activity]

10.58* Sucrose and D-glucose rotate plane-polarized light to the right; D-fructose rotates light to the left. When sucrose is hydrolyzed, the glucose–fructose mixture rotates light to the left. (a) What does this indicate about the relative degrees of rotation of light of glucose and fructose? (b) Why do you think that the mixture is called invert sugar? [App: Polarized Light and Optical Activity]

10.59 What generalization can you make about the direction and degree of rotation of light by enantiomers? [App: Polarized Light and Optical Activity]

10.60* Our bodies don't have the enzyme required to digest cellulose, yet it is a necessary addition to a healthy diet. Why? [App: Carbohydrates in the Diet]

10.61 What are two types of soluble fiber and two sources of this type of fiber? [App: Carbohydrates in the Diet]

10.62* What hormone deficiency gives rise to diabetes? What purpose does this hormone serve? [App: Glucose in the Blood and Urine]

10.63 Why is Benedict's test not infallible in testing for glucose? [App: Glucose in the Blood and Urine]

10.64 Briefly describe the enzymatic process for determination of glucose. [App: Glucose in the Blood and Urine]

10.65 Describe what happens when incompatible blood types are mixed. Why does this occur? [App: Cell Surface Carbohydrates]

10.66* What is the role of an antigenic determinant on blood cell? [App: Cell Surface Carbohydrates]

10.67 Look at the structures of the blood group antigenic determinants. What groups do all blood types have in common? Why do you think that blood type O is the "universal donor"? [App: Cell Surface Carbohydrates]

10.68* Sugar substitutes have caloric value, yet are still prescribed for diabetics and for lowering dietary calorie consumption. Look at the sweetness table and propose a justification for this substitution. [Int: Sweetness]

Additional Problems

10.69 What is the relationship between D-ribose and L-ribose? What generalizations can you make about D-ribose and L-ribose with respect to (a) melting point (b) rotation of plane-polarized light (c) density (d) solubility in water (e) chemical reactivity?

10.70* What is the relationship between D-ribose and D-xylose (Problem 10.19). What generalizations can you make about D-ribose and D-xylose with respect to (a) melting point (b) rotation of light (c) density (d) solubility in water (e) chemical reactivity?

10.71 L-Sorbose, which is used in the commercial production of vitamin C, differs from D-fructose only at carbon 5. Draw the open-chain structure of D-sorbose.

10.72 D-Fructose can form a six-membered cyclic hemiacetal as well as the more prevalent five-membered cyclic form. Draw the α isomer of D-fructose in the six-membered ring.

10.73 Are the α and β forms of monosaccharides enantiomers? Why or why not?

10.74 Raffinose, found in sugar beets, is the most prevalent trisaccharide. It is formed by an α-1,6 linkage of D-galactose to the glucose portion of sucrose. Draw the structure of raffinose.

10.75 Does raffinose (Problem 10.74) have a hemiacetal grouping? An acetal grouping? Is raffinose a reducing sugar?

10.76* When you chew a cracker for several minutes, it begins to taste sweet. What do you think that the saliva in your mouth does to the starch in the cracker?

CHAPTER

11

Carbohydrate Metabolism

The tiny surface projections (villi), where digested food is absorbed, are clearly visible in this cross section of the small intestine. In this chapter we'll begin the story of what happens to food molecules during digestion.

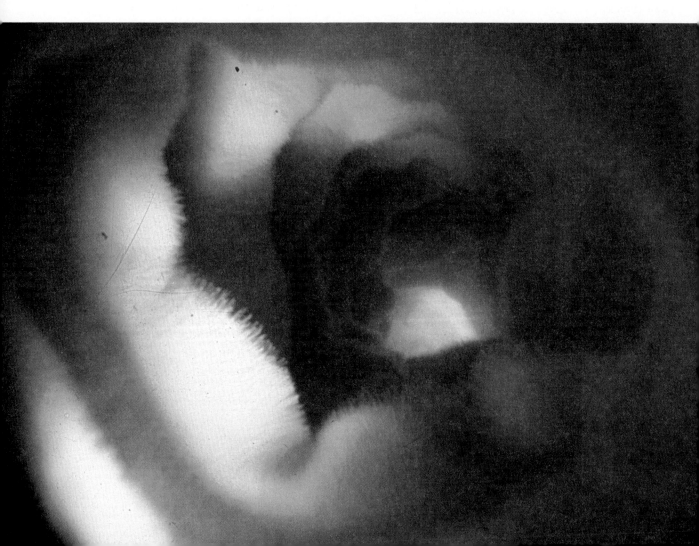

The story of carbohydrate metabolism is essentially the story of glucose: how it is broken down to pyruvate for entrance into the citric acid cycle, how it is stored and then released, and how it is synthesized when carbohydrates are in short supply. Because of the importance of glucose in metabolism, the body has several alternative strategies for maintaining glucose's concentration in blood and providing it to cells that depend on it. In this chapter, we'll answer the following questions about carbohydrate metabolism:

1. **What happens during digestion of carbohydrates?** The goal: Be able to describe where carbohydrates are digested and what the major products are.

2. **What are the major pathways in the metabolism of glucose?** The goal: Be able to identify the alternative pathways available for the synthesis and breakdown of glucose and describe their interrelationships.

3. **What is glycolysis?** The goal: Be able to list the major products of glycolysis and give an overview of the pathway.

4. **What happens to pyruvate once it is formed?** The goal: Be able to describe when each of the alternative pathways occurs and what its product is.

5. **What is the result of complete catabolism of glucose?** The goal: Be able to list the major products of the catabolism of glucose and explain what determines the total amount of ATP produced.

6. **Which hormones influence glucose metabolism?** The goal: Be able to name three hormones that influence glucose metabolism and describe their roles.

7. **What are the roles of glycogen and the pentose phosphate pathway in metabolism?** The goal: Be able to describe the anabolism and catabolism of glycogen, and the major products of the pentose phosphate pathway.

11.1 DIGESTION OF CARBOHYDRATES

Digestion A general term for the breakdown of food into small molecules.

The first stage in catabolism, previously summarized in Figure 9.6, is known as **digestion,** a catch-all term used to describe the breakdown of bulk food into individual small molecules. Digestion entails the physical grinding, softening, and mixing of food, as well as the enzyme-catalyzed hydrolysis of carbohydrates, proteins, and fats. Digestion begins in the mouth when foods are chewed; it continues in the stomach; and it concludes in the small intestine (Figure 11.1a).

The products of food digestion—glucose, fatty acids, glycerol, and amino acids—are mostly relatively small molecules that are absorbed from the intestinal tract. The absorption happens through millions of tiny hair-like projections called *microvilli* (Figure 11.1b) that provide a total surface area as big as a

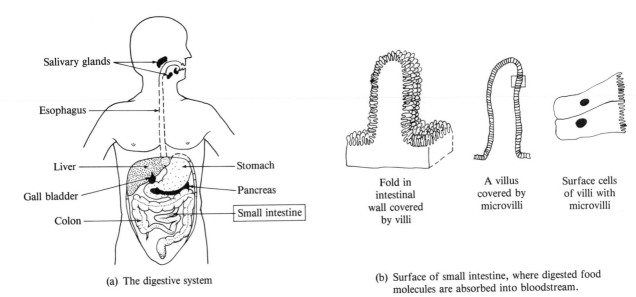

(a) The digestive system

(b) Surface of small intestine, where digested food molecules are absorbed into bloodstream.

Figure 11.1
The digestive system and details of the small intestine surface. The end products of digestion are absorbed into the bloodstream at the microvilli. Enzymes from the mucous membrane of the small intestine complete hydrolysis of disaccharides to monosaccharides.

Figure 11.2
Digestion of carbohydrates.

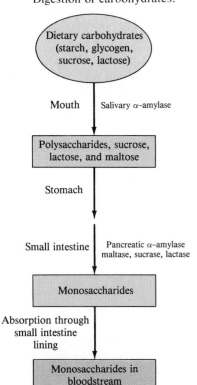

football field. Once in the bloodstream, the small molecules are transported into target cells where many are further broken down to produce energy. Some of the fragments produced by cellular breakdown of food molecules are exhaled as carbon dioxide, some are excreted, and some are used as building blocks to synthesize new biomolecules.

The digestion of carbohydrates, summarized in Figure 11.2, begins in the mouth as the α-amylase in saliva catalyzes hydrolysis of the glycosidic bonds in sugar molecules. Bacteria in the mouth do the same thing, usually not to our benefit (Figure 11.3). Starch from plants and glycogen from meat are hydrolyzed by α-amylase into smaller polysaccharides and the disaccharide maltose. Salivary α-amylase continues to act on dietary polysaccharides in the stomach until, after an hour or so, it is inactivated by stomach acid. No further carbohydrate digestion takes place in the stomach.

$$\text{(Glucose)}_n + H_2O \xrightarrow{\alpha\text{-amylase}} \text{(glucose)}_{<n} + \text{(glucose)}_2$$

Starch and glycogen   Maltose

α-Amylase is also secreted by the pancreas and enters the small intestine, where conversion of polysaccharides to maltose continues. Other enzymes from the mucous lining of the small intestine hydrolyze maltose and the dietary disaccharides sucrose and lactose to the monosaccharides glucose, fructose, and galactose, which are then transported across the intestinal wall.

A non-life-threatening but troublesome condition known as *lactose intolerance* develops in many individuals when the enzyme lactase ceases to be

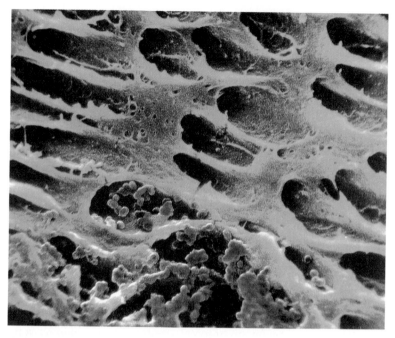

Figure 11.3
Dentine and bacteria. A cavity gets started when bacteria (orange
particles) in dentine (blue) below the outer tooth enamel hydrolyze
sucrose and then make the polysaccharide dextran from the resulting
glucose. The tooth becomes coated with plaque, which is composed
of dextran, microorganisms, and proteins from saliva. Further
breakdown of sugars within the plaque produces acids that attack
the tooth enamel.

made in the small intestine after the age of about 4. Lactose then proceeds
unchanged to the large intestine, where it increases the osmolarity of the
intestine contents and causes diarrhea. Bacterial attack on lactose in the intestine also produces methane and other gases that can cause nausea and vomiting.

Practice Problem   **11.1**   Complete the following word equations:

(a) lactose $+ \ H_2O \xrightarrow{\text{lactase}}$

(b) sucrose $+ \ H_2O \xrightarrow{\text{sucrase}}$

11.2 GLUCOSE METABOLISM: AN OVERVIEW

The principal role of glucose is as a fuel to yield the energy carried by ATP. All
cells require a certain amount of glucose, and there are several metabolic
strategies for maintaining a normal glucose concentration in the blood. Some
cells, notably red blood cells and cells in the brain, central nervous system, and
muscles, rely on glucose for energy but can't synthesize it.

In Chapter 9, we described the final stages of ATP production beginning
with acetyl SCoA, a common intermediate in the catabolism of all foods. Now

we'll go back and further examine the central position of glucose in metabolism, as summarized in Figure 11.4. Look back at this diagram and at Table 11.1 often as you read this chapter for help in sorting out the pathways that have similar names.

When glucose enters a cell, it is immediately converted in an irreversible reaction to glucose 6-phosphate. At this point, several pathways are available. Its most likely fate when energy is needed is *glycolysis,* the pathway leading to pyruvate, acetyl SCoA, and subsequent energy production. When cells are already well supplied with glucose, some is converted to glycogen for storage, a pathway known as *glycogenesis.* A supply of glucose beyond what can be stored as glycogen is diverted to the synthesis of fatty acids, which we'll describe in Chapter 13. In addition, and depending on the types of cells, varying amounts of glucose enter the *pentose phosphate pathway,* which supplies NADPH and sugars needed for biosynthesis. These pathways are summarized in Table 11.1.

The pyruvate produced in glycolysis also has several alternative fates awaiting it. Most likely, the pyruvate will be converted to acetyl SCoA. This pathway, however, is short-circuited in some tissues, especially where there's not enough oxygen. Under these **anaerobic** conditions, pyruvate is instead converted to lactate. The lactate may, in turn, be cycled back to pyruvate and reconverted to glucose by *gluconeogenesis,* a pathway that also allows glucose

Anaerobic In the absence of oxygen.

Figure 11.4
Glucose metabolism. Synthetic pathways (anabolism) are shown in blue, pathways that break down biomolecules (catabolism) are shown in yellow, and connections to lipid and protein metabolism are shown in green.

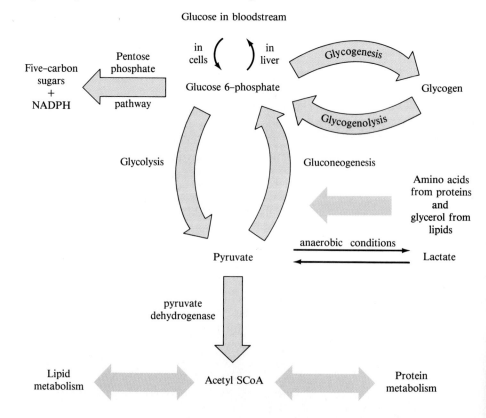

Table 11.1   Metabolic Pathways of Glucose

Name	Derivation of Name	Function
Glycolysis	*glyco-*, glucose (from the Greek, meaning "sweet") *-lysis*, decomposition	Conversion of glucose to pyruvate
Gluconeogenesis	*gluco-*, glucose *-neo-*, new *-genesis*, creation	Synthesis of glucose from amino acids, pyruvate, and other noncarbohydrates
Glycogenesis	*glyco (gen)-*, glycogen *-genesis*, creation	Synthesis of glycogen from glucose
Glycogenolysis	*glycogen-*, glycogen *-lysis*, decomposition	Breakdown of glycogen to glucose
Pentose phosphate pathway	*pentose*, a five-carbon sugar *phosphate*	Conversion of glucose to five-carbon sugar phosphates

to be synthesized from pyruvate, amino acids, or glycerol when the body is starved for glucose. In yeasts, pyruvate is metabolized differently than in human beings, and ethyl alcohol is the end product.

Practice Problem   11.2   Identify each of the following pathways:
(a) pathway for release of glucose from glycogen
(b) pathway for synthesis of glucose from lactate
(c) pathway for synthesis of glycogen

11.3   Name the synthetic pathways that have glucose as their first reactant.

11.3 GLYCOLYSIS

Glycolysis The biochemical pathway that breaks down a molecule of glucose into two molecules of pyruvate plus energy.

Glycolysis is a series of 10 enzyme-catalyzed reactions that break down each glucose molecule into two pyruvate molecules. The steps of glycolysis, also called the *Embden-Meyerhoff pathway* after its discoverers, are summarized in Figure 11.5, where the reactions and the structures of intermediates should be looked at as you proceed through the description of glycolysis in the following paragraphs. Almost all organisms carry out glycolysis; in humans it occurs in the cytosol of all cells, in contrast with the further oxidation of pyruvate, which occurs in mitochondria.

Steps 1–3 of Glycolysis: Phosphorylation   Glucose from the breakdown of food is carried in blood to cells that it enters by facilitated diffusion across cell membranes. As soon as it enters the cell, glucose is phosphorylated in *step 1* of glycolysis, which requires an energy investment from ATP. From here on, all intermediates in glycolysis are phosphates and are trapped within the cells because phosphates do not cross cell membranes. Glucose 6-phosphate, the product of step 1, inhibits hexokinase, an enzyme that plays an important role in the elaborate and delicate control of glucose metabolism.

Figure 11.5
The glycolysis pathway
for converting glucose
to pyruvate.

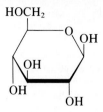

Glucose

Exergonic and
not reversible **1** ⌠—ATP
            ⌡→ADP

Step 1. Glucose undergoes reaction with ATP to yield
glucose 6-phosphate plus ADP in a reaction
catalyzed by *hexokinase*.

$^{2-}O_3POCH_2$

Glucose
6-phosphate

2 ↓

Step 2. Isomerization of glucose 6-phosphate yields
fructose 6-phosphate. The reaction is catalyzed by the
mutase enzyme, *phosphohexoseisomerase*.

$^{2-}O_3POCH_2$ OH

HO CH_2OH

OH

Fructose
6-phosphate

Exergonic and
not reversible **3** ⌠—ATP
            ⌡→ADP

Step 3. Fructose 6-phosphate reacts with a second
molecule of ATP to yield fructose 1,6-bisphosphate
plus ADP. *Phosphofructokinase*, the
enzyme for step 3, is a major control
point in glycolysis.

$^{2-}O_3POCH_2$ OH

HO $CH_2OPO_3^{2-}$

OH

Fructose
1,6-bisphosphate

4 ↓

Step 4. The six-carbon chain of fructose 1,6-bisphosphate
is cleaved into two three-carbon pieces by the enzyme
aldolase. (Continued on next page.)

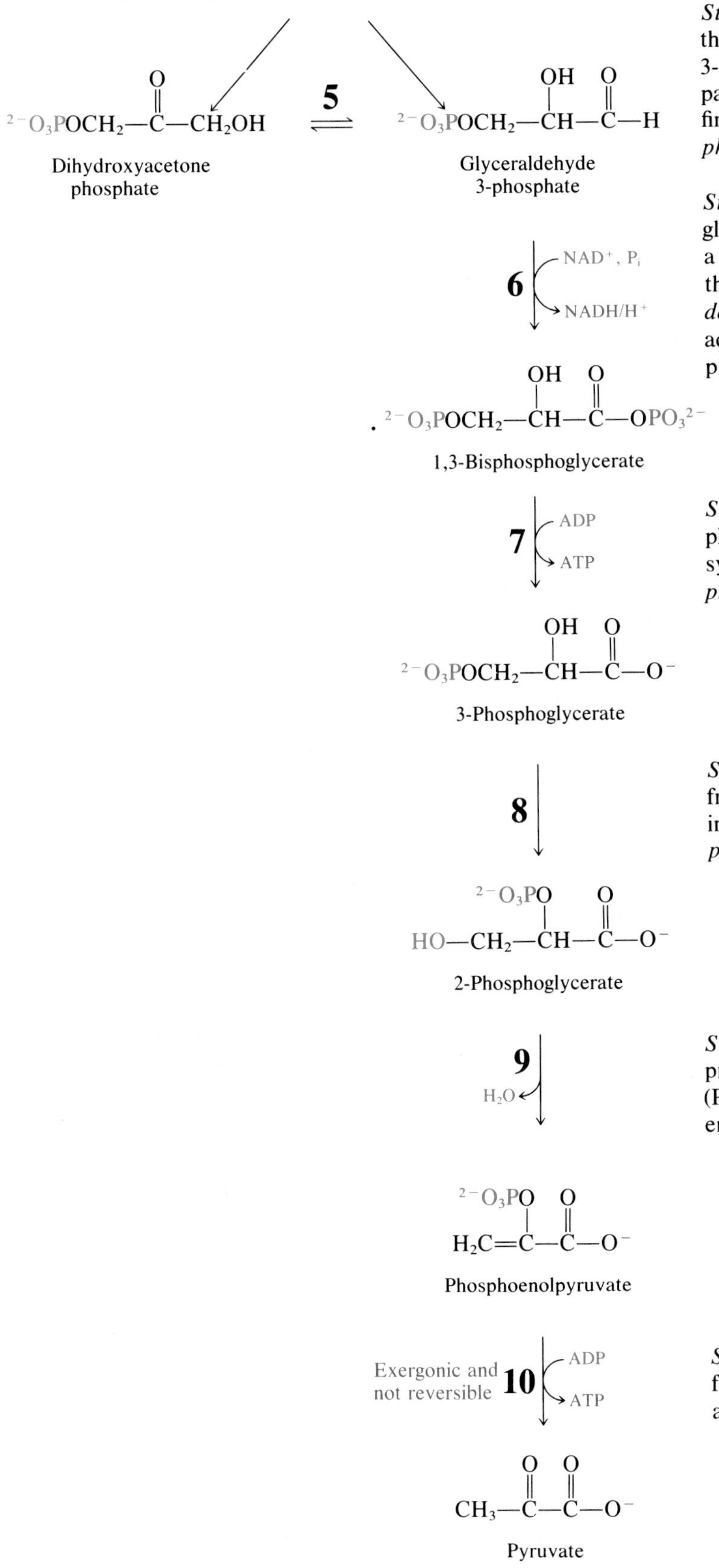

$$^{2-}O_3POCH_2{-}\overset{\overset{\displaystyle O}{\|}}{C}{-}CH_2OH \;\rightleftharpoons\; {^{2-}}O_3POCH_2{-}\overset{\overset{\displaystyle OH}{|}}{CH}{-}\overset{\overset{\displaystyle O}{\|}}{C}{-}H$$

Dihydroxyacetone phosphate 　　　　　Glyceraldehyde 3-phosphate

5

Step 5. The two products of step 4 are both three-carbon sugars, but only glyceraldehyde 3-phosphate can continue in the glycolysis pathway. Dihydroxyacetone phosphate must first be isomerized by the enzyme *triose phosphate isomerase*.

6 ⟨ NAD$^+$, P$_i$ → NADH/H$^+$

Step 6. Two reactions occur as glyceraldehyde 3-phosphate is first oxidized to a carboxylic acid and then phosphorylated by the enzyme *glyceraldehyde 3-phosphate dehydrogenase*. The coenzyme nicotinamide adenine dinucleotide (NAD) and inorganic phosphate ion (P$_i$) are required.

$$.\;{^{2-}}O_3POCH_2{-}\overset{\overset{\displaystyle OH}{|}}{CH}{-}\overset{\overset{\displaystyle O}{\|}}{C}{-}OPO_3{}^{2-}$$

1,3-Bisphosphoglycerate

7 ⟨ ADP → ATP

Step 7. A phosphate group from 1,3-bisphosphoglycerate is transferred to ADP, resulting in synthesis of ATP, and catalyzed by *phosphoglycerate kinase*.

$$^{2-}O_3POCH_2{-}\overset{\overset{\displaystyle OH}{|}}{CH}{-}\overset{\overset{\displaystyle O}{\|}}{C}{-}O^-$$

3-Phosphoglycerate

8

Step 8. A phosphate group is next transferred from carbon 3 to carbon 2 of phosphoglycerate in a step catalyzed by the enzyme *phosphoglyceromutase*.

$$HO{-}CH_2{-}\overset{\overset{\displaystyle ^{2-}O_3PO}{|}}{CH}{-}\overset{\overset{\displaystyle O}{\|}}{C}{-}O^-$$

2-Phosphoglycerate

9 H$_2$O

Step 9. Loss of water from 2-phosphoglycerate produces phosphoenolpyruvate (PEP). The dehydration is catalyzed by the enzyme *enolase*.

$$H_2C{=}\overset{\overset{\displaystyle ^{2-}O_3PO}{|}}{C}{-}\overset{\overset{\displaystyle O}{\|}}{C}{-}O^-$$

Phosphoenolpyruvate

Exergonic and not reversible **10** ⟨ ADP → ATP

Step 10. Transfer of the phosphate group from phosphoenolpyruvate to ADP yields pyruvate and generates ATP, catalyzed by *pyruvate kinase*.

$$CH_3{-}\overset{\overset{\displaystyle O}{\|}}{C}{-}\overset{\overset{\displaystyle O}{\|}}{C}{-}O^-$$

Pyruvate

311

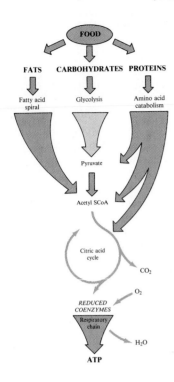

Step 2 is an isomerization that converts glucose 6-phosphate to fructose 6-phosphate. As the following open-chain formulas show, the reaction is conversion of an aldose to a ketose, which prepares the molecule for addition of a second phosphate group on the end carbon.

$$
\begin{array}{ccc}
\begin{array}{l}
\text{H}\diagdown\!\!\!\diagup\text{O} \\
\ \ \text{C} \\
\text{H}-\text{C}-\text{OH} \\
\text{HO}-\text{C}-\text{H} \\
\text{H}-\text{C}-\text{OH} \\
\text{H}-\text{C}-\text{OH} \\
\text{CH}_2\text{OPO}_3{}^{2-}
\end{array}
&
\rightleftharpoons
&
\begin{array}{l}
\text{CH}_2\text{OH} \\
\text{C}=\text{O} \\
\text{HO}-\text{C}-\text{H} \\
\text{H}-\text{C}-\text{OH} \\
\text{H}-\text{C}-\text{OH} \\
\text{CH}_2\text{OPO}_3{}^{2-}
\end{array}
\end{array}
$$

Glucose 6-phosphate (an aldohexose)   Fructose 6-phosphate (a ketohexose)

Step 3 makes a second energy investment as fructose 6-phosphate is converted to fructose 1,6-bisphosphate by reaction with ATP. ("Bis-" means "two.") Step 3 is another major control point for glycolysis. When the cell is short of energy, ADP (adenosine diphosphate) and AMP (adenosine monophosphate) concentrations build up and activate the step 3 enzyme, phosphofructokinase. When energy is in good supply, ATP and citrate build up and inhibit the enzyme. The outcome of steps 1–3 is formation of a molecule ready to be split into the two three-carbon intermediates that will ultimately become two molecules of pyruvate.

Steps 4 and 5 of Glycolysis: Cleavage and Isomerization   *Step 4* converts the six-carbon bisphosphate from step 3 into two three-carbon monophosphates, one an aldose and one a ketose. The bond between carbons 3 and 4 in fructose 1,6-bisphosphate breaks, and a C=O group is formed. If you look back at Section 5.8, you'll see that this is the reverse of an aldol reaction.

$$
\begin{array}{l}
\text{CH}_2\text{OPO}_3{}^{2-} \\
\text{C}=\text{O} \\
\text{HO}-\text{C}-\text{H} \\
\text{H}-\text{C}-\text{OH} \\
\text{H}-\text{C}-\text{OH} \\
\text{CH}_2\text{OPO}_3{}^{2-}
\end{array}
\quad\rightleftharpoons\quad
\begin{array}{l}
\text{CH}_2\text{OPO}_3{}^{2-} \\
\text{C}=\text{O} \\
\text{CH}_2\text{OH} \\
\ \\
+ \\
\ \\
\text{H}\diagdown\!\!\!\diagup\text{O} \\
\ \ \text{C} \\
\text{H}-\text{C}-\text{OH} \\
\text{CH}_2\text{OPO}_3{}^{2-}
\end{array}
$$

Dihydroxyacetone phosphate

Glyceraldehyde 3-phosphate

Fructose 1,6-bisphosphate

The two three-carbon sugars produced in step 4 are isomers that are interconvertible in an aldose–ketose equilibrium (*step 5* in Figure 11.5), but only glyceraldehyde 3-phosphate can continue on the glycolysis pathway. The overall result of steps 4 and 5 is therefore the production of *two* glyceraldehyde 3-phosphate molecules.

Steps 1–5 are sometimes referred to as the *energy investment* part of glycolysis. So far, two ATPs have been invested and no income has been earned, but the stage has been set for a small profit. Note that since one glucose molecule gives two glyceraldehyde 3-phosphates that each pass separately down the rest of the pathway, *steps 6–10 of glycolysis each take place twice for every glucose molecule that enters at step 1.*

Steps 6–10 of Glycolysis: Energy Generation   The second half of glycolysis is devoted to alternately generating molecules with high-energy phosphate groups that can be removed in exergonic reactions and then capturing this energy in ATP.

Step 6 is the combined oxidation of glyceraldehyde 3-phosphate to a carboxylic acid and phosphorylation of the acid. NAD^+ is the oxidizing agent, making this the first energy-generating step of glycolysis. Some of the energy from the exergonic oxidation is captured in NADH, and some is devoted to forming the high-energy phosphate.

Step 7 generates the first ATP of glycolysis by transferring a phosphate group to ADP from 1,3-bisphosphoglycerate. Because this step occurs twice for each glucose molecule, the energy balance sheet in glycolysis is even after step 7. Two ATPs were spent in steps 1–5, and now they've been replaced.

Steps 8 and 9 generate the second high-energy phosphate by an isomerization followed by dehydration to give phosphoenolpyruvate. Transfer of the phosphate group of this molecule to ADP in *step 10* then generates ATP in a highly exergonic reaction. The two ATPs formed by the two occurrences of step 10 are pure profit, and the overall results of glycolysis are as follows:

Net Result of Glycolysis

$$C_6H_{12}O_6 + 2\ NAD^+ + 2\ P_i + 2\ ADP \longrightarrow 2\ CH_3\!-\!\overset{\overset{\displaystyle O}{\|}}{C}\!-\!\overset{\overset{\displaystyle O}{\|}}{C}\!-\!O^- + 2\ NADH + 2\ ATP + 2\ H_2O + 2\ H^+$$

Glucose   Pyruvate

- Conversion of glucose to two pyruvates
- Production of two ATPs
- Production of two reduced coenzyme molecules, NADH.

Practice Problems   **11.4**   Identify the two pairs of steps in glycolysis in which high-energy phosphate intermediates are synthesized and their energy then harvested in ATP.

11.5   Identify each step in glycolysis that is an isomerization.

11.4 ENTRY OF OTHER SUGARS INTO GLYCOLYSIS

Fructose from fruits and sucrose, galactose from the lactose in milk, and mannose from plant polysaccharides are the major simple sugars other than glucose produced by digestion. Each eventually joins the glycolysis pathway.

Fructose is converted to glycolysis intermediates in two ways: In muscle, it is phosphorylated to fructose 6-phosphate, and in the liver it is converted to glyceraldehyde 3-phosphate. Mannose, like glucose, is converted by hexose kinase to a 6-phosphate, which then undergoes a multistep, enzyme-catalyzed rearrangement and enters glycolysis as fructose 6-phosphate. A five-step pathway is needed to convert galactose to glucose 6-phosphate for entry into the glycolysis pathway. In a genetic disease known as *galactosemia,* one of the enzymes of this pathway is missing. The resulting buildup of galactose can cause cataracts, mental retardation, liver damage, and death. The condition can be treated with a lactose- and galactose-free diet, which means strict, lifelong avoidance of all milk and milk products.

Practice Problems **11.6** Use the curved arrow symbolism to write an equation for the conversion of fructose to fructose 6-phosphate by ATP. At what step does fructose 6-phosphate enter glycolysis?

11.7 Compare glucose and galactose (see Section 10.5) and tell how their structures differ.

11.5 THREE REACTIONS OF PYRUVIC ACID

The breakdown of glucose to pyruvate is a central metabolic pathway in most living systems. The further reactions of pyruvate, however, depend on conditions and on the nature of the organism.

Under normal oxygen-rich (**aerobic**) conditions, pyruvate is converted into acetyl SCoA as part of the central pathway of glucose metabolism. Under oxygen-poor (anaerobic) conditions, it is converted into lactate, a reaction that takes place in bacteria, as when milk turns sour, and in muscle tissue of higher organisms.

When pyruvate undergoes **fermentation** by yeast, it is converted into ethanol plus carbon dioxide, the process used to produce beer, wine, and other alcoholic drinks (Figure 11.6). Fermentation is also used to make bread: The carbon dioxide causes the bread to rise, and the alcohol evaporates. The first leavened, or raised, bread was probably made by accident when airborne yeasts got into the dough.

Aerobic In the presence of oxygen.

Fermentation The breakdown of glucose to ethanol plus carbon dioxide by the action of yeast enzymes.

aerobic conditions

$$CH_3-\overset{\overset{\textstyle O}{\|}}{C}-\overset{\overset{\textstyle O}{\|}}{C}-O^-$$

Pyruvate

Pyruvate dehydrogenase

NADH

Anaerobic conditions

NAD$^+$

Fermentation with yeast

$$CH_3-\overset{\overset{\textstyle O}{\|}}{C}-S-CoA \quad + \quad CO_2$$

Acetyl SCoA

$$CH_3-\overset{\overset{\textstyle OH}{|}}{C}H-\overset{\overset{\textstyle O}{\|}}{C}-O^-$$

Lactate

$$CH_3-CH_2-OH + CO_2$$

Ethanol

Oxidation of Pyruvate to Acetyl SCoA   Under aerobic conditions, pyruvate formed in the cytosol crosses the mitochondrial membrane and enters the matrix, where acetyl SCoA is produced with the aid of an enzyme embedded in the mitochondrial membrane. The overall reaction is simple in appearance:

Net Result of Pyruvate Oxidation

$$\text{Pyruvate} + NAD^+ + \text{H-S-CoA} \xrightarrow[\text{dehydrogenase}]{\text{pyruvate}} \text{acetyl SCoA} + CO_2 + NADH$$

- Transformation of two carbon atoms from pyruvate to acetyl SCoA
- Release of one molecule of CO_2
- Production of one reduced coenzyme molecule (NADH) in the mitochondrion.

Pyruvate dehydrogenase complex, the enzyme responsible for pyruvate oxidation, contains about 60 subunits of three different enzymes with a total molecular weight of over 4 million, and the conversion of pyruvate to acetyl SCoA is actually far from simple. It requires NAD^+, CoA, FAD, and two other coenzymes (lipoic acid and thiamine pyrophosphate) derived from thiamine (vitamin B_1). The enzyme subunits, which are adjacent to each other, swing into position one after the other as pyruvate loses CO_2 and is converted to an acetyl group which is then transferred to coenzyme A.

Figure 11.6
Fermentation of grapes. Inside these barrels, yeast is at work, converting the sugar in grape juice into wine.

Practice Problem   ***11.8***   Complete oxidation of glucose produces six molecules of carbon dioxide. Describe the stages of catabolism at which all six are formed.

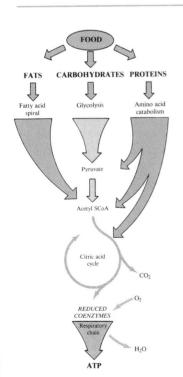

Conversion of Pyruvate to Lactate   Why does pyruvate take an alternate pathway in the absence of oxygen? Since no oxygen has been used in glucose catabolism thus far, what's the connection? The problem lies with the NADH formed in step 6 of glycolysis (Figure 11.5). Under aerobic conditions, NADH is continually reoxidized during electron transport, and fresh NAD^+ is in good supply. If the respiratory chain shuts down because of lack of oxygen, however, NADH builds up in concentration, no NAD^+ is available, and glycolysis can't continue. An alternative way to reoxidize NADH is therefore essential, because without the ATP from oxidative phosphorylation, glycolysis is needed as a source of fresh ATP.

The reduction of pyruvate to lactate solves the problem. NADH serves as the reducing agent and is reoxidized to NAD^+. Red blood cells have no mitochondria and therefore always form lactate as the end product of glycolysis. We'll pursue the fate of the lactate a bit later in this chapter.

11.6 ENERGY OUTPUT IN COMPLETE CATABOLISM OF GLUCOSE

The total energy output from oxidation of glucose is the combined result of (1) glycolysis, (2) conversion of pyruvate to acetyl SCoA, (3) the passage of the two acetyl groups through the citric acid cycle, and (4) the passage of reduced coenzymes from each of these pathways through the respiratory chain.

To determine the total number of ATPs generated from one glucose molecule, we can first add up the net equations for each pathway. Since each glucose yields two pyruvates and two acetyl SCoAs, the net equations for pyruvate oxidation and the citric acid cycle must be multiplied by 2:

Net Result of Catabolism of One Glucose Molecule

Glycolysis

$$\text{Glucose} + 2\text{ NAD}^+ + 2\text{ P}_i + 2\text{ ADP} \longrightarrow 2\text{ pyruvate} + 2\text{ NADH} + 2\text{ ATP} + 2\text{ H}_2\text{O} + 2\text{ H}^+$$

Pyruvate oxidation

$$2\text{ Pyruvate} + 2\text{ NAD}^+ + 2\text{ HSCoA} \longrightarrow 2\text{ acetyl SCoA} + 2\text{ CO}_2 + 2\text{ NADH}$$

Citric acid cycle

$$2\text{ Acetyl SCoA} + 6\text{ NAD}^+ + 2\text{ FAD} + 2\text{ ADP} + 2\text{ P}_i + 4\text{ H}_2\text{O} \longrightarrow$$
$$2\text{ HSCoA} + 6\text{ NADH} + 6\text{ H}^+ + 2\text{ FADH}_2 + 2\text{ ATP} + 4\text{ CO}_2$$

$$\text{Glucose} + 10\text{ NAD}^+ + 2\text{ FAD} + 2\text{ H}_2\text{O} + 4\text{ ADP} + 4\text{ P}_i \longrightarrow$$
$$10\text{ NADH} + 8\text{ H}^+ + 2\text{ FADH}_2 + 4\text{ ATP} + 6\text{ CO}_2$$

For each NADH that passes through electron transport in the mitochondrion, 3 ATPs are produced, and for each $FADH_2$, 2 ATPs are produced (Section 9.10). Translating the net equation for glucose catabolism into ATPs according to these relationships gives a total of 38 ATPs, the maximum number that can be produced from each glucose molecule.

$$10\text{ NADH}\left(\frac{3\text{ ATP}}{\text{NADH}}\right) + 2\text{ FADH}_2\left(\frac{2\text{ ATP}}{\text{FADH}_2}\right) + 4\text{ ATP} = 38\text{ ATP}, \text{ maximum per glucose}$$

There's one step in the pathway from glucose to ATP that we've not mentioned yet, but that in some cells reduces the maximum number of ATPs produced by 2—from 38 to 36. The additional step arises because the two NADHs from glycolysis are produced in the *cytosol*, but oxidative phosphorylation takes place in the *mitochondrion* and NADH can't cross the inner mitochondrial membrane. This problem is solved by shuttle mechanisms that carry hydrogen atoms from NADH across the membrane, producing $FADH_2$ and decreasing the ATP yield.

Practice Problem   11.9   Why does the FAD-requiring shuttle decrease the ATP yield from 38 to 36?

11.7 REGULATION OF GLUCOSE METABOLISM AND ENERGY PRODUCTION

Hypoglycemia Lower-than-normal blood glucose concentration.

Normal blood glucose concentration ranges roughly from 65 to 110 mg/dL. When departures from normal occur, we're in trouble (Figure 11.7). Low blood glucose (**hypoglycemia**) causes weakness, sweating, and rapid heartbeat, and in

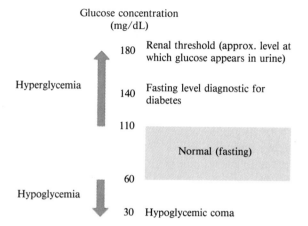

Glucose concentration
(mg/dL)

180	Renal threshold (approx. level at which glucose appears in urine)
140	Fasting level diagnostic for diabetes
110	
	Normal (fasting)
60	
30	Hypoglycemic coma

Hyperglycemia

Hypoglycemia

Figure 11.7
Blood glucose
concentrations.

severe occurrences, mental confusion, convulsions, coma, and death. At a
blood glucose level of 30 mg/dL consciousness is impaired or lost, and prolonged hypoglycemia can cause permanent dementia. Excess blood glucose
(**hyperglycemia**) causes increased urine flow as water is drawn from cells because of the higher osmolarity of urine containing glucose. Prolonged hyperglycemia can cause low blood pressure, coma, and death.

   Two hormones from the pancreas have the major responsibility for blood
glucose regulation (Figure 11.8). The first and most important is *insulin*, which
is released when blood glucose concentration rises (Figure 11.9). Although

Hyperglycemia   Higher-
than-normal blood glucose
concentration.

Figure 11.8
Pancreatic islets. These cell
clusters contain β cells that
produce insulin (shown by
the stain at the left) and α
cells that produce glucagon
(shown by the stain at the
right).

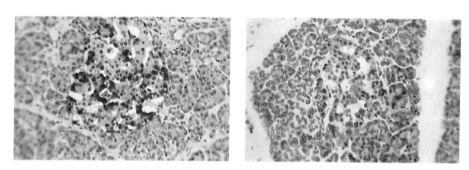

Figure 11.9
Regulation of glucose
concentrations by insulin and
glucagon from the pancreas.

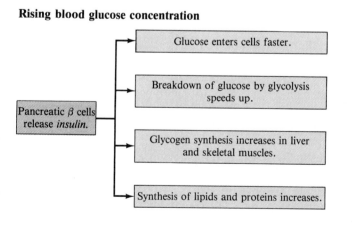

Rising blood glucose concentration

Pancreatic β cells release *insulin*.

- Glucose enters cells faster.
- Breakdown of glucose by glycolysis speeds up.
- Glycogen synthesis increases in liver and skeletal muscles.
- Synthesis of lipids and proteins increases.

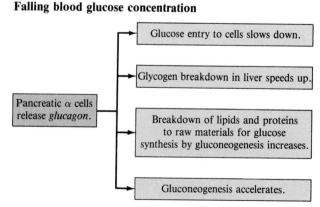

Falling blood glucose concentration

Pancreatic α cells release *glucagon*.

- Glucose entry to cells slows down.
- Glycogen breakdown in liver speeds up.
- Breakdown of lipids and proteins to raw materials for glucose synthesis by gluconeogenesis increases.
- Gluconeogenesis accelerates.

AN APPLICATION: GLYCOGEN STORAGE DISEASES

A group of genetic diseases known as *glycogen storage diseases* result from defects in enzymes associated with glycogen metabolism. Clinical symptoms are of two types: those due to hypoglycemia because of reduced availability of glucose from glycogen, and those due to accumulation of excessive amounts of glycogen in liver and muscle tissue.

von Gierke's disease results from a deficiency of glucose-6-phosphatase, the enzyme that removes phosphate from glucose 6-phosphate so that glucose can be released from the liver into the bloodstream. Symptoms include (1) a greatly enlarged liver containing excessive normal glycogen and (2) dangerously low blood sugar after nothing has been eaten for a few

hours. Management of von Gierke's disease requires maintenance of adequate blood glucose levels around the clock by techniques such as nasogastric feeding in infants, drugs that inhibit uptake of glucose by the liver, or overnight delivery of carbohydrates directly to the stomach through a gastrostomy tube.

In *McArdle's disease,* an individual develops normally but as a young adult experiences rapid fatigue and muscle cramps on exertion. The defect is in the enzyme for breakdown of muscle glycogen to glucose. Moderate buildup of normal muscle glycogen occurs, but cramps arise because insufficient ATP is available from glycolysis due to a shortage of glucose 6-phosphate.

insulin has multiple physiological effects, its primary role is to aid the passage of blood glucose across cell membranes. When insulin concentration rises, cells produce energy from glucose and excess glucose is converted to glycogen, proteins, or lipids.

The second, *glucagon,* is released when blood glucose concentration drops. In a reversal of insulin's effects, glycogen is broken down in the liver rather than synthesized, and proteins and lipids are broken down into simple molecules that can be converted to glucose in the liver by gluconeogenesis. Epinephrine, the fight-or-flight hormone, also accelerates the breakdown of glycogen, but its primary target is muscle tissue where energy is needed for quick action, rather than the liver. A closer look at glycogen metabolism and gluconeogenesis is provided in Sections 11.10 and 11.11.

11.8 METABOLISM IN FASTING AND STARVATION

Imagine that you're lost in the woods. You've had no carbohydrates and very little else to eat for hours, and you're exhausted. The glycogen stored in your liver and muscles will soon be used up, but your brain must still rely on glucose to keep functioning. What happens next? Fortunately, your body's not yet ready to give up. The gluconeogenesis pathway can make glucose from proteins. And if you're lost for a long time, there's a further backup system that extracts energy from compounds other than glucose.

The metabolic changes in the absence of food begin with a gradual decline in blood glucose concentration accompanied by an increased release of glucose from glycogen. All cells contain glycogen, but most is stored in liver cells (about 70 g) and muscle cells (about 200 g). Free glucose and glycogen represent less than 1% of our energy reserves and are used up in 15–20 hr of normal activity (3 hr in a marathon). Fats are our largest energy reserve, but adjusting to dependence on them for energy takes time. In the meantime, as glucose and glycogen are exhausted, metabolism turns to breakdown of proteins in muscles and glucose production from amino acids via gluconeogenesis.

During the first few days of starvation, protein is used up at a rate as high as 75 g/day. Meanwhile, lipid catabolism is being mobilized and acetyl SCoA molecules are being manufactured from fats at an ever-increasing rate. Eventually, the citric acid cycle is overloaded and can't degrade acetyl SCoA as rapidly as it's produced. Acetyl SCoA therefore builds up inside cells and begins to be removed by a new series of metabolic reactions that transform it into substances called *ketone bodies,* which we'll describe further in our discussion of lipid metabolism in Chapter 13.

The ketone bodies enter the bloodstream, and, as starvation continues, the brain and other tissues are able to switch over to producing up to 50% of their ATP from ketone bodies instead of glucose. By the fortieth day of starvation, metabolism has stabilized at the use of about 25 g of protein and 180 g of fat each day, a condition in which glucose and protein are conserved as much as possible. An average person can survive in this state for several months—those with more fat can survive longer.

11.9 METABOLISM IN DIABETES MELLITUS

Diabetes mellitus A chronic condition due to either insufficient insulin or failure of insulin to cross cell membranes.

Metabolic disease A disease due to disruption of metabolism caused by a genetically determined enzyme defect.

Diabetes mellitus, one of the most common **metabolic diseases,** is estimated to affect up to 4% of the population. Diabetes is not a single disease but is divided into two major types, *insulin-dependent* and *non-insulin-dependent*. The insulin-dependent type, also called *juvenile-onset* or Type I diabetes because it often appears in childhood, is caused by an insufficient supply of insulin due to damage to the pancreatic cells that produce it. By contrast, in the non-insulin-dependent type, also called *adult-onset* or Type II diabetes because it usually occurs in obese individuals over about 40 years of age, insulin is in good supply but fails to promote the passage of glucose across cell membranes. Although often thought of only as a disease of glucose metabolism, diabetes affects protein and fat metabolism as well, and in some ways the metabolic response resembles starvation.

The symptoms by which diabetes is usually detected—excessive thirst accompanied by frequent urination, abnormally high glucose concentrations in urine and blood, and wasting of the body despite a good diet—result when available glucose does not enter cells where it is needed. Glucose builds up in the blood, causing the symptoms of hyperglycemia and spilling over into the urine. In untreated diabetes, metabolism responds to the glucose shortage within cells by proceeding through the same stages from depletion of glycogen stores to breakdown of proteins and fats.

Adult-onset diabetes is thought to result when cell membrane receptors fail to recognize insulin. Drugs that increase insulin levels are an effective treatment because more of the undamaged receptors are put to work. To treat juvenile-onset diabetes, the missing insulin must be supplied by injection. Commercially available insulin (Figure 11.10) is now produced by bacteria modified by recombinant DNA techniques (Chapter 15) and is the first product of genetic engineering approved for use in humans.

Figure 11.10
Synthetic insulin. The drop of insulin pictured is one of the first to be made by genetic engineering techniques (Chapter 15). Before the availability of human insulin from genetically engineered bacteria, diabetic individuals had to rely on insulin extracted from the pancreas of cows, which is expensive and differs in three amino acids from human insulin.

AN APPLICATION: GLUCOSE TOLERANCE TEST

The glucose tolerance test is among the clinical laboratory tests usually done to pin down a diagnosis of diabetes mellitus. The patient must fast for 10–16 hr, must avoid a diet high in carbohydrates prior to the fast, and must not be taking any of a long list of drugs that can interfere with the test.

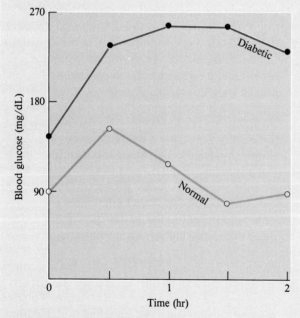

Blood glucose concentration in glucose tolerance test for normal and diabetic individuals.

First, a blood sample is drawn for determination of the fasting glucose concentration. Then, the patient drinks a solution containing 75 g of glucose, and further blood samples are taken at regular intervals after ingesting the glucose. The accompanying figure compares the changes in diabetic and nondiabetic individuals. In both, there is a rise in blood glucose concentration in the first hour. The difference becomes apparent after 2 hr, when the concentration in a nondiabetic individual has dropped to close to the fasting level but that in a diabetic individual remains high.

As listed below, a fasting blood glucose concentration of 140 mg/dL or higher, and/or a glucose tolerance test concentration that remains above 200 mg/dL beyond 1 hr, are considered diagnostic criteria for diabetes. For a firm diagnosis, the glucose tolerance test is usually given more than once.

Key Diagnostic Features of Diabetes Mellitus

Classic symptoms such as frequent urination, excessive thirst, rapid weight loss

Random blood glucose concentration > 200 mg/dL

Fasting blood glucose ≥ 140 mg/dL

Sustained blood glucose concentration ≥ 200 mg/dL after glucose administration in glucose tolerance test

An insulin-dependent diabetic is at risk for two types of medical emergencies: *Acidosis* as the result of buildup of acidic ketones in the late stages of uncontrolled diabetes may lead to coma and diminished brain function but can be reversed by timely insulin administration. Hypoglycemia or "insulin shock," by contrast, may be due to an overdose of insulin or failure to eat. If untreated, diabetic hypoglycemia can cause nerve damage or death. The arrival at the emergency room of a diabetic patient in a coma requires quick determination of whether the condition is due to ketoacidosis or excess insulin. Observations are backed up by bedside tests for glucose and ketones in blood and urine.

11.10   GLYCOGEN METABOLISM: GLYCOGENESIS AND GLYCOGENOLYSIS

Glycogen, the storage form of glucose in animals, is a branched polymer of glucose (Section 10.8). In the muscles, it is a source of energy for the muscle cells themselves; in the liver, it is a source of glucose needed to maintain normal blood glucose concentrations.

Glycogenesis   The biochemical pathway for synthesis of glycogen.

To synthesize glycogen via **glycogenesis,** a strategy like that used for the transfer of phosphate or acetyl groups is needed. Glucose 6-phosphate from the first step of glycolysis is converted to glucose 1-phosphate, which then bonds through its phosphate group to form the high-energy diphosphate of uridine (UDP). One by one, UDP carries glucose molecules to the growing glycogen chain (Figure 11.11).

Figure 11.11
Glycogenesis and glycogenolysis. The pathways to and from glycogen, (glucose)$_n$, are shown schematically.

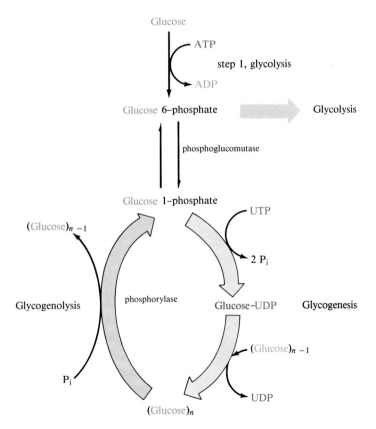

Glucose-UDP, the activated carrier of glucose in glycogen synthesis

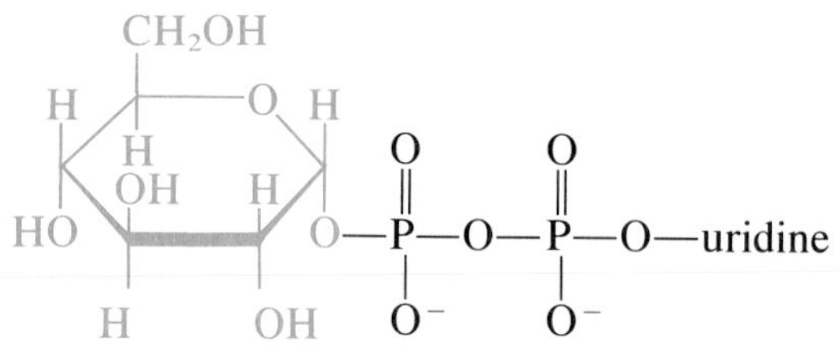

$$
\begin{array}{c}
\text{CH}_2\text{OH} \\
\text{H} \quad \text{O} \quad \text{H} \\
\text{H} \\
\text{OH} \quad \text{H} \\
\text{HO} \qquad \text{O—P—O—P—O—uridine} \\
\text{H} \quad \text{OH} \quad \text{O}^- \quad \text{O}^-
\end{array}
$$

Glycogenolysis The biochemical pathway for breakdown of glycogen to free glucose.

Glycogenolysis, the release of glucose from glycogen, is not the reverse of glycogen synthesis, because it does not require uridine triphosphate (UTP). First, the enzyme phosphorylase catalyzes the conversion of a glucose at the end of the glycogen chain to glucose 1-phosphate by reaction with a phosphate ion. The glucose 1-phosphate is then converted to glucose 6-phosphate in the reverse of the equilibrium by which it is formed in glycogenesis.

Glycogenolysis in muscle cells occurs when there is an immediate need for energy from glucose and may be triggered by release of epinephrine, the fight-or-flight hormone. The glucose 6-phosphate produced in muscle cells goes directly into glycolysis. Glycogenolysis in the liver occurs when there is a need for glucose in the blood. Because phosphates can't cross cell membranes, liver cells contain the enzyme for hydrolysis to yield free glucose for release into the bloodstream.

$$\text{Glucose 6-phosphate} + \text{H}_2\text{O} \longrightarrow \text{glucose} + \text{P}_i$$

Practice Problem 11.10 The following reaction is a good example of coupling of endergonic and exergonic reactions:

$$\text{UTP} + \text{glucose 1-phosphate} + \text{H}_2\text{O} \longrightarrow \text{glucose-UDP} + 2\,\text{P}_i$$

The first of the coupled reactions is formation of glucose-UDP and pyrophosphate ion (PP$_i$) by combination of UTP and glucose 1-phosphate, for which $\Delta G = +1.1$ kcal. The second is hydrolysis of the pyrophosphate ion to give two phosphate ions (2 P$_i$), for which $\Delta G = -8.0$ kcal. What is the common intermediate in these coupled reactions? What is ΔG for the coupled reactions? Is the change favorable or unfavorable?

11.11 GLUCONEOGENESIS: GLUCOSE FROM NONCARBOHYDRATES

Gluconeogenesis The biochemical pathway for the synthesis of glucose from noncarbohydrates, such as lactate, amino acids, or glycerol.

Gluconeogenesis in the liver is a pathway for making glucose from a variety of small molecules. For one, there's lactate, which is in especially abundant supply when muscles are working hard and which is also the normal product of glycolysis in red blood cells. The lactate is converted to pyruvate, which is the substrate for the first step of gluconeogenesis. The cycling of lactate from muscles and red blood cells to the liver and the return of glucose via the bloodstream to muscle cells is shown in Figure 11.12. Amino acids and glycerol from lipid catabolism are also able to enter the gluconeogenesis pathway for conversion to glucose 6-phosphate.

Figure 11.12
Major pathways of carbohydrate metabolites. Lactate produced in muscle during exercise or in red blood cells is sent in the bloodstream to the liver, where it is converted back into glucose 6-phosphate that can be broken down to give energy, stored as liver glycogen, or returned to the bloodstream. The cycle marked in red is known as the *Cori cycle* after the husband-and-wife team Carl and Gerti Cori who first recognized it and who received the Nobel Prize for this work.

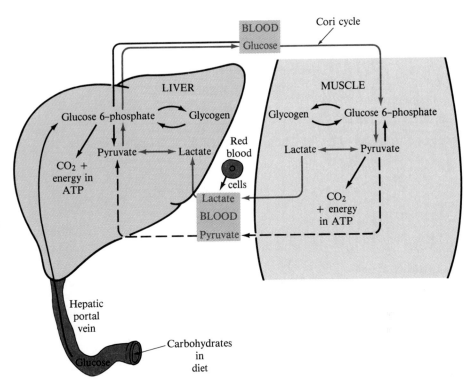

We noted earlier that, for metabolic pathways to be favorable, they must be exergonic. As a result, most are not reversible because the amount of energy required by the reverse, endergonic pathway is too large to be supplied by cellular metabolism. Glycolysis and gluconeogenesis provide a good example of this relationship and of the way around it.

Three reactions in glycolysis—steps 1, 3, and 10 in Figure 11.5—are too exergonic to be reversed. Gluconeogenesis therefore uses different enzymes and coenzymes, with different energy requirements, to bypass these three reactions. The other 7 reactions of the 10-step gluconeogenesis pathway are the reverse of those in glycolysis and are catalyzed by the same enzymes running in reverse.

We're not going to examine all the reactions of gluconeogenesis in detail, but the first provides an interesting example. In what adds up to a reversal of the last step of glycolysis, pyruvate is first converted to oxaloacetate and then to phosphoenolpyruvate. Recall that phosphoenolpyruvate (PEP) is a high-energy compound whose conversion to the more stable pyruvate in glycolysis releases about 6 kcal. To bypass the endergonic reversal of this reaction requires the investment of two energy-carrying molecules, ATP and GTP (guanosine triphosphate, a high-energy phosphate similar in structure to ATP), in a process that releases about 5 kcal.

Oxaloacetate is another significant intermediate in gluconeogenesis because it provides a connection to the first step of the citric acid cycle. If, for example, energy is needed, oxaloacetate from pyruvate can go directly into the citric acid cycle instead of continuing on to phosphoenolpyruvate and the production of glucose.

Practice Problems 11.11   The net reaction for gluconeogenesis is

$$\text{2 Pyruvate} + \text{4 ATP} + \text{2 GTP} + \text{2 NADH} + \text{2 H}^+ + \text{2 H}_2\text{O} \longrightarrow$$
$$\text{glucose} + \text{4 ADP} + \text{2 GDP} + \text{6 P}_i + \text{2 NAD}^+$$

Compare this result with the net reaction for glycolysis in Section 11.3. If glucose is converted to two pyruvates and both pyruvates are recycled back to glucose, is there an overall energy cost or profit?

Practice Problems 11.12   Which would be the appropriate coenzyme in conversion of lactate to pyruvate, NADH, NAD$^+$, or FADH$_2$? Is this oxidation or reduction?

$$\underset{\text{Lactate}}{CH_3-\underset{\underset{OH}{|}}{CH}-\underset{\underset{O}{\|}}{C}-O^-} \quad \overset{?}{\longrightarrow} \quad \underset{\text{Pyruvate}}{CH_3-\underset{\underset{O}{\|}}{C}-\underset{\underset{O}{\|}}{C}-O^-}$$

11.12   THE PENTOSE PHOSPHATE PATHWAY, AN ALTERNATIVE TO GLYCOLYSIS

Pentose phosphate pathway   The biochemical pathway that produces ribose (a pentose), NADPH, and other sugar phosphates from glucose; an alternative to glycolysis.

Depending on the type of cell, varying amounts of glucose enter the **pentose phosphate pathway** as an alternative to glycolysis. One major function of this alternative is to provide NADPH, a coenzyme needed in the synthesis of lipids. Thus, the pathway occurs mainly where lipids are produced: in fatty tissue, the liver, the adrenal cortex, and mammary glands. Another major function is production of ribose, a pentose needed for the synthesis of nucleic acids. When the need for NADPH or ribose is low, intermediates from the pentose phosphate pathway can continue on through glycolysis and the citric acid cycle.

Glucose 6-phosphate formed in step 1 of glycolysis is the starting material for the pentose phosphate pathway, which, like glycolysis, takes place in the cytosol. The pathway is divided into two stages. The first, the *oxidative stage,* results in the conversion of glucose 6-phosphate to ribulose 5-phosphate and the reduction of two NADP$^+$s to two NADPHs:

Oxidative stage of the pentose phosphate pathway

Glucose 6-phosphate $+$ 2 NADP$^+$ $+$ H$_2$O $\longrightarrow$
$$\text{ribulose 5-phosphate} + CO_2 + \text{2 NADPH} + \text{2 H}^+$$

$NAD^+/NADH$ and $NADP^+/NADPH$ differ in structure by only one phosphate group (Figure 9.10), but they differ significantly in function and are not interchangeable. The first of these coenzymes is needed mainly in its oxidized form, NAD^+, as an oxidizing agent in the exergonic reactions of catabolism. The second is needed mainly in its reduced form, NADPH, as a reducing agent in the endergonic reactions of biosynthesis.

Next after the oxidative phase, ribulose 5-phosphate is isomerized to ribose 5-phosphate, the sugar phosphate essential for nucleic acid and coenzyme synthesis.

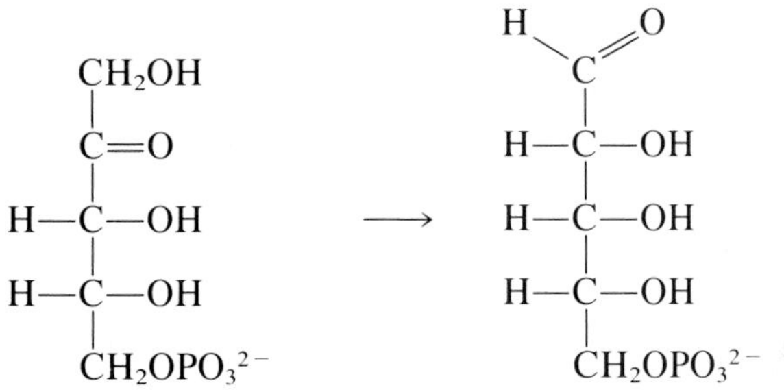

$$
\begin{array}{ccc}
\text{CH}_2\text{OH} & & \text{H}\diagdown\,\diagup\text{O} \\
| & & \text{C} \\
\text{C}{=}\text{O} & & | \\
| & & \text{H}{-}\text{C}{-}\text{OH} \\
\text{H}{-}\text{C}{-}\text{OH} & \longrightarrow & \text{H}{-}\text{C}{-}\text{OH} \\
| & & | \\
\text{H}{-}\text{C}{-}\text{OH} & & \text{H}{-}\text{C}{-}\text{OH} \\
| & & | \\
\text{CH}_2\text{OPO}_3{}^{2-} & & \text{CH}_2\text{OPO}_3{}^{2-}
\end{array}
$$

Ribulose 5-phosphate   Ribose 5-phosphate

At this point, ribose might be converted to one or more of several three-, four-, five-, or seven-carbon sugars that are also intermediates in various biosyntheses. Ultimately, glyceraldehyde 3-phosphate and fructose 6-phosphate are formed:

Net result of the pentose phosphate pathway

$$3 \text{ Glucose 6-phosphate} + 6\,\text{NADP}^+ \longrightarrow 2 \text{ fructose 6-phosphate} +$$
$$3\text{ CO}_2 + \text{glyceraldehyde 3-phosphate} + 6 \text{ NADPH} + 6 \text{ H}^+$$

The pentose phosphate pathway is remarkably versatile. The details aren't important to us here, but consider some of the possible variations: When NADPH demand is high, intermediates are recycled to glucose 6-phosphate for further NADPH production by the oxidative phase. When ATP demand is high, fructose 6-phosphate and glyceraldehyde 3-phosphate can enter glycolysis, and the overall result is complete oxidation of glucose 6-phosphate to CO_2 and H_2O. When nucleic acid synthesis is the priority, most of the nonoxidative phase is skipped and ribose 5-phosphate is the major product.

Practice Problem   11.13   Eighteen carbon atoms enter the pentose phosphate pathway in three glucose 6-phosphate molecules. Describe the fate of these carbon atoms according to the net result of the pathway.

INTERLUDE: BIOCHEMISTRY OF RUNNING

A runner is tense and ready, waiting for the signal to start the race. With the training period over, biochemistry is in control of providing the energy needed to run. The flood of epinephrine and the fight-or-flight reaction take control first, but only during the opening seconds of the race. The next energy resource to be tapped is creatine phosphate, a high-energy amino acid phosphate that stands ready in muscles for instant reaction. Although it's quick because it goes in one step to

ATP, there's enough creatine phosphate to last no more than 30 seconds.

Next, muscles start to use their stores of glycogen, freeing glucose for glycolysis, pyruvate production, and under these anaerobic conditions, lactate formation instead of acetyl SCoA formulation (see Fig. 11.4). In a 100-m sprint, probably all the energy comes from creatine phosphate and anaerobic glycolysis, but anaerobic breakdown of glucose from glycogen suffices for

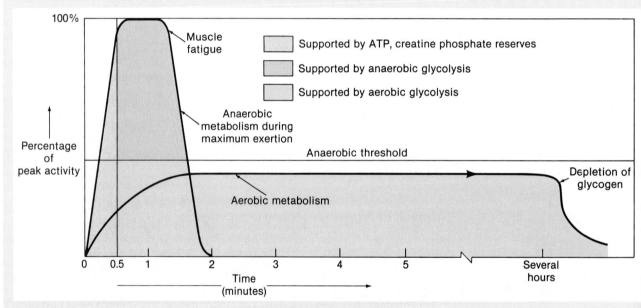

Changes in metabolism with muscle activity. The anaerobic threshold varies with factors such as age, weight, and level of athletic training.

SUMMARY

Carbohydrate **digestion** begins with α-amylase in the mouth, continues with pancreatic α-amylase and other enzymes in the small intestine, and yields monosaccharides that enter the bloodstream by absorption through the intestinal wall. The major route for glucose catabolism is **glycolysis** to yield pyruvate, followed by pyruvate oxidation to acetyl SCoA, which then yields ATP via the citric acid cycle and oxidative phosphorylation. For the com-

plete catabolism of one glucose molecule, 38 (or 36) ATPs are produced.

Glycolysis is a 10-step pathway in which two ATPs are first invested to form active phosphate intermediates that ultimately yield two pyruvates, two ATPs, and two NADHs for each glucose. Under **anaerobic** conditions, NADH serves as the reducing agent for conversion of pyruvate to lactate, thereby providing the NAD$^+$ needed

only a minute or two of maximum exertion since a buildup of lactate causes muscle fatigue (see the curve, p. 326). Beyond this, other pathways come into action. As breathing and heart rate speed up and oxygen-carrying blood flows more quickly to muscles, the aerobic pathway is activated, and ATP is once again generated by oxidative phosphorylation. The trick is to run at a speed just under that relying on anaerobic ATP generation, the "anaerobic threshold."

Now the question is, what fuel will metabolism rely on, carbohydrate or fat? Burning fatty acids from fats is more efficient: More than twice as many calories are generated by burning a gram of fat than by burning a gram of carbohydrate. When we're sitting quietly, in fact, our muscle cells are burning mostly fat, and the fat in storage could support the exertion of marathon running for several days. By contrast, glycogen alone can

provide enough glucose to fuel only 2–3 hr of such running.

The difficulty is that fatty acids can't be delivered to muscle cells fast enough to maintain the ATP level needed for running, so metabolism compromises and glycogen remains the limiting factor for the marathon runner. Once glycogen is gone, extreme exhaustion and mental confusion set in—the condition known as "hitting the wall." Running speed is then limited to that sustainable by fats only. To delay this point as long as possible, glycogen synthesis is encouraged by diets high in carbohydrates prior to and during a race. In the hours just before the race, however, carbohydrates are avoided. Their effect of triggering insulin release is undesirable a this point because the resulting faster use of glucose will hasten depletion of glycogen.

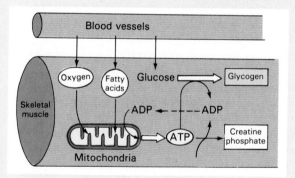

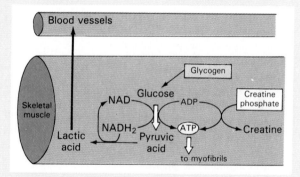

for glycolysis to continue producing ATP. The pancreatic hormones insulin and glucagon regulate blood glucose concentrations and help to guard against the ill effects of **hypoglycemia** and **hyperglycemia.**

Glucose is stored as glycogen made by **glycogenesis,** especially in liver and muscle tissue, and is released via **glycogenolysis.** In the liver, glycogenolysis occurs when glucose is needed in the blood. In muscle cells that need quick energy, glycogenolysis is triggered by epinephrine. The synthesis of glucose from lactate and other noncarbohydrates, **glyconeogenesis,** is a pathway of especial importance in starvation or glucose shortage within cells because of **diabetes mellitus.** Seven of the reactions of glycolysis run in reverse in gluconeogenesis, while alternate enzymes carry out the reverse of the three glycolysis reactions that are too exergonic to be reversed.

Varying amounts of glucose are also metabolized in the **pentose phosphate pathway,** which provides ribose (a pentose), NADPH, and other sugars needed for biosynthesis.

REVIEW PROBLEMS

Digestion and Metabolism

11.14* Where in the body does digestion occur, and what kinds of chemical reactions does it involve?

11.15 What is meant by the words *aerobic* and *anaerobic?*

11.16* What three products are formed from pyruvate under aerobic, anaerobic, and fermentation conditions?

11.17 Complete the following word equation:

$$\text{Maltose} + H_2O \longrightarrow ? + ?$$

Where in the digestive system does this process occur?

11.18* Four of the processes in glucose metabolism have similar-sounding names. Differentiate among glycolysis, gluconeogenesis, glycogenesis, and glycogenolysis.

Glycolysis

11.19 By what other name is the glycolysis pathway known?

11.20* What is the name and structure of the final product of carbohydrate glycolysis?

11.21 Where in the cell does glycolysis occur? How does that compare to where the citric acid cycle occurs?

11.22* Look at the 10 steps in glycolysis and then answer these questions:
(a) Which steps involve phosphorylation?
(b) Which step is an oxidation?
(c) Which step is a dehydration?

11.23 Lactate can be converted into pyruvate by the enzyme lactate dehydrogenase and the coenzyme NAD^+. Write the reaction in the standard biochemical format, using a curved arrow to show the involvement of NAD^+.

11.24* How many moles of ATP are produced by
(a) glycolysis of one mole of glucose?
(b) aerobic conversion of one mole of pyruvate to one mole of acetyl SCoA?
(c) catabolism of one mole of acetyl SCoA in the citric acid cycle?

11.25 How many moles of ATP are released by complete catabolism of one mole of glucose with the FAD shuttle in operation in glycolysis?

11.26* If fructose is isomerized to yield glucose prior to glycolysis, how many moles of ATP would you expect to obtain from complete catabolism of 1.0 mol of sucrose?

11.27 How many moles of acetyl SCoA are produced by catabolism of 1.0 mol of sucrose (Problem 11.26)?

11.28* How many moles of CO_2 are produced by the complete catabolism of sucrose (Problem 11.27)?

11.29 How many moles of ATP are produced by complete catabolism of one mole of maltose to yield CO_2?

11.30* What is the purpose of the formation of lactate under anaerobic conditions?

Glucose Levels in the Blood

11.31 Differentiate between blood sugar levels and resulting symptoms in hyperglycemia and hypoglycemia.

11.32* What is the effect of epinephrine on the glucose metabolic pathways?

11.33 From what molecules is glucose initially produced during starvation or fasting?

11.34* As starvation continues, to what is acetyl SCoA converted to prevent buildup in the cells?

11.35 What is the difference between juvenile- and adult-onset diabetes?

11.36* Where is most of the glycogen in the body stored?

11.37 Of what use is UDP in the formation of glycogen from glucose?

11.38* Why is glycogenolysis not the exact reverse of the process of glycogenesis?

Glucose Anabolism

11.39 What is the name of the anabolic pathway for making glucose?

11.40* What two molecules serve as starting materials for glucose synthesis?

11.41 To what molecule is pyruvate initially converted in the anabolism of glucose? To what substance is that molecule in turn converted?

11.42* Why can't pyruvate be converted to glucose in an exact reverse of the glycolysis pathway?

11.43 What is the major purpose of the pentose phosphate pathway?

11.44* Depending on the body's needs, into what type of compounds is glucose converted in the pentose phosphate pathway?

Applications

11.45 What genetic deficiency gives rise to von Gierke's disease and what are the symptoms? [App: Glycogen Storage Diseases]

11.46* What deficiency gives rise to McArdle's disease and what are the symptoms? [App: Glycogen Storage Diseases]

11.47 How do fasting levels of glucose in a diabetic person compare to those in a nondiabetic person? [App: Glucose Tolerance Test]

11.48 Discuss the differences in the response of a diabetic person compared to those of a nondiabetic person after drinking a glucose solution. [App: Glucose Tolerance Test]

11.49 Why is creatine phosphate a better source of quick energy than glucose? [Int: Biochemistry of Running]

11.50* Why is it not possible for a person to sprint for miles? [Int: Biochemistry of Running]

Additional Questions and Problems

11.51 When NADH is reoxidized in the citric acid cycle, 3 mol of ATP is produced. In most cells, only 2 mol of ATP is produced from the reoxidation of the NADH produced during glycolysis. What causes this difference in energy yield?

11.52* What are the symptoms and cause of the genetic disease galactosemia?

11.53 Why can pyruvate cross the mitochondrial membrane while no other molecule after step 1 in glycolysis can?

11.54* Look at the glycolysis pathway. With what type of process are kinase enzymes usually associated?

11.55 Is the same net production of ATP observed in the complete combustion of fructose as is observed in glucose? Why or why not?

11.56* Explain why one more ATP is produced when glucose is obtained from glycogen rather than used directly from the blood?

CHAPTER

12 Lipids

Soap bubbles and cell membranes are both composed of lipids, and some of their properties are quite similar.

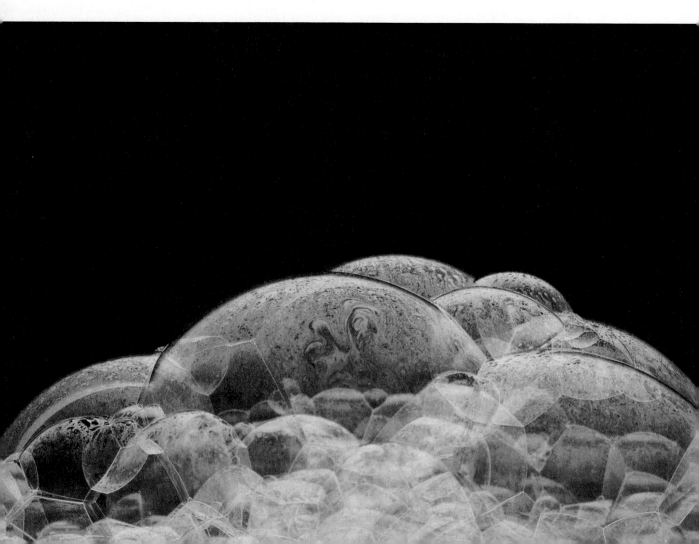

Lipids are less well known to most people than carbohydrates and proteins, yet they're just as essential to our diet and well-being. Lipids have many important biological roles, serving as sources of fuel, as a protective coat around many plants and insects, and as a major component of the membranes that surround every living cell.

Chemically, *lipids* are naturally occurring organic molecules that dissolve in nonpolar organic solvents when a sample of plant or animal tissue is crushed or ground. For example, if a plant is placed in a kitchen blender, finely ground, and then extracted with ether, anything that dissolves in the ether is called a lipid, while anything that remains insoluble (including carbohydrates, proteins, and inorganic salts) is not a lipid.

In this chapter, we'll answer the following questions about lipids:

1. **How are lipids classified?** The goal: Be able to recognize the names and structures of the different kinds of lipids.

2. **What are fats, waxes, and oils?** The goal: Be able to describe the general structures and general properties of waxes, fats, and oils.

3. **What reactions do triacylglycerols undergo?** The goal: Be able to list the most important reactions of these lipids and, given the reactants, predict the products.

4. **What are cell membranes?** The goal: Be able to give the general structure of cell membranes and describe the lipids they contain.

5. **What are steroids, prostaglandins, and leukotrienes?** The goal: Be able to describe the general structures of these biomolecules, describe some of their functions, and recognize specific examples.

12.1 STRUCTURE AND CLASSIFICATION OF LIPIDS

Lipid A naturally occurring molecule found in plant or animal sources that is soluble in nonpolar organic solvents.

Look at the structures of some representative **lipids** shown in Figure 12.1. lipids are defined by solubility in nonpolar solvents (a physical property) rather than by chemical structure, it's not surprising that there are a great many different kinds. Note in all the structures shown that the molecules contain large hydrocarbon portions and not many polar groups, which accounts for their solubility behavior.

Figure 12.2 summarizes the important classes of lipids that we'll be looking at in this chapter. *Waxes* are esters of fatty acids; *triacylglycerols* and *glycerophospholipids* are esters of glycerol; and *sphingolipids* are derivatives of an 18-carbon amino alcohol. *Steroids* (cyclic hydrocarbon derivatives) and *prostaglandins* and *leukotrienes* (both derivatives of a 20-carbon acid) differ from the other lipids in having no ester groups that can be hydrolyzed. Most of the lipid class names in Figure 12.2 are probably unfamiliar to you at this point, so you might want to check back occasionally as you read this chapter.

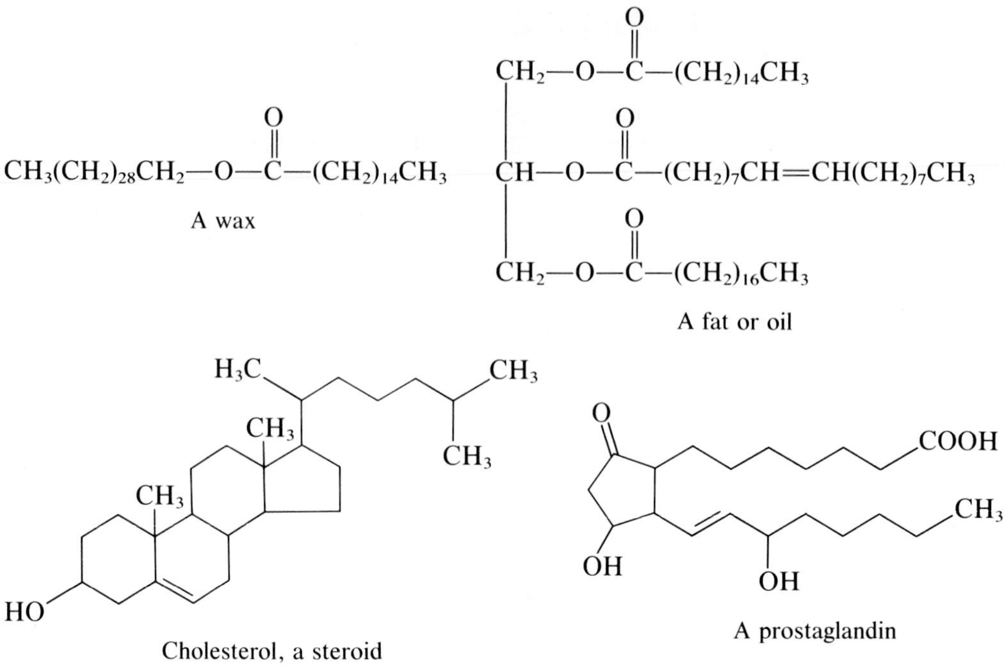

$$CH_3(CH_2)_{28}CH_2-O-\overset{\overset{\displaystyle O}{\|}}{C}-(CH_2)_{14}CH_3$$

A wax

$$CH_2-O-\overset{\overset{\displaystyle O}{\|}}{C}-(CH_2)_{14}CH_3$$
$$CH-O-\overset{\overset{\displaystyle O}{\|}}{C}-(CH_2)_7CH=CH(CH_2)_7CH_3$$
$$CH_2-O-\overset{\overset{\displaystyle O}{\|}}{C}-(CH_2)_{16}CH_3$$

A fat or oil

Cholesterol, a steroid

A prostaglandin

Figure 12.1
Structures of some representative lipids. Compounds that can be isolated from plant and animal tissue by extraction with nonpolar organic solvents are classified as lipids.

Figure 12.2
Some important families of lipids. Red shaded boxes show those derived from glycerol, and blue shaded boxes show those derived from sphingosine, an amino alcohol. Phospholipids are marked with an asterisk, and boxes outlined in red show lipids that are not esters and cannot by hydrolyzed into components.

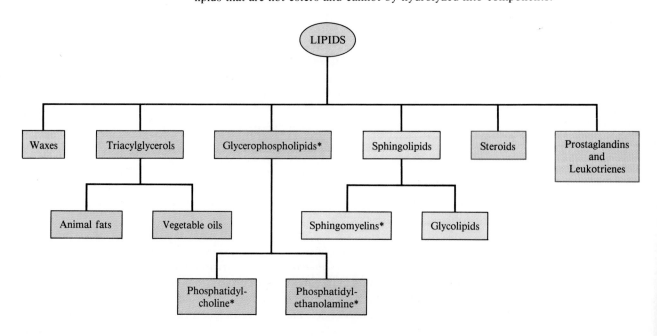

12.2 FATTY ACIDS, WAXES, FATS, AND OILS

Fatty acid A long-chain carboxylic acid; those in animal fats and vegetable oils often have 12–22 carbon atoms.

Saturated Containing only single bonds between carbon atoms and thus unable to accommodate additional hydrogen atoms.

Unsaturated Containing one or more double or triple bonds between carbon atoms and thus able to add more hydrogen atoms.

Polyunsaturated fatty acid (PUFA) A long-chain fatty acid that has two or more carbon–carbon double bonds.

The carboxylic acids present in naturally occurring fats and oils are referred to as **fatty acids.** All consist of long, straight hydrocarbon chains with a carboxylic acid group at one end, and most have even numbers of carbon atoms. Fatty acids may or may not contain double bonds. Those without double bonds are known as **saturated** fatty acids; those with double bonds are known as **unsaturated** fatty acids. If double bonds are present, they are usually cis rather than trans.

$$\begin{array}{c} \text{A fatty acid} \\ (n \text{ is usually an even number}) \end{array} \qquad \underset{\displaystyle CH_3(CH_2)_n\overset{\textstyle O}{\overset{\textstyle \|}{C}}OH}{}$$

About 40 different fatty acids are known to occur naturally, with some of the common ones listed in Table 12.1. Palmitic acid (16 carbons) and stearic acid (18 carbons) are the most abundant saturated acids; oleic and linoleic acids (both with 18 carbons) are the most abundant unsaturated ones. Oleic acid is *monounsaturated,* because it has only one carbon–carbon double bond, but linoleic, linolenic, and arachidonic acids are **polyunsaturated fatty acids** (called **PUFAs**), because they have more than one carbon–carbon double bond.

Two of the polyunsaturated fatty acids, linoleic and linolenic, are essential in the human diet because the body can't synthesize them. Infants grow poorly and develop severe skin lesions if fed a diet of nonfat milk, which lacks these acids; adults usually have sufficient reserves of body fat to avoid such problems. Arachidonic acid also plays an essential role in metabolism (Section 12.11), but the body is able to synthesize it from linolenic acid.

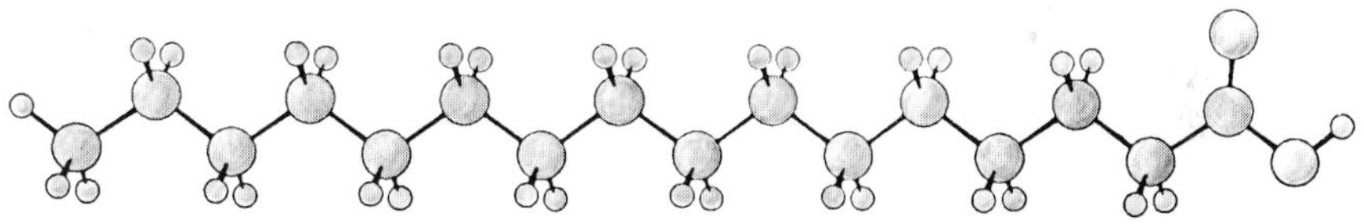

$$CH_3CH_2CH_2CH_2CH_2CH_2CH_2CH_2CH_2CH_2CH_2CH_2CH_2CH_2CH_2\overset{\textstyle O}{\overset{\textstyle \|}{C}}{-}OH$$

Palmitic acid—a saturated fatty acid

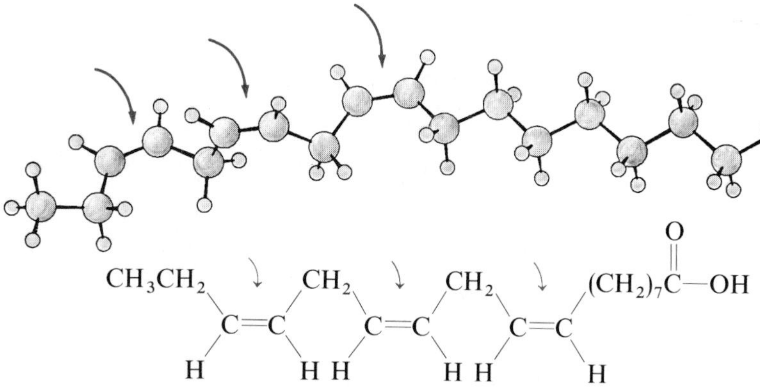

$$CH_3CH_2 \quad CH_2 \quad CH_2 \quad (CH_2)_7\overset{\textstyle O}{\overset{\textstyle \|}{C}}{-}OH$$

Linolenic acid—a polyunsaturated fatty acid (PUFA)

Table 12.1 Structures of Some Common Fatty Acids

Name	Number of Carbons	Number of Double Bonds	Structure	Melting Point (°C)
Saturated				
Lauric	12	0	$CH_3(CH_2)_{10}COOH$	44
Myristic	14	0	$CH_3(CH_2)_{12}COOH$	58
Palmitic	16	0	$CH_3(CH_2)_{14}COOH$	63
Stearic	18	0	$CH_3(CH_2)_{16}COOH$	70
Unsaturated				
Oleic	18	1	$CH_3(CH_2)_7CH{=}CH(CH_2)_7COOH$ (cis)	4
Linoleic	18	2	$CH_3(CH_2)_4CH{=}CHCH_2CH{=}CH(CH_2)_7COOH$ (all cis)	−5
Linolenic	18	3	$CH_3CH_2CH{=}CHCH_2CH{=}CHCH_2CH{=}CH(CH_2)_7COOH$ (all cis)	−11
Arachidonic	20	4	$CH_3(CH_2)_4(CH{=}CHCH_2)_4CH_2CH_2COOH$ (all cis)	−50

Wax A mixture of esters of long-chain carboxylic acids with long-chain alcohols.

Most natural **waxes** are mixtures of esters, formed in nature by reaction of fatty acids with long-chain alcohols. The carboxylic acid portion usually has an even number of carbons from 16 through 36, while the alcohol portion contains an even number of carbons in the range from 24 through 36. For example, one of the major components in beeswax is triacontyl hexadecanoate, the ester formed from the 30-carbon alcohol (triacontanol) and the 16-carbon acid (hexadecanoic acid, commonly known as palmitic acid). The waxy protective coatings on most fruits, berries, leaves, and animal furs have similar structures. Aquatic birds have a water-repellent waxy coating on their feathers and are in danger when caught in an oil spill because the waxy coating dissolves (Figure 12.3).

A wax:

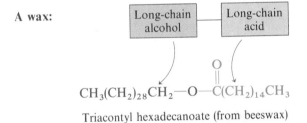

$$CH_3(CH_2)_{28}CH_2{-}O{-}\overset{\displaystyle O}{\overset{\|}{C}}(CH_2)_{14}CH_3$$

Triacontyl hexadecanoate (from beeswax)

Figure 12.3
Bird caught in a petroleum spill. The oil-soaked feathers have lost their waxy coating and also their insulating ability.

Fat A mixture of triacylglycerols that is solid because it contains a high proportion of saturated fatty acids.

Oil A mixture of triacylglycerols that is liquid because it contains a high proportion of unsaturated fatty acids.

Triacylglycerol (triglyceride) A triester of glycerol with three fatty acids.

Animal **fats** and vegetable **oils** are the most plentiful lipids in nature. Although they appear different—animal fats like butter and lard are usually solid, while vegetable oils like corn, olive, soybean, and peanut oil are liquid—their structures are closely related. All fats and oils are **triacylglycerols,** or *triglycerides;* that is, they are esters of glycerol (1,2,3-propanetriol, *glycerin*) with three fatty acids.

A triacylglycerol

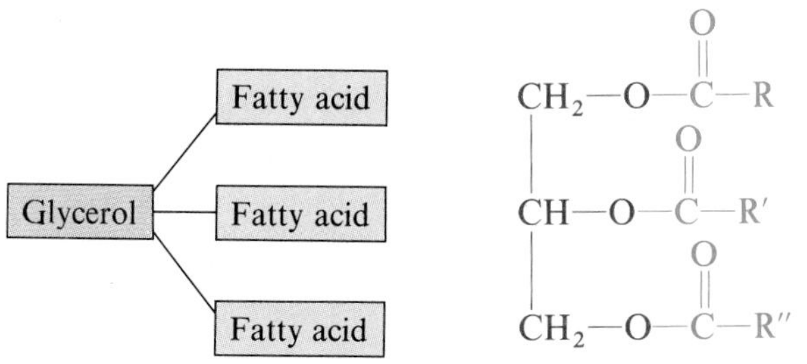

A fat or oil

For example, a typical triglyceride might have the following structure:

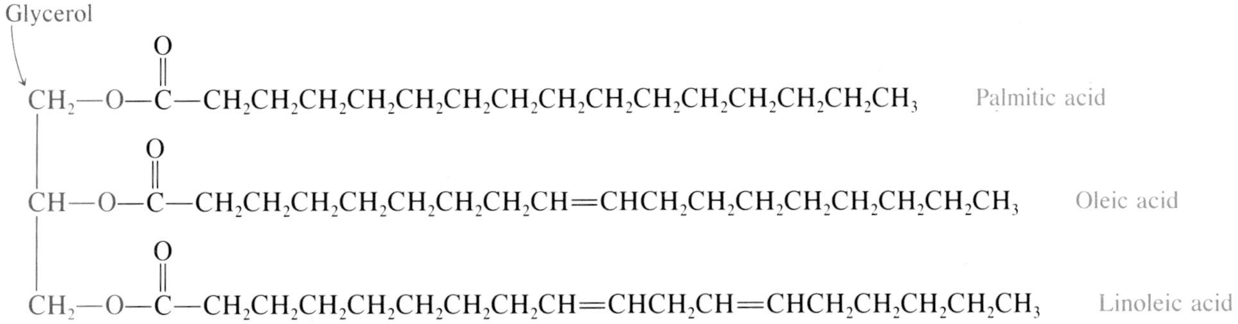

As shown in the triacylglycerol structure above, the three fatty acids of any specific fat or oil molecule are not necessarily the same. Furthermore, the fat or oil from a given natural source is a complex mixture of many different triacylglycerols. Table 12.2 lists the average composition of fats and oils from several different sources. Note particularly in Table 12.2 that vegetable oils consist almost entirely of unsaturated fatty acids, while animal fats contain a much larger percentage of saturated fatty acids. We'll see in the next section that this difference in composition is the primary reason for the different melting points of fats and oils.

Practice Problems **12.1** One of the constituents of the carnauba wax used in floor and furniture polish is an ester of a C_{32} straight-chain alcohol with C_{20} straight-chain carboxylic acid. Draw the structure of this ester.

12.2 Show the structure of glyceryl trioleate, a fat molecule whose components are glycerol and three oleic acid units.

Table 12.2   Approximate Composition of Some Common Fats and Oils[a]

Source	Saturated Fatty Acids (%)				Unsaturated Fatty Acids (%)	
	C_{12} Lauric	C_{14} Myristic	C_{16} Palmitic	C_{18} Stearic	C_{18} Oleic	C_{18} Linoleic
Animal Fat						
Lard	—	1	25	15	50	6
Butter	2	10	25	10	25	5
Human fat	1	3	25	8	46	10
Whale blubber	—	8	12	3	35	10
Vegetable Oil						
Corn	—	1	8	4	46	42
Olive	—	1	5	5	83	7
Peanut	—	—	7	5	60	20
Soybean	—	—	7	4	34	53

[a]Where totals are less than 100%, small quantities of several other acids are present.

12.3   PROPERTIES OF FATS AND OILS

The melting points of the various fatty acids listed in Table 12.1 show that as a general rule, the more double bonds a fatty acid has, the lower its melting point. Thus, the saturated C_{18} acid (stearic) melts at 70°C, the monounsaturated C_{18} acid (oleic) melts at 4°C, and the diunsaturated C_{18} acid (linoleic) melts at −5°C. This same trend also holds true for triacylglycerols: The more highly unsaturated a triacylglycerol, the lower its melting point. The difference in melting points between fats and oils is a consequence of differences in their chemical structures. As shown in Table 12.2, vegetable oils generally have a higher proportion of unsaturated to saturated fatty acids than animals fats.

How do the double bonds make such a significant difference in properties? Compare the structures of palmitic and linoleic acids shown in Section 12.2, for example. The hydrocarbon chains in saturated acids like palmitic are uniform in shape with identical angles of each carbon atom, allowing the chains to nestle easily together in a crystal. By contrast, the carbon chain in an unsaturated acid like linoleic has a kink wherever it passes through a cis double bond. The kinks make it difficult for such chains to fit next to each other in a uniform fashion to form crystals. The more double bonds there are in a triacylglycerol, the harder it is for it to crystallize, an effect illustrated in Figure 12.4 with molecular models.

We are accustomed to the characteristic yellow color and flavors of cooking oils, but these are contributed by natural materials carried along during production of the oils from plants; pure oils are colorless and odorless. Overheating, or exposure to air, oxidizing agents, or water, causes decomposition to products with unpleasant odors or flavors, creating what we call a *rancid* oil.

Practice Problem   12.3   Draw the full structure of the arachidonic acid (Table 12.1) in a way that shows the cis stereochemistry of its four double bonds.

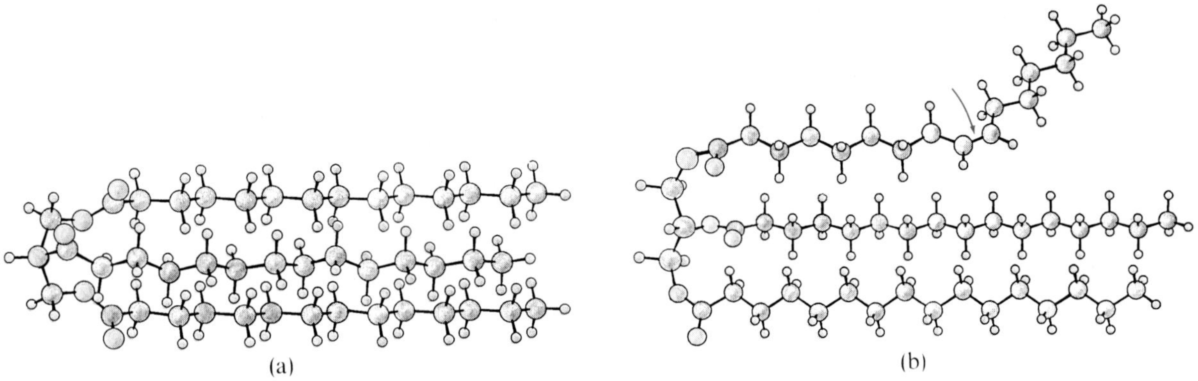

Figure 12.4
Molecular models of (a) a saturated and (b) an unsaturated triacylglycerol. The double bond (red arrow) in (b) prevents the molecule from adopting a regular shape and crystallizing easily. Fats are solids because of their high proportion of saturated hydrocarbon chains, and oils are liquids because of their high proportion of unsaturated hydrocarbon chains.

12.4 CHEMICAL REACTIONS OF TRIACYLGLYCEROLS

Hydrogenation The carbon–carbon double bonds in vegetable oils can be hydrogenated to yield saturated fats in the same way that any alkene can react with hydrogen to yield an alkane (Section 2.6). Margarine and solid cooking fats like Crisco are produced commercially by hydrogenation of vegetable oils to give a product chemically similar to that found in animal fats.

Part-structure of an unsaturated vegetable oil

Part-structure of a saturated fat

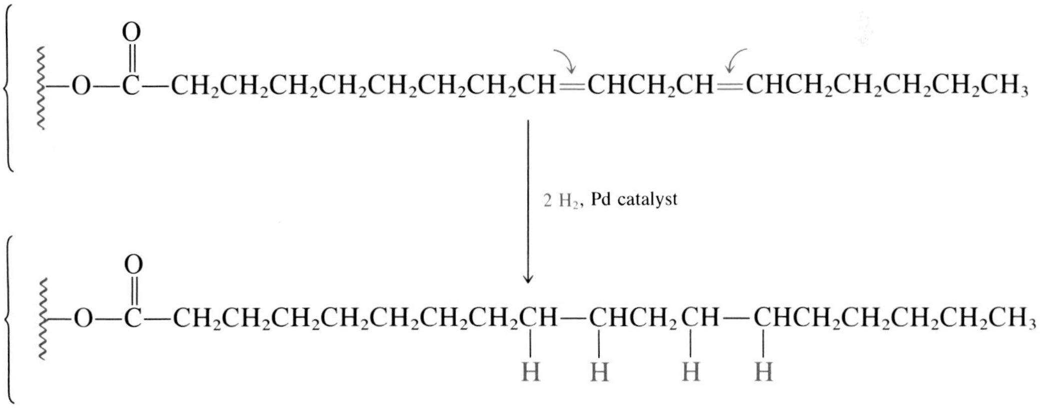

By carefully controlling the exact extent of hydrogenation, the final product can be made to have any desired consistency. Margarine, for example, is prepared so that only about two-thirds of the double bonds present in the starting vegetable oil are hydrogenated. The remaining double bonds are left unhydrogenated, so the margarine has exactly the right consistency to keep soft in the refrigerator and to melt on warm toast.

AN APPLICATION: LIPIDS IN THE DIET

The major recognizable sources of fats and oils in our diet are butter and margarine, lard and vegetable oils, the visible fat in meat, and chicken skin. In addition, there are "invisible" lipids in milk and milk products, meat, fish, poultry, nuts and seeds, whole-grain cereals, and prepared foods. As illustrated in Table 12.2, different foods have characteristic distributions of unsaturated and saturated triacylglycerols. In meat, poultry, fish, dairy products, and eggs, the triacylglycerols are mostly saturated and are accompanied by small quantities of cholesterol. Vegetable oils have a higher unsaturated fatty acid content and contain no cholesterol. They never have, so the salad oil labels that proclaim, "No cholesterol," are stating a fact, not announcing something special the manufacturer has done for us.

Fats and oils are a popular component of our diet: they taste good, give a pleasant texture to food, and, because they are digested slowly, give a feeling of satisfaction after a meal. The percentage of calories from fats and oils in the average U.S. diet in the early 1980s was 40–45%, considerably more than needed. Remember that only two fatty acids are designated as essential because our bodies can't make them. Fats and oils in our diet beyond what we burn are mostly stored as fat in *adipose tissue* that lies in layers under the skin and around internal organs.

Concern for the relationships among saturated fats, cholesterol levels, and various diseases caused a modest decrease to 37% average calories from fats and oils in the U.S. diet by 1990. We have significantly decreased our consumption of butter, eggs, beef, and whole milk (all containing relatively high proportions of saturated fat and cholesterol) but at the same time have increased our consumption of cheese and the oils in salad dressings and baked goods. An interesting development, largely the result of public pressure, has been the agreement by several large commercial bakeries to decrease their

---

Practice Problem **12.4** Write an equation showing the complete hydrogenation of glyceryl trioleate (Practice Problem 12.2) to yield glyceryl tristearate.

Hydrolysis The breakdown of a compound by reaction with water.

Saponification Reaction of a fat or oil with aqueous hydroxide ion to yield glycerol and carboxylate salts of fatty acids.

Hydrolysis: Soap Triacylglycerols, like all esters, can be hydrolyzed to yield carboxylic acid and alcohol components. In the body, this **hydrolysis** is carried out by enzymes as the first step in lipid metabolism. In the laboratory, the same reaction is usually carried out with aqueous NaOH or KOH, where it is called **saponification** (Section 6.7). The initial products of fat (or oil) saponification are one molecule of glycerol and three molecules of fatty acid carboxylate salts.

A fat or oil →(NaOH / H₂O)→ Glycerol + Fatty acid salts (soap)

use of the "tropical oils" palm oil and coconut oil. Although called "oils" because of their plant origin, these substances are solids at room temperature and contain significant proportions of saturated fats.

Several organizations concerned with health maintenance have recommended a diet with no more than 30% of its calories from fats and oils. Furthermore, it's recommended that no more than 10% of daily calories come from saturated fat and no more than 300 mg of cholesterol per day be included. In a daily diet of 3000 Calories, 30% from fats and oils is about 100 g, the amount in a bit more than 1/4 lb of butter.

Triacylglycerols from the seeds of these sunflowers will soon appear on supermarket shelves as vegetable oil and margarine.

Soap The mixture of carboxylate salts of fatty acids formed on saponification of animal fat.

Hydrophilic Water-loving; a hydrophilic substance dissolves in water.

Hydrophobic Water-fearing; a hydrophobic substance does not dissolve in water.

Micelle A spherical cluster formed by the aggregation of soap molecules (or other molecules with hydrophilic and hydrophobic ends) in water.

The complex mixture of fatty acid salts produced by saponification of animal fat with NaOH or KOH is known as **soap.** Crude soap curds, which contain glycerol and excess alkali as well as soap, are first purified by boiling with water and adding NaCl to precipitate the pure sodium carboxylate salts. The smooth soap that precipitates is then dried, perfumed, and pressed into bars for household use. Dyes are added for colored soaps; antiseptics are added for medicated soaps; pumice is added for scouring soaps; and air is blown in for soaps that float.

Soaps act as cleaning agents because the two ends of a soap molecule are so different. The sodium-salt end of the long-chain molecule is ionic and therefore **hydrophilic** (water-loving); it tends to dissolve in water. The long organic chain portion of the molecule, however, is nonpolar and therefore **hydrophobic** (water-fearing); it tends to avoid water and to dissolve in grease. Because of these two opposing tendencies, soaps are attracted to *both* grease and water.

When soap molecules are dispersed in water, very few of them actually dissolve. Instead, large numbers of soap molecules aggregate together, with their long, hydrophobic hydrocarbon tails clustering in a ball to exclude water. At the same time, their hydrophilic ionic heads on the surface of the cluster stick out into the water. These clusters, called **micelles,** are diagrammed in Figure 12.5. Grease and dirt are made soluble in water when they are coated by the nonpolar tails of the soap molecules in the center of micelles. Once solubilized, the grease and dirt can be washed away.

Figure 12.5
A soap micelle. The hydrophobic ends of the soap molecules extend into the grease particle, while the hydrophilic ends keep the micelle soluble in water. Lipids are transported in the bloodstream in similar micelles, as described in Chapter 13.

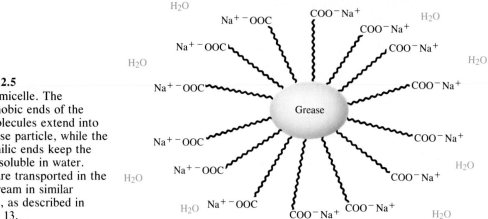

Practice Problems   **12.5**   In hard water, which contains iron, magnesium, and calcium ions, soap often leaves a scummy residue composed of precipitated acid salts. Draw the structure of calcium oleate, one of the components of bathtub scum.

12.6   Write the complete equation for the hydrolysis of a triacylglycerol in which the fatty acids are two molecules of stearic acid and one of oleic acid (see Table 12.1).

AN APPLICATION: DETERGENTS

Strictly speaking, anything that washes away dirt is a *detergent*. The term is usually applied, however, to synthetic materials made from petroleum chemicals that began to replace soaps in the 1950s. The goal was to overcome the problems caused by precipitation of soap scum in hard water, which include waste of soap, residues left in washed clothing, and difficulty in cleaning bathtubs and washing machines.

Like soaps, detergent molecules have hydrophobic hydrocarbon tails and hydrophilic heads, and they cleanse by the same mechanism as soap—forming micelles around greasy dirt. Materials that function in this manner are described as *surface-active agents* or *surfactants*. The hydrophilic heads may be anionic, cationic, or nonionic. Anionic detergents are commonly used in home laundry products.

An anionic detergent

$$Na^+\,^-O-\underset{\underset{\displaystyle O}{\|}}{\overset{\overset{\displaystyle O}{\|}}{S}}-\!\!\!\!\bigcirc\!\!\!\!-CH_2CH_2CH_2CH_2CH_2CH_2CH_2CH_2CH_2CH_2CH_2CH_3$$

Sodium dodecylbenzenesulfonate

12.5 MEMBRANE LIPIDS

The major function of lipids in the body, other than energy storage, is as components of cell membranes. Lipids are ideal for this function because membranes must separate the aqueous solutions inside cells from the aqueous extracellular fluid. The molecules of membrane lipids are similar to soap and detergent molecules in having hydrophobic tails and hydrophilic heads, but they differ in having two tails instead of one.

Membrane lipids are built up from alcohols, amino alcohols, fatty acids, and phosphate groups. Some are esters of glycerol, like the triacylglycerols, but with a phosphate ester in place of one fatty acid ester group. Others are esters of the amino alcohol *sphingosine*.

$$
\begin{array}{l}
1\ CH_2OH \\
|\\
2\ CHOH \\
|\\
3\ CH_2OH
\end{array}
\qquad
\begin{array}{l}
1\ CH_2OH \\
|\\
2\ CHNH_2 \\
|\\
3\ CHOH \\
|\\
4\ CH{=}CH(CH_2)_{12}CH_3
\end{array}
$$

Glycerol Sphingosine

The general structures and names of the major types of membrane lipids are summarized in Figure 12.6. After describing these membrane lipids, we'll take a look at how cell membranes are organized and function.

Cationic detergents are used in fabric softeners and in disinfectant soaps.

A cationic detergent

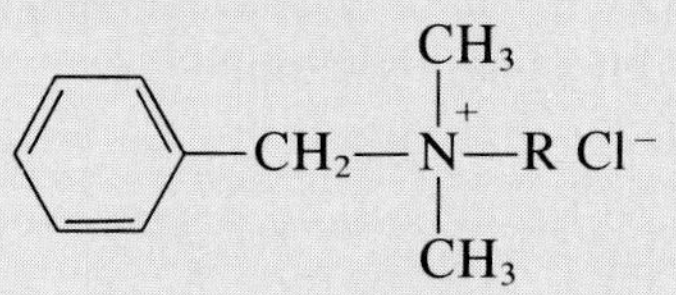

$$\text{A benzalkonium chloride; } R = C_8H_{17} \text{ to } C_{18}H_{37}$$

Nonionic detergents are low-sudsing and are effective at low temperatures.

A nonionic detergent

$$CH_3(CH_2)_{10}CH_2OCH_2CH_2(OCH_2CH_2)_7OH$$

A polyether

Note that the hydrocarbon chains in these detergents are unbranched. Some of the first detergents contained branched-chain hydrocarbons, but this was soon discovered to be a mistake. The bacteria in natural waters and sewage treatment plants are slow to eat branched-chain hydrocarbons, so detergents containing them went undecomposed and produced suds in streams and lakes.

The average washing machine consumes about 20 lb of detergents per year, but the use of surfactants goes way beyond the home laundry. Of the 7.6 billion pounds of surfactants used in 1988, 29% went into household products, 54% went into industrial uses, and 17% went into personal care products such as shampoos and cosmetics.

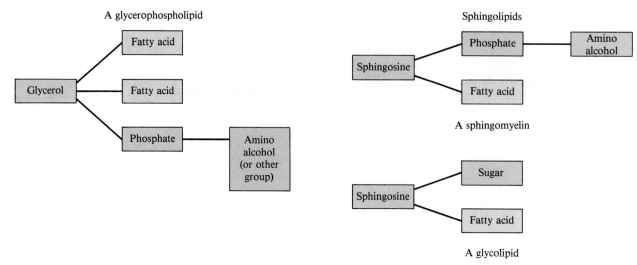

A glycerophospholipid

Glycerol — Fatty acid
— Fatty acid
— Phosphate — Amino alcohol (or other group)

Sphingolipids

Sphingosine — Phosphate — Amino alcohol
— Fatty acid

A sphingomyelin

Sphingosine — Sugar
— Fatty acid

A glycolipid

Figure 12.6
The major membrane lipids. The polar head groups are shown in the green boxes.

12.6 GLYCEROPHOSPHOLIPIDS

Glycerophospholipid (phosphoglyceride) A lipid in which glycerol is linked by ester bonds to two fatty acids and one phosphate, which is in turn linked by another ester bond to an amino alcohol (or other alcohol); a derivative of phosphatidic acid.

Glycerophospholipids, also known as **phosphoglycerides,** resemble triacylglycerols in having a glycerol unit linked by ester bonds to two fatty acids (the two hydrophobic tails). They differ, however, at the third position of glycerol, which is linked to a phosphate instead of a fatty acid. (Recall from Section 6.12 how phosphoric acid can form phosphate esters with alcohols.) The fatty acids may be any of the C_{12} to C_{20} acids normally present in fats, with the acid bonded to C1 of glycerol usually saturated, while that at C2 is usually unsaturated. The phosphate group, which is always bonded to C3, is also connected by a separate ester link to another compound, often a small amino alcohol. The glycerophospholipids, some of which are listed in Table 12.3, are named as derivatives of *phosphatidic acid.* For example, phosphatidylcholine:

Phosphatidic acid structure:

$$CH_2-O-\overset{\overset{\displaystyle O}{\|}}{C}-R$$
$$CH-O-\overset{\overset{\displaystyle O}{\|}}{C}-R'$$
$$CH_2-O-\overset{\overset{\displaystyle O}{\|}}{P}-O-H$$
$$O-H$$

Phosphatidic acid

Phosphatidylcholine structure:

Glycerol — A saturated fatty acid
$$CH_2-O-\overset{\overset{\displaystyle O}{\|}}{C}-R$$
An unsaturated fatty acid
$$CH-O-\overset{\overset{\displaystyle O}{\|}}{C}-R'$$
A phosphate group
$$CH_2-O-\overset{\overset{\displaystyle O}{\|}}{P}-O-CH_2CH_2\overset{+}{N}(CH_3)_3$$
$$O^-$$
Choline

Phosphatidylcholine (a lecithin)

Table 12.3 Some Glycerophospholipids

General Structure	X =	X—OH Name	Type of Glycerophospholipid
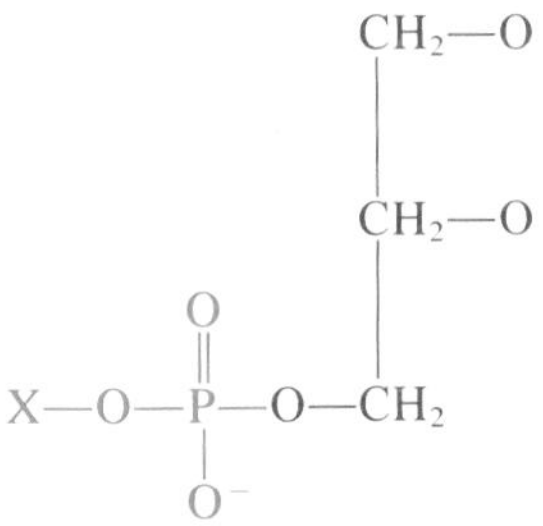	H—	Water	Phosphatidic acid (parent compound for glycerophospholipids)
	$H_3\overset{+}{N}CH_2CH_2$—	Ethanolamine	Phosphatidylethanolamine
	$(CH_3)_3\overset{+}{N}CH_2CH_2$—	Choline	Phosphatidylcholine (lecithin; most abundant cell membrane phospholipid)
	$^-OOCCHCH_2$— (with NH_3^+)	Serine	Phosphatidylserine (found in most tissues)
	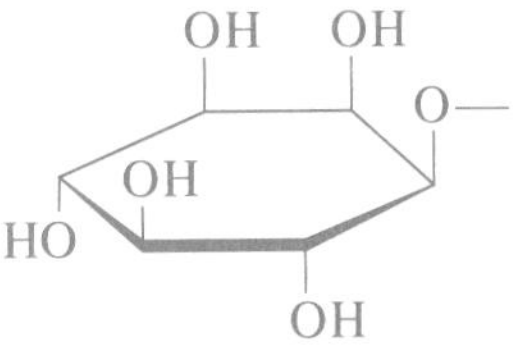 Inositol ring (OH, OH, OH, HO, OH, O—)	Inositol	Phosphatidylinositol (relays chemical signals from outside cell to inside cell)

General Structure:

$$CH_2-O$$
$$CH_2-O$$
$$X-O-\overset{\overset{O}{\|}}{\underset{O^-}{P}}-O-CH_2$$

Phosphatidylcholine (lecithin) A glycerophospholipid containing choline.

The **phosphatidylcholines,** also known as **lecithins,** are major components of cell membranes in all higher organisms and are also important as a source of the choline needed in transmission of nerve signals. In solution at physiological pH, the hydrophilic head group in a phosphatidylcholine has a positive charge on the quaternary nitrogen in choline and a negative charge on the ionized phosphate group, as shown in green in the structure above. A compound referred to in the singular as phosphatidycholine or lecithin is often a mixture of molecules with different R and R′ tails.

Phosphatidylcholine and other glycerophospholipids are *emulsifying agents,* substances that can coat droplets of nonpolar liquids and hold them in suspension in water. You can find lecithin, usually obtained from soybean oil, listed as an ingredient in chocolate bars and other foods in which it serves as an emulsifier. It's also the lecithin in egg yolk that emulsifies the oil droplets in mayonnaise. Lecithin can be purchased at many vitamin counters and health food stores, and it has been touted as able to lower blood pressure and improve memory, although firm evidence for these effects is yet to be presented.

Phosphatidylethanolamine and phosphatidylserine are next after phosphatidylcholine in occurrence in cell membranes and are particularly abundant in brain tissue. Note in Table 12.3 that the hydrophilic head groups in these glycerophospholipids (shown in green) also contain ionic charges in solution at physiological pH.

Practice Problem **12.7** Show the structure of the glycerophospholipid that contains a stearic acid unit, an oleic acid unit, and a phosphate bonded to ethanolamine.

12.7 SPHINGOLIPIDS

Sphingolipid A lipid derived from the amino alcohol sphingosine (or related amino alcohols) rather than glycerol.

Sphingomyelin A sphingolipid with a fatty acid bonded to the C2 —NH_2 and a phosphate bonded to the C1 —OH group of sphingosines.

Phospholipid A lipid that has an ester link between phosphoric acid and an alcohol.

Sphingolipids are derivatives of the amino alcohol *sphingosine* rather than glycerol. The most common sphingolipids are the **sphingomyelins,** which contain phosphorus and are therefore also **phospholipids.** In sphingomyelins, a long-chain fatty acid is bonded to the C2 —NH_2 group by an amide link, and a phosphoric acid unit is bonded to the C1 —OH group. The phosphate group is also linked by an ester bond to choline. The sphingomyelins are a major constituent of the coating around nerve fibers (the *myelin sheath;* Figure 12.7) and are found in large quantities in brain tissue. A diminished amount of sphingomyelins in the brain myelin is associated with multiple sclerosis, a disease whose cause is not understood.

Choline — Phosphate group — Amide link

$$(CH_3)_3\overset{+}{N}CH_2CH_2{-}O{-}\underset{\underset{O^-}{|}}{\overset{\overset{O}{\|}}{P}}{-}O{-}\underset{2}{\overset{1}{C}}H_2$$

$$\underset{2}{CH}{-}NH{-}\overset{\overset{O}{\|}}{C}(CH_2)_{14}CH_3 \quad \text{— A fatty acid}$$

$$\underset{3}{CH}{-}OH$$

Sphingosine — $\underset{4}{CH}{=}CH(CH_2)_{12}CH_3$

Sphingomyelin
(a sphingolipid)

Glycolipid A sphingolipid with a fatty acid bonded to the C2 —NH_2 and a sugar bonded to the C1 —OH group of sphingosines.

Yet another class of sphingosine-based lipids are the **glycolipids,** or *sphingoglycolipids:* compounds that contain both carbohydrate and lipid parts but no phosphorus. Glycolipids differ from sphingomyelins in having a sugar at C1 instead of a phosphate ester group. In the *cerebrosides,* for example, an amide link connects a fatty acid and the —NH_2 group at C2, and a glycoside link connects a monosaccharide and the —OH group at C1. Cerebrosides are particularly abundant in nerve cell membranes in the brain, where the sugar unit is D-galactose. They are also found in nonneural cell membranes, where the sugar unit is D-glucose.

Figure 12.7
Myelin sheath (orange) around the human auditory nerve. The myelin forms within cells that surround the axons of nerve cells.

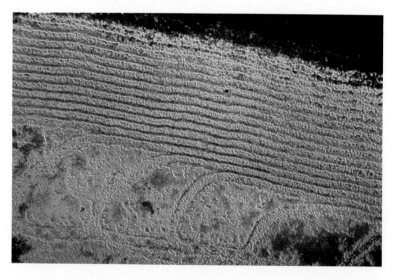

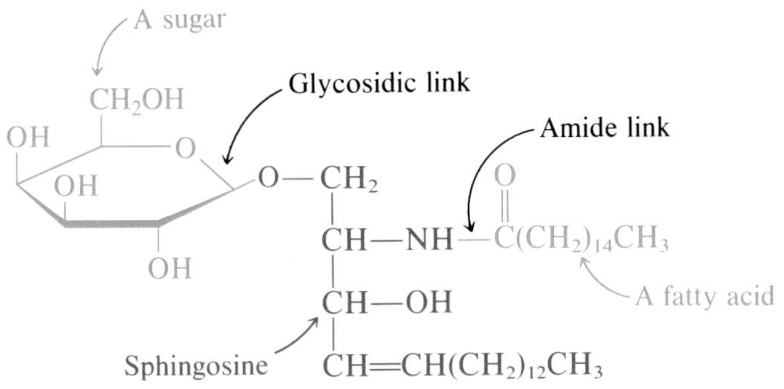

Closely related to the cerebrosides are the *gangliosides:* glycolipids in which the carbohydrate component linked to sphingosine is a small polysaccharide rather than a monosaccharide. Over 60 gangliosides are known. Their saccharide portions extend out of the cell membrane and apparently play important roles as receptors for hormones and neurotransmitters and in recognizing toxins and other cells.

Tay-Sachs disease, a fatal genetic disorder, is due to a defect in lipid metabolism that causes a greatly elevated concentration of a particular ganglioside in the brain. An infant born with this defect suffers mental retardation and liver enlargement and often succumbs by age 3. Tay-Sachs is one of a group of *sphingolipid storage diseases,* metabolic diseases that result from deficiencies in the supply of enzymes that break down sphingolipids.

Solved Problem 12.1   A class of membrane lipids known as *plasmalogens* has the general structure shown here. Identify the component parts of this lipid and choose the terms that apply to it: phospholipid, glycerophospholipid, sphingolipid, glycolipid. Is it most similar to a phosphatidylethanolamine, a phosphatidylcholine, a cerebroside, or a ganglioside?

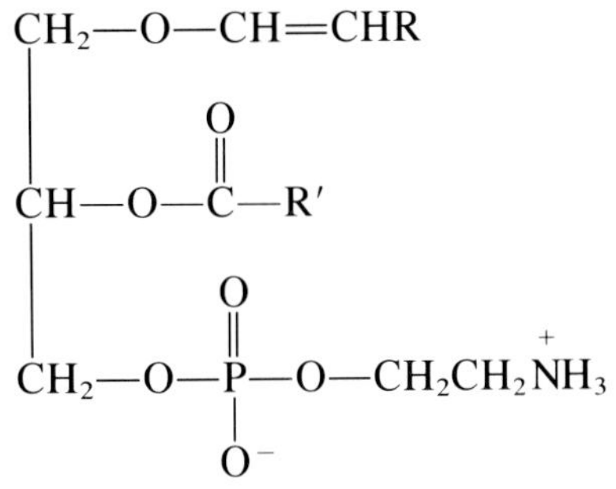

Solution   The molecule contains a phosphate group and thus is a *phospholipid,* The glycerol backbone of three carbon atoms bonded to three oxygen atoms is also present, so the compound is a *glycerophospholipid,* but one in which there is an ether linkage (CH_2—O—CH=CHR) instead of one ester linkage. The phosphate group is bonded to ethanolamine ($HOCH_2CH_2NH_2$). The compound is not a sphingolipid or a glycolipid because it is not derived from sphingosine; for the same reason it is not a cerebroside or a ganglioside. Except for the ether group in place of an ester group, the compound has the same structure as a *phosphatidylethanolamine.*

Practice Problem   *12.8*   Show the structure of the sphingomyelin that contains a myristic acid unit. Identify the hydrophilic head group and the hydrophobic tails in this molecule.

12.8   CELL MEMBRANES

Phospholipids form a major part of the membranes that surround all living cells. The lipid membranes serve not only to separate the interior of cells from their outside environment but also act as a kind of semipermeable barrier to allow the selective passage of nutrients and other components into and out of cells.

Phospholipids aggregate in a closed, sheet-like membrane structure called a **lipid bilayer** (Figure 12.8). Two parallel layers of lipids are oriented in the bilayer so that ionic head groups protrude into the aqueous environment on the inner and outer faces of the bilayer. The nonpolar tails cluster together in the middle of the bilayer where they can interact with each other and avoid water.

Cell membranes are a good deal more complex than the simple bilayer picture indicates, leading to an overall membrane structure called the *fluid-mosaic model* shown in Figure 12.9. One complexity, for example, is that the inner and outer faces of cell membranes are not identical. In human red blood cells (erythrocytes), for example, the outer layer consists primarily of sphingomyelin and phosphatidylcholine, while the inner layer is mainly phosphatidylethanolamine and phosphatidylserine.

Another complexity is that cell membranes contain many other kinds of molecules embedded in the bilayer. Glycolipids and cholesterol are present, and at least 20% of the weight of a membrane consists of protein globules (visible in Figure 12.10). Some proteins are only partially embedded (*peripheral* proteins), but others extend completely through the membrane (*integral* proteins). Lipid components of the membrane serve a primarily structural purpose, while proteins act to mediate the interactions of the cell with outside agents. Some proteins, for example, act as channels to allow specific molecules or ions to enter or leave the cell. The carbohydrate portions of glycolipids and glycoproteins (proteins bonded to carbohydrates) interact with molecules in the extracellular fluid.

Lipid bilayer The basic structural unit of cell membranes; composed of two parallel sheets of membrane lipid molecules arranged tail to tail.

Figure 12.8
Aggregation of phospholipids into the bilayer structure of a cell membrane.

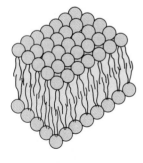

Where ⧸ is

$$CH_2{-}CH{-}CH_2{-}O{-}\overset{\overset{\displaystyle O}{\|}}{P}{-}O{-}CH_2CH_2\overset{+}{N}(CH_3)_3$$

$$O \qquad O \qquad\qquad O^-$$

Hydrophilic head

$$O{=}C \quad O{=}C$$

$$CH_2 \quad CH_2$$
$$CH_2 \quad CH_2$$
$$CH_2 \quad CH_2$$ Hydrophobic tails
$$CH_2 \quad CH_2$$
$$CH_2 \quad CH_2$$

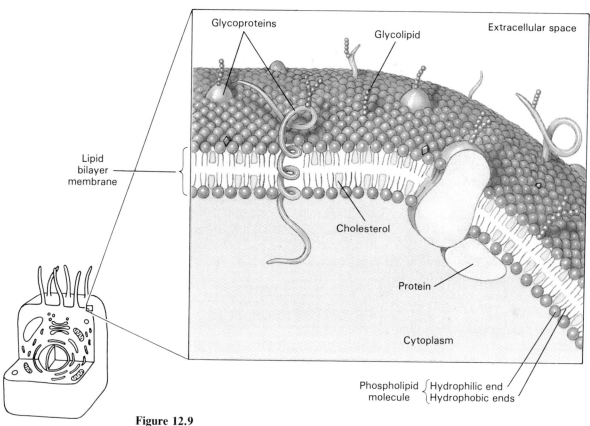

Glycoproteins

Glycolipid

Extracellular space

Lipid bilayer membrane

Cholesterol

Protein

Cytoplasm

Phospholipid molecule { Hydrophilic end Hydrophobic ends

Figure 12.9
Fluid mosaic model of a cell membrane. Cholesterol forms part of the membrane, proteins are embedded in the bilayer, and the carbohydrate chains of glycoproteins and glycolipids extend into the extracellular space, where they interact with other substances.

Figure 12.10
A fractured cell membrane. The lipid bilayer has been split between the two lipid layers. The smoother, hydrophilic side of the top layer—the outside of the cell—is at the top in the photo. The rougher, hydrophobic interior side of the bottom layer—the inside of the bilayer—is on the bottom. The globular particles (6–9 nm in diameter) are membrane proteins.

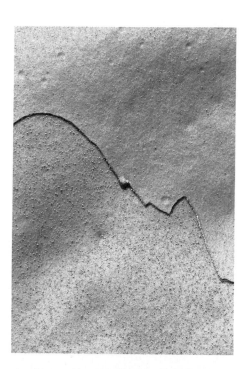

AN APPLICATION: TRANSPORT ACROSS CELL MEMBRANES

On the one hand, the membrane surrounding a living cell can't be impermeable, because nutrients must enter and waste products must leave the cell. On the other hand, the membrane can't be completely permeable, or substances would move back and forth until their concentrations were equal on both sides. In actuality, some substances move across the membranes freely in what is known as *passive transport,* and others do so only by using energy, a process known as *active transport*.

Transport across cell membranes is not yet fully understood, but three major types of transport have been identified.

Simple Diffusion Some solutes *diffuse* in and out of cells, which means they simply wander by normal molecular motion from areas of high concentration to areas of low concentration. Simple diffusion is one of the two types of passive transport. To cross the cell membrane by simple diffusion, a solute must leave the aqueous solution on one side of the membrane, dissolve in the nonpolar region in the bilayer, and return to the aqueous solution on the other side. The lipid bilayer is just about impermeable to ions and polar molecules, which aren't soluble in the nonpolar hydrocarbon region, but oxygen molecules, for example, easily diffuse through.

Facilitated Diffusion Like simple diffusion, facilitated diffusion is passive transport and requires no energy input. The difference is that in facilitated diffusion solutes are helped across the membrane by proteins, allowing their transport at a faster rate than could be accomplished by simple diffusion. Glucose is believed to be transported into red blood cells by facilitated diffusion. One possible way facilitated diffusion might occur is diagrammed at the bottom of this box.

Active Transport It's essential to life that the concentrations of some solutes be different inside and outside cells. Such differences are opposite the natural tendency of solutes to move about until their concentration equalizes and therefore can be maintained only by the expenditure of energy. A notable example of active transport is the constant movement of Na^+ and K^+ ions across cell membranes. Only in this manner is it possible to maintain low Na^+ concentrations within cells and high Na^+ concentrations in extracellular fluids, and the opposite for K^+. It's thought that energy is used to change the shape of an integral protein, simultaneously bringing three Na^+ ions into the cell and moving two K^+ ions out of the cell.

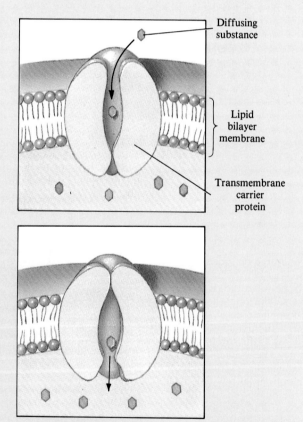

Diffusing substance

Lipid bilayer membrane

Transmembrane carrier protein

Facilitated diffusion. Molecules from the region of higher concentration diffuse into a carrier protein, which then changes conformation and releases molecules into the cell. Facilitated diffusion requires no energy and can only move species from higher- to lower-concentration regions.

One of the central features of the fluid-mosaic model is that the membrane is *fluid* rather than rigid. As a consequence, the membrane is not easily ruptured, because the lipids in the bilayer simply flow back together to repair any small hole or puncture. The effect is similar to what's observed in cooking when a thin film of oil or melted butter lies on top of water. The film can be punctured and broken, but it immediately flows back together when left alone. Still other consequences of bilayer fluidity are that small nonpolar molecules can move easily through the membrane and that individual lipid or protein molecules can diffuse rapidly from place to place within the membrane.

12.9 STEROIDS: CHOLESTEROL

Steroid A lipid whose structure is based on a tetracyclic (four-ring) carbon skeleton.

In addition to triacylglycerols and phospholipids, plant and animal cells also contain steroids. A **steroid** is a lipid whose structure is based on the following tetracyclic (four-ring) system. Three of the rings are six-membered, while the fourth is five-membered. The rings are designated A, B, C, and D, beginning at the lower left-hand corner, and the carbon atoms are numbered from 1 to 19, beginning in ring A.

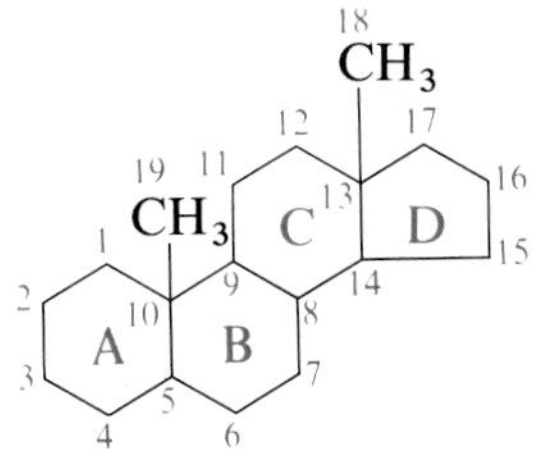

Steroid skeleton

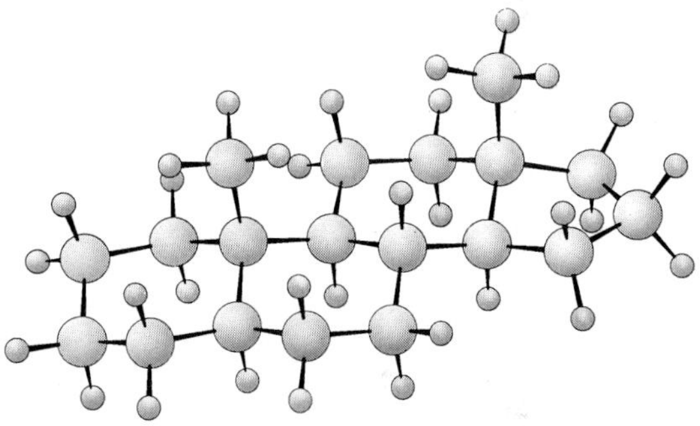

Steroids have many diverse roles throughout both the plant and animal kingdoms. Some, such as the plant steroid digitoxigenin isolated from the purple foxglove (*Digitalis purpurea*), are used in medicine as heart stimulants; others are hormones; and still others, such as cholic acid, which is synthesized in the liver from cholesterol, act as emulsifying agents to aid in the digestion of fats.

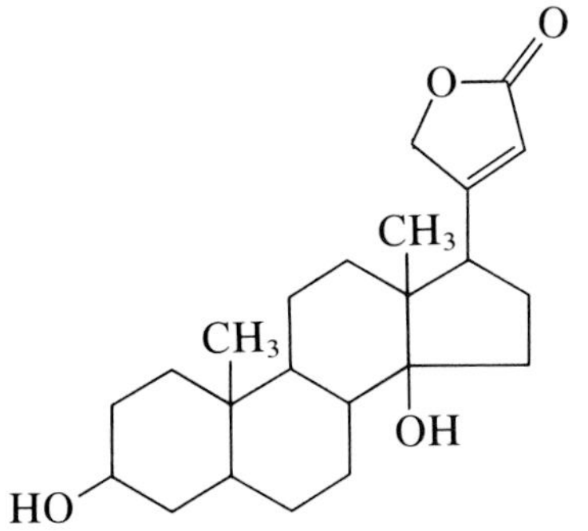

Digitoxigenin
(from purple foxglove)

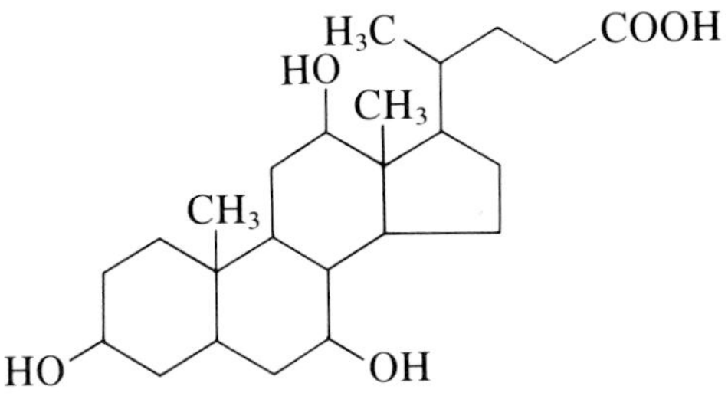

Cholic acid, a bile acid
(from liver bile)

Cholesterol, an unsaturated alcohol, is the most abundant animal steroid. It's been estimated that a 130 lb person has a total of about 175 g of cholesterol distributed throughout the body. Much of this cholesterol is esterified with fatty acids, but some is also found as the free alcohol. Gallstones, for example, are nearly pure cholesterol.

Cholesterol, an unsaturated steroidal alcohol

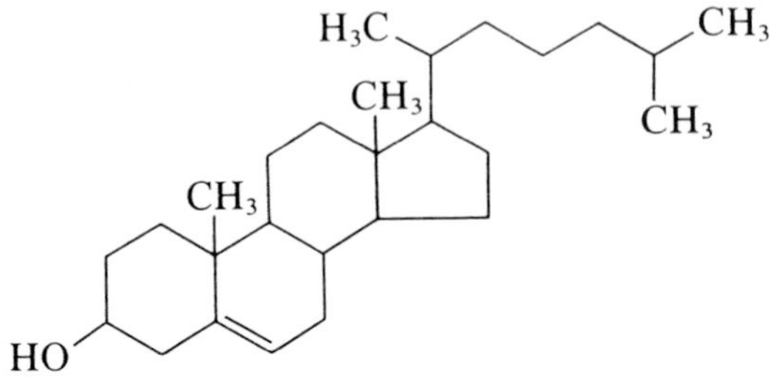

Cholesterol serves two important functions in the body. First, it is a component of cell membranes, where it acts as a regulator of membrane fluidity. Cholesterol is able to fit in between the fatty acid chains of the lipid bilayer as shown in Figure 12.9, restricting their motion and making the bilayer more rigid. Second, cholesterol serves as the body's starting material for the synthesis of all other steroids, including all the sex hormones and bile acids.

The human body obtains its cholesterol both by synthesis in the liver and by ingestion of food. Even on a strict no-cholesterol diet, the body of an adult is able to synthesize approximately 800 mg of cholesterol per day.

12.10 STEROID HORMONES

Hormones are chemical messengers that mediate biochemical events in target tissues (Chapter 8). There are two major classes of steroid hormones: the sex hormones, which control maturation and reproduction, and the adrenocortical hormones, which regulate a variety of metabolic processes.

Sex Hormones   *Testosterone and androsterone* are the two most important male sex hormones, or *androgens*. Androgens are responsible both for the development of male secondary sex characteristics during puberty and for promoting tissue and muscle growth. Both are synthesized in the testes from cholesterol and to a lesser extent in the adrenal cortex.

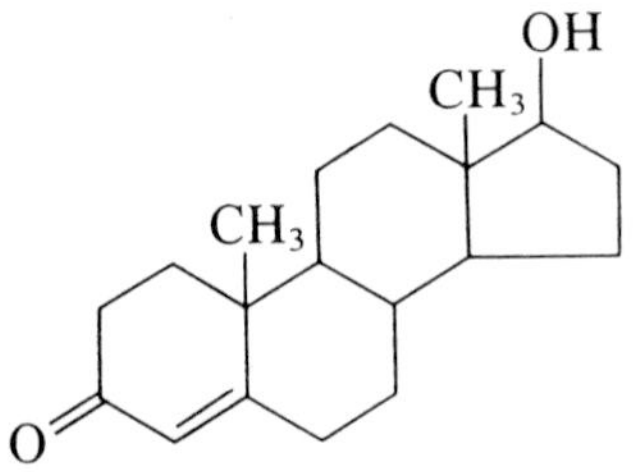

Testosterone
(an androgen)

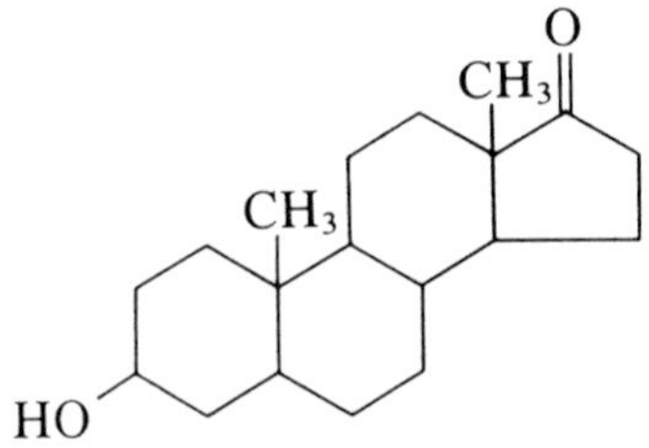

Androsterone
(an androgen)

Estrone and *estradiol* are the two most important female sex hormones, or *estrogens*. These substances, which are synthesized in the ovaries from testosterone and also to a small extent in the adrenal cortex, are responsible for the development of female secondary sex characteristics and for regulation of the menstrual cycle. Note that both have a benzene-like aromatic A ring. In addition, another kind of sex hormone called a *progestin* is essential for preparing the uterus for implantation of a fertilized ovum during pregnancy. *Progesterone* is the most important progestin.

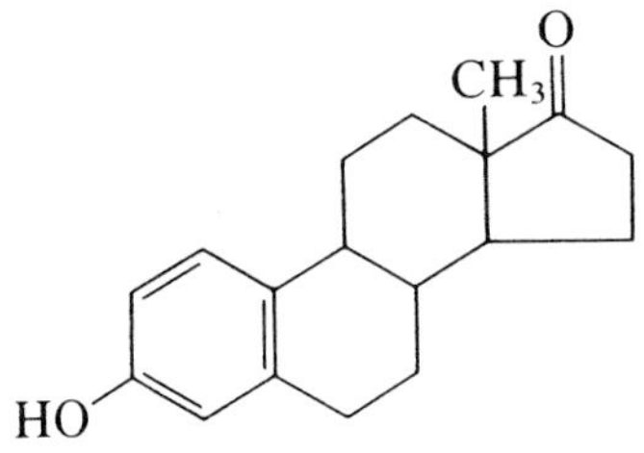

Estrone

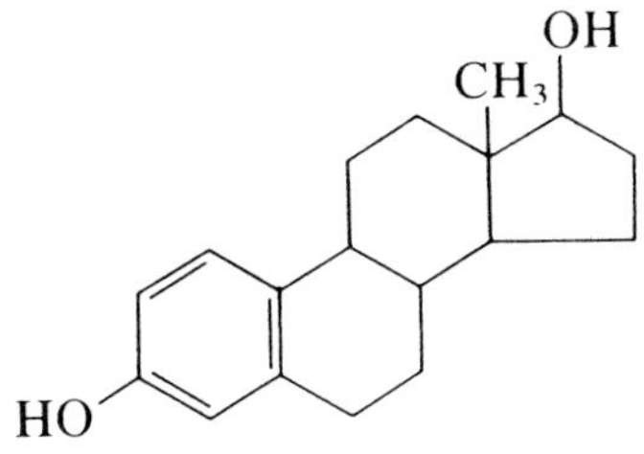

Estradiol

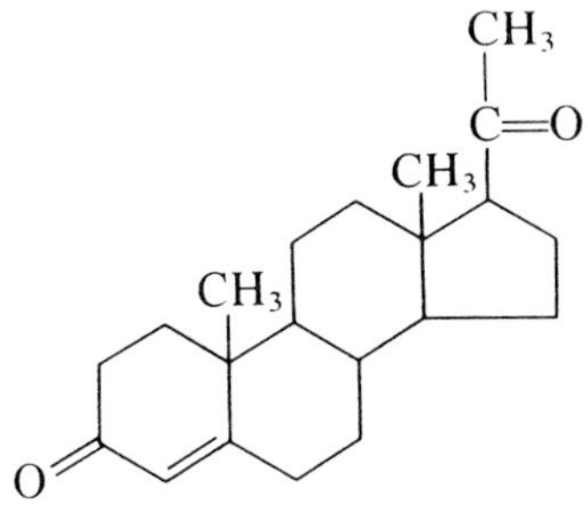

Progesterone (a progestin)

(estrogens)

Adrenocortical Hormones   Steroid hormones are secreted from the cortex (outer part) of the adrenal glands, small organs about half the size of a thumb, located near the upper end of each kidney. These steroids are of three types: *mineralocorticoids, glucocorticoids,* and the sex hormones mentioned above. Mineralocorticoids, such as *aldosterone,* function to control tissue swelling by regulating the delicate cellular salt balance between Na$^+$ and K$^+$ (hence the "mineral" in their name). Glucocorticoids, such as *hydrocortisone* and its close relative *cortisone,* are involved in the regulation of glucose metabolism and in the control of inflammation. You may well have used an antiinflammatory ointment containing cortisone to reduce the swelling and itching if you've ever been exposed to poison oak or poison ivy.

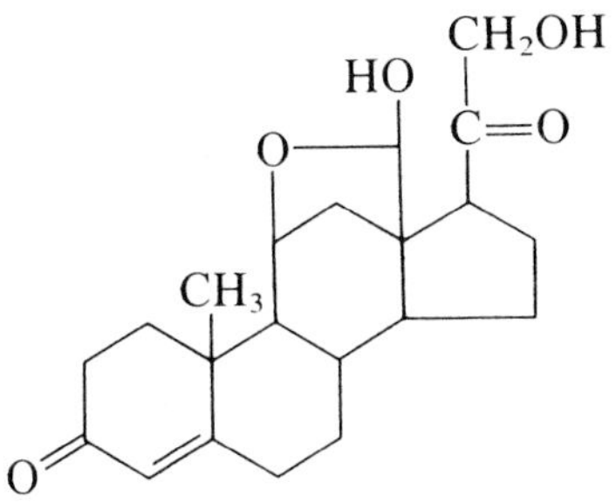

Aldosterone
(a mineralocorticoid)

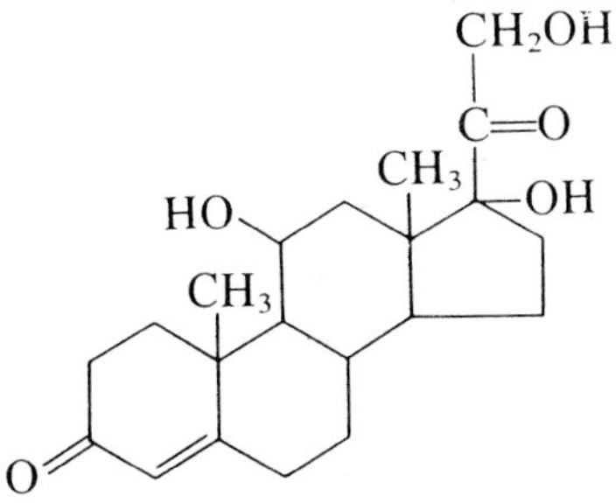

Hydrocortisone
(a glucocorticoid)

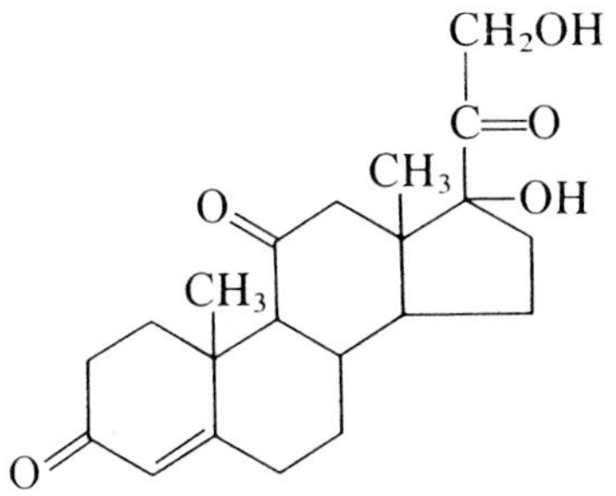

Cortisone

Synthetic Steroids   In addition to the several hundred known steroids isolated from plants and animals, a great many more have been synthesized in the laboratory in the search for new drugs. The general idea is to start with a natural hormone, carry out chemical modifications of the structure, and then see what biological properties the modified steroid might have.

    Among the best known of the synthetic steroids are oral contraceptive and anabolic agents. Most birth control pills are a mixture of two compounds, a synthetic estrogen such as *ethynylestradiol* and a synthetic progestin such as *norethindrone*. These synthetic steroids appear to function by tricking the body into thinking it's pregnant and therefore temporarily infertile. Anabolic steroids such as stanozolol, detected in several athletes during the 1988 Olympics, are synthetic androgens that mimic the tissue-building effect of testosterone.

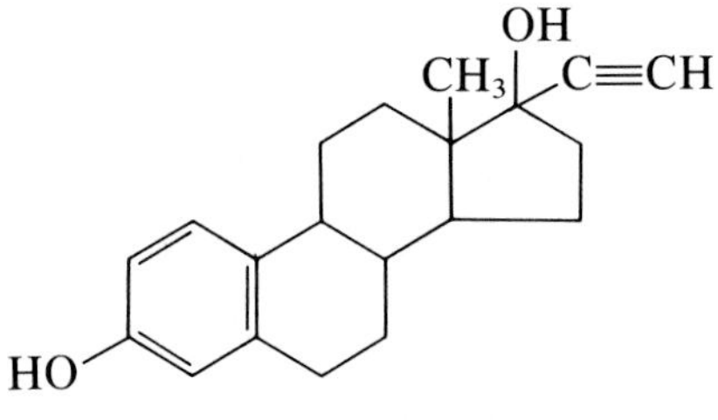

Ethynylestradiol
(a synthetic estrogen)

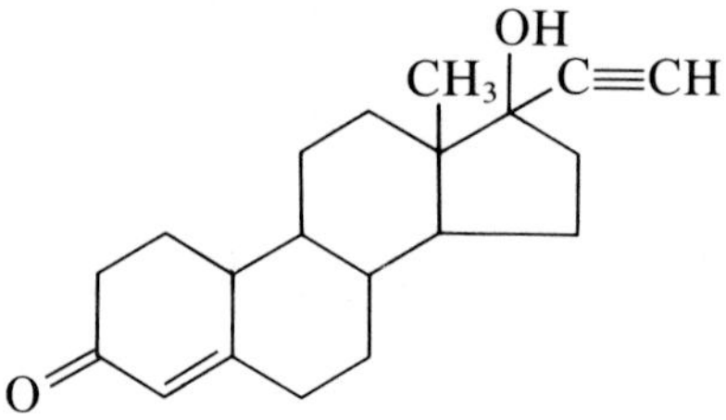

Norethindrone
(a synthetic progestin)

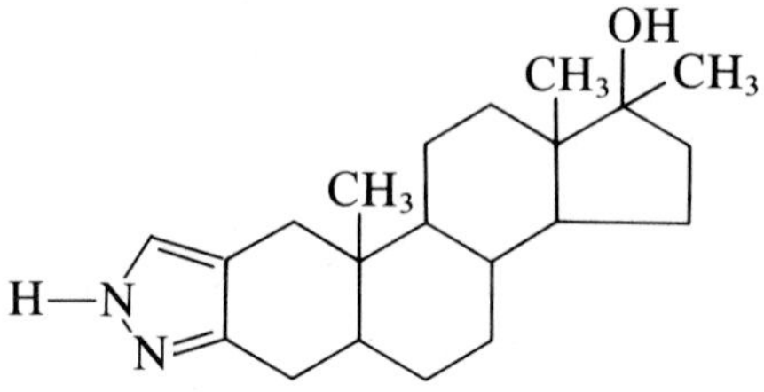

Stanozolol
(a synthetic androgen)

Practice Problems **12.9** Look at the structure of progesterone and identify the functional groups in the molecule.

12.10 Look carefully at the structures of estradiol and ethynylestradiol and point out the differences. What common structural feature do they share that makes both estrogens?

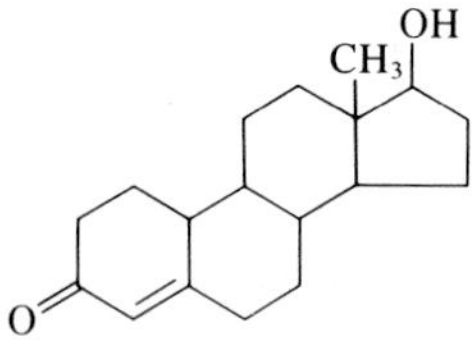

Nandrolone,
an anabolic steroid

12.11 *Nandrolone* is an anabolic, or tissue-building, steroid sometimes taken by athletes seeking to build muscle mass. Among its side effects is a high level of androgenic activity. Compare the structures of nandrolone and testosterone and point out their structural similarities.

12.11 PROSTAGLANDINS AND LEUKOTRIENES

Few groups of compounds have caused as much excitement among medical researchers in the past 15 years as the prostaglandins and related compounds. Although the name *prostaglandin* reflects the original isolation of these compounds from the male prostate gland, they have subsequently been found in small amounts in all body tissues and fluids.

Prostaglandin A lipid derived from a C_{20} carboxylic acid and containing a cyclopentane ring with two long side chains.

Chemically, *prostaglandins* are C_{20} carboxylic acids that contain a five-membered ring with two long side chains. They differ only in the number of oxygen atoms they contain and in the number of double bonds present. Prostaglandin E_1 (abbreviated PGE_1) and prostaglandin $F_{2\alpha}$ are examples:

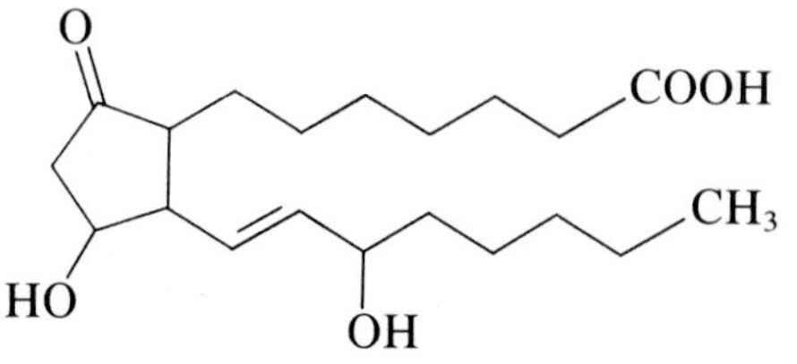

Prostaglandin E_1 (PGE_1)

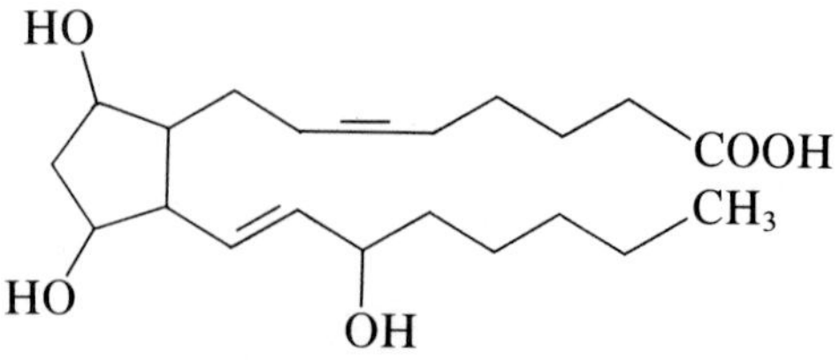

Prostaglandin $F_{2\alpha}$ ($PGF_2\alpha$)

AN APPLICATION: RESEARCH REPORT ON A LEUKOTRIENE INHIBITOR

An article appeared recently about a new drug that apparently inhibits leukotriene synthesis. The story provides a good illustration of how drug development proceeds.

In a clinical trial, a new compound prepared by chemists at a pharmaceutical company had given "significant" relief to 48 victims of ulcerative colitis, an inflammatory disease that causes pain and bloody diarrhea and can eventually require removal of the colon. There is at present no cure for ulcerative colitis, nor any understanding of its cause. The new drug is also undergoing tests for treating asthma, hay fever, psoriasis (a skin disease), and rheumatoid arthritis.

The researchers began their studies by focusing on known information about arachidonic acid and the compounds synthesized from it. It was known that aspirin reacts as illustrated here with an enzyme that initiates the series of reactions leading from arachidonic acid to the prostaglandins. Transfer of the acetyl group from aspirin inactivates the enzyme.

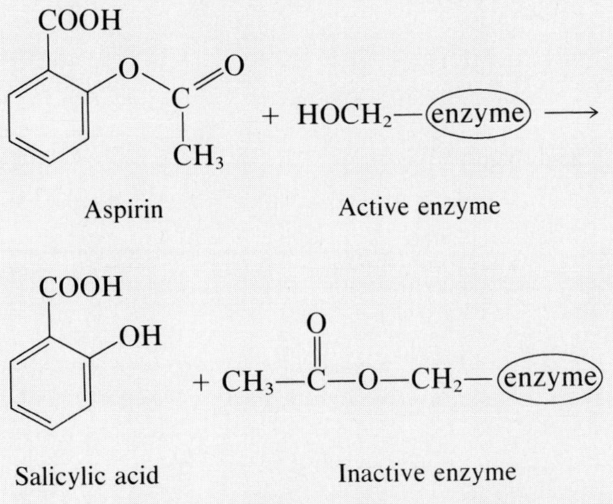

$$\text{Aspirin} + \text{HOCH}_2 - \text{enzyme} \longrightarrow$$

Aspirin   Active enzyme

$$\text{Salicylic acid} + \text{CH}_3 - \overset{\text{O}}{\underset{}{\text{C}}} - \text{O} - \text{CH}_2 - \text{enzyme}$$

Salicylic acid   Inactive enzyme

A similar series of reactions, catalyzed by a different set of enzymes, leads from arachidonic acid to leukotrienes, which are potent initiators of allergic and inflammatory responses. The researchers set their sights on finding a compound that would shut down leukotriene synthesis in the same manner that aspirin prevents prostaglandin synthesis. They appear to have been successful—their experimental drug is reported to bind solely to the enzyme for the first step in leukotriene synthesis.

The clinical results, though preliminary, are exciting because they provide the first link between leukotrienes and inflammatory bowel disease. If all goes well, in a few years we may have a drug that's effective against a most unpleasant disease that formerly had no cure. The drug might even be more broadly useful and alleviate the symptoms of many diseases, as does aspirin. On the other hand, this particular drug may not pass the rigorous testing needed to bring it into general clinical use. Should that happen, you can be sure that structure modifications designed to lead to a safe drug with the same leukotriene-locking action will be under way.

Everyone working with science and technology must keep in touch with new developments in their specialty.

Leukotriene A simple lipid derived from a C_{20} carboxylic acid like a prostaglandin but with no cyclopentane ring.

The several dozen known prostaglandins have an extraordinary range of biological effects. They can lower blood pressure, influence platelet aggregation during blood clotting, stimulate uterine contractions, and lower the extent of gastric secretions. In addition, they are responsible for some of the pain and swelling that accompanies inflammation.

In addition to the prostaglandins themselves, closely related compounds have still other important physiological actions. Interest has centered particularly on the **leukotrienes,** which differ from prostaglandins by the absence of the ring. Leukotriene release in the body has been found to trigger the asthmatic response, severe allergic reactions, and inflammation. Like all hormones, prostaglandins and leukotrienes control the body's responses to various stimuli. Unlike most other hormones, however, these compounds are produced in the cells where they act.

Prostaglandins and leukotrienes are synthesized in the body from the 20-carbon unsaturated fatty acid arachidonic acid. Arachidonic acid, in turn, is synthesized in the body from linolenic acid, helping to explain why linolenic is one of the two essential fatty acids. To illustrate the relationships among arachidonic acid, the prostaglandins, and the leukotrienes, we've shown arachidonic acid and a leukotriene so that they're "bent" into similar shapes.

20 19 18 17 16 15 14 13 12 11 10 9 8 7 6 5 4 3 2 1

$$CH_3CH_2CH_2CH_2CH_2CH{=}CHCH_2CH{=}CHCH_2CH{=}CHCH_2CH{=}CHCH_2CH_2CH_2COOH$$

Arachidonic acid

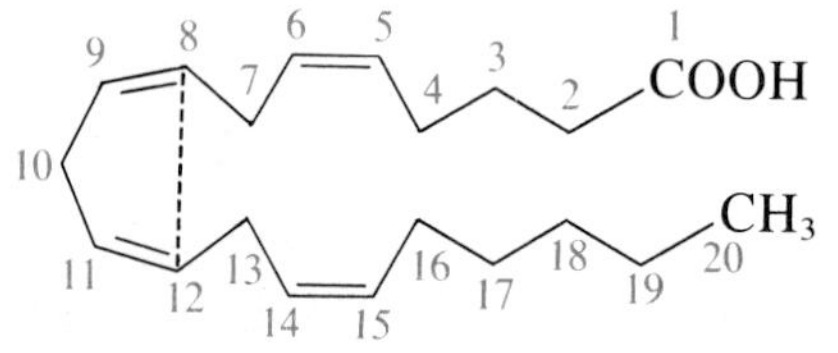
COOH
CH3

Arachidonic acid (bent)

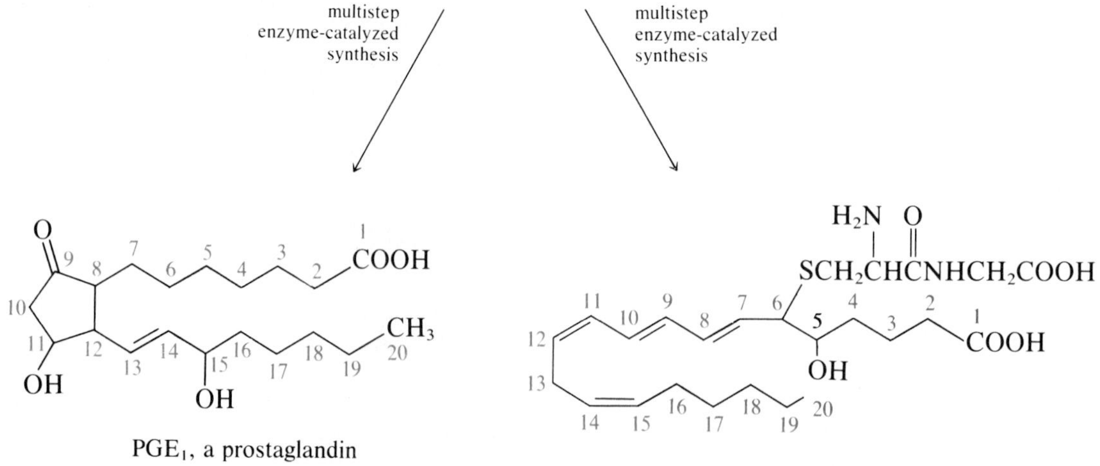
multistep
enzyme-catalyzed
synthesis
multistep
enzyme-catalyzed
synthesis
COOH
CH3
OH
OH
H2N O
SCH2CHCNHCH2COOH
COOH
OH

PGE$_1$, a prostaglandin

Leukotriene D$_4$

INTERLUDE: CHEMICAL COMMUNICATION

To be honest about it, a lot of scientific work is done for the fun of it, without great concern for immediate practical application. Practical applications often *do* follow, but the applications are not themselves the initial goal. Take the study of insects, for example. Have you ever wondered how ants are able to follow a precise path from their nests to a food source, or how the male and female of an insect species find each other for mating? Such questions are interesting to the curious, but it's not clear what value the answers might have.

Insects and many other organisms deal with each other and with their surroundings primarily by sending and receiving chemical messages using substances called *semiochemicals.* Among the most important semiochemicals are *pheromones,* chemicals that are released by one member of a species to evoke a specific response in another member of the same species. Ants and termites, for example, lay down trail pheromones that mark a path from nest to food. Similarly, butterflies, moths, and other flying insects release sex pheromones to indicate their location to other interested parties.

A good semiochemical must be volatile, so that it disappears after its job has been done, and must be structurally unique, so that it is species-specific. After all, release of a sex pheromone wouldn't be of much use to a butterfly if it attracted moths, bees, houseflies, and every other flying insect within range. Low-molecular-weight lipids are ideally suited for use as insect semiochemicals, as shown by the examples given at the bottom of this box.

These mating gypsy moths probably didn't meet just by chance. Volatile lipids used as pheromones enable insects to attract members of the opposite sex, often from great distances.

What about practical applications? An understanding of the chemical messages used by insects offers hope for developing environmentally safe, species-specific means of insect control. For example, gypsy moth infestations in hardwood forests of the northeast United States are now being fought with *pheromone traps* rather than with broad-range insecticides. A tiny amount of the gypsy moth sex pheromone is used to lure the moths into a trap without harming insects of any other species.

Finally, what about *human* sex pheromones? That's another story, but one that is actively being worked on.

A termite trail pheromone

Honeybee alarm pheromone

Gypsy moth sex pheromone

SUMMARY

Lipids are the naturally occurring organic molecules that dissolve when a plant or animal sample is extracted with a nonpolar solvent. Because they are defined by a physical property rather than by structure, there are a great many different kinds of lipids.

Waxes are esters between long-chain alcohols and long-chain carboxylic acids, whereas fats and oils are **triacylglycerols**—triesters of glycerol with three long-chain **fatty acids.** The fatty acids are unbranched; most have an even number of carbon atoms; and they may be either **saturated** or **unsaturated.** Vegetable oils contain a higher proportion of unsaturated fatty acids than animal fats and have consequently lower melting points.

Fats and oils can be **saponified** by treatment with aqueous NaOH or KOH to yield **soap,** a mixture of long-chain fatty acid carboxylate salts. Soaps act as cleansers because their two ends are so different. The negatively charged carboxylate end is ionic and **hydrophilic,** while the hydrocarbon chain end is nonpolar and **hydrophobic.** Thus, a soap molecule is attracted to both grease and water.

Phospholipids are lipids that contain a phosphate group. **Glycerophospholipids** are derived from glycerol by esterification with two fatty acids and one phosphoric acid. The phosphate group, in turn, is also bonded through an ester link to another molecule such as choline, ethanolamine, or serine. **Sphingolipids,** the second main class of phospholipids, are based on the alcohol sphingosine rather than on glycerol. The sphingolipids known as **sphingomyelins** and **glycolipids** are important components of cell membranes.

Phospholipids oriented into a **lipid bilayer** form a major part of the membranes that surround living cells. The ionic phosphate head groups orient on the outside of the bilayer toward the aqueous environment, while the nonpolar hydrocarbon chains cluster in the middle of the bilayer to avoid water.

Steroids are lipids based on a structure with four rings joined together, whereas **prostaglandins** are C_{20} carboxylic acids that have a single five-membered ring connected to two long side chains. **Leukotrienes** are like prostaglandins but have no ring.

REVIEW PROBLEMS

Waxes, Fats and Oils

12.12* What is a lipid?

12.13 Why are there so many different kinds of lipids?

12.14* What is a fatty acid?

12.15 What does it mean to say that fats and oils are triacylglycerols?

12.16 Draw the structure of glycerol trilaurate made from glycerol and three lauric acid molecules.

12.17 How does animal fat differ chemically from vegetable oil?

12.18* What function does a fat serve in an animal?

12.19 What function does a wax serve in a plant or animal?

12.20* Spermaceti, a fragrant substance isolated from sperm whales, was commonly used in cosmetics until it was banned in 1976 to protect the whales from extinction. Chemically, spermaceti is cetyl palmitate, the ester of palmitic acid with cetyl alcohol (the straight-chain C_{16} alcohol). Show the structure of spermaceti.

12.21 What kind of lipid is spermaceti (Problem 12.20)—a fat, a wax, or a steroid?

12.22* There are two isomeric fat molecules whose components are glycerol, one palmitic acid, and two stearic acid units. Draw the structures of both and explain how they differ.

12.23 One of the two molecules in Problem 12.22 is chiral (Section 10.3). Which molecule is chiral and why?

12.24* Write the structures of these molecules:
(a) sodium palmitate   (b) decyl oleate
(c) glyceryl palmitodioleate

Chemical Reactions of Lipids

12.25 How would you convert a vegetable oil like soybean oil into a solid cooking fat?

12.26* Draw the structures of all products you would obtain by saponification of the following lipid with aqueous KOH. What are the names of the products?

$$CH_2-O-\overset{\overset{O}{\|}}{C}(CH_2)_{16}CH_3$$
$$CH-O-\overset{\overset{O}{\|}}{C}(CH_2)_7CH=CH(CH_2)_7CH_3$$
$$CH_2-O-\overset{\overset{O}{\|}}{C}(CH_2)_7CH=CHCH_2CH=CHCH_2CH=CHCH_2CH$$

12.27 Draw the structure of the product you would obtain on hydrogenation of the lipid in Problem 12.26. What is its name?

12.28* Would the product in Problem 12.27 have a higher or a lower melting point than the original lipid? Why?

12.29 What products would you obtain by treatment of oleic acid with these reagents?
(a) Br_2 (b) H_2, Pd catalyst
(c) CH_3OH, HCl catalyst

Phospholipids, Glycolipids, and Cell Membranes

12.30* What is the difference between a fat and a phospholipid?

12.31 How do sphingomyelins and cerebrosides differ structurally?

12.32* Why is it that phosphoglycerides rather than triacylglycerols are found in cell membranes?

12.33 Why are phosphoglycerides more soluble in water than triacylglycerols?

12.34* How does a soap micelle differ from a membrane bilayer?

12.35 What constituents besides phospholipids are present in a cell membrane?

12.36* What are the names of the two different kinds of sphingosine-based lipids?

12.37 Show the structure of a cerebroside made up of D-galactose, sphingosine, and myristic acid.

12.38 Draw the structure of a phosphoglyceride that contains palmitic acid, oleic acid, and the phosphate bonded to propanolamine.

12.39 Draw the structure of a sphingomyelin that contains a stearic acid unit.

12.40* *Cardiolipin,* a compound found in heart muscle, has the following structure. What products would be formed if all ester bonds in the molecule were saponified by treatment with aqueous NaOH?

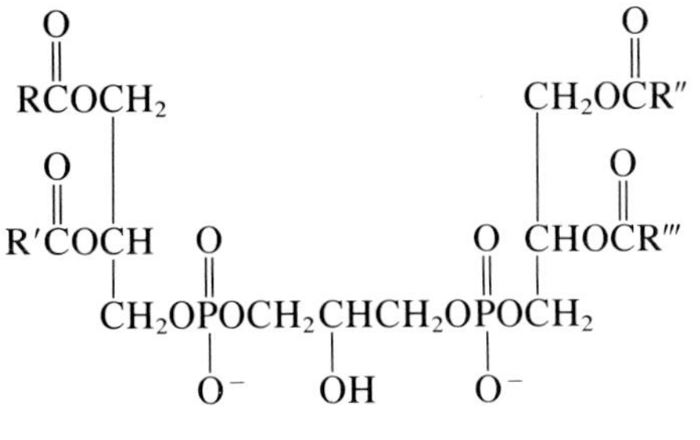

Cardiolipin

Steroids and Prostaglandins

12.41 What functional groups are present in the two sex hormones estradiol and testosterone?

12.42* How do the sex hormones estradiol and testosterone differ structurally?

12.43 The female sex hormone estrone has four chiral centers. Identify them.

12.44* What function does cholesterol serve in cell membranes?

12.45 Draw the products you would expect from treatment of cholesterol with these reagents:
(a) Br_2 (b) H_2, Pd catalyst
(c) [O], an oxidizing agent

12.46 Diethylstibestrol (DES) exhibits estradiol-like hormonal activity even though it is not itself a steroid. Once used widely as an animal food additive, DES has been implicated as a causative agent in several types of cancer. Show how DES can be drawn so that it is structurally similar to estradiol.

Diethylstilbestrol

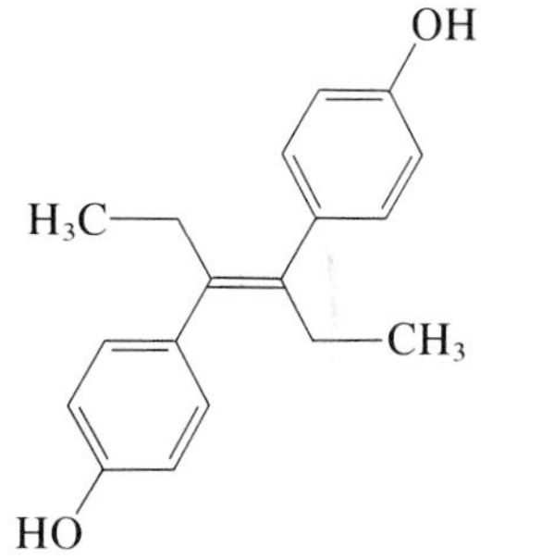

12.47 Thromboxane A_2 is a lipid involved in the blood-clotting process. To what category of lipids does thromboxane A_2 belong?

Thromboxane A_2

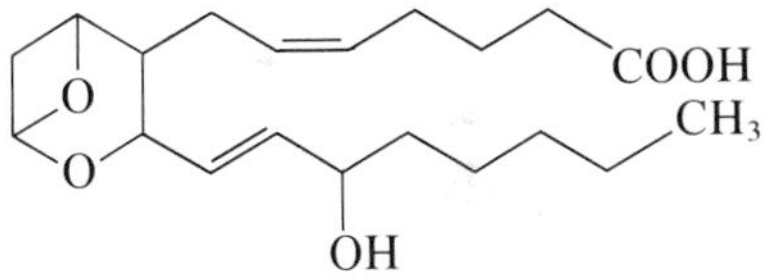

12.48* What fatty acid do you think serves as a biological precursor of thromboxane A_2? (Problem 12.47.)

Applications

12.49 What is the advantage of a detergent relative to a soap as a cleaning agent? [App: Detergents]

12.50 Describe the mechanism by which soaps and detergents provide cleaning action. [App: Detergents]

12.51 Cationic detergents are rarely used for cleaning. For what purposes are they used? [App: Detergents]

12.52* Why are branched-chain hydrocarbons no longer used for detergents? [App: Detergents]

12.53 Why must cells be semipermeable? [App: Transport Across Cell Membranes]

12.54* Which process requires energy—passive or active transport? Why is energy sometimes required to move solute across the cell membrane? [App: Transport Across Cell Membranes]

12.55 Why is it desirable to inhibit the production of leukotrienes? [App: Leukotriene Inhibitor]

12.56* How does aspirin function to inhibit the formation of prostaglandins from arachidonic acid? [App: Leukotriene Inhibitor]

12.57 What is a semiochemical? [Int: Chemical Communication]

12.58* What are the advantages of using pheromone traps for insect control compared to using insecticides? [Int: Chemical Communication]

Additional Questions and Problems

12.59 Which of the following are saponifiable lipids?
(a) prostaglandin E_1 (b) a lecithin
(c) progesterone   (d) a sphingomyelin
(e) a cerebroside   (f) glyceryl trioleate

12.60* Identify the component parts of each saponifiable lipid listed in Problem 12.59.

12.61 Draw the structure of the fat made from two molecules of linolenic acid and one molecule of myristic acid.

12.62* Would the fat described in Problem 12.61 have a higher or lower melting point than the fat made from one molecule each of linolenic, myristic, and stearic acids?

12.63 Why is cholesterol not saponifiable?

12.64* Jojoba wax, used in candles and cosmetics, is partially composed of the ester of stearic acid and a straight-chain 22-carbon alcohol. Draw the structure of this wax component.

12.65 What three types of lipids are particularly abundant in brain tissue?

12.66* What is the general purpose of a hormone?

12.67 What are some of the functions that prostaglandins serve in the body?

12.68* Lecithins are often used as food additives to provide emulsification. How do they accomplish this purpose?

12.69 If the average molecular weight of a sample of soybean oil is 1500, how many grams of NaOH are needed to saponify 5.0 g of the oil?

12.70* The concentration of cholesterol in the blood serum of a normal adult is approximately 200 mg/dL. How much total blood cholesterol does a person with a blood volume of 5.75 L have?

CHAPTER

13

Lipid Metabolism

Cholesterol! It's a relatively simple molecule (carbons in green, hydrogens in blue, and oxygen in red), but understanding how it affects body chemistry is not so simple.

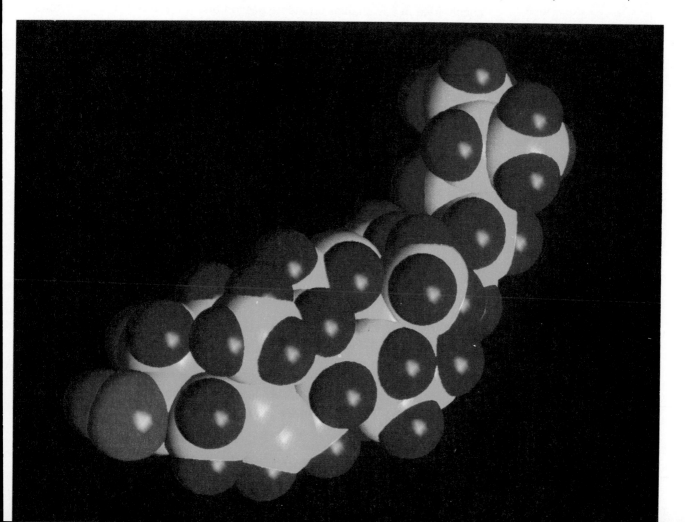

In the kitchen, we make a distinction between fats and oils based on melting point. In biochemistry, however, this distinction is unimportant because both are *triacylglycerols*—triesters of glycerol with a variety of fatty acids. Triacylglycerols make up 98–99% of the lipids in our diet, and about 90% of the lipids in our bodies.

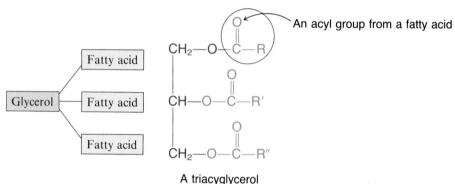

A triacyglycerol

Having introduced the chemistry of triacylglycerols in Chapter 12, we'll concentrate in this chapter on their digestion and metabolism. The numerous other types of lipids mentioned in Chapter 12 follow similar metabolic pathways. Some questions we'll answer include

1. *What happens during the digestion of triacylglycerols?* The goal: Be able to list the sequence of events in the digestion of triacylglycerols and their transport into the bloodstream.

2. *What are the major pathways in the metabolism of triacylglycerols?* The goal: Be able to name the major pathways for the synthesis and breakdown of triacylglycerols and fatty acids and be able to show their interrelationships.

3. *How are triacylglycerols moved in and out of storage in adipose tissue?* The goal: Be able to explain the regulation and reactions of the storage and mobilization of triacylglycerols.

4. *How are fatty acids oxidized?* The goal: Be able to show what happens to a fatty acid from its entry into a cell to its conversion to acetyl SCoA.

5. *What is the energy yield from fatty acid catabolism?* The goal: Be able to calculate the number of molecules of ATP produced from a given fatty acid.

6. *What is the function of ketogenesis?* The goal: Be able to identify ketone bodies, describe their synthesis, and explain their role in metabolism.

7. *How are fatty acids synthesized?* The goal: Be able to compare the pathways for fatty acid synthesis and oxidation, and be able to describe the reactions of the synthesis pathway.

13.1 DIGESTION OF TRIACYLGLYCEROLS

Chyme The semisolid mixture produced by digestion in the stomach.

Bile Fluid secreted by the liver and released into the small intestine during digestion; contains bile salts, bicarbonate ion, and other electrolytes.

Bile acids Steroid acids derived from cholesterol that are secreted in bile.

Figure 13.1
Digestion of triacylglycerols.

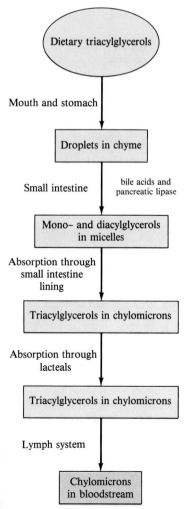

When eaten, triacylglycerols (TAGs) pass through the mouth unchanged and enter the stomach, where their physical breakdown takes place (Figure 13.1). The heat and churning action of the stomach break lipids into smaller droplets, a process that takes longer than the physical breakdown and digestion of other foods. To be sure that there's time for this breakdown, the presence of lipids in consumed food slows down the rate at which the mixture of partially digested foods known as **chyme** leaves the stomach. One of the reasons lipids are a pleasing part of the diet is that the stomach feels "full" for a longer time after a fatty meal.

When chyme leaves the stomach, it enters the upper end of the small intestine, where its arrival triggers the release of *pancreatic lipases,* enzymes for the hydrolysis of lipids (Figure 13.2a). (The pancreas also produces enzymes for carbohydrate and protein hydrolysis, and bicarbonate ion for neutralization of stomach acid in the chyme.) The gallbladder simultaneously releases **bile,** which is manufactured in the liver and stored in the gallbladder until needed (Figure 13.2b). Enzymes can't attack the lipids inside water-insoluble droplets, and it's the job of bile to emulsify them. Steroids called **bile acids,** or *bile salts,* solubilize lipid droplets by forming micelles much like the soap micelle shown previously in Figure 12.5. You can see from the structure of cholic acid, the major bile acid, that it resembles soaps and detergents in containing both hydrophilic and hydrophobic regions. When freed of the TAGs they transport during digestion, bile acids are mostly recycled. The small amount that escapes recycling is excreted.

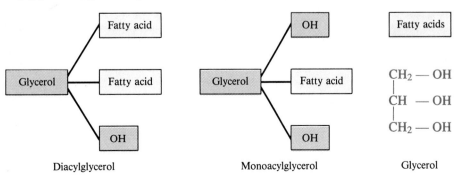

Once micelles have formed, the digestion of triacylglycerols proceeds. Within the small intestine, pancreatic lipase acts at the micelle surface to partially hydrolyze the triacylglycerols, producing mainly mono- and diacylglycerols, plus fatty acids and a small amount of glycerol.

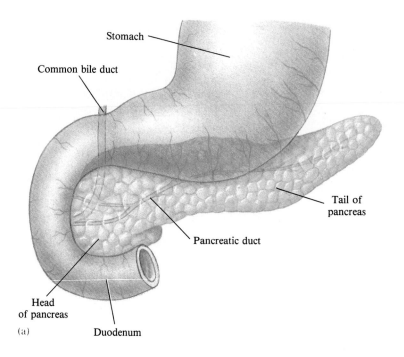

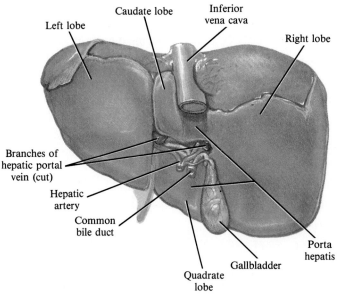

Figure 13.2

Anatomy of the pancreas and liver. (a) The pancreas secretes enzymes for the digestion of lipids, carbohydrates, and proteins, plus bicarbonate ion. Note the bile duct through which bile from the gallbladder enters the small intestine at its upper end (the duodenum). (The pancreas also secretes the hormones insulin and glucagon, from a different group of cells.) (b) Blood carrying metabolites from the digestive system enters the liver through the hepatic portal vein. The gallbladder is the site for storage of bile.

The smaller fatty acids and glycerol are water-soluble and are absorbed directly through the surface of the villi that line the small intestine. Within the villi, they enter the bloodstream through capillaries (Figure 13.3a) and are carried to the liver via the hepatic portal vein. The larger acylglycerols and fatty acids are resolubilized in the intestine by the bile salts. The resulting micelles carry these water-insoluble molecules to the intestinal lining. There, the acylglycerols and fatty acids are released from the micelles and absorbed.

Next, within the intestinal lining, the still-insoluble fatty acids and acylglycerols are repackaged by a roundabout route to enter the bloodstream (Figure 13.3b). First, the fatty acids and acylglycerols are reconverted to triacylglycerols. Then, the triacylglycerols are bound into **chylomicrons,** one of several kinds of **lipoproteins** that aid in lipid transport. The general structure of a lipoprotein is a globule of lipid surrounded by a phospholipid-based membrane with the hydrophilic regions toward the outside (Figure 13.4).

Chylomicron Spherical, low-density lipoprotein particles that transport triacylglycerols

Lipoprotein A lipid–protein complex that transports lipids.

Figure 13.3
Site of lipid absorption in the small intestine. Each villus (a) is lined with huge numbers of microvilli, which provide the large surface area needed for absorption. After absorption (b), smaller molecules go directly into the bloodstream via capillaries. Larger lipid molecules are repackaged into chylomicrons for entry into lacteals and delivery to the bloodstream via the lymphatic system.

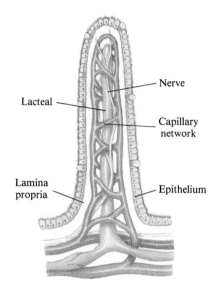

(a) A single villus

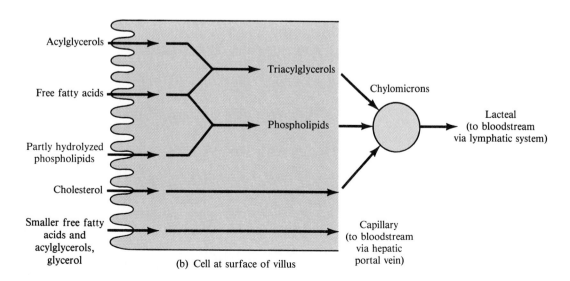

(b) Cell at surface of villus

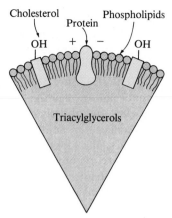

Figure 13.4
The structure of a TAG-carrying lipoprotein. The lipoproteins that transport cholesterol and other lipids have similar structures.

Because chylomicrons are too large to pass through capillary walls, they instead enter the lymphatic system by absorption into lacteals (Figure 13.3a). The chylomicrons are then carried to the thoracic duct (just below the collarbone), where the lymphatic system empties into the bloodstream. At this point, the lipids from foods are ready to be put to use.

13.2 TRIACYLGLYCEROL METABOLISM: AN OVERVIEW

Triacylglycerols and fatty acids enter metabolism from three different sources: the diet, storage in adipose tissue, and synthesis in the liver. Whatever their origin, these lipids must be made soluble by association with water-soluble proteins. The various ways this is accomplished are summarized in Figure 13.5.

The chylomicrons that carry dietary TAGs through the lymphatic system into the blood are the lowest density lipoproteins because they carry the highest ratio of lipids to proteins. The slightly more dense ones, the *VLDLs* (very-low-density lipoproteins), carry TAGs from the liver where they have been synthesized to tissues where they will be utilized or stored. When TAGs stored in adipose tissue are needed for energy, they are first hydrolyzed and then the resulting fatty acids are carried in the bloodstream by proteins known as *albumins*.

As summarized in Figure 13.6, the first step in utilizing TAGs is complete hydrolysis to glycerol and fatty acids:

$$\text{Triacylglycerol (TAG)} \xrightarrow[\text{H}_2\text{O}]{\text{lipase}} \text{Glycerol} + \text{Fatty acids}$$

Figure 13.5
Transport of TAGs and fatty acids. The carriers are shown in blue.

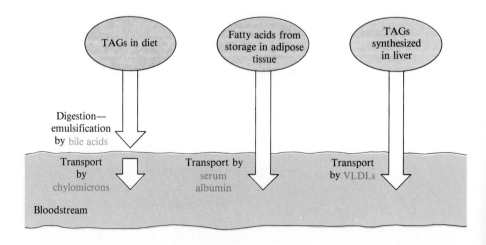

Mobilization (of triacylglycerols)
Hydrolysis of triacylglycerols in adipose tissue and release of fatty acids into bloodstream.

TAGs that are released from storage, or **mobilized,** are hydrolyzed in fat cells known as *adipocytes* by a lipase enzyme that is activated by epinephrine, glucagon, and several other hormones. TAGs from digested foods are hydrolyzed when the lipoproteins carrying them in the bloodstream encounter the *lipoprotein lipase,* which is anchored in capillary walls in adipose tissue and elsewhere. The resulting free fatty acids then travel through the bloodstream in association with albumins and enter cells for further metabolism.

The glycerol from TAG hydrolysis is carried in the bloodstream to the liver or kidneys where it is converted in the following series of reactions to dihydroxyacetone phosphate (DHAP). Isomerization of DHAP gives glyceraldehyde 3-phosphate, which can enter either the glycolysis pathway or the glyconeogenesis pathway. Thus, dihydroxyacetone phosphate represents one of several points at which carbohydrate and lipid metabolism are connected.

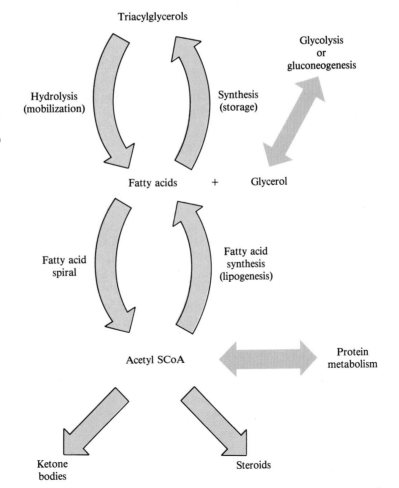

Figure 13.6
Metabolism of triacylglycerols. Pathways that break down molecules (catabolism) are shown in yellow, and synthetic pathways (anabolism) are shown in blue. Connections to other pathways or intermediates of metabolism are shown in green.

AN APPLICATION: CHOLESTEROL, LIPOPROTEINS, AND HEART DISEASE

A debate about the relationship between cholesterol and heart disease has been going on for over three decades. Newspaper and television ads trumpet the idea that cholesterol is a killer and that we'd all live much healthier lives if only we would switch our brand of cooking oil.

What are the facts? Several points seem clear. The first is that a diet rich in saturated animal fats generally leads to an increase in blood-serum cholesterol. The second is that a diet lower in saturated fat and higher in unsaturated fat leads to a lowering of the serum cholesterol level. Since the body manufactures much of its own cholesterol, however, the serum level can't be controlled entirely by diet. Remember also that since cholesterol is an essential component of cell membranes and the precursor of all other steroids in the body, a certain amount of serum cholesterol is necessary for life.

The third point is that a level of serum cholesterol greater than 200 mg/dL (a normal value is 150–200 mg/dL) has been correlated with *atherosclerosis,* a condition in which yellowish deposits composed of cholesterol and other lipid-containing materials form within the larger arteries, notably those of the brain and heart. Numerous studies have also correlated high cholesterol levels with the risk of heart attack brought on by blockage of blood flow to heart muscles.

It's recently been found that a better indication of a person's risk of heart disease comes from a measurement of blood lipoprotein levels. As shown in Figure 13.4, lipoproteins are complex assemblages of lipids and proteins that serve to transport water-soluble lipids throughout the body. They can be somewhat arbitrarily divided into four major types distinguishable by their density, as shown in the accompanying table. Since lipids are generally less dense than proteins, the chylomicrons and the very-low-density lipoproteins (VLDLs) are richer in lipid and lower in protein than the low-density lipoproteins (LDLs) and high-density lipoproteins (HDLs).

The four lipoprotein fractions have different roles. Chylomicrons and VLDL act primarily as carriers of triacylglycerols as described in Sections 13.1 and 13.2. LDL and HDL act as carriers of cholesterol to and from the liver. Present evidence suggests that LDL transports cholesterol as its linolenate ester *to* peripheral tissues where it is deesterified and can accumulate in the deposits of atherosclerosis. HDL transports choles-

The fatty acids from TAG hydrolysis have two possible fates. When energy is in good supply, fatty acids are converted back to TAGs for storage. When cells are in need of energy, carbon atoms are removed two at a time from fatty acids via the *fatty acid spiral*. The result is complete conversion of fatty acids with even numbers of carbon atoms to acetyl SCoA, and subsequent ATP production via the citric acid cycle and oxidative phosphorylation. In resting muscle, fatty acids are the major energy source.

Acetyl SCoA, which is at a crossroads in metabolism (Figure 13.6), serves as the starting material for the biosynthesis of fatty acids (*lipogenesis*) in the liver. Another pathway available to acetyl SCoA is *ketogenesis,* the production from acetyl SCoA of the three compounds known as *ketone bodies*. Normally acetyl SCoA takes this route to only a small extent. Under the stress of glucose shortage, however, ketogenesis becomes increasingly important because ketone bodies can be used as fuel instead of glucose. Acetyl SCoA is also the starting material for the synthesis of cholesterol, from which all other steroids are made.

terol as its stearate ester *from* dead or dying cells back to the liver where it is converted to bile acids and excreted.

If LDL delivers more cholesterol than is needed, and if not enough HDL is present to remove it, the excess is deposited in cells and arteries. Thus, the higher the HDL level, the less the likelihood of deposits and the lower the risk of heart disease. A rule of thumb is that a person's risk drops about 24% for each increase of 5 mg/dL in HDL concentration. Normal HDL values are about 45 mg/dL for men and 55 mg/dL for women, perhaps helping to explain why women are generally less susceptible than men to heart disease.

If high serum levels of HDL are good, then how do we get them? The answer is simple enough: As common sense tells you, the most important factor is a generally healthy lifestyle. Obesity, smoking, lack of exercise, and heavy drinking appear to lead to low HDL levels, while regular exercise, a prudent diet, and moderate alcohol consumption lead to high HDL levels. Runners and other endurance athletes, in particular, have HDL levels nearly 50% higher than those of the general population.

Serum Lipoproteins

Name	Density (g/mL)	% Lipid	% Protein
Chylomicrons	<0.94	98	2
VLDL (very-low-density)	0.940–1.006	90	10
LDL (low-density)	1.006–1.063	75	25
HDL (high-density)	1.063–1.210	60	40

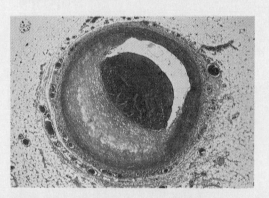

This coronary artery is severely narrowed by atherosclerosis, the buildup of cholesterol deposits in vessel walls (pink). The resulting roughening of artery walls also promotes the formation of clots (dark red) that can cut off blood flow completely, causing a heart attack.

Practice Problem   **13.1**   Look back at Figure 11.5 and tell where dihydroxyacetone phosphate can enter the glycolysis or gluconeogenesis pathway.

13.3   STORAGE AND MOBILIZATION OF TRIACYLGLYCEROLS

We've noted that adipose tissue is the storage depot for TAGs and that TAGs are our primary energy storage form. TAGs don't just sit unused until needed, however. The passage of fatty acids in and out of storage in adipose tissue is a continuous process essential to maintaining homeostasis.

To see how storage and mobilization are regulated, look back at Figure 11.9 which shows the effects of the hormones insulin and glucagon on metabolism. After a meal, blood glucose levels are high, insulin levels rise, and glucagon levels drop. Glucose is entering cells, and glycolysis is proceeding actively. In addition, insulin stimulates the synthesis of TAGs for storage.

The reactants in TAG synthesis are glycerol 3-phosphate and fatty acyl derivatives of coenzyme A. The glycerol 3-phosphate is made from dihydroxyacetone phosphate (DHAP), which you've seen is an intermediate in glycolysis and also a product of glycerol metabolism in the liver. Since adipocytes don't have the enzyme needed to convert glycerol to glycerol 3-phosphate, they can't make triacylglycerols unless some glucose is undergoing glycolysis and producing glycerol 3-phosphate.

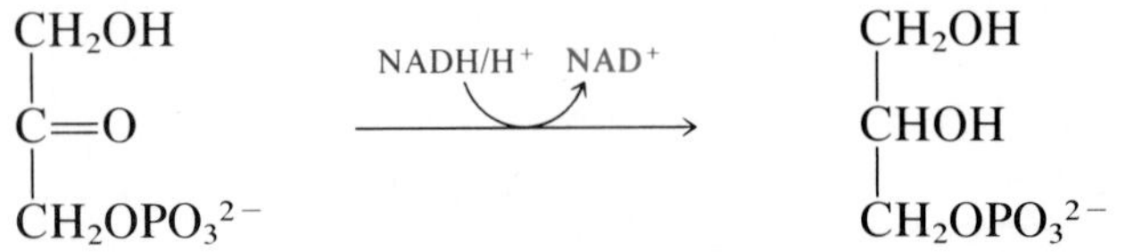

$$
\begin{array}{ccc}
\underset{\substack{|\\ \text{C}=\text{O}\\ |\\ \text{CH}_2\text{OPO}_3{}^{2-}}}{\text{CH}_2\text{OH}}
& \xrightarrow{\text{NADH/H}^+ \quad \text{NAD}^+}
& \underset{\substack{|\\ \text{CHOH}\\ |\\ \text{CH}_2\text{OPO}_3{}^{2-}}}{\text{CH}_2\text{OH}} \\
\text{DHAP} & & \text{Glycerol 3-phosphate}
\end{array}
$$

Fatty acyl SCoAs consist of a fatty acid acyl group (like an acetyl group, but with a long carbon chain) bonded to coenzyme A. The fatty acids may have come from digestion or from biosynthesis. TAG synthesis begins with the conversion of glycerol 3-phosphate to phosphatidic acid by addition of two fatty acid groups:

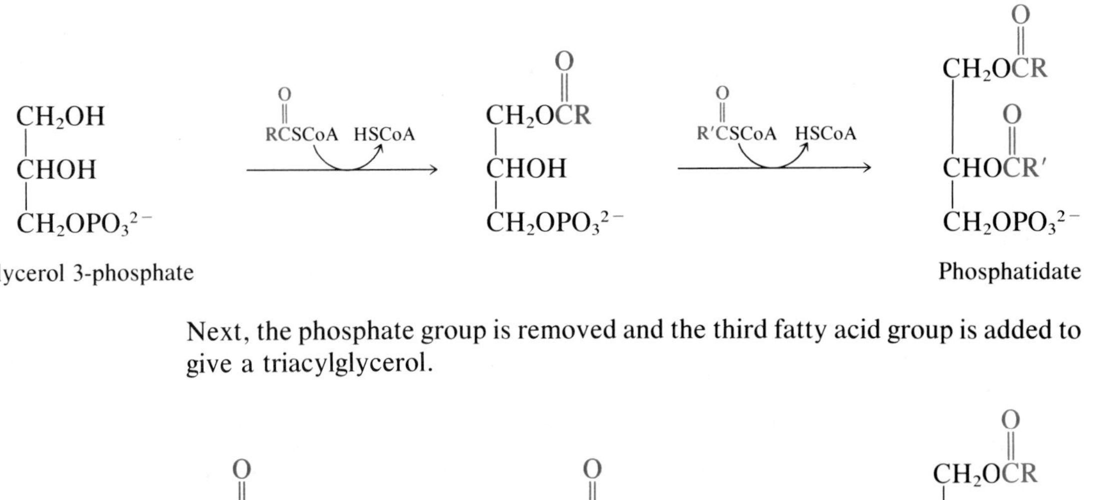

Glycerol 3-phosphate → (RCSCoA, HSCoA) → intermediate → (R'CSCoA, HSCoA) → Phosphatidate

Next, the phosphate group is removed and the third fatty acid group is added to give a triacylglycerol.

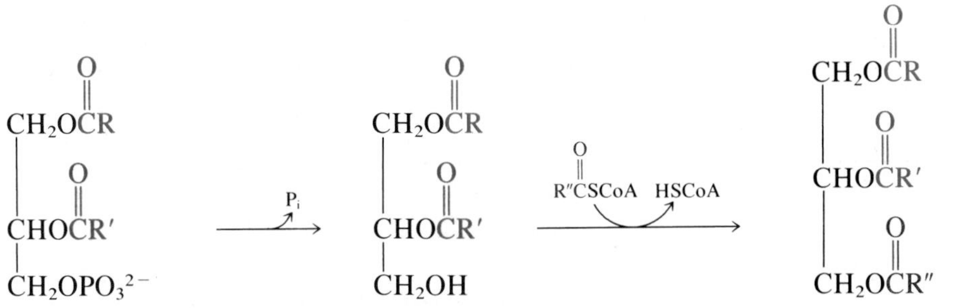

Phosphatidate → (P_i) → intermediate → (R"CSCoA, HSCoA) → A triacylglycerol

After a meal has been digested, blood glucose levels are low so that insulin levels drop and glucagon levels rise, activating the enzymes in adipose tissue responsible for mobilizing TAGs. Also, with glycerol 3-phosphate in short supply, fewer fatty acids can be recycled into storage as TAGs. Now, fatty acids and glycerol are produced in adipose tissue and released into the bloodstream.

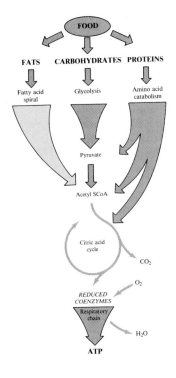

13.4 ACTIVATION, MEMBRANE TRANSPORT, AND OXIDATION OF FATTY ACIDS

Once a fatty acid crosses the cell membrane into the cytosol of a cell that needs energy (Figure 13.7), three successive processes must occur.

1. The fatty acid must be *activated* by conversion to fatty acyl SCoA.

2. The fatty acyl SCoA must be *transported* into the mitochondrial matrix.

3. The fatty acyl SCoA must be *oxidized* by enzymes in the matrix to produce acetyl SCoA plus reduced coenzymes ($FADH_2$ and NADH).

Fatty Acid Activation   Activation of a fatty acid serves the same purpose as the first few steps in oxidation of glucose by glycolysis. That is, some energy must be initially invested in conversion of the fatty acid to a form that can be broken down more easily.

The reaction of a fatty acid with coenzyme A to form a fatty acyl SCoA is endergonic. In order to take place, the reaction is coupled with the exergonic hydrolysis of ATP to yield pyrophosphate ion ($P_2O_7^{4-}$) and adenosine monophosphate (AMP), and with the even more exergonic subsequent hydrolysis of the pyrophosphate, to yield hydrogen phosphate ion (HPO_4^{2-}). The overall reaction breaks one high-energy phosphate bond in ATP and one in pyrophosphate, which is equivalent to "spending" two ATPs.

$$\underset{\substack{\\ \\}}{R-\overset{\overset{\displaystyle O}{\|}}{C}-O^-} \;+\; HSCoA \;+\; ATP \;+\; H_2O \;\longrightarrow$$

$$R-\overset{\overset{\displaystyle O}{\|}}{C}-SCoA \;+\; AMP \;+\; 2\,HPO_4^{2-} \;+\; H^+$$

Figure 13.7
A glandular cell. The nucleus and its contents appear in yellow, and the mitochondria in dark red. The mitochondria are the site of energy production: Fatty acid oxidation, amino acid breakdown, the citric acid cycle, and oxidative phosphorylation occur there. The cytosol is the site of fatty acid biosynthesis, glycolysis, the pentose phosphate pathway, and gluconeogenesis.

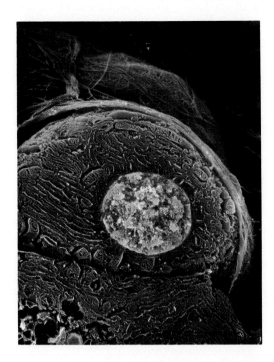

Transport Into Mitochondrion   The fatty acyl SCoA molecule is too large to cross the inner mitochondrial membrane and must be transferred instead by a shuttle mechanism that uses carnitine as a temporary carrier molecule. The fatty acyl SCoA reacts with carnitine to yield a fatty acyl carnitine; transport across the membrane occurs; the fatty acyl group is picked up by HSCoA; and carnitine crosses back to begin another cycle.

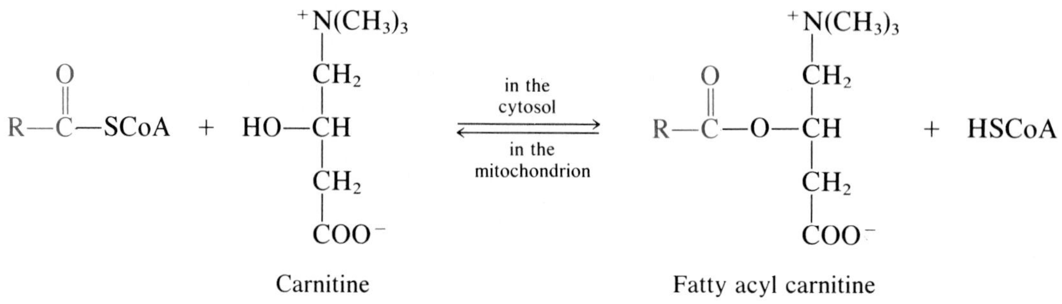

$$
\begin{array}{ccc}
\underset{R-\overset{\overset{\displaystyle O}{\|}}{C}-SCoA}{} +
\underset{\text{Carnitine}}{HO-\overset{\overset{\displaystyle ^+N(CH_3)_3}{|}\ \overset{\displaystyle CH_2}{|}}{\underset{\underset{\displaystyle COO^-}{|}\ \underset{\displaystyle CH_2}{|}}{CH}}}
& \underset{\substack{\text{in the}\\\text{mitochondrion}}}{\overset{\substack{\text{in the}\\\text{cytosol}}}{\rightleftharpoons}}
& \underset{\text{Fatty acyl carnitine}}{R-\overset{\overset{\displaystyle O}{\|}}{C}-O-\overset{\overset{\displaystyle ^+N(CH_3)_3}{|}\ \overset{\displaystyle CH_2}{|}}{\underset{\underset{\displaystyle COO^-}{|}\ \underset{\displaystyle CH_2}{|}}{CH}}} + HSCoA
\end{array}
$$

Fatty Acid Oxidation   Fatty acids are oxidized in the mitochondrial matrix by the series of four reactions shown in Figure 13.8 that make up the **fatty-acid spiral.** Each reaction cycle results in cleavage of a two-carbon acetyl group from the end of a fatty acyl group. The pathway is referred to as a *spiral* because a long-chain fatty acyl group must continue to return to the pathway until each pair of carbon atoms is removed as acetyl SCoA. The alternative name for the pathway, **β oxidation,** reflects the oxidation in two of the steps of the carbon atom β to the SCoA group.

Fatty acid spiral (β oxidation)   A repetitive series of biochemical reactions that degrade fatty acids to acetyl SCoA.

Figure 13.8
The fatty acid spiral. Passage of acyl SCoA through these four steps leads to the cleavage of one acetyl group from the end of the fatty acid chain.

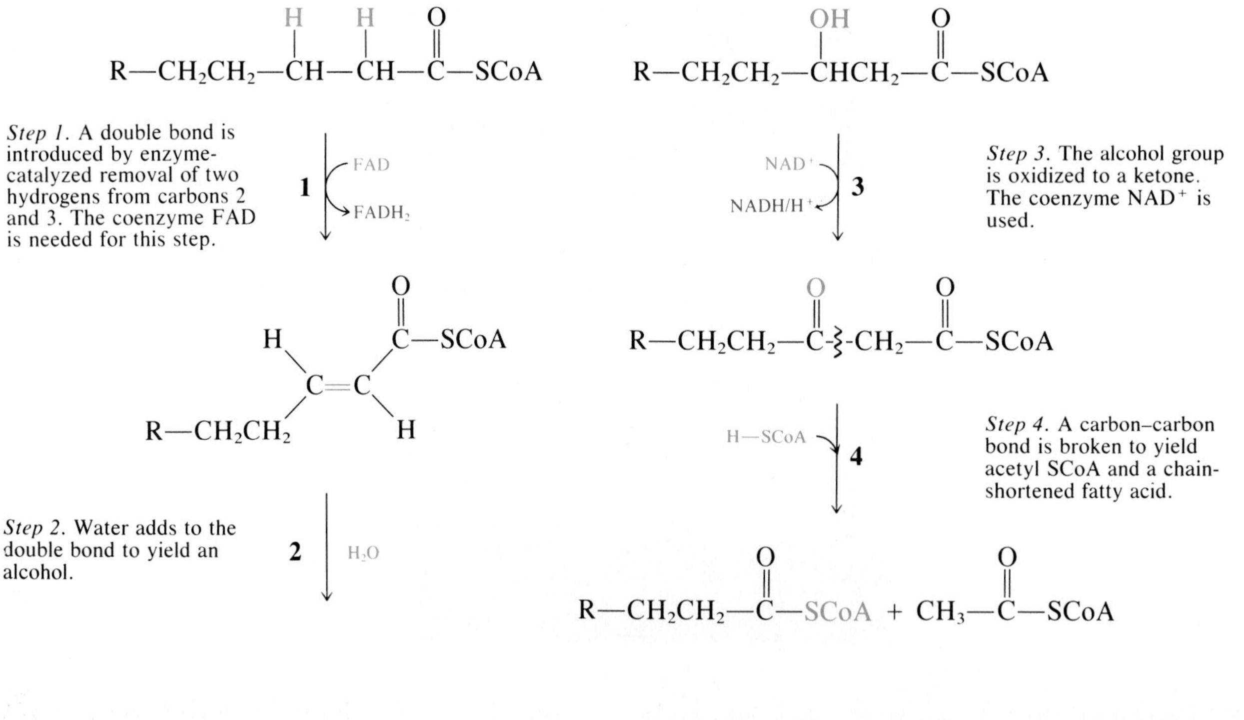

Step 1. A double bond is introduced by enzyme-catalyzed removal of two hydrogens from carbons 2 and 3. The coenzyme FAD is needed for this step.

Step 2. Water adds to the double bond to yield an alcohol.

Step 3. The alcohol group is oxidized to a ketone. The coenzyme NAD⁺ is used.

Step 4. A carbon–carbon bond is broken to yield acetyl SCoA and a chain-shortened fatty acid.

Steps 1 and 2 of fatty acid oxidation: The first β oxidation and hydrolysis. In step 1, the oxidizing agent FAD removes two hydrogen atoms from the carbon atoms α and β to the C=O group in the fatty acyl SCoA, forming a double bond. In step 2, H_2O adds across this newly created double bond to give an alcohol with the —OH group on the β carbon.

Step 3 of fatty acid oxidation: The second β oxidation. NAD^+ serves as an oxidizing agent for conversion of the β —OH group to a carbonyl group.

Step 4 of fatty acid oxidation: Cleavage to remove an acetyl group. In the final step of the fatty acid spiral, an acetyl group is split off and attached to a new coenzyme A molecule, leaving behind an acyl SCoA that is two carbon atoms shorter. If you look carefully, you might recognize this step as the reverse of a Claisen condensation reaction (Section 6.8).

To see how the fatty acid spiral works, look at the catabolism of palmitic acid shown in Figure 13.9. One turn of the spiral converts the 16-carbon palmitoyl SCoA into the 14-carbon myristyl SCoA plus acetyl SCoA; a second turn of the spiral converts myristyl SCoA into the 12-carbon lauryl SCoA plus acetyl SCoA; a third turn converts lauryl SCoA into the 12-carbon capryl SCoA plus acetyl SCoA; and so on.

You can predict how many molecules of acetyl SCoA will be obtained from a given fatty acid simply by counting the number of carbon atoms and dividing by 2. For example, the 16-carbon palmitic acid yields eight molecules of acetyl SCoA after seven turns of the spiral. The number of turns of the spiral is always one less than the number of acetyl SCoA molecules produced, because the last turn cleaves a 4-carbon chain into two acetyl SCoAs.

Figure 13.9
Passage of palmitoyl SCoA (16 C atoms) through the fatty acid spiral. Each repetition of the spiral cleaves two carbons from the end of the chain, yielding a molecule of acetyl CoA and a fatty acid chain shortened by two carbon atoms.

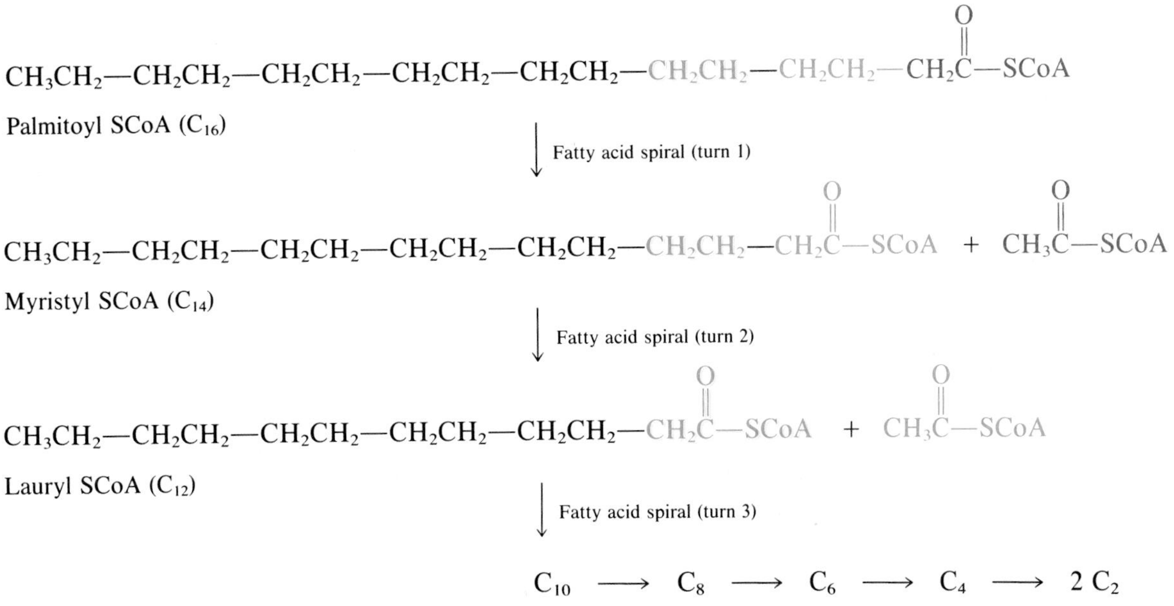

$$CH_3CH_2-CH_2CH_2-CH_2CH_2-CH_2CH_2-CH_2CH_2-CH_2CH_2-CH_2CH_2-CH_2\overset{O}{\overset{\|}{C}}-SCoA$$

Palmitoyl SCoA (C_{16})

Fatty acid spiral (turn 1)

$$CH_3CH_2-CH_2CH_2-CH_2CH_2-CH_2CH_2-CH_2CH_2-CH_2CH_2-CH_2\overset{O}{\overset{\|}{C}}-SCoA \ + \ CH_3\overset{O}{\overset{\|}{C}}-SCoA$$

Myristyl SCoA (C_{14})

Fatty acid spiral (turn 2)

$$CH_3CH_2-CH_2CH_2-CH_2CH_2-CH_2CH_2-CH_2CH_2-CH_2\overset{O}{\overset{\|}{C}}-SCoA \ + \ CH_3\overset{O}{\overset{\|}{C}}-SCoA$$

Lauryl SCoA (C_{12})

Fatty acid spiral (turn 3)

$$C_{10} \longrightarrow C_8 \longrightarrow C_6 \longrightarrow C_4 \longrightarrow 2\ C_2$$

Since fatty acids usually have an even number of carbon atoms, none are left over. Additional oxidation steps deal with fatty acids with odd numbers of carbon atoms and those with double bonds, ultimately also releasing all carbon atoms for further oxidation in the citric acid cycle.

Practice Problems

13.2 How many molecules of acetyl SCoA are produced by catabolism of these fatty acids, and how many turns of the fatty acid spiral are needed?
(a) lauric acid, $CH_3(CH_2)_{10}COOH$
(b) arachidic acid, $CH_3(CH_2)_{18}COOH$

13.3 Write the equations for the next three turns of the fatty acid spiral following those in Figure 13.9.

13.4 Look back at the reactions of the citric acid cycle (Figure 9.11) and identify the three reactions that are similar to the first three reactions of the fatty acid spiral.

AN APPLICATION: THE LIVER, CLEARINGHOUSE FOR METABOLISM

The liver is the largest reservoir of blood in the body and is the largest internal organ, making up about 2.5% of the body's mass. Blood carrying the end products of digestion enters the liver through the hepatic portal vein before going on into general circulation, and the liver is therefore ideally situated to regulate the concentrations of nutrients and other substances in the blood.

Various functions of the liver have been previously described in scattered sections of this book, but it's only by taking an overview that the central role of the liver in metabolism can be appreciated. Among its many functions, the liver synthesizes glycogen from glucose, glucose from noncarbohydrates, triacylglycerols from mono- and diacylglycerols, fatty acids from acetyl SCoA, phospholipids, cholesterol, bile acids, and the majority of the serum proteins. In addition, liver cells can catabolize glucose, fatty acids, and amino acids to yield carbon dioxide and energy stored in ATP. The *urea cycle,* by which nitrogen from amino acids is converted to urea for excretion, also takes place in the liver.

Because reserves of glycogen, certain lipids and amino acids, iron, and fat-soluble vitamins are held in storage in the liver, it is able to release them as needed to maintain homeostasis. In addition, only liver cells have the enzyme needed to convert glucose 6-phosphate from glycogenolysis and gluconeogenesis to glucose that can enter the bloodstream.

Given its central role in metabolism, the liver is subject to a number of pathologic conditions based on excessive accumulation of various metabolites. One example is *cirrhosis,* the development of fibrous tissue that is preceded by excessive triacylglycerol buildup. Cirrhosis occurs in alcoholism, uncontrolled diabetes, and metabolic conditions in which the synthesis of lipoproteins from triacylglycerols is blocked. Another example is *Wilson's disease,* a genetic defect in copper metabolism. In Wilson's disease, copper accumulates in the liver rather than being excreted or recycled for use in a number of coenzymes. Chronic liver disease, as well as brain damage and anemia, are symptoms of Wilson's disease. The disease is treated by a low-copper diet and drugs that enhance the excretion of copper.

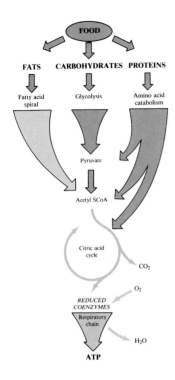

13.5   ENERGY OUTPUT DURING FATTY ACID CATABOLISM

The total energy output during fatty acid catabolism, like that from glucose catabolism, is measured by the total number of ATPs produced. In the case of fatty acids, this total is the sum of the ATPs from acetyl SCoA and those from reduced coenzymes from fatty acid oxidation. Recall from Section 9.8 that for each acetyl SCoA that passes through the citric acid cycle, the overall reaction is

$$\text{Acetyl SCoA} + 3\,\text{NAD}^+ + \text{FAD} + \text{ADP} + \text{P}_i + 2\,\text{H}_2\text{O} \longrightarrow$$
$$\text{HSCoA} + 3\,\text{NADH} + 3\,\text{H}^+ + \text{FADH}_2 + \text{ATP} + 2\,\text{CO}_2$$

Since in electron transport each NADH generates 3 ATPs and each FADH_2 generates 2 ATPs, the total number of ATPs produced by each acetyl SCoA is 12. Thus, a fatty acid that yields n molecules of acetyl SCoA releases $12n$ ATPs, where n is one-half the number of carbon atoms in a given fatty acid.

In addition, each turn of the fatty acid spiral (Figure 13.8) releases two reduced coenzymes: one FADH_2 and one NADH. The FADH_2 coenzyme generated in step 1 yields two molecules of ATP when it is recycled, and the NADH coenzyme generated in step 3 yields three ATPs. The total is thus five ATPs for each turn of the spiral, and the spiral turns $n - 1$ times for a given acid.

To arrive at the final total of ATPs, however, we have to subtract the two molecules of ATP that are required for the initial bonding of the free fatty acid with coenzyme A. Thus the overall energy scorecard looks like this:

	Change in Number of ATPs per Fatty Acid
Initial bonding to coenzyme A	-2
(Molecules of acetyl SCoA) $\times \dfrac{12\ \text{ATP}}{\text{acetyl SCoA}}$	$+12n$
(Number of turns of spiral) $\times \dfrac{5\ \text{ATP}}{\text{turn}}$	$+5(n-1)$
Net change for complete fatty acid catabolism	$17n - 7$

where

$$n = 1/2\ \text{number of carbons in the fatty acid chain}$$
$$= \text{number of molecules of acetyl SCoA produced}$$
$$n - 1 = \text{number of turns of the fatty acid spiral}$$

Comparing the amount of ATP produced by catabolism of a fatty acid to the amount produced by catabolism of glucose illustrates why our bodies use triacylglycerols rather than carbohydrates for long-term energy storage. One mole of glucose (180 g) generates 36 mol of ATP, whereas one mole of lauric acid (200 g) generates 95 mol of ATP. Thus, fatty acids yield more than twice as much energy per gram as carbohydrates (and do twice as much damage to diets). In addition, stored fats have a greater "energy density" because they are hydrophobic. Glycogen is hydrophilic, however, and each gram of stored glycogen is associated with about 2 g of water.

Solved Problem 13.1   Calculate the number of molecules of ATP produced during the catabolism of lauric acid, $CH_3(CH_2)_{10}COOH$.

Solution   First, count the number of carbons in the lauric acid chain and then divide by 2 to find how many molecules of acetyl SCoA are produced:

$$CH_3(CH_2)_{10}COOH \longrightarrow C_{12} \longrightarrow 12/2 = 6 \text{ molecules of acetyl SCoA}$$

Next, find how many turns of the fatty acid spiral are needed to degrade lauric acid by subtracting 1 from the number of acetyl SCoA molecules:

$$6 \text{ Acetyl SCoA} - 1 \longrightarrow 5 \text{ turns of the fatty acid spiral}$$

Finally, do the arithmetic:

Initiation	= − 2 ATP
6 Acetyl SCoA × (12 ATP/acetyl SCoA) =	72 ATP
5 Turns × (5 ATP/turn)	= 25 ATP
Net change	= 95 ATP

Checking shows that with $n = 6$, $(17 \times 6) - 7 = 95$ ATPs.

Practice Problems   **13.5**   Calculate the amount of ATP produced by catabolism of one mole of palmitic acid, $CH_3(CH_2)_{14}COOH$.

13.6   How many grams of ATP (mol wt 507 amu) are produced by catabolism of 1.0 g of palmitic acid (mol wt 256 amu)?

Ketone bodies Compounds produced in the liver that can be used as fuel by muscle and brain tissue; β-hydroxybutyrate, acetoacetate, and acetone.

Ketogenesis The synthesis of ketones bodies from acetyl SCoA.

13.6 PRODUCTION OF KETONE BODIES; KETOACIDOSIS

Glucose and fatty acids are our major fuel molecules, but small amounts of alternative fuels called **ketone bodies** are also produced from acetyl SCoA in liver mitochondria, a process known as **ketogenesis**. The first step in the production of ketone bodies is the combination of two acetyl SCoAs to give acetoacetyl SCoA. Hydrolysis then yields acetoacetate, and further reactions yield acetone and 3-hydroxybutyrate.

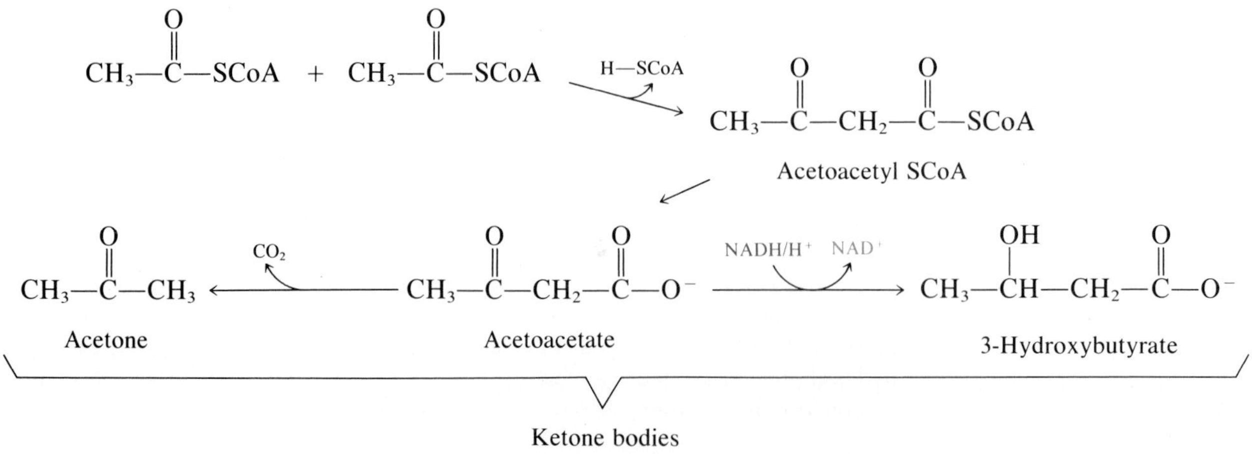

Acetoacetyl SCoA

Acetone   Acetoacetate   3-Hydroxybutyrate

Ketone bodies

Under well-fed, healthy conditions, skeletal and heart muscles derive a small portion of their daily energy needs from one of the ketone bodies, acetoacetate, which is converted back to two acetyl SCoAs for oxidation via the citric acid cycle. But consider the situation when glucose levels are abnormally low. The body must then respond by providing other sources for energy. For example, the synthesis of glucose by gluconeogenesis and the breakdown of stored fat to give acetyl SCoA are accelerated. Meanwhile, however, glucose metabolism and the citric acid cycle are slowing down. The result is production from fatty acids of more acetyl SCoA than can be processed by the citric acid cycle. This condition leads to diversion of more and more acetyl SCoA into ketone bodies that can be used to produce energy. During the early stages of starvation, heart and muscle tissues burn larger quantities of acetoacetate, thereby preserving glucose for use in the brain. In prolonged starvation, even the brain can switch to ketone bodies to meet up to 75% of its energy needs.

Ketoacidosis Lowered blood pH due to accumulation of ketone bodies.

The condition in which ketone bodies are produced faster than they are utilized (*ketosis*) is recognized by the smell of acetone on the breath and the presence of ketone bodies in the urine (*ketonuria*). Because two of the ketone bodies are carboxylic acids, continued ketosis such as might occur in untreated diabetes leads to the potentially serious condition known as **ketoacidosis:** acidosis resulting from increased concentrations of ketone bodies in the blood. The blood's buffers are overwhelmed, and blood pH drops. An individual experiences dehydration due to increased urine flow, labored breathing because acidic blood is a poor oxygen carrier, and depression. Ultimately, the condition can lead to coma and death.

Practice Problem   13.7   Which of the following classifications apply to the formation of 3-hydroxybutyrate from acetoacetyl SCoA: (i) hydrolysis, (ii) oxidation, (iii) reduction, (iv) condensation?

13.7 BIOSYNTHESIS OF FATTY ACIDS

Lipogenesis The biochemical pathway for synthesis of fatty acids from acetyl SCoA.

The biosynthesis of fatty acids from acetyl SCoA, a process known as **lipogenesis,** provides a link between carbohydrate, lipid, and protein metabolism. Because the catabolism of both carbohydrates and amino acids produces acetyl SCoA, fatty acid synthesis followed by triacylglycerol synthesis allows excess carbohydrates and amino acids to be stored in adipose tissue. As you've seen for glycolysis and gluconeogenesis, biochemical pathways that run in reverse directions don't usually occur by reverse routes because the reverse of an energetically favorable pathway is likely to be energetically unfavorable. This principle applies to oxidation of fatty acids by the fatty acid spiral and to its reverse, the biosynthesis of fatty acids (Table 13.1).

Acyl carrier protein Protein that carries acyl groups during fatty acid synthesis.

Fatty acid synthesis begins when an acetyl SCoA is converted to malonyl SCoA, and the malonyl group is then transferred to the **acyl carrier protein (ACP).** One malonyl SACP is required for each turn of the biosynthesis spiral, because it carries the carbon atoms to be added to the growing chain. At the same time, the acetyl group from another acetyl SCoA is bonded to a second acyl carrier protein to generate acetyl SACP.

Table 13.1 Comparison of Fatty Acid Oxidation and Synthesis

Oxidation	Synthesis
Occurs in mitochondria.	Occurs in cytosol.
Enzymes different from synthesis.	Enzymes different from oxidation.
Intermediates carried by coenzyme A.	Intermediates carried by acyl carrier protein.
Coenzymes: PAD, NAD$^+$.	Coenzyme: NADPH.
Spiral pathway, carbon atoms removed two at a time.	Spiral pathway, carbon atoms added to at a time.

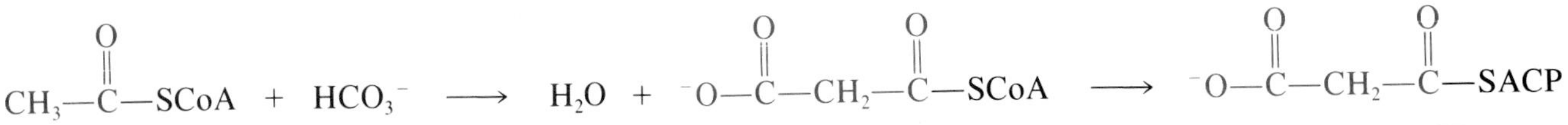

$$CH_3-\overset{\overset{\displaystyle O}{\|}}{C}-SCoA \ + \ HCO_3^- \longrightarrow H_2O \ + \ ^-O-\overset{\overset{\displaystyle O}{\|}}{C}-CH_2-\overset{\overset{\displaystyle O}{\|}}{C}-SCoA \longrightarrow \ ^-O-\overset{\overset{\displaystyle O}{\|}}{C}-CH_2-\overset{\overset{\displaystyle O}{\|}}{C}-SACP$$

Acetyl SCoA   Malonyl SCoA   Malonyl SACP

$$CH_3-\overset{\overset{\displaystyle O}{\|}}{C}-SCoA \longrightarrow CH_3-\overset{\overset{\displaystyle O}{\|}}{C}-SACP$$

Acetyl SCoA   Acetyl SACP

Once malonyl SACP and acetyl SACP have been generated, a repeating cycle of four reactions takes place to lengthen the growing fatty acid chain by two carbon atoms with each repetition (Figure 13.10).

Step 1 of fatty acid synthesis: Condensation. The malonyl group from malonyl ACP transfers to acetyl SACP with the loss of CO_2, resulting in addition of two carbon atoms. Note that malonyl SACP is the source of the added carbon atoms.

Steps 2–4 of fatty acid synthesis: Reduction, dehydration, and reduction. These three reactions accomplish the reverse of steps 3, 2, and 1 in the fatty acid spiral (Figure 13.8). The carbonyl group is reduced to an —OH group, dehydration yields a carbon–carbon double bond, and the double bond is reduced by addition of hydrogen.

The result of the first cycle in fatty acid synthesis is the addition of 2 carbon atoms to an acetyl group to give a 4-carbon acyl group. The next cycle then adds 2 more carbon atoms to give a 6-carbon acyl group.

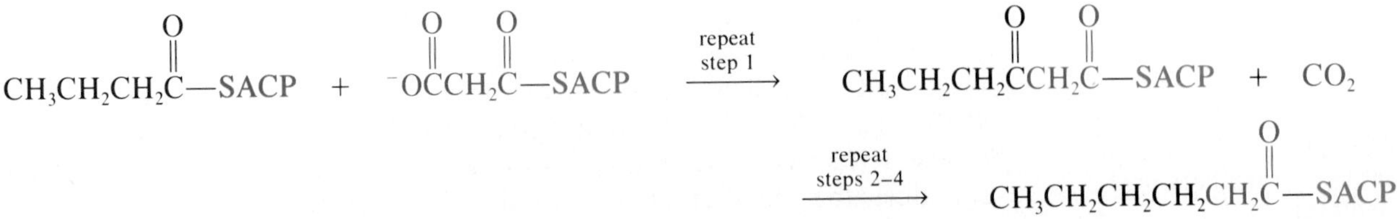

$$CH_3CH_2CH_2\overset{\overset{\displaystyle O}{\|}}{C}-SACP \ + \ ^-O\overset{\overset{\displaystyle O}{\|}}{C}CH_2\overset{\overset{\displaystyle O}{\|}}{C}-SACP \ \xrightarrow[\text{step 1}]{\text{repeat}} \ CH_3CH_2CH_2\overset{\overset{\displaystyle O}{\|}}{C}CH_2\overset{\overset{\displaystyle O}{\|}}{C}-SACP \ + \ CO_2$$

$$\xrightarrow[\text{steps 2–4}]{\text{repeat}} \ CH_3CH_2CH_2CH_2CH_2\overset{\overset{\displaystyle O}{\|}}{C}-SACP$$

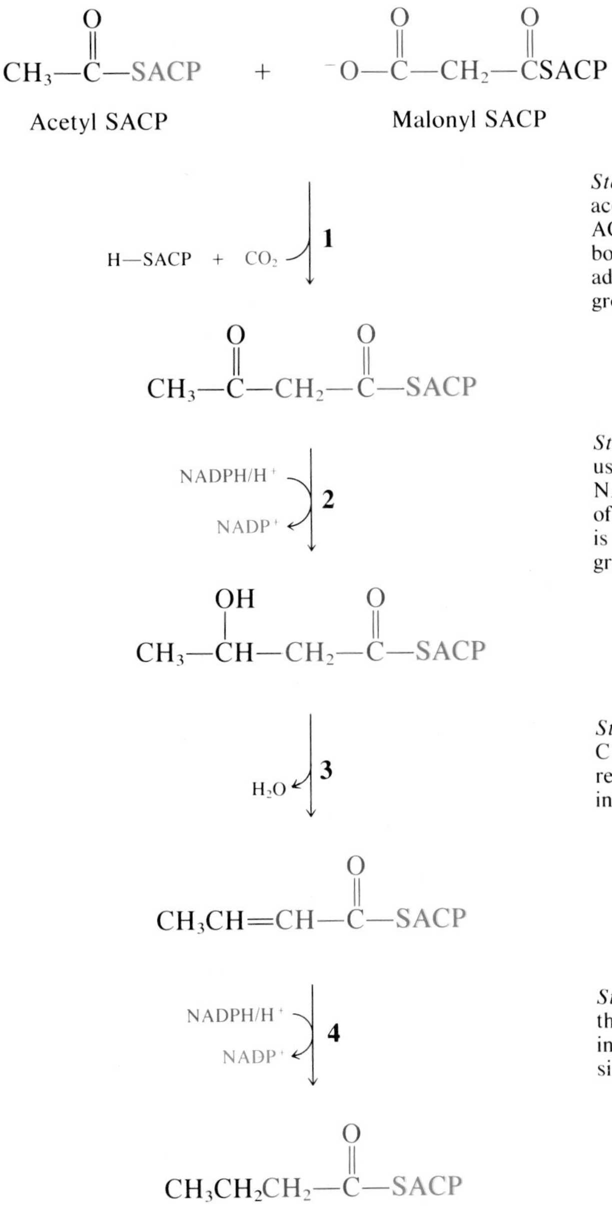

Step 1. Acetyl groups from acetyl ACP and malonyl ACP are joined by a C—C bond, with loss of the CO_2 added to form the malonyl group.

Step 2. In this reduction using the coenzyme NADPH, the carbonyl group of the original acetyl group is reduced to a hydroxyl group.

Step 3. Dehydration at the C atoms α and β to the remaining carbonyl group introduces a double bond.

Step 4. In another reduction, the double bond introduced in step 3 is converted to a single bond.

Figure 13.10
Chain elongation in the biosynthesis of fatty acids. The steps shown begin with acetyl acyl carrier protein (acetyl SACP), the reactant in the first cycle of palmitic acid synthesis. Each new pair of carbon atoms is carried into the next cycle by a fresh malonyl SACP. The growing chain remains attached to the carrier protein from the original acetyl SACP.

After seven trips through the elongation cycle, a 16-carbon palmitoyl group has been produced. Hydrolysis then breaks the thioester bond to give palmitic acid, $CH_3(CH_2)_{14}COOH$. All other fatty acids are synthesized from palmitic acid.

The major transport and biochemical pathways of acylglycerols and fatty acids are summarized in Figure 13.11.

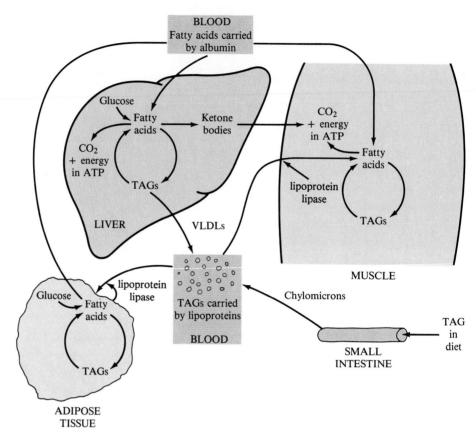

Figure 13.11
Major pathways of lipid metabolites. Note that synthesis of fatty acids from glucose takes place via acetyl SCoA mainly in adipose tissue and the liver.

SUMMARY

The digestion and metabolism of triacylglycerols (TAGs) and other lipids is complicated by their insolubility in water. **Bile acids** therefore emulsify dietary TAGs in the small intestine, where they are partially hydrolyzed by pancreatic lipase and absorbed through the intestinal lining. After absorption of the hydrolysis products, TAGs are resynthesized and repackaged in **lipoproteins** known as **chylomicrons** that are carried in the lymphatic system to the bloodstream. Lipoproteins known as VLDLs similarly transport TAGs synthesized in the liver, and albumins in blood serum transport fatty acids released from storage in adipose tissue.

The major catabolic pathway of TAGs is hydrolysis to fatty acids plus glycerol, followed by oxidation of fatty acids in the **fatty acid spiral** to yield acetyl SCoA. The spiral proceeds by removal of two carbon atoms at a time as acetyl SCoA. A fatty acid undergoes $n - 1$ turns of the spiral and yields n acetyl SCoAs, where n is one-half the number of carbon atoms in the fatty acid. The acetyl SCoAs may then either enter the citric acid cycle for conversion into energy or be used as starting material for the synthesis of steroids and **ketone bodies.**

The **lipogenesis** pathway synthesizes fatty acids from acetyl SCoA. Fatty acid synthesis occurs in the cytosol, and the reactants are acyl groups bonded to **acyl carrier protein.** Carbon atoms are added two at a time to produce palmitic acid, from which other fatty acids are synthesized. TAG synthesis also requires dihydroxyacetone phosphate, which is made in the liver from glycerol or comes from glycolysis. Because the acetyl SCoA needed for lipogenesis is a product of carbohydrate and protein metabolism, lipogenesis followed by synthesis of TAGs is a pathway for converting excess protein and carbohydrate to fat.

Ketogenesis in the liver produces ketone bodies from acetyl SCoA. When glucose is in short supply, ketogenesis accelerates and ketones are burned for energy. An excess of ketone bodies, two of which are acids, produces the dangerous condition known as **ketoacidosis.**

INTERLUDE: MAGNETIC RESONANCE IMAGING (MRI)

The recent development in body imaging known as *magnetic resonance imaging (MRI)* is creating considerable excitement in the medical community because of its advantages over techniques that use X rays or radioactive materials. In MRI, the patient is not exposed to any ionizing radiation, there is no need to introduce contrast materials into the patient's body, and soft-tissue structures that are obscured by bone in X-ray images are visible.

Essentially, MRI takes advantage of the magnetic properties of certain nuclei atomic and looks at the location of those nuclei in the body. By placing the nuclei in a magnetic field and irradiating them with electromagnetic radiation in the radiofrequency range, the nuclei are made to absorb and release specific amounts of energy, which are measured. Body images are produced by computer processing of the resulting data.

Most MRI images currently look at hydrogen nuclei, present in abundance wherever there is water or fat in the body. The signals produced vary with the density of hydrogen atoms and with the nature of their surroundings, allowing identification of different types of tissue and the visualization of motion. For example, the volume of blood leaving the heart in a single stroke can be measured, allowing observation of the heart in motion. Several types of atoms in addition to hydrogen can also be detected by MRI, and the applications of images based on the phosphorus atoms in ATP and other phosphates are actively being explored. The technique holds great promise for studies of metabolism.

MRI has quickly become the method of choice for detecting brain tumors not seen with other techniques; stories abound of dramatic cures of patients with motor difficulties caused by tiny brain tumors. The technique is also valuable in diagnosing knee damage because it is a painless alternative to arthroscopy, in which an endoscope is physically introduced into the knee joint.

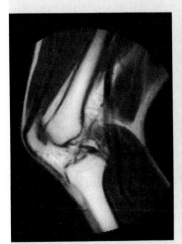 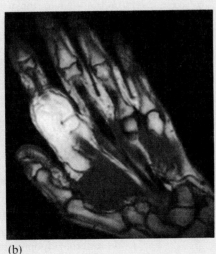 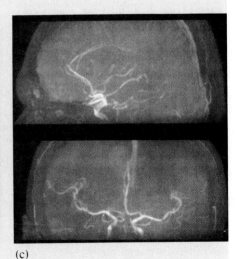

(a)   (b)   (c)

MRI images of (a) a normal knee, (b) a hand with a bone abnormality on the index finger, and (c) a normal brain from the side (top image) and front (bottom image) showing blood flow in arteries. In (a) and (b) the dark lines are blood vessels and the white areas are bone. In (a) the cruciate ligaments, often injured in sporting accidents, show faintly as dark lines between the two bones. The signal intensity in (c) has been modified to emphasize blood flow; such an image might be used to detect an aneurism or observe blood flow to a tumor.

REVIEW PROBLEMS

Digestion and Catabolism of Lipids

13.8* Where does digestion of lipids occur?

13.9 Why do lipids make one feel full for a long time after a meal?

13.10* What is the purpose of bile in lipid catabolism?

13.11 Lipases break down triacylglycerols by catalyzing hydrolysis. What are the products of this hydrolysis?

13.12* What are chylomicrons, and how are they involved in lipid metabolism?

13.13 What is the origin of the triacylglycerols transported by very-low-density lipoproteins?

13.14* How are the fatty acids from adipose tissue transported?

13.15 The glycerol derived from digestion of dietary triacylglycerols is converted into glyceraldehyde 3-phosphate, which then enters into step 6 of the glycolysis pathway. What further transformations are necessary to convert glyceraldehyde 3-phosphate into pyruvate?

13.16* If the conversion of glycerol to glyceraldehyde 3-phosphate releases one molecule of ATP, how many molecules of ATP are released during the conversion of glycerol to pyruvate? (Problem 13.15.)

13.17 How many molecules of ATP are released in the overall catabolism of glycerol to acetyl SCoA?

13.18* How many molecules of ATP are released in the complete catabolism of glycerol to CO_2 and H_2O? (Problem 13.17.)

13.19 What is an adipocyte?

13.20* Where in the cell does the fatty acid spiral occur?

13.21 What is the alternative name of the fatty acid spiral? Why is this name appropriate?

13.22* What initial chemical transformation takes place on a fatty acid to activate it for catabolism?

13.23 What is the function of carnitine in fatty acid catabolism?

13.24* Why do you suppose the sequence of reactions that catabolize fatty acids is called the fatty acid *spiral* rather than the fatty acid *cycle?*

13.25 How many moles of ATP are produced by one turn of the fatty acid spiral?

13.26* Arrange these four molecules in order of their biological energy content per mole:
(a) glucose
(b) capric acid, $CH_3(CH_2)_8COOH$
(c) sucrose
(d) myristic acid, $CH_3(CH_2)_{12}COOH$

13.27 Show the products of each step in the following fatty acid spiral on hexanoic acid:

(a)
$$CH_3CH_2CH_2CH_2CH_2CSCoA \xrightarrow[\substack{\text{acetyl SCoA} \\ \text{dehydrogenase}}]{\text{FAD} \quad \text{FADH}_2} \ ?$$

(b) product of (a) + H_2O $\xrightarrow[\text{hydratase}]{\text{enoyl SCoA}}$?

(c) product of (b) $\xrightarrow[\substack{\beta\text{-hydroxyacyl SCoA} \\ \text{dehydrogenase}}]{\text{NAD}^+ \quad \text{NADH/H}^+}$?

(d) product of (c) + HSCoA $\xrightarrow[\text{transferase}]{\text{acetyl SCoA}}$?

13.28* Write the equation for the final step in the catabolism of any fatty acid with an even number of carbons.

13.29 How many molecules of acetyl SCoA result from complete catabolism of these compounds?
(a) caprylic acid, $CH_3(CH_2)_6COOH$
(b) myristic acid, $CH_3(CH_2)_{12}COOH$

13.30* How many turns of the fatty acid spiral are necessary to completely catabolize caprylic and myristic acids? (Problem 13.29.)

Fat Anabolism

13.31 What is the name of the anabolic pathway for synthesizing fatty acids?

13.32* What is the starting material for fatty acid synthesis?

13.33 Why can't the fatty acid spiral be run backward to produce triacylglycerols?

13.34* Why are fatty acids generally composed of an even number of carbons?

13.35 What is the fatty acid from which all other fatty acids are synthesized?

13.36* How many rounds of the lipigenesis cycle are needed to synthesize palmitic acid, $C_{15}H_{31}COOH$?

Applications

13.37 What is a normal blood cholesterol range? [App: Cholesterol, Lipoproteins, and Heart Disease]

13.38* What is atherosclerosis? [App: Cholesterol, Lipoproteins, and Heart Disease]

13.39 What is the difference in the roles of LDL and HDL? [App: Cholesterol, Lipoproteins, and Heart Disease]

13.40* Which is better, a high or low HDL/LDL level? Why? [App: Cholesterol, Lipoproteins, and Heart Disease]

13.41   What is cirrhosis of the liver and what can trigger it? [App: Liver, Clearinghouse for Metabolism]

13.42*   What type of atomic nucleus is primarily monitored in MRI? [Int: Magnetic Resonance Imaging (MRI)]

13.43   What is the advantage of MRI compared to radiation techniques of imaging the body? [Int: Magnetic Resonance Imaging (MRI)]

Additional Questions and Problems

13.44*   Consuming too many carbohydrates causes deposition of fats in adipose tissue. How can this happen? Why aren't the molecules stored in glycogen instead?

13.45   Are any of the intermediates in the fatty acid spiral chiral? Explain.

13.46*   What three compounds are classified as ketone bodies? Why are they so designated? What process in the body produces them? Why do they form?

13.47   What is ketosis? What condition results from prolonged ketosis? Why is it dangerous?

13.48*   How many molecules of acetyl SCoA result from catabolism of glyceryl trimyristate?

13.49   How many ATPs are released in the complete catabolism of glyceryl trimyristate? (Problem 13.48.)

13.50*   Compare fats and carbohydrates as energy sources in terms of the amount of energy released per mole and account for the observed energy difference.

CHAPTER

14

Protein and Amino Acid Metabolism

Urea, an end product of protein metabolism, forms beautiful crystals, as seen here under polarized light. Nitrogen not needed for making new amino acids must be excreted as urea, which is formed by a metabolic pathway described in this chapter.

To understand how we function, we need to understand proteins. The complexities of their structure and their role as enzymes have already been discussed. Their biosynthesis is determined by the genetic code and is such a broad subject that it requires an entire chapter itself. What remains for this chapter is a look at the metabolism of amino acids. We'll answer the following questions:

1. *What happens during the digestion of proteins?* The goal: Be able to list the sequence of events in the digestion of proteins and the transport of amino acids into the bloodstream.
2. *What are the major strategies in the catabolism of proteins?* The goal: Be able to explain the general catabolic pathways of amino acids.
3. *What is transamination?* The goal: Be able to describe where transamination occurs in amino acid metabolism and be able to draw the structures of the products of transamination.
4. *What is the urea cycle?* The goal: Be able to list the major reactants and products of the reactions in the urea cycle and explain how the cycle is linked to the citric acid cycle.
5. *What are glucogenic and ketogenic amino acids?* The goal: Be able to define these terms and tell how these kinds of amino acids are metabolized.
6. *What is the strategy for amino acid synthesis?* The goal: Be able to describe in general the compounds used in amino acid synthesis.

14.1 PROTEIN DIGESTION

The end result of protein digestion is simple—the hydrolysis of all peptide bonds to produce a collection of amino acids:

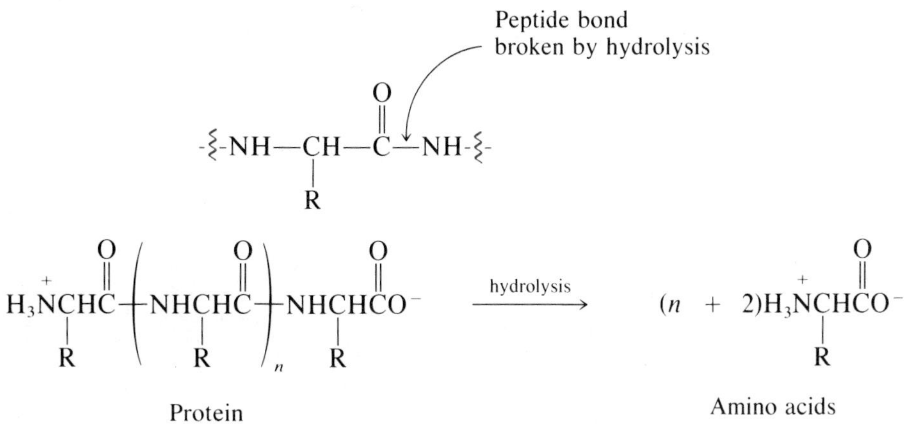

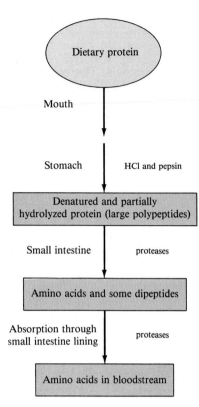

Figure 14.1
Digestion of proteins.

The digestion of dietary proteins, as summarized in Figure 14.1, begins with their denaturation in the strongly acidic environment of the stomach (pH 1–2). In addition to hydrochloric acid, gastric secretions include *pepsinogen*, a zymogen that is activated by acid to give the enzyme *pepsin* (Figure 14.2). Some protein peptide bonds made accessible by the unfolding of denaturation are hydrolyzed by pepsin, giving large polypeptides that then enter the small intestine.

As the polypeptides enter the small intestine, the pH is raised to about 7–8, pepsin is inactivated, and a group of pancreatic zymogens is secreted. A small amount of *trypsinogen* is activated by an enzyme from the intestinal lining to give *trypsin*, one of the **protease** enzymes that carry out peptide hydrolysis. This trypsin then activates the remaining trypsinogen and other pancreatic zymogens to give *chymotrypsin, carboxypeptidase*, and *elastase*, proteases that further hydrolyze polypeptides.

The combined action of the pancreatic proteases in the small intestine and other proteases in the intestinal lining converts all polypeptides to amino acids. After transport across cell membranes lining the intestine, the amino acids are absorbed directly into the bloodstream.

14.2   AMINO ACID METABOLISM: AN OVERVIEW

The entire collection of free amino acids throughout the body—the **amino acid pool**—occupies a central position in protein and amino acid metabolism (Figure 14.3). Since living organisms are dynamic rather than static, all tissues and biomolecules in the body are constantly being degraded, repaired, and replaced.

Figure 14.2
Digestion in cheese making. (a) The first step in cheese making is the addition of bacteria and enzymes that sour the milk and denature the milk protein (casein). (b) After excess water is removed, the remaining solid curd is shaped and stored while enzymes from microorganisms hydrolyze proteins, lactose, and triacylglycerols.

(a)

(b)

Protease   An enzyme for hydrolysis of the peptide bonds in proteins.

Amino acid pool   The entire collection of free amino acids in the body.

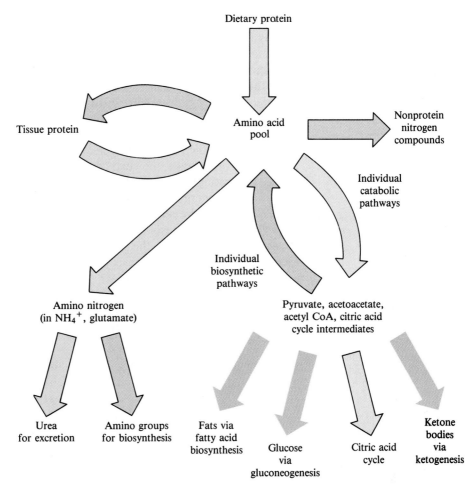

Figure 14.3
Protein and amino acid metabolism. Synthetic pathways are shown in blue, pathways that break down molecules are shown in yellow, and connections to fat and glucose metabolism are shown in green. The amino groups are used in the biosynthesis of amino acids and nonprotein nitrogen compounds.

Thus, amino acids enter the pool not only by digestion of proteins and synthesis from other compounds, but also from breakdown of tissues. In fact, a normal adult breaks down about 2% of their protein, or approximately 400 g of protein, every day.

The enzymes for hydrolysis of waste protein to form amino acids are located in cells throughout the body. Also, cells throughout the body call on amino acids from the pool for protein synthesis. In addition, as the principal nitrogen-containing compounds of the body, amino acids are needed for the synthesis of nonprotein nitrogen-containing compounds.

Amino acid catabolism is quite complex since each of the 20 different amino acids is degraded through its own unique pathway. The general idea of all the pathways, however, is that the amino nitrogen atom is removed and the amino acids are converted into compounds that can enter the citric acid cycle. For some amino acids, a single step is sufficient; for others, a multistep pathway is needed.

In keeping track of amino acid metabolism, it helps to think of each amino acid as having two parts—the amino group and the carbon atoms—that take separate courses. The amino group may be used in the synthesis of other nitrogen-containing compounds, including nucleic acids and other amino acids. Since there is no storage form for nitrogen compounds equivalent to glycogen or triacylglycerols, however, any nitrogen not immediately needed must be excreted by conversion to urea, as discussed in the next section.

Once the carbon atoms of amino acids are converted to compounds associated with the citric acid cycle, they are available for several alternative pathways. For instance, they can continue through the cycle in the body's main energy-generating pathway to give CO_2 and energy stored in ATP. About 10–20% of our energy is normally produced from amino acids. If not needed for energy, however, the carbon-carrying intermediates produced from amino acids enter the gluconeogenesis pathway for glucose synthesis (Section 11.11), the ketogenesis pathway for the synthesis of ketone bodies from acetyl SCoA (Section 13.6), or the lipogenesis pathway for the synthesis of fatty acids from acetyl SCoA (Section 13.7).

14.3 AMINO ACID CATABOLISM: REMOVAL OF THE AMINO GROUP

Transamination The interchange of the amino group of an amino acid and the keto group of an α-keto acid.

Transamination   As the first step in their catabolism, most amino acids undergo **transamination,** a reaction in which the amino group of the amino acid and the keto group of an α-keto acid change places.

$$R'-\underset{\underset{NH_3^+}{|}}{CH}-COO^- \;+\; R''-\overset{\overset{O}{\|}}{C}-COO^- \;\underset{\xrightarrow{\text{transaminase}}}{\rightleftharpoons}\; R'-\overset{\overset{O}{\|}}{C}-COO^- \;+\; R''-\underset{\underset{NH_3^+}{|}}{CH}-COO^-$$

Amino acid   α-Keto acid   α-Keto acid   Amino acid

There are a number of *transaminase* enzymes. Most are specific for α-ketoglutarate as the amino group acceptor and work with several different amino acid donors, resulting in formation of an α-keto acid and glutamate as diagrammed in Figure 14.4. For example, alanine is converted to pyruvate by transamination. The enzyme for this conversion, *alanine aminotransferase (ALT),* is especially abundant in the liver, and above-normal ALT concentrations are taken as an indication of liver damage, which allows ALT to leak into the bloodstream.

Figure 14.4
Major pathway of nitrogen from an amino acid to urea. The red pathway shows the complete catabolism of amino acid nitrogen to give urea. The green pathway shows the reverse route used in the biosynthesis of some amino acids.

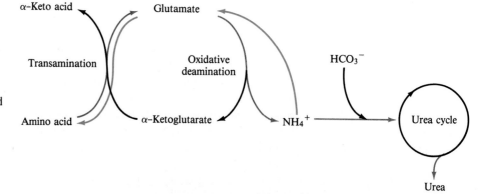

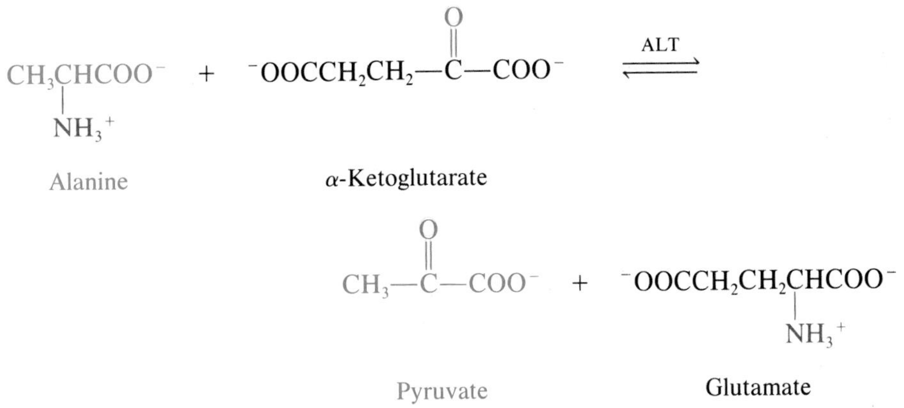

$$\text{CH}_3\text{CHCOO}^- \quad + \quad {}^-\text{OOCCH}_2\text{CH}_2 \overset{\overset{\displaystyle O}{\|}}{-\text{C}} -\text{COO}^- \quad \underset{\longleftarrow}{\overset{\text{ALT}}{\longrightarrow}}$$
$$\underset{\displaystyle \text{NH}_3^+}{|}$$

Alanine α-Ketoglutarate

$$\text{CH}_3 \overset{\overset{\displaystyle O}{\|}}{-\text{C}} -\text{COO}^- \quad + \quad {}^-\text{OOCCH}_2\text{CH}_2\text{CHCOO}^-$$
$$\underset{\displaystyle \text{NH}_3^+}{|}$$

Pyruvate   Glutamate

Each transaminase uses the coenzyme pyridoxal phosphate, a derivative of vitamin B$_6$ (pyridoxine) that bonds to amino acids as shown in Figure 14.5.

Transamination is a key part of many biochemical pathways involving amino acids and α-keto acids, interconverting amino groups and carbonyl groups as necessary. The transamination reactions are equilibria that go easily in either direction depending on the concentrations of the reactants. In this way, they regulate amino acid concentrations by keeping synthesis and breakdown in balance.

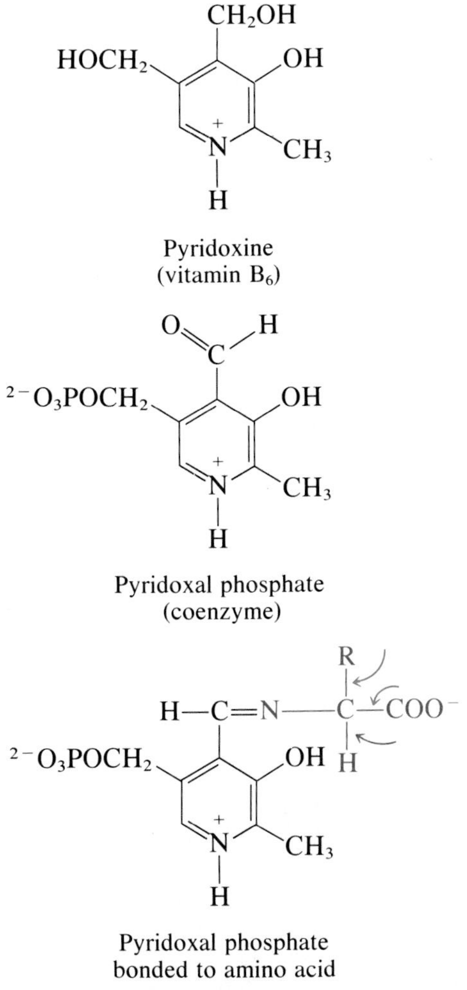

Figure 14.5
Pyridoxal phosphate (PLP), a vitamin B$_6$ derivative and a coenzyme for amino acid reactions. In transamination, the bond marked by the red arrow breaks, the resulting intermediate rearranges, and the coenzyme structure is restored after the products have formed. PLP is a coenzyme for other types of amino acid reactions and also facilitates breaking of the bonds marked with blue arrows.

Pyridoxine
(vitamin B$_6$)

Pyridoxal phosphate
(coenzyme)

Pyridoxal phosphate
bonded to amino acid

Solved Problem 14.1 The serum concentration of the transaminase from heart muscle, *aspartate aminotransferase (AST)*, is used in the diagnosis of heart disease because the enzyme escapes into the serum from damaged heart cells. AST catalyzes transamination of aspartate with α-ketoglutarate. What are the products of this reaction?

Solution The reaction is the interchange of an amino group from aspartate with the keto group from α-ketoglutarate. We know that α-ketoglutarate always gives glutamate in transamination, so one product is glutamate. The product from the amino acid will have a keto group instead of the amino group. Consulting Table 7.1 shows that the structure of aspartate is

$$\text{Aspartate} \qquad ^-OOCCH_2\overset{\alpha}{\underset{|}{CH}}COO^-$$
$$\underset{NH_3^+}{}$$

Removing the $-NH_3^+$ and $-H$ groups bonded to the α carbon and replacing them by a $=O$ gives the desired α-keto acid, oxaloacetate, a compound you've seen in the citric acid cycle.

$$\text{Oxaloacetate} \qquad ^-OOC-CH_2-\overset{\overset{O}{\|}}{C}-COO^-$$

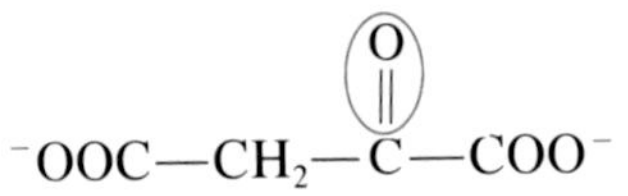

The reaction is

$$\text{Aspartate} + \alpha\text{-ketoglutarate} \xrightarrow{\text{AST}} \text{oxaloacetate} + \text{glutamate}$$

Practice Problems **14.1** What are the structure and name of the α-keto acid formed by transamination of the amino acid leucine?

14.2 What is the product of the following reaction?

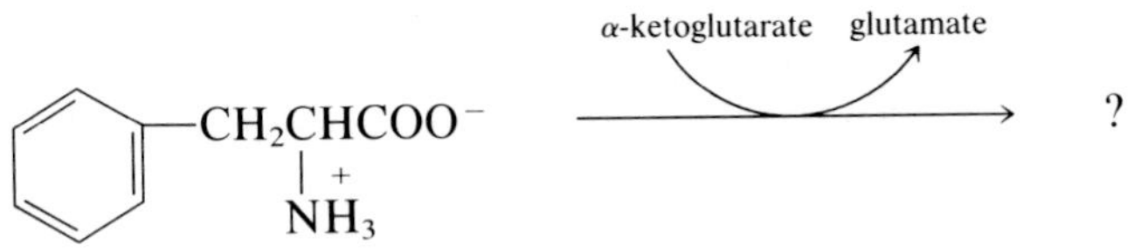

Oxidative deamination
Conversion of an amino acid $-NH_2$ group to an α-keto group, with removal of NH_4^+.

Oxidative Deamination The glutamate produced from transamination of α-ketoglutarate acts as an amino group carrier. Although it's a reactant in the synthesis of several nonessential amino acids (Section 14.6), most of it undergoes recycling to fresh α-ketoglutarate plus ammonia that is eliminated (see Figure 14.4). In a process known as **oxidative deamination,** the glutamate amino group is removed as ammonia (*deamination*) and replaced by a carbonyl group (*oxidation*) to give back α-ketoglutarate. The enzyme for this reaction, *glutamate dehydrogenase*, is unique in being able to use either NAD⁺ or NADP⁺ as its oxidizing coenzyme. Oxidative deamination, like transamination, is a reaction that is used in the reverse direction (called *reductive amination*) in biosynthesis.

$$^-OOCCH_2CH_2CH-COO^- \quad + \quad H_2O$$
$$\overset{|}{NH_3^+}$$

Glutamate

$$NAD^+(NADP^+) \qquad NADH(NADPH)$$
$$\xrightarrow[\substack{\text{glutamate} \\ \text{dehydrogenase}}]{}$$

$$NH_4^+ \quad + \quad ^-OOCCH_2CH_2\overset{\displaystyle O}{\overset{\|}{C}}-COO^-$$

α-Ketoglutarate

14.4 THE UREA CYCLE

Urea cycle The cyclic biochemical pathway that produces urea for excretion.

Ammonia is highly toxic to living things and must be eliminated efficiently. Fish are able to excrete ammonia into their watery surroundings without undergoing any harm (Figure 14.6), but mammals must first convert the ammonia to nontoxic urea by the **urea cycle.** By so doing, they avoid elimination of ammonia in the urine, which could only be done safely if it were dissolved in a large volume of water.

The conversion of ammonia to urea takes place in the liver, from which it is transported to the kidney and transferred to urine for excretion. Like many other biochemical pathways, urea formation begins with an energy investment. Ammonia from oxidative deamination (actually, ammonium ion at physiological pH), carbon dioxide from the citric acid cycle (in solution as bicarbonate ion), and ATP combine to form carbamoyl phosphate. Two ATPs are invested in this reaction, and one high-energy phosphate is transferred to the carbamoyl phosphate.

$$NH_4^+ \quad + \quad HCO_3^- \quad + \quad 2\,ATP \quad \longrightarrow \quad H_3\overset{+}{N}-\overset{\displaystyle O}{\overset{\|}{C}}-O-PO_3^{2-} \quad + \quad 2\,ADP \quad + \quad HPO_4^{2-} \quad + \quad H_2O$$

Carbamoyl phosphate

Carbamoyl phosphate next reacts in the first step of the four-step urea cycle, shown in Figure 14.7.

Figure 14.6
An orange-fin anemone fish in its natural environment in the Fiji Islands. Because fish are surrounded by water, their metabolism need not bother to convert ammonia to urea.

Figure 14.7
The urea cycle.

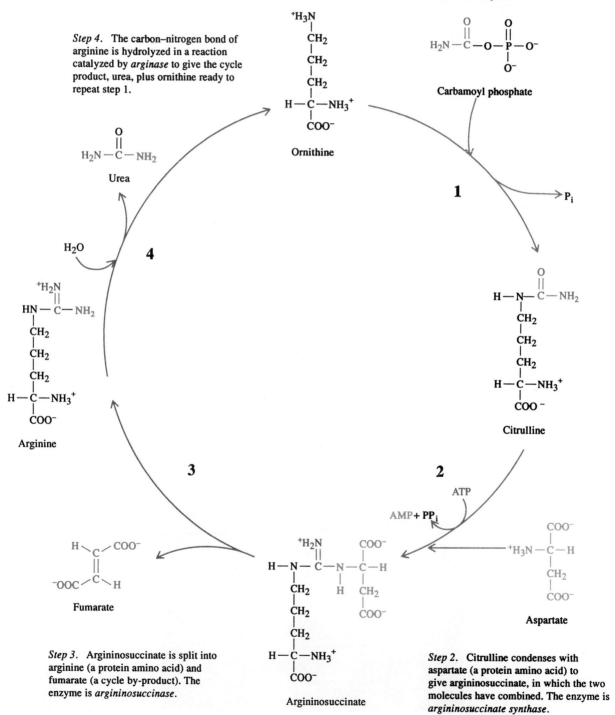

Step 1. **Carbamoyl phosphate transfers its** $H_2NC=O$ **group to ornithine (a nonprotein amino acid) to give citrulline in a reaction catalyzed by** *ornithine transcarbamoylase.*

Step 4. The carbon–nitrogen bond of arginine is hydrolyzed in a reaction catalyzed by *arginase* to give the cycle product, urea, plus ornithine ready to repeat step 1.

Carbamoyl phosphate

Ornithine

Urea

1

4

2

Citrulline

Arginine

3

Fumarate

ATP

AMP + PP$_i$

Aspartate

Step 3. Argininosuccinate is split into arginine (a protein amino acid) and fumarate (a cycle by-product). The enzyme is *argininosuccinase.*

Argininosuccinate

Step 2. **Citrulline condenses with aspartate (a protein amino acid) to give argininosuccinate, in which the two molecules have combined. The enzyme is** *argininosuccinate synthase.*

Steps 1 and 2 of the urea cycle: Building up a reactive intermediate. The first step of the urea cycle transfers the carbamoyl group, $H_2NC{=}O$, from carbamoyl phosphate to *ornithine*, an amino acid not found in proteins, to give *citrulline*, another nonprotein amino acid. Note that the bond broken in carbamoyl phosphate is an energetic anhydride link. The reaction introduces the first nitrogen of the end-product urea into the cycle.

Next, a molecule of aspartate, a standard protein amino acid, condenses with citrulline in a reaction driven by conversion of ATP to AMP and pyrophosphate (PP_i), followed by the additional exergonic hydrolysis of pyrophosphate. Both of what will become the urea nitrogen atoms are now bonded to the same carbon atom in argininosuccinate.

Steps 3 and 4 of the urea cycle: Cleavage and hydrolysis of the step 2 product. Step 3 cleaves argininosuccinate into two pieces: arginine, an amino acid, and fumarate, an intermediate in the citric acid cycle. Now all that remains, in step 4, is the hydrolysis of arginine to give urea and to restore ornithine, the reactant in step 1 of the cycle.

Net result of the urea cycle

$$CO_2 \; + \; NH_4^+ \; + \; 3\,ATP \; + \; {}^-OOC{-}CH_2{-}\underset{\underset{\displaystyle NH_3^+}{|}}{CH}{-}COO^- \; + \; 2\,H_2O \; \longrightarrow$$

Aspartate

$$\underset{\text{Urea}}{H_2N{-}\overset{\overset{\displaystyle O}{\|}}{C}{-}NH_2} \; + \; 2\,ADP \; + \; AMP \; + \; 4\,P_i \; + \; \underset{\text{Fumarate}}{{}^-OOC{-}CH{=}CH{-}COO^-}$$

- Elimination as urea of carbon from CO_2, nitrogen from NH_4^+, and nitrogen from the amino acid aspartate
- Breaking of four high-energy phosphate bonds to provide energy
- Production of the citric acid cycle intermediate, fumarate.

Fumarate links the citric acid cycle with the urea cycle. Fumarate travels around the citric acid cycle to give oxaloacetate, which can be diverted to form aspartate by a reverse transamination, thereby replenishing this reactant for step 2 of the urea cycle.

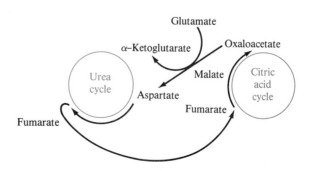

Practice Problem   14.3   Look at the citric acid cycle (Figure 9.11) and then write the structure of each reaction product in the pathway connecting fumarate and aspartate and describe the structural change in each reaction.

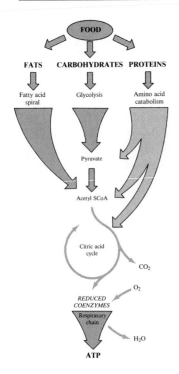

14.5   AMINO ACID CATABOLISM: FAT CARBON ATOMS

The 20 amino acids found in proteins can be grouped according to the carbon-containing products of their catabolism. The amino acids listed in the blue boxes in Figure 14.8 can, instead of being completely catabolized, be converted to glucose. Recall that the first two reactions in glucose synthesis from noncarbohydrates (gluconeogenesis, Section 11.11) convert pyruvate to phosphoenolpyruvate, the high-energy intermediate needed to continue on the pathway.

$$\text{Pyruvate} \rightarrow \text{oxaloacetate} \rightarrow \text{phosphoenolpyruvate}$$

All amino acids catabolized to pyruvate or to citric-acid-cycle intermediates that precede oxaloacetate can enter gluconeogenesis via oxaloacetate. They are referred to as **glucogenic amino acids.**

Amino acids that are catabolized to acetyl SCoA or acetoacetyl SCoA can be used in the synthesis of ketone bodies and are therefore referred to as **ketogenic amino acids.** Ketone bodies, you may recall, are made from acetoacetyl SCoA synthesized from acetyl SCoA (Section 13.6). The ketogenic amino acids are also able to enter fatty acid biosynthesis via acetyl SCoA (Section 13.7). Of the ketogenic amino acids, there are only two—leucine and lysine—that aren't also glucogenic.

A few of the carbon containing end products of amino acid catabolism result from simple transformations. Alanine, for example, gives pyruvate on transamination, and serine is converted to pyruvate in a single step that results in loss of the amino group.

Glucogenic amino acid An amino acid catabolized to a compound that can enter the citric acid cycle for conversion to glucose.

Ketogenic amino acid An amino acid catabolized to a compound that can be converted to ketone bodies (also fatty acids).

$$\underset{\text{Serine}}{\text{HOCH}_2\underset{\underset{\text{NH}_3{}^+}{|}}{\text{CH}}\text{COO}^-} \xrightarrow[\text{dehydratase}]{\text{serine}} \underset{\text{Pyruvate}}{\text{CH}_3\overset{\overset{\text{O}}{\|}}{\text{C}}\text{COO}^- \;+\; \text{NH}_4{}^+}$$

Asparagine is hydrolyzed to aspartate, followed by transamination of aspartate to give oxaloacetate.

$$\underset{\text{Asparagine}}{\text{H}_2\text{N}\overset{\overset{\text{O}}{\|}}{\text{C}}\text{CH}_2\underset{\underset{\text{NH}_3{}^+}{|}}{\text{CH}}\text{COO}^-} \;+\; \text{H}_2\text{O} \longrightarrow \underset{\text{Aspartate}}{{}^-\text{OOCCH}_2\underset{\underset{\text{NH}_3{}^+}{|}}{\text{CH}}\text{COO}^-} \;+\; \text{NH}_4{}^+$$

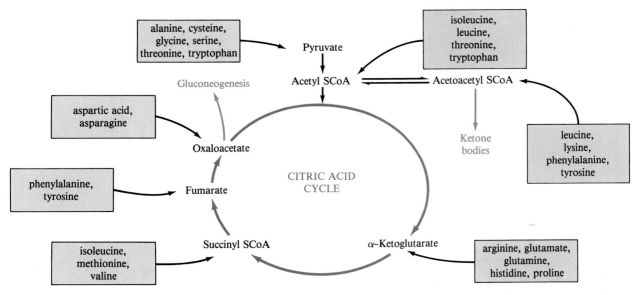

Figure 14.8
Amino acid catabolism. The carbon atoms of different amino acids enter the citric acid cycle as the indicated compounds. Glucogenic amino acids are listed in blue boxes, and ketogenic amino acids are listed in pink boxes.

The amino acids that enter the citric acid cycle as α-ketoglutarate are first transformed by one- to four-step pathways to glutamate, which is then converted by transamination to α-ketoglutarate containing the amino acid carbon atoms. The larger amino acids arrive at their end products through more complex pathways.

Practice Problem   14.4   Arginine is converted to ornithine in the last step of the urea cycle (Figure 14.7). To carry arginine carbon atoms to the citric acid cycle, ornithine undergoes transamination at its terminal amino group to give an aldehyde, followed by oxidation to glutamate and conversion to α-ketoglutarate. Write the structures of the five molecules in the pathway beginning with arginine and ending with α-ketoglutarate. Circle the region of structural change in each.

14.6 BIOSYNTHESIS OF NONESSENTIAL AMINO ACIDS

Nonessential amino acid
One of 10 amino acids that are synthesized in the body and are therefore not necessary in the diet.

Essential amino acid One of 10 amino acids that cannot be synthesized by the body and so must be obtained in the diet.

We said in Section 7.2 that humans are able to synthesize only 10 of the 20 amino acids found in proteins, known as the **nonessential amino acids** because they don't have to be in our diets. The remaining 10 are **essential amino acids** that must be obtained from dietary sources (Table 14.1). In evolutionary terms, it seems that when the diet could provide an amino acid that would otherwise have to be synthesized by a complex route, our metabolism gave up on making it. The nonessential amino acids can all be made in 1–3 steps, but the essential amino acids would require 7–10 steps based on observations of their synthesis in microorganisms.

Table 14.1   Amino Acids Essential and Nonessential in the Diet

Essential Amino Acids	Nonessential Amino Acids
Histidine	Alanine
Isoleucine	Arginine
Leucine	Asparagine
Lysine	Aspartate
Methionine	Cysteine
Phenylalanine	Glutamate
Threonine	Glutamine
Tryptophan	Glycine
Tyrosine[a]	Proline
Valine	Serine

[a] Sometimes classified as nonessential because it is synthesized from phenylalanine.

AN APPLICATION: NITROGEN BALANCE AND KWASHIORKOR

A healthy adult is normally at *nitrogen equilibrium*, meaning that the amount of nitrogen taken in each day is equal to the amount excreted. The recommended daily dietary allowance for the average 70 kg adult is 56 g of protein per day, equivalent to 0.8 g of protein per kilogram of ideal body weight. Total protein needs are increased when the diet is low in calories from fats and carbohydrates, when the foods consumed have a poor distribution of essential amino acids, or when physical activity increases. The average protein intake in the United States is about 110 g/day, well above what most of us need.

Infants and children, pregnant women, those recovering from starvation, and those with healing wounds are usually in *positive nitrogen balance*, that is, they're excreting less nitrogen than they consume, a condition to be expected when new tissue is growing. The reverse condition, *negative nitrogen balance*, occurs when more nitrogen is excreted than consumed. This happens when protein intake is inadequate, during starvation, and in a number of pathologic conditions including malignancies, malabsorption syndromes, and kidney disease.

Health and nutrition professionals group all disorders caused by inadequate protein intake as *protein energy malnutrition (PEM)*. Children, because of their higher protein needs, suffer most from such a diet. The problem is rampant where meat and milk are in short supply and where the dietary staples are vegetables or grains.

Protein deficiency alone is rare, and its symptoms are often accompanied by those of vitamin deficiencies, infectious diseases, and starvation. At one end of the spectrum is *kwashiorkor*, in which protein is deficient although caloric intake may be adequate. Children with kwashiorkor have edema (swelling due to water retention), have an enlarged, fatty liver, and are underdeveloped. The word "kwashiorkor" is from the language of Ghana and is translated as "the sickness the older gets when the next child is born," which is the time when weaning from mother's milk means conversion to a high-carbohydrate, low-protein diet. At the other end of the spectrum is *marasmus*, a condition caused mainly by starvation. As distinguished from kwashiorkor, marasmus is identified with severe muscle wasting, below-normal stature, and poor response to treatment. As many as 1 billion people worldwide suffer from PEM, most in the poor populations of developing countries.

Reductive amination
Conversion of an α-keto acid to an amino acid by reaction with NH_4^+.

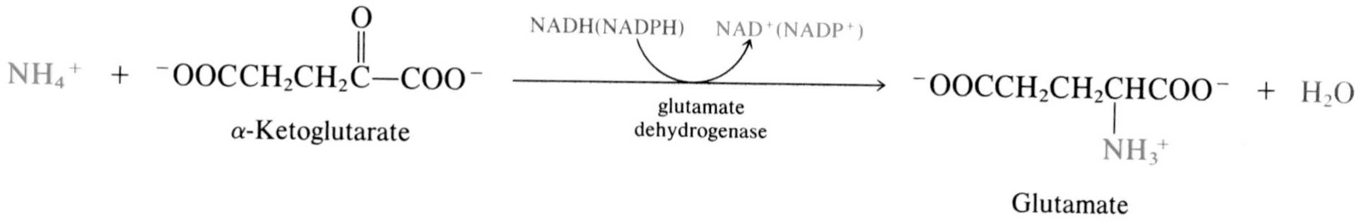

All of the 10 nonessential amino acids derive their amino groups from glutamate. The glutamate itself can be made from ammonia and α-ketoglutarate by **reductive amination,** the reverse of the oxidative deamination so common in amino acid catabolism (Section 14.3). The same glutamate dehydrogenase enzyme carries out the reaction.

$$NH_4^+ \;+\; {}^-OOCCH_2CH_2\overset{\displaystyle O}{\overset{\|}{C}}-COO^- \xrightarrow[\substack{glutamate\\dehydrogenase}]{\substack{NADH(NADPH) \quad NAD^+(NADP^+)}} {}^-OOCCH_2CH_2\underset{\underset{NH_3^+}{|}}{C}HCOO^- \;+\; H_2O$$

α-Ketoglutarate

Glutamate

The following four common metabolic intermediates, which you've seen by now in many roles, are the precursors for synthesis of the nonessential amino acids. As examples of how the intermediates are converted to amino acids, a few syntheses are shown in Table 14.2.

Table 14.2   Some Examples of Amino Acid Biosynthesis

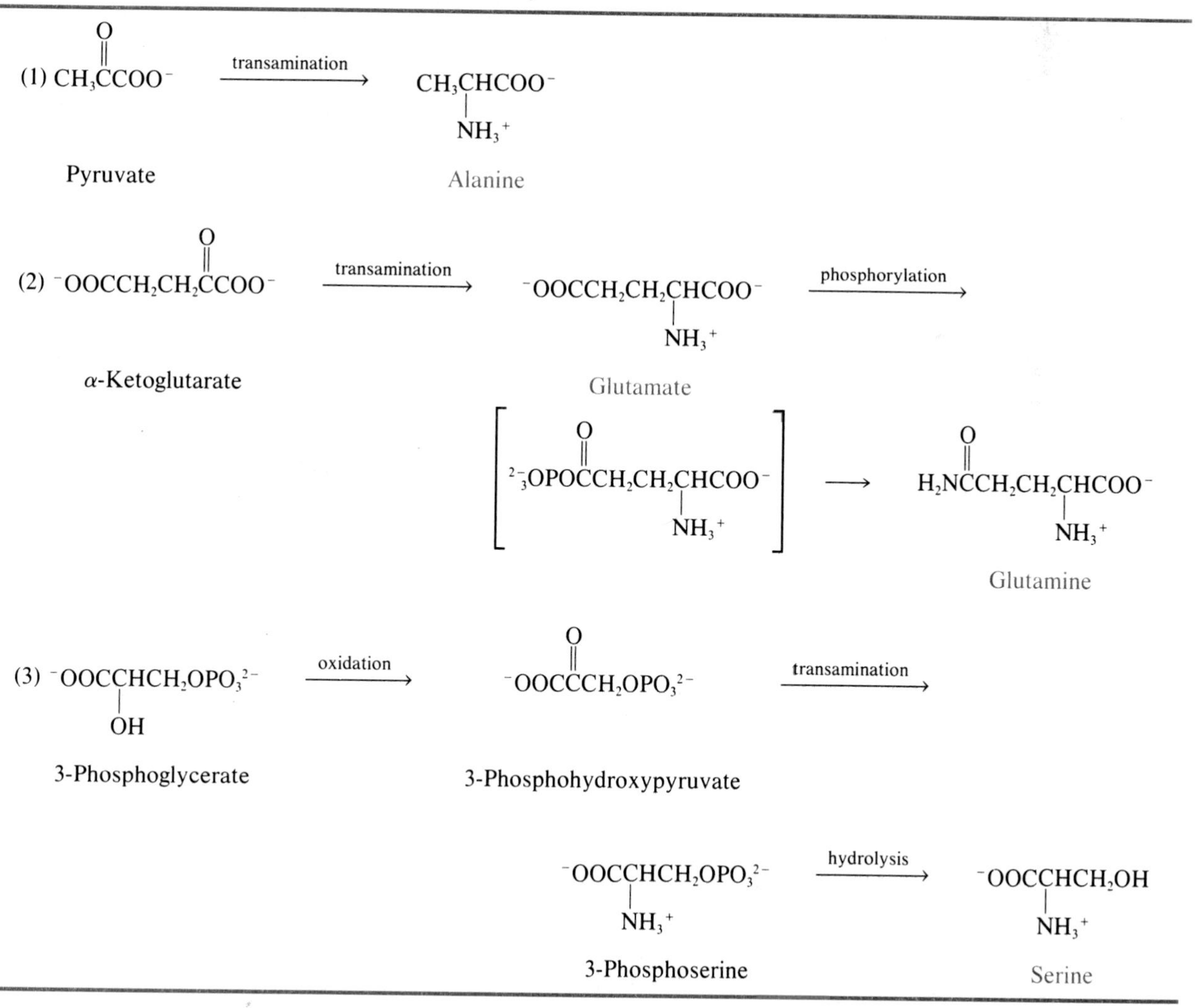

(1) $CH_3\overset{\displaystyle O}{\overset{\|}{C}}COO^- \xrightarrow{\text{transamination}} CH_3\underset{\underset{NH_3^+}{|}}{C}HCOO^-$

Pyruvate — Alanine

(2) ${}^-OOCCH_2CH_2\overset{\displaystyle O}{\overset{\|}{C}}COO^- \xrightarrow{\text{transamination}} {}^-OOCCH_2CH_2\underset{\underset{NH_3^+}{|}}{C}HCOO^- \xrightarrow{\text{phosphorylation}}$

α-Ketoglutarate — Glutamate

$\left[{}^{2-}{}_3OPO\overset{\displaystyle O}{\overset{\|}{C}}CH_2CH_2\underset{\underset{NH_3^+}{|}}{C}HCOO^- \right] \longrightarrow H_2N\overset{\displaystyle O}{\overset{\|}{C}}CH_2CH_2\underset{\underset{NH_3^+}{|}}{C}HCOO^-$

Glutamine

(3) ${}^-OOC\underset{\underset{OH}{|}}{C}HCH_2OPO_3^{2-} \xrightarrow{\text{oxidation}} {}^-OOC\overset{\displaystyle O}{\overset{\|}{C}}CH_2OPO_3^{2-} \xrightarrow{\text{transamination}}$

3-Phosphoglycerate — 3-Phosphohydroxypyruvate

${}^-OOC\underset{\underset{NH_3^+}{|}}{C}HCH_2OPO_3^{2-} \xrightarrow{\text{hydrolysis}} {}^-OOC\underset{\underset{NH_3^+}{|}}{C}HCH_2OH$

3-Phosphoserine — Serine

Precursors in syntheses of nonessential amino acids

$$CH_3\overset{\overset{\displaystyle O}{\|}}{C}-COO^-$$

$$^-OOC-\overset{\overset{\displaystyle O}{\|}}{C}CH_2COO^-$$

$$^-OOCCH_2CH_2\overset{\overset{\displaystyle O}{\|}}{C}-COO^-$$

$$^{2-}O_3POCH_2\underset{\underset{\displaystyle OH}{|}}{C}HCOO^-$$

Pyruvate   Oxaloacetate α-Ketoglutarate   3-Phosphoglycerate

The amino acid tyrosine is sometimes classified as nonessential rather than essential because it is synthesized from phenylalanine. Whatever the classification, we have a high nutritional requirement for phenylalanine and

INTERLUDE: XENOBIOTICS

How do our bodies tackle a chemical compound they've never met before? The compound might be a drug, a pesticide, or an environmental pollutant. It could be addictive, it could be a carcinogen, or it could be a harmless compound that need only be excreted. In medical terminology, such compounds are known as *xenobiotics*— chemical compounds foreign to the body.

A xenobiotic is likely to encounter a two-pronged attack when it enters the body. First, it undergoes chemical reactions that make it hydrophilic and water-soluble, so that it doesn't accumulate in fatty tissues. Second, enzymes further modify its structure, converting it to a compound that is even more water-soluble and is likely to be excreted.

In the first reaction for many xenobiotics, a hydrogen atom is replaced by a hydroxyl group in a *hydroxylation* reaction carried out by a *monooxygenase* enzyme complex known as *cytochrome P-450*. (The enzyme name derives from a peak in its absorption spectrum at 450 nm.) The xenobiotic (RH), oxygen, and NADPH react as follows:

$$RH + O_2 \xrightarrow{\quad NADPH \quad NADP^+ \quad} R-OH + H_2O$$

About half of all drugs are metabolized by cytochrome P-450. For example, hydroxylation

of phenobarbital solubilizes it for excretion. As you might expect, this protective action takes place mainly in the liver as blood passes through on its way to general circulation.

The second phase of xenobiotic metabolism is usually bonding through an oxygen, nitrogen, or sulfur atom to an even more polar group, in many cases the glucose derivative glucuronic acid or the amino acid derivative glutathione:

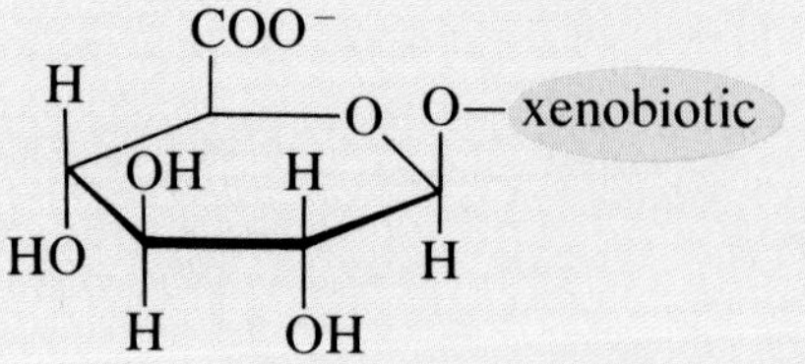

Glucuronide group

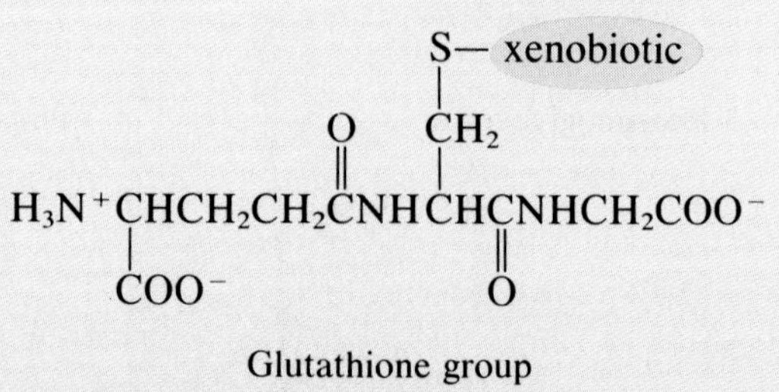

$$H_3N^+\underset{\underset{\displaystyle COO^-}{|}}{C}HCH_2CH_2\overset{\overset{\displaystyle O}{\|}}{C}NH\underset{\underset{\displaystyle}{|}}{C}H\overset{\overset{\displaystyle}{}}{C}NHCH_2COO^-$$

Glutathione group

several metabolic diseases are associated with defects in the enzymes needed to convert it to tyrosine. The best known of these diseases is *phenylketonuria (PKU)*, the first inborn error of metabolism for which the biochemical cause was recognized. In 1947 it was found that failure of the tyrosine synthesis leads to PKU.

$$\text{Phenylalanine} \longrightarrow \text{Tyrosine}$$

Phenylalanine: $\text{—CH}_2\text{CHCOO}^-$ with NH_3^+

Tyrosine: $\text{HO—}\bigcirc\text{—CH}_2\text{CHCOO}^-$ with NH_3^+

Cytochrome P-450 has both good and bad effects: good in detoxifying some poisons; bad in converting some compounds, especially polycyclic aromatic hydrocarbons like those in cigarette smoke, to carcinogenic compounds that react with and damage DNA. Like many biomolecules, the cytochrome is *inducible*, meaning that when more of it is needed, more of it is made.

It's well known that smokers have more cytochrome P-450 than nonsmokers. In fact, the wide variability of xenobiotic enzyme activity with age, sex, genetic makeup, and previous exposure to xenobiotics is thought to play a role in the wide variability in individual responses to carcinogens.

Smoke swirling down the windpipe and into the lungs. This person probably has more cytochrome P-450 than a nonsmoker.

PKU results in elevated blood serum and urine concentrations of phenylalanine, phenylpyruvate, and several other metabolites produced from excess phenylalanine. Undetected PKU causes mental retardation by the second month of life. Estimates are that, prior to the 1960s, 1% of those institutionalized for mental retardation were PKU victims. The only defense against PKU and similar treatable metabolic disorders that take their toll early in life is widespread screening of newborn infants. In the 1960s a test for PKU was introduced, and virtually all hospitals in the United States now routinely screen for PKU. Treatment consists of a restricted phenylalanine diet, which is maintained in infants with special formulas and in older individuals by eliminating meat and using low-protein grain products.

Practice Problem   **14.5**   Look at the reactions in Table 14.2 and identify those that are
(a) transaminations   (b) hydrolyses   (c) oxidations

SUMMARY

The digestion of dietary proteins results in complete enzyme-catalyzed hydrolysis of the proteins to yield amino acids. Digestion takes place in the stomach, where the enzyme is *pepsin,* and in the small intestine under the influences of several other **proteases.** The catabolism of amino acids is complex because different pathways are needed for each of the 20 amino acids in proteins. The pathways have in common the initial removal of the amino group, followed by conversion of the carbon atom part of the molecule to a compound that can enter the citric acid cycle.

The amino group is usually removed by **transamination,** in which the amino group of an amino acid and the keto group of α-ketoglutarate are interchanged to give glutamate and a new α-keto acid. In **oxidative deamination** the amino group of glutamate is removed as ammonium ion for reaction with bicarbonate ion to form carbamoyl phosphate. The carbamoyl phosphate is then converted to urea for excretion via the **urea cycle.** The urea cycle requires aspartate and is connected to the citric acid cycle by the production of fumarate. Fumarate is converted in the citric acid cycle to oxaloacetate from which the necessary aspartate is produced by transamination.

Glucogenic amino acids are those that are catabolized to pyruvate or to intermediates in the citric acid cycle. **Ketogenic amino acids** are catabolized to acetyl SCoA or acetoacetyl SCoA, from which ketone bodies can be synthesized. Ketogenic amino acids can also enter fatty acid synthesis via acetyl SCoA.

Nine of the 10 nonessential amino acids are biosynthesized from four precursor molecules (pyruvate, oxaloacetate, α-ketoglutarate, and 3-phosphoglycerate); their amino groups come from glutamate. Tyrosine is synthesized from the essential amino acid phenylalanine.

REVIEW PROBLEMS

Amino Acid Pool

14.6*   In what part of the digestive tract does the digestion of proteins begin?

14.7   What is the body's amino acid pool?

14.8*   What are the sources of the body's amino acid pool?

14.9   What are the fates of amino acids in the amino acid pool?

Amino Acid Catabolism

14.10*   What is meant by transamination?

14.11   How is vitamin B_6 associated with the process of transamination?

14.12*   What is the structure of the α-keto acid formed by transamination of the following amino acids with α-ketoglutarate? (See Table 7.1 for structures.)
(a) isoleucine   (b) cysteine   (c) phenylalanine

14.13 Pyruvate and oxaloacetate, as well as α-keto-glutarate, can be used as receptors for the amino group in transamination. Write the structures of the products formed when these two substances react in the transamination process.

14.14* What is meant by an oxidative deamination reaction?

14.15 What coenzymes are associated with oxidative deamination?

14.16* What type of enzyme is associated with oxidative deamination?

14.17 Write the structures of the α-keto acids produced by oxidative deamination of
(a) phenylalanine   (b) tryptophan

14.18* What is the other product formed in oxidative deamination besides an α-keto acid?

14.19 What is a glucogenic amino acid?

14.20* What is a ketogenic amino acid?

The Urea Cycle

14.21 Why does the body convert NH_4^+ to urea for excretion?

14.22* Why might the urea cycle be called the ornithine cycle?

14.23 What are the sources of the carbon and each of the two nitrogens in urea?

Amino Acid Anabolism

14.24* How do essential and nonessential amino acids differ in the number of steps required for their synthesis?

14.25 What substance serves as the starting material for amino acid anabolism?

14.26* What is the name of the process by which amino acids are made from common nonnitrogen metabolites? What process is this the reverse of?

14.27 How is tyrosine made in the body?

14.28* What causes PKU, and what are its symptoms?

Applications

14.29 What is positive nitrogen balance? [App: Nitrogen Balance and Kwashiorkor]

14.30* What are some conditions that require a positive nitrogen balance? [App: Nitrogen Balance and Kwashiorkor]

14.31 What is negative nitrogen balance? [App: Nitrogen Balance and Kwashiorkor]

14.32* What are some conditions that cause a negative nitrogen balance to occur? [App: Nitrogen Balance and Kwashiorkor]

14.33 How do marasmus and kwashiorkor differ in cause and in symptoms? [App: Nitrogen Balance and Kwashiorkor]

14.34* What does the term "xenobiotic" mean? [Int: Xenobiotics]

14.35 In what two ways does a body respond to a xenobiotic agent? [Int: Xenobiotics]

Additional Questions and Problems

14.36* Why is the formation of urea an energy-expensive process?

14.37 How can amino acids "pay" for the disposing of extra nitrogen?

14.38* Write the equation for the transamination reaction that occurs between valine and pyruvate. What do you think is the name of the associated enzyme?

14.39 Detail the five points at which amino acids can enter the citric acid cycle.

14.40* Can an amino acid be both glucogenic and ketogenic? Why or why not?

14.41 Briefly explain how the carbons from amino acids can end up in adipose tissue.

14.42* Consider all the metabolic processes that we have studied. Why do we say that tissue is dynamic? Detail some examples of dynamic relationships.

CHAPTER

15

Nucleic Acids and Protein Synthesis

We've had Superman and Superwoman in comic strips. Maybe, thanks to genetic engineering, someday we'll have real Supercattle that produce low-fat milk and low-cholesterol beef.

How does a seed "know" what kind of plant to become? How does a fertilized ovum know how to grow into a human being? And how does a cell know what part of the body it's in, whether brain or big toe, so that it can produce the right chemicals necessary for sustaining life? The answers to these and myriad other questions about all living organisms reside in the biological molecules called *nucleic* (nu-**clay**-ic) *acids.*

The nucleic acids, *deoxyribonucleic acid (DNA)* and *ribonucleic acid (RNA),* are the chemical carriers of an organism's genetic information. Coded in an organism's DNA is all the information that determines the nature of the organism, whether dandelion, goldfish, or human being. Coded also in DNA are all the instructions needed for cellular functioning, and all the directions needed for producing the tens or hundreds of thousands of different proteins required by the organism. In this chapter, we'll answer the following questions about nucleic acids:

1. *What is the composition of the nucleic acids, DNA and RNA?* The goal: Be able to describe and identify the components of nucleosides, nucleotides, DNA, and RNA.
2. *What is the structure of DNA?* The goal: Be able to describe the double helix and base pairing in DNA.
3. *How is DNA reproduced?* The goal: Be able to explain the process of DNA replication.
4. *What are the functions of RNA?* The goal: Be able to list the types of RNA, their locations in the cell, and their functions.
5. *How do organisms synthesize RNA?* The goal: Be able to explain the process of transcription.
6. *How does RNA control protein synthesis?* The goal: Be able to explain the genetic code and describe each step in translation.
7. *What are mutations?* Be able to describe mutations, how they might occur, and their role in hereditary disease.

15.1   DNA, CHROMOSOMES, AND GENES

The terms "chromosome" and "gene" were coined long before the chemical nature of these cell components was understood. A "chromosome" was a structure in the cell nucleus thought to be the carrier of *genetic information*—all the information needed by an organism to duplicate itself. A "gene" was the portion of a chromosome that controlled a specific inheritable trait such as brown eyes or red hair. We've come a long way since these terms were introduced, with especially dramatic progress since the 1970s thanks to a number of new techniques we'll discuss shortly.

When a cell is not actively dividing, its nucleus is occupied by *chromatin,* a tangle of fibers composed of protein and **DNA (deoxyribonucleic acid),** the

DNA (deoxyribonucleic acid)   The nucleic acid that stores the genetic information.

Chromosome A complex between proteins and a DNA molecule that is visible during cell division.

Gene All segments of DNA needed to direct the synthesis of a protein with a specific function.

Nucleic acid A biological polymer made by the linking together of nucleotide units.

Nucleotide A building block for nucleic acid synthesis, consisting of a five-carbon sugar bonded to a cyclic amine base and to phosphoric acid.

Heterocycle A ring that contains nitrogen or some other atom in addition to carbon.

RNA (ribonucleic acid) Nucleic acid responsible for putting the genetic information to use in protein synthesis.

nucleic acid molecule that stores genetic information. Just prior to cell division, the DNA is duplicated so that each new cell can receive a complete copy. During cell division chromatin organizes itself into **chromosomes,** each of which contains a DNA molecule (Figure 15.1). Each DNA molecule, in turn, is made up of many **genes**—individual segments of the DNA that contain the instructions needed to direct the synthesis of a protein with a specific function.

Different organisms differ in their complexity and therefore have different numbers of chromosomes and genes. A frog, for example, has 26 chromosomes (13 pairs), whereas a human has 46 chromosomes (23 pairs). Each chromosome contains a single immense molecule of DNA that, in humans, has a molecular weight of up to 150 *billion* amu and a length of up to *12 centimeters* when stretched out. The 23 pairs of human chromosomes are estimated to include about 100,000 genes.

15.2   COMPOSITION OF NUCLEIC ACIDS

Like proteins and carbohydrates, **nucleic acids** are polymers made up of individual molecules linked together in long chains. Proteins are polypeptides, carbohydrates are polysaccharides, and nucleic acids are poly*nucleotides*. A **nucleotide** can itself be further broken down by hydrolysis to yield three components: a simple aldopentose sugar, a **heterocyclic** amine base, and phosphoric acid.

$$\text{Nucleic acid} \xrightarrow{\text{hydrolysis}} \begin{array}{c}\text{many} \\ \text{nucleotides}\end{array} \xrightarrow{\text{hydrolysis}} \begin{array}{c}\text{aldopentose sugars} \\ + \\ \text{amine bases} \\ + \\ \text{H}_3\text{PO}_4\end{array}$$

There are two types of nucleic acids: DNA and RNA. DNA stores genetic information, and **RNA (ribonucleic acid)** makes possible the use of that information. Before we discuss the overall structure of the nucleic acids, it's important to understand how their component parts are joined together and how DNA and RNA differ.

Figure 15.1
Human chromosomes. Chromosomes are roughly X-shaped during cell division and have four banded arms connected at a centromere.

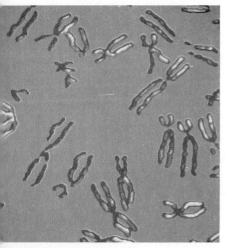

Nucleosides   In RNA, the sugar component is D-ribose, as indicated by the name "*ribo*nucleic acid." In DNA, or *deoxy*ribonucleic acid, the sugar component is 2-D-*deoxy*ribose. (The prefix "2-deoxy-" means that an oxygen atom is missing from the C2 position of ribose.)

Ribose

2-Deoxyribose ⸺ Oxygen missing

Nucleoside A compound consisting of a five-carbon sugar bonded to a cyclic amine base; like a nucleotide but missing the phosphate group.

The compounds formed by bonding of either of these sugars to a heterocyclic amine are known as **nucleosides.** As illustrated in the following general formula, the sugar and the amine are connected by a β-glycosidic bond between a heterocyclic ring nitrogen and the sugar's anomeric carbon atom (C1).

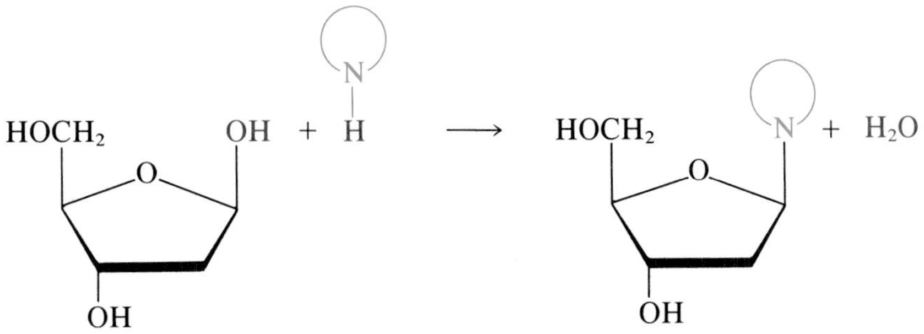

2-Deoxyribose   Heterocyclic   A nucleoside
(or ribose)   amine

Five heterocyclic amine bases are found in DNA and RNA. Two are derivatives of purine, and three are derivatives of pyrimidine (Section 4.3). The hydrogen atoms lost in the formation of nucleosides are shown in color in the following structures. Adenine, guanine, and cytosine are included in both DNA and RNA, thymine is part of only DNA molecules, and uracil is part of only RNA molecules.

Bases in DNA and RNA

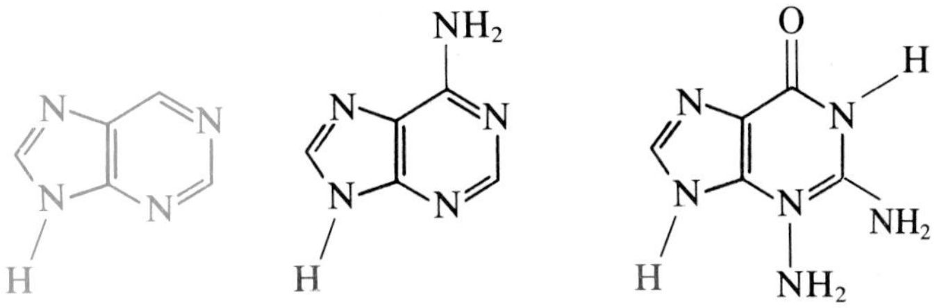

Purine   Adenine   Guanine
                 (DNA, RNA)   (DNA, RNA)

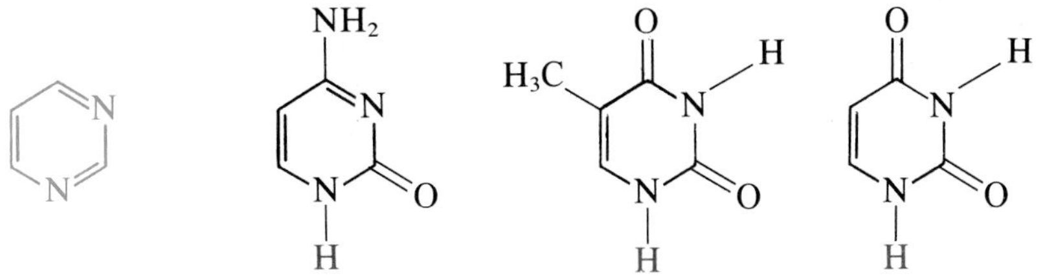

Pyrimidine   Cytosine   Thymine   Uracil
                (DNA, RNA)   (DNA)   (RNA)

Nucleosides are named with the base name modified by the ending -*osine* for the purine bases and -*idine* for the pyrimidine bases. No prefix is used for names of nucleosides containing ribose, but the prefix "deoxy-" is added for those that contain deoxyribose. To distinguish between them, numbers without primes are used for the nitrogen base, and numbers with primes are used for the sugar ring.

Two nucleosides

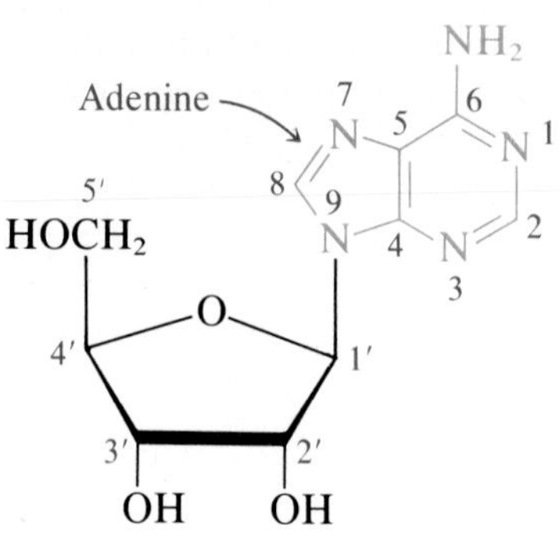

Adenosine 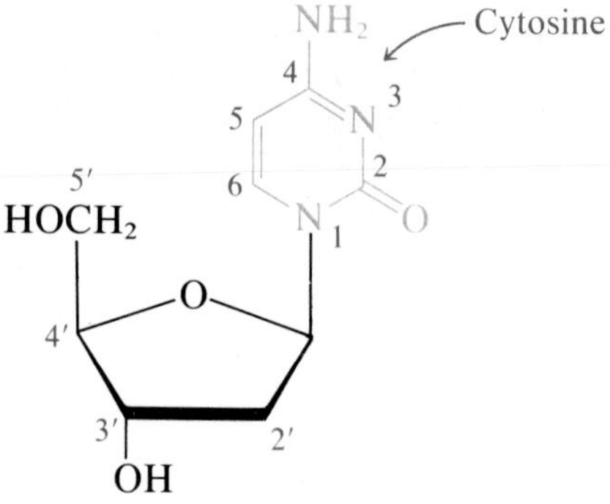   Deoxycytidine

Nucleotides   The nucleotides that are the building blocks of nucleic acids are monophosphate esters of nucleosides, formed by reaction of phosphoric acid with the alcohol —OH group on C5′ of the sugar. The phosphate group is ionized in body fluids and is often represented as —OPO_3^{2-}.

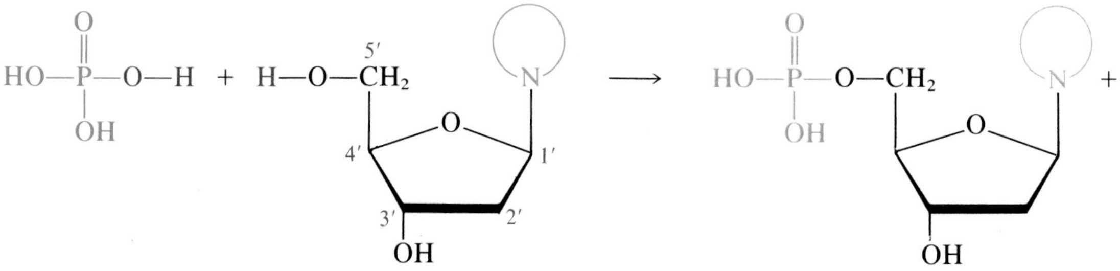

Phosphoric   A deoxyribonucleoside   A deoxyribonucleotide
acid

Nucleotides are named by adding *5′-monophosphate* to the name of the nucleoside. The nucleotides corresponding to adenosine and deoxycytidine are thus adenosine 5′-monophosphate (AMP) and deoxycytidine 5′-monophosphate (dCMP). Nucleotides containing D-ribose are known as **ribonucleotides,** and those containing D-deoxyribose are known as **deoxyribonucleotides.**

Ribonucleotide   A nucleotide containing D-2-ribose.

Deoxyribonucleotide   A nucleotide containing D-deoxyribose.

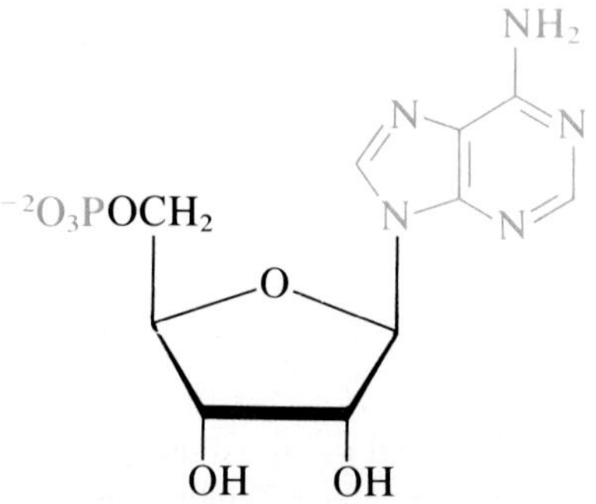

Adenosine 5′-monophosphate (AMP)   Deoxycytidine 5′-monophosphate (dCMP)
(a ribonucleotide)   (a deoxyribonucleotide)

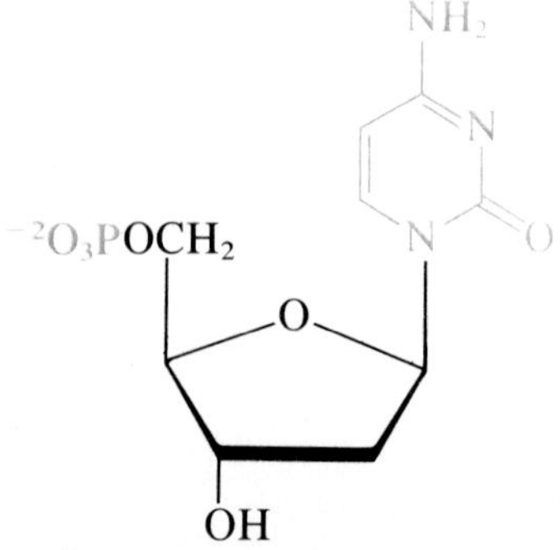

The names of the bases, nucleosides, and nucleotides are summarized in Table 15.1 together with their abbreviations. We rely heavily on abbreviations in biochemistry, so it's important to be familiar with them. Any of the nucleosides or nucleotides can add phosphate groups to form diphosphate or triphosphate esters like the guanosine phosphates. As you've seen on numerous occasions, a number of these phosphates—particularly adenosine triphosphate, ATP—play essential roles in biochemistry.

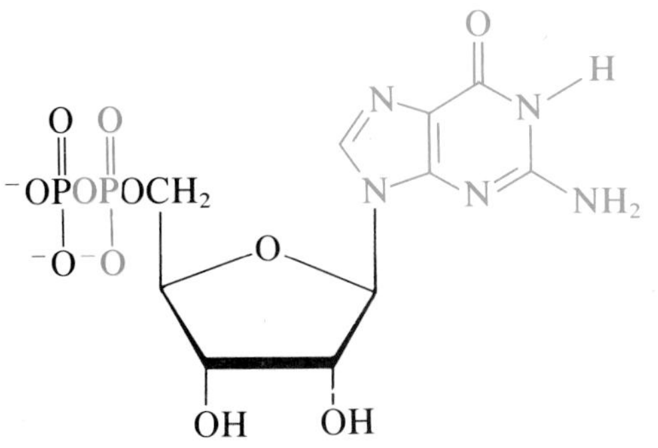

Guanosine diphosphate (GDP)

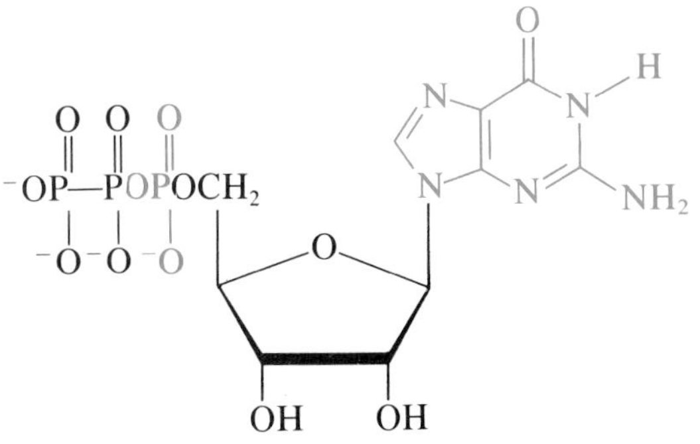

Guanosine triphosphate (GTP)

Summary of nucleic acid composition

DNA (deoxyribonucleic acid)

- Sugar is D-2-deoxyribose.
- DNA is a polymer of deoxyribonucleotides.
- Bases are adenine, guanine, cytosine, and *thymine*.

RNA (ribonucleic acid)

- Sugar is D-ribose.
- RNA is a polymer of ribonucleotides.
- Bases are adenine, guanine, cytosine, and *uracil* (instead of thymine).

Table 15.1   Names of Bases, Nucleosides, and Nucleotides in DNA and RNA

Bases	Nucleosides	Nucleotides	
DNA			
Adenine (A)	Deoxyadenosine	Deoxyadenosine 5-monophosphate	dAMP
Guanine (G)	Deoxyguanosine	Deoxyguanosine 5′-monophosphate	dGMP
Cytosine (C)	Deoxycytidine	Deoxycytidine 5′-monophosphate	dCMP
Thymine (T)	Deoxythymidine	Deoxythymidine 5′-monophosphate	dTMP
RNA			
Adenine (A)	Adenosine	Adenosine 5-monophosphate	AMP
Guanine (G)	Guanosine	Guanosine 5′-monophosphate	GMP
Cytosine (C)	Cytidine	Cytidine 5′-monophosphate	CMP
Uracil (U)	Uridine	Uridine 5′-monophosphate	UMP

Solved Problem 15.1 Classify the following structure as that of a nucleoside or a nucleotide, identify its sugar and base components, name it, and give its abbreviation.

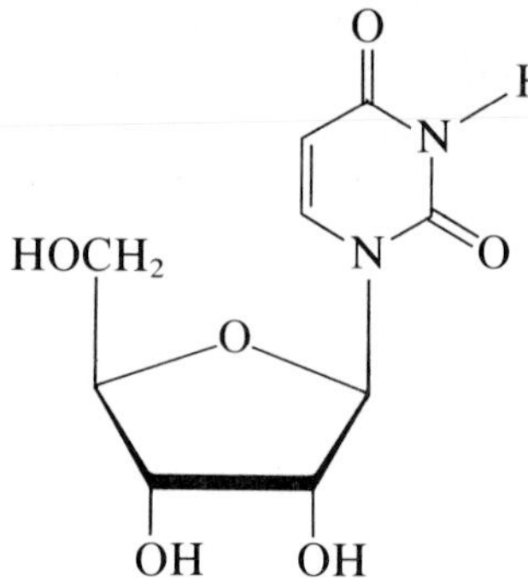

Solution The compound contains a sugar and a nitrogen base, but no phosphate group, so it is a *nucleoside*. The sugar has an —OH in the 2′ position and is therefore *ribose*. Checking the base structures given earlier shows that this one is *uracil*, a pyrimidine base. The nucleoside is therefore named *uridine*.

Practice Problems **15.1** Draw the structure of 2′-deoxythymidine 5′-phosphate.

15.2 Draw the structure of the triphosphate of deoxycytidine.

15.3 Write the full names of dTMP, AMP, ADP, and ATP.

15.3 THE STRUCTURE OF NUCLEIC ACID CHAINS

Nucleotides are joined together in DNA and RNA by phosphate ester bonds between the phosphate component of one nucleotide and the carbon 3′ —OH group on the sugar component of the next nucleotide:

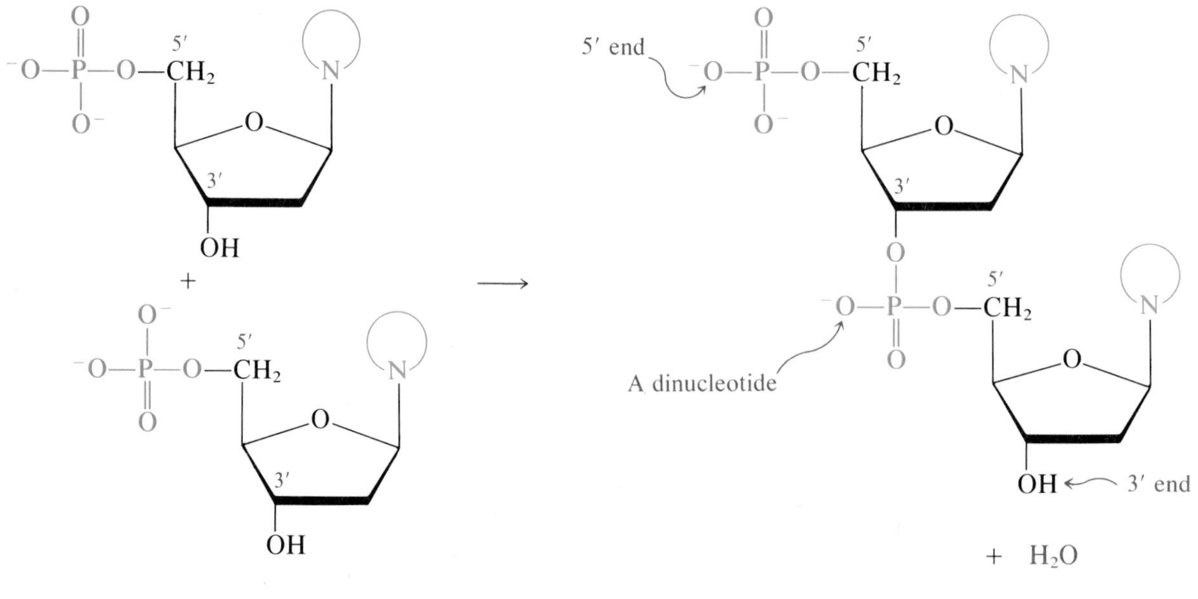

Additional nucleotides join by formation of still more phosphate ester bonds until ultimately an immense chain is formed (Figure 15.2). Regardless of how long a polynucleotide chain is, one end of the nucleic acid molecule always has a free —OH group at C3' (called the *3' end*), and the other end of the molecule always has a phosphoric acid group at C5' (the *5' end*).

Just as the exact structure of a protein depends on the sequence in which the individual amino acids are connected (Section 7.6), the exact structure of a nucleic acid molecule depends on the sequence in which individual nucleotides are connected. To carry the analogy even further, just as a protein has a polyamide backbone with different side chains attached to it at regular intervals, a nucleic acid has an alternating sugar-phosphate backbone with different heterocyclic amine bases attached to it at regular intervals.

The sequence of nucleotides in a chain is described by starting at the 5' end and identifying the bases in order of occurrence. Rather than write the full name of each nucleotide or each base, however, it's more convenient to use the simple one-letter abbreviations of the bases: A for adenine, G for guanine, C for cytosine, T for thymine (and U for uracil in RNA). Thus, a typical sequence in DNA might be written as -T-A-G-G-C-T-.

Figure 15.2
Generalized structure of a nucleic acid chain. Phosphate ester bonds link the individual nucleotides together.

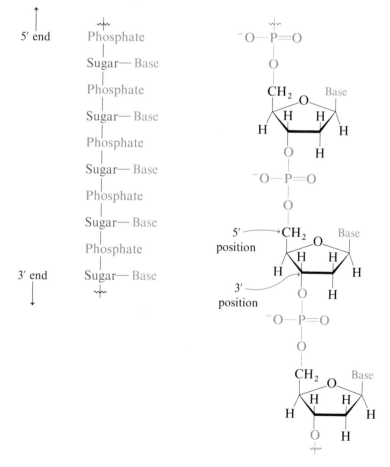

Comparison of protein and nucleic acid backbones and side chains

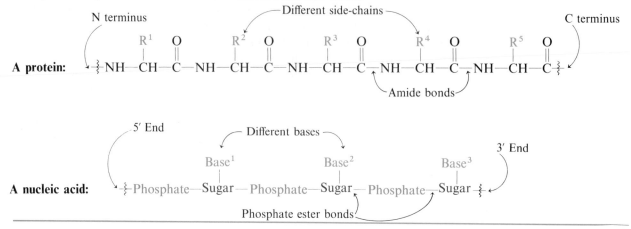

Practice Problem **15.4** Write the full structure of the DNA dinucleotide A-G. Identify the 5' and 3' ends of the dinucleotide.

15.4 BASE PAIRING IN DNA: THE WATSON-CRICK MODEL

DNA samples from different cells of the same species have the same proportions of the four heterocyclic bases, but samples from different species can have quite different proportions of bases. For example, human DNA contains about 30% each of adenine and thymine, and 20% each of guanine and cytosine. The bacterium *Escherichia coli,* however, contains only about 24% each of adenine and thymine, and 25% each of guanine and cytosine. Note that in both cases, the bases occur in *pairs*. A and T are usually present in equal amounts, as are G and C. Why should this be?

In 1953, James Watson and Francis Crick proposed a structure for DNA that not only accounts for the pairing of bases but also provides a simple method for the storage and transfer of genetic information. According to the *Watson-Crick model* (Figure 15.3), a DNA molecule consists of *two* polynucleotide strands coiled around each other in a helical, screw-like fashion. The sugar-

Figure 15.3
The double-helical structure of DNA. The sugar-phosphate backbone (blue) runs along the outside of the helix, while amine bases (purines in red and pyrimidines in green) hydrogen-bond to one another on the inside. A section of the wider major groove is at the bottom in the picture, with a section of the minor groove immediately above it.

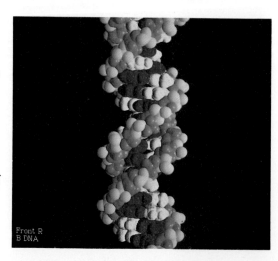

Double helix Two strands coiled around each other in a screw-like fashion, as in the two polynucleotide strands in DNA.

phosphate backbone is on the outside of the so-called **double helix,** and the heterocyclic bases are on the inside, so that a base on one strand points directly toward a base on the second strand (Figure 15.4).

The two strands of the DNA double helix run in opposite directions, one in the $5' \rightarrow 3'$ direction, the other in the $3' \rightarrow 5'$ direction, as shown in Figure 15.4. The strands are held together by hydrogen bonds between bases. This hydrogen bonding is not random, but rather is the basis for the structure of DNA, just as hydrogen bonding interactions determine the secondary structure of proteins. In the double helix, adenine and thymine form two hydrogen bonds to each other but not to cytosine or guanine. Similarly, cytosine and guanine form three hydrogen bonds to each other in the double helix, but not to adenine or thymine.

Hydrogen-bonding base pairs in DNA

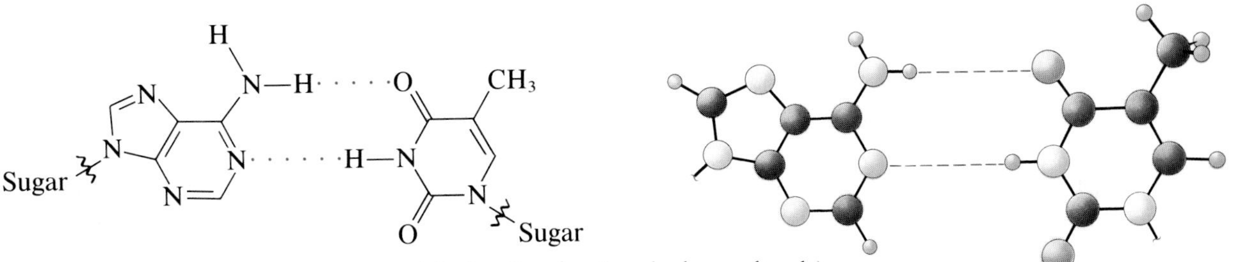

Adenine-thymine (two hydrogen bonds)

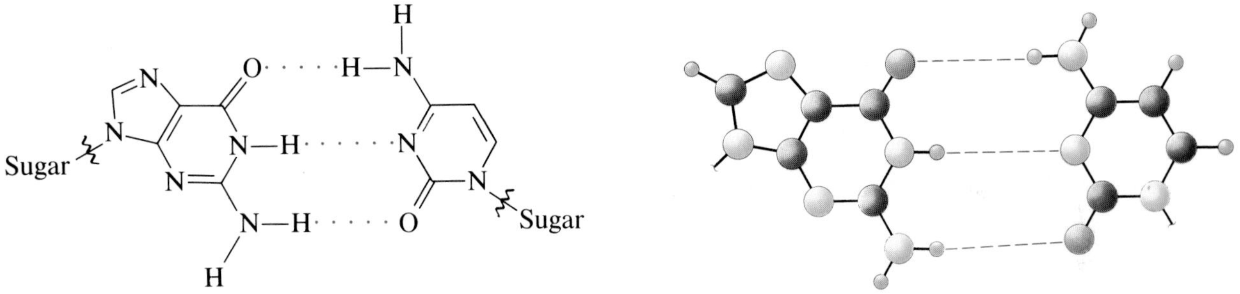

Guanine-cytosine (three hydrogen bonds)

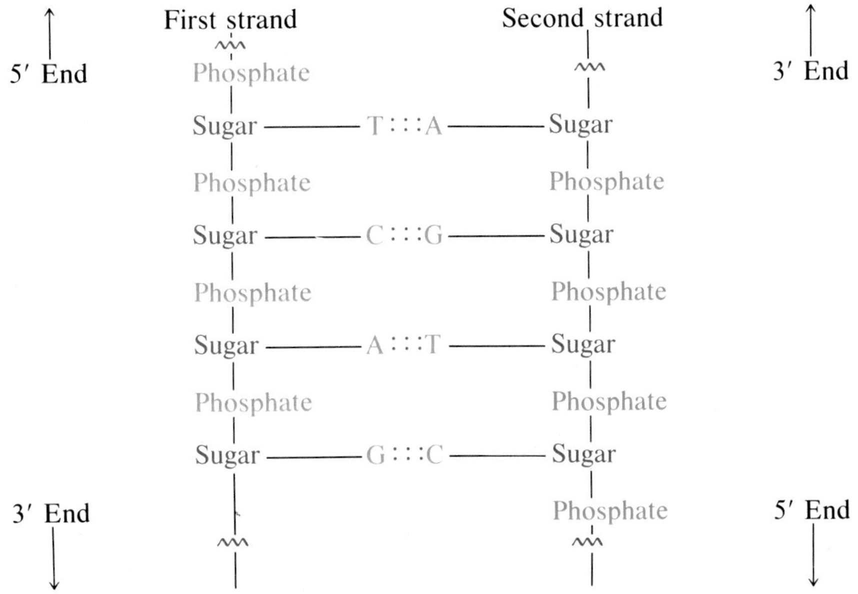

Figure 15.4
Complementary base pairing in the DNA double helix.

The two strands of the DNA double helix are *complementary* rather than identical. Whenever an A base occurs in one strand, a T base occurs opposite it in the other strand; when a C base occurs in one strand, a G base occurs in the other. This complementary *base pairing* in the two strands explains why A/T and C/G always occur in equal amounts.

As indicated in Figure 15.3, the DNA double helix is like a twisted ladder, with the sugar-phosphate backbone making up the sides and the hydrogen-bonded base pairs the rungs. X-ray measurements show that the helix is 2.0 nm wide, that there are exactly 10 nucleotides in each full turn, and that each turn is 3.4 nm long. A helpful way to remember the base pairing in DNA is to memorize the phrase "Pure silver taxi."

Pure	Silver	Taxi
Pur	**AG**	**TC**
The purines	A and G	pair with T and C

Notice in Figure 15.3 that the two strands of the double helix coil in such a way that two kinds of "grooves" result, a *major groove* and a *minor groove*. The major groove is slightly deeper than the minor groove, and both are lined by potential hydrogen-bond donors and acceptors. Thus, a variety of molecules are able to *intercalate*, or fit, into one of the grooves between the strands. A large number of cancer-causing and cancer-preventing agents are thought to function by interacting with DNA in this way.

Solved Problem 15.2   What sequence of bases of one strand of DNA is complementary to the sequence T-A-T-G-C-A-G on the other strand?

Solution   Remembering that A and G (silver) bond to T and C (taxi), respectively, go through the original sequence replacing each A by T, each G by C, each T by A, and each C by G.

Original: T-A-T-G-C-A-G
                : : : : : : :
Complement: A-T-A-C-G-T-C

Practice Problems   15.5   What sequences of bases on one DNA strand are complementary to these sequences on another strand?
(a) G-C-C-T-A-G-T   (b) A-A-T-G-G-C-T-C-A

15.6   Why doesn't DNA include any adenine-guanine base pairs?

15.7   Draw the structures of adenine and uracil (which replaces thymine in RNA) and show the hydrogen bonding between them.

15.5 NUCLEIC ACIDS AND HEREDITY

How do organisms use nucleic acids to store and express genetic information? If the information is to be preserved and passed on from generation to generation, a mechanism must exist for copying DNA. If the information is to be used, mechanisms must exist for decoding the information and for carrying out the instructions therein.

According to what has been called the *central dogma of molecular genetics,* the function of DNA is to store information and pass it on to RNA, while the function of RNA is to read, decode, and use the information received from DNA to make proteins. Each of the hundreds or thousands of individual genes on a DNA molecule contains the instructions necessary to make a specific protein that is in turn needed for a specific biological purpose. The primary structure of each enzyme, for example, is coded for by a different gene. By decoding the right genes at the right time in the right place, an organism can use genetic information to synthesize the many thousands of proteins necessary to carry out the biochemical reactions required for smooth functioning.

Three fundamental processes take place in the transfer and use of genetic information:

$$\text{DNA} \xrightarrow{\text{replication}} \text{DNA} \xrightarrow{\text{transcription}} \text{mRNA} \xrightarrow{\text{translation}} \text{Proteins}$$

Replication The process by which copies of DNA are made in the cell.

Transcription The process by which the information in DNA is read and used to synthesize mRNA.

Translation The process by which mRNA directs protein synthesis.

1. Replication is the process by which a replica, or identical copy, of DNA is made. Replication occurs every time a cell divides so that information can be preserved and handed down to offspring.

2. Transcription is the process by which the genetic messages contained in DNA are ''read,'' or transcribed. The product of transcription, known as *messenger RNA* (mRNA), leaves the cell nucleus and carries the message to the sites of protein synthesis.

3. Translation is the process by which the genetic messages carried by mRNA are decoded and used to build proteins.

In the following sections, we'll look at these important processes, focusing on what happens to DNA, the RNAs, and the proteins being synthesized. The complex processes of replication, transcription, and translation proceed with great accuracy and require participation by many other types of molecules. Energy-supplying nucleoside triphosphates and many different enzymes play essential roles, some understood, some not yet fully understood, and most likely some not yet discovered. At this point, however, our goal is to present an overview of how the genetic information is duplicated and put to work.

15.6 REPLICATION OF DNA

The Watson-Crick double-helix model of DNA does more than just explain base pairing, it also provides for an explanation of the ingenious way that DNA molecules reproduce exact copies of themselves. DNA replication begins with a partial unwinding of the double helix (Figure 15.5). As the two DNA strands separate and the bases are exposed, the enzyme *DNA polymerase* moves into

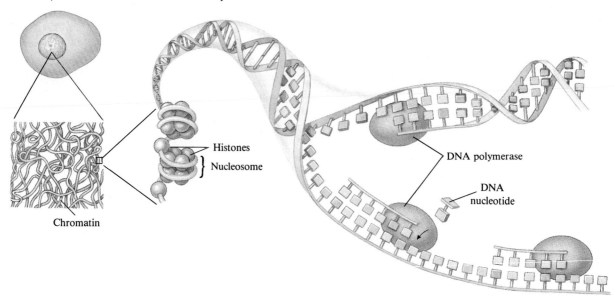

Figure 15.5
DNA replication. In a nondividing cell, chromatin consists of the DNA double helix coiled around *histones* (proteins) in a series of beads known as *nucleosomes*. When replication is about to occur, the DNA unwinds from the histones and the hydrogen bonds of the double helix open up, allowing DNA polymerase to move along the strands.

position at the point where synthesis will begin. One by one, nucleotide triphosphates approach and hydrogen-bond to the bases in each strand in an exactly complementary manner, A to T, and G to C. DNA polymerase then catalyzes bond formation between each arriving nucleotide and its neighbors, accompanied by removal of the extra phosphate groups, and the two new strands begin to grow.

Since each new strand is complementary to its old template strand, two identical new copies of the DNA double helix are produced during replication. In each new helix, one strand is the old template and the other is newly synthesized, a result described by saying that the replication is *semi-conservative*. The process is shown schematically in Figure 15.6.

Crick probably described the DNA replication process best when he described the fitting together of two DNA strands as being like a hand in a glove. The hand and glove separate, a new hand forms inside the old glove, and a new glove forms around the old hand. Two identical copies now exist where only one existed before.

DNA polymerase catalyzes the reaction between the 5′ phosphate on an incoming nucleotide and the free 3′ OH on the growing polynucleotide. As a result, the new DNA strands can grow only in the 5′ → 3′ direction, and strand growth must begin at the 3′ end of the template. Because the original DNA strands are complementary, though, only one new strand can begin at the 3′ end and grow continuously as the point of replication (the *replication fork*) moves along the original DNA. The other strand must grow in the opposite direction. The result is production of a series of short sections of DNA, as shown schematically in Figure 15.7. To complete formation of this strand, the sections are joined by the action of a *DNA ligase* enzyme.

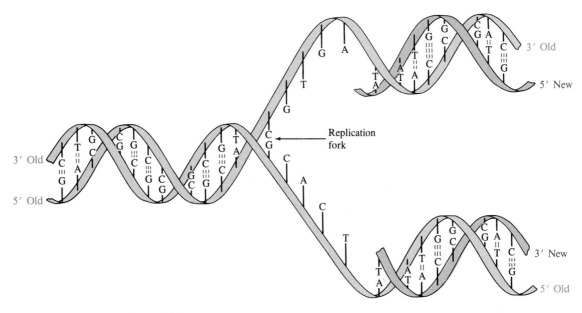

Figure 15.6
A schematic representation of DNA replication. The original DNA double helix partially unzips at a point called the *replication fork,* and complementary new nucleotides line up on each strand. When the new nucleotides are joined by DNA polymerase, two identical new DNA molecules result.

Figure 15.7
Replication at a fork. Because the polynucleotide must grow in the $5' \rightarrow 3'$ direction, one strand (top) grows continuously toward the replication fork and the other grows in segments as the fork moves. The segments are later joined by a ligase enzyme.

Genome The sum of all genes in an organism.

It's difficult to conceive of the magnitude of the replication process. The sum of all genes in a human cell—the human **genome**—is estimated to be approximately 3 billion base pairs, and a single DNA chain might contain up to 250 million pairs of bases. Regardless of the size of these enormous molecules, their base sequence is faithfully copied during replication, and an error occurs only about once in each 10–100 billion bases. The complete copying process in human cells takes several hours. To replicate such huge molecules as human DNA at this speed requires not one, but many replication forks, producing many segments of DNA strands that ultimately are united to form the new double helix.

Growth of new DNA strands from numerous replication forks

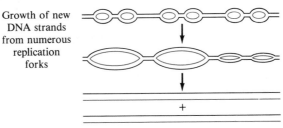

AN APPLICATION: THE HUMAN GENOME PROJECT

To "map the human genome" means to identify the location on the chromosomes of our 100,000 genes (a few thousand have been located so far) and to find the complete sequence of the 3 billion base pairs that make up human DNA. Ever since techniques to make this possible began to appear, debate has raged about whether the mapping should be done. Is it worth the huge expense of time and money? Who should do it? How should it be done? The technology is changing so rapidly, it's hard to make plans.

Perhaps the toughest question is, What will be done with the results? The benefits in understanding the molecular basis of inherited diseases are clear, for improved diagnosis and treatment will certainly result. Identification of individuals especially at risk to environmental factors that cause cancer or heart disease will also become possible. Many scientists look forward to intellectual benefits in their varied fields. But tough emotional and ethical problems are bound to arise when an individual can know many years in advance that he or she is destined to develop a degenerative and costly disease. Should this information be available to potential employers, insurers, or spouses?

Whatever the outcomes, the course toward mapping the human genome seems to have been set in the late 1980s. The Human Genome Project was given initial funding by the U.S. government, and James Watson, codiscoverer of the DNA double helix 36 years earlier, was appointed to coordinate efforts at research centers throughout the United States and, perhaps, the world. The project is predicted to take 15 years and cost at least $3 billion, or $1 per nucleotide. Genome mapping is a topic that's bound to show up with regularity in newspapers and magazines for many years to come.

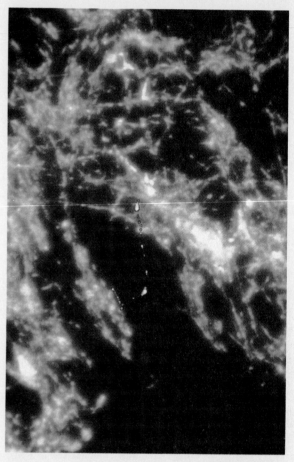

The picture shows in blue the DNA released onto a microscope slide from a single human sperm cell. Fluorescently labeled DNA sequences complementary to those in genes under study have been added to the slide, a technique that will be used to map the human genome. The bright dots identify locations where the complementary fluorescent DNA probes have attached to the genes in question. The highlighted genes are on chromosome 19, are expressed during embryo development, and have been implicated in formation of certain types of tumors.

15.7   STRUCTURE AND FUNCTION OF RNA

RNA is structurally similar to DNA—both are sugar-phosphate polymers and both have heterocyclic bases attached—but there are important differences. For one thing, they differ in composition: The sugar in RNA is ribose rather than the 2-deoxyribose in DNA, and the base uracil is present in RNA instead of thymine. They also differ in size and structure—RNA molecules are smaller than DNA molecules, and RNA is single-stranded, not double-stranded like DNA.

Another difference between RNA and DNA is in function. DNA has only one function—storing genetic information—but there are three main kinds of ribonucleic acid, each of which has a specific function.

Ribosome   A structure in the cell where protein synthesis occurs.

Ribosomal RNA (rRNA) The type of RNA complexed with proteins in ribosomes.

Messenger RNA (mRNA) The RNA whose function is to carry genetic messages transcribed from DNA and to direct protein synthesis.

Transfer RNA (tRNA)   The RNA whose function is to transport specific amino acids into position for protein synthesis.

● **Ribosomal RNAs** Outside the nucleus but within the cytoplasm of a cell are the **ribosomes,** small, granular structures where protein synthesis takes place. Each ribosome is a complex consisting of about 60% **ribosomal RNA (rRNA)** and 40% protein and has a total molecular weight of approximately 5,000,000 amu.

● **Messenger RNAs** The nucleic acids that record information from DNA in the cell nucleus and carry it to the ribosomes are known as **messenger RNAs (mRNA).** Protein synthesis can be controlled by variation of the rates at which mRNAs are synthesized and then destroyed once the protein they produce is no longer needed.

● **Transfer RNAs** The function of **transfer RNAs (tRNA)** is to deliver amino acids one by one to protein chains growing at ribosomes.

15.8   RNA SYNTHESIS: TRANSCRIPTION

The process of converting the information contained in a DNA segment into proteins begins with the synthesis of mRNA molecules containing anywhere from several hundred to several thousand ribonucleotides, depending on the size of the protein to be made. Each of the 100,000 or so proteins in the human body is synthesized from a different mRNA that has been transcribed from a specific gene on DNA.

Messenger RNA is synthesized in the cell nucleus by *transcription* of DNA, a process similar to DNA replication. As in replication, a small section of the DNA double helix unwinds, and the bases on the two strands are exposed. RNA nucleotides (*ribonucleotides*) line up in the proper order by hydrogen-bonding to their complementary bases on DNA, the nucleotides are joined together by an *RNA polymerase* enzyme, and mRNA results.

Unlike what happens in DNA replication, where both strands are copied, only one of the two DNA strands is transcribed into mRNA. The DNA strand that is transcribed is called the *template strand,* while its complement is called the *informational strand.* Since the template strand and the informational strand are complementary, and since the template strand and the mRNA molecule are also complementary, it follows that *the messenger RNA molecule produced during transcription is a copy of the DNA informational strand.* The only difference is that the mRNA molecule has a U base everywhere that the DNA informational strand has a T base. The transcription process is shown schematically in Figure 15.8, and an electron micrograph of transcription taking place is shown in Figure 15.9.

Transcription of DNA by the process just discussed raises as many questions as it answers. How does the DNA know where to unwind? Where along the chain does one gene stop and the next one start? How do the ribonucleotides know exactly the right place along the template strand to begin lining up and the right place to stop? These and many other questions about both replication and

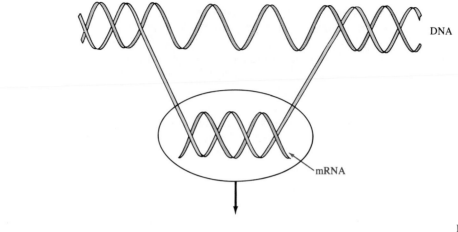

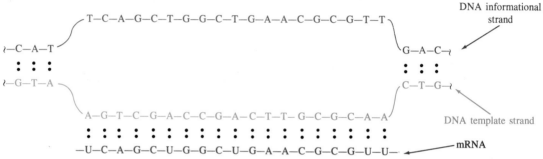

Figure 15.8

The transcription of DNA to synthesize mRNA. The mRNA molecule produced is complementary to the template strand from which it is transcribed and is identical to the informational strand except for the replacement of T by U.

Figure 15.9
Electron micrograph of DNA undergoing transcription. The backbone of the feather-like structure running down the image is the DNA strand, and the clusters along the backbone are the mRNA molecules, which increase in length as they move along the DNA.

transcription are extremely difficult to answer. The current picture is that a DNA chain contains certain base sequences called *promoter sites,* which bind to the RNA polymerase enzyme that actually carries out RNA synthesis, thus signaling the beginning of a gene. Similarly, there are other base sequences at the end of a gene that signal a stop to mRNA synthesis.

Another part of the picture is the recent discovery that genes are not necessarily continuous segments of the DNA chain. Often a gene will begin in one small section of DNA called an **exon,** then be interrupted by a seemingly nonsensical section called an **intron** that does not code for any part of the protein to be synthesized, and then take up again farther down the chain in another exon. The final mRNA molecule to be released from the nucleus results only after the nonsense sections are cut out by appropriate enzymes and the remaining pieces are spliced together. Current evidence is that up to 90% of human DNA is made up of introns and only about 10% of DNA actually contains genetic instructions. The possible functions of introns and how they came to be there in the first place are, as you might guess, lively areas of study and speculation.

Exon A DNA segment in a gene that codes for part of a protein molecule.

Intron A seemingly nonsensical segment of DNA that does not code for part of a protein.

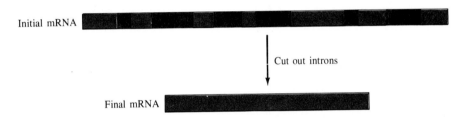

rRNA and tRNA are synthesized in the nucleus in essentially the same manner as mRNA, and before leaving the nucleus, all three types of RNA molecules are modified in various ways needed for their different functions.

Practice Problem   ***15.8***   What mRNA base sequences are complementary to each of the following DNA template sequences?
(a) -G-A-T-T-A-C-C-G-T-A-   (b) -T-A-T-G-G-C-T-A-G-G-C-A-

15.9 THE GENETIC CODE

Since all biological processes are ultimately regulated by proteins, the primary function of mRNA is to direct the biosynthesis of the thousands of diverse peptides and proteins required by an organism. The ribonucleotide sequence in an mRNA chain acts like a long, coded sentence to specify the order in which different amino acid residues should be joined to form a protein. Each "word" or **codon** in the mRNA sentence consists of a series of three ribonucleotides that is specific for a particular amino acid. For example, the series cytosine-uracil-guanine (C-U-G) on an mRNA chain is a codon directing incorporation of the amino acid leucine into a growing protein. Similarly, the sequence guanine-adenine-uracil (G-A-U) codes for aspartic acid.

Codon A sequence of three ribonucleotides in the RNA chain that codes for a specific amino acid.

AN APPLICATION: VIRUSES AND AIDS

Viruses are small structures containing a piece of nucleic acid wrapped in a protective coat of protein. The nucleic acid may be either DNA or RNA, it may be either single-stranded or double-stranded, and it may consist either of a single piece or of several pieces. Many hundreds of different viruses are known, each of which can infect a particular plant or animal cell.

Viruses occupy the gray area between living and nonliving. By itself, a virus has none of the cellular machinery necessary for replication. Once it enters a living cell, though, a virus can take over the host cell and force it to produce virus copies. Some of the infected cells eventually die, but others continue to produce viral copies, which then leave the host and spread the infection to other cells.

Viral infection begins when a virus particle enters a host cell, loses its protein coat, and releases its nucleic acid. What happens next de-

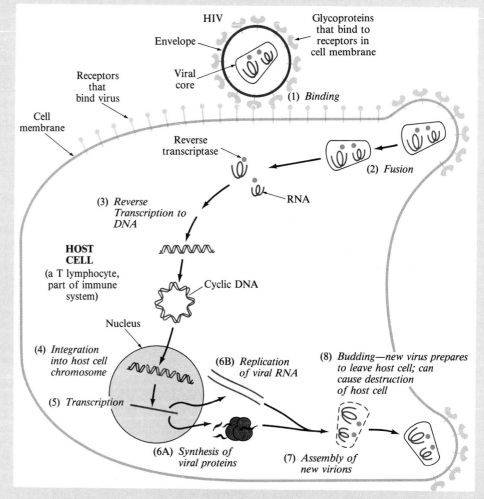

Interaction of HIV virus with host cell, a T lymphocyte. The parenthetical numbers show the events beginning with binding and cell entry by the virus, proceeding to reproduction of the viral RNA and proteins by reverse transcription, and ending with release of a new virus particle from the host cell.

pends on whether the virus is based on DNA or RNA. If the infectious agent is a DNA virus, the host cell first replicates the viral DNA and then decodes it in the normal way. The viral DNA is transcribed to produce RNA, the RNA is translated to synthesize viral protein coating, and the newly formed virus copies are released from the cell. If the infectious agent is an RNA virus, however, a problem exists. Either the cell must transcribe and produce proteins directly from the viral *RNA* template, or else it must first produce DNA from the viral RNA by a process called *reverse transcription*. Viruses that follow the reverse transcription route are called *retroviruses*. The human immunodeficiency virus (HIV-1) responsible for most cases of AIDS is an example; its mode of action is shown in the accompanying diagram.

Viral infections, whether HIV infections or the common cold, are difficult to treat with chemical agents. The challenge is to design a drug that can act on viruses within cells without damaging the cells and their genetic machinery. AZT, the first drug to be licensed for treatment of AIDS, is similar enough in structure to the nucleoside thymidine to halt the process of reverse transcription. Though not a cure for full-blown AIDS, AZT is effective at slowing the progress of the

disease in individuals infected with HIV but as yet without symptoms. Another approach to AIDS therapy is the search for a chemical agent that could inhibit binding of the virus to the cell membrane (step 1 in the diagram).

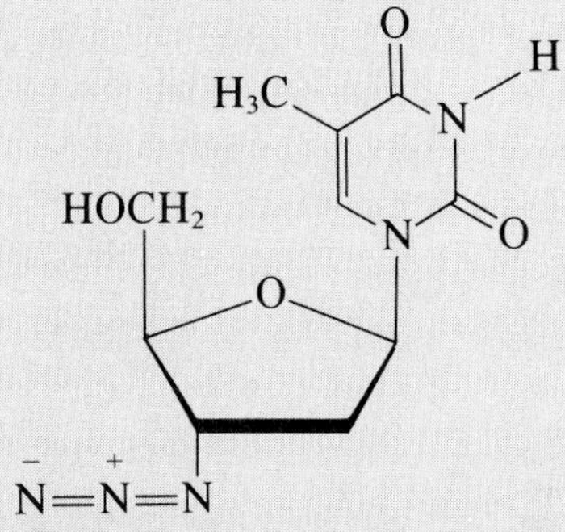

AZT (3'-azido-3'-deoxythymidine)

One possible defense against viral infections is the use of vaccines, which cause the human immune system to produce antibodies that will inactivate a virus. Advances have been made toward developing an HIV vaccine, but the task is especially challenging because HIV undergoes frequent mutations—variations in its nucleotide structure. The result is that an HIV vaccine active against one strain of HIV might soon encounter a mutant it can't react with.

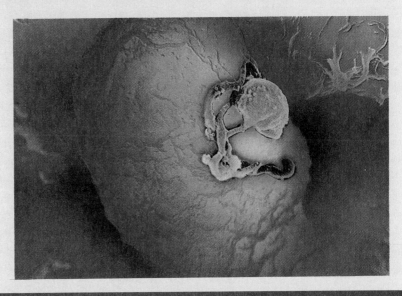

An electron micrograph of a cell that has been invaded by HIV-1 and is about to release a newly formed virus particle (green).

Of the 64 possible three-base combinations in RNA, 61 code for specific amino acids, and 3 code for chain termination (the "stop codes"). The meaning of each codon—the **genetic code** universal to all but a few living organisms—is shown in Table 15.2. Note that most amino acids are specified by more than one codon and that codons are always written in the $5' \rightarrow 3'$ direction.

Genetic code The complete assignment of the 64 mRNA codons to specific amino acids (or stop signals).

Practice Problems

15.9 List possible codon sequences for the following amino acids:
(a) Ala   (b) Phe   (c) Leu   (d) Val   (e) Tyr

15.10 What amino acids do the following sequences code for?
(a) A-U-U   (b) G-C-G   (c) C-G-A   (d) A-A-C

15.10 RNA AND PROTEIN BIOSYNTHESIS: TRANSLATION

The message contained in mRNA is decoded by tRNA molecules. Each different tRNA acts as a carrier to transport a specific amino acid into place on the ribosome so that it can be incorporated into the growing protein chain. A typical tRNA molecule is a single chain held together by regions of base-pairing in a structure something like a cloverleaf (Figure 15.10a) and containing about 70–100 ribonucleotides.

In three dimensions, a tRNA is L-shaped, as shown in Figures 15.10b and c. Every tRNA is bonded to its specific amino acid by an ester linkage between the —COOH of the amino acid and the free —OH group at the C3' position of ribose on the 3' end of the tRNA chain. At the other end of the L, near the

Table 15.2   Codon Assignments of Base Triplets

First Base (5' end)	Second Base	Third Base (3' end)			
		U	C	A	G
U	U	Phe	Phe	Leu	Leu
	C	Ser	Ser	Ser	Ser
	A	Tyr	Tyr	Stop	Stop
	G	Cys	Cys	Stop	Trp
C	U	Leu	Leu	Leu	Leu
	C	Pro	Pro	Pro	Pro
	A	His	His	Gln	Gln
	G	Arg	Arg	Arg	Arg
A	U	Ile	Ile	Ile	Met
	C	Thr	Thr	Thr	Thr
	A	Asn	Asn	Lys	Lys
	G	Ser	Ser	Arg	Arg
G	U	Val	Val	Val	Val
	C	Ala	Ala	Ala	Ala
	A	Asp	Asp	Glu	Glu
	G	Gly	Gly	Gly	Gly

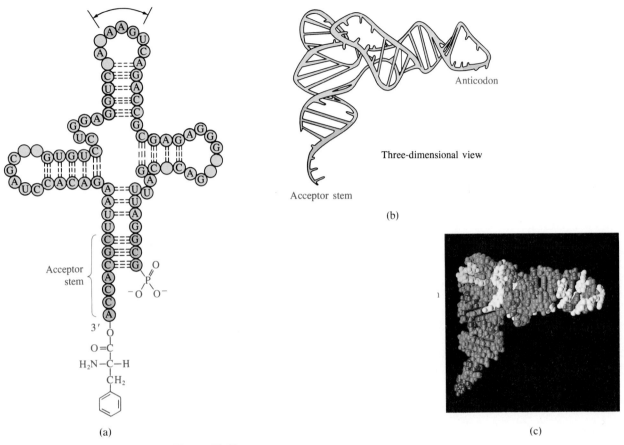

Figure 15.10
Structure of a tRNA molecule. (a) Schematic, flattened tRNA molecule. The cloverleaf-shaped tRNA contains an anticodon triplet on one "leaf" and a covalently bonded amino acid at its 3′ end. The example shown is a yeast tRNA that codes for phenylalanine. (The nucleotides not specifically identified are slightly altered analogs of the four normal ribonucleotides.) (b) The three-dimensional shape (the tertiary structure) of a tRNA molecule. Note how the anticodon is at one end of the molecule and the "acceptor stem" where the amino acid bonds is at the other end. (c) A computer-generated space-filling model of a tRNA.

Anticodon A sequence of three ribonucleotides on tRNA that recognizes the complementary sequence on mRNA.

molecule's halfway point, each tRNA contains a section called an **anticodon,** a sequence of three ribonucleotides complementary to certain codon sequences on mRNA. For example, the codon sequence C-U-G on mRNA can match up with the complementary anticodon sequence G-A-C on a leucine-bearing tRNA.

The three main steps in protein synthesis are initiation, elongation, and termination.

Initiation   Protein synthesis is initiated when an mRNA, a ribosome, and the first tRNA come together (1, Figure 15.11). The first A-U-G codon on the 5′ end of mRNA acts as a "start" signal for the translation machinery and codes for the introduction of a methionine unit. Initiation is completed when the methionine tRNA occupies one of the two binding sites on the ribosome. Since this is the site where the growing peptide will reside, it's known as the *active site,* or

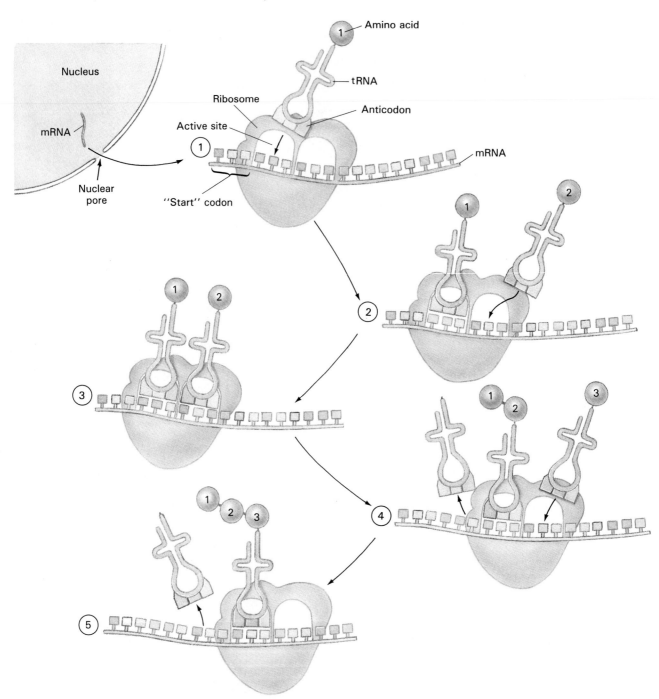

Figure 15.11
Initiation and elongation steps in protein synthesis. *Initiation:* An mRNA molecule joins with the two ribosomal subunits and a methionine tRNA at the active, or P binding site (①). *Elongation:* The second tRNA approaches the ribosome and binds at the A site (②, ③). A peptide bond then forms between the first two amino acids. Then translocation shifts the entire ribosome one codon further down the mRNA chain as the first tRNA is freed from the P site, the peptide-carrying tRNA previously at the A site shifts to the P site, and the A site is available for the next tRNA (④). With the second peptide bond formed, the ribosome translocates again and is ready for the cycle to repeat as the peptide chain grows (⑤).

the *P binding site*. (Clearly, not all proteins have methionine at one end—if it's not needed, the methionine from chain initiation is removed before the new protein goes to work.)

Elongation   Next to the active site on the ribosome is the *A binding site*, the point where the next codon on mRNA is exposed and where the tRNA carrying the next amino acid will be attached (hence, the "A" binding site). All available tRNA molecules can approach and try to fit, but only one with an appropriate anticodon sequence can bind (2, Figure 15.11). Let's assume that the second codon in our mRNA is C-U-G, which codes for leucine, and a leucine-bearing tRNA binds to the A site (3, Figure 15.11). Next, a peptide bond forms between methionine and leucine, and the bond to the first tRNA breaks.

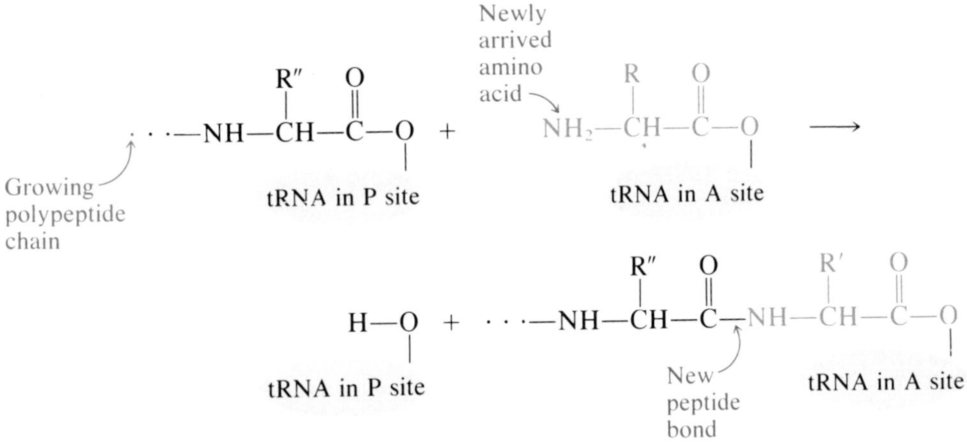

Newly arrived amino acid

$$\cdots-NH-\underset{\underset{\text{tRNA in P site}}{}}{\overset{R''}{CH}}-\overset{O}{C}-O \;+\; \underset{\underset{\text{tRNA in A site}}{}}{NH_2-\overset{R}{CH}-\overset{O}{C}-O} \longrightarrow$$

Growing polypeptide chain

$$H-O \;+\; \cdots-NH-\underset{}{\overset{R''}{CH}}-\overset{O}{C}-NH-\underset{}{\overset{R'}{CH}}-\overset{O}{C}-O$$

tRNA in P site   New peptide bond   tRNA in A site

Formation of peptide bond between amino acids during translation

With the first peptide bond formed, the entire ribosome shifts three positions (one codon) along the mRNA chain, a process called *translocation*. At the same time, the empty methionine tRNA previously at the P site is freed from the ribosome, the leucine tRNA bearing the new polypeptide chain is moved from the A site to the active site, and the A site is opened up to accept the tRNA carrying the next amino acid (4, Figure 15.11).

The three elongation steps now repeat:

1. The next appropriate tRNA binds to the A site.

2. Peptide bond formation occurs to attach the newly arrived amino acid to the growing chain.

3. Translocation takes place to free the A site for the next tRNA (5, Figure 15.11).

A single RNA molecule can be "read" simultaneously by many ribosomes, as shown schematically in the following diagram. The growing polypeptides increase in length as the ribosomes move down the mRNA strand.

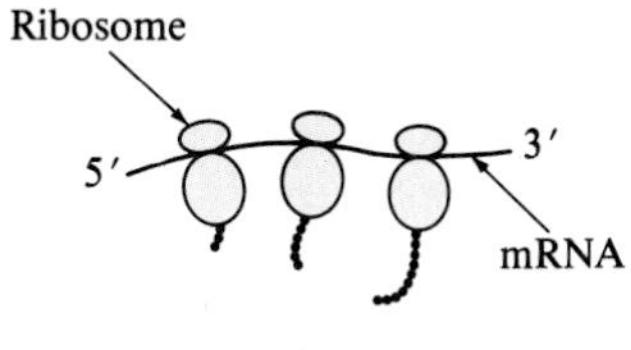

Ribosome

5'   3'

mRNA

Termination   When synthesis of the proper protein is completed, a "stop" codon (U-A-A, U-G-A, or U-A-G) signals the end of the process. An enzyme called a *releasing factor* then frees the polypeptide chain from the last tRNA, and the mRNA molecule is released from the ribosome.

In our discussion in this and the preceding sections, we've left many questions about replication, transcription, and translation unanswered. What tells a cell when to start replication? Since each cell contains the entire genome and since cells differ widely in their function, what keeps genes for unneeded proteins turned off? What turns on genes for proteins needed by the cell? Are there mechanisms to repair damaged DNA or correct errors in its structure? Exactly how do mRNA and a ribosome unite in the proper orientation? These are just a few examples of further questions. Some can be answered with details beyond what we can cover here. Others can't yet be fully answered, but they may be soon, because molecular genetics is one of today's most active areas of scientific research.

Solved Problem 15.3   What amino acid sequence is coded for by the mRNA base sequence AUC-GGU?

Solution   Table 15.1 indicates that AUC codes for isleucine and that GGU codes for glycine. Thus, AUC-GGU codes for Ile-Gly.

Practice Problems   15.11   What amino acid sequence is coded for by the following mRNA base sequence?
CUU-AUG-GCU-UGG-CCC-UAA

15.12   What anticodon sequences of tRNAs can match the codons in the mRNA in Problem 15.11?

15.13   What was the base sequence in the original DNA template strand on which the mRNA sequence in Problem 15.11 was made?

15.11   GENE MUTATION AND HEREDITARY DISEASE

The base-pairing mechanism of DNA replication and RNA transcription is an extremely efficient and accurate method for preserving and using genetic information, but it's not perfect. Occasionally an error occurs, resulting in the incorporation of an incorrect base at some point.

If an occasional error occurs during the transcription of an mRNA molecule, the problem is not too serious. After all, large numbers of mRNA molecules are continually being produced, and an error that occurs only one out of a million or so times would hardly be noticed. If an occasional error occurs during the replication of a *DNA* molecule, however, the consequences can be far more damaging. Each chromosome in a cell contains only *one* DNA molecule, and if that molecule is miscopied during replication, then the error is passed on when the cell divides.

Mutation An error in base sequence occurring during DNA replication.

Mutagen A substance that causes mutations.

An error in base sequence that occurs during DNA replication is called a **mutation.** Some mutations occur spontaneously, others are caused by exposure to ionizing radiation such as cosmic rays and gamma rays, and still others are caused by exposure to certain chemicals called **mutagens.** Some, perhaps most, mutations are harmless, but others can be devastating. Imagine, for example, that the sequence A-T-G on the informational strand of DNA is miscopied as A-C-G during replication. The mRNA transcribed from the corresponding template strand will then have the incorrect codon sequence A-C-G rather than the correct sequence A-U-G. But since A-C-G codes for threonine, whereas A-U-G codes for methionine, an incorrect amino acid will be inserted into the corresponding protein during translation. Furthermore, *every* copy made of the protein will have the same error.

The biological effects of incorporating an incorrect amino acid into a protein can range from negligible to catastrophic. If the incorrect amino acid occurs at some unimportant site, there may be little or no change in the biological properties of the protein. If, however, the error occurs at an important point, the biological activity of the protein can be completely changed.

Somatic cell Any cell other than a reproductive one.

Germ cell A reproductive (sperm or egg) cell.

When mutation occurs in a **somatic cell,** meaning any cell other than a sperm or egg cell, that cell might undergo uncontrolled growth leading to cancer. When mutation occurs in a **germ cell** (sperm or egg), then the genetic alteration is passed on to the offspring where it might show up as a hereditary disease. There are some 2000 known hereditary diseases that affect humans, with consequences that are sometimes fatal. Some of the more common ones are listed in Table 15.3.

The biochemical nature of the genetic defects in some hereditary diseases is well understood, but the nature of many others has eluded investigation. The key to understanding hereditary diseases is to identify the defective gene and its protein product. Cystic fibrosis, a hereditary disease that affects one of every 1600 Caucasian children, resisted the search for its cause until 1989. After narrowing in on a particular chromosome, researchers scanned the DNA in that chromosome and finally located the faulty gene. Its mutation results in production of a protein that normally aids in transporting sodium and chloride ions across cell membranes but is missing one phenylalanine. The next step toward a treatment will be to discover exactly how this defective protein causes the disease symptoms. A more immediate benefit is development of a screening test to identify carriers of this gene.

Table 15.3   Some Common Hereditary Diseases and Their Causes

Name	Nature and Cause of Defect
Phenylketonuria	Brain damage in infants caused by the defective enzyme phenylalanine hydroxylase
Albinism	Absence of skin pigment caused by the defective enzyme tyrosinase
Tay-Sachs disease	Mental retardation caused by a defect in production of the enzyme hexosaminidase A
Cystic fibrosis	Bronchopulmonary, liver, and pancreatic obstructions by thickened mucus; defective gene and protein identified
Sickle-cell anemia	Anemia and obstruction of blood flow caused by a defect in hemoglobin

Practice Problems   15.14   If the sequence T-G-G in a gene on a DNA informational strand mutated and became T-G-A, the protein encoded by that gene would stop short of completion. Explain.

15.15   What changes would result in the proteins made from genes in which the following mutations had occurred?
(a) A-T-C becomes A-T-G   (b) C-C-T becomes C-G-T

INTERLUDE: RECOMBINANT DNA

The revolution in molecular biology that began in the late 1970s owes its existence to the discovery of powerful techniques for manipulating the transfer and expression of genetic information. Using what has come to be called **recombinant DNA** technology, it is actually possible to cut a specific gene out of one organism and recombine it into the genetic machinery of a second organism. Normally, a specific gene from a higher or-

ganism is spliced into the DNA of a bacterium, and the bacterium then manufactures the specified protein, perhaps insulin or some other valuable material.

Bacterial cells, unlike the cells of higher organisms, contain part of their DNA in circular pieces called *plasmids,* each of which carries just a few genes. Plasmids are extremely easy to isolate, several copies of each plasmid are present in a cell, and each plasmid replicates through the normal base-pairing pathway.

To prepare a plasmid for insertion of a foreign gene, the plasmid is first cut open by use of a restriction endonuclease—one of over 200 enzymes known to cleave a DNA molecule between two nucleotides where a specific base sequence occurs. For example, the much-used restriction endonuclease *Eco*RI is capable of cutting a plasmid between G and A in the sequence G-A-A-T-T-C. The cut is not straight, however, but is slightly offset so that both DNA strands are left with a few unpaired bases on each end—so-called *sticky ends.*

A schematic representation for preparing recombinant DNA. (a) A bacterial plasmid and a chromosome of a higher organism are cut with the same restriction enzyme, leaving "sticky ends." (b) The opened plasmid and the desired gene fragment are then joined by DNA ligase enzyme to yield (c) an altered plasmid that is reinserted into a bacterial cell.

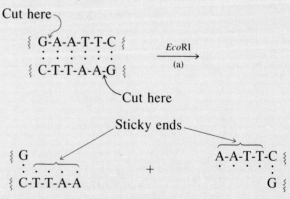

SUMMARY

Deoxyribonucleic acid (DNA) and **ribonucleic acid (RNA)** are the carriers of an organism's genetic information. DNA and protein are combined in chromatin in cell nuclei and during cell division in higher organisms are visible as **chromosomes**. Each chromosome contains one DNA molecule, which contains several thousand **genes**—sections of the DNA chain that encode instructions for constructing a single protein.

Nucleic acids are polymers of **nucleotides.** Each nucleotide consists of a **heterocyclic amine base** linked to C1 of an aldopentose sugar, with the sugar in turn linked through its C5 —OH group to phosphoric acid. The sugar component in RNA is ribose, while that in DNA is 2-deoxyribose. DNA contains two purine amine bases (adenine and guanine) and two pyrimidine bases (cytosine and thymine); RNA contains adenine, guanine, cytosine, and

Once the plasmid has been cut, the specific gene to be inserted is then cut from its chromosome using the same restriction endonuclease so that the sticky ends on the gene fragment are complementary to the sticky ends on the opened plasmid. The gene fragment and opened plasmid are then mixed in the presence of a DNA ligase enzyme that joins them together and reconstitutes the now altered plasmid. The entire sequence of events is shown schematically in the accompanying figure.

Once the altered plasmid has been made, it can be inserted back into a bacterial cell where the normal processes of transcription and translation take place to synthesize the protein encoded by the inserted gene. Since bacteria multiply rapidly, there are soon a large number of them, all containing the altered plasmid and all manufacturing the desired protein. In a sense, the bacterium has been harnessed and put to work as a protein factory.

Among the present commercial uses of recombinant DNA technology are the syntheses of human insulin, human growth hormone, and interferon, an antiviral protein that is being tested for cancer treatment and has already proven effective in protecting individuals with an immune system disorder (chronic granulomatous disease) from microbial infections.

A landmark in the rapid growth of biotechnology was reached in the early 1990s when gene therapy, the injection into humans of genetically altered copies of their own cells, had its first clinical trials. A child with a rare hereditary immune disorder was treated with her own white blood cells to which the missing gene had been added and her condition apparently improved. In a larger trial, white blood cells altered to produce tumor necrosis factor, which kills cancerous cells, were administered to a group of skin cancer patients. These trials are just the beginning of what promises to be a long series of studies leading to new therapies based on manipulation of DNA.

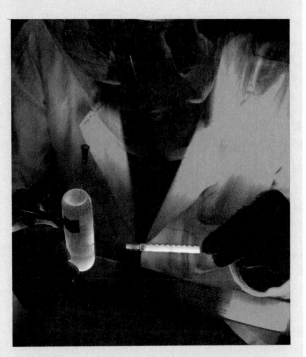

The scientist is withdrawing a sample of human DNA from a tube containing centrifuged lymphocytes marked with a dye that fluoresces yellow under ultraviolet light.

a second pyrimidine base called uracil in place of thymine. Nucleotides join together to yield nucleic acid chains by formation of an ester link between the phosphate component of one nucleotide and the —OH group at C3′ on the sugar of the neighboring nucleotide.

Molecules of DNA consist of two polynucleotide strands held together by hydrogen bonds between bases on the two strands and coiled into a **double helix.** Adenine (A) and thymine (T) form hydrogen bonds only to each other, as do cytosine (C) and guanine (G). Thus, the two strands are complementary rather than identical. Whenever an A occurs in one strand, a T occurs opposite it in the other strand; when a G occurs in one strand, a C occurs opposite it.

Three main processes take place in the transfer of genetic information:

1. Replication of DNA is the process by which identical copies of DNA are made and genetic information is preserved. The DNA double helix partially unwinds, complementary deoxyribonucleotides line up in order, DNA polymerase enzyme forms bonds between them, and two new DNA molecules are produced.
2. Transcription is the process by which RNA is produced. A segment of the DNA double helix unwinds, complementary ribonucleotides line up on the **template strand** of DNA, and an RNA polymerase enzyme forms bonds between them. The RNA produced is an exact copy of the **informational strand** of the DNA.

There are three types of RNA: **Ribosomal RNA (rRNA)** combines with proteins to form **ribosomes,** structures in the cytoplasm that are the sites of protein synthesis. **Messenger RNA (mRNA)** duplicates the information from DNA needed for synthesis of a given protein, carries it out of the nucleus to ribosomes, and directs protein synthesis. **Transfer RNA (tRNA)** delivers amino acids in the correct order for protein synthesis.
3. Translation is the process by which mRNA directs protein synthesis. Each mRNA has three-base segments called **codons** along its chain. Each codon is recognized by the **anticodon** of the tRNA that delivers the appropriate amino acid to the ribosome during protein synthesis.

Mutations occur when an error is made during DNA replication, an incorrect base is placed into the DNA chain, and a protein containing an incorrect amino acid sequence is produced. If the genetic alteration occurs in a sperm or egg cell, the mutation is passed on to offspring.

REVIEW PROBLEMS

Structure and Function of Nucleic Acids

15.16* What is a nucleotide, and what three kinds of components does it contain?

15.17 What are the names of the sugars in DNA and RNA, and how do they differ?

15.18* What does the prefix *deoxy-* mean when used in DNA?

15.19 What are the names of the four heterocyclic bases in DNA?

15.20* What are the names of the four heterocyclic bases in RNA?

15.21 Where in the cell is most DNA found?

15.22* What are the three main kinds of RNA, and what is the function of each?

15.23 What is meant by these terms?
(a) base pairing
(b) replication (of DNA)
(c) translation
(d) transcription

15.24* What is the difference between a gene and a chromosome?

15.25 What genetic information does a single gene contain?

15.26* Approximately how many genes are in human DNA?

15.27 What kind of bonding holds the DNA double helix together?

15.28* What does it mean to speak of bases as being ''complementary''?

15.29 Rank the following kinds of nucleic acids in order of size: DNA, mRNA, tRNA.

15.30* What is the difference between a nucleotide's 3′ end and its 5′ end?

15.31 Show by drawing structures how the phosphate and sugar components of a nucleic acid are joined.

15.32 Show by drawing structures how the sugar and heterocyclic base components of a nucleic acid are joined.

15.33 Draw the complete structure of deoxycytidine 5′-phosphate, one of the four deoxyribonucleotides.

15.34 Draw the full structure of the RNA dinucleotide U-C. Identify the 5′ and 3′ ends of the dinucleotide.

Nucleic Acids and Heredity

15.35 What are the names of the two DNA strands, and how do they differ?

15.36* What is a codon, and on what kind of nucleic acid is it found?

15.37 What is an anticodon, and on what kind of nucleic acid is it found?

15.38* What are the general shape and structure of a tRNA molecule?

15.39 What are exons and introns? Which of the two carries genetic information? Which of the two is more abundant in DNA?

15.40* How can the phrase "Pure silver taxi" help you remember the nature of base pairing in DNA?

15.41 The DNA from sea urchins contains about 32% A and about 18% G. What percentages of T and C would you expect in sea urchin DNA? Explain.

15.42* Look at Table 15.2 and find codons for these amino acids:
(a) Pro   (b) Lys   (c) Met

15.43 What amino acids are specified by these codons?
(a) A-C-U   (b) G-G-A   (c) C-U-U

15.44* What anticodon sequences on tRNA are complementary to the codons in Problem 15.43?

15.45 If the sequence T-A-C-C-G-A appeared on the information strand of DNA, what sequence would appear opposite it on the template strand?

15.46* What sequence would appear on the mRNA molecule transcribed from the DNA in Problem 15.45?

15.47 What dipeptide would be synthesized from the gene sequence in Problem 15.45?

15.48* What is a mutation, and how can it be caused?

15.49 Why does a mutation in RNA have a much smaller effect on an organism than a mutation in DNA?

15.50* If the gene sequence A-T-T-G-G-C-C-T-A on the informational strand of DNA mutated and became A-C-T-G-G-C-C-T-A, what effect would the mutation have on the sequence of the protein produced?

15.51 What kind of cell must undergo mutation in order for the genetic error to be passed down to future generations?

15.52* What problem would you foresee if codons were made of only two nucleotides rather than three? How many codons are possible using combinations of two bases?

15.53 In general terms, what is the cause of hereditary diseases?

15.54* A substantial percentage of chemical agents that cause mutations also cause cancer. Explain in general terms why this might be.

15.55 Metenkephalin is a small peptide having morphine-like properties and found in animal brains. Give an mRNA sequence that will code for the synthesis of metenkephalin:

Tyr-Gly-Gly-Phe-Met

15.56* Give a DNA gene sequence that will code for metenkephalin (Problem 15.55).

Applications

15.57 Do you think that the presence of introns will make the process of genome mapping more difficult? Why or why not? [App: Human Genome Project]

15.58 Discuss the implications of knowing your own genetic map. [App: Human Genome Project]

15.59 How do viruses differ from normal living organisms? [App: Viruses and AIDS]

15.60* Explain the process of reverse transcription. What is the name given to viruses that use this process? [App: Viruses and AIDS]

15.61 How do vaccines work? Why will it be difficult to design a vaccine to counteract the virus responsible for AIDS? [App: Viruses and AIDS]

15.62* Give two reasons why bacterial cells are used for recombinant DNA techniques. [Int: Recombinant DNA]

15.63 Discuss the general procedure for recombinant DNA techniques, including the purposes of plasmids, restriction endonucleases, and DNA ligases. [Int: Recombinant DNA]

15.64* What types of compounds are produced by genetically altered bacteria? [Int: Recombinant DNA]

Additional Questions and Problems

15.65 Discuss how human genome mapping might be combined with recombinant DNA techniques to alleviate or eliminate many hereditary diseases.

15.66* Normal hemoglobin has a glutamic acid at the same site in which sickle-cell hemoglobin has a valine. List a possible mRNA sequence that could be present in each type of hemoglobin.

15.67 List a DNA sequence that could account for the amino acids in normal and sickle-cell hemoglobin (Problem 15.66).

15.68* Would you expect a mutation in the DNA from TAA to TAG to produce a hazardous effect? Why or why not? What about a change from TAA to GAA?

15.69 Human and horse insulin are both composed of two polypeptide chains with one chain containing 21 amino acids and the other containing 30 amino acids. How many nitrogen bases are present in the DNA coding for each chain?

15.70* Human and horse insulin (Problem 15.69) differ in primary structure at two amino acids: at the ninth position in one chain (human has serine and horse has glycine) and in the thirtieth position on the other chain (human has threonine and horse has alanine). How must the DNA differ to account for this?

15.71 Explain how a tRNA delivers the correct amino acid to the growing protein chain.

15.72* The initiation code for all proteins is AUG. What happens if a protein does not include methionine as its first amino acid?

C H A P T E R

16

Body Fluids

Blood is a water solution (the plasma) that also contains (left to right) red cells
(5 billion/mL), white cells (7 million/mL), and platelets (250 million/mL). In this chapter
you'll discover the functions of these four components of blood.

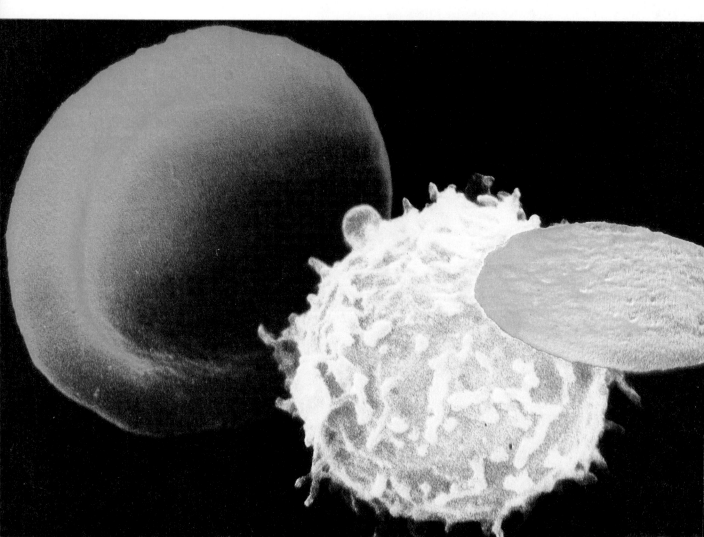

Just about every major aspect of chemistry you've studied so far applies to the subject of this chapter—body fluids. Electrolytes, nutrients and waste products, metabolism intermediates, and chemical messengers flow through the body in blood, and wastes exit in the urine. The chemical composition of blood and urine therefore mirrors chemical reactions throughout the body. Fortunately, samples of these fluids are easily collected. Many advances in understanding biological chemistry have been based on information from blood and urine analysis. In addition, studies of blood and urine chemistry provide information essential for the diagnosis and treatment of disease.

The questions answered in this chapter demonstrate the central role of chemistry in understanding physiology:

1. *How are body fluids classified?* The goal: Be able to describe the major categories of body fluids, their general composition, and the exchange of solutes between them.

2. *What are the roles of blood in maintaining homeostasis?* The goal: Be able to explain the composition and functions of blood.

3. *How do red blood cells participate in the transport of blood gases?* The goal: Be able to explain the relationships among O_2 and CO_2 transport, acid–base balance, and temperature.

4. *How do blood components participate in the body's defense mechanisms?* The goal: Be able to identify and describe the roles of blood components that participate in inflammation, the immune response, and blood clotting.

5. *How does the kidney function?* The goal: Be able to describe the process of urine formation.

6. *How does the composition of urine vary?* The goal: Be able to explain the influences on urine volume and acid elimination.

16.1 BODY WATER AND ITS SOLUTES

Intracellular fluid Fluid inside cells.

Extracellular fluid Fluid outside cells.

Blood plasma Liquid portion of blood; an extracellular fluid.

Interstitial fluid Fluid surrounding cells; an extracellular fluid.

Water is the solvent in all body fluids, and the water content of the human body averages about 60% (Figure 16.1). Physiologists describe body water as occupying two different "compartments"—the intracellular and the extracellular compartments. Thus far we have looked mainly at chemical reactions occurring in the **intracellular fluid,** the fluid inside cells, which includes about two-thirds of all body water. The remaining one-third of body water is **extracellular fluid,** which includes mainly **blood plasma** (the fluid portion of blood) and **interstitial fluid** (the fluid that fills the spaces between cells).

To be soluble in water a substance must be an ion, a gas, a small polar molecule, or a large molecule with many polar or ionic groups on its surface. All four types of solutes are present in body fluids. The major components of body fluids are inorganic ions and ionized biomolecules, mainly proteins, as shown in

431

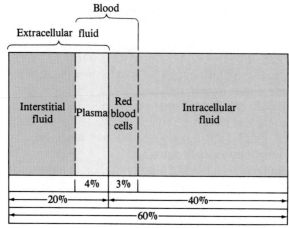

Figure 16.1
Body water as a percent of total body weight. The body averages 60% water by weight, which is divided between extracellular and intracellular fluid.

Electrolyte A substance that conducts electricity when dissolved in water; specifically, ions in body fluids.

Osmolarity The number of moles of dissolved solute particles (ions or molecules) per liter of solution.

the comparison of blood plasma, interstitial fluid, and intracellular fluid in Figure 16.2.

Inorganic ions, known collectively as **electrolytes,** have many specific roles in the reactions of metabolism, are major contributors to the **osmolarity** of body fluids, and move about as necessary to maintain charge balance inside and outside cells. Water-soluble proteins, which are negatively charged at physiological pH, make up a large proportion of the mass of solutes in blood plasma and intracellular fluid. One hundred milliliters of blood contains about 7 g of protein. Blood proteins transport lipids and other molecules, and play essential roles in the protection afforded by blood clotting (Section 16.6) and the immune response (Section 16.5). The blood gases oxygen and carbon dioxide, along with

Figure 16.2
The distribution of cations and anions in body fluids. Outside cells, Na^+ is the major cation and Cl^- is the major anion. Inside cells, K^+ is the major cation and HPO_4^{2-} is the major anion. Note that at physiological pH, proteins are negatively charged.

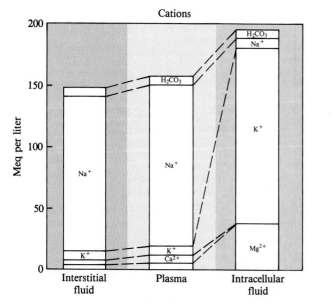

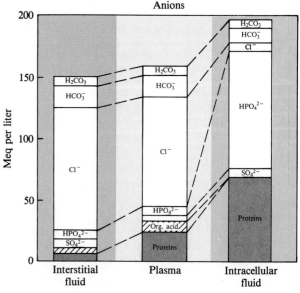

REVIEW Osmosis is the passage of solvent molecules across semipermeable membranes such as cell membranes, from a more dilute solution to a more concentrated solution. The osmolarity of fluids inside and outside body cells must be kept in balance to prevent water flow that would cause cells to shrivel or expand.

Active transport Energy-requiring transport of molecules or ions across cell membranes against a concentration gradient.

glucose, amino acids, and the nitrogen-containing by-products of protein catabolism, are the major small molecules in body fluids.

Blood travels through peripheral tissue in a network of tiny, hair-like capillaries that connect the arterial and venous parts of the circulatory system (Figure 16.3a). It is here that the exchange of nutrients and end products of metabolism occurs. Capillary walls consist of a single layer of loosely spaced cells (Figure 16.3b). Water and many small solutes move freely across the walls in response to differences in fluid pressure and concentration.

On the arterial ends of capillaries (connected to *arteries,* which carry blood from the heart), blood pressure is higher than interstitial fluid pressure and pushes solutes and water into interstitial fluid. On the venous ends (connected to *veins,* which carry blood to the heart), blood pressure is lower, and water and solutes reenter the plasma. Solutes that can cross membranes passively by diffusion move from regions of high concentration to regions of low concentration. The combined result of water and solute exchange at capillaries is that, except for proteins, blood plasma and interstitial fluid are very similar in composition (see Figure 16.2).

In addition to blood capillaries, peripheral tissue is networked with lymph capillaries that terminate in blind pockets. The lymphatic system collects excess interstitial fluid, debris from cellular breakdown, and proteins and lipid droplets too large to pass through capillary walls. Interstitial fluid and the substances that accompany it into the lymphatic system are referred to as *lymph,* and the walls of lymph capillaries are constructed so that lymph cannot return to the surrounding tissue. Ultimately lymph enters the bloodstream at the thoracic duct.

Solutes in the interstitial fluid and the intracellular fluid are exchanged by crossing cell membranes (Figure 16.4). Here, major differences in concentration are maintained by **active transport** against concentration gradients (from regions of low concentration to regions of high concentration) and by the impermeability of cell membranes to certain solutes, notably Na$^+$.

Figure 16.3
(a) The capillary network. Solute exchange between blood and interstitial fluid occurs across capillary walls. (b) Cross section of a blood capillary. The walls are made of individual cells lying parallel to the capillary blood channel. The nucleus of a single cell appears as a green semicircle on the right. This falsely colored electron micrograph enlarges the original tissue more than 2500 times.

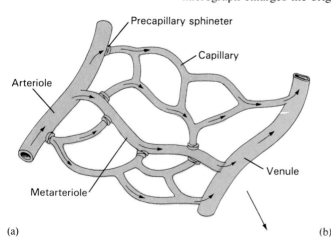

(a)

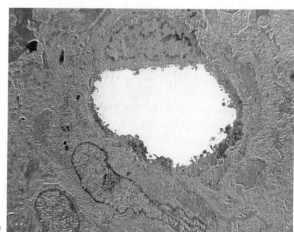

(b)

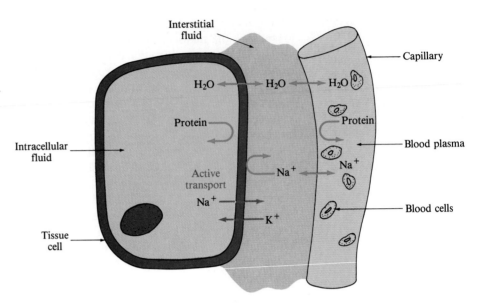

Figure 16.4
Exchange among body fluids. Water exchanges freely in most tissues, with the result that the osmolarity of blood plasma, interstitial fluid, and intracellular fluid is the same. Large proteins cross neither capillary walls nor cell membranes, leaving the interstitial fluid protein concentration low. Concentration differences between interstitial fluid and intracellular fluid are maintained by active transport.

AN APPLICATION: ORGAN CRYOPRESERVATION

Organ cryopreservation is the freezing of live tissue for storage and subsequent use. To make this type of preservation successful, the molecular mechanisms essential to life in the frozen state must be understood. Freezing unprotected body fluids causes formation of ice crystals that rip through cell membranes, damage subcellular organelles, and allow the cell contents to spill out. Much of what has been attempted in medical cryopreservation is based on strategies observed in the survival of insects, mammals, and fish in subfreezing temperatures.

Successful cryopreservation of human sperm frozen in a glycerol solution first occurred in 1949. Since then, techniques have been developed for freezing many single-cell suspensions (sperm, red and white blood cells, platelets) and simple tissues (embryos, skin, cornea, pancreatic islets). Cryopreservation of larger organs for transplantation and the return of the frozen organs to a fully viable state is more difficult and is not yet possible.

To survive freezing, tissues must satisfy three biochemical conditions. First, ice formation in extracellular fluids must be controlled so that freezing occurs slowly and ice crystals are small.

To promote small crystal formation, freeze-tolerant animals use blood proteins as nuclei around which large numbers of very small crystals form. Since small ice crystals are thermodynamically unstable, however, and like to form larger and larger crystals (much as sizable ice crystals appear in ice cream kept too long after opening), antifreeze compounds that slow crystal growth are also needed.

Secondly, cell structure and the cell membrane must be protected because water tends to exit cells by osmosis when the extracellular fluids freeze. Freeze-tolerant animals use low-molecular-weight compounds (trehalose, proline, glycerol, sorbitol, glucose) to prevent the injuries that would result during freezing.

Finally, cell viability must be maintained, which means that metabolism cannot cease. Although the lowered temperature automatically lowers the metabolism rate, there must be some production of ATP. (In freeze-tolerant animals, it's as low as 1–10% of the normal resting rate.) Since there is no blood supply, the fuel must be present in the cells. Also, since there isn't any oxygen supply, the metabolism must be anaerobic, and the metabolites must not be toxic be-

16.2 FLUID BALANCE

The body is kept in overall fluid balance by roughly equal daily intake and output of water as summarized in the following table. In a hot environment or when doing strenuous work, the intake of drinking water and loss in sweat and exhaled gases both increase greatly.

Water intake (mL/day)		Water output (mL/day)	
Drinking water	1200	Urine	1400
Water from food	1000	Skin	400
Water from metabolic		Lungs	400
oxidation of food	300	Sweat	100
		Feces	200
Total	2500		2500

cause there is no mechanism for their elimination.

Large-organ cryopreservation must meet the same conditions as the survival of freeze-tolerant animals. Ice-nucleating compounds and/or antifreeze compounds must be provided (and then removed while the tissue is thawed if the compounds are potentially toxic). Also, membranes must be stabilized and metabolism inhibited, but not so much so that ATP production ceases. Further studies of freeze-tolerant animals are expected to contribute to the eventual successful cryopreservation of organs for transplantation.

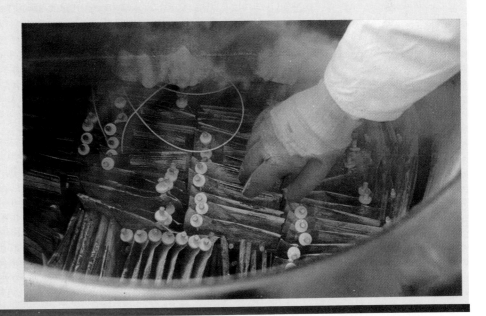

Bags of cryopreserved blood for transfusions or for controlled thawing that allows separation of useful components such as a clotting agent for treatment of hemophilia.

One important role of the kidneys is to keep water and electrolytes in balance by increasing or decreasing the amounts eliminated. The intake and output of water are, in turn, controlled by hormones. Receptors in the hypothalamus monitor the concentration of solutes in blood plasma, and as little as a 2% change in osmolarity can cause an adjustment in hormone secretion. For example, when a rise in osmolarity indicates an increased concentration of solutes in blood and therefore a shortage of water, secretion of *antidiuretic hormone* (also known as vasopressin) increases. In the kidney, antidiuretic hormone causes a decrease in the water content of the urine. At the same time, the thirst mechanism is activated to cause increased water intake.

A long list of abnormal conditions cause what physicians refer to as the *syndrome of inappropriate antidiuretic hormone secretion (SIADH),* the result of excess secretion of the hormone. When this occurs, the kidney excretes too little water, the water content of body compartments increases, and serum concentrations of electrolytes drop to dangerously low levels. Among the causes of SIADH are regional low blood volume due to decreased return of blood to the heart (caused by, for example, asthma, pneumonia, pulmonary obstruction, or heart failure) and misinterpretation by the hypothalamus of osmolarity (due, for example, to central nervous system disorders, barbiturates, or morphine).

The reverse problem, inadequate secretion of antidiuretic hormone, often a result of injury to the hypothalamus, causes *diabetes insipidus* (unrelated to diabetes mellitus). In this condition, up to 15 L of dilute urine is excreted each day. Administration of synthetic hormone controls the problem.

16.3   BLOOD

Blood flows through the body in the circulatory system, which in the absence of trauma or disease is an essentially closed system. About 55% of blood is plasma, which contains the solutes shown in Figure 16.5, and the remaining 45% is a mixture of red blood cells (erythrocytes), platelets, and white blood cells suspended in the plasma.

Whole blood   Blood plasma plus blood cells.

The plasma and cells together make up **whole blood** (Figure 16.6), which is usually collected for clinical laboratory analysis directly into evacuated tubes. An anticoagulant must be added to whole blood to prevent clotting, which otherwise will occur in 20–60 min at room temperature. Often the anticoagulant is supplied right in the collection tube. *Heparin,* a common anticoagulant, is a natural polysaccharide that interferes with the action of enzymes needed for clotting. Other anticoagulants, such as citrate ion and oxalate ion, form precipitates with calcium ion, which is also needed for blood clotting. Plasma is separated from blood cells by spinning the sample in a centrifuge.

Blood serum   Fluid portion of blood remaining after clotting has occurred.

Many laboratory analyses are performed on **blood serum,** the fluid remaining after blood has completely clotted. When a serum sample is desired, whole blood is collected in the presence of an agent that hastens clotting, often thrombin, a natural component of the clotting system (Section 16.6). Centrifugation separates the clot and cells to leave behind the serum.

The functions of blood fall into three major categories:

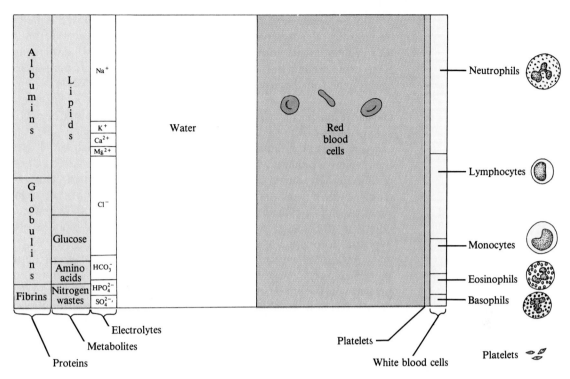

Figure 16.5
The composition of whole blood.

Figure 16.6
Whole blood for a blood count analysis. The number of red cells, white cells, and platelets per liter will be determined by an electronic cell counter.

Erythrocytes   Red blood cells.

● **Transport**   The circulatory system is the body's equivalent of the interstate highway network, transporting materials from where they are produced to where they are used or disposed of.

● **Regulation**   Blood redistributes body heat as it flows along, thereby participating in the regulation of body temperature. It also picks up or delivers water and electrolytes as they are needed to maintain fluid and electrolyte balance. In addition, blood buffers are essential to the maintenance of acid–base balance.

● **Defense**   Blood carries the molecules and cells needed for two major defense mechanisms: (1) blood clotting, which prevents loss of blood and begins healing of wounds; (2) the immune response, which destroys foreign invaders.

16.4   RED BLOOD CELLS AND BLOOD GASES

Red blood cells, or **erythrocytes,** have one major purpose: to transport blood gases. Erythrocytes have no nuclei or ribosomes and cannot replicate themselves. In addition, they have no mitochondria or glycogen and must obtain glucose from the surrounding plasma. Their enormous number—about 250 million in a single drop of blood—and their large surface area provide for rapid exchange of gases throughout the body. Because they are small and flexible, erythrocytes can squeeze through the tiniest capillaries one at a time.

Ninety-five percent of the protein in an erythrocyte is hemoglobin, the transporter of oxygen and carbon dioxide. Hemoglobin (Hb) is composed of four protein chains with the quaternary structure shown earlier in Figure 7.16. Each protein chain has a heme molecule embedded in its nonpolar interior, and each of the four hemes can combine with one O_2 molecule.

Oxygen Transport   The iron(II) ion, Fe^{2+}, in the center of each heme molecule is the site of the action in transporting oxygen, which is held to the iron by bonding through an unshared electron pair. In contrast to the cytochromes of the respiratory chain, where iron cycles between Fe^{2+} and Fe^{3+}, heme iron must remain in the reduced Fe^{2+} state or it loses its oxygen-carrying ability. Hemoglobin carrying four oxygens (*oxyhemoglobin*) is bright red, hemoglobin that has lost one or more oxygens (*deoxyhemoglobin*) is dark red-purple and accounts for the darker color of venous blood, and dried blood is brown because exposure to oxygen has oxidized the heme iron.

At normal physiological conditions, the percentage of heme molecules that carry oxygen, known as the *percent saturation,* is dependent on the **partial pressure** of oxygen in surrounding tissues as shown in Figure 16.7. The relation is not a simple equilibrium, as shown by the S shape of the curve, but is determined by allosteric interaction (Section 8.9). Binding each O_2 causes changes in hemoglobin structure that enhance binding of the next O_2, and releasing each oxygen enhances release of the next. As a result, oxygen is more readily released to tissue where the partial pressure of oxygen is low. The average oxygen partial pressure in peripheral tissue is 40 mm Hg, a pressure at which Hb remains 75% saturated by oxygen, leaving a large amount of O_2 in reserve for emergencies. Note, however, the rapid drop in the curve between 40 mm Hg and 20 mm Hg, the oxygen pressure in tissue where metabolism is occurring rapidly.

Partial pressure   The contribution to total gas pressure by each individual component of a mixture of gases.

Carbon Dioxide Transport, H^+ Transport, and Temperature Effects   Carbon dioxide from metabolism in peripheral cells diffuses into interstitial fluid and then into capillaries, where it is transported in three ways: (1) dissolved, (2) bonded to Hb, or (3) as HCO_3^-. About 7% of the CO_2 dissolves in blood plasma. The rest enters erythrocytes, where some of it bonds to hemoglobin, not in the heme portion of the molecule but by reaction with nonionized amino acid —NH_2 groups,

Figure 16.7
Oxygen saturation of hemoglobin at normal physiologic conditions. Oxygen pressure is about 100 mm Hg in arteries and 20 mm Hg in active muscles. Note the big release of oxygen between 40 mm Hg and 20 mm Hg.

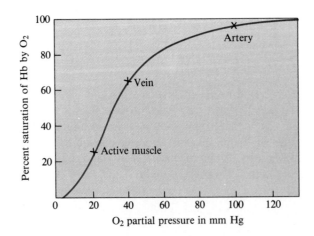

$$Hb-NH_2 + CO_2 \longrightarrow Hb-NHCOO^- + H^+$$

Most of the CO_2 is rapidly converted to bicarbonate ion within erythrocytes, which contain a large concentration of carbonic anhydrase:

$$H_2O + CO_2 \xrightarrow{\text{carbonic anhydrase}} H^+ + HCO_3^-$$

Without some compensating change, the result of the two reactions of carbon dioxide shown above would be an unacceptably large increase in blood acidity. To cope with the increased acidity, hemoglobin responds by reversibly binding hydrogen ions:

$$Hb \cdot 4\,O_2 + 2\,H^+ \rightleftharpoons Hb \cdot 2\,H^+ + 4\,O_2$$

The release of oxygen is enhanced by allosteric effects when the hydrogen ion binding increases, and oxygen is held more firmly when the hydrogen ion binding decreases. The result of changes in H^+ concentration on the oxygen affinity of hemoglobin is shown in the following diagram. Where H^+ concentration is high in peripheral tissues, Hb releases O_2 and picks up H^+. The Hb then carries H^+ to the lungs, where it is released as O_2 is picked up.

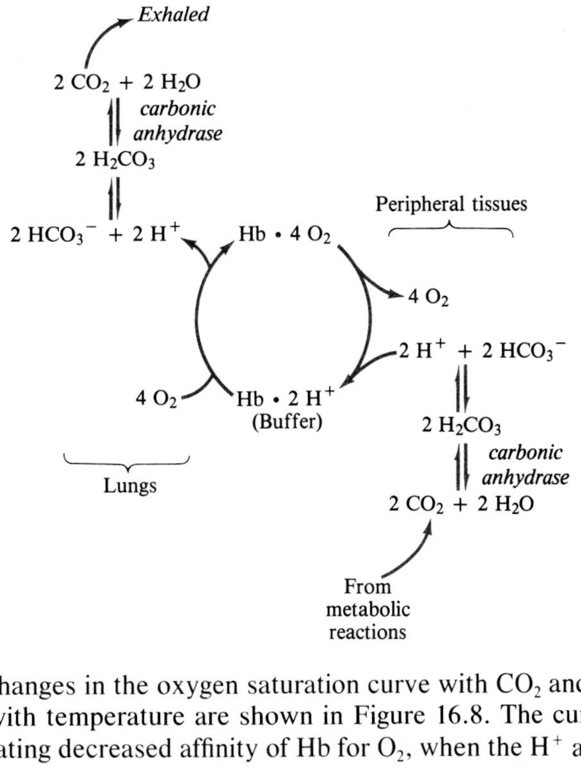

The changes in the oxygen saturation curve with CO_2 and H^+ concentrations and with temperature are shown in Figure 16.8. The curve shifts to the right, indicating decreased affinity of Hb for O_2, when the H^+ and CO_2 concentrations increase and when the temperature increases. These are exactly the conditions in muscles that are working hard and need more oxygen. The curve shifts to the left, indicating increased affinity of Hb for oxygen, under the opposite conditions of decreased H^+ and CO_2 concentrations and lower temperature.

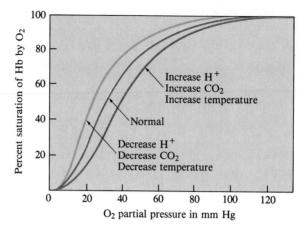

Figure 16.8
Changes in oxygen affinity of hemoglobin with changing conditions. The normal curve of Figure 27.7 is shown in red.

REVIEW   Buffers that help to maintain acid-base balance in body fluids are (1) the bicarbonate buffer system, $[HCO_3^-]/[CO_2]$; (2) the phosphate buffer system, $[HPO_4^{2-}]/[H_2PO_4^-]$; (3) acidic and basic proteins acting as buffers. (Appendix B.10.)

The intimate relationships among H^+ and HCO_3^- concentrations and O_2 and CO_2 partial pressures are essential to maintaining electrolyte and acid–base balance. Clinical laboratory measurements of these parameters are often used in diagnosis. Table 16.1 lists a few common disorders that cause acidosis and their effects on blood concentrations.

16.5   PLASMA PROTEINS, WHITE BLOOD CELLS, AND IMMUNITY

Having examined the function of red blood cells, we now want to take a look at the roles of plasma proteins and other types of blood cells (Table 16.2) in the *defense* functions of blood: the immune response and blood clotting. Each of these processes is complex, and we can only take an overview of them. In this section we will focus on the immune function, and in the next section we will do the same for blood clotting.

Antigen   A substance that stimulates the immune response.

A foreign invader, known as an **antigen,** is any substance recognized by the body as not part of itself. The invader might be a foreign molecule or a microorganism. Antigens can also be small molecules known as *haptens* that are recognized as antigens only after they have bonded to carrier proteins. Haptens include some antibiotics, environmental pollutants, and allergens from plants and animals.

Table 16.1   Effects on Blood Gas Concentrations of Some Common Disorders That Cause Acidosis

Disorder	H^+ Concentration	CO_2 Partial Pressure	O_2 Partial Pressure	HCO_3^- Concentration
Kidney failure	Increase	Initially normal, then decrease	Normal	Decrease
Diabetic ketoacidosis	Increase	Initially normal, then decrease	Normal	Decrease
Emphysema	Increase	Increase	Decrease	Normal
Respiratory distress syndrome	Increase	Increase	Decrease	Normal

Table 16.2   Protein and Cellular Components of Blood

Blood Component	Function
Proteins	
Albumins	Transport lipids, hormones, drugs; major contributor to plasma osmolarity
Globulins	
Immunoglobulins (γ-globulins, antibodies)	Identify antigens (microorganisms and other foreign invaders) and initiate their destruction
Transport globulins	Transport lipids and metal ions
Fibrinogen	Forms fibrin, the basis of blood clots
Blood cells	
Red blood cells	Transport O_2, CO_2, H^+
White blood cells	
Lymphocytes (T cells and B cells)	Defend against specific pathogens and foreign substances
Neutrophils, eosinophils, and monocytes	Carry out phagocytosis—engulf foreign invaders
Basophils	Release histamine during inflammatory response of injured tissue
Platelets	Help to initiate blood clotting

Inflammation A nonspecific defense mechanism triggered by antigens or tissue damage; includes swelling, redness, warmth, and pain.

Figure 16.9
Red blood cells squeezing through capillary wall (top left to bottom right). When inflammation reduces the permeability of a capillary wall, red and white cells, plus plasma, cross into the surrounding tissue.

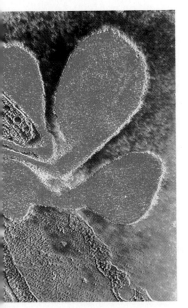

The recognition of an antigen can initiate three different responses, the first, inflammation, being nonspecific and the other two being specific to a given antigen.

Inflammatory Response   Cell damage due to infection or injury initiates **inflammation.** For example, the swollen, painful, red bump that develops around a splinter in your finger is an inflammation. Chemical messengers released at the site of the injury direct the inflammatory response. One important such messenger is *histamine,* which is synthesized from the amino acid histidine.

$$\underset{\text{Histidine}}{\begin{array}{c}N\diagup\diagdown\\ \quad\quad-CH_2CH-CO^-\\ \underset{H}{N}\quad\quad ^+NH_3\end{array}} \xrightarrow{\substack{\text{histidine}\\ \text{decarboxylase}}} \underset{\text{Histamine}}{\begin{array}{c}N\diagup\diagdown\\ \quad\quad-CH_2CH_2-NH_3^+\\ \underset{H}{N}\end{array}} + CO_2$$

Histamine sets off dilation of capillaries and increases the permeability of capillary walls (Figure 16.9). The resulting increased blood flow into the damaged area reddens and warms the skin, and swelling occurs as plasma carrying blood-clotting factors and defensive proteins enters the intercellular space. At the same time, white blood cells cross capillary walls to attack invaders.

Bacteria or other antigens at the site of inflammation are attacked by **phagocytosis:** White blood cells (*phagocytes*) surround the bacteria and destroy them by enzyme-catalyzed hydrolysis reactions (Figure 16.10). Phagocytes that

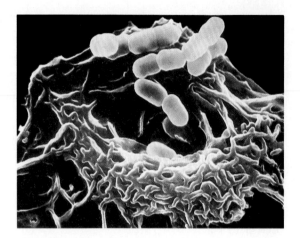

Figure 16.10
Phagocytosis. The green
Escherichia coli bacteria,
normal residents of the
human intestine that can
cause infection under some
conditions, are about to be
engulfed by a white blood
cell.

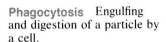

Phagocytosis Engulfing
and digestion of a particle by
a cell.

have encountered antigens also emit a variety of chemical messengers that help
to direct the inflammatory response and initiate specific immune responses.

Yet another defense system consisting of a large family of plasma proteins
known as the *complement* system is activated by inflammation. The complement proteins assemble into a *membrane attack complex* able to destroy bacteria directly by creating holes in their cell walls.

An inflammation caused by a wound will heal completely only if all
infectious agents have been removed, with dead cells and other debris absorbed
into the lymph system. The inflammatory response is dependent upon normal
numbers of white blood cells, and if the white blood cell count falls below 1000
per mL of blood, any infection can be life-threatening.

Specific Immune Responses   As opposed to the inflammatory response, an
immune response depends on recognition of specific invaders. At the molecular
level, an antigen is detected by an interaction very much like that between an
enzyme and its substrate. Noncovalent attraction allows a spatial fit between
the antigen, or more likely a small region in an antigen, and an **antibody** specific
to it. The antigen and the antibody may be independent molecules or they may
be portions of proteins or glycoproteins protruding from the surfaces of cell
membranes.

Before an immune response is triggered, an antigen often undergoes
phagocytosis in such a manner that a segment of the antigen molecule is
transported to the surface of the phagocyte. At the surface, it will be recognized
by a lymphocyte, a white blood cell responsible for one of the two antigen-
specific responses: the antibody-directed or the cell-directed immune response.
This recognition can take place anywhere in the body but often occurs in lymph
nodes, the tonsils, or the spleen, which have large concentrations of lymphocytes.

Immune response The
recognition of and attack on
antigens, including viruses,
bacteria, toxic substances,
and infected cells.

Antibody (immunoglobulin)
Protein molecule that
identifies antigens.

Antibody-directed Immune Response   In the antibody-directed immune response, antigens are first detected by binding to antibodies on the surfaces of B
lymphocytes, known as B cells. The B cells divide into *plasma cells* that
produce identical non-cell-bound antibodies specific for the antigen they have
encountered.

AN APPLICATION: THE BLOOD–BRAIN BARRIER

Nowhere in human beings is the maintenance of a constant internal environment more important than in the brain. If the brain were exposed to the fluctuations in concentrations of hormones, amino acids, neurotransmitters, and potassium that occur elsewhere in the body, inappropriate nervous activity would result. Therefore, the brain must be rigorously isolated from variations in blood composition.

How can the brain receive nutrients from the blood in capillaries and yet be protected? The answer lies in the unique structure of the *endothelial* cells that form the walls of brain capillaries. Unlike the cells in other capillaries, those in brain capillaries form a series of continuous tight junctions so that nothing can pass between them. To reach the brain, therefore, a substance must cross this *blood–brain barrier* by crossing the endothelial cell membranes.

The brain, of course, can't be completely isolated or it would die from lack of nourishment. Glucose, the main source of energy for brain cells, and certain amino acids the brain can't manufacture are recognized and brought across the cell membranes by transporters specific to each nutrient. Similar specific transporters move surplus substances out of the brain.

An asymmetric (one-way) transport system exists for glycine, a small amino acid that is a potent neurotransmitter. Glycine inhibits rather than activates transmission of nerve signals, and its concentration must be held at a lower level in the brain than in the blood. To accomplish this, there is a glycine transport system in the cell membrane closest to the brain, but no matching transport system on the other side. Thus, glycine can be transported out of the brain but not into it.

The brain is also protected by the "metabolic" blood–brain barrier. In this case, a compound that gets into an endothelial cell is converted there to a metabolite that is unable to enter the brain. A striking demonstration of the metabolic brain barrier is provided by dopamine, a neurotransmitter that is deficient in Parkinson's

disease, and L-dopa, a metabolic precursor of dopamine.

L-Dopa

Dopamine

L-Dopa can both enter and leave the brain because it is recognized by an amino acid transporter. The brain is protected from an entering excess of L-dopa, however, by its conversion to dopamine *within* the endothelial cell. Dopamine, which is also produced from L-dopa within the brain (thereby allowing treatment of Parkinson's disease with L-dopa), can, like glycine, leave the brain but cannot enter it.

Since crossing the endothelial cell membrane is the necessary route into the brain, substances soluble in the membrane lipids readily breach the blood–brain barrier. Among such substances are nicotine, caffeine, codeine, diazepam (Valium, a tranquilizer), and heroin. Heroin differs from morphine in having two nonpolar acetyl groups where the morphine molecule has polar hydroxyl groups. The resulting difference in lipid solubility allows heroin to enter the brain much more efficiently than morphine. Once heroin is inside the brain, enzymes remove the acetyl groups to give morphine, in essence trapping it in the brain. When developing therapeutic drugs, it is possible to take advantage of such trapping so that a carefully controlled dosage can have a potent effect.

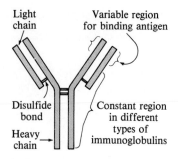

Figure 16.11
The basic structure of an immunoglobulin.

Antibodies, known also as **immunoglobulins,** make up 20% of the plasma proteins. The body may contain up to 10,000 different immunoglobulins at any one time, and we have the capacity to make more than one hundred million others. The immunoglobulins are glycoproteins that have in common two *heavy* polypeptide chains and two *light* polypeptide chains joined by disulfide bonds, as shown schematically in Figure 16.11. The antigen-binding sites are composed of antigen-specific sequences of amino acids. Once synthesized, antibodies spread out to mark their antigens for destruction by phagocytosis or by the complement system.

B-cell division also yields *memory cells* that remain on guard and quickly produce more plasma cells if the same antigen reappears in the future.

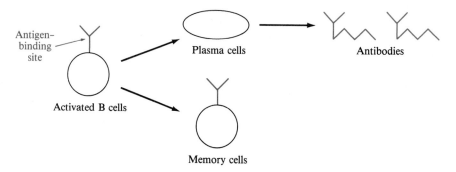

The long-lived memory cells are responsible for long-term immunity to diseases after the first illness or after a vaccination.

Immunoglobulin G antibodies (known as gamma globulins) protect against viruses and bacteria. Allergies and asthma are caused by an oversupply of *immunoglobulin E.* Numerous disorders result from the mistaken identification of normal body constituents as foreign and the overproduction of antibodies to combat them. These *autoimmune diseases* include attack on connective tissue at joints in rheumatoid arthritis, attack on pancreatic islet cells in some forms of diabetes mellitus, and a generalized attack on nucleic acids and blood components in systemic lupus erythematosus.

Cell-directed Immune Response   In the cell-directed immune response, antigenic regions on the surface of infected body cells are detected by compatible regions on the surface of T lymphocytes, or T cells. The T cells themselves destroy the infected cells by releasing a toxic protein that kills by perforating the cell membranes. Cell-directed immunity is responsible for protection against cancer cells and virus-infected cells and also causes the rejection of transplanted organs.

16.6 BLOOD CLOTTING

Fibrin Insoluble protein that forms the fiber framework of a blood clot.

A blood clot consists of blood cells trapped in a mesh of the insoluble fibrous protein known as **fibrin** (Figure 16.12). Formation of a clot is a many-step process requiring participation of 12 clotting factors. The calcium ion is one such clotting factor, and vitamin K is necessary for synthesis in the liver of other clotting factors, most of which are glycoproteins. Therefore, a deficiency of vitamin K, the presence of a competitive inhibitor of vitamin K, or a defi-

Figure 16.12
Red blood cells trapped in a fibrin clot.

Hemostasis The stopping of bleeding.

Blood clotting Formation of a structure of fibrin and blood cells to close the rupture of a blood vessel.

Zymogen A compound that becomes an active enzyme after undergoing a chemical change.

ciency of a clotting factor can cause excessive bleeding, sometimes even from minor tissue damage. *Hemophilia* is a disorder caused by an inherited genetic defect that results in the absence of one or another of the clotting factors.

The body's mechanism for halting blood loss from even the tiniest capillary is referred to as **hemostasis.** The first events in hemostasis are constriction of surrounding blood vessels and formation at the site of tissue damage of a plug composed of the blood cells known as *platelets.*

Next, **blood clotting,** which may be triggered by two different pathways, swings into action. One pathway (the *intrinsic pathway*) begins when blood makes contact with the negatively charged surface of the fibrous protein collagen exposed at the site of tissue damage. Glass is also negatively charged, and clotting is activated in exactly the same manner when blood is placed in a glass tube. The other pathway (the *extrinsic pathway*) begins with release by tissue damage of an integral membrane glycoprotein known as *tissue factor.*

The result in either case is a cascade of reactions in which several inactive clotting factors that are zymogens are activated one after another by cleavage of polypeptides to give active clotting factors, many of which are enzymes that then catalyze activation of the next factor in the cascade. The two pathways merge and, in the final step of the *common pathway,* the factor *thrombin* catalyzes cleavage of small polypeptides from the soluble plasma protein *fibrinogen* (Figure 16.13). Negatively charged groups in these polypeptides make

Figure 16.13
Schematic drawing of fibrinogen (mol wt 340,000) and a fibrin clot, which is essentially a polymer of fibrin. In fibrinogen there are three globular protein regions connected by fibrous protein. Removal of ionic polypeptides from the central region gives fibrin, in which central sites are free to associate with compatible sites on the ends of the molecule, forming a half-staggered arrangement which is then stabilized by covalent cross-links.

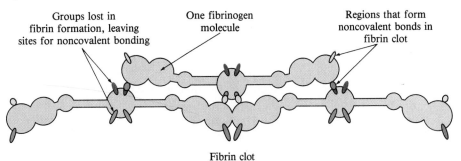

Groups lost in fibrin formation, leaving sites for noncovalent bonding

One fibrinogen molecule

Regions that form noncovalent bonds in fibrin clot

Fibrin clot

fibrinogen soluble and keep the molecules apart. Once these polypeptides are removed, the resulting insoluble fibrin molecules immediately associate with each other by noncovalent interactions. Then, they are bound into fibers by formation of peptide bond cross-links between lysine and glutamine side chains in a reaction catalyzed by another of the clotting factors:

$$\text{Gln}-\text{CH}_2\text{CH}_2-\overset{\overset{\text{O}}{\|}}{\text{C}}-\text{NH}_2 \quad + \quad {}^+\text{H}_3\text{NCH}_2\text{CH}_2\text{CH}_2\text{CH}_2-\text{Lys} \qquad \text{Protein chain}$$

$$\downarrow$$

$$\text{Gln}-\text{CH}_2\text{CH}_2-\overset{\overset{\text{O}}{\|}}{\text{C}}-\text{NHCH}_2\text{CH}_2\text{CH}_2\text{CH}_2-\text{Lys} \quad + \quad \text{NH}_4{}^+$$

Cross-link between protein chains

The final stage of hemostasis is breakdown of the clot by hydrolysis of peptide bonds after it has done its job of preventing blood loss and binding together damaged surfaces as they heal.

16.7 THE KIDNEY AND URINE FORMATION

The kidneys bear the major responsibility for maintaining a constant internal environment in the body. By managing the elimination of appropriate amounts of water, electrolytes, hydrogen ions, and nitrogen-containing wastes, the kidneys respond to changes in health, diet, and physical activity.

About 25% of the blood pumped from the heart goes directly to the kidneys, where the functional units are the *nephrons*. Each kidney contains over a million of them. Blood enters a nephron at a *glomerulus* (Figure 16.14), a tangle of capillaries surrounded by a fluid-filled space. **Filtration,** the first of three essential kidney functions, occurs here. The pressure of blood pumped into the glomerulus directly from the heart is high enough to push plasma and all its solutes except large proteins across the capillary membrane into the surrounding fluid, the **glomerular filtrate.** The filtrate flows from the capsule into the tubule that makes up the rest of the nephron, and the blood enters the network of capillaries intertwined with the tubule.

Since about 125 mL of filtrate per minute enters the kidneys, they produce a total of 180 L of filtrate per day. This filtrate contains not only waste products but also many solutes the body can't afford to lose, such as glucose and electrolytes. Since we excrete only about 1.4 L of urine each day, you can see that another important function of the kidneys is **reabsorption:** the recapture of water and essential solutes by moving them *out of* the tubule. Reabsorption alone, however, is not sufficient to provide the kind of control over urine composition that is needed. Suppose, for example, more of a particular solute

Filtration, kidney Filtration of blood plasma through a glomerulus and into a kidney nephron.

Glomerular filtrate Fluid that enters the nephron from the glomerulus; filtered blood plasma.

Reabsorption, kidney Movement of solutes out of filtrate in the kidney tubule.

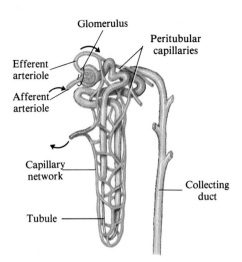

Figure 16.14
Structure of a nephron.

Secretion Movement of
solutes into filtrate in a
kidney tubule.

must be excreted than is present in the filtrate. This situation is dealt with by
secretion: the transfer of solutes *into* the kidney tubule.

Reabsorption and secretion require the transfer of solutes and water
among the filtrate, the interstitial fluid surrounding the tubule, and blood in the
capillaries. Solutes may cross the tubule and capillary membranes by passive
diffusion in response to concentration or ionic charge differences, or by active
transport. Water moves in response to differences in the osmolarity of the fluids
on the two sides of the membranes. Solute and water movement is also controlled by hormone-directed variations in the permeability of the tubule
membrane. The changes in filtrate composition during urine formation are
diagrammed in Figure 16.15.

16.8 URINE COMPOSITION AND FUNCTION

Urine contains the products of glomerular filtration, minus the substances
reabsorbed in the tubules, plus the substances secreted in the tubules. The
actual concentration of these substances in urine at any time is determined by
the amount of water being excreted, which can vary significantly with water
intake, exercise, temperature, and state of health.

About 50 g of solids in solution are excreted every day. The major
components of these solids are about 20 g of electrolytes and 30 g of nitrogen-
containing wastes, which include urea and ammonia from amino acid catabolism, creatinine from breakdown of creatine phosphate in muscles, and uric acid
from purine catabolism. Normal urine composition is usually reported as
amount excreted per day, and laboratory urinalysis often requires collection of
a 24-hr urine sample.

The following paragraphs briefly describe a few of the mechanisms that
control the composition of urine.

Fluid and Na$^+$ Balance   The higher the concentration of solutes in urine, the
less water is reabsorbed and the greater the volume of urine. The elevated
concentration of glucose in the urine of individuals with diabetes mellitus, for
example, is responsible for their increased urine volume.

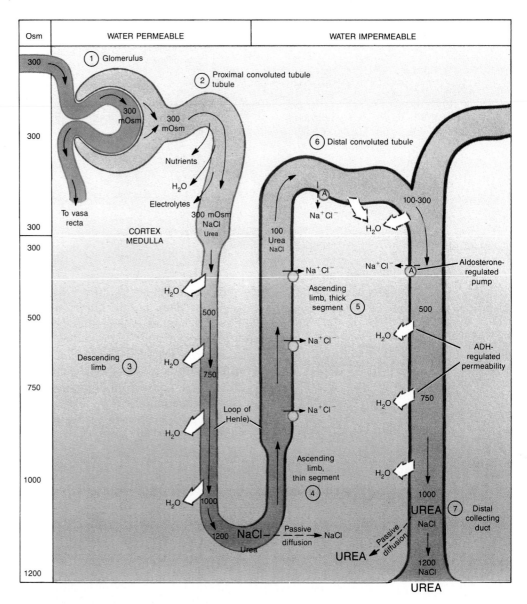

Figure 16.15

Formation of urine in the kidney. The numerical scale at the left shows the osmolarity of the outer (cortex) and inner (medulla) regions of the kidney. The high osmolarity of the medulla is maintained by removal of water via the capillary network (not shown). After filtration in the *glomerulus* (①), about 80% of the Na^+, 100% of the glucose, and significant amounts of other electrolytes and nutrients are reabsorbed by active transport in the *proximal convoluted tubule* (②). Next, in the *descending limb* of the loop of Henle (③), which is not permeable to Na^+ and carries out no active transport, water passively exits the tubule because the surrounding interstitial fluid has a higher osmolarity. Then, in the *ascending limb,* the situation reverses. The tubule here is not very permeable to water, but Na^+ departs by passive diffusion at the lower end (④) and active transport at the upper end (⑤). To maintain charge balance, Cl^- moves with Na^+. Finally, in the *distal convoluted tubule* (⑥) and *collecting duct,* the filtrate concentration receives its final adjustment.

The amount of water reabsorbed is in turn dependent on both the antidiuretic-hormone-controlled permeability of the collecting duct membrane and the amount of sodium actively reabsorbed. Increased sodium reabsorption means higher interstitial osmolarity, greater water reabsorption, and decreased urine volume. In the opposite condition of decreased sodium reabsorption, less water is reabsorbed and urine volume increases. Diuretic drugs such as furosemide (Endural or Lasix), which is used in treating hypertension and congestive heart failure, act by inhibiting the active transport of Na^+ out of the loop of Henle. Caffeine has a similar effect.

The reabsorption of Na^+ is normally under the control of the steroid hormone aldosterone. The arrival of chemical messengers signaling a decrease in total blood plasma volume accelerates the secretion of aldosterone. The result is increased Na^+ reabsorption in the kidney tubules accompanied by increased water reabsorption.

Acid–base Balance   Respiration, buffering of body fluids, and excretion of hydrogen ions in the urine combine to maintain acid–base balance. Metabolism normally produces an excess of hydrogen ions, 50–100 mEq of which must be excreted each day to prevent acidosis. In the process of eliminating H^+, the kidneys also regulate the HCO_3^- concentration of body fluids. Very little free hydrogen ion exists in blood plasma, and therefore very little enters the glomerular filtrate. Instead, the H^+ to be eliminated is produced from CO_2 in the cells lining the proximal and distal convoluted tubules.

$$CO_2 + H_2O \xrightarrow{\text{carbonic anhydrase}} H^+ + HCO_3^-$$

The HCO_3^- ions return to the bloodstream, and the H^+ ions enter the filtrate.

The urine must carry away the necessary quantity of H^+ without becoming excessively acid. To accomplish this, the H^+ is tied up by reaction with HPO_4^{2-} that is absorbed at the glomerulus or by reaction with NH_3 produced in the distal tubule cells by deamination of glutamate.

$$H^+ + HPO_4^{2-} \longrightarrow H_2PO_4^-$$
$$H^+ + NH_3 \longrightarrow NH_4^+$$

When acidosis occurs, the kidney responds by synthesizing more ammonia, thereby increasing the quantity of H^+ eliminated.

A further outcome of H^+ production in tubule cells is the net reabsorption of HCO_3^- that entered the filtrate at the glomerulus. The body cannot afford to lose HCO_3^-: The result would be production of additional acid from carbon dioxide by reaction with water. Instead, H^+ secreted into the tubules combines with HCO_3^- in the filtrate to produce CO_2 and water. Upon returning to the bloodstream, the CO_2 is reconverted to HCO_3^-.

In summary, acid–base reactions in the kidneys have the following results:

- Secreted H^+ is eliminated in the urine as NH_4^+ or $H_2PO_4^-$.
- Secreted H^+ combines with filtered HCO_3^-, producing CO_2 that returns to the bloodstream and again is converted to HCO_3^-.

INTERLUDE: AUTOMATED CLINICAL LABORATORY ANALYSIS

What happens when a physician orders chemical tests of blood, urine, or spinal fluid? The sample goes to a clinical chemistry laboratory, often in a hospital, where most tests are done by automated clinical chemistry analyzers. There are basically two types of analysis, one for the quantity of a chemical component and the other for the quantity of an enzyme with a specific metabolic activity.

Many chemical components are measured by mixing a reagent with the sample—the analyte—and determining the quantity of a colored product with a *photometer*, a unit that measures the absorption of light of a wavelength specific to the product. For each test specified, a portion of the sample is mixed with the appropriate reagent and the photometer is adjusted to the exact wavelength necessary.

When it's not possible to directly form a product that can be seen by a photometer, other types of reactions are used. Since most analytes are important in body fluids, they are usually substrates for enzyme-catalyzed reactions that can often be used in analysis. Glucose is determined in this manner by a pair of enzyme-catalyzed reactions: The glucose is converted to glucose 6-phosphate by the hexokinase-catalyzed reaction with ATP, the phosphate is then oxidized by $NADP^+$, and the quantity of NADPH produced is measured photometrically.

The second type of analysis, determination of the quantity of a specific enzyme or the ratio of two or more enzymes, is valuable in detecting organ damage that allows enzymes to leak into body fluids. For example, elevation of both ALT (alanine aminotransferase) and AST (aspartate aminotransferase) with an AST/ALT ratio greater than 1.0 is characteristic of liver disease. If, however, the AST is greatly elevated and the AST/ALT ratio is higher than 1.5, a myocardial infarction (heart attack) is likely to have occurred. When the analyte is an enzyme, advantage is often taken of its action on a substrate. ALT, for example, is determined by photometrically monitoring the disappearance of NADH in the following pair of coupled reactions (where LD = lactate dehydrogenase):

$$\text{L-Alanine} + \alpha\text{-ketoglutarate} \xrightarrow{\text{ALT}} \text{pyruvate} + \text{L-glutamate}$$

SUMMARY

Two-thirds of body fluid is **intracellular fluid** and one-third is **extracellular fluid,** mainly **blood plasma** and **interstitial fluid.** Daily water intake and output are kept in balance by hormonal control of water in urine and thirst. At capillaries throughout the body, nutrients and metabolism end products carried in blood are exchanged by passage across capillary walls, through interstitial fluid, and across cell membranes of peripheral tissue.

Blood functions fall into three major categories: transport, regulation, and defense. Oxygen and carbon dioxide are transported in **erythrocytes** by hemoglobin, with the amounts of bound gases under allosteric control such that more oxygen is released to tissues where metabolism is active.

Plasma proteins and white blood cells are responsible for defense mechanisms. Infection or cell damage initiates **inflammation,** a nonspecific response that includes release of histamine, increased blood flow, and phagocytosis of antigens by white blood cells. In the antibody-directed and cell-directed **immune responses,** specific antigens are recognized by noncovalent interactions between protein amino acid sequences like those in enzyme-substrate interaction. The antibody-directed immune response occurs when B lymphocytes detect an antigen and divide into plasma cells that produce **immunoglobulins,** glycoprotein **antibodies** that mark that antigen for destruction. B cell division also yields memory cells that produce antibody if their antigen reappears. The cell-directed immune response occurs when T lymphocytes detect antigenic regions on the surface of infected body cells and release a toxic protein that perforates the cell membrane and destroys the infected cells.

Hemostasis, the halt of blood loss, begins with formation of a platelet plug at the site of tissue damage. **Blood clotting,** a process requiring stepwise activation of a series of 12 clotting factors, follows. The end result is removal

$$\text{Pyruvate} + \text{NADH/H}^+ \xrightarrow{\text{LD}} \text{lactate} + \text{NAD}^+$$

Automated analyzers rely on premixed reagents and automatic division of a fluid sample into small portions for each test. A low-volume analyzer that can provide rapid results for a few tests accepts a bar-coded serum or plasma sample cartridge (about the size of a small cassette tape cartridge) followed by bar-coded reagent cartridges. The instrument software reads the bar codes and directs an automatic pipettor to transfer the appropriate volume of sample to each test cartridge. The instrument then moves the test cartridge along as the sample and reagents are mixed, the reaction takes place for a measured amount of time, and the photometer reading is taken and converted to the test result.

A high-volume analyzer with more-complex software can randomly access 40 or more tests, can do STAT tests in less than 5 minutes, and can run over 400 tests per hour at a cost of less than 10 cents per test. The end result is a printed report on each sample listing the types of tests, the sample values, and a normal range for each test.

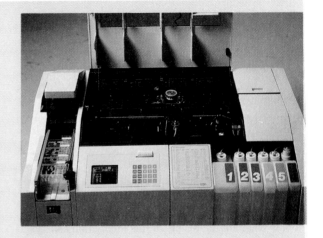

The small automated analyzer shown is used mainly for specialized tests such as those for cardiac enzymes, drugs of abuse, and therapeutic drugs. Eighty-two different analyses are available and up to 10 can be chosen for a single blood or urine sample and completed in about 30 minutes. The sample cup and a reagent pack for each test are loaded at the left. All results are read by a photometer and printed out on a slip that emerges just above the push-button control panel. The numbered bottles at the right contain water automatically dispensed to dissolve and dilute reagents.

by thrombin of solubilizing polypeptides from fibrinogen, followed by clot formation by cross-linking of the resulting **fibrin** fibers.

The kidney maintains homeostasis by controlled elimination of water, electrolytes, hydrogen ions, and nitrogen-containing wastes. The three essential kidney functions are **filtration** of all but large proteins from blood, **reabsorption** of water and essential solutes, and **secretion** of solutes to be eliminated. These functions are controlled by passive diffusion between regions of different **osmolarity, active transport,** and hormone-directed variation in the permeability of the kidney tubule membrane. The water and Na$^+$ content of urine are interrelated and controlled by antidiuretic hormone and aldosterone. Acid–base balance is maintained by elimination of H$^+$ as NH$_4^+$ or H$_2$PO$_4^-$ and net reabsorption of HCO$_3^-$.

REVIEW PROBLEMS

Body Fluids

16.1 What are the three principal body fluids?

16.2* Which body fluid contains the majority of the body's water?

16.3 What characteristics are needed for a substance to be soluble in body fluids?

16.4* How does blood pressure compare to interstitial fluid pressure in arterial capillaries? In venous capillaries? What effects do these presure diferences have on solute crossing cell membranes?

16.5 What is the purpose of the lymphatic system?

16.6* What is another name for antidiuretic hormone, and what is the purpose of this hormone?

16.7 What is plasma?

16.8* What are the three main types of blood cells?

16.9 What is the major function of each type of blood cell?

16.10* Heparin used for commercial purposes is commonly isolated from beef lung or pork intestines. Why do these tissues contain this substance?

16.11 What is blood serum?

16.12* How many O_2 molecules can be bonded by a hemoglobin molecule?

16.13 What must be the oxidation state of the iron in hemoglobin for it to perform its function?

16.14* How do deoxyhemoglobin and oxyhemoglobin differ in color?

16.15 How does the degree of saturation of hemoglobin vary with the partial pressure of O_2 in the tissues?

16.16* Oxygen has an allosteric interaction with hemoglobin. What are the results of this interaction as oxygen is bonded and as it is released?

16.17 What are the three ways of transporting CO_2 in the body?

16.18* Use Figure 16.7 to estimate the partial pressure at which hemoglobin is 50% saturated with oxygen.

16.19 Explain two ways in which the CO_2 released by oxidation of glucose tends to raise the H^+ concentration of the blood?

16.20* Does each of the following effects cause hemoglobin to release more O_2 to the tissues or to absorb more O_2?

(a) raising the temperature

(b) production of CO_2

(c) increase of the H^+ concentration

16.21 What are the three types of responses the body makes upon exposure to an antigen?

16.22* Antihistamines are often prescribed to counteract the effects of allergies. Propose one way in which these drugs might work to stop production of histamine from histidine.

16.23 How are specific immune responses similar to the enzyme–substrate interaction?

16.24* What types of cells are associated with antibody-directed immune response and how do they work?

16.25 What are memory cells?

16.26* T cells are often discussed in conjunction with the disease AIDS, in which a virus destroys these cells. How do T cells work to combat disease?

16.27 What is a blood clot?

16.28* What vitamin and what mineral are specifically associated with the clotting process?

16.29 What two pathways trigger blood clotting?

16.30* Why do you suppose that many of the enzymes involved in blood clotting are secreted by the body as zymogens?

16.31 Kidneys are often referred to as filters that purify the blood. What other two essential functions do the kidneys perform to help maintain homeostasis?

16.32* Write the reactions by which HPO_4^{2-} and HCO_3^- absorb excess H^+ from the urine before elimination.

Applications

16.33 What is cryopreservation? [App: Organ Cryopreservation]

16.34* Why do cells rupture as the water in them freezes? [App: Organ Cryopreservation]

16.35 What two steps are taken in cryopreservation to ensure that small, rather than large, ice crystals form and that the crystals do not merge to make larger crystals upon standing? [App: Organ Cryopreservation]

16.36* What problems are associated with the need to maintain metabolism while tissues are frozen? [App: Organ Cryopreservation]

16.37 How do endothelial cells in brain capillaries differ from those in other capillary systems? [App: Blood–Brain Barrier]

16.38* What is meant by an asymmetric transport system? Give one specific example of such a system. [App: Blood–Brain Barrier]

16.39 What type of substance is likely to breach the blood–brain barrier? Would ethanol be likely to cross this barrier? Why or why not? [App: Blood–Brain Barrier]

16.40* How are photometers used in automated analysis? [Int: Automated Clinical Chemistry Analysis]

16.41 Why is it useful to test for enzyme levels in body fluids? [Int: Automated Clinical Chemistry Analysis]

16.42 What are some advantages of using automated analyzers compared to using technicians? [Int: Automated Clinical Chemistry Analysis]

Additional Questions

16.43 Why is ethanol soluble in blood?

16.44* Nursing mothers are said to impart immunity to their infants. Why do you think this is so?

16.45 Many people find they retain water, with fingers and ankles swelling after eating salty food. Explain this in terms of the method of operation of the kidneys.

16.46* How does active transport differ from osmosis?

16.47 When is active transport necessary to move substances through cell walls?

16.48* Discuss the importance of the CO_2/HCO_3^- equilibrium in blood and in urine.

16.49 Name the three principal functions of blood and give an example of each type of function.

Chemical Bonds

Clearly, there must be some force that holds atoms together in chemical compounds. Otherwise, the atoms would simply fly apart, and no compounds could exist. The forces that hold atoms together are called **chemical bonds** and are of two types—*ionic bonds* and *covalent bonds*. As a general rule, ionic bonds occur between a metal and a nonmetal, whereas covalent bonds occur between two nonmetallic elements.

A.1   TYPES OF CHEMICAL BONDS

Ionic Bonds

Sodium is a soft, silvery metal that reacts violently with water, and chlorine is a corrosive, poisonous green gas. When atoms of the two elements react with one another, they produce our familiar table salt, sodium chloride. The salt is formed when a sodium atom transfers its valence $3s$ electron to a chlorine atom to form a positively charged sodium cation (Na^+) and a negatively charged chloride anion (Cl^-). Since opposite electrical charges attract each other, the positive sodium ion and negative chloride ion are held together by an **ionic bond.** Indicating the valence electrons by dots, the reaction can be written in the following way:

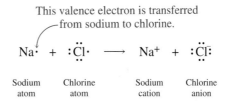

When a vast number of sodium atoms transfer electrons to an equally vast number of chlorine atoms, the result is a visible crystal of sodium chloride in which equal numbers of sodium and chloride ions are packed together in a regular arrangement. Each positively charged sodium ion is surrounded by six negatively charged chloride ions, and each chloride ion is surrounded by six sodium ions (Figure A.1). This packing arrangement allows each ion to be stabilized by the attraction of unlike charges on its six nearest-neighbor ions, while being as far as possible from like-charged ions.

Covalent Bonds: Molecules

A **covalent bond** results, not from the complete transfer of an electron between atoms, but from the *sharing* of electrons (usually two) between atoms. The

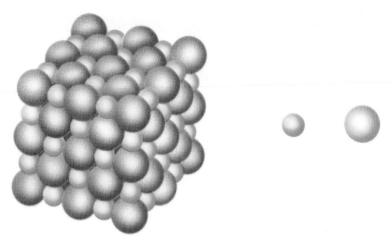

Figure A.1
The arrangement of Na$^+$ ions and Cl$^-$ ions in a crystal of sodium chloride. Each Na$^+$ ion is surrounded by six Cl$^-$ ions, and each Cl$^-$ ion is surrounded by six Na$^+$ ions. The crystal is held together by ionic bonds—the electrical attraction between oppositely charged ions.

unit of matter that results when atoms are joined by covalent bonds is called a **molecule.** A water molecule (H_2O) results when two hydrogen atoms each share two electrons with a single oxygen atom, an ammonia molecule (NH_3) results when three hydrogen atoms each share two electrons with a nitrogen atom, a methane molecule (CH_4) results when four hydrogen atoms each share two electrons with a carbon atom, and so on. To visualize these and other molecules, it helps to imagine the individual atoms as spheres that stick together to give a discrete molecular unit with a specific three-dimensional shape (Figure A.2).

Even some *elements* exist as molecules rather than as atoms. Thus, hydrogen, oxygen, nitrogen, fluorine, chlorine, and bromine all exist as diatomic (two-atom) molecules held together by covalent bonds. We therefore have to write them as such when using any of these elements in a chemical equation.

H_2	O_2	N_2	F_2	Cl_2	Br_2
Hydrogen	Oxygen	Nitrogen	Fluorine	Chlorine	Bromine

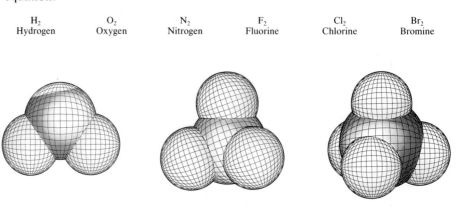

Figure A.2
Visualizing some simple molecules. The individual atoms, represented by spheres, are joined together in specific arrangements and with specific geometries by sharing electrons in a covalent bond.

A.2 HOW COVALENT BONDS FORM

The gaseous element hydrogen provides the simplest example of how a covalent bond forms. A hydrogen atom has one valence electron in a $1s$ orbital. When two hydrogen atoms approach each other, the electrons pair up in the region between the two nuclei and hold the atoms together by a covalent bond. In effect, the negative charges of the electrons act as a kind of "glue" to hold the positively charged hydrogen nuclei together.

Shared electron pair

$$H\cdot \ + \ \cdot H \ \longrightarrow \ H\!:\!H$$

Two hydrogen atoms   A hydrogen molecule

As a further example of covalent bonding, look at the chlorine molecule, Cl_2. Each individual chlorine atom has seven valence electrons in the $3s^2\,3p^5$ electronic configuration. Two of the three $3p$ orbitals are filled by two electrons each, while the third $3p$ orbital holds only one electron. When two chlorine atoms approach each other, the unpaired $3p$ electrons from each are shared by both atoms in a covalent bond. Such bond formation can be pictured as the head-on overlap of the p orbitals containing the single electrons, with the resulting formation of a region of high electron density between the atoms.

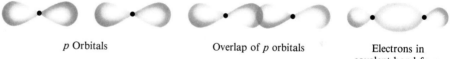

p Orbitals   Overlap of p orbitals   Electrons in covalent bond form region of high electron density between atoms

A.3 COVALENT BONDS AND THE OCTET RULE

As a general rule, an atom will share as many of its valence-shell electrons as possible, either until it has no more to share or until it obtains eight valence electrons (the **octet rule**) and thereby reaches the electron configuration of a noble gas. For the second-row elements commonly found in organic molecules, the following guidelines apply:

● Group 3A elements such as boron have three valence electrons and form three electron-pair bonds, as in borane, BH_3.

Only 6 electrons around boron

$$\cdot \dot{B}\cdot \ + \ 3\ H\cdot \ \longrightarrow \ H\!:\!\overset{\displaystyle H}{\underset{\displaystyle \cdot\cdot}{B}}\!:\!H \qquad \text{Borane}$$

● Group 4A elements such as carbon have four valence electrons and form four bonds as in methane, CH_4.

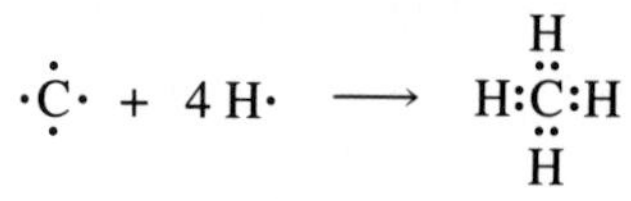

$$\cdot\overset{\cdot}{\underset{\cdot}{C}}\cdot \ + \ 4\ H\cdot \ \longrightarrow \ H\!:\!\overset{H}{\underset{H}{\overset{\cdot\cdot}{C}}}\!:\!H \qquad \text{Methane}$$

● Group 5A elements such as nitrogen have five valence electrons and form three bonds as in ammonia, NH_3.

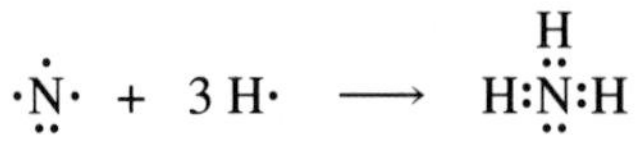

$$\cdot\overset{\cdot}{\underset{\cdot\cdot}{N}}\cdot \ + \ 3\ H\cdot \ \longrightarrow \ H\!:\!\overset{H}{\underset{\cdot\cdot}{N}}\!:\!H \qquad \text{Ammonia}$$

● Group 6A elements such as oxygen have six valence electrons and form two bonds as in water, H_2O.

$$\cdot\overset{\cdot\cdot}{\underset{\cdot}{O}}\cdot \ + \ 2\ H\cdot \ \longrightarrow \ H\!:\!\overset{\cdot\cdot}{\underset{\cdot\cdot}{O}}\!:\!H \qquad \text{Water}$$

● Group 7A elements (halogens) such as fluorine have seven valence electrons and form one bond, as in hydrogen fluoride, HF.

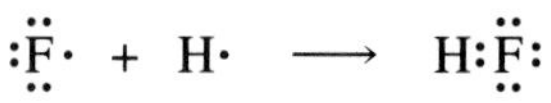

$$:\!\overset{\cdot\cdot}{\underset{\cdot\cdot}{F}}\!\cdot \ + \ H\cdot \ \longrightarrow \ H\!:\!\overset{\cdot\cdot}{\underset{\cdot\cdot}{F}}\!: \qquad \text{Hydrogen fluoride}$$

● Group 8A elements (noble-gases) such as neon rarely form covalent bonds because they already have valence-shell octets.

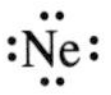

$$:\!\overset{\cdot\cdot}{\underset{\cdot\cdot}{Ne}}\!: \quad \text{(Does not form covalent bonds)}$$

These conclusions are summarized in Table A.1.

Table A.1   Covalent Bonding for Second-Row Elements

Group	Number of Bonds	Example
3A	3	BH_3
4A	4	CH_4
5A	3	NH_3
6A	2	H_2O
7A	1	HF
8A	0	—

A.4   MULTIPLE COVALENT BONDS

Not all molecules contain only single bonds like those in methane, ammonia, and water. In molecules such as O_2, N_2, and many others, the atoms share more than one pair of electrons, leading to the formation of *multiple* covalent bonds. The only way, for instance, that both oxygen atoms in the O_2 molecule can have valence-shell octets is if they share four electrons (two pairs), giving a **double bond.** Similarly, the nitrogen atoms in the N_2 molecule share six electrons (three pairs), giving a **triple bond.**

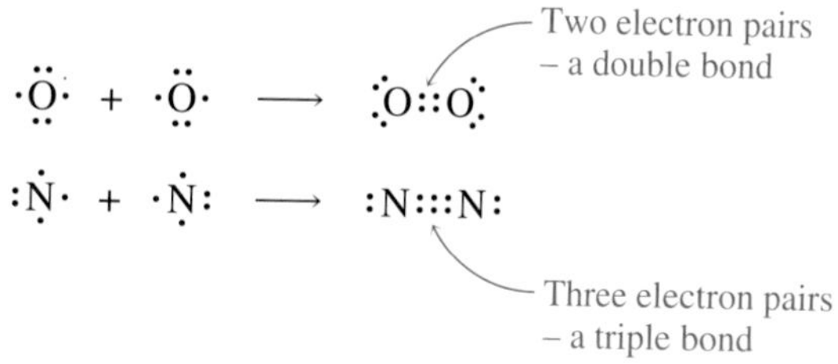

$$\cdot \overset{..}{O} \cdot \; + \; \cdot \overset{..}{O} \cdot \; \longrightarrow \; \overset{..}{:} O :: O \overset{..}{:}$$

Two electron pairs
– a double bond

$$: \overset{.}{N} \cdot \; + \; \cdot \overset{.}{N} : \; \longrightarrow \; : N ::: N :$$

Three electron pairs
– a triple bond

Multiple covalent bonding is particularly common for the second-row elements in organic molecules. For example, ethylene, a simple compound used commercially as a plant hormone to induce ripening in fruit, has the formula C_2H_4. The only way that all six atoms can have octets is for the two carbon atoms to share four electrons in a carbon–carbon double bond:

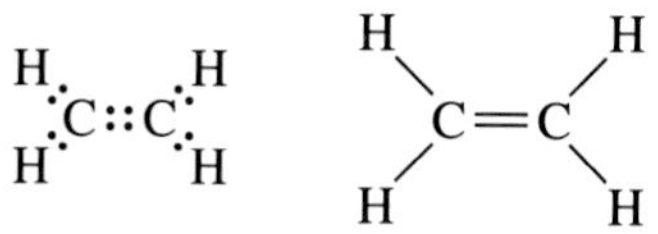

Ethylene—the carbon atoms share four electrons in a double bond.

Acetylene, the gas used in welding, has the formula C_2H_2. Thus, the two acetylene carbon atoms must share six electrons in a carbon–carbon triple bond:

$$H : C ::: C : H \qquad H - C \equiv C - H$$

Acetylene—the carbon atoms share six electrons in a triple bond.

Note that in compounds with multiple bonds like ethylene and acetylene, each carbon atom still forms a total of four covalent bonds.

Carbon, nitrogen, and oxygen are the elements most often present in multiple bonds. Carbon and nitrogen form both double and triple bonds; oxygen forms double bonds, frequently with carbon but also with phosphorus and sulfur.

Some common kinds of multiple covalent bonds

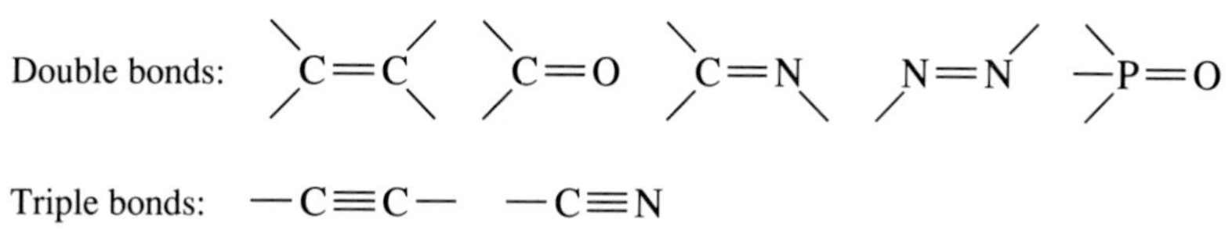

Double bonds: $\overset{\diagdown}{\underset{\diagup}{C}} = \overset{\diagup}{\underset{\diagdown}{C}} \quad \overset{\diagdown}{\underset{\diagup}{C}} = O \quad \overset{\diagdown}{\underset{\diagup}{C}} = N \overset{\diagdown}{\underset{\diagup}{}} \quad \overset{\diagdown}{\underset{\diagup}{N}} = N \overset{\diagup}{} \quad - \overset{}{\underset{\diagup}{P}} = O$

Triple bonds: $- C \equiv C - \quad - C \equiv N$

A.5   STRUCTURAL FORMULAS

Formulas such as H_2O and CH_4, which show the numbers and kinds of atoms in one molecule, are called *molecular formulas*. More useful, though, are structural formulas, which show how the atoms in a molecule are connected. For example, water, ammonia, and methane are represented as follows:

Lewis structures

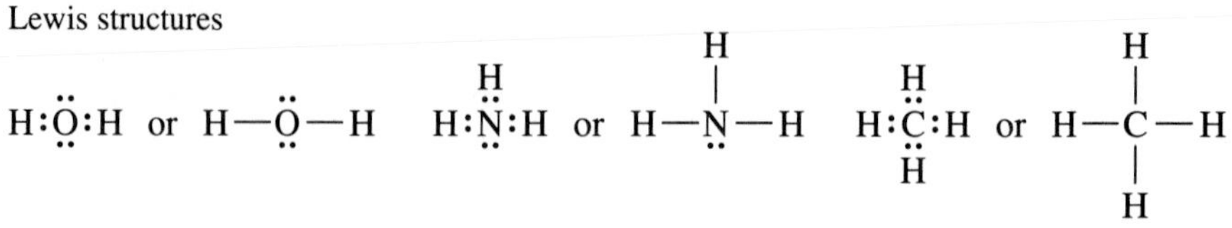

$$H\!:\!\ddot{O}\!:\!H \quad \text{or} \quad H\!-\!\ddot{O}\!-\!H \qquad H\!:\!\ddot{N}\!:\!H \quad \text{or} \quad H\!-\!\underset{|}{\overset{H}{N}}\!-\!H \qquad H\!:\!\overset{H}{\underset{H}{C}}\!:\!H \quad \text{or} \quad H\!-\!\underset{|}{\overset{|}{\underset{H}{C}}}\!-\!H$$

Structural formulas that represent valence electrons by dots are called **Lewis structures,** after G. N. Lewis of the University of California, who devised them. More commonly, though, pairs of electrons in covalent bonds are indicated by a line drawn between atoms, and only those pairs of electrons not used for bonding are shown as dots. In a water molecule, for example, the oxygen atom is surrounded by eight electrons, four of them represented by two lines to hydrogen atoms (two covalent O—H bonds) and four others that are not used in bonding and are shown as dots. Such nonbonding pairs of electrons are known as **lone pairs.**

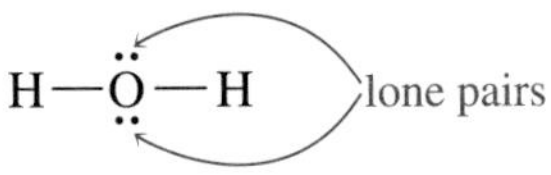

$$H\!-\!\ddot{O}\!-\!H \quad \rangle \text{lone pairs}$$

A.6 DRAWING STRUCTURAL FORMULAS OF ORGANIC MOLECULES

To draw a structural formula, the arrangement of atoms must be known. Sometimes the arrangement is obvious: Water, for example, can only be H—O—H because only oxygen can be in the middle and form two covalent bonds. Usually, however, the use of complex analytical instruments is needed to gather the necessary structural information.

Once the arrangement of atoms is known, structural formulas of organic molecules can be generated by knowing some common bonding patterns:

- Carbon atoms are often bonded to each other.
- Carbon atoms form four covalent bonds (and have no lone pairs in neutral molecules).
- Nitrogen atoms form three covalent bonds and have one lone pair.
- Oxygen atoms form two covalent bonds and have two lone pairs.
- Halogen atoms (X = F, Cl, Br, I) usually form one covalent bond and have three lone pairs.
- Hydrogen atoms form one covalent bond.

Knowing these common bonding patterns simplifies the drawing of structural formulas for organic molecules. In ethane, C_2H_6, for example, three of the four covalent bonds of each carbon atom are used in bonds to hydrogen, and the fourth is a carbon–carbon bond. There is no other arrangement in which all eight atoms can have their usual bonding patterns.

$$6 \; H\cdot \atop 2 \; \cdot\ddot{C}\cdot \Bigg\} \Longrightarrow$$

$$\begin{array}{cc} H & H \\ H\!:\!\!\overset{\displaystyle \cdot\cdot}{\underset{\displaystyle \cdot\cdot}{C}}\!:\!\overset{\displaystyle \cdot\cdot}{\underset{\displaystyle \cdot\cdot}{C}}\!:\!H \\ H & H \end{array} \quad \text{or} \quad H\!-\!\overset{\displaystyle H}{\underset{\displaystyle H}{C}}\!-\!\overset{\displaystyle H}{\underset{\displaystyle H}{C}}\!-\!H$$

Ethane, C_2H_6

Even structural formulas soon become awkward for most organic molecules, and **condensed structures** are often used. In its condensed form, ethane is written as CH_3CH_3, meaning that each carbon atom has three hydrogen atoms bonded to it (CH_3) and that the two CH_3 units are bonded to each other.

A.7 THE SHAPES OF MOLECULES

What determines the overall shape of a molecule? Why, for example, are the three atoms in water connected at an angle of 104.5° rather than in a straight line? Like so many properties, molecular shapes are related to the numbers and locations of the valence electrons in covalent bonds and lone pairs.

Predicting molecular shape by analyzing how bonds and electron pairs surround individual atoms is done by applying what is called the **valence-shell electron-pair repulsion (VSEPR) model.** Pairs of valence electrons in bonds and in lone pairs occupy negatively charged electron clouds that stay as far apart as possible, causing molecules to assume specific shapes. Molecular shapes can be predicted in the following way:

● *Identify the atom whose bonding geometry is of interest.*

● *Count the number of electron-charge clouds surrounding the atom of interest in the Lewis structure.* The number of charge clouds is simply the total number of bonds and lone pairs. Multiple bonds count the same as single bonds, because we're interested only in the *number* of charge clouds, not in how many electrons there are in each cloud.

● *Determine the geometry around the atom according to the number of electron-charge clouds as summarized in Table A.2.* An atom with two charge clouds has linear geometry, an atom with three charge clouds has planar triangular geometry, and an atom with four charge clouds has tetrahedral geometry.

As indicated in Table A.2, an atom with only two charge clouds, such as the carbon atom in CO_2, has the clouds farthest apart when they point in opposite directions. Thus, CO_2 is a linear molecule with bond angles of 180°.

Carbon dioxide, CO_2 $\ddot{O}\!=\!C\!=\!\ddot{O}$ —180°—

When an atom is surrounded by three charge clouds, as occurs on the carbon atom in formaldehyde, CH_2O, the clouds can be farthest apart if they lie in a plane and point to the corners of an equilateral triangle. Thus, a formaldehyde molecule is planar triangular, with H—C—O and H—C—H bond angles of 120°. For similar reasons, an ozone molecule (O_3) is bent, with

Table A.2 Geometry Around Atoms With Two, Three, and Four Charge Clouds

Number of Bonds	Number of Lone Pairs	Number of Charge Clouds on Atom	Shape of Molecule	Example
2	0	2	 Linear	$O{=}C{=}O$
3	0	3	 Planar triangular	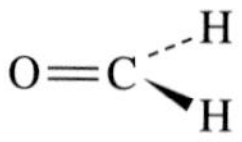 $O{=}C\begin{smallmatrix}H\\H\end{smallmatrix}$
2	1		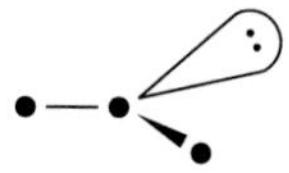 Bent	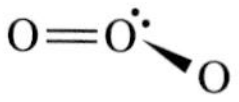 $O{=}\ddot{O}{-}O$
4	0	4	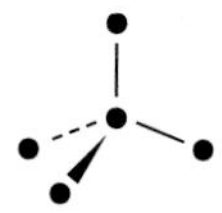 Tetrahedral	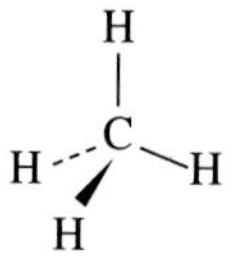 $H{-}C(H)(H){-}H$
3	1		 Pyramidal	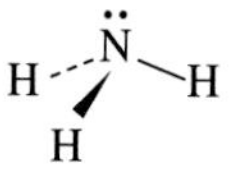 $H{-}\ddot{N}(H){-}H$
2	2		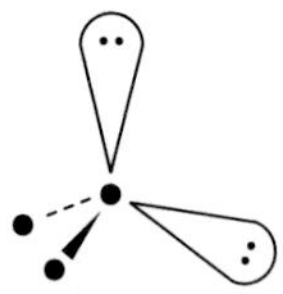 Bent	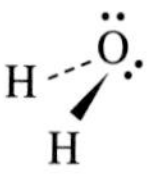 $H{-}\ddot{\ddot{O}}{-}H$

an O—O—O angle of approximately 120°. The central oxygen atom in ozone has three charge clouds, one of which is a lone pair:

Formaldehyde, CH₂O   O=C⟨H H

Side view

Top view

120° 120° 120° 120°

Ozone, O₃   O=C⟨H

Side view

Top view

120°

When there are four charge clouds, as occurs on the central atom in CH_4, NH_3, and H_2O, the clouds can be farthest apart if they extend to the corners of a regular tetrahedron. As illustrated in Figure A.3, a regular tetrahedron is a geometric figure whose four identical faces are equilateral triangles. The atom of interest lies at the center of the tetrahedron, the charge clouds point to the four corners, and the angle between two lines drawn from the center to any two corners is 109.5°.

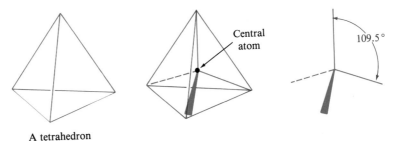

A tetrahedron

Central atom

109.5°

Figure A.3
Tetrahedral molecular geometry. An atom is located at the center of the tetrahedron, and the four electron charge clouds are oriented toward the corners. The angle between any two charges clouds is 109.5°. Water, ammonia, methane, and many other molecules have geometries based on the tetrahedron.

Because valence-shell octets are so common, a great many molecules have geometries based on the tetrahedron. In methane (CH_4), for example, the carbon atom has tetrahedral geometry with H—C—H bond angles of exactly 109.5°. In ammonia (NH_3), the nitrogen atom has a tetrahedral arrangement of its four charge clouds, but one point of the tetrahedron is occupied by a lone pair, resulting in an overall pyramidal shape for the molecule. Similarly, water has two points of the tetrahedron occupied by lone pairs and has an overall bent shape.

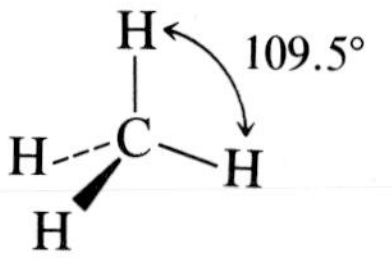

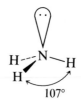

A methane molecule is tetrahedral, with bond angles of 109.5°.

An ammonia molecule is pyramidal, with bond angles of 107°.

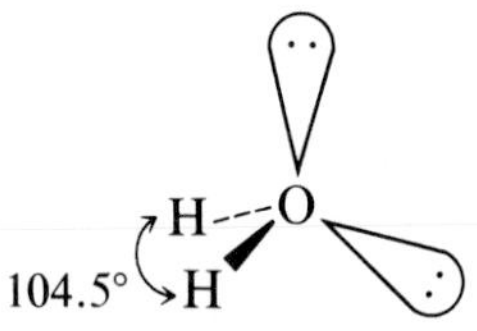

A water molecule is angular, with a bond angle of 104.5°.

Note that the H—N—H bond angle in ammonia (107°) and the H—O—H bond angle in water (104.5°) are close to, but not exactly at, the 109.5° tetrahedral value. The angles are diminished somewhat from their ideal values because the lone-pair charge clouds spread out and compress the rest of the molecule.

The geometry around individual atoms in larger molecules also derives from the geometries shown in Table A.2. For example, each of the two carbon atoms in ethylene ($H_2C=CH_2$) is attached to three other atoms, giving rise to planar triangular geometry. The molecule as a whole is planar, with H—C—C and H—C—H bond angles of approximately 120°.

The ethylene molecule is planar, with bond angles of 120°

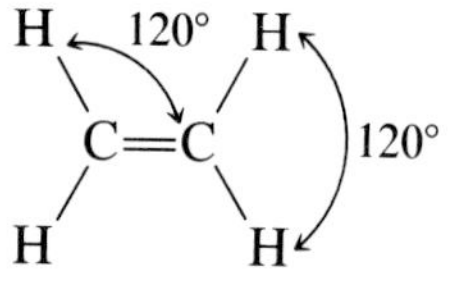

Top view

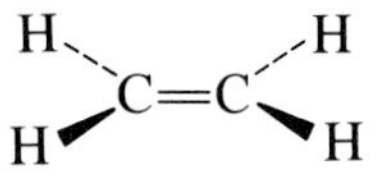

Side view

Carbon atoms bonded to four other atoms are each at the center of a tetrahedron, as shown here for ethane, $H_3C—CH_3$:

The ethane molecule has tetrahedral carbon atoms, with bond angles of 109.5°

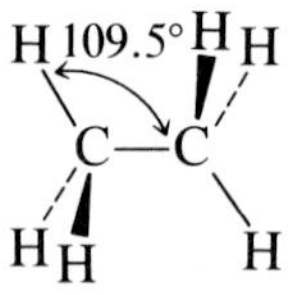

A.8 ELECTRONEGATIVITY

Electrons in a covalent bond occupy the space between the bonded nuclei. If the atoms are identical, as in H_2 and Cl_2, the electrons are equally attracted to both and are shared equally. If the atoms are not identical, though, the electrons are usually not equally attracted to both and are not shared equally.

Smaller atoms with higher nuclear charges, such as fluorine, oxygen, chlorine, and nitrogen attract electrons strongly and are said to be highly *electronegative*. Fluorine is the most highly electronegative of all elements, followed by oxygen, nitrogen, and chlorine. In a covalent bond such as that in hydrogen fluoride (HF), electrons spend more time at the fluorine end of the bond than at the hydrogen end, creating a **polar covalent bond** that has some ionic character to it. Although the molecule as a whole is neutral, the fluorine end is more negative than the hydrogen end. The unequal charge distribution is represented by placing a δ^- (Greek lowercase delta) at the more negative end of the bond and a δ^+ at the more positive end of the bond to indicate *partial* charges. The direction of bond polarity is often represented by an arrow pointing toward where the electrons are most strongly attracted. The arrow is pointed at the negative end and crossed at the positive end.

$$\overset{\delta^+}{H}\!-\!\overset{\delta^-}{F}$$

←———— Direction of electron movement

The ability of an atom in a covalent bond to attract electrons is called the atom's **electronegativity.** On the numerical electronegativity scale shown in Figure A.4, fluorine is assigned a value of 4 and less electronegative atoms are assigned lower values. Electronegativity increases as atoms get smaller on going from left to right across the periodic table, and decreases as atoms get larger on going down a group in the table. Metals in general have lower electronegativities than nonmetals.

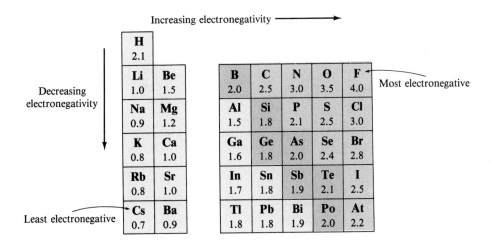

Increasing electronegativity ————→

H	
2.1	

Decreasing electronegativity ↓

Li	**Be**
1.0	1.5
Na	**Mg**
0.9	1.2
K	**Ca**
0.8	1.0
Rb	**Sr**
0.8	1.0
Cs	**Ba**
0.7	0.9

B	**C**	**N**	**O**	**F**
2.0	2.5	3.0	3.5	4.0
Al	**Si**	**P**	**S**	**Cl**
1.5	1.8	2.1	2.5	3.0
Ga	**Ge**	**As**	**Se**	**Br**
1.6	1.8	2.0	2.4	2.8
In	**Sn**	**Sb**	**Te**	**I**
1.7	1.8	1.9	2.1	2.5
Tl	**Pb**	**Bi**	**Po**	**At**
1.8	1.8	1.9	2.0	2.2

Most electronegative

Least electronegative

Comparing the electronegativity values of two bonded atoms makes it possible to predict the polarity of a bond. Oxygen (electronegativity 3.5), for example, is more electronegative than nitrogen (3.0), and both are more electronegative than carbon (2.5). Both carbon–oxygen and carbon–nitrogen bonds are therefore polar, with the positive ends at the carbon atoms. The larger difference in electronegativity values shows that the carbon–oxygen bond is the more polar of the two.

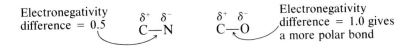

Electronegativity difference = 0.5 $\overset{\delta^+}{C}\!-\!\overset{\delta^-}{N}$ $\overset{\delta^+}{C}\!-\!\overset{\delta^-}{O}$ Electronegativity difference = 1.0 gives a more polar bond

A.9 POLAR MOLECULES

Just as individual bonds can be polar, entire molecules can also be polar if electrons are attracted more strongly to one part of the molecule than to another. Overall molecular polarity depends both on the nature of the bonds in the molecule and on the shape of the molecule. The chloromethane (CH_3Cl) molecule is polar, for example, because it contains a strongly polar C—Cl bond whose polarity is not canceled by the less polar C—H bonds.

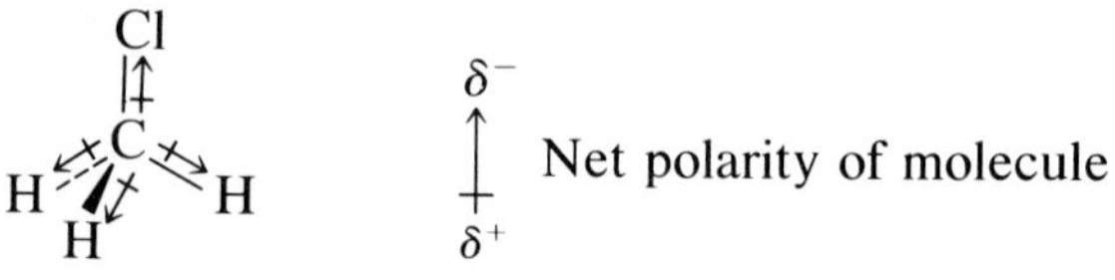

$$CH_3Cl$$
polar molecule

δ^-
$\downarrow$ Net polarity of molecule
δ^+

By contrast with chloromethane, neither carbon dioxide (CO_2) nor tetrachloromethane (CCl_4) molecules are polar, despite their polar C—O and C—Cl bonds. The overall symmetry of both molecules is such that the individual bond polarities cancel out.

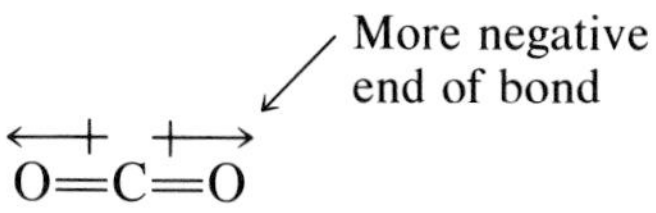

$$O=C=O$$

More negative end of bond

CO_2, nonpolar because individual bond polarities cancel

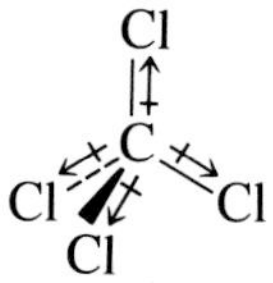

CCl_4, nonpolar because individual bond polarities cancel

A.10 HYDROGEN BONDS

Just as opposite electrical charges cause *ions* to be attracted to one another (an ionic bond), opposite partial charges can cause polar *molecules* to be attracted to one another. The most important example of such an intermolecular attraction is the *hydrogen bond,* an interaction that is responsible for many of the unique properties of water and for holding huge biomolecules in the shapes needed to play their essential roles in biochemistry.

When a hydrogen atom is covalently bonded to a strongly electronegative atom such as O, N, or F, electron density is pulled away from the hydrogen atom toward the electronegative O, N, or F. The hydrogen is thus left with a partial positive charge, while the electronegative O, N, or F atom is left with a partial negative charge. A **hydrogen bond** results when the positive hydrogen atom on one molecule is attracted to an unshared electron pair on the electronegative atom of a neighboring molecule. In water, for example, the hydrogen atom of one molecule is attracted to the oxygen atom of another molecule:

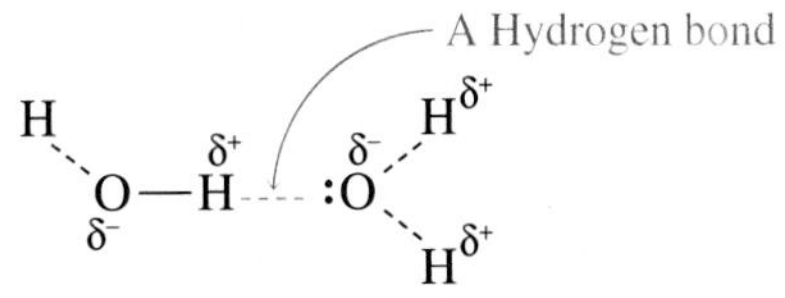

A Hydrogen bond

$$H \diagdown \underset{\delta^-}{O} - \underset{\delta^+}{H} - - - : \underset{}{\overset{\delta^-}{O}} \diagup \overset{H^{\delta^+}}{\diagdown} \underset{H^{\delta^+}}{}$$

Hydrogen bonding can occur in any molecule that has an O—H or an N—H bond. There are many such examples throughout organic and biological chemistry.

Acids and Bases

B.1   ACID DEFINITION

The classical definition of an acid is as a substance that gives hydrogen ions (H^+) when dissolved in water solution. According to a broader definition proposed in 1923 by Johannes Brønsted and Thomas Lowry, however, an acid is a substance that is able to give away H^+ to another molecule or ion, whether or not water is present. Since a hydrogen *atom* consists of a proton and an electron, a hydrogen *ion*, H^+, is simply a proton. Thus, we refer to acids as *proton donors*.

Different acids can have different numbers of hydrogen atoms available for giving away. Acids with one proton to donate, such as HCl and HNO_3, are **monoprotic acids**; H_2SO_4 is a **diprotic acid** since it has two protons; and H_3PO_4 is a **triprotic acid** since it has three protons.

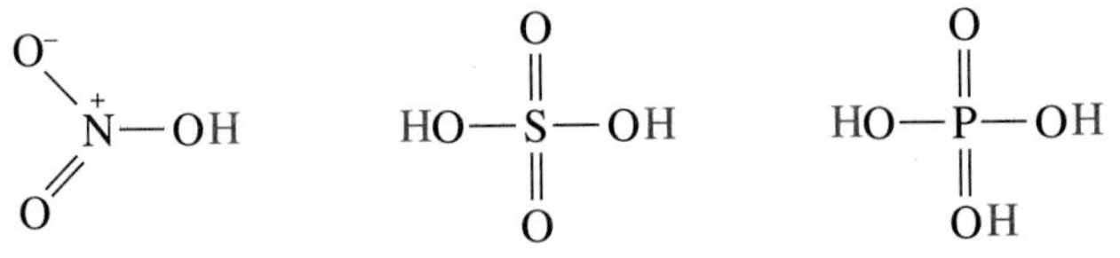

| Nitric acid (monoprotic) | Sulfuric acid (diprotic) | Phosphoric acid (triprotic) |

Acetic acid (CH_3COOH), an example of an organic acid, actually has four hydrogens, but only the one attached to oxygen is acidic. The three hydrogens bonded to carbon are not given up easily. Most organic acids are similar: They contain many hydrogen atoms, but only the one in the —COOH group is acidic.

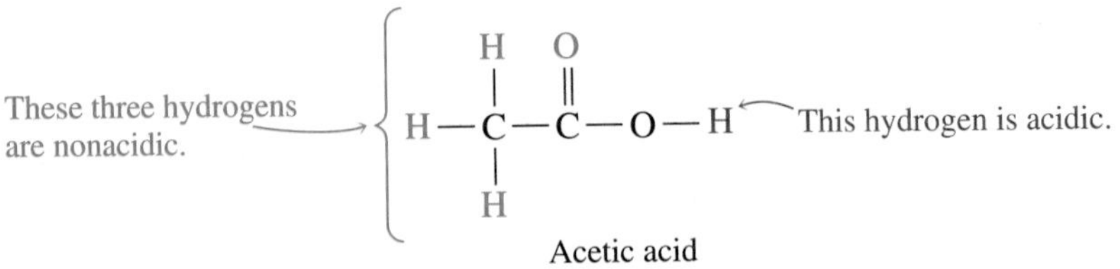

These three hydrogens are nonacidic.   This hydrogen is acidic.

Acetic acid

B.2   BASE DEFINITION

The classical definition of a base is as a compound like NaOH that gives hydroxide ions (OH^-) when it dissolves in water. According to the broader and more useful Brønsted-Lowry definition, however, a base is any substance that accepts a proton from an acid. Thus, during an acid–base reaction, a proton is transferred from the acid to the base.

The general reaction between proton-donors (acids) and proton-acceptors (bases) can be represented as

Electrons from the base form
the new B — H bond.

$$H—A + \ B: \ \rightleftharpoons \ B—H^+ + A^-$$

$$H—A + \ B:^- \ \rightleftharpoons \ B—H + A^-$$

where H—A represents an acid and B: or B:$^-$ represents a base. Notice that, when a bond forms to H$^+$ in an acid–base reaction, both electrons in the resulting B—H bond come from the base. In fact, a base *must* have such a lone pair of electrons in order to accept H$^+$ from an acid.

A base can be either neutral (B:) or negatively charged (B:$^-$). If the base is neutral, such as NH$_3$, then the protonated product has a positive charge (B—H$^+$); if the base is negatively charged, such as OH$^-$, then the protonated product is neutral (B—H):

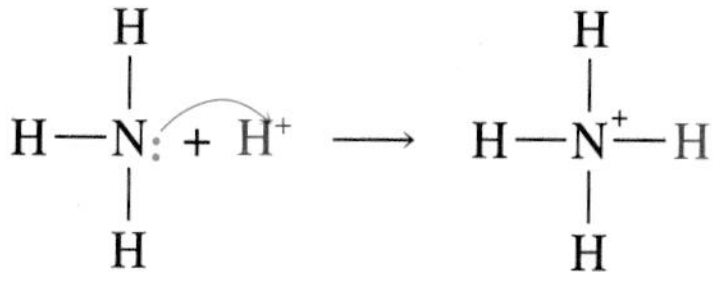

$$H—\overset{\displaystyle H}{\underset{\displaystyle H}{N}}: + \ H^+ \ \longrightarrow \ H—\overset{\displaystyle H}{\underset{\displaystyle H}{\overset{+}{N}}}—H$$

Ammonia   Ammonium ion
(neutral)   (positively charged)

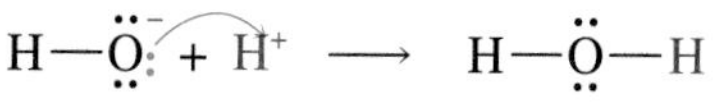

$$H—\ddot{\underset{..}{O}}:^- + \ H^+ \ \longrightarrow \ H—\ddot{\underset{..}{O}}—H$$

Hydroxide ion   Water
(negatively charged)   (neutral)

B.3 WATER AS AN ACID AND A BASE

Water is not an acid in the classical sense because it does not contain appreciable concentrations of H$^+$. Nevertheless, the Brønsted-Lowry definition includes water as an acid because it can react as a proton donor. In its reaction with acetate ion, for example, water donates a proton to the acetate ion to form acetic acid:

$$CH_3COO^- + H_2O(l) \ \rightleftharpoons \ CH_3COOH(aq) + OH^-$$

Base   Water as an acid

Water also does not meet the classical definition of a base because it does not contain substantial concentrations of OH$^-$. Nevertheless, water reacts as a Brønsted-Lowry base when it accepts a proton from an acid and forms the hydronium ion (H_3O^+). If HCl gas is dissolved in water, for example, a solution of H_3O^+ and Cl^- ions results:

Hydrochloric acid

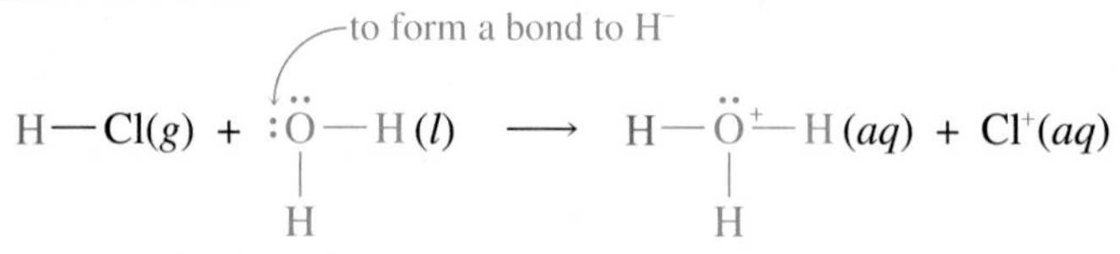

$$H—Cl(g) \ + \ :\overset{..}{O}—H\,(l) \ \longrightarrow \ H—\overset{..}{\overset{+}{O}}—H\,(aq) \ + \ Cl^+(aq)$$

to form a bond to H^-

(Acid) (Base) (Aqueous solution of ions)

Although it's sometimes convenient to write a chemical equation that portrays the splitting apart (**dissociation**) of an acid into a proton and an anion, this dissociation doesn't occur spontaneously in practice. A bare proton is too unstable to exist by itself, and an acid therefore gives up a proton only when a base such as water is present to accept it.

$$H—A \ \overset{}{\not\longrightarrow} \ H^+ \ + \ A^-$$

The dissociation of pure acid doesn't occur in isolation.

$$H—A \ + \ Base \ \rightleftharpoons \ H—Base^+ \ + \ A^-$$

Instead, a base is necessary to accept the proton from the acid.

B.4 REACTIONS OF ACIDS WITH BASES

Reactions of Acids with Metal Hydroxides   Acids react with metal hydroxide bases such as KOH to yield water and a *salt*, an ionic compound derived from the cation of the base and the anion of the acid:

$$HCl(aq) \ + \ KOH(aq) \ \rightarrow \ H_2O(l) \ + \ KCl(aq)$$

An acid   A base   Water   A salt

All such reactions are written with a single arrow because they are driven virtually to completion by formation of water. The net ionic equation for all such reactions (unless an insoluble salt is formed) is:

$$H^+(aq) \ + \ OH^-(aq) \rightarrow H_2O(l)$$

Reactions of Acids with Metal Bicarbonates and Metal Carbonates   The reaction of an acid with a metal bicarbonate yields carbonic acid (H_2CO_3), an unstable substance that decomposes to give carbon dioxide and water:

$$HBr(aq) \ + \ NaHCO_3(aq) \rightarrow H_2O(l) \ + \ CO_2(g) \ + \ NaBr(aq)$$

Dropping the spectator ions, Br^- and Na^+, from the equation shows that the bicarbonate ion is indeed a base (a proton acceptor) and that the reaction is an acid–base process:

$$H^+(aq) \ + \ HCO_3^-(aq) \rightarrow \ [H_2CO_3(aq)] \ \rightarrow H_2O(l) + \ CO_2(g)$$

| An acid | Bicarbonate ion (a base) | Carbonic acid (unstable) | Water | Carbon dioxide |

Acids react with metal carbonates such as Na_2CO_3 in exactly the same way they react with metal bicarbonates, except that a carbonate ion (CO_3^{2-}) can neutralize twice as much acid as a bicarbonate ion (HCO_3^-):

$$2\,HCl(aq) \ + \ Na_2CO_3(aq) \rightarrow H_2O(l) \ + \ CO_2(g) \ + \ 2\,NaCl(aq)$$

Although most metal carbonates are insoluble in water, they nevertheless react easily with aqueous acid. In fact, geologists often test for carbonate-bearing rocks such as limestone ($CaCO_3$) simply by putting a few drops of aqueous HCl on the rock and watching to see if bubbles of CO_2 form.

Reactions of Acids with Ammonia   Aqueous solutions of ammonia are sometimes labeled "ammonium hydroxide," but this is really a misnomer. Ammonia is such a weak base that only about 1% of the ammonia molecules in aqueous solution are protonated by water to yield NH_4^+ and OH^- ions; the remaining 99% are unprotonated.

$$NH_3(aq) + H{-}O{-}H \ \rightleftharpoons \ NH_4^+ \ (aq) \ + \ OH^- \ (aq)$$

99% of ammonia molecules are unprotonated.

1% of ammonia molecules are protonated by water.

Acids react with ammonia to yield ammonium salts such as ammonium chloride, NH_4Cl, most of which are water-soluble.

$$HCl(aq) \ + \ NH_3(aq) \rightarrow NH_4Cl(aq)$$

Living organisms contain a group of compounds called *amines*, which contain ammonia-like nitrogen atoms and undergo neutralization reactions with acids just as ammonia does. Methylamine, for example, reacts with hydrochloric acid:

$$HCl(aq) \ + \ CH_3NH_2(aq) \rightarrow CH_3NH_3Cl(aq)$$

B.5   ACID AND BASE STRENGTH

Different acids differ in their ability to give up a proton. At the left in Table B.1, a number of common acids are listed in order of decreasing *acid strength*, or ability to give up a proton. The six acids at the top of the table are *strong*

Table B.1   Relative Acid and Base Strengths

		Acid			Base		
Increasing acid strength	Strong acids: 100% ionized	Perchloric acid	$HClO_4$	ClO_4^-	Perchlorate ion	Little or no reaction as bases	Increasing base strength
		Sulfuric acid	H_2SO_4	HSO_4^-	Hydrogen sulfate ion		
		Hydriodic acid	HI	I^-	Iodide ion		
		Hydrobromic acid	HBr	Br^-	Bromide ion		
		Hydrochloric acid	HCl	Cl^-	Chloride ion		
		Nitric acid	HNO_3	NO_3^-	Nitrate ion		
		Hydronium ion	$\mathbf{H_3O^+}$	$\mathbf{H_2O}$	**Water**		
	Weak acids	Hydrogen sulfate ion	HSO_4^-	SO_4^{2-}	Sulfate ion	Very weak bases	
		Phosphoric acid	H_3PO_4	$H_2PO_4^-$	Dihydrogen phosphate ion		
		Nitrous acid	HNO_2	NO_2^-	Nitrite ion		
		Hydrofluoric acid	HF	F^-	Fluoride ion		
		Acetic acid	CH_3COOH	CH_3COO^-	Acetate ion		
	Very weak acids	Carbonic acid	H_2CO_3	HCO_3^-	Bicarbonate ion	Weak bases	
		Dihydrogen phosphate ion	$H_2PO_4^-$	HPO_4^{2-}	Hydrogen phosphate ion		
		Ammonium ion	NH_4^+	NH_3	Ammonia		
		Hydrocyanic acid	HCN	CN^-	Cyanide ion		
		Bicarbonate ion	HCO_3^-	CO_3^{2-}	Carbonate ion		
		Hydrogen phosphate ion	HPO_4^{2-}	PO_4^{3-}	Phosphate ion		
		Water	$\mathbf{H_2O}$	$\mathbf{OH^-}$	**Hydroxide ion**		
	Little or no reaction as acids	Methyl alcohol	CH_3OH	CH_3O^-	Methoxide ion	Strong bases: 100% reaction with H_2O to give OH^-	
		Ammonia	NH_3	NH_2^-	Amide ion		

acids, meaning that they are completely ionized in aqueous solution. Those remaining are *weak acids,* meaning that they establish equilibria between ionized and non-ionized forms in aqueous solution.

Note that diprotic acids such as sulfuric acid undergo two stepwise dissociations in water. The first dissociation of H_2SO_4 to yield HSO_4^- (hydrogen sulfate ion) occurs to the extent of nearly 100%. The second dissociation of HSO_4^- to yield SO_4^{2-} (sulfate ion) takes place to a much lesser extent because it involves loss of a positively charged H^+ from an ion that already carries one negative charge.

$$H_2SO_4(aq) + H_2O(l) \rightarrow H_3O^+ + HSO_4^-$$
Hydrogen sulfate ion

$$HSO_4^- + H_2O(l) \rightleftharpoons H_3O^+ + SO_4^{2-}$$
Sulfate ion

The most striking feature of Table B.1 is the relationship between acid and base strength. The acid anions shown on the right of the table are all bases and are listed in order of increasing base strength. Thus, the weaker an acid, the stronger the base it forms when it loses a proton.

The pairs on the left and right in Table B.1 are referred to as **conjugate acid/base pairs**—acid/base pairs that are related by the gain or loss of a proton. For example, HCl is the conjugate acid of Cl^-, and Cl^- is the conjugate base of HCl.

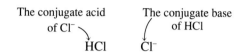

Why is there a relationship between acid and base strength? Think about what it means for an acid or base to be strong or weak. The fact that a strong acid gives up a proton readily means that the anion of a strong acid has little affinity for the proton (otherwise the acid wouldn't have given up the proton in the first place). But this is exactly the definition of a weak base—a substance that has little affinity for a proton.

$$\text{H—A} + H_2O \longrightarrow H_3O^+ + A^-$$

If this is a strong acid because it gives up the proton readily . . .   . . . then this anion is a weak base because it has little affinity for the proton

For example:

$$\text{HCl} + H_2O \longrightarrow H_3O^+ + Cl^-$$

(HCl—a strong acid)   (Cl^-—a very weak base)

In the same way, a weak acid is one that holds a proton tightly and gives it up with difficulty. The resulting anion is therefore a strong base.

$$A^- + H_2O \rightleftharpoons \text{H—A} + OH^-$$

If this anion has a high affinity for a proton . . .   . . . then this is a weak acid because it does not readily give up the proton.

B.6 PROTON-TRANSFER REACTIONS

The list of acids and bases in Table B.1 is extremely valuable because it can be used to make predictions about proton-transfer reactions. As a general rule, *an acid–base proton-transfer equilibrium always favors formation of the weaker acid and base.*

To try out this rule, let's compare the reactions of acetic acid with water and with hydroxide ion. The general procedure for predicting the direction of an acid-base reaction is to write the equation, identify the acid and base on each side of the arrow, and then decide which of each pair is stronger and weaker. Acetic acid reacts with water, for example, to give hydronium ion and acetate ion. Writing the reaction and looking at Table B.1 shows that the reaction is favored in the *reverse* direction to that written, because acetic acid and water are the weaker acid and base, respectively.

$$CH_3COOH(aq) + H_2O(l) \rightleftharpoons CH_3COO^- + H_3O^+ \quad \text{Reverse reaction favored}$$

| Weaker acid | Weaker base | Stronger base | Stronger acid |

On the other hand, the reaction of acetic acid with hydroxide ion is favored in the *forward* direction, because acetate ion and water are the weaker base and acid:

$$CH_3COOH(aq) + OH^- \rightleftharpoons CH_3COO^- + H_2O(l) \quad \text{Forward reaction favored}$$

| Stronger acid | Stronger base | Weaker base | Weaker acid |

B.7 WEAK-ACID DISSOCIATION CONSTANTS

As with any chemical equilibrium, the reaction of a weak acid HA with water can be described quantitatively by an equilibrium constant expression. The equilibrium constant K_a for the reaction of a weak acid HA with water is called an **acid dissociation constant (K_a)** and is defined as the product of the hydrogen-ion concentration $[H_3O^+]$ times the anion concentration $[A^-]$ divided by the concentration of undissociated acid $[HA]$ (all concentrations are expressed in *molarity (M)*, or mol/L):

$$HA(aq) + H_2O(l) \rightleftharpoons H_3O^+(aq) + A^-(aq)$$

$$K_a = \frac{[H_3O^+][A^-]}{[HA]}$$

The stronger the acid, the larger the value of its K_a. In fact the K_a values of strong acids such as HCl that are 100% dissociated are extremely large and difficult to measure.

The K_a values of some common acids are given in Table B.2 and illustrate several important points:

● Weak acids dissociate to a small extent in water and have K_a less than 1.
● Donation of each successive H^+ from a polyprotic acid is more difficult, as indicated by successively smaller values of K_a.
● Most organic acids containing the —COOH group are weak acids and have values of K_a near 10^{-5}.

B.8 THE IONIZATION OF WATER AND pH

Pure water is very slightly dissociated into ions. Since each dissociation yields one H_3O^+ ion and one OH^- ion, the concentrations of the two ions are the

Table B.2   Some Acid Dissociation Constants, K_a (at 25°C)

Acid	K_a	Acid	K_a
Hydrofluoric acid (HF)	3.5×10^{-4}	*Polyprotic acids*	
Hydrocyanic acid (HCN)	4.9×10^{-10}	Sulfuric acid	
Ammonium ion (NH_4^+)	5.9×10^{-10}	H_2SO_4	large
		HSO_4^-	1.2×10^{-2}
Organic Acids			
Formic acid (HCOOH)	1.8×10^{-4}	Phosphoric acid	
Acetic acid (CH_3COOH)	1.8×10^{-5}	H_3PO_4	7.5×10^{-3}
Propanoic acid		$H_2PO_4^-$	6.2×10^{-8}
(CH_3CH_2COOH)	1.3×10^{-5}	HPO_4^{2-}	2.2×10^{-13}
Ascorbic acid	7.9×10^{-5}	Carbonic acid	
(vitamin C)		H_2CO_3	4.3×10^{-7}
		HCO_3^-	5.6×10^{-11}

same—1.00×10^{-7} mol/L at 25°C.

$$2\ H_2O(l) \rightleftharpoons H_3O^+(aq) + OH^-(aq)$$

$$[H_3O^+] = [OH^-] = 1.00 \times 10^{-7}\ \text{mol/L}$$

The expression for the equilibrium constant K is simplified by dropping the water concentration to give an expression for the **ion product constant of water, K_w.**

$$K = \frac{[H_3O^+][OH^-]}{[H_2O]^2}$$

$$K_w = [H_3O^+][OH^-] = (1.00 \times 10^{-7})(1.00 \times 10^{-7}) = 1.00 \times 10^{-14}$$

The ion product constant K_w has the same value for every aqueous solution, regardless of whether an acid or a base is present. If an acid is present, then $[H_3O^+]$ is larger than 1.00×10^{-7} and $[OH^-]$ is smaller than 1.00×10^{-7}. If a base is present, then $[H_3O^+]$ is smaller than 1.00×10^{-7} and $[OH^-]$ is larger than 1.00×10^{-7}.

In medicine, chemistry, and many other fields, it's often necessary to know the exact concentration of H_3O^+ and OH^- ions in a solution. Rather than use molarity, these concentrations are normally expressed using *pH*, a number between 0 and 14 that indicates a solution's acid strength. A pH less than 7 corresponds to an acidic solution, a pH larger than 7 corresponds to a basic solution, and a pH of exactly 7 corresponds to a neutral solution.

In more precise terms, **pH** is defined as the negative common logarithm of the H_3O^+ concentration in mol/L:

$$pH = -\log[H_3O^+]$$

If you've studied logarithms, you know that the common logarithm of a number is the power to which 10 must be raised to equal the number. The pH definition can therefore be restated as:

$$[H_3O^+] = 10^{-pH}$$

For example, in neutral water (at 25°C) where $[H_3O^+] = 10^{-7}$, the pH is 7; in a strong acid solution, where $[H_3O^+] = 10^{-1}$, the pH is 1; and in a strong base solution, where $[H_3O^+] = 10^{-14}$, the pH is 14.

- Acidic solution: pH < 7   ($[H_3O^+] > 1 \times 10^{-7}$)
- Neutral solution: pH = 7 ($[H_3O^+] = 1 \times 10^{-7}$)
- Alkaline solution: pH > 7 ($[H_3O^+] < 1 \times 10^{-7}$)

To find pH, $[H_3O^+]$ is first expressed in scientific notation. For instance, if you want to know the pH of a 0.0001 M solution of HCl, for which $[H_3O^+]$ = 0.0001:

$$[H_3O^+] = 0.0001 = 1 \times 10^{-4}$$
$$pH = 4$$

It's important to keep in mind that the pH scale covers an enormous range of acidities because it involves powers of 10 (Figure B.1). A change of only 1 pH unit means a 10-fold change in $[H_3O^+]$; a change of 2 pH units means a 100-fold change in $[H_3O^+]$; and a change of 12 pH units, from a strong acid solution with pH 1 to a strong base solution with pH 13, means a change of 10^{12} (a million million) in $[H_3O^+]$.

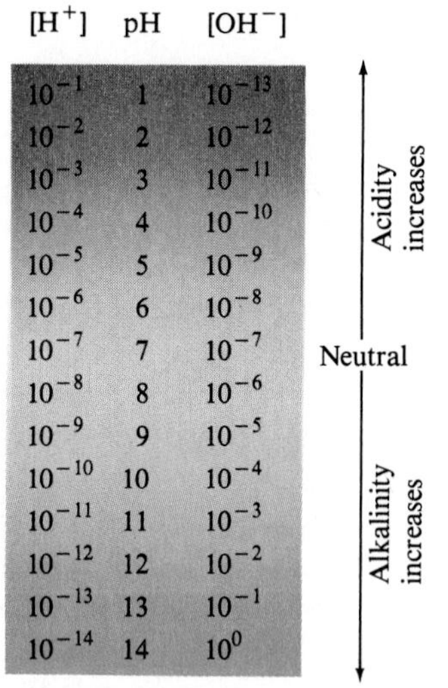

Figure B.1
Relationship of the pH scale to $[H^+]$ and $[OH^-]$ concentrations. Note that as one concentration increases, the other concentration decreases, so that the sum of the $[H^+]$ and $[OH^-]$ exponents is always −14.

B.9   WORKING WITH pH

Converting between pH and $[H_3O^+]$ is simple when the pH is a whole number. How, though, do you find $[H_3O^+]$ of blood, which has a pH of 7.4, or the pH of a solution with $[H_3O^+] = 4.6 \times 10^{-3}$? Sometimes it's sufficient to make an estimate. For example, since the pH of blood (7.4) is between 7 and 8, the H_3O^+ concentration of blood must be between 10^{-7} M and 10^{-8} M. To be exact about using pH values requires either a calculator or a log table. Most likely, you'll use a calculator.

Converting H_3O^+ concentration to pH on most scientific calculators is done with a "log" key and an "expo" or "e" key plus a "(−)" key for indicating negative exponents of 10. The number may be entered before or after pushing the "log" key. Consult your calculator instructions if you're not sure how to use these keys. The number given by the calculator must be converted to pH by changing the sign from minus to plus.

Finding the pH equivalent to $[H_3O^+] = 4.6 \times 10^{-3}$, for example, gives the following result:

$$pH = -[\log(4.6 \times 10^{-3})] = -(-2.337242) = 2.34$$

A note about significant figures: The log is reported with the same number of digits *after* the decimal point as in the whole original number: log *4.6 × 10⁻³ = 2.34.*

B.10 BUFFER SOLUTIONS

The pH of a solution can be kept relatively constant by using a **buffer,** a combination of substances that act together to prevent a drastic change in the solution's pH. Most common buffer solutions are mixtures of a weak acid and the anion of that acid, for example, a mixture of acetic acid (CH_3COOH) and acetate ion (CH_3COO^-). When acid is added to the solution, it reacts with the basic anion:

$$H_3O^+ \, (aq) + CH_3COO^-(aq) \rightarrow CH_3COOH(aq) + H_2O(l)$$

When base is added to the solution, it reacts with the acid:

$$OH^-(aq) + CH_3COOH(aq) \rightarrow CH_3COO^-(aq) + H_2O(l)$$

As a result, both added acid and added base are removed from solution, and the pH of the solution remains nearly constant.

An effective buffer solution contains roughly equal concentrations of the acid and its anion. The 0.1 M acetic acid–0.1 M acetate ion buffer is a good example. At equilibrium in this buffer solution, the pH is 4.74. Rearranging the equilibrium constant expression for acetic acid shows that the H_3O^+ concentration, and therefore the pH, depends on the ratio of acid to anion concentrations:

$$K_a = \frac{[H_3O^+][CH_3COO^-]}{[CH_3COOH]}$$

$$[H_3O^+] = K_a \frac{[CH_3COOH]}{[CH_3COO^-]}$$

When acid is added to the buffer system, the equilibrium is shifted so that the concentration of CH_3COOH increases and that of CH_3COO^- decreases. But as long as the changes in $[CH_3COOH]$ and $[CH_3COO^-]$ are small relative to their original values, the value of $[CH_3COOH]/[CH_3COO^-]$ doesn't change much and there is little change in the pH. When base is added to the buffer system, the equilibrium is shifted so that the concentration of CH_3COOH decreases and that of CH_3COO^- increases. Here too, though, as long as the concentration changes are small, there is little change in the pH.

The dramatic effect of a buffer solution is illustrated by comparing the pH changes that occur when 0.01 mol of H^+ or OH^- is added to water and to the buffer solution:

	pH of 1 L of Water	**pH of 1 L of 0.1 M CH_3COOH – 0.1 M CH_3COO^- Buffer**
	pH 7	pH 4.74
Plus 0.01 mol H^+ gives	pH 2	pH 4.68
Plus 0.01 mol OH^- gives	pH 12	pH 4.85

The pH range in which a buffer is effective is controlled by both the choice of acid HA and the concentration ratio $[HA]/[A^-]$. To make this apparent, we can take the negative logarithm of the general equation:

$$[H_3O^+] = K_a \frac{[HA]}{[A^-]}$$

$$pH = pK_a - \log\left(\frac{[HA]}{[A^-]}\right)$$

(In the above expression, pK_a is the negative logarithm of the K_a.) Rearranging the result gives what is known as the **Henderson-Hasselbalch equation,** which makes it possible to calculate the pH of a buffer solution if the pK_a of the acid and the $[A^-]/[HA]$ ratio are known:

Henderson-Hasselbalch equation: $pH = pK_a + \log\left(\frac{[A]}{[HA]}\right)$

When the $[A^-]/[HA]$ ratio is equal to 1, the pH of a buffer solution is equal to the pK_a of the acid. Taking the acetic acid–acetate buffer, for example, the K_a of acetic acid is 1.8×10^{-5}, and the pK_a is 4.74:

$$K_a = 1.8 \times 10^{-5}$$

$$pK_a = -\log K_a = -[\log(1.8 \times 10^{-5})] = -[-4.744727] = 4.74$$

Thus, when the $[CH_3COO^-]/[CH_3COOH]$ ratio is equal to 1, the pH of the buffer is 4.74:

$$pH = pK_a + \log\left(\frac{[A^-]}{[HA]}\right) = 4.74 + \log(1) = 4.74$$

(Note that the logarithm of 1 is 0.)

In general, a buffer is effective within a pH range of its initial pH plus or minus 1. The buffering effect is overcome when such large quantities of acid or base are added that the $[A^-]/[HA]$ ratio of a buffer undergoes a large change.

B.11 EQUIVALENTS OF ACIDS AND BASES: NORMALITY

For some purposes, particularly in clinical chemistry, it's more useful to think in terms of **acid** or **base equivalents** rather than molar concentrations. One equivalent of an acid is equal to the molar mass of the acid divided by the number of H^+ ions produced per formula unit. Thus, one equivalent of an acid is the weight in grams that can donate one mole of H^+ ions. Similarly, one equivalent of a base is the weight in grams that can produce one mole of OH^- ions:

$$\text{One equivalent of acid} = \frac{\text{molar mass}}{\text{number of } H^+ \text{ ions produced}}$$

$$\text{One equivalent of base} = \frac{\text{molar mass}}{\text{number of } OH^- \text{ ions produced}}$$

One equivalent of the monoprotic acid HCl is 36.5 g, the molar mass of the acid, but one equivalent of the diprotic acid H_2SO_4 is 49 g, the molar mass of the acid (98 g) divided by 2.

$$\text{One equivalent } H_2SO_4 = \frac{\text{molar mass } H_2SO_4}{2} = \frac{98 \text{ g}}{2} = 49 \text{ g}$$

Since H_2SO_4 is diprotic . . .   divide by 2

Using acid–base equivalents has two practical advantages: First, it's useful when only the acidity or alkalinity of a solution, rather than the identity of the acid or base, is of interest. Second, it shows quantities that are chemically equivalent: 36.5 g of HCl and 49 g of H_2SO_4 are chemically equivalent quantities, for example, because they each react with one equivalent of base. *One equivalent of any acid neutralizes one equivalent of any base.*

Because acid–base equivalents are so useful, acid and base concentrations are sometimes expressed in *normality* (*N*) rather than in molarity. The **normality** of an acid or base solution is defined as the number of equivalents of acid or base per liter of solution. For example, a solution made by dissolving 1.0 equivalent (36.6 g) of HCl in 1.0 L of water has a concentration of 1.0 Eq/L, or 1.0 N.

$$\text{Normality (N)} = \frac{\text{equivalents of acid or base}}{1 \text{ L of solution}}$$

The values of molarity (M) and normality (N) are the same for monoprotic acids such as HCl but are not the same for diprotic or triprotic acids. A solution made by dissolving 1.0 equivalent (49 g, 0.50 mol) of the diprotic acid H_2SO_4 in 1.0 L of water has a *normality* of 1.0 Eq/L, or 1.0 N, but has a *molarity* of 0.50 mol/L, or 0.50 M. For any acid (or base), normality is always equal to molarity times the number of H^+ (or OH^-) ions produced per formula unit.

Normality of acid = (molarity of acid) × (number of H^+ ions produced per mol)

Normality of base = (molarity of base) × (number of OH^- ions produced per mol)

A P P E N D I X   C

Exponential Notation

Numbers that are either very large or very small are usually represented in *exponential notation* as a number between 1 and 10 multiplied by a power of 10. In this kind of expression, the small raised number to the right of the 10 is the *exponent*.

Number	Exponential Form	Exponent
1,000,000	1×10^6	6
100,000	1×10^5	5
10,000	1×10^4	4
1,000	1×10^3	3
100	1×10^2	2
10	1×10^1	1
1		
0.1	1×10^{-1}	-1
0.01	1×10^{-2}	-2
0.001	1×10^{-3}	-3
0.000 1	1×10^{-4}	-4
0.000 01	1×10^{-5}	-5
0.000 001	1×10^{-6}	-6
0.000 000 1	1×10^{-7}	-7

Numbers greater than 1 have *positive* exponents, which tell how many times a number must be *multiplied* by 10 to obtain the correct value. For example, the expression 5.2×10^3 means that 5.2 must be multiplied by 10 three times:

$$5.2 \times 10^3 = 5.2 \times 10 \times 10 \times 10 = 5.2 \times 1000 = 5200$$

Note that doing this means moving the decimal point three places to the right:

$$5\ 2\ 0\ 0.$$

1 2 3

The value of a positive exponent indicates *how many places to the right the decimal point must be moved* to give the correct number in ordinary decimal notation.

Numbers less than 1 have *negative exponents,* which tell how many times a number must be *divided* by 10 (or multiplied by one-tenth) to obtain the correct value. Thus, the expression 3.7×10^{-2} means that 3.7 must be divided by 10 two times:

$$3.7 \times 10^{-2} = \frac{3.7}{10 \times 10} = \frac{3.7}{100} = 0.037$$

Note that doing this means moving the decimal point two places to the left:

$$0.\ 0\ 3\ 7$$

$$2\ 1$$

The value of a negative exponent indicates *how many places to the left the decimal point must be moved* to give the correct number in ordinary decimal notation.

CONVERTING DECIMAL NUMBERS TO EXPONENTIAL NOTATION

To convert a number greater than 1 from decimal notation to exponential notation, first move the decimal point to the *left* until there is only a single digit to the left of the decimal point. The *positive* exponent needed for exponential notation is the same as *the number of places the decimal point was moved.*

$$6\ 3\ 5\ 7\ 8\ 1.\ = 6.35781 \times 10^5$$

$$5\ 4\ 3\ 2\ 1$$

To convert a number smaller than 1 from decimal notation to exponential notation, first move the decimal point to the *right* until there is *a single nonzero digit* to the left of the decimal point. The *negative* exponent needed for exponential notation is the same as *the number of places the decimal point was moved.*

$$0.\ 0\ 0\ 0\ 4\ 2\ 6 = 4.26 \times 10^{-4}$$

$$1\ 2\ 3\ 4$$

MULTIPLYING EXPONENTIAL NUMBERS

To multiply two numbers in exponential form, the exponents are *added.* For example:

$$(3.5 \times 10^3) \times (4.2 \times 10^4) = 3.5 \times 4.2 \times 10^{(3\ +\ 4)}$$
$$= 14.7 \times 10^7$$
$$= 1.47 \times 10^8 = 1.5 \times 10^8 \text{ (rounded off)}$$
$$(5.2 \times 10^4) \times (4.6 \times 10^{-3}) = 5.2 \times 4.6 \times 10^{[4\ +\ (-3)]}$$
$$= 23.92 \times 10^1$$
$$= 2.392 \times 10^2 = 2.4 \times 10^2 \text{ (rounded off)}$$

DIVIDING EXPONENTIAL NUMBERS

To divide two numbers in exponential form, the exponents are *subtracted*. For example:

$$\frac{4.1 \times 10^4}{6.2 \times 10^6} = \frac{4.1}{6.2} \times 10^{(4-6)} = 0.6613 \times 10^{-2} = 6.6 \times 10^{-3}$$

$$\frac{6.6 \times 10^3}{8.4 \times 10^{-2}} = \frac{6.6}{8.4} \times 10^{[3-(-2)]} = 0.7857 \times 10^5 = 7.9 \times 10^4$$

Conversion Factors

Length   SI Unit: Meter (m)

1 meter = 0.001 kilometer (km)

= 100 centimeters (cm)

= 1.0936 yards (yd)

1 centimeter = 10 millimeters (mm)

= 0.3937 inch (in.)

1 nanometer = 1×10^{-9} meter

1 Angstrom (Å) = 1×10^{-10} meter

1 inch = 2.54 centimeters

1 mile = 1.6094 kilometers

Volume   SI Unit: Cubic meter (m^3)

1 cubic meter = 1000 liters (L)

1 liter = 1000 cubic centimeters (cm^3)

= 1000 milliliters (mL)

= 1.056710 quarts (qt)

1 cubic inch = 16.4 cubic centimeters

Temperature   SI Unit: Kelvin (K)

0 K = $-273.15°C$

= $-459.67°F$

°F = $(9/5)°C + 32°$; $(1.8 \times °C) + 32°$

°C = $(5/9)(°F - 32°)$; $\dfrac{(°F - 32°)}{1.8}$

K = $°C + 273.15°$

Mass   SI Unit: Kilogram (kg)

1 kilogram = 100 grams (g)

= 2.205 pounds (lb)

1 gram = 1000 milligrams (mg)

= 0.03527 ounce (oz)

1 pound = 453.6 grams

1 atomic mass unit = 1.66054×10^{-24} gram

Pressure   SI Unit: Pascal (Pa)

1 pascal = 9.869×10^{-7} atmospheres

1 atmosphere = 101,325 pascals

= 760 mm Hg (Torr)

= 14.70 lb/in^2

Energy   SI Unit: Joule (J)

1 joule = 0.23901 calorie (cal)

1 calorie = 4.184 joules

GLOSSARY

Note: This glossary is alphabetized according to the letter-by-letter method, as are most dictionaries. All the letters of compound terms up to the first comma are taken into account in the alphabetization. For example, "Acid, Brønsted-Lowry" comes before "Acid anhydride," which comes before "Acid–base indicator."

Acetal A compound that has two —OR groups bonded to the same carbon atom.

Acetyl coenzyme A (acetyl SCoA) Acetyl-substituted coenzyme A that is the common intermediate carrying acetyl groups into the citric acid cycle.

Acetyl group The CH_3C=O group.

Achiral The opposite of chiral; not having (right or left) handedness.

Acid, Brønsted-Lowry A substance that is able to donate a hydrogen ion, H^+.

Acid anhydride A compound that has a carbonyl group bonded to a carbon substituent and an —OCOR group, RCO_2COR.

Acid–base reaction, Brønsted-Lowry The transfer of a proton from an acid to a base.

Acidic solution A solution in which pH is less than 7 and hydronium ion concentration is greater than hydroxide ion concentration.

Activation of an enzyme Any process that initiates or increases the action of an enzyme.

Active site A small, three-dimensional portion of an enzyme with the specific shape and structure necessary to bind a substrate.

Active transport Energy-requiring transport of molecules or ions across cell membranes against a concentration gradient.

Acyclic alkane An alkane that contains no rings.

Acyl carrier protein Protein that carries acyl groups during fatty acid synthesis.

Addition reaction, aldehydes and ketones Addition of an alcohol or other reagent to the carbon–oxygen double bond to give a carbon–oxygen single bond.

Addition reaction, alkenes and alkynes Addition of a reactant of the general form X-Y to the multiple bond of an unsaturated compound to yield a saturated product.

Adenosine triphosphate (ATP) The triphosphate of adenosine; the "energy currency" molecule of metabolism.

Aerobic In the presence of oxygen.

Alcohol A compound that contains an —OH functional group covalently bonded to an alkane-like carbon atom, R—OH.

Aldehyde A compound that has a carbonyl group bonded to one carbon atom and one hydrogen atom, RCHO.

Aldol reaction The reaction of a ketone or aldehyde to form a hydroxy ketone product on treatment with base catalyst.

Aldose A monosaccharide that contains an aldehyde carbonyl group.

Alkaloid A naturally occurring nitrogen-containing compound isolated from a plant; usually basic, bitter, and poisonous.

Alkane A compound that contains only carbon and hydrogen and has only single bonds.

Alkene A compound that contains only carbon and hydrogen and has a carbon–carbon double bond.

Alkoxy group An RO— group.

Alkyl group The part of an alkane that remains when one hydrogen atom is removed.

Alkyl halide A compound that contains an alkyl group bonded to a halogen atom, RX.

Alkyne A compound that contains only carbon and hydrogen and has a carbon–carbon triple bond.

Allosteric control Cooperative interaction by which binding a substrate or a regulator at one site in an enzyme influences shape and therefore binding at other sites.

Allosteric enzyme Enzyme with two or more protein chains (quaternary structure) and two or more interactive binding sites for substrate and regulators.

Alpha (α) amino acid An amino acid in which the amino group is bonded to the carbon atom next to the —COOH group.

Alpha (α) helix A common secondary protein structure in which a protein chain wraps into a coil stabilized by hydrogen bonds between peptide links in the backbone.

Amide A compound that has a carbonyl group bonded to one carbon atom and one nitrogen atom group, $RCONR_2'$, where R′ may be an H atom.

Amide bond The bond between a carbonyl group and the nitrogen atom in an amide.

Amine A compound that has one or more organic groups bonded to nitrogen, RNH_2, R_2NH, or R_3N.

Amino acid A molecule that contains both an amino group and a carboxylic acid functional group; proteins are polymers of amino acids.

Amino acid pool The entire collection of free amino acids in the body.

Amino group The $-NH_2$ functional group.

Ammonium salt An ionic substance formed by reaction of ammonia or an amine with an acid; an ammonium ion, NH_4^+, and an anion or an amine salt.

Amphoteric Able to react as both an acid and a base.

Anabolism Metabolic reactions that build larger biological molecules from smaller pieces.

Anaerobic In the absence of oxygen.

Anion A negatively charged ion.

Anomeric carbon atom The hemiacetal C atom in a cyclic sugar; the C atom bonded to an —OH group and an —OR group (R is the ring).

Anomers Cyclic sugars that differ only in positions of substituents at the hemiacetal carbon (the anomeric carbon); the α form has the —OH on the opposite side the $-CH_2OH$; the β form has the —OH on the same side as the $-CH_2OH$.

Antibody (immunoglobulin) Protein molecule that identifies antigens.

Anticodon A sequence of three ribonucleotides on tRNA that recognizes the complementary sequence (the codon) on mRNA.

Antigen A substance that stimulates the immune response.

Antiseptic An agent that can be used to destroy or prevent the growth of harmful microorganisms on or in the body.

Apoenzyme The protein portion of an enzyme.

Aromatic compound A compound that contains a six-membered ring of carbon atoms with three double bonds or a similarly stable ring.

Aryl group General name for any substituent derived from an aromatic compound.

Basal metabolic rate The minimum amount of energy required per unit time to stay alive.

Base, Brønsted-Lowry A substance that can accept a hydrogen ion (H^+) from an acid.

Benedict's reagent A reagent (Cu^{2+} in basic solution) that converts an aldehyde into a carboxylic acid and yields a brick-red precipitate of Cu_2O.

β-Pleated sheet A common secondary protein structure in which segments of a protein chain fold back on themselves to form parallel strands held together by hydrogen bonds between peptide links in the backbone.

Bile Fluid secreted by the liver and released into the small intestine during digestion; contains bile salts, bicarbonate ion, and other electrolytes.

Bile acids Steroid acids derived from cholesterol that are secreted in bile.

Blood clotting Formation of a structure of fibrin and blood cells to close the rupture of a blood vessel.

Blood plasma Liquid portion of blood; an extracellular fluid.

Blood serum Fluid portion of blood remaining after clotting has occurred.

Branched-chain alkane An alkane that has a branching connection of carbon atoms along its chain.

C-Terminal amino acid The amino acid with the free —COOH group at the end of a polypeptide.

Carbocation A polyatomic ion with a positively charged carbon atom.

Carbohydrate A member of a large class of naturally occurring polyhydroxy ketones and aldehydes.

Carbonyl group A functional group that has a carbon atom joined to an oxygen atom by a double bond, C=O.

Carbonyl-group substitution reaction A reaction in which a new group X replaces (substitutes for) a group Y attached to a carbonyl-group carbon.

Carboxylate anion The anion that results from dissociation of a carboxylic acid, $RCOO^-$.

Carboxyl group The —COOH functional group.

Carboxylic acid A compound that has a carbonyl group bonded to one carbon substituent and one —OH group, RCOOH.

Carboxylic acid salt An ionic compound containing a carboxylate ion.

Carcinogenic Cancer-causing.

Catabolism Metabolic reactions that break down molecules into smaller pieces.

Catalyst A substance that speeds up a chemical reaction without itself undergoing any permanent chemical change.

Catalytic cracking A procedure for converting the alkane mixture in kerosene into the smaller, branched-chain alkanes needed for gasoline.

Cation A positively charged ion.

Cell membrane (plasma membrane) The membrane surrounding a cell (a lipid bilayer).

Chain-growth polymer Polymer formed by the addition of monomer molecules one by one to the end of a growing chain.

Chemiosmotic hypothesis An explanation of how the establishment of a pH difference across the mitochondrial membrane by the respiratory chain is coupled to the synthesis of ATP.

Chiral Having (right or left) handedness; able to have two different mirror-image forms.

Chiral carbon atom A carbon atom bonded to four different groups.

Chromosome A complex between proteins and a DNA molecule that is visible during cell division.

Chyme The semisolid mixture produced by digestion in the stomach.

Chylomicron Spherical, low-density lipoprotein particles that transport triacylglycerols.

Cis isomer Isomer with a specific pair of atoms or groups on same side of a double bond.

Citric acid cycle The series of biochemical reactions that break down acetyl groups to carbon dioxide and energy stored in reduced coenzymes.

Claisen condensation reaction A reaction that joins two ester molecules together to yield a keto ester product.

Clinical chemistry Chemical analysis of body tissues and fluids.

Codon A sequence of three ribonucleotides in the RNA chain that codes for a specific amino acid.

Coenzyme A small, organic molecule that acts as an enzyme cofactor.

Coenzyme A Coenzyme that functions as a carrier of acetyl groups and other acyl (R—C=O) groups.

Cofactor A small, nonprotein part of an enzyme that is essential to the enzyme's catalytic activity.

Combustion A chemical reaction in which heat and often light are produced; usually refers to burning in the presence of oxygen.

Competitive enzyme inhibition Enzyme regulation in which an inhibitor competes with a similarly shaped substrate for binding to the enzyme active site.

Condensed structure A structure in which central atoms and the atoms bonded to them are written as groups, e.g., $CH_3CH_2CH_3$.

Conformation The exact three-dimensional shape of a molecule at any given instant.

Conjugated protein A protein that yields one or more other substances in addition to amino acids when hydrolyzed.

Coupled reactions Combination of an energy-requiring and energy-releasing reaction that occur together to give an overall energy-releasing reaction.

Cycloalkane An alkane that contains a ring of carbon atoms.

Cytochromes Heme-containing coenzymes that function as electron carriers.

Cytoplasm The region between the cell membrane and the nuclear membrane in a eukaryotic cell.

Cytosol The contents of the cytoplasm surrounding the organelles.

D sugar Monosaccharide with —OH group on chiral carbon atom farthest from the carbonyl group pointing to the right in the Fischer projection.

Dehydration Loss of water, as from an alcohol to yield an alkene.

Denaturation The loss of secondary, tertiary, or quaternary protein structure due to disruption of noncovalent interactions that leaves peptide bonds intact.

Deoxyribonucleotide A nucleotide containing D-deoxyribose.

Diabetes mellitus A chronic condition due to either insufficient insulin or failure of insulin to cross cell membranes.

Diastereomers Stereoisomers that are not mirror images of each other.

Digestion A general term for the breakdown of food into small molecules.

Dimer A unit formed by the joining together of two identical molecules.

Disaccharide A carbohydrate that yields two monosaccharides on hydrolysis.

Disinfectant An agent that can be used to destroy or prevent the growth of harmful microorganisms on inanimate objects only.

Distillation The process by which a mixture is separated according to boiling points.

Disulfide A compound that contains a sulfur–sulfur single bond, R—S—S—R.

Disulfide bridge An S—S bond formed between two cysteine residues that can join two peptide chains together or cause a loop in a peptide chain.

DNA (deoxyribonucleic acid) The nucleic acid that stores the genetic information.

Double helix Two strands coiled around each other in a screw-like fashion, as in the two polynucleotide strands in DNA.

Electrolyte A substance that conducts electricity when dissolved in water; specifically, ions in body fluids.

Endergonic Describes a process that absorbs free energy from the surroundings (positive ΔG).

Endocrine system A system of specialized cells, tissues, and ductless glands that excretes hormones and shares with the nervous system the responsibility for maintaining constant internal body conditions and responding to changes in the environment.

Enzyme A protein or other molecule that acts as a catalyst for biological reactions.

Enzyme–substrate complex A complex of enzyme and substrate in which the two are bound together by noncovalent interactions with the substrate in position to react.

Erythrocytes Red blood cells.

Essential amino acid An amino acid that cannot be synthesized by the body and so must be obtained in the diet.

Ester A compound that has a carbonyl group bonded to one carbon atom and one —OR group, RCOOR'.

Esterification reaction The reaction between an alcohol and a carboxylic acid to yield an ester plus water.

Ether A compound that has an oxygen atom bonded to two carbon atoms, R—O—R.

Ethyl group —CH$_2$CH$_3$, the alkyl group derived from ethane.

Eukaryotic cell Cell with a membrane-enclosed nucleus; found in all higher organisms.

Exergonic Describes a process that releases free energy to the surroundings (negative ΔG).

Exon A DNA segment in a gene that codes for part of a protein molecule.

Extracellular fluid Fluid outside cells.

Fat A mixture of triacylglycerols that is solid because it contains a high proportion of saturated fatty acids.

Fatty acid A long-chain carboxylic acid; those in animal fats and vegetable oils often have 12 to 22 carbon atoms.

Fatty-acid spiral (β-oxidation) A repetitive series of biochemical reactions that degrade fatty acids to acetyl SCoA.

Feedback control Activation or inhibition of the first reaction in a sequence by a product of the sequence.

Fermentation The breakdown of glucose to ethanol plus carbon dioxide by the action of yeast enzymes.

Fibrin Insoluble protein that forms the fiber framework of a blood clot.

Fibrous protein A tough, insoluble protein whose peptide chains are arranged in long filaments.

Filtration, kidney Filtration of blood plasma through glomerulus and into kidney nephron.

Fischer projection Structure that represents a chiral carbon atom as the intersection of two lines. The horizontal line represents bonds pointing out of the page, and the vertical line represents bonds pointing behind the page. For sugars, the aldehyde or ketone is at the top.

Flavin adenine dinucleotide (FAD) Coenzyme that functions as an oxidizing agent and forms FADH$_2$.

Free energy change (ΔG) A quantity that is a function of energy change and entropy change and is the criterion for spontaneous change (negative ΔG; $\Delta G = \Delta H - T\Delta S$).

Free radical An atom or group that has an unpaired electron.

Functional group A part of a larger molecule composed of an atom or group of atoms that has characteristic chemical behavior.

Gene All segments of DNA needed to direct the synthesis of a protein with a specific function.

Genetic code The complete assignment of the 64 mRNA codons to specific amino acids (or stop signals).

Genetic enzyme control Regulation of enzyme activity by control of the synthesis of enzymes.

Genetic mutation An error in the code that determines inherited traits.

Genome The sum of all genes in an organism.

Germ cell A reproductive (sperm or egg) cell.

Globular protein A water-soluble protein that adopts a compact, coiled-up shape.

Glomerular filtrate The fluid that enters the nephron from the glomerulus; filtered blood plasma.

Glucogenic amino acid An amino acid catabolized to a compound that can enter the citric acid cycle for conversion to glucose.

Gluconeogenesis The biochemical pathway for the synthesis of glucose from noncarbohydrates such as lactate, amino acids, and glycerol.

Glycerophospholipid (phosphoglyceride) A lipid in which glycerol is linked by ester bonds to two fatty acids and one phosphate, which is in turn linked by another ester bond to an amino alcohol (or other alcohol); a derivative of phosphatidic acid.

Glycogenesis The biochemical pathway for synthesis of glycogen.

Glycogenolysis The biochemical pathway for breakdown of glycogen to free glucose.

Glycol A dialcohol; a compound that contains two —OH groups.

Glycolipid A sphingolipid with a fatty acid bonded to the C2 —NH_2 and a sugar bonded to the C1 —OH group of sphingosine.

Glycolysis The biochemical pathway that breaks down a molecule of glucose into two molecules of pyruvate plus energy.

Glycoside A cyclic acetal formed by reaction of a monosaccharide with an alcohol, accompanied by loss of H_2O.

Glycosidic bond Bond between the anomeric carbon atom of a monosaccharide and an —OR group.

Grain alcohol A common name for ethyl alcohol, CH_3CH_2OH.

Halogenation, alkane The substitution of one or more hydrogen atoms in an alkane by halogen atoms.

Halogenation, alkene The reaction of an alkene with a halogen (Cl_2 or Br_2) to yield a 1,2-dihaloalkane product.

Hemiacetal A compound that has both an alcohol-like —OH group and an ether-like —OR group bonded to the same carbon atom.

Hemostasis The stopping of bleeding.

Heterocycle A ring that contains nitrogen or some other atom in addition to carbon.

Holoenzyme The combination of apoenzyme and cofactor that is active as a biological catalyst.

Homeostasis Maintenance of unchanging internal conditions by living things.

Hormone A chemical messenger, usually secreted by an endocrine gland and transported through the bloodstream to elicit response from a specific target tissue.

Hydration The surrounding of a solute ion or molecule by water molecules.

Hydration, alkene The reaction of an alkene with water to yield an alcohol.

Hydrocarbon A compound that contains only carbon and hydrogen.

Hydrogenation The reaction of an alkene (or alkyne) with H_2 to yield an alkane product.

Hydrogen bonding Intermolecular force due to attraction between an electronegative atom (O, N, or F) and a hydrogen atom bonded to an electronegative atom (usually O or N).

Hydrohalogenation The reaction of an alkene with HCl or HBr to yield an alkyl halide product.

Hydrolysis The breakdown of a compound, such as an acetal, by reaction with water; the H's and O from water add to the atoms in the broken bond.

Hydrophilic Water-loving; a hydrophilic substance dissolves in water.

Hydrophobic Water-fearing; a hydrophobic substance does not dissolve in water.

Hydroxyl group A name for the —OH group in an organic compound.

Hyperglycemia Higher than normal blood glucose concentration.

Hypoglycemia Lower than normal blood glucose concentration.

Immune response The recognition of and attack on antigens, including viruses, bacteria, toxic substances, and infected cells.

Induced-fit model A model that pictures an enzyme with a conformationally flexible active site that can change shape to accommodate a range of different substrate molecules.

Inflammation A nonspecific defense mechanism triggered by antigens or tissue damage; includes swelling, redness, warmth, and pain.

Inhibition of an enzyme Any process that slows down or stops the action of an enzyme.

Interstitial fluid Fluid surrounding cells; an extracellular fluid.

Intracellular fluid Fluid inside cells.

Intron A seemingly nonsensical segment of DNA that does not code for part of a protein.

Irreversible enzyme inhibition A mechanism of enzyme deactivation in which an inhibitor forms covalent bonds to the active site.

Isoelectric point The pH at which a large sample of amino acid molecules has equal numbers of + and − charges.

Isomers, cis–trans Alkenes that have the same formula and connections between atoms but differ in having pairs of groups on opposite sides of the double bond.

Isomers, constitutional Compounds with the same molecular formula but with different connections between atoms.

Isomers, optical The two mirror-image forms of a chiral molecule.

Isopropyl group —$CH(CH_3)_2$, the alkyl group derived by removing a hydrogen atom from the central carbon of propane.

Ketoacidosis Lowered blood pH due to accumulation of ketone bodies.

Ketogenesis The synthesis of ketone bodies from acetyl SCoA.

Ketogenic amino acid An amino acid catabolized to a compound that can be converted to ketone bodies (also fatty acids).

Ketone A compound that has a carbonyl group bonded to two carbon atoms, $R_2C{=}O$.

Ketone bodies Compounds produced in the liver that can be used as fuel by muscle and brain tissue; β-hydroxybutyrate, acetoacetate, and acetone.

Ketose A monosaccharide that contains a ketone carbonyl group.

L sugar Monosaccharide with —OH group on chiral carbon atom farthest from the carbonyl group pointing to the left in the Fischer projection.

Leukotriene A simple lipid derived from a C_{20} carboxylic acid like a prostaglandin but with no cyclopentane ring.

Line structure A shorthand way of representing ring structures as polygons without showing individual carbon atoms.

1,4 Link An acetal link between the hydroxyl group at C1 of one sugar and the hydroxyl group at C4 of another sugar.

Lipid A naturally occurring molecule found in plant and animal sources that is soluble in nonpolar organic solvents.

Lipid bilayer The basic structural unit of cell membranes; composed of two parallel sheets of membrane lipid molecules arranged tail to tail.

Lipogenesis The biochemical pathway for synthesis of fatty acids from acetyl SCoA.

Lipoprotein A lipid–protein complex that transports lipids.

Lock-and-key model A model for enzyme specificity that pictures an enzyme as a large molecule with a cleft into which only specific substrate molecules can fit.

Markovnikov's rule In the addition of HX to an alkene, the H becomes attached to the carbon that already has the most H's, and the X becomes attached to the carbon that has fewer H's.

Mercaptan An alternate name for a thiol, R—SH.

Messenger RNA (mRNA) The RNA whose function is to carry genetic messages transcribed from DNA and direct protein synthesis.

Meta Indicates 1,3 substituents on a benzene ring.

Metabolic disease A disease due to disruption of metabolism caused by a genetically determined defect.

Metabolism The overall sum of the many reactions taking place in an organism.

Methyl group —CH_3, the alkyl group derived from methane.

Micelle A spherical cluster formed by the aggregation of soap molecules (or other molecules with hydrophilic and hydrophobic ends) in water.

Mirror image The reverse image produced when an object is reflected in a mirror.

Mitochondrial matrix The space surrounded by the inner membrane.

Mitochondrion An egg-shaped organelle where small molecules are broken down to provide the energy to power an organism.

Mobilization (of triacylglycerols) Hydrolysis of triacylglycerols in adipose tissue and release of fatty acids into bloodstream.

Monomer Small molecule combined to form a polymer.

Monosaccharide (simple sugar) A carbohydrate that can't be chemically broken down into smaller sugars by hydrolysis with aqueous acid.

Mutagen A substance that causes mutations.

Mutarotation Change in rotation of plane-polarized light resulting from the equilibrium between cyclic anomers and the open-chain form of a sugar.

Mutation An error in base sequence occurring during DNA replication.

_n_-Propyl group —$CH_2CH_2CH_3$, the alkyl group derived by removing a hydrogen atom from an end carbon of propane.

N-Terminal amino acid The amino acid with the free —NH_2 group at the end of a protein.

Natural gas A gaseous mixture of small alkanes, primarily methane.

Neurotransmitter A chemical messenger that transmits a nerve impulse between neighboring nerve cells.

Nicotinamide adenine dinucleotide (NAD⁺) Coenzyme that functions as an oxidizing agent and forms NADH/H^+.

Nitrate ester A compound formed by reaction of an alcohol with nitric acid.

Nomenclature A system for naming chemical compounds.

Noncompetitive enzyme inhibition Enzyme regulation in which an inhibitor binds to an enzyme elsewhere than at the active site, thereby changing the shape of the enzyme's active site.

Nonessential amino acid One of 10 amino acids that are synthesized in the body and are therefore not necessary in the diet.

Nucleic acid A biological polymer made by the linking together of nucleotide units.

Nucleoside A compound consisting of a five-carbon sugar bonded to a cyclic amine base; like a nucleotide but missing the phosphate group.

Nucleotide A building block for nucleic acid synthesis, consisting of a five-carbon sugar bonded to a cyclic amine base and to phosphoric acid.

Octane number A measure of the antiknock properties of a fuel.

Oil A mixture of triacylglycerols that is liquid because it contains a high proportion of unsaturated fatty acids.

Optical isomers (enantiomers) The two mirror-image forms of a chiral molecule.

Organelle A small, organized unit in the cell that performs a specific function.

Ortho Indicates 1,2 substituents on a benzene ring.

Oxidation The loss of electrons or increase in oxidation number of a reactant in a chemical reaction; in organic chemistry, the removal of hydrogen from a molecule or the addition of oxygen to a molecule.

Oxidative deamination Conversion of an amino acid —NH_2 group to an α-keto group, with removal of NH_4^+.

Oxidative phosphorylation The synthesis of ATP from ADP using energy released in the respiratory chain.

Para Indicates 1,4 substituents on a benzene ring.

Paraffin A mixture of waxy alkanes having 20 to 36 carbon atoms.

Pentose phosphate pathway The biochemical pathway that produces ribose (a pentose), NADPH, and other sugar phosphates from glucose; an alternative to glycolysis.

Peptide bond An amide bond that links two amino acids together.

Petroleum A complex mixture of hydrocarbons, primarily of marine origin.

pH A number usually between 0 and 14 that describes the acidity of an aqueous solution; mathematically, the negative common logarithm of a solution's H_3O^+ concentration.

Phagocytosis Engulfing and digestion of a particle by a cell.

Phenol A compound that has an —OH functional group bonded directly to an aromatic, benzene-like ring.

Phenyl group The name of the C_6H_5— unit when a benzene ring is considered a substituent group.

Phosphate ester A compound formed by reaction of an alcohol with phosphoric acid.

Phosphate group A —PO_3^{2-} group in an organic molecule.

Phosphatidylcholine (lecithin) A glycerophospholipid containing choline.

Phospholipid A lipid that has an ester link between phosphoric acid and an alcohol.

Phosphoric anhydride A —P—O—P— group, like those in ADP and ATP.

Phosphorylation A reaction that results in addition of a phosphate group (—PO_3^{2-}).

Polycyclic aromatic compound A substance that has two or more benzene-like rings fused together along their edges.

Polymer Very large molecule composed of identical repeating units and formed by combination of small molecules.

Polymerization A reaction in which monomers combine to form a polymer.

Polypeptide A molecule composed of roughly 10 to 100 amino acids linked by peptide bonds.

Polysaccharide (complex carbohydrate) A carbohydrate composed of many monosaccharides bonded together.

Polyunsaturated fatty acid (PUFA) A long-chain fatty acid that has two or more carbon–carbon double bonds.

Precursor A compound necessary for the synthesis of another compound.

Primary alcohol An alcohol in which the OH-bearing carbon atom is bonded to one other carbon (and two hydrogens), RCH_2OH.

Primary amine An amine that has one organic group bonded to nitrogen, RNH_2.

Primary (1°) carbon A carbon atom that is bonded to one other carbon atom.

Primary protein structure The sequence in which amino acids are linked together in a protein.

Prokaryotic cell Cell that has no nucleus; found in bacteria and algae.

Prostaglandin A lipid derived from a C_{20} carboxylic acid and containing a cyclopentane ring with two long side chains.

Protease An enzyme for hydrolysis of the peptide bonds in proteins.

Protein A large biological molecule made of many amino acids linked together through amide bonds.

Quaternary ammonium salt An ammonium salt with four organic groups bonded to the nitrogen atom.

Quaternary (4°) carbon A carbon atom that is bonded to four other carbons.

Quaternary protein structure The way in which two or more protein chains aggregate to form large, ordered structures.

R— The general symbol for an alkyl group.

Reabsorption, kidney Movement of solutes out of filtrate in kidney tubule.

Reaction mechanism A complete description of how a reaction occurs, including the details of each individual step in the overall process.

Reagent A reactant used to bring about a specific chemical reaction.

Reducing sugar A carbohydrate that reacts with an oxidizing agent such as Benedict's reagent.

Reduction The gain of electrons or decrease in oxidation number of a reactant in a chemical reaction; in organic chemistry, the addition of hydrogen or the removal of oxygen.

Reductive amination Conversion of an α-keto acid to an amino acid by reaction with NH_4^+.

Refining The process by which petroleum is converted into gasoline and other useful products.

Replication The process by which copies of DNA are made in the cell.

Residue, amino acid An alternative name for an amino acid unit in a polypeptide or protein.

Resonance The existence of a molecule in a single structure intermediate among two or more correct possible double-bond-containing structures that can be drawn.

Respiratory chain (electron transport chain) The series of biochemical reactions that passes electrons from reduced coenzymes to oxygen and is coupled to ATP formation.

Ribonucleotide A nucleotide containing D-ribose.

Ribosomal RNA (rRNA) The type of RNA complexed with proteins in ribosomes.

Ribosome A structure in the cell where protein synthesis occurs.

RNA (ribonucleic acid) Nucleic acid responsible for putting the genetic information to use in protein synthesis.

Saponification Reaction of a fat or oil with aqueous hydroxide ion to yield glycerol and carboxylate salts of fatty acids.

Saponification reaction The reaction of an ester with aqueous hydroxide ion to yield an alcohol and the metal salt of a carboxylic acid.

Saturated Containing only single bonds between carbon atoms and thus unable to accommodate additional hydrogen atoms.

Secondary alcohol An alcohol in which the OH-bearing carbon atom is bonded to two other carbons (and one hydrogen), R_2CHOH.

Secondary amine An amine that has two organic groups bonded to nitrogen, R_2NH.

Secondary (2°) carbon A carbon atom that is bonded to two other carbons.

Secondary protein structure The way in which nearby segments of a protein chain are oriented into a regular pattern, for example, an α-helix or a β-pleated sheet.

Secretion, kidney Movement of solutes into filtrate in kidney tubule.

Simple protein A protein that yields only amino acids when hydrolyzed.

Soap The mixture of carboxylate salts of fatty acids formed on saponification of animal fat.

Somatic cell Any cell other than a reproductive one.

Specificity The extent to which an enzyme reacts with different substrates.

Sphingolipid A lipid derived from the amino alcohol sphingosine (or related amino alcohols), rather than glycerol.

Sphingomyelin A sphingolipid with a fatty acid bonded to the C2 —NH_2 and a phosphate bonded to the C1 —OH group of sphingosine.

Stereoisomers Isomers that have the same molecular and structural formulas but different arrangements of their atoms in space.

Steroid A lipid whose structure is based on a tetracyclic (four-ring) carbon skeleton.

Straight-chain alkane An alkane that has all its carbon atoms connected in a row.

Structural formula A formula that shows how atoms are connected to each other.

Substituent A group attached to a root compound.

Substitution reaction An organic reaction in which two reactants exchange atoms or groups, AB + XY → AY + XB.

Substitution reaction, aromatic Substitution of an atom or group for one of the hydrogens on an aromatic ring.

Substrate The reactant in an enzyme-catalyzed reaction.

Substrate-level phosphorylation Formation of ATP by transfer of a phosphate to ADP from another phosphate-containing compound.

Symmetry plane An imaginary plane cutting through the middle of an object so that one half of the object is a mirror image of the other half.

Synapse Narrow gap between nerve cells across which a signal is carried by a neurotransmitter.

Tertiary alcohol An alcohol in which the OH-bearing carbon atom is bonded to three other carbons (and no hydrogens), R_3COH.

Tertiary amine An amine that has three organic groups bonded to nitrogen, R_3N.

Tertiary (3°) carbon A carbon atom that is bonded to three other carbons.

Tertiary protein structure The way in which an entire protein chain is coiled and folded into its specific three-dimensional shape.

Thiol A compound that contains the —SH functional group, R—SH.

Tollens' reagent A reagent ($AgNO_3$ in aqueous NH_3) that converts an aldehyde into a carboxylic acid and deposits a silver mirror on the inside surface of the reaction flask.

Transamination The interchange of the amino group of an amino acid and the keto group of an α-keto acid.

Transcription The process by which the information in DNA is read and used to synthesize mRNA.

Transfer RNA (tRNA) The RNA whose function is to transport specific amino acids into position for protein synthesis.

Trans isomer Isomer with a specific pair of atoms or groups on opposite sides of double bond.

Translation The process by which mRNA directs protein synthesis.

Triacylglycerol (triglyceride) A triester of glycerol with three fatty acids.

Triple helix The secondary protein structure of tropocollagen in which three protein chains coil around each other to form a rod.

Turnover number The number of substrate molecules acted on by one molecule of enzyme per unit time.

Unsaturated Containing one or more double or triple bonds between carbon atoms and thus able to add more hydrogen atoms.

Urea cycle The cyclic biochemical pathway that produces urea for excretion.

Vinyl monomer A compound with the structure $CH_2{=}CHZ$ that undergoes polymerization to give a polymer with the repeating unit $-CH_2-CHZ-$.

Vitamin A small, organic molecule that must be obtained in the diet and that is essential in trace amounts for proper biological functioning.

Wax A mixture of esters of long-chain carboxylic acids with long-chain alcohols.

Whole blood Blood plasma plus blood cells.

Wood alcohol A common name for methyl alcohol, CH_3OH.

Zwitterion A neutral dipolar compound that contains both + and − charges in its structure.

Zymogen (proenzyme) A compound that becomes an active enzyme after undergoing a chemical change.

A N S W E R S

Selected Answers
to Problems

Answers are given for most in-chapter problems and even-numbered end-of-chapter problems.

Chapter 1

1.1 (a) alcohol, carboxylic acid (b) double bond, ester **1.2** (a) CH_3CHO (b) CH_3CH_2COOH **1.3** $CH_3CH_2CH_2CH_2CH_2CH_2CH_3$ **1.6** Structures (a) and (c) are identical and are isomers of structure (b). **1.9** (a) 2,6-dimethyloctane (b) 2,2-diethylheptane **1.11** CH_3's are primary, CH_2's are secondary, CH's are tertiary, and C's are quaternary. **1.12** (a) 2-methylbutane (b) 2,2,3-trimethylbutane **1.13** $2 C_2H_6 + 7 O_2 \rightarrow 4 CO_2 + 6 H_2O$ **1.15** (a) 1-ethyl-4-methylcyclohexane (b) 1-ethyl-3-isopropylcyclopentane **1.20** Add more water and see which is miscible. **1.22** (a) CH_3OH (b) CH_3NH_2 (c) CH_3COOH (d) CH_3OCH_3 **1.26** A straight-chain alkane has all its carbons in a row. **1.28** No. Isomers must have the same formula. **1.30** Carbon forms only four bonds. **1.32** $CH_3CH_2CH_2OH$, $CH_3CH(OH)CH_3$, and $CH_3CH_2OCH_3$ **1.34** There are too many hydrogens. **1.36** identical: (a); isomers: (b), (d), (e); unrelated: (c) **1.38** (a) First and second are identical. (b) First and second are identical. **1.42** hexane, 2-methylpentane, 3-methylpentane, 2,2-dimethylbutane, 2,3-dimethylbutane **1.44** (a) 1-isopropyl-1-methylcyclopentane (b) 1,1,3,3-tetramethylcyclopentane (c) propylcyclohexane (d) 4-butyl-1,1-dimethylcyclohexane (e) ethylcyclooctane (f) 1,2-diethyl-3-methylcyclopropane (g) 2-ethyl-1-methyl-3-propylcyclopentane **1.46** heptane, 2-methylhexane, 3-methylhexane, 2,2-dimethylpentane, 2,3-dimethylpentane, 2,4-dimethylpentane, 3,3-dimethylpentane, 3-ethylpentane, 2,2,3-trimethylbutane **1.48** $2 C_8H_{18} + 25 O_2 \rightarrow 16 CO_2 + 18 H_2O$ **1.50** hydrogen **1.54** A semisynthetic compound is one derived by laboratory manipulation of a naturally occurring substance. **1.56** Shape affects reactivity and biological properties. **1.58** branched-chain alkanes **1.60** On a percentage basis, the molecular weights of hexadecane and heptadecane are closer than those of methane and ethane. **1.62** CH_3's are primary, CH_2's are secondary, CH's are tertiary, and C's are quaternary. **1.64** Compounds with similar structures are often miscible. **1.68** ethylcyclopropane, 1,1-dimethylcyclopropane, 1,2-dimethylcyclopropane, methylcyclobutane, cyclopentane.

Chapter 2

2.1 (a) 2-methyl-3-heptene (b) 2-methyl-1,5-hexadiene **2.3** Compounds (a) and (c) can exist as cis–trans isomers. **2.5** (a), (b), (c) butane; (d) methylcyclohexane **2.8** (a) 3-ethyl-2-pentene (b) 2,3-dimethyl-2-butene or 2,3-dimethyl-1-butene **2.10** 2-ethyl-1-butene or 3-methyl-2-pentene **2.11** $(CH_3)_3C^+$ **2.13** (a) *ortho*-bromochlorobenzene (b) butylbenzene (c) *ortho*-bromomethylbenzene or *ortho*-bromotoluene **2.18** The word "aromatic" refers only to chemical structure. **2.20** alkene: -*ene*; alkyne: -*yne*; aromatic: -*benzene* **2.22** (a) 1-pentene (b) 5-methyl-2-pentyne (c) 2,3-dimethyl-2-butene (d) 2-ethyl-3-methyl-1,3-pentadiene (e) 4-ethyl-3,5-dimethylcyclohexene (f) 3,3-diethylcyclobutene **2.24** 1-pentyne, 2-pentyne, 3-methyl-1-butyne **2.30** 1,2-pentadiene, 1,3-pentadiene, 1,4-pentadiene, 3-methyl-1,2-butadiene, 2-methyl-1,3-butadiene **2.32** A triple bond is linear. **2.34** 2-pentene **2.36** identical: (a), (b) **2.44** $2 C_2H_2 + 5 O_2 \rightarrow 4 CO_2 + 2 H_2O$ **2.50** (b) bromobenzene **2.52** cyclohexane **2.56** The cis double bond in rhodopsin isomerizes to trans. **2.60** ultraviolet **2.62** They have no double bonds. **2.64** Cyclohexene reacts with Br_2. **2.70** 2-bromopentane and 3-bromopentane **2.74** hydration of 3,4,4-trimethyl-2-pentene or of 2-ethyl-3,3-dimethyl-1-butene

Chapter 3

3.1 (a) alcohol (b) alcohol (c) phenol (d) alcohol (e) ether (f) ether **3.2** A hydroxyl group is a part of a larger molecule.   **3.4** (a) 3-pentanol (b) 2-ethyl-1-pentanol (c) 5-bromo-2-ethyl-1-pentanol (d) 4,4-dimethylcyclohexanol   **3.5** 14.3: tertiary (a); secondary (b), (c), (d); primary (e); 14.4: secondary (a), (d); primary (b), (c)   **3.6** (a)   **3.7** (b)   **3.8** (a) propene (b) cyclohexene (c) 4-methyl-1-pentene or 4-methyl-2-pentene   **3.9** (a) 2,3-dimethyl-2-butanol (b) 1-butanol or 2-butanol   **3.11** (a) 2-propanol (b) cycloheptanol (c) 3-methyl-1-butanol   **3.13** (a) p-chlorophenol (b) 4-bromo-3-methylphenol   **3.16** A primary alcohol has one, a secondary alcohol has two, and a tertiary alcohol has three organic R groups attached to the OH-bearing carbon.   **3.18** phenol   **3.20** phenol, ether   **3.24** (a) o-ethylphenol (b) isopropyl methyl ether (c) methyl p-nitrophenyl ether (d) cyclopentyl methyl ether (e) o-butylphenol (f) dipropyl ether   **3.26** (a) < (c) < (b)   **3.28** a ketone   **3.30** aldehyde or carboxylic acid   **3.32** Phenols dissolve in aqueous NaOH; alcohols don't.   **3.38** odor   **3.42** Alcohols can form hydrogen bonds; thiols and alkyl chlorides can't.   **3.44** depressant   **3.48** alcohol dehydrogenase   **3.50** sterilant, anesthetic   **3.52** vitamin E   **3.54** Chlorofluorocarbons destroy ozone.   **3.58** Neither is similar to water.   **3.60** Antiseptics can be used on living tissue.   **3.62** flammability   **3.66** CH_3COOH **3.68** Sugar hydroxyl groups form hydrogen bonds with water.

Chapter 4

4.1 primary (a), (c); secondary (b), (d); tertiary (e)   **4.2** (a) propylamine (b) dimethylamine (c) N-ethylaniline   **4.4** $(CH_3)_2NH + H_2O \rightleftharpoons (CH_3)_2NH_2{}^+ + OH^-$ **4.5** (a) $CH_3CH_2CH(CH_3)NH_2 + HBr(aq) \rightleftharpoons CH_3CH_2CH(CH_3)NH_3{}^+Br^-(aq)$ (b) $C_6H_5NH_2 + HCl(aq) \rightleftharpoons C_6H_5NH_3{}^+Cl^-(aq)$ (c) $CH_3CH_2NH_2 + CH_3COOH(aq) \rightleftharpoons CH_3CH_2NH_3{}^+CH_3COO^-(aq)$ (d) $CH_3NH_3{}^+Cl^- + NaOH(aq) \rightleftharpoons CH_3NH_2 + H_2O + NaCl(aq)$ **4.6** (a) sec-butylammonium bromide (b) anilinium chloride (c) ethylammonium acetate   **4.7** (a) ethylamine (b) triethylamine   **4.9** (a) tertiary (b) primary   **4.10** N-ethyl-N,N-dimethylcyclohexyl ammonium chloride   **4.12** -amine   **4.16** Problem 15.14: primary (a), (f); secondary (b), (c), (e); tertiary (d). Problem 15.15: primary (a), (b) secondary (c); tertiary (d).   **4.18** weaker   **4.20** (a) $CH_3CH_2NH_2 + H_2O \rightleftharpoons CH_3CH_2NH_3 + OH^-$; ethylamine is predominant (b) ethylammonium ion is predominant   **4.24** diethylamine   **4.26** tertiary   **4.28** Quinine dissolves in aqueous acid.   **4.30** does

not react   **4.34** The salts are more soluble in body fluids.   **4.36** $C_{17}H_{20}N_2S$ **4.40** Decylamine has a larger organic part.   **4.44** (a) purine   **4.48** Amines are (a) stronger smelling, (b) more basic, (c) lower-boiling.

Chapter 5

5.1 (a) ester, carboxylic acid (b) ketone (c) aldehyde (d) ketone (e) aldehyde (f) ester   **5.3** (a) pentanal (b) 3-pentanone (c) 4-methylhexanal   **5.5** (a) 2-methyl-1-propanol (b) m-chlorobenzyl alcohol (c) cyclopentanol   **5.6** (a) 4,4-dimethylcyclohexanone (b) 4-methylpentanal (c) 2-methylpentanal   **5.9** hemiacetal: (b); acetal (a), (d)   **5.10** (a) $C_6H_5CH_2COCH_2CH_3$, methanol (b) formaldehyde, propanol   **5.12** (a), (b)   **5.14** -al, -one   **5.16** (a) cyclopentanone (b) octanal (c) $CH_3COCH_2CH_2CH_2CHO$ (d) 2-hydroxycyclopentanone   **5.18** (a), (b), (d), (e)   **5.22** (a) 2-methylbutanal (b) 2-methylpentanal (c) 2,2-dimethylpropanal (d) 3-methyl-2-nitrobenzaldehyde (e) 3-ethyl-3-methylpentanal (f) 3,3-dibromobutanal   **5.24** (a) butanal (b) butanal (c) 2-butanone (d) impossible structure (e) 2-butanone   **5.26** acetal   **5.28** (a) $CH_3COCH_3 \rightarrow CH_3CH(OH)CH_3$ (b) $CH_3CHO \rightarrow CH_3COOH$ (c) $CH_3CHO + CH_3OH \rightarrow CH_3CH(OCH_3)_2$ **5.30** (a) cyclopentanol (b) 1-hexanol (c) 2-ethyl-1-pentanol (d) benzyl alcohol (e) 2-butanol (f) 2,2-dichloroethanol   **5.32** A Tollens' test is positive for pentanal.   **5.34** (a) p-methylbenzyl alcohol (b) 2-ethyl-4-methyl-1-pentanol (c) 2-buten-1-ol (d) m-hydroxybenzyl alcohol   **5.38** (a) butanal, methanol, and ethanol (b) acetone and ethylene glycol (c) cyclohexanone and methanol (d) benzaldehyde and ethanol   **5.40** $HOCH_2CH_2CH_2CH_2CHO$ and CH_3OH **5.42** The hemiacetal and aldehyde forms of glucose are in equilibrium.   **5.46** (a) cyclohexanone (b) 3-cyclohexenol (c) 3,4-dibromocyclohexanone   **5.50** 2-methylpropanal   **5.56** HCN is a gas that reacts with NaOH.   **5.58** HCN is generated outside the insect.   **5.60** no   **5.64** (a) 2-methyl-3-pentanone (b) 1,5-hexadiene (c) m-bromotoluene (d) 4,5,5-trimethyl-3-hexanone (e) o-methoxyisopropylbenzene (f) 5,5-diethyl-3-heptyne

Chapter 6

6.1 (b) > (c) > (a)   **6.2** highest (a); lowest (c)   **6.3** (a) $CH_3CH_2CONH_2$ (b) $CH_3CH_2CH_2COOH$ (c) $(CH_3)_2CHCOOH$ **6.4** (a) 4-methylpentanoic acid (b) isopropyl butanoate (c) *N*-methyl-*p*-chlorobenzamide **6.6** (a) potassium butanoate (b) barium 2-methylpentanoate **6.7** (a) 2 HCOO⁻Ca²⁺ (b) H_2C=CHCOO⁻Na⁺ **6.8** HCOOCH₂CH(CH₃)₂ **6.9** (a) cyclohexanol and 4-methylpentanoic acid (b) 2-propanol and pentanoic acid **6.10** (a) 2-propanol and 2-methylpropanoic acid (b) ethanol and 2-butenoic acid (c) 1-propanol and *p*-bromobenzoic acid **6.13** acetic acid and *p*-ethoxyaniline **6.14** (a) 2-butenoic acid and methylamine (b) *p*-chlorobenzoic acid and diethylamine **6.15** (a) amide; acetic acid and ammonia (b) phosphate ester; ethanol and phosphoric acid (c) ester; propanoic acid and methanol **6.18** The carbonyl carbon is bonded to an electronegative atom. **6.22** (a–b) CH₃CH₂COOH (c) CH₃CH₂COO⁻ **6.24** (a) potassium 3-ethylpentanoate (b) ammonium benzoate (c) calcium propanoate **6.26** heptanoic acid, 2-methylhexanoic acid, 3-methylhexanoic acid **6.30** (a) 9.34 g (b) 28.1 g **6.32** 190 mL **6.34** chloroacetic acid **6.36** (a) 3-methylbutyl acetate (b) methyl 4-methylpentanoate (c) ethyl 2,2-dimethylpropanoate (d) ethyl benzoate (e) cyclopentyl propanoate **6.38** (a) methanol and pentanoic acid (b) 2-propanol and 2-methylbutanoic acid (c) cyclohexanol and acetic acid (d) phenol and *o*-hydroxybenzoic acid **6.40** (a) 2-ethylbutanoic acid and ammonia (b) benzoic acid and aniline (c) formic acid and dimethylamine (d) propanoic acid and isopropylamine **6.42** (a) 3-methylpentanoic acid and ammonia (b) acetic acid and aniline (c) methylethylamine and benzoic acid (d) 2,3-dibromohexanoic acid and ammonia **6.44** *o*-aminobenzoic acid and methanol **6.46** HOCH₂CH₂CH₂COOH **6.50** H₂NCH₂CH₂CH₂CH₂CH₂COOH **6.52** acetic anhydride and *p*-ethoxyaniline **6.54** a hydrogen on carbon next to the ester **6.58** They react with water to give phosphoric acid. **6.60** (a) addition (b) addition (c) elimination **6.62** molds and fungi **6.64** CH₃COSH **6.70** Propanamide forms hydrogen bonds. **6.72** ethyl butanoate **6.74** (a) 2-chloro-3,4-dimethyl-3-hexene (b) *N*-methyl-*N*-phenylpropanamide (c) phenyl 2,2-diethylbutanoate (d) *N*-ethyl-*o*-nitrobenzamide

Chapter 7

7.1 aromatic ring: phenylalanine, tryptophan, tyrosine; sulfur: cysteine, methionine; alcohols: serine, threonine; alkyl groups: alanine, isoleucine, leucine, valine **7.3** low pH, protonated; isoelectric point, neutral; high pH, deprotonated **7.4** hydrophilic **7.5** (a), (c) **7.6** handed: shoe, shirt, coin; not handed: pencil, paper clip, drinking glass **7.7** 2-aminobutane has four different groups attached to C2 **7.8** (b), (c) **7.9** isoleucine, threonine **7.10** Val-Cys, Cys-Val **7.11** Val-Tyr-Gly, Val-Gly-Tyr, Tyr-Gly-Val, Tyr-Val-Gly, Gly-Val-Tyr, Gly-Tyr-Val **7.12** (a) Leu-Asp (b) Tyr-Ser-Lys **7.14** (a) hydrogen bonds (b) hydrophobic (c) salt bridge (d) hydrophobic (e) hydrogen bonds **7.15** eight amino acids **7.16** a molecule with both an amino group and a carboxylic acid group **7.18** (a) serine (b) threonine (c) proline (d) phenylalanine (e) cysteine **7.20** (a) valine (b) threonine (c) cysteine (d) tyrosine **7.22** (a) basic (b) neutral (c) acidic (d) neutral **7.26** *Chiral* means handed; *a-chiral* means not handed. **7.28** (a), (c), (e), (f) **7.30** (a) is chiral **7.32** a large amino acid polymer **7.34** A simple protein contains only amino acids; a conjugated protein contains other kinds of compounds as well. **7.36** A fibrous protein is thread-like and water-insoluble; a globular protein is coiled up and water-soluble. **7.38** (a) amino acid sequence (b) helical or pleated sheet segments (c) overall three-dimensional coiling (d) aggregate structure of several chains **7.40** It forms disulfide links. **7.42** (a) Hydrophobic amino acids stay on inside of globular proteins. (b) Salt bridges hold charged groups near each other. (c) Hydrogen bonding holds groups together. **7.44** Tertiary structure is disrupted. **7.46** N-terminus: Asp; C-terminus: Phe **7.50** Proteins are hydrolyzed in the stomach. **7.52** N-terminus: Tyr; C-terminus: Met **7.54** Glycine is an internal salt. **7.56** Val-Gly-Ser-Met-Ala-Asp **7.58** They have no net charge at this pH. **7.60** Asp, Phe **7.62** 94% **7.64** one that does not contain all essential amino acids **7.66** identity and amounts of amino acids in a protein **7.68** Enzyme is denatured during processing. **7.70** Tropocollagen has a complex quaternary structure. **7.72** basic pH, perhaps 9.0

Chapter 8

8.2 (a) hydration of fumaric acid (b) oxidation of squalene (c) phosphorylation of glucose (d) hydrolysis of cellulose **8.3** ligase **8.4** Vitamin A is a hydrocarbon; vitamin C has many hydroxyl groups. **8.6** An enzyme is a biological catalyst, usually a protein. **8.8** by lowering the activation energy

8.10 (a) breaking a bond by addition of water **(b)** isomerization of a substrate **(c)** addition or elimination of a small molecule **8.12** A cofactor is a small, nonprotein part of an enzyme; a coenzyme is an organic cofactor. **8.14** Enzymes bind substrates in their active sites and hold them in the correct position for reaction. **8.16 (a)** hydrolysis of a protein **(b)** bond formation in DNA **(c)** transfer of a methyl group **8.18** competitive inhibition **8.20** hydrolases **8.24** by acting as a noncompetitive inhibitor for bacterial enzymes **8.26** Reaction rates increase with substrate concentration and then level off. **8.28 (a)** decrease **(b)** probable decrease **(c)** probable stop **(d)** increase **8.30** competitive, noncompetitive, irreversible **8.32** competitive and noncompetitive: noncovalent bonds; irreversible: covalent bonds **8.34** Papain hydrolyzes peptide bonds in meat. **8.36** The product of an enzyme-catalyzed reaction series is an inhibitor for an earlier step in the series. **8.38** A positive regulator changes the shape of an active site so that it accepts substrate more readily. A negative regulator changes the shape of an active site so that it accepts substrate less readily. **8.40** fewer calories **8.42** A hormone is a chemical messenger produced in the endocrine glands. A neurotransmitter transmits chemical messages in the nervous system. **8.44** vitamins **8.46** Vitamin C is not stored in the body because it is water-soluble. **8.48** hypothalamus **8.52** Enzymes are usually proteins; hormones can have any structure. **8.54** cAMP is a second messenger released in a cell's interior. **8.56** central and peripheral **8.58** cholinergic and adrenergic **8.60** autonomic: unconsciously controlled functions; somatic: consciously controlled functions **8.62** CK(MB), AST, and LDH$_1$ **8.64** Water-soluble vitamins are not stored in the body. **8.66** by preventing cell wall synthesis **8.70** Pyroglutamic acid is a cyclic amide of glutamic acid. **8.72** Sulfa drugs mimic PABA.

Chapter 9

9.1 exergonic **(a)**, **(c)**; endergonic **(b)**; **(a)** proceeds furthest **9.3** What is favorable in one direction is unfavorable in the other direction. **9.4** Acetyl phosphate + ADP → acetate + ATP; favorable by −3.0 kcal **9.5 (a)** NAD$^+$ **(b)** FAD **9.7** Citric acid, isocitric acid **9.8** Acetyl SCoA + 2 O$_2$ → 2 CO$_2$ + HSCoA + H$_2$O **9.9** A stable carboxylate ion is produced. **9.10** NADH/H$^+$, FMNH$_2$, FeSP

(red), CoQ **9.11** It has a large hydrocarbon portion. **9.12** exergonic: releases energy; endergonic: absorbs energy **9.14** Both ΔH and ΔS are important. **9.16** exergonic: **(a)**, **(c)**; endergonic: **(b)**; furthest toward products: **(a)** **9.18** Prokaryotic: bacteria and algae; eukaryotic: higher organisms **9.20** cytoplasm: everything between the cell membrane and the nuclear membrane in a eukaryotic cell; cytosol: material that fills the interior of the cell **9.22** ATP is produced there. **9.24** digestion: breakdown of bulk food; metabolism: total of all cellular reactions **9.26** acetyl SCoA **9.28** adenosine triphosphate **9.30** Energy is released when a phosphate group is transferred. **9.34** exergonic by 4.5 kcal **9.36** No. Reaction is unfavorable by 4.0 kcal/mol. **9.38** in mitochondria **9.40** oxaloacetic acid **9.42** formation of ATP by transfer of a phosphate to ADP from another phosphate-containing compound such as GTP **9.44** 3 NADH, 1 FADH$_2$ **9.46** the electron transport system **9.48** NADH or FADH$_2$ is oxidized; ADP is phosphorylated. **9.50** H$_2$O and energy **9.52** iron **9.54** FADH$_2$, CoQ, 2 Fe^{3+} **9.56** It would stop. **9.58** the minimum energy expenditure to keep functioning **9.60** Energy is needed above the minimum. **9.62** They prevent electron transport. **9.64** 2,4-dinitrophenol **9.68** oxidoreductases **9.70** cytochrome oxidase

Chapter 10

10.1 (a) aldopentose **(b)** ketotriose **(c)** aldotetrose **10.4** eight **10.5 (a)** D-ribose **(b)** L-mannose **10.8** β anomer **10.12** reducing because the right-hand sugar has a hemiacetal linkage **10.13** a β 1,4 glycoside link **10.14** two β-D-glucoses **10.15** nonreducing **10.16** a polyhydroxy aldehyde or ketone **10.18** aldose: aldehyde; ketose: ketone **10.22** glucose: in most food, fruits, and vegetables; galactose: in brain tissue and lactose; fructose: in honey and fruits **10.24** Enantiomers are mirror-image stereoisomers; diastereomers are non-mirror-image stereoisomers. **10.26** no **10.28** The product from erythrose has a plane of symmetry. **10.30** It reduces oxidizing agents like Benedict's reagent. **10.32** α form **10.40** A new chiral center is formed when fructose is reduced. **10.42** A hemiacetal has a carbon atom bonded to one —OH group and one —OR group. An acetal has a carbon bonded to two —OR groups. **10.46** Maltose occurs in fermenting grains; lactose occurs in milk; sucrose occurs in many plants.   **10.48** Both are α-glucose polymers; amylopectin has branches.   **10.50** Both monosaccharide units are acetals.   **10.52** Gentiobiose is reducing and has a hemiacetal linkage on the right-hand sugar.   **10.54** Trehalose is nonreducing and has an acetal linkage.   **10.56** one that rotates plane-polarized light   **10.58** D-Fructose rotates light more strongly than D-glucose.   **10.60** It is dietary fiber.   **10.62** Insulin regulates blood sugar.   **10.66** It causes the production of antibodies.   **10.68** Their caloric value is low relative to their sweetness.   **10.70** They are diastereomers; their properties have no relation.   **10.76** Saliva hydrolyzes the glycoside links, yielding glucose.

Chapter 11

11.1 (a) glucose + galactose **(b)** glucose + fructose   **11.2 (a)** glycogenolysis **(b)** gluconeogenesis **(c)** glycogenesis   **11.3** glycolysis, glycogenesis, pentose phosphate pathway   **11.4** steps 6, 7, and 9, 10   **11.5** steps 2, 5, and 8   **22.6** at step 3   **11.7** They differ in the stereochemistry of the hydroxyl group at C4.   **11.8** Glucose → 2 pyruvate → 2 CO_2 + 2 acetyl SCoA; 2 acetyl SCoA → 4 CO_2 **11.9** Because NADH can't cross the inner mitochondrial membrane but must be converted into $FADH_2$ which produces only 2 ATPs.   **11.10** PP_i; $\Delta G = -6.9$ kcal; favorable   **11.11** energy cost of 2 ATP + 2 GTP   **11.12** NAD^+; oxidation   **11.13** 2 fructose 6-phosphate + 3 CO_2 + glyceraldehyde 3-phosphate   **11.14** mouth, stomach, and small intestines; hydrolysis of carbohydrates, fats, and proteins   **11.16** anaerobic: lactate; aerobic: acetyl SCoA; fermentation: ethanol   **11.18** glycolysis: breakdown of glucose to pyruvate; gluconeogenesis: synthesis of glucose from pyruvate; glycogenesis: synthesis of glycogen from glucose; glycogenolysis: breakdown of glycogen to glucose   **11.20** pyruvate   **11.22 (a)** steps 1, 3, 6 **(b)** step 6 **(c)** step 9   **11.24(a)** 6 mol **(b)** 2 × 3 mol **(c)** 12 mol   **11.26** 72 mol   **11.28** 12 mol   **11.30** allows synthesis of glucose from fat or protein   **11.32** triggers breakdown of glycogen   **11.34** ketone bodies   **11.36** liver   **11.38** Some steps are too exergonic to be reversed.   **11.40** lactate, glycerol   **11.42** Some steps are too exergonic to be reversed.   **11.44** fructose 6-phosphate, glyceraldehyde 3-phosphate, or ribose 5-phosphate   **11.46** deficiency of enzyme for breakdown of muscle glycogen   **11.50** Energy for sprinting comes from creatine phosphate. **11.52**

Enzyme for galactose catabolism is missing.   **11.54** phosphorylation   **11.56** Glucose from glycogen is phosphorylated directly with inorganic phosphate rather than with ATP.

Chapter 12

12.1 $CH_3(CH_2)_{18}COOCH_2(CH_2)_{30}CH_3$
12.5 $[CH_3(CH_2)_7CH{=}CH(CH_2)_7CO_2{}^-]_2Ca^{2+}$
12.9 double bond, ketone   **12.10** An aromatic A ring makes them estrogens.   **12.12** a naturally occurring molecule that dissolves in nonpolar solvents   **12.14** A long, straight-chain carboxylic acid   **12.18** insulation, long-term fuel storage   **12.20** $CH_3(CH_2)_{14}COOCH_2(CH_2)_{14}CH_3$
12.22 The two stearic acids can be bonded to adjacent or nonadjacent oxygens.
12.24 (a) $CH_3(CH_2)_{14}COO^-Na^+$
(b) $CH_3(CH_2)_7CH{=}CH(CH_2)_7COOCH_2(CH_2)_8CH_3$
12.26 glycerol, stearic acid, oleic acid, linolenic acid   **12.28** higher   **12.30** A fat has three fatty acids bonded to glycerol; a phospholipid has a phosphate group.   **12.32** The charged head group is necessary.   **12.34** A soap micelle is a monolayer.   **12.36** sphingomyelins and sphingoglycolipids   **12.40** four fatty acids, two phosphoric acids, three glycerols   **12.42** The A ring of estradiol is a phenol; the A ring of testosterone has a double bond and a ketone.   **12.44** regulates membrane fluidity   **12.48** arachidonic acid   **12.52** They are not biodegradable.   **12.54** active transport   **12.56** It inhibits an enzyme.   **12.58** They are species-specific.   **12.60 (b)** choline, glycerol, phosphoric acid, two fatty acids **(d)** sphingosine, fatty acid, phosphoric acid, choline **(e)** sphingosine, fatty acid, monosaccharide **(f)** three oleic acids, glycerol   **12.62** lower melting   **12.64** $CH_3(CH_2)_{16}COOCH_2(CH_2)_{20}CH_3$
12.66 to send chemical messages in the body   **12.68** Like soaps, lecithins are attracted to both water and fats.   **12.70** 11–12 g

Chapter 13

13.1 step 5 of glycolysis   **13.2 (a)** 6 acetyl CoA, 5 turns **(b)** 10 acetyl CoA, 9 turns   **13.4** steps 6–8 of the citric acid cycle   **13.5** 129 ATP   **13.6** 260 g   **13.7** (i), (iii)   **13.8** small intestine   **13.10** to emulsify lipids   **13.12** lipoproteins that aid in lipid transport   **13.14** in association with albumins   **13.16** 5 ATP   **13.18** 20 ATP   **13.20** in mitochondrial matrix   **13.22** A thiol ester is formed with HSCoA.   **13.24** A different product results from each turn. **13.26** (d) > (b) > (c) > (a) **13.28** CH₃COCH₂COSCoA + HSCoA → 2 CH₃COSCoA **13.30** (a) three turns (b) six turns **13.32** Acetyl SCoA **13.34** They are made from a two-carbon unit. **13.36** seven rounds **13.38** deposition of cholesterol in the arteries **13.40** high HDL/LDL **13.42** ^{1}H **13.44** Carbohydrates are metabolized to yield acetyl SCoA, which is converted into lipids. **13.46** acetone, 3-hydroxybutyrate, acetoacetate **13.48** 22 **13.50** Fatty acids yield more than twice as much energy per gram as carbohydrates do.

Chapter 14

14.1 $(CH_3)_2CHCH_2COCOO^-$
14.2 $C_6H_5CH_2COCOO^-$ **14.3** fumarate → malate → oxaloacetate → aspartate **14.5** transaminations: (1), (2), (3); hydrolysis: (3); oxidation: (3) **14.6** in the stomach **14.8** food, tissue breakdown **14.10** An amino group of an amino acid exchanges with the keto group of an α-keto acid. **14.12** (a) $CH_3CH_2CH(CH_3)COCOO^-$ (b) $HSCH_2COCOO^-$ (c) $C_6H_5CH_2COCOO^-$ **14.14** An amino group is replaced by a keto group. **14.16** a dehydrogenase **14.18** ammonia **14.20** an amino acid that can be converted into a ketone body **14.22** Ornithine is the reactant in step 1 and the product of step 4. **14.24** Nonessential amino acids require 1–3 steps; essential amino acids require 7–10 steps. **14.26** reductive amination; oxidative deamination **14.28** a defect in tyrosine biosynthesis **14.30** growing children, pregnant women, those with healing wounds **14.32** inadequate protein intake, kidney disease **14.34** a compound that is foreign to the body **14.36** Two molecules of ATP are consumed. **14.38** $(CH_3)_2CHCH(NH_2)COOH$ + $CH_3COCOO^- → (CH_3)_2CHCOCOO^-$ + $CH_3CH(NH_2)COOH$; valine aminotransferase **14.40** yes **14.42** Tissue is constantly undergoing degradation and resynthesis.

Chapter 15

15.3 deoxythymidine 5′-monophosphate, adenosine 5′-monophosphate, adenosine 5′-diphosphate, adenosine 5′-triphosphate
15.5 (a) C-G-G-A-T-C-A (b) T-T-A-C-C-G-A-G-T **15.6** A and G don't hydrogen-bond to each other. **15.8** (a) C-U-A-A-U-G-G-C-A-U (b) A-U-A-C-C-G-A-U-C-C-G-U **15.9** (a) GCU, GCC, GCA, GCG (b) UUU, UUC (c) UUA, UUG,

CUU, CUC, CUA, CUG (d) GUU, GUC, GUA, GUG (e) UAU, UAC **15.10** (a) Ile (b) Ala (c) Arg (d) Asn **15.11** Leu-Met-Ala-Trp-Pro
15.12 GAA, UAC, CGA, ACC, GGG, AUU
15.13 GAA-TAC-CGA-ACC-GGG-ATT
15.14 TGA codes for UGA, which is a stop codon. **15.15** (a) Stop becomes Tyr. (b) Gly becomes Ala. **15.16** A nucleotide consists of a heterocyclic amine base, an aldopentose, and a phosphoric acid. **15.18** oxygen atom missing **15.20** adenine, cytosine, guanine, and uracil **15.22** Messenger RNA carries genetic message from DNA to ribosomes; ribosomal RNA makes up ribosomes; transfer RNA transports amino acids to ribosomes. **15.24** A chromosome is a large molecule of DNA; a gene is a part of the chromosome that codes for a single protein. **15.26** approximately 100,000 **15.28** They hydrogen-bond to each other. **15.30** The 5′ end has a free phosphoric acid group; the 3′ end has a free —OH group. **15.36** a sequence of three nucleotides on mRNA that codes for a specific amino acid **15.38** cloverleaf; contains 70–100 nucleotides **15.40** The purines A and G pair with T and C, respectively. **15.42** (a) CCU, CCC, CCA, CCG (b) AAA, AAG (c) AUG
15.44 (a) UGA (b) CCU (c) GAA
15.46 U-A-C-C-G-A **15.48** an error in base sequence that occurs during DNA replication **15.50** Thr replaces Ile. **15.52** If codons were made up of 2 nucleotides, only $4^2 = 16$ combinations would be possible, and it would not be possible to code for all 20 amino acids. **15.54** A mutation changes the biological activity of a protein and can lead to uncontrollable cell growth.
15.56 TAT-GGT-GGT-TTT-ATG-TAA
15.60 Reverse transcriptases give DNA from RNA. **15.62** They are small and contain plasmids. **15.64** proteins **15.66** Glu: GAA, GAG; Val: GUU, GUC, GUA, GUG **15.68** TAA → TAG would have no effect; TAA → GAA would change Ile to Leu. **15.70** At least two bases in DNA must differ. **15.72** It is removed.

Chapter 16

16.2 intracellular fluid **16.4** higher in arterial; lower in venous **16.6** vasopressin; to control water–electrolyte balance **16.8** red blood cells, platelets, white blood cells **16.10** Clotting would be dangerous in these organs. **16.12** four **16.14** Oxyhemoglobin is red; deoxyhemoglobin is purple.

16.16 The binding and release of each successive oxygen is enhanced.   **16.18** approximately 30 mm Hg   **16.20** releases O_2: **(a)**, **(b)**, **(c)**   **16.22** by inhibiting the decarboxylase enzyme   **16.24** lymphocytes cause phagocytosis   **16.26** T cells release a toxin that destroys infected cells.   **16.28** Ca^{2+}, vitamin K   **16.30** so that they do not trigger unwanted clots   **16.32** $H^+ + HPO_4^{2-} \rightarrow H_2PO_4^-$; $H^+ + HCO_3^- \rightarrow H_2O + CO_2$ **16.34** Ice crystals disrupt cell membranes.   **16.36** ATP production must continue, and metabolism must be anaerobic.   **16.38** Transport occurs in only one direction.   **16.40** to measure amounts of colored products   **16.44** Antibodies are passed from mother to child.   **16.46** Active transport involves an expenditure of energy to carry a substance across a membrane. Osmosis is spontaneous.   **16.48** The equilibrium is necessary for pH control.

CREDITS

Photo/Illustration Credits

2, Paul Lowe/Matrix; 5, Sygma; 13, Gary Retherford/ Photo Researchers (vitamins) and Any Levin/Photo Researchers (vegetables); 16, image generated on the Evans & Sutherland PS 390 computer graphics system using Tripos® SYBYL computational chemistry software; 33, Hans Reinhard/Okapia/Photo Researchers; 45, © Joel Gordon, Courtesy West Publishing Company/CHEMISTRY, I/E by Radel and Navidi; 50, Luiz Claudio Marigo/ Peter Arnold; 51, © Christo, 1983; 52, Richard Megna/ Fundamental Photographs; 53, Michael English/Medical Images Inc.; 58, Dr. E.R. Degginger; 61, Tom Bochsler; 66, Philippe Plailly/Science Photo Library/Photo Researchers; 70, Donald Clegg and Roxy Wilson; 71, Richard Megna/Fundamental Photographs; 81, Donald Clegg and Roxy Wilson; 82, John Bova/Photo Researchers; 86, Anaquest Pharmaceutical Products; 90, NASA/Science Source/Photo Researchers; 95, Susan Lapides/Woodfin Camp & Associates.

100, reproduced from *The Merck Index*, Eleventh Edition (1989), S. Budavari, M.J. O'Neil, A. Smith, P.E. Heckelman, Eds., by permission of the copyright owner, Merck & Co., Inc., Rahway, NJ, U.S.A. © Merck & Co., Inc., 1989; 101, reproduced with permission of Medical Economics Co., Oradell, N.J. (No Doz) and Dean: Lange's Handbook of Chemistry, 13/E, © 1987, McGraw-Hill. Reprinted with permission of McGraw-Hill; 95, 108, Donald Clegg and Roxy Wilson; 110, Jerome Wexler/Photo Researchers; 111, Dr. E.R. Degginger (opium poppy) and Hans Pfletschinger/Peter Arnold (mosquito); 115, Will and Deni McIntyre/Photo Researchers; 119, T. Eisner and D. Aneshanslet, Cornell University; 125, J.P. Ferrero Jacana/Photo Researchers; 129, S. Varnedoe; 130, © Joel Gordon, Courtesy West Publishing Company/CHEMISTRY, I/E by Radel and Navidi; 141, adapted with permission of Academic Press Inc., San Diego, Calif.; 146, Bridgeman/Art Resource; 152, Tom Bochsler/Photography Unlimited; 159, Nobel Proctor/Photo Researchers; 163, The Granger Collection; 164, CNRI/Science Photo Library/Photo Researchers; 176, Tom Bochsler/Photography Unlimited; 182, Jean Marc Barey/Agence Vandystadt/Photo Researchers. 197, Bill Longcore/Photo Researchers; 198, 199, adapted from Annino and Grese, *Clinical Chemistry*, 4th Ed., copyright 1976, Little, Brown & Company, used with permission. 203, William Ober and Claire Garrison; 206, after

Jane Richardson (Fig. 18.14); 207, Science Photo Library/ Photo Researchers; 216, CNRI/Science Photo Library/ Photo Researchers; 219, © Richard Megna/Fundamental Photographs; 225, from Mathews and van Holde, *Biochemistry*, copyright 1990, with permission of The Benjamin/Cummings Publishing Company; 239, Dr. E.R. Degginger; 240, William Ober and Claire Garrison (nerve cell); 243, Dr. Edward J. Bottone, Dept. of Microbiology, Mount Sinai Hospital, New York; 247, Globus Brothers/ The Stock Market; 249, Michael English/Medical Images; 255, Richard Megna/Fundamental Photographs; 258, Peter G. Aitken/Photo Researchers; 269, image generated on the Evans & Sutherland PS 390 computer graphics system using Tripos® SYBYL computational chemistry software; 272, Norman R. Lightfoot/Photo Researchers; 276, CNRI/ Science Photo Library/Photo Researchers; 283, James H. Carmichael, Jr./Photo Researchers; 288, Bonnie Rauch/ Photo Researchers; 292, Don W. Fawcett/Ito/Photo Researchers; 293, Rainer Berg/Okapia/Science Source/Photo Researchers; 296, George Haling/Photo Researchers; 297, M.I. Walker/Photo Researchers.

300, United Nations Photo; 304, Lennart Nilsson: *BEHOLD MAN*; 307, CNRI/Science Photo Library/Photo Researchers; 308, from Mathews and van Holde, *Biochemistry*, copyright 1990, with permission of The Benjamin/Cummings Publishing Company; 315, Jack Fields/Photo Researchers; 317, Frederick Martini; 320, Eli Lilly and Company; 330, Peter Aprahamian/Science Library/Photo Researchers; 334, F. Durand/Sygma; 339, Patrick Donehue/Photo Researchers; 344, CNRI/Science Photo Library/Photo Researchers; 347, William Ober and Claire Garrison (cell membrane) and Don W. Fawcett/ Photo Researchers; 348, William Ober and Claire Garrison; 353, SIU/Photo Researchers; 355, John Serrao/Photo Researchers; 359, Tripos Associates, St. Louis, MO, U.S.A.; 362, Craig Luce (liver); 363, William Ober and Claire Garrison; 367, Biophoto Associates, Science Source/Photo Researchers; 369, Lennart Nilsson from *The Body Victorious*, © Delacorte Press; 379, Toshiba America Medical Systems, Inc.; 382, M.I. Walker/Photo Researchers; 384, John Colwell from Grant Heilman; 389, Andrew J. Martinez/Photo Researchers; 397, Lennart Nilsson: *BEHOLD MAN*.

400, Land O'Lakes, Inc.; 402, Howard Sochurek/ Medical Images; 408, Richard Feldman/National Institute

I N D E X

The page references given in boldface refer to pages where terms are defined.

A-50

SOME IMPORTANT FAMILIES OF ORGANIC MOLECULES

Family Name	Functional Group Structure[a]	Simple Example	Name Ending
Alkane	(contains only C—H and C—C single bonds)	CH_3CH_3 ethane	*-ane*
Alkene	$\diagdown C{=}C \diagup$	$H_2C{=}CH_2$ ethylene	*-ene*
Alkyne	—C≡C—	H—C≡C—H acetylene (ethyne)	-yne
Arene	(benzene ring structure)	(benzene ring structure) benzene	none
Alkyl halide[b]	—C—X	CH_3—Cl methyl chloride	none
Alcohol	—C—O—H	CH_3—OH methyl alcohol (methanol)	*-ol*
Ether	—C—O—C—	CH_3—O—CH_3 dimethyl ether	none
Amine	—N—H, —N—H, —N—	CH_3—NH_2 methylamine	*-amine*
Aldehyde	—C(=O)—H	CH_3—C(=O)—H acetaldehyde (ethanal)	*-al*
Ketone	—C—C(=O)—C—	CH_3—C(=O)—CH_3 acetone	*-one*
Carboxylic acid	—C(=O)—OH	CH_3—C(=O)—OH acetic acid	*-ic acid*
Anhydride	—C(=O)—O—C(=O)—	CH_3—C(=O)—O—C(=O)—CH_3 acetic anhydride	none
Ester	—C(=O)—O—	CH_3—C(=O)—O—CH_3 methyl acetate	*-ate*
Amide	—C(=O)—NH_2, —C(=O)—N—H, —C(=O)—N—	CH_3—C(=O)—NH_2 acetamide	*-amide*

[a] The bonds whose connections aren't specified are assumed to be attached to carbon or hydrogen atoms in the rest of the molecule.
[b] X = F, Cl, Br, or I.